Innovation Management for Disruptive Maturity in Competitive Scenarios

George Leal Jamil
Informacoes em Rede Consultoria e Treinamento Ltda, Brazil

Liliane Carvalho Jamil
Independent Researcher, Brazil

Francisco Carlos Paletta
University of São Paulo, Brazil

Nuno Geada
ISCTE, University Institute of Lisboa, Portugal

Augusto Alves Pinho Vieira
Independent Researcher, Brazil

IGI Global
Scientific Publishing
Publishing Tomorrow's Research Today

Vice President of Editorial Melissa Wagner
Director of Acquisitions Mikaela Felty
Director of Book Development Jocelynn Hessler
Production Manager Mike Brehm
Cover Design Jose Rosado

Published in the United States of America by
IGI Global Scientific Publishing
701 East Chocolate Avenue
Hershey, PA, 17033, USA
Tel: 717-533-8845
Fax: 717-533-7115
Website: https://www.igi-global.com E-mail: cust@igi-global.com

Library of Congress Cataloging-in-Publication Data

Names: Jamil, George Leal, 1959- editor. | Jamil, Liliane, 1988- editor. | Paletta, Francisco Carlos, 1963- editor. | Geada, Nuno, 1983- editor. | Vieira, Augusto, 1988- editor.
Title: Innovation management for disruptive maturity in competitive scenarios / edited by George Leal Jamil, Liliane Carvalho Jamil, Francisco Carlos Paletta, Nuno Geada, Augusto Alves Pinho Vieira.
Description: Hershey, PA : IGI Global, [2026] | Includes bibliographical references and index. | Summary: "Considering any competitive scenario is essential for organizations to perfect value aggregation chains with solutions from innovative sources, such as business model design, new technologies insertion and dynamic processes reviews"-- Provided by publisher.
Identifiers: LCCN 2024052830 (print) | LCCN 2024052831 (ebook) | ISBN 9798369340301 (hardcover) | ISBN 9798369351536 (paperback) | ISBN 9798369340318 (ebook)
Subjects: LCSH: Strategic planning. | Disruptive technologies.
Classification: LCC HD30.28 .I5397 2025 (print) | LCC HD30.28 (ebook) | DDC 658.4/012--dc23/eng/20250121
LC record available at https://lccn.loc.gov/2024052830
LC ebook record available at https://lccn.loc.gov/2024052831

British Cataloguing in Publication Data
A Cataloguing in Publication record for this book is available from the British Library.

The editors George, Liliane and Augusto dedicate this book to their beloved daughter and granddaughter Sofia. There is no innovation such as her!

Editorial Advisory Board and List of Reviewers

Editorial Advisory Board

List of Reviewers

João Fernandes Siqueira, *Centro de Pós-Graduação e Pesquisas em Administração, Universidade Federal de Minas Gerais, Brazil*
Rafael Ignacio Peres Uribe, *University La Salle, Colombia*
Artur Alves Pinho Vieira, *Independent Researcher, Brazil*

Table of Contents

Detailed Table of Contents

Ethiopia. The study used an explanatory research design and a quantitative methodology. To gather data from a sample of 451, the researchers used a probability sampling technique, more precisely a multi-stage clustering sampling. Based on information from a pilot study with 95 respondents, an exploratory factor analysis (EFA) was carried out to evaluate the internal consistency and reliability of the study variables. Confirmatory factor analysis (CFA) and structural equation modeling (SEM) were used to analyse the gathered data. The findings show that entrepreneurship orientation and innovation mediation as a combination have a big impact on organisational success.

This work aims to present a specific case, in internal control, and to report the actions taken in the planning, creation and implementation of the Internal Control Advisory of the Institute of Technology in Pharmaceuticals - Farmanguinhos. For the development of this article, firstly, the organizational charts of Oswaldo Cruz Foundation (Fundação Oswaldo Cruz – Brazilian name) / Fiocruz and Farmanguinhos were brought into context, to briefly situate the reader in the complex environment of the activities developed in both institutions. That done, the steps adopted by the Institute's Executive Board were described step by step to structure the Internal Control Advisory - ICA, in view of the need to meet the demands arising during the internal audits carried out by Audin (Fiocruz's auditing body) / Fiocruz. Finally, the performance of ICA presents the interaction with the several sectors inside and outside the Institute of Drugs Technology (Farmanguinhos/Fiocruz)

The increasing demand for greater efficiency in public administration has driven the adoption of innovative human resource management approaches, such as the Management and Performance Program (PGD). The PGD modernizes public administration by replacing traditional attendance-based control with a productivity-driven model focused on results. Its implementation at the Brazilian Administrative Council for Economic Defense (CADE) has fostered greater flexibility in task execution without compromising service quality. This chapter employs a case study methodology, enabling an in-depth analysis of the PGD's implementation and its organizational impact. In 2025, the primary challenge will be to consolidate this new model while ensuring its sustainability and securing leadership engagement. The establishment of a dedicated team to oversee the PGD at CADE strengthens structured governance, which is crucial for the modernization of the federal public administration.

Preface

The rapidly evolving landscape of global markets presents both challenges and opportunities for businesses, governments, and societies alike. In this age of disruption, innovation is more crucial than ever, shaping competitive strategies, business models, and national economies. The edited book *Innovation Management for Disruptive Maturity in Competitive Scenarios* provides an insightful exploration into how organizations, institutions, and nations can manage innovation to remain competitive in an ever-changing environment.

This book brings together a diverse range of perspectives from scholars and professionals across multiple disciplines and geographies, providing a comprehensive overview of the theories, methodologies, and practices that are driving innovation in the digital and post-digital economy. From strategic innovation to digital transformation, entrepreneurship, and sustainable development, each chapter in this volume contributes to a holistic understanding of how innovation management can guide organizations and individuals toward achieving disruptive maturity.

CHAPTER OVERVIEW

Chapter 1, *Quintuple Helix and Impact Economy: The Bases for Generating a Disruptive Innovative Model*, introduces the Quintuple Helix model, emphasizing the convergence of the digital economy with the impact economy. It explores how innovation, supported by government, businesses, universities, civil society, and the environment, can create scalable and sustainable solutions for social and environmental challenges.

Chapter 2, *Turning Maintenance into Competitive Advantage: Innovation in the Analytical Chemistry Industry*, shifts the focus to the analytical chemistry sector, where innovation in leadership, management, and digital solutions has reconfigured business models, turning maintenance into a strategic advantage for companies.

Chapter 3, *Six Steps to Implement Innovation in Companies: Innovation Hubs, Internal Hackathons, Innovation Communities*, provides practical insights into implementing innovation strategies within organizations, stressing the importance of mindset shifts, agile methodologies, and exposure to new ideas as critical steps for fostering a culture of innovation.

Chapter 4, *Analyzing the Impact of Managerial Innovation Dimensions on the Financial Performance of Moroccan Listed Companies*, presents a systematic analysis of how various managerial innovation dimensions—process, organizational, strategic, and technological—impact the financial performance of companies, offering valuable insights into the role of innovation in driving financial success.

In Chapter 5, *Information Science as the Base for Multi, Inter, and Transdisciplinary Studies: A More Robust Way to Innovate Through Digital Transformation*, explores how Information Science, through its multi-, inter-, and transdisciplinary approaches, can provide a solid scientific foundation for digital transformations that foster innovation and success in complex projects.

In Chapter 6, *Impact of Innovation Dimension Mediation on the Performance of Ethiopian SMEs with an Entrepreneurship Orientation*, the authors investigate how entrepreneurship orientation and innovation mediation affect the performance of small and medium-sized enterprises (SMEs) in Ethiopia, offering insights into the factors driving success in emerging markets.

Chapter 7, *Internal Control Advisory to Support Innovation in the Corporate Governance Diplomacy: A Case Study of a Brazilian Science and Technology Institution (Oswaldo Cruz Foundation – Fiocruz)*, presents a case study of the internal control advisory system at Fiocruz, demonstrating how internal control mechanisms can support innovation in research and development institutions.

The focus shifts to the public sector in Chapter 8, *Performance Management Program as a Disruption in Public Management: The Case of the Administrative Council for Economic Defense (CADE)*, which explores how innovative human resource management practices, such as the Management and Performance Program (PGD), are transforming public administration in Brazil.

Chapter 9, *Innovation Management at the Fuzzy Front End: AI, Business Cases, and Product Portfolios*, examines the strategic management of the Fuzzy Front End (FFE) in innovation, highlighting how AI can enhance early-stage decision-making and business model innovation, ultimately shaping competitive advantage.

In Chapter 10, *Disruptive Maturity Through the Incorporation of Strategic Innovation in Ethiopia Higher Education*, the authors analyze how strategic alignment, human capital, and technological capabilities contribute to institutional revitalization and the development of strategic innovation within higher education institutions in Ethiopia.

Chapter 11, *Challenges to the Adoption of the Balanced Scorecard as a Managerial Innovation in Moroccan SMEs*, delves into the barriers preventing the adoption of the Balanced Scorecard (BSC) in Moroccan SMEs, offering insights into how these enterprises can overcome these challenges to improve performance and sustainability.

In Chapter 12, *Business Models' Effect on Innovation with Mediation of Open Innovation Analysis in Africa*, the authors examine the relationship between business models and innovation, with a particular focus on the role of open innovation in enhancing outcomes for African industries, providing evidence for the importance of business model design in driving innovation.

Chapter 13, *Artificial Intelligence and the Application of Augmented Reality Techniques as Disruptive Innovation: A Thinking for Process Management in the Pharmaceutical Industry*, explores the use of AI and augmented reality in pharmaceutical process management, specifically addressing regulatory compliance, and highlighting how these technologies can drive efficiency and precision in highly regulated industries.

In Chapter 14, *Digital Risk Management in Digital Transformation Contexts*, the authors discuss the emerging risks associated with digital transformation, emphasizing the need for comprehensive risk management strategies that integrate data protection, information security, and broader governance considerations in a rapidly evolving digital landscape.

Chapter 15, *The Impact of Technological and Non-technological Innovation on Export Performance for SMEs in Marrakech-Safi Region in Morocco: A Structural Equation Analysis*, investigates how different types of innovation influence the export performance of SMEs in Morocco, offering insights into the role of innovation in improving international competitiveness for SMEs.

Chapter 16, *Leveraging AI for Disruptive Green Human Resource Management: Challenges and Opportunities to Strategic People Management*, explores the role of AI in transforming Green Human Resource Management (GHRM). The authors discuss how AI can drive sustainable practices, enhance employee development, and improve corporate social responsibility. The chapter also addresses the ethical, legal, and social challenges AI brings to HRM, emphasizing the need for responsible and strategic integration.

Chapter 17, *Innovation and Its Relation to Strategic Positioning with Mediation of Market Competition in Horn of Africa*, investigates the relationship between innovation, market competition, and strategic positioning in the Horn of Africa. It highlights how innovation drives a competitive edge and how competition mediates strategic positioning, offering recommendations for managers to foster innovation in dynamic market conditions.

Chapter 18, *Innovation Management for Disruptive Maturity with Mediation of Competitive Scenario in Horn of Africa*, focusing on innovation management, this chapter explores how firms in the Horn of Africa can leverage innovation to achieve disruptive maturity. It presents a model linking innovation management with competitive pressures and the need for firms to adapt to market changes in order to thrive in increasingly competitive environments.

Chapter 19, *Variables That Explain the Way of Innovating in Colombian Companies: A Perspective Under the RISE Model*, delves into the factors influencing innovation in Colombian companies through the RISE (Route for Innovative and Sustainable Enterprises) model. It identifies key variables like ecodesign and work environment that impact how organizations innovate and achieve their sustainability goals.

Chapter 20, *Efficacy of Deep Neural Embeddings-based Approach vs Traditional Machine Learning: An Experimental CASE Study*, presents an experimental case study comparing deep neural embeddings with traditional machine learning techniques. The authors explore how embedding techniques can offer advantages in areas like medicine and agriculture, providing an alternative to conventional machine learning models, especially when large, annotated datasets are scarce.

Chapter 21, *Innovation Management in Health Technology Scenarios: A Reflection on Immunosuppressant Drugs for Organ Transplants*, in the context of health technology, this chapter reflects on innovation management in the development of immunosuppressant drugs for organ transplants. It examines how advancements in immunosuppression therapy and related technologies can improve transplant success rates and overall healthcare outcomes.

Chapter 22, *Innovation Positioning Panorama Influence on Innovation Source Diversity in Healthcare Sector Through R&D Investment in Ethiopia*, investigates the role of R&D investment in the healthcare sector in Ethiopia and its influence on innovation positioning. It highlights the importance of diverse innovation sources and demonstrates the critical role of R&D expenditure in converting innovative positioning into tangible outcomes.

Chapter 23, *Leveraging Blockchain in High Frequency Trading for Shaping Market Behavior and Enhancing Portfolio Returns,* explores the integration of blockchain in high-frequency trading (HFT) and its impact on market behavior and portfolio returns. The authors analyze how AI, machine learning, and blockchain can help reduce emotional interference in trading, leading to more consistent results in volatile capital markets.

In Chapter 24, *Strategic Innovation Management Using Emerging Technologies for Building Resilience in Competitive Scenarios*, organizations looking to build resilience must incorporate strategic innovation management, especially through emerging technologies like AI, blockchain, and IoT. This chapter focuses on how these technologies can be leveraged for long-term growth and adaptability in competitive and volatile markets, with an emphasis on sustainability and ethical practices.

Chapter 25, *The Dynamic Model for Management Improvement Projects,* introduces a dynamic model for management improvement projects, emphasizing flexibility and adaptability in the face of challenges such as complexity and speed. Using the star diagram methodology, the authors provide a framework for organizations to enhance project performance, impact, and sustainability across various sectors.

The chapters in this collection highlight how innovation, when strategically managed and enhanced by emerging technologies, can foster growth, adaptability, and competitive advantage in various sectors. Each contribution provides a comprehensive exploration of how organizations can use innovation to navigate the challenges of today's dynamic business environment, while ensuring long-term sustainability and resilience.

This edited volume not only illuminates the multifaceted nature of innovation management but also offers actionable insights for practitioners and policymakers looking to navigate the complexities of innovation in competitive and disruptive environments. Through diverse case studies, theoretical frameworks, and empirical analyses, the book sets the stage for further exploration of how organizations can build disruptive maturity through strategic innovation.

Acknowledgement

We, the book editors, would like to congratulate and thank all authors for their efforts in providing such interesting, updated and thoroughly researched texts, which are the real components of this project delivery. Since we had to restart it over, several authors kept their trust on our editorial work, resulting in a real contribution to our readers. One mention to Dr. Shashi Kant, who wrote chapters, invited other authors to contribute and was always available to help us over the reviews of this book, becoming a partner for this editorial project.

It is also needed to thank our reviewers for their contributions, examining each production sent, caring for the essential perfection of our project. They provided an additional scrutiny and balance to encompass the complex topics we addressed in our coverage.

Finally, a special thanks to our IGI Global team, who, as usual, dedicated a time on guiding us to improve the quality, productivity and responsiveness throughout this remarkable work.

Chapter 1
Quintuple Helix and Impact Economy:
The Bases for a Generative Disruptive Innovative Model

Frederico Cesar Mafra Pereira
https://orcid.org/0000-0002-1971-8069
Federal University of Minas Gerais, Brazil & Federal University of São Carlos, Brazil

Rogério Aparecido Sá Ramalho
https://orcid.org/0000-0002-8491-3514
Federal University of São Carlos, Brazil

Aline Laureano Suave
Federal University of São Carlos, Brazil

ABSTRACT

The digital economy has reshaped interactions between governments, businesses, universities, and society, driving innovations geared toward social and environmental impact. This article investigates the convergence between the digital economy and the impact economy, using the Quintuple Helix Model as an analytical framework. The study examines the interaction between government, businesses, universities, civil society, and the environment, highlighting how this dynamic can foster sustainable innovation. The methodology includes a theoretical review on digital transformation, impact entrepreneurship, and tools such as Design Thinking, Canvas Models, and ESG, SDG, and SROI indicators. The results indicate that the adoption of collaborative models is essential to maximize positive impact and promote scalable solutions. It is concluded that the impact economy, supported by helical structures, represents a strategic model for aligning economic growth with sustainability, offering guidelines for responsible and inclusive innovation.

DOI: 10.4018/979-8-3693-4030-1.ch001

INTRODUCTION

Digital transformation is driving an unprecedented revolution in relationships between individuals, organizations and governments, reshaping the way economic, social and cultural resources are distributed and used globally. This revolution has been enabled by a set of disruptive technologies including big data, artificial intelligence (AI), blockchain, the Internet of Things (IoT) and cloud computing, which have not only driven economic growth but also created a new paradigm for social and economic organization.

One of the central features of this 'digital economy' is its ability to connect people, organizations and governments on an exponential and global scale. Through digital platforms, individuals can interact with complex systems that were previously inaccessible, including access to financial services, education, healthcare and labour markets. On the other hand, digitalization has also exposed deep inequalities, creating significant challenges for greater (and 'intended') social and economic inclusion. Regions with low internet access and reach, for example, face barriers to fully participating in this new economy, which intensifies the need for solutions that prioritize digital inclusion as a basic right.

This technological revolution also brings with it ethical and structural challenges that need to be addressed to ensure sustainable and equitable development. Among the main challenges are the ethical dilemmas of AI (Floridi, 2023), involving data privacy (since, with the massive collection of information, concerns about the security and consent of the use of personal data grow), algorithmic bias (since many AI systems reflect prejudices embedded in training data, leading to discriminatory decisions), transparency (since AI models often operate as 'black boxes', making it difficult to explain their decisions), accountability in autonomous decisions and possible errors committed by these systems, the need for policies that encourage intersectoral collaboration and the resistance of traditional industries to the adoption of collaborative models. Transparency in AI systems and adequate clear regulation, as well as an audit of algorithms, can help minimize such negative impacts, ensuring truly ethical and trustworthy AI.

Another significant impact of digital transformation is related to the speed and accuracy of decision-making. By collecting and analyzing large volumes of data in real time, governments can implement more effective policies, companies can optimize operations, and individuals can make more informed decisions about consumption and investment. In the meantime, for innovation resulting from digital transformation to be inclusive and effective, public policies must encourage the promotion of a collaborative ecosystem that ensures that its benefits reach all levels of society. Some of these approaches include, for example: (i) tax incentives for companies that invest in joint research; (ii) public-private partnerships that facilitate access to new technologies for small businesses and underserved communities; (iii) digital education programs to train professionals and reduce the technology skills gap; and (iv) flexible regulation that allows experimentation with new technologies without compromising security and privacy.

A new model of capitalism (Cohen, 2022) emerges in this context as a response to global social and environmental demands, proposing a mode of development that harmonizes the creation of economic value with social and environmental benefits. Unlike the traditional capitalist model, which ultimately prioritizes profit maximization, this new model (which in this article will be considered as "impact economy") seeks to meet the needs of all stakeholders of a business, including marginalized communities and the environment. This is a new paradigm that is particularly relevant in a world where climate crises, social inequalities, and economic instability threaten the collective well-being and long-term sustainability of the entire planet.

For Gabriel, Osorio and Morel (2023), impact can be defined as "the long-term, positive and negative, primary and secondary effects produced by a development intervention, whether direct or indirect, intended or not", and corroborates Cohen (2022), for whom impact "is the measure of the benefit created by a given action for people and the planet". Studies by Carayannis and Campbell (2022) highlight that the ongoing transition to an impact economy requires a broader integration between technological innovation and social objectives. Digital technologies provide the tools to scale sustainable solutions, while the impact economy offers the ethical and social framework to ensure that these solutions benefit society as a whole. It is clear, therefore, that the relationship between the digital economy and the impact economy is not just a convergence of ideas, but a pragmatic need to face the challenges of the 21st century, an approach that implies the adoption of tools and methods that allow us to measure and maximize positive impacts, such as, for example, the use of blockchain to track supply chains and ensure sustainability, the use of digital platforms to democratize access to microcredit in vulnerable communities or the use of AI algorithms to develop public health solutions that meet the needs of populations in remote areas or to optimize natural resources.

Some concrete examples of the adoption of new technologies to enhance the socioeconomic development of underserved regions around the world include: (i) Brazil – Amazon Project 4.0[1] (use of digital technology for environmental monitoring and training of riverside communities); (ii) India – Aadhaar[2] (digital identification system that allows access to financial and government services for marginalized populations); (iii) Africa – Digital Education[3] (initiatives that bring online education and connectivity to remote areas, reducing educational inequality, such as Project Hello World[4] and Projecs for all[5]). These cases demonstrate how technology can be used to promote social inclusion and economic development.

Another key element of the impact economy is its alignment with the Sustainable Development Goals (SDGs) set by the UN (2015). These goals provide a global framework for addressing issues such as poverty, inequality and climate change. Companies and governments are increasingly using the SDGs as a guide to align their impact strategies, promoting a more integrated and coherent approach to sustainable development.

Finally, the impact economy represents a fundamental trend in redefining the business models of the future, offering a pragmatic and visionary response to global challenges, promoting a more ethical, inclusive and sustainable capitalism. The adoption of disruptive business models that incorporate impact principles is already transforming entire sectors. One example is the emergence of 'B Corps' (B Labs, 2006), which operate with a clearly defined social purpose and measure their success not only in financial terms, but also in relation to the impact generated. In addition, startups focused on renewable energy, precision agriculture and the circular economy, for example, have already demonstrated that it is possible to align profitability with sustainability and social impact. A model that supports this impact economy should encourage the adoption of collaborative practices that involve governments, companies, universities and civil society, exploring the theory of helix innovation systems, towards a unified theory of helical architectures (Carayannis & Campbell, 2022).

This article therefore explores how the impact economy, a new paradigm that combines economic value generation with social and environmental benefits, aligns with the dynamics of the digital economy. Based on the Quintuple Helix Model (Carayannis, Barth & Campbell, 2012), it seeks to analyze how the integration of its five spheres — government, business, academia, civil society and the environment — can foster impact-oriented innovations.

To this end, this article is structured into 7 sections. In addition to this introduction, section 2 addresses how digital transformation, by expanding the capacity to generate data and connecting individuals, organizations, and governments, has contributed to the emergence of an environment favorable to impact-oriented innovation. Section 3 addresses the concept of impact entrepreneurship, highlighting its role as a catalyst for structural transformations in the market. Section 4 explores some of the main methodological approaches and tools used to design and implement impact businesses. Section 5 presents the Quintuple Helix Model as an approach to be used in the modeling and implementation of innovative businesses with positive impact. Section 6 addresses the importance of measuring the impact generated by impact businesses, presenting a review of the main indicators used, such as the UN SDGs, and measurement methodologies that combine quantitative and qualitative approaches. Finally, in the final considerations, the main lessons learned and reflections derived from the topics covered are consolidated, offering an integrated and prospective vision on the impact economy and the application of the Quintuple Helix model as an enabling structure.

FROM DIGITAL TRANSFORMATION TO IMPACT REVOLUTION

According to Oliveira, Mafra Pereira and Fialho (2023), the development and proliferation of new digital technologies have made their adoption an increasingly critical aspect for organizations, aiming at their survival in increasingly dynamic and intensive markets. As a result of the challenges imposed by this new context, digital transformation emerged (Rogers, 2016; Schwab, 2017; Vial, 2019; Verhoef et al., 2021), representing not only a technological change, but a multidimensional process that has profoundly changed the way individuals, organizations and governments interact and create value. This context was made possible by the convergence of technologies such as big data, the Internet of Things (IoT), blockchain and artificial intelligence (AI) which, together, created an ecosystem of global connections and data-driven decision-making. This scenario expanded the capacity to collect, store and analyze information, allowing for more agile, personalized and effective solutions.

It is worth noting again that the adoption of the various technologies highlighted and collaborative models faces political and cultural challenges. Such barriers include resistance to innovation by outdated government regulations that do not encourage interoperability between different sectors, cultural differences in the implementation of new technologies (in particular, rigid organizational cultures and fear of competition), the lack of leadership training to encourage an innovation mindset, the lack of financial incentives for the adoption of new technologies, and the lack of promotion of partnerships between established companies and startups for the exchange of knowledge. In short, the need to adopt a more open approach to innovation (Chesbrough, 2006; 2019), both by the government and by traditional companies, can benefit the implementation of innovation without compromising organizational competitiveness.

Below is a brief history that will allow us to understand how digital transformation has not only expanded the capacity for global connection and data generation, but has also paved the way for an impact economy that prioritizes innovation for the sake of collective well-being and sustainability.

Data Generation and the Global Connection

The expansion of Internet access and the adoption of connected devices have brought an explosion of data never before seen in the context of global society, creating opportunities for governments, businesses and civil society to exploit the information and knowledge generated to develop innovative solutions. Data has become the most valuable resource of the 'new' digital economy, providing insights that have enabled the transformation of organizational processes and improved the efficiency of public and private systems (when used for such purposes).

Some examples confirm such advances. In smart cities, IoT sensors monitor traffic, energy consumption and air quality in real time, and this data allows city managers to implement immediate and evidence-based solutions, such as optimizing public transportation or promoting renewable energy. These initiatives not only improve the quality of life of the population, but also generate positive environmental impacts, illustrating the transformative role of technology in achieving sustainability goals.

The Contribution of Innovation to Positive Impact

The integration of digital transformation with impact-oriented innovation creates a virtuous cycle where technologies catalyze solutions that seek to address global challenges, such as climate change and social inequality, and not just focused on the development of specific products or services for a given company, or a particular economic sector.

Thus, it is necessary to mobilize innovative capital to sustain the impact economy, combining public and private investments that support startups and growing companies whose business models are impact-oriented. According to the Global Impact Investing Network (GIIN, 2022), the global impact investment market was estimated at US$ 1.164 trillion in 2022, highlighting investors' appetite for solutions that balance profitability and socio-environmental responsibility, and reflecting the increased awareness of climate issues and the demand for more transparent reporting standards. These funds play the role of connecting investors with projects that promote social and environmental transformation, directing capital to initiatives focused on clean energy, affordable healthcare and digital education, some of humanity's greatest challenges in this century.

Companies like Tesla exemplify this transition by combining digital technologies with a commitment to sustainability, revolutionizing not only the automotive sector, but also demonstrating that impact-focused business models can be profitable and scalable. Another example is the use of blockchain in supply chains, which allows the origin of products to be traced and ensures that ethical and sustainable practices are adopted at all levels of production.

Another example is the use of blockchain, which has proven to be a powerful tool for ensuring transparency, traceability, and security in digital processes, as it is based on a decentralized and immutable technology, whose main applications include: (i) corporate governance and smart contracts (with the immutable record of decisions and automated contractual processes, reducing bureaucracy and ensuring greater reliability in commercial transactions); (ii) data protection and cybersecurity (with decentralized models that reduce the risk of leakage of sensitive information, secure storage, and transparent sharing of operational data, reducing the risk of manipulation and fraud); (iii) supply chains (allowing secure and reliable tracking of the product, from the raw material to the end consumer, ensuring authenticity, quality, and compliance with environmental regulations). The adoption of blockchain not only improves operational efficiency, but also strengthens trust in digital systems and between companies, consumers, and regulators (OECD, 2019; Islam & Meher, 2023; Calandra, Pinna & Pinna, 2024; Vladucu et al., 2024).

From Digital Transformation to Impact Economy

Therefore, this aforementioned evolution is not just a trend, but a necessity for a more inclusive and sustainable future. Digital transformation has provided the foundations for large-scale innovation, while the impact economy has redefined how success is measured, going beyond financial profit to include social and environmental values. For Carayannis and Campbell (2022), society in general (including individuals, companies, governments) is witnessing a historic inflection point, where emerging technologies are integrated into business models that prioritize positive impact (with a social and environmental focus). This convergence therefore reflects a global movement to reconcile economic growth with the ecological limits of the planet and the needs of the most vulnerable communities.

For some authors (Ávila & Gomes, 2021; Sott, Baum & Bender, 2022; Fernandes, 2022), this moment marks a new stage in digital transformation, focusing on the humanization of technology and collaboration between humans ('Society 5.0') and AI. A new concept of 'Industry 5.0' thus proposes a more humane, sustainable and collaborative model, in which technological progress goes hand in hand with respect for human beings and the environment. The objective is to combine innovation with well-being, sustainability and social responsibility, aspects that align precisely with the concept of the "impact economy".

Case studies show practical applications of this new context around the world: (i) Japan uses collaborative robots in manufacturing; (ii) the European Union is investing in the circular economy; (iii) Brazil is adopting technologies such as sensors and blockchain in agriculture; and (iv) India is automating healthcare services with AI. These examples demonstrate the diversity of applications of Industry 5.0 for social and environmental benefits.

However, as mentioned above, this progress brings ethical challenges, such as data privacy, impacts on employment, responsibility for autonomous decisions and inequality in access to technology. Addressing these challenges requires cooperation between governments, companies and civil society. To this end, it is essential to develop public policies that encourage sustainable technologies, promote professional retraining, make regulations more flexible and encourage public-private partnerships.

The next section will highlight the role of impact entrepreneurship as a catalyst for structural transformations in the market and the idea of the impact economy, presenting practical examples and detailed reflections on its contribution to building a more sustainable and inclusive future.

IMPACT ENTREPRENEURSHIP AND DISRUPTIVE BUSINESS MODELS

Conceptualizing Impact Entrepreneurship

Impact entrepreneurship represents an innovative approach that places the resolution of social and environmental problems at the center of its activities. More than just a business model, it is a philosophy that challenges traditional market structures, seeking to create shared value among all stakeholders. In other words, the essence of this type of entrepreneurship lies in the integration of ethical and social purposes with market logic. While traditional entrepreneurship prioritizes profit as the main indicator of success, impact entrepreneurship evaluates its performance based on three pillars: economic, social and environmental (Elkington, 1997; Cohen, 2022).

One of the most striking characteristics of impact entrepreneurship is its ability to catalyze structural transformations in the market. Impact entrepreneurs not only challenge established models, but also introduce innovative solutions that address global problems in a creative and effective way. A prime example is the Grameen Bank (in Bangladesh), founded by Muhammad Yunus, which popularized the concept of microcredit (Bello Filho, 2010), offering small loans to low-income entrepreneurs without the need for collateral, allowing marginalized communities to develop small businesses. This financial innovation not only empowered millions of vulnerable individuals, but also demonstrated that it is possible to combine social impact with economic sustainability, having proven to be a highly effective model in reducing poverty and replicated in several countries.

Another example of transformative impact is the renewable energy sector. Companies like Tesla have not only revolutionized the automotive market with electric vehicles, but have also encouraged the adoption of sustainable practices throughout the production chain (Friebel, 2021). This disruptive model inspires other industries to rethink their processes, contributing to the construction of a greener and more resilient economy (Cohen, 2022).

In other words, impact entrepreneurship enables (i) double or triple bottom line, through which organizations evaluate success based on social, environmental and economic impact, (ii) scalability and innovation, whose business models can be expanded to generate greater impact, (iii) financial sustainability, whose ventures, unlike NGOs, seek their own revenues to ensure their continuity, (iv) measurement and transparency, based on the use of metrics to evaluate the real impact of the initiative and (vi) intersectoral collaboration, making partnerships between the private sector, governments and civil society tangible to enhance impact.

Disruptive Business Models and Social Innovation

From impact entrepreneurship, new business models, with innovative and disruptive characteristics, can be associated with their ability to challenge paradigms and introduce solutions that meet needs previously unmet by traditional models and systems. And these innovations play a fundamental role in solving problems, mainly of a social and environmental focus.

Some examples of how disruptive business models are being applied to promote positive impact can be observed: (i) Sharing Economy Platforms - companies such as Airbnb[6] and BlaBlaCar[7] leverage digital technology to create marketplaces that maximize the use of underutilized resources (in this case, real estate and automobiles). Although they began as market solutions, they have evolved to include initiatives that promote responsible tourism and reduce carbon emissions; (ii) Precision Agriculture - startups such as Indigo Agriculture[8] are using AI and IoT sensors to optimize the use of resources in agriculture, reducing waste and promoting sustainability. This approach not only increases productivity, but also minimizes environmental impact. In Brazil, the startup Agrosmart[9] uses AI and IoT to help farmers optimize the use of water and agricultural inputs, promoting more efficient and sustainable agriculture. Its technology allows small and large producers to make data-driven decisions, reducing waste and increasing productivity; (iii) Affordable Health Technologies - startups such as Zipline[10] use drones to deliver medicines to remote areas, ensuring that vulnerable communities have access to essential treatments. This solution combines technological innovation with clear social purposes; (iv) Clean Energy – Solar Sister[11] (Sub-Saharan Africa) empowers women entrepreneurs to sell solar energy solutions in communities without access to electricity. The impact goes beyond clean energy, promoting economic inclusion and female empowerment, ensuring more families have access to sustainable lighting and reducing dependence on fossil fuels; (v) Manufacturing – Fairphone[12] develops modular and sustainable smartphones, ensuring that their components are ethically sourced and easily replaceable. By focusing on supply chain transparency, the company encourages responsible production and promotes better working conditions in mining and electronics manufacturing; (vi) Apparel – Outdoor clothing brand Patagonia[13] is an example of a company that combines financial success with environmental impact. The company donates part of its profits to environmental causes, invests in recycled materials and encourages conscious consumption practices, such as repairing products instead of replacing them.

According to Cohen (2022), there were around three thousand businesses certified as B Corps (B Labs, 2006), in 150 different sectors, distributed across 64 countries around the world, redefining a new corporate format, called "benefit corporation", from 2010 onwards, in the United States. This corporate format is exempt from the obligation to maximize its profits, allowing it to dedicate itself equally to the search for impact, without fear of legal action by its shareholders.

Challenges and Opportunities in Impact Entrepreneurship

While impact entrepreneurship offers promising solutions to global problems, it also faces significant challenges. These include the difficulty of accessing financing, the (still) cultural resistance to adopting new models, and the lack of clear standards for measuring impact. According to Carayannis et al. (2018), overcoming these obstacles requires the creation of collaborative ecosystems that integrate government, the private sector, and civil society. At the same time, the opportunities are vast. The growing demand for sustainable and responsible solutions paves the way for impact entrepreneurs to gain prominence. The advancement of technologies such as blockchain and AI offers new tools to measure and amplify the impact of initiatives.

Thus, the future of impact entrepreneurship is intrinsically linked to the ability to combine technological innovation with social and environmental responsibility, correlating precisely with the concept of the impact economy. The emergence of more inclusive and sustainable business models represents an opportunity to redefine current market rules in favor of building a more ethical and collaborative capitalism. For Carayannis et al. (2022), impact entrepreneurship will play a central role in this transition to a more resilient economy oriented towards collective well-being, based on the integration of technological and social innovations. By expanding the understanding of inherent concepts and presenting real cases, it becomes easier to engage non-experts and encourage new initiatives aligned with the principles of disruptive maturity.

In the next section, we will present some of the most relevant methodological approaches and tools used in the implementation of impact businesses, which seek to balance innovation, sustainability and inclusion, such as the Triple Bottom Line (TBL), Design Thinking and Canvas Models adapted for impact (such as the C Model). These practices have helped entrepreneurs and organizations to effectively align their economic, social and environmental objectives.

APPROACHES AND TOOLS FOR IMPACT BUSINESS

Triple Bottom Line (TBL): Performance on Three Pillars

The Triple Bottom Line (TBL) concept, introduced by Elkington (1997), proposes an approach to assessing organizational performance that goes beyond financial profit, considering three fundamental dimensions: economic (profit), social (people) and environmental (planet). The economic pillar assesses the financial sustainability of the business, ensuring that it is viable in the long term. The social pillar analyzes the benefits generated for the communities served, such as job creation, improved health and education. The environmental pillar examines the impacts of activities on the environment, including reducing carbon emissions and responsible use of natural resources. For Elkington (1997), the interaction between organizations and the community supports the three pillars, representing how the impact of business relates to sustainable growth and whether actions do not compromise and contribute to future generations.

This framework is widely used as a conceptual-methodological basis by impact businesses to measure and communicate their performance more comprehensively. An example of an organization that adopts TBL in its strategy is Patagonia[14], which not only manufactures sustainable clothing, but also invests in environmental conservation projects and promotes ethical working conditions in its supply chain. This integrated approach reflects, in practice, the commitment to the three pillars of TBL.

Corroborating this evolution of digital transformation towards an impact economy, aiming to meet a growing demand and help companies issue their reports with information on the impacts generated by their business model, Ciasca (2021) points out new initiatives that have emerged throughout the first decades of the 21st century, such as the European Eco-Management and Audit Scheme – EMAS, the Global Reporting Initiative – GRI, the International Organization for Standardization – ISO 14001, The Prince's Accounting for Sustainability – A4S and International Integrated Reporting – IIRC.

For example, the IIRC (2011) noted that its framework resulted in a broader view in relation to other sustainability reports, consolidating the concept of financial and non-financial capital as fundamental to reporting information ranging from the company's strategy to the allocation of scarce resources in its business. Thus, the three original pillars of the TBL became interrelated in six capitals: financial, manufactured, intellectual, human, social and relationship, and natural (Ciasca, 2011).

Later, Elkington (2018), in an interview with the Harvard Business Review about the 25 years of the TBL concept, highlighted that some companies have moved in this direction, citing the Danish Novo Nordisk[15], the Brazilian Natura[16] and the North American Danone[17], which not only used the concepts he introduced in 1997, but also adopted IR as a disclosure model, including the six capitals in its framework, and confirming the progress in the conceptual framework on the pillars of sustainability.

Design Thinking for User-Centered Solutions

Design Thinking is a creative and iterative approach that places the user at the center of the innovation process (Brown, 2009). Initially developed for product design, this methodology has been adapted to solve complex problems in diverse areas, including impact businesses. According to Brown (2009), Design Thinking is based on three fundamental pillars: empathy, experimentation, and collaboration, and its process is divided into five main stages: 1) empathy, where qualitative research is carried out to deeply understand the target audience; 2) problem definition, which involves analyzing the information collected to identify central challenges; 3) ideation, which promotes the generation of creative solutions; 4) prototyping, which transforms ideas into tangible models; and 5) testing, which assesses the feasibility and effectiveness of the proposed solutions (Xu et al., 2021).

In the context of impact businesses, Design Thinking enables a deep understanding of beneficiaries' needs, facilitating the co-creation of solutions that truly address social and environmental demands. IDEO.org, for example, used Design Thinking to develop sanitation solutions in vulnerable communities in Africa, resulting in the creation of affordable and sustainable products.

Canvas Templates Adapted for Impact

Canvas models are visual tools that help entrepreneurs map and structure the main elements of a business model, with the Business Model Canvas (Osterwalder & Pigneur, 2010) being the precursor to all other derived canvas models. In the case of impact businesses, these models have been adapted to include social and environmental dimensions, allowing a holistic view of operations. Among these adaptations, Model C (ICE, 2018; Polman & Winston, 2022) stands out, an evolution of the traditional Business Model Canvas.

Figure 1. C-Model

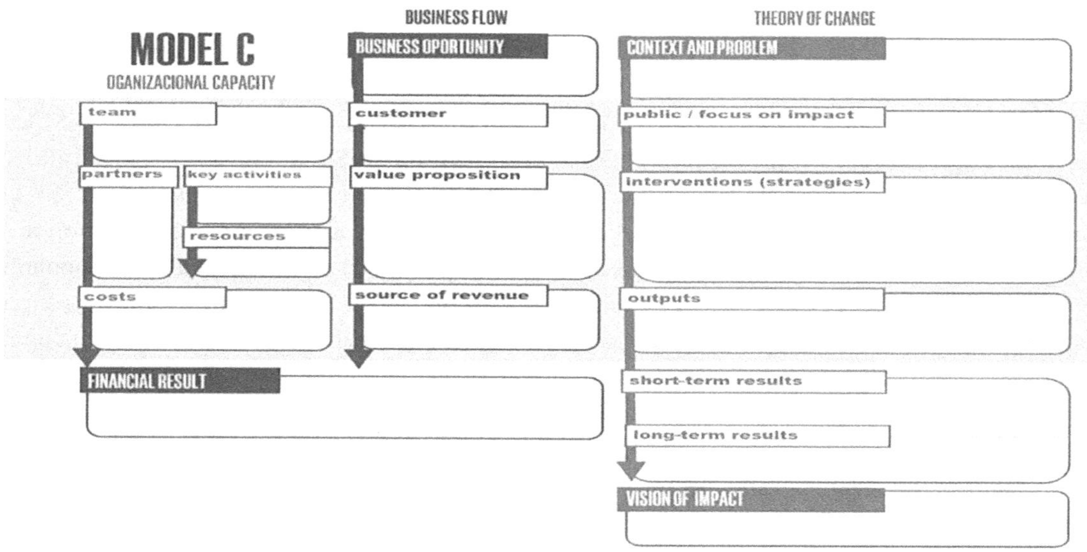

Source: Lima, Correia & Oliveira (2024)

The C-Model was developed to support social entrepreneurs in designing initiatives that generate shared value, and includes specific elements such as "expected social impact", "community partnerships" and "impact indicators". This framework allows entrepreneurs to analyze how their activities directly contribute to collective well-being and environmental sustainability. As an example of an application, an organization that promotes access to clean water in rural areas can use the C-Model to map partnerships with local communities, identify sustainable sources of financing and measure the social impact generated. This not only increases transparency, but also attracts investors interested in supporting impact initiatives.

Integration of Approaches and Tools

While each of the three approaches has its own focus area, integrating them is essential for the success of impact businesses. Design Thinking provides the framework for deeply understanding challenges and needs, while Canvas Models help structure sustainable solutions. TBL, in turn, ensures that performance is assessed comprehensively, promoting transparency and commitment to impact objectives.

As an example of an integrated application of these approaches and tools, a clean energy startup could use Design Thinking to identify barriers to accessing sustainable technologies in remote communities. It could then structure its business model using the C-Model, ensuring that local partnerships are considered and that impact indicators are established. Finally, the TBL can be used to evaluate the effectiveness of the initiative, considering economic, social and environmental impacts.

Thus, the aforementioned approaches and tools, among others, are fundamental to designing and implementing impact businesses that truly contribute to sustainable development. By combining creativity, strategic planning and comprehensive assessment, these practices allow entrepreneurs to address global challenges in an innovative and effective way.

However, it is important to recognize that adopting these approaches requires a change in mindset. Entrepreneurs and organizations must be willing to prioritize social and environmental impact, even when this requires adjustments to traditional business models. As Carayannis and Campbell (2011) argue, this transition is essential to building a more inclusive and sustainable capitalism.

The next section will present the Quintuple Helix Model, a significant conceptual evolution in relation to the Triple and Quadruple Helix frameworks, which integrates five key actors — government, business, academia, civil society and the environment — into a collaborative ecosystem focused on innovation and positive impact. It is inferred that this model reflects the need for more inclusive and sustainable solutions to global challenges, such as climate change, social inequalities and the transition to a greener economy.

THE QUINTUPLE HELIX MODEL: A FRAMEWORK FOR INNOVATIVE AND SUSTAINABLE BUSINESSES

Helix innovation models are characterized by their multi-structural, multi-functional, and non-linear nature. In recent decades, these models have evolved from the triple helix model to the quadruple and quintuple helix models of innovation (Gabriel, Osorio & Morel, 2023).

The Triple Helix concept, proposed by Etzkowitz and Leydesdorff (2000), was a milestone in understanding the interaction between universities, industry and government to promote innovation. This approach highlighted the role of universities as drivers of knowledge, while government and business acted as facilitators in the practical application of this knowledge. The Triple Helix explicitly recognizes the importance of higher education and emphasizes the production of knowledge for innovation in the economy, being compatible with the concept of a knowledge economy (Carayannis et al., 2012), which, together with a knowledge-based society, depends on the institutional spheres of university, industry and government (Gabriel et al., 2023).

Subsequently, the Quadruple Helix expanded the model by including civil society as a fourth essential actor, recognizing the importance of citizen participation in the co-creation of innovative solutions and emphasizing the role of media, culture and democracy in this innovation process, highlighting that sustainable development requires a co-evolution between knowledge economy, knowledge society and knowledge democracy (Carayannis & Campbell, 2016). In this model, society is central and aims to integrate the public, media and culture in the innovation process, recognizing that reality is constructed and communicated by the media and influenced by values and cultures. Such evolution, in the authors' view, was essential to incorporate more diverse perspectives and increase the legitimacy of initiatives.

Thus, the Quadruple Helix model proposes a collective interaction, where knowledge is exchanged through four subsystems: educational system (universities and human capital), economic system (industry and economic capital), political system (laws and political capital) and civil system (culture, media and values) (Carayannis & Grigoroudis, 2016). The model suggests the participation of citizens not only as collaborators, but also with autonomy to propose new types of innovation, connecting users with industry, academia and government. The role of other actors is to support civil society in its innovative activities, allowing the industrial sector and the public sector to benefit from the innovations developed by citizens (Carayannis & Grigoroudis, 2016).

The quadruple helix model emphasizes cooperation in innovation, especially in the dynamics of coopetition (collaborative competition), coevolution and co-specialization in regional or sectoral innovation ecosystems, and the European Commission recognizes this model in its smart specialization strategies (RIS3), proposing it as an evolution of the classic triple helix model (Carayannis & Grigoroudis, 2016).

The Quintuple Helix Model

In the current academic debate, it is widely accepted that an adequate solution or response to the challenge of global warming can only be found by utilizing the asset of human knowledge. And for Carayannis and Campbell (2022), the key to success lies in the use of available and newly created "knowledge" in correspondence with the Quintuple Helix Model. This model adds the environment as another central element, reflecting the urgency of aligning economic development with environmental sustainability, since the environment is not only a beneficiary of initiatives, but also an active actor that influences policy decisions and business practices.

For Carayannis et al. (2012), the Quintuple Helix Model is interdisciplinary and transdisciplinary at the same time, since the complexity of the five-helix structure implies a complete analytical understanding of all of them and requires the continuous involvement of the entire disciplinary spectrum, ranging from the natural sciences (due to the natural environment) to the social sciences and humanities (due to society, democracy and the economy). The natural environment is particularly important for the process of knowledge production and creation of new innovation, as it contributes to the preservation, survival and revitalization of humanity, enabling the creation of new green technologies; and, ultimately, humanity must learn more from nature (especially in times of climate change). With the Natural Environment Helix, "sustainable development" and "social ecology" become components of social (societal) innovation and knowledge production.

Thus, it can be considered that the Quintuple Helix model outlines what sustainable development can mean and imply for 'eco-innovation' and 'eco-entrepreneurship' in the current situation and for the future, which aligns with the concept and idea of impact entrepreneurship.

Table 1. Comparison of the three helical innovation models

Aspect	Triple Helix	Quadruple Helix	Quintuple Helix
Main Proponents	Etzkowitz & Leydesdorff (2000)	Carayannis & Campbell (2016); Carayannis & Grigoroudis (2016)	Carayannis & Campbell (2022); Carayannis et al. (2012)
Involved Helices	University – Industry – Government	University – Industry – Government – Civil Society	University – Industry – Government – Civil Society – Environment
Main Focus	Knowledge-based innovation and institutional collaboration	Inclusion of society in co-creation and innovation legitimacy	Environmental sustainability integrated with innovation and knowledge production
Core Concept	Knowledge economy and society	Knowledge democracy and citizen participation	Eco-innovation, eco-entrepreneurship, and alignment with sustainable development
Role of Civil Society	Implicit or secondary	Active and autonomous in innovation (initiator and co-creator)	Essential part, also focusing on cultural values and ecological responsibility
Role of the Environment	Not considered as a helix	Not formally included yet	Active actor influencing policies, business practices, and knowledge generation
Disciplinary Approach	Multidisciplinary (mainly social and economic sciences)	Multidisciplinary with inclusion of cultural and communication dimensions	Interdisciplinary and transdisciplinary (integrating natural, social, and human sciences)
Purpose of Innovation	Knowledge-based economic development	Inclusive and culturally informed social innovation	Sustainability, environmental regeneration, and innovation aimed at survival and ecological balance
Highlighted Applications	Innovation policies and regional development	Smart specialization strategies (RIS3) recognized by the European Commission	Solutions to global challenges such as climate change, based on knowledge and aligned with sustainability

Source: Prepared by the authors (2025)

In order to address global environmental problems, the production and circulation of knowledge is crucial for smart, sustainable and inclusive solutions, integrating the natural environment with the architecture of innovation (Barth, 2011; Carayannis et al., 2022). Knowledge, as a resource, is created through creative processes, combinations and productions in "knowledge models" or "innovation models" and thus becomes available to society.

According to Carayannis et al. (2012), there are six models of knowledge creation and creativity in innovation: (i) Mode 1, which focuses on the traditional role of university research in a linear model of innovation, where success is defined by quality approved by hierarchically established peers; (ii) Mode 2, which is characterized by five principles: (1) knowledge produced in the context of application, (2) transdisciplinarity, (3) heterogeneity and organizational diversity, (4) social responsibility and reflexivity, and (5) quality control; (iii) Triple Helix, which offers a model to explain mode 2 as an emerging historical structure for the production of scientific knowledge and its relationship with mode 1, functioning as a model of trilateral networks and hybrid organizations between university, industry and government; (iv) Mode 3, which encourages interdisciplinary thinking and the transdisciplinary application of interdisciplinary knowledge, promoting the coexistence and co-evolution of different paradigms of knowledge and innovation; (v) Quadruple Helix, which builds on the Triple Helix model, adding as a fourth helix the "public", specifically the media and culture-based society, promoting the diffusion of knowledge and supporting a knowledge society; and (vi) Quintuple Helix, expanding the Quadruple Helix, adds the natural environment, incorporating social ecology and sustainable development as an integral part of innovation processes.

These models, taken together, suggest a knowledge society where knowledge is continually processed, disseminated and used to promote sustainable development. In the Quintuple Helix model, this exchange of knowledge between subsystems—education, economy, environment, media and cultural society, and political system—seeks to harmonize human progress with nature, emphasizing the role of socio-ecological transition in sustainable development (Carayannis et al., 2012).

Figure 2. The evolution of knowledge creation models

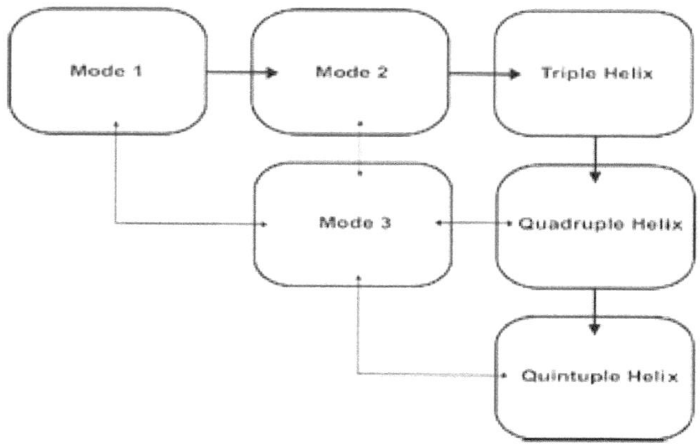

Source: Carayannis et al. (2012)

According to Gabriel et al. (2023), cross-sector collaborations inherent in the triple/quadruple/quintuple helix models are essential for successful innovation, as they allow a comprehensive understanding of diverse perspectives, facilitate knowledge exchange, offer additional opportunities (such as financing, projects, products), reduce redundant knowledge and stimulate innovation and economic performance.

The Role of Each Helix in the Collaborative Ecosystem

As already highlighted, the most important element of the Quintuple Helix model — besides the active "human agents" — is the resource of "knowledge," which, through its circulation between social subsystems, is transformed into innovation and know-how in a society and for the economy. Thus, the interaction and collective exchange of knowledge in a state (or nation-state) is visualized through the following five subsystems (or helices): (1) educational system, (2) economic system, (3) natural environment, (4) media-based public and culture (also civil society), and (5) political system (Carayannis et al., 2012). For Machado, Sartori and Rosa (2024), these five subsystems imply an analytical and continuous understanding in disciplines ranging from the natural sciences to the social sciences and humanities, through which knowledge is transformed into innovation, creating sustainable knowledge that is reflected in environmental performance and quality, based on democracy and ecology.

Figure 3. The subsystems of the Quintuple Helix model

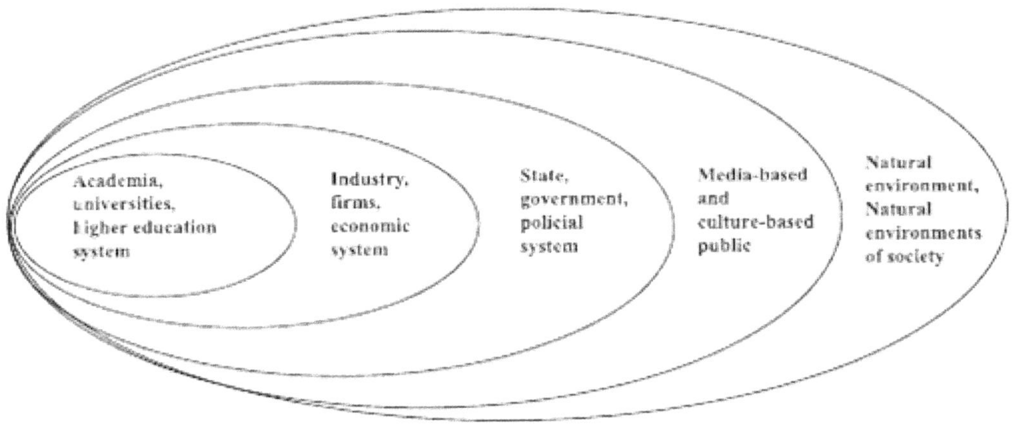

Source: Carayannis et al. (2012)

The education system refers to academia, universities, higher education systems and schools, through which the necessary "human capital" (students, teachers, scientists, academic entrepreneurs) of a state is formed through the dissemination and research of knowledge. The economic system comprises industries, companies, services and banks, concentrating and focusing the "economic capital" (entrepreneurship, machines, products, technology, money) of a state. The natural environment is crucial for sustainable development and provides people with "natural capital" (resources, plants, animal diversity). The media-based public and culture integrate two forms of "capital": on the one hand, the social capital of the cultural society (tradition, values) and, on the other, the information capital of the media-based public. The political system is important in formulating the will of where the state is headed in the present and in the future, defining, organizing and managing the general conditions. Thus, it has "political and legal capital" (ideas, laws, plans, politicians) (Carayannis et al., 2012).

To understand the dynamics of the Quintuple Helix model, Carayannis et al. (2012) consider the following steps (Figure 4):

- Step 1: When more investment is directed to the education system helix to promote sustainable development from the perspective of global warming, it creates new impulses and suggestions for knowledge creation in the education system. Targeted investment produces new equipment, creates jobs for scientists and teachers, and increases research opportunities. Thus, a greater volume of scientific and research innovations can be achieved. At the same time, education and training can improve their effectiveness. Investment in education should have a positive impact on human capital, because with more resources, education and training become more effective, allowing human capital to better seize opportunities and direct its actions. The resulting output of this human capital for greener or more sustainable development is also, in turn, an input to the economic system helix;
- Step 2: Through the input of new knowledge provided by human capital into the helix of the economic system, the value of an advanced knowledge economy increases. Through the valorization of knowledge, important production facilities and development opportunities for a future-oriented

(future-sensitive) green economy based on knowledge creation can be stimulated and realized. Not only does this knowledge create new types of jobs, green products and green services in the economic system, but it also enables new and decisive impulses for greener economic growth. In this subsystem, new values (such as corporate social responsibility) are demanded, enabling and supporting a new output of know-how and innovations by the economic system. Thus, Barth (2011) states that "*the economic know-how capital in this context is sustainability. Here, the output of economic know-how will be a high-quality and sustainable economy, but in fact the special know-how that the economic system now implies is probably a new harmony between human beings and nature*";

- Step 3: This new sustainability, as an output of the economic system, will be a new knowledge input into the natural environment helix. This new knowledge "communicates" to nature that it will be increasingly protected, as there will be less exploitation, destruction, contamination and waste. The natural environment can thus regenerate itself and strengthen its natural capital, and humanity can also learn again and even more from nature by creating knowledge. The aim of this helix should be to live in balance with nature, develop regenerative technologies and use available and finite resources sustainably and with a sensitive approach. In particular, the natural science disciplines come into play, to form new green know-how for humans and as an output of the natural environment subsystem, providing more environmental protection and a higher quality of life for people. The development of new environmentally friendly technologies can reduce CO_2 emissions more effectively and help to reduce climate change. In short, in the context of Barth (2011), the natural environment helix is "*the output of the natural environment, therefore it is green know-how*";
- Step 4: The output of the natural environment is followed by an input of new knowledge about nature and a "greener" lifestyle to the media- and culture-based audience subsystem. In this helix, it is crucial to communicate and live this lifestyle. The media-based audience is given a new and important function (information capital), which consists in disseminating, through the media, information about a new environmental awareness and a new human lifestyle. This capital should provide incentives on how a green lifestyle can be implemented in a simple, accessible and conscious way (knowledge creation), which fosters the necessary social capital of the culture-based audience, on which a society depends for sustainable development. Social capital should therefore transmit information about citizens' desires, needs, problems or satisfactions as output to the political system;
- Step 5: The knowledge input into the political system is the know-how coming from the media- and culture-based public and also represents the collective knowledge of the other three subsystems of society. Discussions about this new knowledge in political systems are necessary impulses for knowledge creation, the goal of which is a "political and legal capital" that makes the Quintuple Helix more effective, of high quality and more sustainable. Consequently, the newly acquired know-how is an output of suggestions, sustainable investments and objectives, knowledge and know-how from the political system that leads back, through the circulation of knowledge, to the educational system, the economic system, the natural environment and the media- and culture-based public.

Figure 4. The Quintuple Helix model and its functions

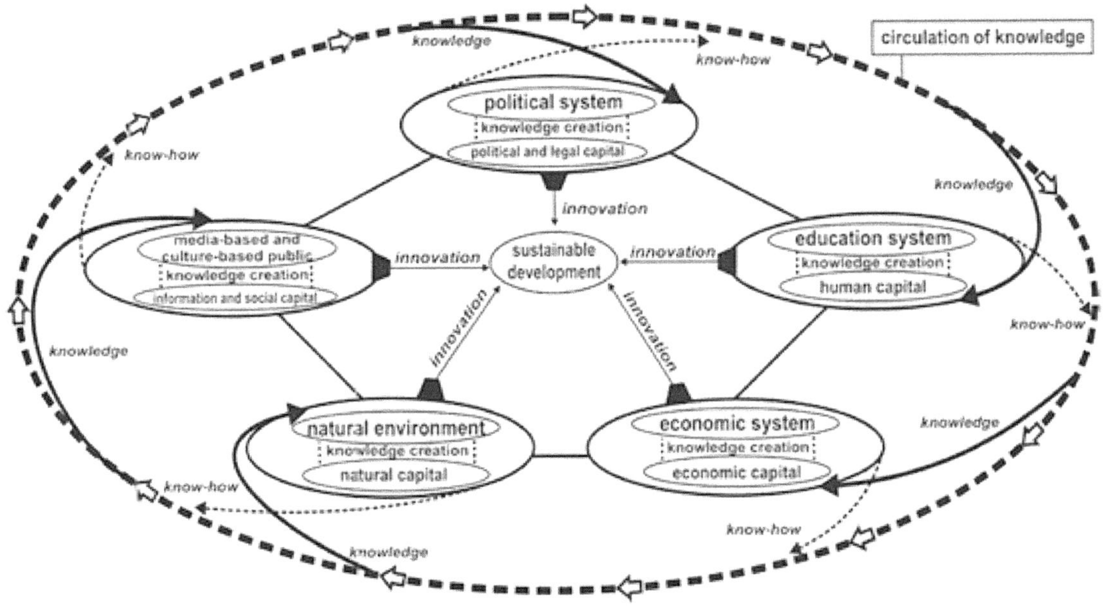

Source: Carayannis et al. (2012)

The Quintuple Helix model thus enables societies to become more resilient, especially in times of crisis, such as climate change and geopolitical tensions. The inclusion of civil society and the environment in the innovation model allows communities to be an active part of the innovation process, ensuring that the solutions developed meet a wide range of social and environmental needs, thus promoting inclusion. One of the main contributions of the Quintuple Helix model would be, precisely, the promotion of what the authors call "knowledge democracy", implying that the advancement of innovation depends on a democratic environment, where the diversity of voices and perspectives can manifest themselves. Knowledge is not considered only an economic resource, but a fundamental pillar for democratic and sustainable governance. And the interaction between the five helices promotes a knowledge governance system, where access to information, transparency and citizen participation become central to the development of innovative and sustainable solutions.

Figure 5. Knowledge production and innovation

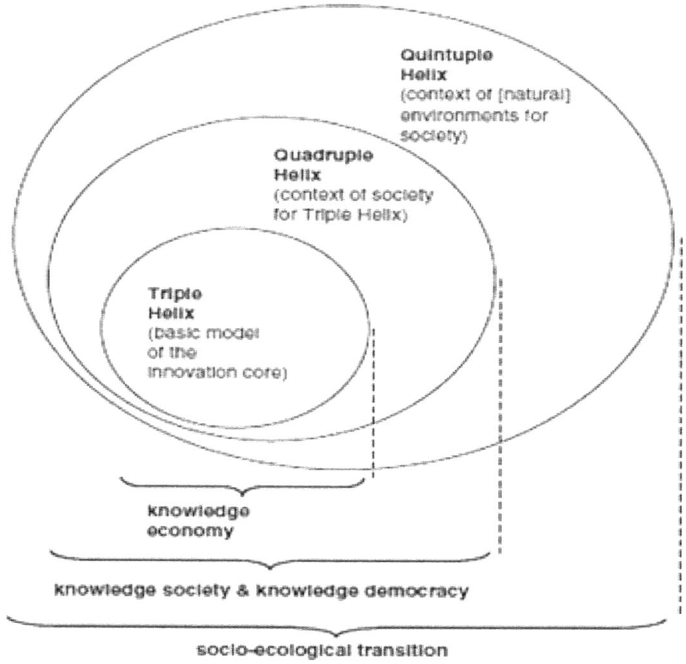

Source: Carayannis et al. (2012)

In order to demonstrate the growing interest among researchers and policy makers regarding, specifically, the Quadruple and Quintuple Helix models, since their creation and publication, Machado, Sartori and Rosa (2024) carried out a bibliometric study (Garcia et al., 2021) on the terms "quadruple helix" and "quintuple helix", resulting in 419 articles (considering the Web of Science, Science Direct, Scopus, Wiley, Emerald and IEEE databases). In total, 57 articles were selected for an integrative literature review (Chueke & Amatucci, 2021). The study identified the main publications, authors and co-authors in the field, and divided the academic production into thematic categories on the impact of these models on the democratization of innovation, regional development, technology transfer, sustainability and social innovation. Challenges and recommendations for future research are also presented, suggesting a deeper exploration of the complexity of the relationships between the actors of these models.

From the Quadruple Helix model, several themes were identified, such as (i) Democratization of Innovation (involving the integration of media and culture into the model, with an emphasis on the role of the arts and open innovation), (ii) Technology Transfer by Companies (discussing innovation in the creative industries and capacity development), (iii) Technology Transfer in Universities (including a literature review on absorptive capacity and technology commercialization challenges), (iv) Development of Localities (exploring the application of the model in medium-sized cities, rural regions and regional innovation systems), (v) Effects on Public Policy Development (analyzing how regional authorities can foster democracy and innovation in public policies) and (vi) Sustainability and Social Innovation (examining the use of the quadruple helix for green economy strategies and sustainability practices).

Regarding the Quintuple Helix model, the themes identified were: (i) Model Proposition (describing the quintuple helix as a model of ecological innovation and eco-entrepreneurship), (ii) Model Configurations (with some studies suggesting specific configurations of the quintuple model, including technology-driven green innovation), (iii) Locality Development (analyzing the use of the circular economy and the application of the model in rural areas and specific sectors, such as aviation), and (iv) Challenges in Knowledge Sharing (exploring the complexities of knowledge sharing among actors in the model).

Machado, Sartori and Rosa (2024) therefore state that the Quintuple Helix model is based on democracy and ecological sensitivity, highlighting the relevance of these factors as fundamental enablers for innovation and knowledge production, stating that a democratic and ecologically conscious society is essential for the continuous advancement of innovation, as it promotes diversity and plurality of ideas, while the environment is seen as an essential driver for sustainable innovation. Furthermore, they demonstrate that this model contrasts with static utopian visions of society, adopting the perspective of a more open, imperfect, constantly evolving society based on democratic processes, that is, a society with socioeconomic, sociotechnical and sociopolitical complexities.

Thus, the integration of the five dimensions of the Quintuple Helix, by facilitating the advancement of digital innovations that are not only efficient but also sustainable and inclusive, promotes a development that considers the environmental and social impacts of technologies, in line with the objectives of Industry 5.0 and Society 5.0, which emphasize a transformation centered on people and ecological sustainability.

Figure 6. Multilevel helical structure of the Quintuple Helix model

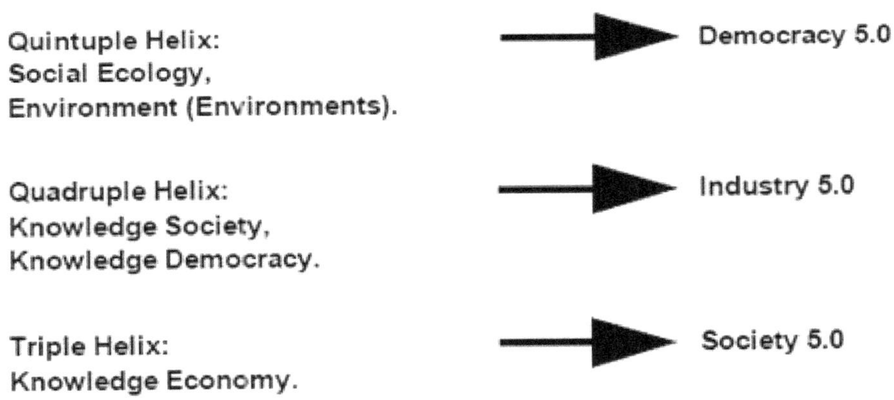

Source: Carayannis & Campbell (2022)

The Industry 5.0 and Society 5.0 models, aligned with the Quintuple Helix model, provide emerging guidelines and policies that prioritize a balanced approach between human-centricity and technology. Carayannis and Campbell (2022) suggest that adopting these guidelines will contribute to the creation of greener, more democratic and resilient economies.

The Emerging Unified Theory of Helical Architectures (EUTOHA)

When comparing the Quintuple Helix model with the Triple Helix model, Carayannis and Campbell (2022) criticize the lack of explicit mentions of democracy and ecology in the latter. It is also worth noting that in publications about the Triple Helix by Etzkowitz and Leydesdorff (2000), democracy is mentioned only in reference to "technological democracy", and the concept of ecology is not discussed, making it clear that the absence of these dimensions limits the effectiveness of the Triple Helix in the context of contemporary challenges, such as those arising from digital transformation and the impact economy.

As already demonstrated, Machado et al. (2024) identified studies that address the Quadruple and Quintuple Helix models, both separately and together. The authors analyzed the contribution of these models to sustainable development, culture, and inclusive innovation in the regions studied, and it was observed that both models promote sustainable growth and the democratization of smart technologies, providing a theoretical framework that strengthens collaboration between ecology, knowledge, and innovation. Among the various contributions arising from the authors researched by Machado et al. (2024), (i) they recommend that future studies examine how relationships evolve and how the roles of actors change over time, in addition to how the entry and exit of actors in innovation networks occurs, (ii) they reinforce the need for studies that deepen the connections and interactions between actors, (iii) they recommend continuous progress in the comparison and dialogue between the two models, seeking to build synergy between them. Another suggested line of research is the understanding of the coevolution of relationships within helices, which can be the subject of longitudinal studies. Finally, considering the small number of studies on the quintuple helix model, future research focusing on this model may be promising to improve the understanding and applicability of this system.

As a way to resolve such criticisms or gaps between the proposals of these models, Carayannis and Campbell (2022) proposed the conception of the Unified Emergent Theory of Helical Architectures (EUTOHA), a unified theory to better structure the use of helical architectures, promoting a digital transformation of knowledge-based economies, aiming to make them more democratic and sustainable. EUTOHA seeks to integrate the triple, quadruple and quintuple innovation structures in a cohesive and effective way, promoting a system where the flow of knowledge between universities, government, industry, civil society and the natural environment is optimized. For the authors, this integration would facilitate the creation of innovative solutions to complex problems, especially in relation to the digital and ecological transformation of economies and societies. Thus, this environment would become essential for the development of a knowledge economy in which innovations could be widely applied and disseminated. One guideline arising from this theory concerns the role of policymakers, who are responsible for creating environments conducive to open innovation, where the public sector and civil society play active roles in creating social and environmental value.

Carayannis and Campbell (2022) also explore the concept of "co-opetition" (competitive cooperation) in the context of the Quadruple and Quintuple Helices. In this model, actors that traditionally compete, such as companies and universities, can collaborate to achieve common goals, especially those related to sustainability and social well-being. This balance between competition and cooperation is seen as an enabler of responsible innovation and collective value creation. In line with this concept, the authors also encourage the formation of cross-sector partnerships that allow the co-creation of value, both economic and social, while simultaneously promoting technological advancement and environmental sustainability.

Figure 7. The five-fold innovation helix

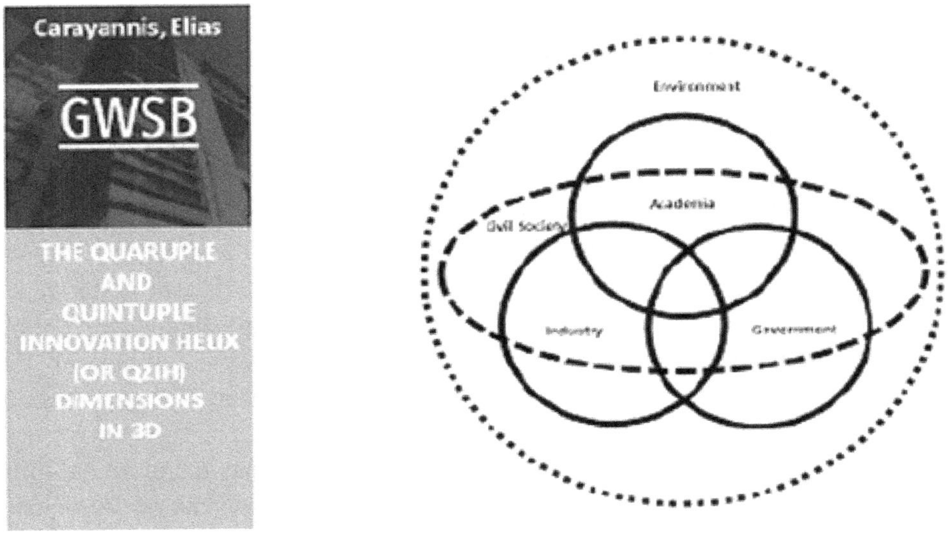

Source: Carayannis & Campbell (2022)

Carayannis and Campbell (2022) therefore propose that the Quintuple Helix model is the core of EUTOHA, serving as an essential integrating device, especially relevant for the digital and ecological transformations needed in modern democratic societies. However, EUTOHA is still an emerging theory and future research is needed to expand and validate its use in different contexts, in addition to being adapted to different regions and sectors, investigating its implications for politics, economy and society.

The Quintuple Helix Model Applied to Impact Businesses

The application of the Quintuple Helix Model is particularly relevant in impact business modeling. By integrating all stakeholders, it is possible to create a collaborative ecosystem that enhances innovation and the implementation of effective solutions.

Some practical examples corroborate its potential applicability, such as: (i) Renewable Energy Projects - initiatives such as the creation of wind farms depend on collaboration between governments (tax incentives), universities (technological research), companies (large-scale implementation), civil society (local support) and the environment (ecosystem preservation); (ii) Smart Cities - the development of solutions for urban mobility requires the integration of technologies, regulations, citizen participation and environmental considerations, such as the use of IoT sensors to optimize public transport; (iii) Sustainable Agriculture - programs that use AI and blockchain to monitor supply chains involve the contribution of all stakeholders to ensure sustainability and social inclusion.

In order to demonstrate the practical application of the Quintuple Helix Model, the following stand out: (i) in Brazil, the Amazon 4.0 Project (already mentioned in this article), an initiative that uses digital technology for environmental monitoring and sustainable development in the Amazon region, and (ii) in the European Union, the Horizon 2020 Program[18], which encourages collaborations between academia, industry and governments for technological and sustainable advancement.

The next section will discuss the need to measure the impact generated by impact businesses, as well as some of the main approaches to this.

INDICATORS AND MEASUREMENT OF IMPACT BUSINESSES

Gabriel et al. (2023) consider that impact evaluation refers to the systematic process of assessing the value or significance of the results of an intervention, such as an activity, program or policy, and that it involves an analysis of the long-term effects on the beneficiaries of the intervention — both positive and negative, primary and secondary — and the understanding of how these interventions contribute to producing an effect, with the aim of extracting valuable lessons. Thus, measuring the results inherent to impact businesses becomes essential to ensure that their activities are aligned with the objectives of creating economic, social and environmental value, allowing the assessment of not only the scope but also the depth of the changes promoted, contributing to strategic decision-making, attracting investors and increasing transparency with stakeholders.

The increased demand for sustainable and responsible initiatives has therefore led to the growing popularity of tools and frameworks to measure impact, such as the Sustainable Development Goals (SDGs) (UN, 2015), SROI (Social Return on Investment) and ESG (Environmental, Social, and Governance) (Cohen, 2022). These approaches combine quantitative and qualitative methods, providing a comprehensive view of the contributions of impact businesses to sustainable development.

Quantitative approaches are based on the collection and analysis of numerical data, offering an objective view of the impact generated, including financial metrics such as SROI, carbon emissions reduction or number of beneficiaries served by an initiative. Qualitative methodologies complement numerical data, providing deeper insights into the experiences and perceptions of stakeholders. Tools such as interviews, case studies and focus groups help to understand social and environmental impacts at the human level.

Some of the most widely used tools for measuring impact include: (i) Triple Bottom Line (TBL) (already presented in this article), which assesses an organization's performance in three dimensions: economic, social and environmental, and is a framework widely adopted by impact businesses to ensure an integrated view of their contributions; (ii) Impact Measurement Project (IMP), a framework that helps organizations identify, measure and manage their impact, ensuring alignment with global standards; (iii) GRI (Global Reporting Initiative), which offers guidelines for preparing sustainability reports, helping organizations communicate their performance in a transparent and reliable manner.

Among the most used indicators for measuring impact, there is:

- *Sustainable Development Goals (SDGs)*

The SDGs are a set of 17 global goals established by the UN (UN, 2015), which serve as a reference for businesses and governments in aligning their initiatives with global goals. Measuring impact through the SDGs allows organizations to identify priority areas and demonstrate how their actions contribute to sustainable development. Companies such as Unilever[19] use the SDGs to align their sustainability goals, such as access to clean water (SDG 6) and the promotion of gender equality (SDG 5). In addition, impact startups can link their innovations to specific SDGs, facilitating the communication of value to social investors.

- *SROI (Social Return on Investment)*

SROI is a methodology that translates the social benefits generated by a project into financial terms, allowing for a cost-benefit analysis. This approach is particularly useful for impact businesses, as it quantifies the economic return of social and environmental initiatives, facilitating the attraction of investments. Organizations that implement professional qualification programs in vulnerable communities can use SROI to measure the impact in terms of increased income and employability of beneficiaries (Cohen, 2022; Polman & Winston, 2022).

- *ESG (Environmental, Social, and Governance)*

ESG is a set of criteria that assesses the environmental, social, and governance performance of companies. This framework is widely used by investors to identify companies committed to responsible and sustainable practices. It is worth noting that companies that adopt ESG criteria often have better access to capital and attract investors aligned with socially responsible values. In addition, ESG allows organizations to continually evaluate their practices and identify areas for improvement (Cohen, 2022; Polman & Winston, 2022).

According to Gabriel et al. (2023), considering specifically the context of helix innovation models, several indices are available. For example, Carayannis, Goletsis & Grigoroudis (2018) introduce composite innovation metrics, which use multicriteria decision analysis techniques, such as the Analytic Hierarchy Process (AHP) and the Technique for Ranking Preferences by Similarity to the Ideal Solution (TOPSIS), to determine the weight of indicators and evaluate alternative options. In addition, Jovanović et al. (2022) present an efficiency index for innovation systems, based on a two-phase Data Envelopment Analysis (DEA) method, which is widely used to assess the efficiency of innovation systems.

In addition to innovation evaluation, there is an emerging field known as Social Impact Assessment (SIA), defined as "the process of identifying the future consequences of a current or proposed action as they relate to individuals, organizations, and large-scale social systems" (Gabriel et al., 2023), and whose goal is to ensure that planned developments or interventions maximize benefits and minimize costs, particularly those that fall on the community.

FINAL CONSIDERATIONS

The impact economy emerges as a comprehensive response to the urgent demands of a changing world. Integrating the generation of economic value with the resolution of social and environmental issues, this paradigm represents a break with the traditional model of capitalism, proposing solutions that are simultaneously profitable, sustainable and inclusive.

Throughout this article, it is clear that the Quintuple Helix Model, by integrating government, companies, universities, civil society and the environment, establishes itself as an essential framework for fostering collaborative innovation, and its ability to promote synergistic interactions between actors offers a solid basis for the modeling and implementation of impact businesses. Thus, its main objective of exploring how the impact economy - a new paradigm that combines the generation of economic value with social and environmental benefits - aligns with the dynamics of the digital economy is fulfilled, considering the Quintuple Helix as an innovation model that can face the existing challenges of global warming through the application of knowledge and know-how, focusing on social exchange and knowledge transfer within its subsystems.

As a non-linear innovation model that combines knowledge, know-how and the natural-environmental system in a single interdisciplinary and transdisciplinary framework, it can provide a step-by-step approach to understanding quality-based management for effective development, restoring balance with nature and enabling future generations to live in a world with plurality and diversity. In short, this article corroborates the thesis of Carayannis et al. (2012) that the Quintuple Helix model represents an adequate model in theory and practice offered to society to understand the link between knowledge and innovation, with the aim of promoting lasting development.

It is worth noting that if investments directed to any of the specific propellers of the model begin to flow, then there will be a positive impact on all other subsystems and on society as a whole, also influencing the way in which impact entrepreneurship can be understood and perceived as a condition for the applicability of the model.

The originality of this article goes beyond the presentation of helical models and the proposition of the Quintuple Helix and EUTOHA as frameworks to be considered by impact businesses. The study also manages to align its theoretical-applied findings with the area of Information Science, both due to its relationship with the determinants of data, information and knowledge generated and shared, as well as with the areas of knowledge demonstrated by Machado et al. (2024), fitting into the percentage of emphasis on Applied Social Sciences.

Figure 8. Articles on helical models, by scientific area

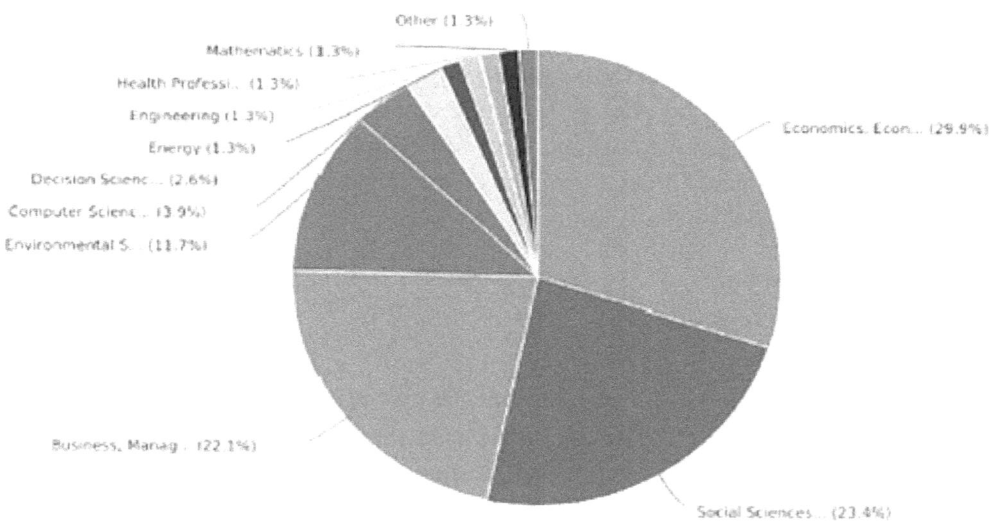

Source: Machado et al. (2024)

From a theoretical perspective, this article contributes to expanding the debate on the transition from digital transformation to the impact economy, exploring how emerging technologies and collaborative models can be used to address global challenges. The discussion on impact entrepreneurship and disruptive business models contributes to enriching the understanding of how organizations can align profitability with social and environmental benefits.

In the practical dimension, it is possible to extract clear guidelines for entrepreneurs, policymakers and other stakeholders who seek to create impact businesses. These guidelines include the adoption of specific tools, the promotion of intersectoral collaborations and the use of robust indicators such as SDGs, ESG and SROI to measure the impact generated. In addition, it is useful for researchers and postgraduate students in the areas of innovation and regional development, as well as a reference for policymakers on innovation, sustainability and territorial management.

In summary, the growing body of research on these models suggests an integrated approach to innovation that takes into account the complexity of a knowledge-based economy and the importance of ecology and democracy in sustainable development. Continued bibliometric studies can also contribute to a broader understanding of the trajectory of this expanding field.

However, there are significant challenges to be overcome. The lack of standardization in impact indicators makes it difficult to compare and evaluate results. In addition, cultural barriers and resistance to the adoption of collaborative models continue to limit the full potential of the impact economy. To mitigate these challenges, it is essential to strengthen education and awareness about the benefits of these models and create platforms that promote the exchange of good practices.

Looking ahead, technological advancement and the strengthening of global innovation networks are expected to further expand the possibilities for positive impact. The transition to a more ethical and sustainable capitalism depends on the ability of governments, businesses, universities and civil society to collaborate in an integrated manner, placing social and environmental impact at the center of economic decisions.

Finally, the relevance of the Quintuple Helix model as an enabling framework for promoting sustainable innovations is reaffirmed. By integrating diverse perspectives and considering the environment as a key actor, this model offers a promising vision for the development of concrete and sustainable solutions. It can be said that the future of the impact economy is not only a hope, but a necessity for building a more just, resilient world that is aligned with the limits of the planet.

REFERENCES

Ávila, A.P.H.L., & Gomes, D.M.O.A. (2021). Challenges and opportunities for digital transformation and society 5.0 in the post-pandemic era. Reason and word, 24(109).

B LABS. (2006). B Corporations. Available at: < https://www.bcorporation.net/en-us/what-are-b-corps/the-b-corp-declaration/>

Barth, T. D. (2011). The idea of a green new deal in a Quintuple Helix model of knowledge, know-how and innovation. *International Journal of Social Ecology and Sustainable Development*, *2*(1), 1–14. DOI: 10.4018/jsesd.2011010101

Bello Filho, W. D. B. (2010). Genesis and reproduction of the belief in the effectiveness of the micro-credit policy in combating poverty. Thesis (Doctorate in Public Policy) – Graduate Program in Public Policy, Federal University of Maranhão. 148f.

Brown, T. (2009). Change by Design: How Design Thinking Transforms Organizations and Inspires Innovation. HarperCollins e-books.

Calandra, D., Pinna, A., & Pinna, M. (2024). The impact of blockchain technology on enhancing sustainability in the agri-food sector: A scoping review. *Journal of Cleaner Production*, *456*, 142379. DOI: 10.1016/j.jclepro.2024.142379

Carayannis, E. G., Barth, T. D., & Campbell, D. F. J. (2012). The Quintuple Helix innovation model: Global warming as a challenge and driver for innovation. *Journal of Innovation and Entrepreneurship*, *1*(1), 2. DOI: 10.1186/2192-5372-1-2

Carayannis, E. G., & Campbell, D. F. J. (2011). Open innovation diplomacy and a 21st century fractal research, education and innovation (FREIE) ecosystem: Building on the quadruple and quintuple helix innovation concepts and the "Mode 3" knowledge production system. *Journal of the Knowledge Economy*, *2*(3), 327–372. DOI: 10.1007/s13132-011-0058-3

Carayannis, E. G., & Campbell, D. F. J. (2016). Quadruple and quintuple helix innovation systems. In S. Blachfellner & T. M. Werener (Eds.), Book of abstracts European meetings on cybernetics and system research Viena (pp. 25–29). Bertalanffy Center of the Study of Systems Science.

Carayannis, E. G., & Campbell, D. F. J. (2022). Towards an Emerging Unified Theory of Helix Architectures (EUTOHA): Focus on the Quintuple Innovation Helix Framework as the Integrative Device. *Triple Helix (Heidelberg)*, *9*(1), 65–75. DOI: 10.1163/21971927-bja10028

Carayannis, E. G., Dezi, L., Gregori, G., & Calo, E. (2022). Smart environments and techno-centric and human-centric innovations for industry and society 5.0: A quintuple helix innovation system view towards smart, sustainable, and inclusive solutions. *Journal of the Knowledge Economy*, *12*(2), 926–955. DOI: 10.1007/s13132-021-00763-4

Carayannis, E.G., Goletsis, Y., & Grigoroudis, E. (2018). Composite innovation metrics: MCDA and the Quadruple Innovation Helix framework. Technol. Forecast. Soc. Change, 131, 4 17. http://doi:10.1016/j.techfore.2017.03.008

Carayannis, E. G., & Grigoroudis, E. (2016). Quadruple innovation helix and smart specialization: Knowledge production and national competitiveness. *Форсайт*, *10*(1), 31–42. DOI: 10.17323/1995-459x.2016.1.31.42

Carayannis, E. G., Grigoroudis, E., Campbell, D. F. J., Meissner, D., & Stamati, D. (2018). The ecosystem as helix: An exploratory theory-building study of regional co-opetitive entrepreneurial ecosystems s Quadruple/Quintuple Helix Innovation Models. *R & D Management*, *48*(1), 148–162. DOI: 10.1111/radm.12300

Chesbrough, H. (2006). *Open Innovation: The New Imperative for Creating and Profiting from Technology*. Harvard Business School Press. DOI: 10.1093/oso/9780199290727.001.0001

Chesbrough, H. (2019). *Open Innovation Results: Going Beyond the Hype and Getting Down to Business.* OUP Oxford. DOI: 10.1093/oso/9780198841906.001.0001

Chueke, G. V., & Amatucci, M. (2021). Methods of systematization of literature in scientific studies bibliometrics, meta-analysis and systematic review. *Internext Electronic Journal of International Business*, *17*(2), 284–292.

Ciasca, D. N. (2021). Do 'triple bottom line' for the six capitals of integrated reporting. Finanças FECAP. *Journal of the Institute of Finance FECAP*, *4*(7), 16–19.

Cohen, R. (2022). *Impact: a new model of capitalism to generate real change in the world.* Matrix.

Elkington, J. (1997). Cannibals With Forks: Triple Bottom Line of 21st Century Business. Capstone Publishing, Org. Oxford.

Elkington, J. (2018). 25 years ago I coined the phrase TBL. Heres why its time to rethink it.

Etzkowitz, H., & Leydesdorff, L. (2000). The dynamics of innovation: From National Systems and "Mode 2" to a Triple Helix of university-industry-government relations. *Research Policy*, *29*(2), 109–123. DOI: 10.1016/S0048-7333(99)00055-4

Fernandes, A. C. R. D. (2022). Detection and development of new business opportunities in the context of society 5.0: A socio-technical approach [Master's thesis, Iscte - University Institute of Lisbon]. Iscte Repository. http://hdl.handle.net/10071/26457

Floridi, L. (2023). *The Ethics of Artificial Intelligence: Principles, Challenges, and Opportunities.* Oxford University Press., DOI: 10.1093/oso/9780198883098.001.0001

Friebel, N. G. (2021). Analysis of the factors that led Tesla to be a world reference in the automobile market. Course Completion Work. Undergraduate Program in Advertising, School of Communication, Arts and Design, Pontifical Catholic University of Rio Grande do Sul.

Gabriel, A., Osorio, R., & Morel, L. (2023). Quadruple-helix in university innovation programs: toward a digital tool for socio-economic impact evaluation. In: IEEE 29th International Conference on Engineering, Technology and Innovation (ICE/ITMC), University of Edinburgh, Jun 2023, Edinburgh, Scotland, United Kingdom. DOI: 10.1109/ICE/ITMC58018.2023.10332436

Garcia, A., Dariz, W. F., Tonietto, R., Torezan, G., & Ramon, M. E. G. (2021). Use of active methodologies in the teaching-learning process in health. *Proceedings of the Teaching Movements Congress*, 1, 44–48.

Global Impact Investing Network. (2022). Annual Impact Investor Survey 2022. GIIN. Available at: https://thegiin.org/research/publication/annualsurvey2022

ICE - Instituto de Cidadania Empresarial. (2018) Model C. Available at: https://www.cmodel.co/en

IIRC - International Integrated Reporting Council. (2011). Towards Integrated Reporting: communicating value in the 21st century. Integrated Reporting, (September), 17. Available at: http://www.ifrs.org/Current-Projects/IASB-Projects/Conceptual-Framework/Documents/May 2015/ED_CF_MAY 2015.pdf%0

Islam, S. M. D., & Meher, J. (2023). Blockchain, sustainability and clean energy transition. *Global Transitions*, *5*, 64–78. DOI: 10.1016/j.glt.2023.04.002

Jovanović, M., Savić, G., Cai, Y., & Levi-Jakšić, M. (2022). Towards a Triple Helix based efficiency index of innovation systems. Scientometrics, 127(5), 2577 2609. http://doi:10.1007/s11192-022-04304-x.

Lima, M. A. C., Correia, S. E. N., & Oliveira, V. M. (2024). Model C in social business modeling. *Revista de Administração Mackenzie*, *25*(4), 1–27.

Machado, H.P.V., Sartori, R., & Rosa, P.F.M. (2024). Beyond the Triple Helix Model: Scientific Production on the Quadruple and Quintuple Helix. Journal of the Knowledge Economy, July.

OECD - ORGANISATION FOR ECONOMIC CO-OPERATION AND DEVELOPMENT. (2019). Blockchain technologies as a digital enabler for sustainable infrastructure. *OECD Environment Policy Papers*, (16). Advance online publication.

Oliveira, B. X., Mafra Pereira, F. C., & Fialho, W. A. (2023). DIGITAL TRANSFORMATION: proposed theoretical-analytical model for value generation. International Journal of Innovation - IJI, São Paulo, 11(2), e23451.

Osterwalder, A., & Pigneur, Y. (2010). *Business model generation: a handbook for visionaries, game changers, and challengers*. John Wiley & Sons.

Polman, P., & Winston, A. (2022). *Positive Impact: How courageous companies thrive by giving more than they take*. Sextante.

Rogers, D. L. (2016). *The Digital Transformation Playbook: Rethink Your Business for the Digital Age*. Columbia Business School Publishing. DOI: 10.7312/roge17544

Schwab, K. (2017). *The Fourth Industrial Revolution*. Portfolio Penguin.

Sott, M.K., Baum, K. da S., & Bender, M. S. (2022). Society 5.0: exploring the dilemmas of the social ecosystem of the future. REVES - Social Relations Journal, 5(4), 14920–01e.

UN – UNITED NATIONS. (2015). Sustainable Development Goals. Available at: https://www.un.org/sustainabledevelopment/

Verhoef, P. C., Broekhuizen, T., Bart, Y., Bhattacharya, A., Dong, J. Q., Fabian, N., & Haenlein, M. (2021). Digital Transformation: A multidisciplinary reflection and research agenda. *Journal of Business Research, 122*, 889–901. DOI: 10.1016/j.jbusres.2019.09.022

Vial, G. (2019). Understanding digital transformation: A review and a research agenda. *The Journal of Strategic Information Systems, 28*(2), 118–144. DOI: 10.1016/j.jsis.2019.01.003

Vladucu, M.-V., Wu, H., Medina, J., Salehin, K. M., Dong, Z., & Rojas-Cessa, R. (2024). Blockchain in Environmental Sustainability Measures: a Survey. arXiv preprint arXiv:2412.15261.

Xu, J., Ye, M., Lu, W., Bao, Z., & Webster, C. (2021). A four-quadrant conceptual framework for analyzing extended producer responsibility in offshore prefabrication construction. *Journal of Cleaner Production, 282*(0), 124540. DOI: 10.1016/j.jclepro.2020.124540 PMID: 33052174

ENDNOTES

1 https://www.gov.br/agricultura/pt-br/assuntos/noticias/amazonia-4-0-sugere-utilizacao-da-tecnologia-para-exploracao-sustentavel-da-biodiversidade

2 https://uidai.gov.in/en/

3 https://resilient.digital-africa.co/en/blog/2024/01/30/the-state-of-digital-education-in-africa/

4 https://www.projecthelloworld.org/

5 https://www.linkedin.com/company/projects-for-all/

6 https://www.airbnb.com.br/

7 https://www.blablacar.com.br/

8 https://www.indigoag.com/

9 https://agrosmart.com.br/

10 https://www.flyzipline.com/

11 https://solarsister.org/

12 https://www.fairphone.com/nl

13 https://www.patagonia.com/home/

14 https://www.patagonia.com/home/

15 https://www.novonordisk.com/

16 https://www.natura.com.br/

17 https://www.danone.com/

[18] https://eur-lex.europa.eu/legal-content/PT/TXT/?uri=LEGISSUM:horizon_2020

[19] https://www.unilever.com/

Chapter 2
Turning Maintenance Into Competitive Advantage:
Innovation in the Analytical Chemistry Industry

Felipe De Sousa Alcantara
https://orcid.org/0009-0008-7965-3595
Dom Cabral Foundation, Brazil

ABSTRACT

The transformation described in this chapter exemplifies how an innovation strategy combined with high level of leadership and management, can completely reconfigure a company's business model. The combination of three-horizons planning, ambidextrous management and digital solutions has created a sustainable cycle of value creation, benefiting both customers and shareholders. This approach not only ensured short-term results but also laid a solid foundation for future evolution, positioning the company as a leader in its industry.

INTRODUCTION

The analytical instrumentation market encompasses a wide range of devices and systems designed to detect, measure, and analyze the physical and chemical properties of substances. These instruments are essential across numerous industrial sectors—including oil and gas, pharmaceuticals, food and beverages, petrochemicals, environmental monitoring, and life sciences, serving as critical tools to ensure quality control, regulatory compliance, safety, and operational efficiency. The market includes technologies such as gas and liquid chromatographs, spectrometers, analyzers, distillers, and various types of sensors and monitoring systems.

Given their pivotal role in industrial processes, analytical instruments must operate with high levels of precision, repeatability, and reliability. The absence of structured maintenance, proper calibration, and routine verification exposes companies to a multitude of risks and operational setbacks.

DOI: 10.4018/979-8-3693-4030-1.ch002

First, there is the risk of non-compliance with national and international regulatory standards, such as those set by ISO, ASTM, or industry-specific regulatory bodies. Inaccurate or uncalibrated equipment may lead to erroneous readings, which in turn can result in the production of non-conforming goods, environmental violations, or failure in safety-critical applications.

Second, the financial impact of undetected equipment degradation is significant. Without preventive or predictive maintenance, organizations are more susceptible to unexpected equipment failures, leading to unplanned downtime, reduced production capacity, and increased operational costs due to emergency repairs and loss of materials or product batches.

Third, inadequate measurement and control can compromise product quality, damaging a company's reputation, reducing customer trust, and ultimately affecting market share. For industries that rely on strict process controls, such as pharmaceutical or petrochemical manufacturing, the consequences of incorrect analytical data can range from loss of entire production lots to the introduction of potentially unsafe products into the market.

Finally, the absence of calibration and traceability impairs decision-making. Data-driven environments depend on the integrity of measurements to guide operational, financial, and strategic actions. Instruments that are not properly calibrated or validated produce data that lacks reliability, hindering continuous improvement efforts and increasing exposure to risk in audits and inspections.

Thus, in the context of digital transformation and growing demands for operational excellence, the implementation of robust maintenance and calibration programs in analytical instrumentation is not merely a technical requirement, it is a strategic imperative. Organizations that recognize the criticality of these practices not only ensure regulatory compliance and operational continuity, but also gain a sustainable competitive advantage by guaranteeing the quality, safety, and efficiency of their processes.

In the business world, especially in projects that involve innovation and transformation, such as the creation of a Research, Development and Innovation (RD&I) area and the change in the business model, it is essential to understand the difference and complementarity between management and leadership. Both competencies, when balanced and well applied, become essential to ensure not only the efficient execution of actions, but also to inspire and mobilize teams towards a greater goal.

As Warren Bennis, a renowned expert on leadership, explains in his book *On Becoming a Leader (Bennis, 2009, p. 42)*:

> "The manager administers; the leader innovates. The manager maintains; the leader develops. The manager relies on systems; the leader relies on people."

Warren G. Bennis (1925–2014) was one of the foremost scholars of leadership in the 20th century and is recognized as a pioneer in the field of modern organizational leadership. A psychologist by training, he earned his doctorate from the MIT Sloan School of Management and served as a professor at the University of Southern California (USC), where he founded the Leadership Institute. Bennis authored several seminal works on the distinction between management and leadership, emphasizing the importance of vision, authenticity, and human development within organizations. His contributions were instrumental in establishing leadership as an independent discipline within the field of management.

Management refers to the ability to plan, organize, and control resources, setting clear goals, monitoring indicators, and ensuring the execution of each step as planned. It is management that brings predictability and efficiency, aligning operational efforts with the expected results. It is responsible for maintaining the focus on performance and time management, ensuring that processes are followed and that deliveries happen within the established parameters.

On the other hand, leadership is the ability to engage people, create a shared purpose, and foster an environment of trust and collaboration. A visionary leader is able to connect the strategic plan to people's "hearts", translating goals into inspirations and transforming challenges into opportunities. While management seeks efficiency and control, leadership promotes engagement and resilience, being an engine of motivation and innovation in times of uncertainty.

In complex projects such as the one described in this chapter, both roles are indispensable. The accurate diagnosis, which identifies points for improvement and opportunities for innovation, requires a careful and pragmatic analysis – a typical role of management. Strategic planning and the implementation of an innovative business model, on the other hand, require leadership capable of leading the team through new horizons, with clarity of purpose and emotional intelligence to deal with resistance and uncertainties.

Throughout this process, success lies not only in achieving the established goals, but in creating a continuous cycle of learning and improvement. This is only possible when the manager understands that his role is not limited to controlling processes, but includes the ability to influence, inspire and transform. Ultimately, a good manager is one who knows how to balance the rigor of management with the inspiration of leadership, creating ambidextrous management that strengthens the culture of innovation and generates sustainable results.

Thus, all the diagnosis, planning, execution and innovation presented in this chapter show that the strength of management lies in the union of these two faculties. It is in this balance that lies the ability to transform strategic visions into concrete reality, ensuring that not only the goals are achieved, but that the trajectory leaves a legacy of continuous improvement and positive impact on the organizational environment.

In the contemporary scenario of the analytical chemistry industry, the search for sustainable competitive advantages has led organizations to rethink their traditional business models. This chapter presents a real case of organizational transformation, in which a leading company in the analytical instrumentation segment implemented an innovative maintenance management strategy, supported by digital technologies and a new service delivery model.

The central objective of this chapter is to demonstrate the power of convergence between theory and practice in innovation management. Through this real case, it is evident how concepts, methodologies and theoretical frameworks, when applied in a structured way and adapted to the specific context, produce concrete and measurable results. This bridge between academic knowledge and practical implementation is particularly relevant in the context of industry 4.0, where digital transformation demands not only theoretical understanding, but also the ability to execute and adapt to organizational realities.

The author, who served as the executive director responsible for this transformation, presents the complete journey: from identifying the initial challenges to implementing innovative solutions that have revolutionized the way the company manages its maintenance operations and relates to its customers.

The company in question, established for more than two decades in the market, specializes in the supply and maintenance of analytical instruments for laboratories and industrial processes. With a diversified portfolio serving several segments, including oil and gas (representing 70% of revenues), food and beverages, pharmaceuticals and agribusiness, the organization has established itself as a reference in its sector, maintaining an international presence in Latin America and its own laboratory structure.

The initial business model was based on the representation of foreign manufacturers, with exclusivity in the regions where it operates. This model, while successful, presented significant challenges:

- Complexity in the technical domain of various products with different specifications;

- Dependence on external suppliers for spare parts;
- High operating cost in corrective maintenance;
- Need for a specialized technical team distributed geographically;
- Long life cycle of the products, with an average of ten years;
- Complex management of more than three hundred different models of equipment.

The transformation began from a fundamental strategic reflection: how to transcend the role of a mere service provider to become a strategic partner of its customers, offering solutions that would ensure greater equipment availability and operational predictability.

This chapter details how the company developed and implemented a new maintenance management model, supported by three fundamental pillars:

- Implementation of advanced maintenance management methodologies (RAM and RCM);
- Development of proprietary predictive monitoring solutions;
- Creation of an RD&I structure focused on innovation and nationalization of components.

The relevance of this case to the innovation management literature lies in its holistic approach, which combines technical, managerial and strategic aspects to create a new business model. The results obtained demonstrate how the integration between maintenance management, technology and innovation can generate significant value for both the company and its customers.

CURRENT BUSINESS SCENARIO

The transformation faced a significant challenge related to organizational culture. With a team composed mostly of longtime employees, the company had developed deeply ingrained paradigms and practices that, while well-intentioned, limited its evolution.

An emblematic example of this dynamic manifested itself in the corporate value's interpretation of "customer passion." Paradoxically, the organization interpreted excellence in care as the ability to respond quickly to problems, rather than their systematic prevention. This view resulted in a vicious cycle: the team prided itself on its ability to "put out fires", unconsciously perpetuating an operating model that allowed these same problems to recur.

This dynamic characterized a self-fulfilling prophecy: the more the organization excelled in emergency care, the less attention was devoted to the root causes of the problems. Issues easily avoided through proactive follow-up and preventive responses continued to occur, fueling the cycle of emergencies.

The implementation of the new business model demanded, therefore, more than technical and operational transformation, it required a profound cultural change. It was necessary to reorient the understanding of "customer focus" to a preventive and proactive model, where true excellence would be measured by the ability to avoid problems, not just solve them quickly.

This cultural transformation proved to be one of the fundamental pillars to raise the business to a new level of excellence, demonstrating that the evolution of mental models precedes and sustains operational and technological changes.

In short, it would be necessary to change, to become scalable and obtain higher standards of excellence. Digital transformation would be one way.

True digital transformation is based on well-structured data and processes. When implemented strategically, it revolutionizes the way companies relate to their customers and manage their operations. In today's landscape, leading companies have come to understand: customers expect real-time interactions, on-demand deliveries, and seamless experiences across all touchpoints. Think Amazon – the simplicity of shopping in a few clicks is only possible because there is a complex digital architecture supporting every interaction. But beware: digitizing poorly defined processes is automating inefficiency. The success of digital transformation requires a solid foundation of optimized processes that, when mirrored in systems, capture essential data for management based on accurate indicators. The result? An exponential leap in the value delivered to customers, a significant reduction in the cost structure and, above all, the ability to scale the business in a sustainable way.

In this company that we use as a real example, the business model is to represent foreign manufacturers, with exclusivity in the regions where it operates. Currently there are more than sixty-five companies represented with more than three hundred different models of equipment. Despite the positive point of having a very broad portfolio, it carries the complexity of mastering several products of different specifications for a strictly technical market, both in sales and maintenance.

The equipment has a long useful life cycle, ten years on average, so we can admit that the competitive advantage is in the after-sales relationship with customers, in the recurring sale of maintenance, parts and services. It is a promising and high-tech market segment, however, even carrying in its values the passions for sustainable growth, excellence and customers first, the company does not explore its full potential for creating and generating value, for distribution to its different stakeholders (Shareholders, Customers, Suppliers, Employees, Society).

Figure 1. Evolution of Digital Transformation in different segments.

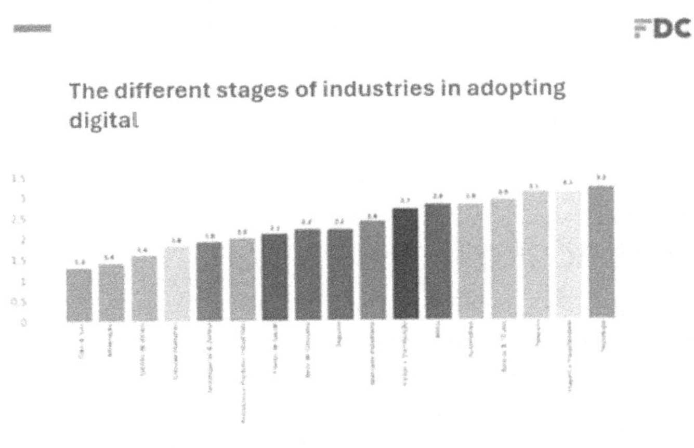

Source: Slides Fundação Dom Cabral.

The Commercial Department, when approaching a customer, uses an argumentative line based on that old maxim: "our equipment has unparalleled quality and robustness", while the most current academic interpretation teaches that the target (in this case the customer) needs, in its judgment, to perceive the proposed value - "*the judgment of the one who perceives the offer of value in fact as value*" (Salum, 2022, p. 38)[1].

Once the equipment has been sold and installed, the Service Department comes into play and an after-sales team focuses on closing a contract for the provision of preventive maintenance services with customers, however, currently, preventive maintenance represents only forty percent of the total net revenue from maintenance performed by the company.

Faced with this scenario, the following challenge was presented: *"What is impossible to do today in your organization, but if it were done it would radically change your business?" (Salum, 2022, p22)*[2].

Considering all the aspects presented in the paragraph above, the company's senior management proposed the adoption of an internal campaign, in order to promote and motivate everyone to participate in the discussion of a new Value Proposition, which seeks to solve a customer problem or satisfy a need, and each value proposition consists of a selected package of products and/or services that meet the needs of a specific customer segment, some proposals may be innovative and represent a new or disruptive offer.

Some important data:

- Fifty percent of the maintenance's cost budget are the technician's travels: air or car travel, hotel and meals;
- Only forty percent of the installed park is active with recurrent maintenance;
- One third of the services are carried out by displacement, that is, by technicians who do not live in the client's cities;
- There is no specific customer service channel, lacking clarity and generating personality in contacts when they need a service. That is, customers know some people in the company and not their process;
- Due to the high rate of correctives and vast portfolio, an assertive stock of parts becomes unfeasible, and due to the import lead time, the SLA's (Service Level Agreement) of service are long;
- Ten percent of the maintenance is done internally by the company, which has its own laboratory.

It was defined that the company's competitive advantage, which supports all strategic planning, will be excellence in services.

THEORETICAL BACKGROUND

We approach as a definitive solution the development of the sensors necessary for equipment connectivity and predictive maintenance through AI (Artificial Intelligence), however, this independent solution is just a good deed while together with the whole plan we believe that it becomes genius. It is valuable for the reader to contextualize some methodological tools used in this process from diagnosis to plan and execution, adaptable and pertinent to any business.

Strategic Maintenance Management

Strategic maintenance planning and management are fundamental pillars to create competitive advantages in the market, ensuring high availability of equipment, low operating costs and efficiency in production processes. Maintenance is often perceived as an unavoidable cost, but in fact, when structured correctly, it becomes a strategic tool that generates value for both companies and customers.

One of the main points addressed is the differentiation between the visible and invisible costs associated with maintenance. While visible costs such as labor, tools, and materials are easy to identify, invisible costs—lost production, declining product quality, trade penalties, and impacts on company image—account for the bulk of financial and operational losses. The absence of adequate planning amplifies these hidden costs, compromising the competitiveness of companies. By structuring preventive or predictive maintenance, it is possible to mitigate these risks, ensuring not only the continuity of processes, but also the predictability of performance and costs.

Another crucial point is the concept of the optimal point of maintenance, where investments in preventive actions balance the costs associated with equipment failures. The material demonstrates that, although the search for zero failure is desirable, it can be financially unfeasible, since excessive maintenance costs can outweigh the benefits generated. Identifying this break-even point is essential to maximize financial return, especially in contracts that offer guarantees of availability and performance to the customer Companies that master this practice are able to align operational efficiency with profitability, standing out in the market.

Predictive maintenance emerges as an important strategic differential. Based on data analysis and real-time monitoring of assets, this approach makes it possible to predict failures before they occur, reducing the need for high inventories and avoiding unexpected downtime. The use of technologies such as IoT sensors and predictive analytics systems significantly extends the ability to proactively predict and fix problems. This not only reduces operational costs, but also creates a higher level of service for customers, who become more trusting of the contracts and solutions offered, increasing revenue recurrence.

The need for customized maintenance policies for different types of equipment is also highlighted. Each asset has its criticality and impact on the production process, which requires a differentiated approach. While less relevant equipment can be managed with corrective maintenance, critical assets require predictive or preventive maintenance to avoid significant financial losses. This customization in maintenance planning is a key strategy for allocating resources efficiently, reducing waste and maximizing operational efficiency.

Finally, a well-structured maintenance plan directly impacts the customer experience, promoting trust and satisfaction. Ensuring the availability of equipment, reducing costs, and offering predictability in services creates a competitive advantage that is difficult to replicate. In addition, customers who perceive value in the reliability and performance guaranteed by maintenance contracts are more likely to establish long-term relationships with the company, strengthening recurring revenue and creating opportunities for new value offerings.

The synthesis on this subject is that maintenance planning and management not only optimize internal processes but also connect directly to the strategic objectives of companies. When well implemented, practices such as predictive maintenance, alignment to the optimal cost-benefit point, and customization of maintenance policies transform maintenance into a competitive advantage, ensuring sustainable results and strengthening the company's position in the market.

RAM and RCM

RAM and RCM are two modern subjects in maintenance engineering. In the broadest sense, addressing technical and managerial aspects. This is clearly a way to professionalize a technical area, acquiring authority and respect on the subject. Let's say that it was like going from being a repair shop to becoming an engineering company. In this context, these matters would need to be studied and implemented.

The integration of RAM (Reliability, Availability, Maintainability) management with Reliability-Centered Maintenance (RCM) provides a solid technical foundation for maximizing the performance of systems and equipment. This approach allows for aligning reliability, availability, and maintainability with the company's strategic objectives, ensuring high performance and resource optimization. While RAM focuses on quantifying and improving reliability metrics, RCM brings a structured analytical methodology to determine the best maintenance plan based on operational risks and impacts.

Technical components of RAM management

- Reliability:
- Measures the probability of a system or component operating flawlessly for a defined period of time, under specific conditions;
- Technique used: Reliability modeling with statistical distributions to predict failures and determine the useful life of assets;
- Practical example: Apply reliability analysis to identify the optimal frequency of preventive maintenance in a pump system.
- Availability:
- Calculates the proportion of time that a system is operational and available for use;
- Technical formula: Availability = (MTBF + MTTR) / MTTR. Where:
    - MTBF = Mean Time Between Failures
    - MTTR = Mean Time to Repair
- Practical example: Reduce maintenance time (MTTR) through standardization of procedures and use of spare parts.
- Manutenibilidade (Maintainability):
- Evaluates the ease and time required to perform maintenance on asset;
- Technique used: Design engineering for quick access and replacement of critical components and standardization of step-by-step maintenance;
- Practical example: Define kits of necessary parts and check list of execution of tasks in preventive maintenance. Or, in corrective cases, standardize the solution to be implemented for the most frequent causes of equipment downtime.

Reliability Centered Maintenance (RCM) Application

Reliability-centered maintenance provides the analytical framework for determining optimal maintenance policies. The process follows clear steps:

- Definition of the operational context:
- Identify the role of the equipment within the system and its critical functions;
- Example: In a refinery, determine that a valve controls the pressure in a gas line critical to the operation.

- Failure Modes and Effects Analysis (FMEA):
- Identify all the ways in which the asset may fail and the associated impacts;
- Example:
    - Failure mode: Corrosion on the valve surface;
    - Effect: Loss of pressure control;
    - Cause: Exposure to aggressive fluids without proper coating.
- Risk Assessment:
- Classify failure modes based on severity, frequency, and detectability;
- Tool used: Risk Matrix or Prioritization of Failure Modes.
- Definition of maintenance policies:
- Selection of preventive, predictive or corrective actions based on the criticality of the asset and the impact of the failure mode;
- Example: For the valve at risk of corrosion, implement quarterly inspections and scheduled replacement after 18 months.
- Implementation and monitoring:
- Integrate tasks into the maintenance plan and continuously monitor performance indicators.

Practical Benefits of RAM and RCM Integration

- Union of the best maintenance engineering techniques;
- Critical Failure Prevention:
- Technical: Identification of latent faults and proactive correction based on FMEA analysis and predictive monitoring.
- Example: The application of IoT sensors to monitor vibration and temperature in critical motors reduced unexpected failures by 40%.
- Asset lifecycle optimization:
- Technical: Reliability modeling to determine the optimal point of replacement of components.
- Example: Replacement of bearings at predictive intervals based on wear analyses.
- Reduction of MTTR (Mean Time to Repair):
- Technical: Use of tools such as Fault Tree Analysis (FTA) to simplify diagnostics and prioritize repairs.
- Example: Reducing the MTTR of a stationary fuel engine from 10 days to 7 days by redesigning the flow of engine disassembly, reassembly, and calibration testing.
- Improved operational safety:
- Technical: Prioritization of critical failures based on health, safety, and environmental (HSE) impact.
- Example: Implement redundancy in security systems to prevent high-severity events.

The combination of RAM management with the RCM methodology allows you to achieve exceptional levels of efficiency, security and reliability. The technical and analytical approach provides proactive asset management, which not only meets but exceeds operational and financial expectations.

By aligning advanced technology, such as IoT and predictive analytics, with robust fundamentals such as RAM and RCM, organizations can build a sustainable competitive advantage, transforming the maintenance of a cost center into a strategic value vector.

Willingness to Pay

In the current scenario, there is unofficial competition with other informal service providers based on price, not quality. Another decisive factor is the response time in the service, in which the company is at a disadvantage against small local assistance that does not burden maintenance with travel costs. By monitoring equipment remotely and anticipating possible downtime with small adjustments, the company will sell availability instead of a quick repair, increasing the perceived value and willingness to pay of customers, at a lower cost for the company and with a solution that the competition does not have.

Strategic Ambidexterity

Planning is phased in horizons and everyone is clear about their responsibilities, with the H3 team separated from the current operation.

Hart's four-quadrant model also helped in this planning of defining sustainable value creation in the short term and sustaining value creation in the long term, as we look at the internal and external environments, today and tomorrow.

The chart below summarizes the strategy for gaining efficiency in the short term, when content and data will be created for greater productivity with more intelligence in management and creation of a disruptive business model in the long term.

Figure 2. Illustrative graph of ambidextrous management for the different execution horizons.

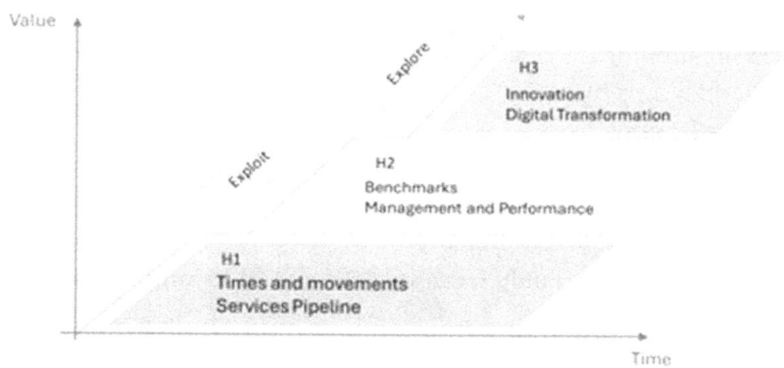

Source: Author.

Organizational Design and Open Innovation

The structure was redefined so that the focus on current results, resource allocation and operational demands would not overlook a team focused on developing the solutions that will bring the long-term competitive advantage. An ambidextrous management structure was adopted.

The RD&I department was created and to enhance its capacity, an open innovation community was also created on the Reddit platform, where problems and challenges that need to be solved are launched, encouraging with content of interest and contests, inviting key people to the community.

The current objective is to design solutions that anticipate equipment failures and therefore has the participation of all electronic technicians who provide service in the company.

Concepts and Stages of Design Thinking

As quoted by Tim Brown in the metaphor of "any child's room" (Brown, 2008), in an organization that uses Design Thinking, prototypes are seen everywhere.

The objective is not to transform the company's culture and management at this time, but to use methodological techniques to solve the proposed problem.

Prototyping does not necessarily require many expensive or technological resources, simple materials can constitute the act of "thinking with your hands". Prototyping is the essence of any creative organization and experiments with all the specifications and planning of a product before it is manufactured.

The success of a product, or a company, is to solve its customers' problems, in a better way than how it is currently executed or in an unprecedented way. In this way, we should fall more in love with the solutions than with the products themselves.

Innovation is not something technologically advanced, totally new and complex. Innovation is perceived value, an innovative product has an impact on people's lives, for this we need to deeply understand our customers, their personas, and co-create solutions. How people behave, think and decide, are culture and values, and for that there are methodologies, such as Design Thinking.

We can summarize the concepts and mindset used in the solutions we are looking for. Understand the problem through facts and data and in active and empathetic listening to people, creatively co-creating with customers, so that the product, in addition to being simple, generates value perceived by the user and trying the solutions early enough to modify or adjust them, avoiding waste.

In the very professional and competitive era we live in, we cannot make the mistake of managing a business by intuition. Design Thinking is the key to the right questions, why not call it the alchemy between pragmatic and analytical management but with a lot of creativity and empiricism.

The principles and fundamentals of Design Thinking are applicable in the approach to problem solving and approach to innovation. Translating into this application, it means meeting customer expectations, generating value for all stakeholders: customers, employees, shareholders and society, creating a new business model with competitive advantages over current market solutions.

Design Thinking, a mental work tool, an approach, a methodology or simply a set of techniques that operate holistically, making the creation process and the resulting product fascinating.

Another methodology used, HCD (Human-Centered Design), created by the author Klauss Krippendorff, which guides design as a way to meet the needs of human beings. This methodology has such importance and relationship with Design Thinking that we consider its conceptual pillars:

- Second-order understanding – when there is a concern with deeply knowing the people for whom you are projecting – persona;
- Stakeholders – consider all people impacted in the processes of creation and implementation of the artifact;
- Interfaces – ensure that the created artifact provides desirable experiences.

In the view of Tim Brown, author of the book Design Thinking and CEO of IDEO, Design Thinking cannot be created without the three lenses of HCD: desire, practicality and feasibility. The artifact, solution, must meet the expectations and solve the problems of the customers, be feasible and financially viable

HCD also stands for the acronym: hear-create-deliver, a methodology we use in the phases of understanding the problem, designing the solutions and prototyping.

The Design Thinking methodology had a great relevance in structuring the solutions described here, whose expression means: "*DESIGN THINKING IS AN APPROACH THAT CATALYZES COLLABORATION, INNOVATION AND THE SEARCH FOR SOLUTIONS THROUGH OBSERVATION AND CO-CREATION, BASED ON THE CONCEPT OF RAPID PROTOTYPING AND THE ANALYSIS OF DIFFERENT REALITIES". (Cavalcanti & Filatro, 2016)*[3]

The *HCD (Human Centered Design) Toolkit* introduces the Design Thinking methodology by presenting three lenses that help find impactful solutions to complex problems, represented below:

Figure 3. HCM diagramming.

Source: HCM article by FDC.

From an operational point of view, HCD proposes three steps: listening (understanding the problem), creating and delivering, which will be commented on in subsequent topics.

STRATEGY AND PLANNING

After leadership meetings, the strategic challenge was defined as follows: How to make the company a reference in availability and reliability, with excellent products and services, transcending the customer experience through technology, generating value for society with the preservation of the environment?

Exploit and Explore, as demonstrated in the graphical illustration of the three execution horizons. The current challenges can be defined in an equation that represents customer happiness, or for the business and value generation, the willingness to pay.

IF = IT+ MTTR + Intangible

Where IT = downtime, MTTR = mean time to repair.

The first two variables represent the finite data of the equation, the core of the company, which means serving the customer better and faster. Therefore, in the execution horizon one, one needs to master the service pipeline, which means mastering the following variables:

- Technical knowledge matrix by region, which impacts local or commuting services, a factor that incurs in the waste of technical hours (on average, two idle days per commuting service);
- Assertive parts inventory. All suppliers are foreign, the average lead time for importing parts is three months, focusing on the customer's inactive time;
- Predictions of the beginning and end of the services, with standardized MTTRs, allowing a prior planning of the agenda and fitting of the new services demanded;
- Definition, together with customers, and a schedule of preventive maintenance provided for in contracts, for prior scheduling.

All these factors incur a negative cycle of high maintenance costs, increasing the rate of corrective maintenance, which, because they are random, make it difficult to predict scheduling and the need for parts.

But even so, no matter how well the company performs, reducing downtime and gaining productivity and standardizing the benchmarks of average repair times by equipment model, it will have a cost and service time limit. The situation will reach an optimal point limit at which, for each point of improvement in meeting the SLAs, the cost will be unfeasible.

Another challenge encountered is the unauthorized and specialized service providers in the market: formed by former employees or former representatives of the brands in Brazil. It is clear that in addition to an environment of unequal competition in the relationship between quality and costs, there will be a limit to the generation of value in this business model.

The intangible part of the equation consists of finding how to increase the customer's perception of value and increase the willingness to pay. To this end, an RD&I area was established within the structure of the Services Board.

Thus, the board defined the strategic guidelines in three horizons of action to ensure competitive advantage.

The finite and immediate goals are to reduce customer service times, comprised of the downtime since the ticket was opened plus the average repair time per equipment and to reduce the average repair cost per equipment.

Another priority action was to adjust the matrix of technical skills, multidisciplinary, to increase local services without travel costs and without wasting technical daily rates, increasing the useful time of the agenda, in addition to increasing the volume of internal maintenance, previously planned, allowing better standardization and control of productivity in the services provided.

A system for digitizing the operational routine and technical reports was implemented, which allowed the qualitative analysis of the attendances, cause and effect of troubleshooting and control of the field team's displacements.

And to become an exponential organization ready to scale, in the new structure of the area an RD&I team is defined, focusing on the development of proprietary products, 3D equipment and parts, and electronic sensors that allow remote monitoring of equipment.

Therefore, the three horizons of action can be summarized as follows:

1. Improved management in the service pipeline, optimizing times and movements;
2. Productivity and quality gains with the standardization of service benchmarks and implementation of a Hotline for centralization of calls, pre-diagnostics and remote service;
3. Digital transformation of the business with on-time telemetry of equipment and remote service by video transmission.

Although the entire package of solutions makes perfect sense together, we will discuss in more detail the development of electronic sensors.

But it is appropriate to briefly discuss the creation of the Hotline, a centralized channel for technical calls.

As part of the process of innovation and continuous improvement in service management, the creation of the Hotline represented an important milestone in the transformation of the company's business model. Prior to this deployment, technical diagnostics were performed in person, which often resulted in unproductive visits, increased downtime for customers' equipment, and increased operating costs. In addition, the absence of a single channel made customers turn directly to the technicians they knew, often calling on professionals who were already allocated to other demands. This compromised the quality of service and made it difficult to manage the workflow, leaving the company without control over the volume and type of occurrences that happened with the equipment.

The implementation of the technical service channel was structured with the use of the Octágora tool, a platform that integrates video, telephone and WhatsApp messages in an efficient remote assistance environment. The Hotline consists of the allocation of an internal technician, responsible for conducting the initial services and following a standardized service trail. With this approach, all calls started to be recorded, diagnosed, and triaged in one place, allowing the technical team to travel to the customer already prepared with the necessary information to carry out resolute and quick interventions.

This process has brought numerous operational and strategic gains. Remote diagnostics have significantly reduced the need for in-person preliminary visits, increasing technicians' availability for critical care and decreasing customer downtime. In addition, the registration of occurrences on the platform provided complete visibility on the history of calls, allowing for more precise control of the number and type of failures presented by the equipment. This level of detail not only improved maintenance decision-making, but also enabled the company to take a predictive approach by identifying failure patterns and predicting future needs.

Another important benefit was the improvement in customer relationship management. Before the centralized channel, it often happened that a technician forgot to return a call or give an opinion on the progress of a call, which was not documented and compromised confidence in the service provided. With the traceability system in place, each interaction was recorded, eliminating communication gaps and ensuring that no customer was left without adequate feedback.

The centralization of services also brought valuable insights to the decision of which maintenance should be performed internally in the company. With remote diagnosis, it was possible to guide the customer to send the equipment only when the analysis indicated a need for more complex intervention.

This reduced unnecessary travel and optimized internal resources, strengthening the company's image as an efficient and reliable technical partner.

The implementation of the centralized channel with the support of the technological tool demonstrated how the intelligent use of innovation can transform the customer experience, increase operational efficiency and generate essential data for strategic management based on facts. This example reinforces that, by combining leadership with structured management, it is possible not only to ensure superior service, but also to create a continuous cycle of improvement that sustains results and reinforces the company's competitiveness in the market.

Returning to the development of electronic sensors, when analyzing the Pareto of the equipment with the highest volume of maintenance, it was identified that the Atmospheric Distiller represented twenty-five percent of all occurrences.

It is the equipment with the largest volume of installed units, it is the majority in the installed park of the company's largest customer, Petrobras, in addition to being in the critical production process of the refineries with a high impact on the financial result of the industry in cases of inadequate shutdown or calibration.

In the refining process, this equipment is in the laboratories for the analysis of the final product and connected directly to the distillation towers, where the process of separating the derivatives takes place. In the towers, the oil is heated at high temperatures until it evaporates. This vapor returns to a liquid state as it cools at different levels within the distillation tower. In each level there is a container that collects a certain by-product of petroleum, diesel, gasoline, naphtha, and kerosene. The equipment qualifies this result according to the physicochemical characteristics of the analysis.

The analysis process of the equipment is similar to the distillation tower, having a resistance that heats the product to its boiling point. It is a standard method in which a volume of a sample evaporates at a constant rate of heating. As the sample begins to evaporate, the temperature value is obtained for each percentage of evaporation of the sample. The gas is captured in a chamber and in the same way the temperature is measured for each percentage of condensation of the sample and the volume of loss for the initial volume measured. As the components of the sample have different boiling and condensation points, the composition of the sample and the fuel is obtained according to the classification of the ASTM D-86 Standard (American Society for Testing and Materials).

The first reflections led to questioning the goals they were seeking to achieve, as they could be assuming that the customers' desires were known for sure. The first answers were: they needed to serve better, with more technical quality, and faster, which would mean greatly reducing the downtime (composed of many variables, see section...) and the average time to repair (RMR). In fact, all this is sensible and necessary, but it is finite. When they reached the limit of efficiency, each percentage of improvement in the service SLA would become very expensive and unfeasible. They needed to listen to customer needs, create innovative ideas together with users, and test them before implementing.

Following the HCD methodology and starting from a strategic decision, we listened to the customers.

Making the Services and After-Sales area a competitive advantage for the company transcends correctively meeting equipment downtime in the shortest possible time.

Meetings were held with some customers in order to understand their business and challenges. Direct questions were not used, keeping in mind what to capture from these conversations to design some practical solution.

It is necessary to keep up with inductive and abductive thoughts. Starting from real cases and customer experiences to reach the conclusions that functional solutions, which do not yet exist, could be created.

Three clients were interviewed:

- Client 1 – Paulínia Refinery: the largest refinery in Latin America and a reference in operations at Petrobras, the company's largest client;
- Client 2: reference laboratory in lubricating oils;
- Customer 3: unit of the Jundiaí plant of the largest beverage company in the world.

Meetings are held at the customers' workplaces, where the equipment works in their processes, to immerse themselves in the context of the operation.

The correct personas were identified, all the relevant characteristics of the operators and decision-makers in the customers, and group meetings were held, encouraging them to tell the stories of their routines in order to fill the empathy map in the following quadrants: say, think, do and feel. By telling stories, people reveal important issues from their daily experiences, being in the work environment. The interview guide was followed to structure the meetings also using the sacrificial concept. Two hypothetical scenarios were created to better understand people's mindset in front of possible solutions: if they had two maintenance packages, one at a lower cost, in which they would have the commitment to act in a previously agreed SLA but not avoiding possible downtime; or, a package with twice the cost, in which the equipment would be remotely monitored, anticipating the dispersion in the measured quantities, interpreting the flags emitted and acting in advance avoiding equipment stoppages.

Of all the content captured in the interviews, two words stood out: availability of equipment and quality of analysis.

There was a pre-existing knowledge scenario, identifying the strategic challenge and talking to the right people.

An interview was conducted with a specialist, a SENAI professor from the schools of mechatronics and electronics engineering, who develops artifacts for remote monitoring of equipment to have a direction and define the best model and proceed with the prototyping of the solution.

New conversations were held with the customers, with a suggestion for a solution, to be in accordance with their wishes.

With the problem clearly in view, it was time to design the solutions.

According to the methodology, three steps are necessary: problem refinement, brainstorming, and evaluation of ideas.

The conclusion of the empathy map is that, by achieving 100% availability in the equipment with a perennial control of calibration measurement, that is, quality assurance in the analyses, a lot of value would be generated for customers and a new commercial business model would be created.

Therefore, the main agenda of brainstorming should answer the question: how can one first read the data from the equipment (through sensors or through the equipment's own software) and, second, interpret the data and relate it to the causes and effects of troubleshooting? Like in a car, the dashboard warns the user to take some actions and prevent the stop.

Many options have emerged and are possible:

- Develop, together with the manufacturers, direct communication of the equipment with a central for data transfer;
- Create your own communication board for the transfer of this data, log files and flags and more sensors that were important to monitor;

- During the organization of previous knowledge, one possibility would be to adapt a sensor that one of the company's engineers developed in his home, to remotely monitor the temperature, pressure and humidity in his aquarium.

Almost like a popular adage: the solutions are inside, and so are the talents.

The first conclusion is that having a proprietary solution generates more value and protects intellectual capital. Furthermore, the company had a team of engineers and electronics technicians, it already had a prototype, so the best thing would be to adapt.

Also, when creating sensors with online communication to measure the necessary parameters, a natural evolution would be to capture and transfer the data from the software as well. The vision that in the future, selling equipment updates and indications for improvement would be a business opportunity.

Now you had the starting point, you needed to know if it would work and promote an iteration cycle of the solution.

Illustration of the refinement of the problem:

Figure 4. Diagram of the refinement of the problem.

Source: Author.

It was decided not to create anything new yet, the starting point would be the sensor board that monitored the employees' aquarium.

Prototype Testing

If you are an engineer or are curious, you will understand in detail how the tests were carried out. If you have little or no technical knowledge, you can move on to the next topic.

The objective was to make it work well, learn from mistakes, resolve disagreements and start conversations about the necessary adaptations to measure the quantities of the equipment.

The research on the possibility of developing a telemetry system began with an internet search on the subject, as they were always observing blogs and forums about programming, Arduino and embedded systems, I knew that there were already solutions for this, the question was to find the simplest and most viable way to carry out a proof of concept.

The research led to an option that seemed quite simple and functional, in the tutorial the example presented sent data via Wi-Fi from a microcontroller to the cloud, using google applications. The solution consists of a program that collects the data and sends it to a google form (Google Forms), this form contains questions that will correspond to the columns of a table similar to Excel (Google Sheets).

The example presented in the tutorial sent only two parameters randomly generated by the microcontroller. To test a system more similar to what we would have in an analytical equipment, it was decided to use sensors as a data source, so a temperature, pressure and humidity sensor was used to generate the data to be sent to Google, and obtaining great results during the test. The prototype was turned on for more than a month, collecting and sending the data to a spreadsheet, which generated variation graphs of all these parameters, creating an overview of the variations and their effects on the climate.

In view of the success of the test, it began to be imagined which equipment could be connected to this system, taking into account the best possible compatibility and applicability, so, based on the installed park, a distiller seemed to be the option that would bring the most impact, in addition, it was known that the most modern version of the Distiller had diagnostic resources that would enable a minimally invasive interface between the equipment and the telemetry system, which makes installation easier and reduces the likelihood of interference from one system to another.

To build a telemetry prototype for the Distiller, it started with the studies of the manuals, in search of options for points and ways to collect data from the equipment, initially the idea was to use a diagnostic resource via serial communication that the equipment has, to collect data from various parameters, however, practical tests showed that this resource needs a manual configuration to be enabled, and does not remain active if the machine is restarted. The manufacturer was even contacted in search of a solution for this loss of configuration, but unfortunately he said that this would not be possible.

Without serial communication, a second possibility was chosen, a side door that the equipment has for connection with a diagnostic module called Dr. Box, this tool was developed by the manufacturer so that the customer could perform basic checks on the operation of the equipment's hardware, such as fuses, valve activation, pumps and resistances without the need to open the housing. It was researched throughout the manual about the specifications of this door, however unfortunately the contents are vague, so it was discovered that it had a unit and a reverse engineering process was started, to understand how Dr. Box worked.

Figure 5. Perspective of the Distiller, indicating the Doctor Box port at point E.

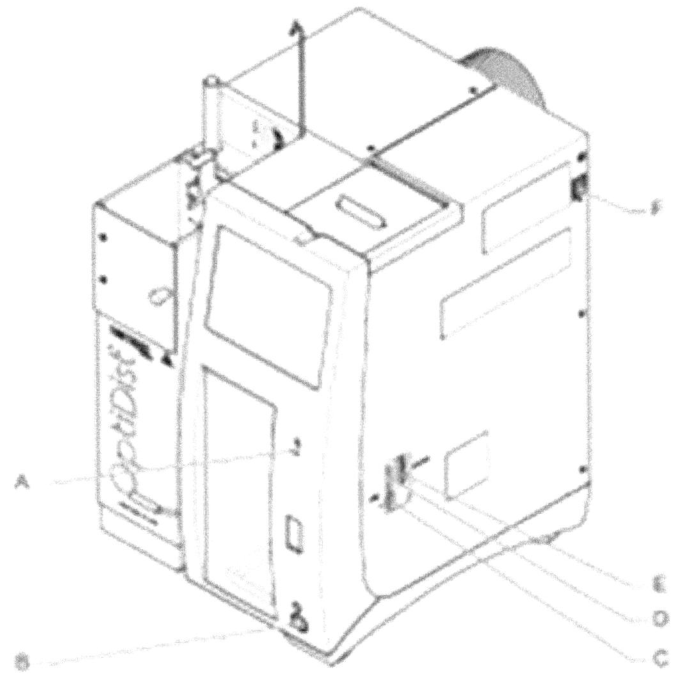

Figure 6. Illustration of the Doctor Box device.

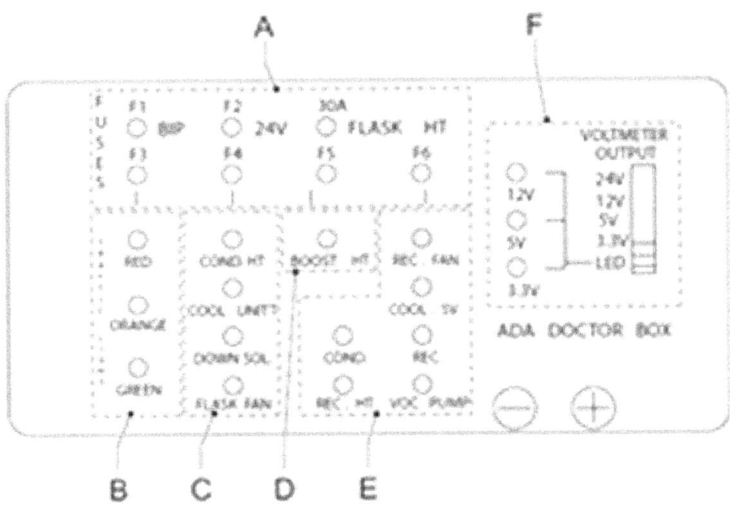

After studying the operation of Dr. Box in practice, it was possible to raise the diagram of the connector pinout on the side of the equipment, and understand how the signals that the equipment made available to indicate failures were, in this way, it was possible to develop a hardware capable of performing the interface between the Distiller and the microcontroller, so that the latter can measure the same points as Dr. Box.

The connector on the side of the equipment has twenty-five pins, through which twenty-four elements or conditions of the equipment are checked, namely:

- F6 Fuse Condition, Receiver Fan Protection, Condenser Tube Solenoid, Condenser Pump, Receiver Pump, Receiver Resistor and VOC Pump;
- F4 fuse condition, capacitor resistance protection, refrigeration unit, elevator solenoid, and balloon fan;
- F3 fuse condition, protection status indication system;
- F5 Fuse Condition, Heater Protection Boost Capacitor;
- F1 fuse condition, internal printer protection;
- Receiver fan activation;
- Activation of the condenser cooling valve;
- Receiver cooling pump activation;
- Condenser cooling pump drive;
- Activation of the receiver's resistance;
- Capacitor Resistance Activation;
- Activation of the capacitor Boost resistance;
- Supply voltage of 5V;
- Supply voltage of 12V;
- Supply voltage of 3.3V;
- 24V supply voltage / F2 fuse, 24V supply protection;
- F30 fuse condition, balloon resistance protection;
- Activation of the refrigeration unit;
- Activation of the balloon elevator solenoid;
- Activation of the balloon fan;
- Activation of the error indication light during distillation;
- Activation of the distillation indication light in progress;
- Activation of the ready for distillation indication light;
- VOC pump drive.

For the initial tests, only the pins related to the hardware items, such as pumps, valves and resistors, were connected, after being able to detect the activation of these items, the distillation status and supply voltages pins were added. To facilitate the process of identifying errors in programming more quickly, a display was also added, so that the variables could be monitored in real time, while the program is executed. The entire assembly was mounted on a standard board to facilitate in case of a future change in hardware, such as replacement of components or changes in the connection scheme of the same.

Figure 7. Image of the prototype under test, connected to the equipment.

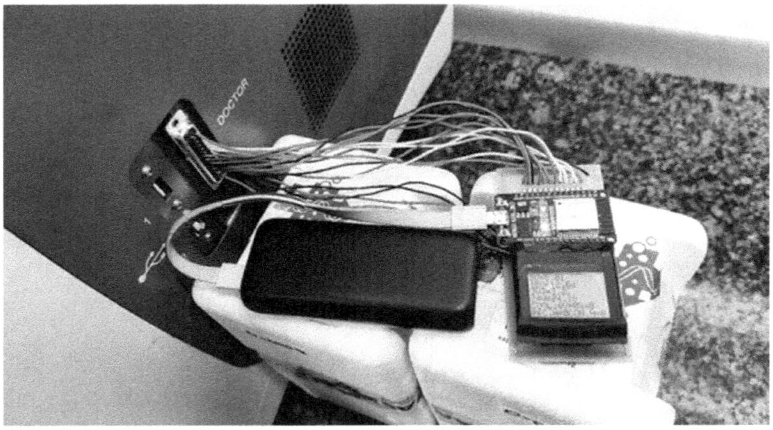

With the prototype hardware partially completed, tests with the software began, and with a few adjustments it was possible to export by Wi-Fi the data of all the parameters measured so far, namely:

- Receiver fan activation;
- Activation of the condenser cooling valve;
- Receiver cooling pump activation;
- Condenser cooling pump drive;
- Activation of the receiver's resistance;
- Capacitor Resistance Activation;
- Activation of the capacitor Boost resistance;
- Supply voltage of 5V;
- Supply voltage of 12V;
- Supply voltage of 3.3V;
- 24V supply voltage / F2 fuse. 24V supply protection;
- Activation of the refrigeration unit;
- Activation of the balloon fan;
- Activation of the error indication light during distillation;
- Activation of the distillation indication light in progress;
- Activation of the ready for distillation indication light.

At this point, the prototype was working exactly like an online version of Dr. Box, indicating the status of items with pumps and valves digitally (on or off), however, in practice, the equipment controls the power of several of these items through a system known as PWM (pulse width modulation) which consists of changing the percentage of time that a certain component remains on in a fixed interval of time (*Imagine, for example, a fixed cycle of 100 milliseconds, if you want a certain component, such as an engine, for example, to work at 50% of its power, you would keep it on for 50ms and off for another 50ms, if you wanted 25% of its power, you would leave it on for 25ms, and off for 75ms, repeating this same proportion consecutively*). With programming, it was possible to measure this pulse width of different components of the equipment hardware and extract as a percentage the power applied to these items (*the resistance of the condenser tube, for example, is activated in this way, and with the programming it was possible to convert the reading in which in the Dr. Box was expressed as on/off into an analog value, such as 23% power for example*).

Figure 8. Distiller software screen.

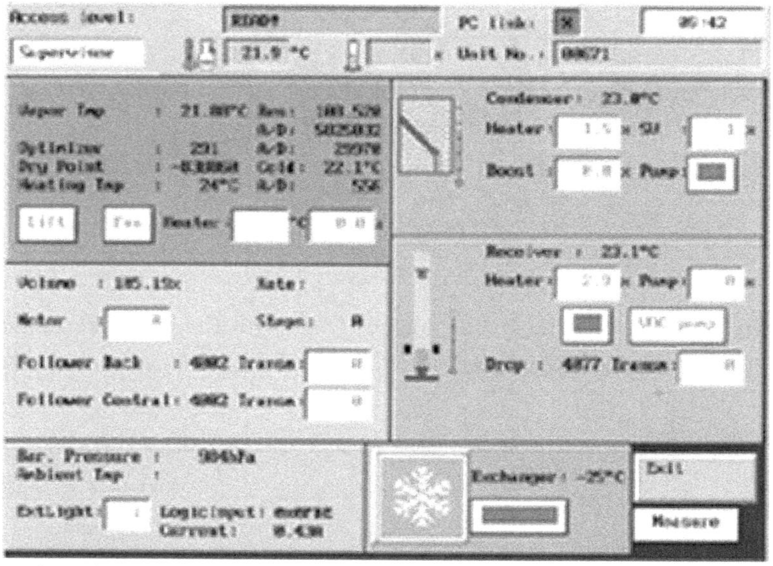

The image above shows the equipment's diagnostic screen, in yellow you have some of the points calculated by the module, which correspond to indicated values.

Figure 9. Monitoring panel created from the data received from the device.

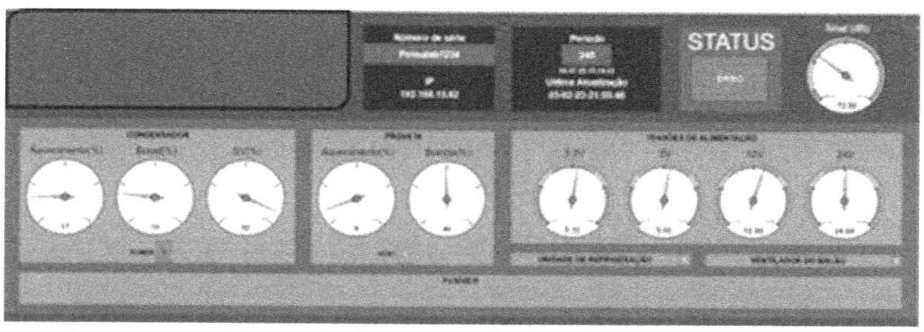

In the image, you can see the upper part of the monitoring panel, with the summary of the monitored parameters.

With the prototype connected to the equipment, tests were carried out following 3 distillations, where data were collected every 30 seconds, generating graphs of the control values of the various monitored variables. The test was satisfactory, but it was possible to notice points for improvement and improvement of the code that extracts the values of the variables and optimization of the frequency of data update, aiming to increase the intervals between readings when the equipment is at rest and increasing the frequency when it is under analysis.

Figure 10. Data received by Google Forms.

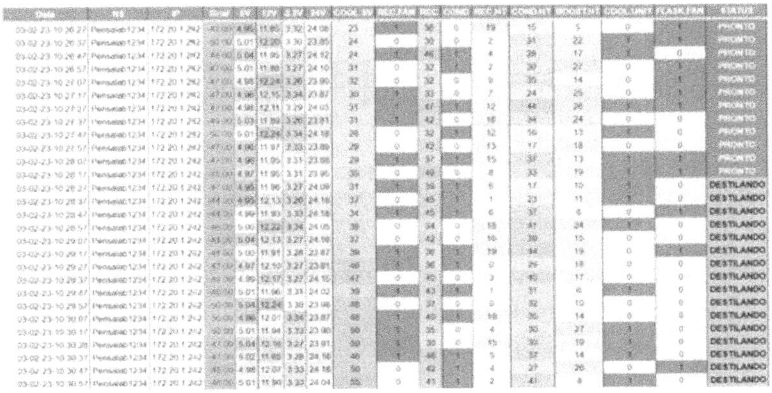

Above is an example of the data received by the form (Google Forms).

Figure 11. Data received in graphic format.

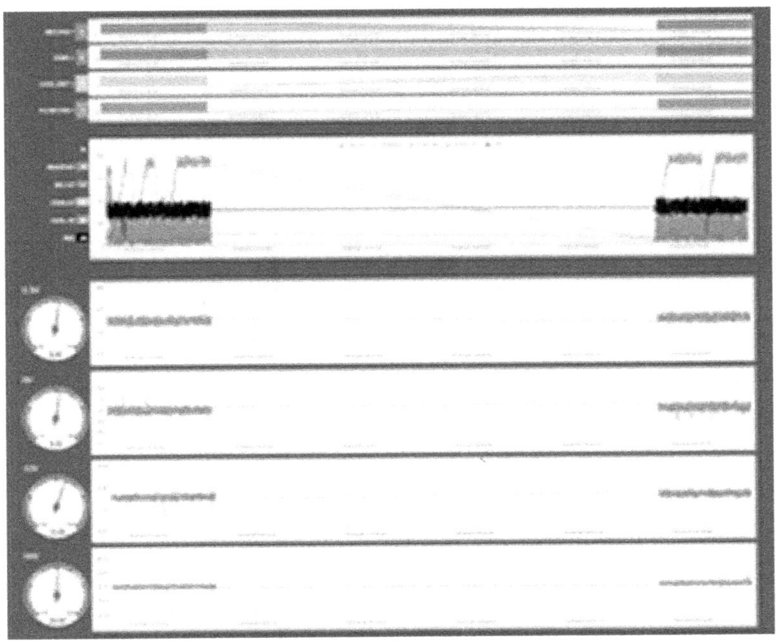

Above is an example of the data received, recorded graphically.

The next steps would be:

- Change the prototype to add fuse monitoring;
- Perform more tests on equipment to validate the correct operation of all aspects of programming, in addition to identifying (at least initially) patterns of behavior of good and problematic equipment through the graphs of the readings obtained
- Consolidate the electrical schematic of the prototype for the manufacture of printed circuit boards and plastic case (probably 3D printed);
- Contact some partner customers to probe the possibility of testing the device in their laboratories
- To "closely" monitor the data collected in the equipment in order to find patterns that indicate the decline in the performance of its components.

Pilot Project and Timelines

The project implementation plan is a document that describes relevant aspects to be considered for the implementation of the solutions created. Some figures below are examples of the tables used in the pilot project, not translated as they are original and real files.

Figure 12. Solution implementation plan.

PLANO DE IMPLEMENTAÇÃO		
Descrição das soluções a serem implementadas		
Pessoas que serão impactadas pela solução	1. Operadores dos clientes	
	2. Responsáveis por budget dos clientes	
	3. Técnicos da Pensalab	
	4. Comercial da Pensalab	
Pessoas envolvidas na implementação e manutenção da solução	Implementação	Manutenção
	1. Operadores dos clientes	1. Técnicos da Pensalab
	2. Técnicos da Pensalab	
Recursos tecnológicos e materiais a serem utilizados	Recursos já disponíveis na instituição	Recursos a serem adquiridos
	Nenhum	1. Micro processador ESP32
		2. PCB PCBWAY
		3. Resistor 10 e 15 Ohms
		4. Capacitores 100uf 16V
		5. Regulador de tensão step down
		6. Led de alto brilho tipo chapéu
		7. Conector DB25 Macho 180o
Possíveis parceiros que poderiam ser convidados a se envolver na implantação e/ou manutenção da solução	Parceiros para implementação	Parceiros para manutenção
	1. Instituto Paulista de Tecnologia	1. Técnicos próprios
	2. UFMG	2.
	3. Replan (Petrobrás)	3.
	4. Embrapa	4.

Fonte: adaptado de IDEO, 2009; D.SCHOOL, 2011

The pilot project plan is the field implementation of the selected solution.

Figure 13. Implementation plan of the Pilot Project.

PLANO DO PROJETO PILOTO		
A) Planejamento		
Descrição das soluções a serem implementadas no projeto piloto	Implantação de um protótipo do sensor de transmissão das informações eletrônicas do equipamento no cliente IPT - Instituto Paulista de Tecnologia. Após os primeiros testes, implantaremos mais 9 dispositivos para ampliar os testes.	
1. Que recursos são necessários para que cada solução que compõe a opção escolhida seja testada em um projeto piloto?	Recursos	Onde conseguir o recurso?
	1. 9 Micro processador ESP32	1. Fornecedores externos
	2. 10 PCB PCBWAY	2.
	3. 100 Resistor 10 e 15 Ohms	3.
	4. 10 Capacitores 100uf 16V	
	5. 10 Regulador de tensão step down	
	6. 50 Led de alto brilho tipo chapéu	
	7. 10 Conector DB25 Macho 180o	
2. A que questões precisamos responder durante a execução do projeto piloto?	1. Existe sinal de internet no local para conectividade?	
	2. Os dados são transmitidos ininterruptamente em intervalos de tempo pré-definidos?	
	3.	
	4.	
3. Como mediremos o sucesso do projeto piloto?	1. Se conseguirmos atualizar a base de dados continuamente	
	2.	
	3.	
	4.	
4. Como dividiremos as tarefas para implementar o projeto piloto?	Design thinker	Tarefa
	Felipe	Negociar o piloto com o IPT
	Tomaz	Instalar o dispositivo
	Tomaz	Tabulação dos dados transmitidos
	Tomaz	Análise dos dados - padrões

Simplified Schedule for the execution of the RD&I area.

Figure 14. Schedule of the execution stages for the RD&I area.

CRONOGRAMA SIMPLIFICADO DO PROJETO			
Nome da tarefa	Atribuído a	Início	Conclusão
Desenvolvimento do software (arduino)		02/01/2023	12/05/2023
Desenvolvimento do circuito		02/02/2023	14/03/2023
Fabricação das placas (PCBWAY)		14/03/2023	03/04/2023
Montagem das placas protótipo		10/04/2023	14/04/2023
Contatar IPT para verificar a possibilidade de testes iniciais		10/04/2023	14/04/2023
Verificar implicações legais para o produto final		13/04/2023	12/05/2023
Testes práticos de funcionamento e performance e montagem de base de dados inicial		24/04/2023	12/05/2023
1a reunião de apresentação do projeto à Emprapii		17/04/2023	17/04/2023
Contatar mais 9 clientes para o teste de validação		01/05/2023	19/05/2023
Montagem da modelagem do relatório técnico modelo para Emprapii - síntese de causa e efeito de troubleshooting do equipamento		18/10/2023	02/05/2023
2a reunião com a Emprapii para apresentação dos dados e avaliação do projeto		02/05/2023	12/02/2023

Cost of devices.

Figure 15. Direct costs of telemetry devices.

CUSTO PARA 10 UNIDADES DO DISPOSITIVO					
Ítem	Quantidade	Descrição	R$ Unit.	R$ Total	Frete
1	9	Micro processador ESP32	R$ 30,00	R$ 270,00	R$ 11,00
2	10	PCB PCBWAY	R$ 7,81	R$ 78,10	
3	100	Resistor 10 Ohms	R$ 0,13	R$ 13,00	R$ 11,00
4	100	Resistor 15 Ohms	R$ 0,13	R$ 13,00	R$ 11,00
5	10	Capacitores 100uf 16V	R$ 1,10	R$ 11,00	R$ 11,00
6	10	Regulador de tensão step down micro	R$ 7,97	R$ 79,70	
7	50	Led de alto brilho tipo chapéu	R$ 0,30	R$ 15,00	R$ 11,00
8	10	Conector DB25 Macho 180o	R$ 5,17	R$ 51,70	R$ 11,00
		Total	R$ 52,61	R$ 531,50	R$ 66,00

FINAL THOUGHTS

In a scenario where business competitiveness becomes increasingly fierce, the innovation area takes a leading role in the development of a new organizational model, focusing on customer experience and value generation, aiming at the perpetuity of the company and its business.

Incorporating this perception into the company's reality, the objective of this project was to implement the concept of ambidextrous organization in the company, aiming to develop a business model with a balance between the culture of innovation and operational excellence, where on the one hand there are processes focused on innovation, and on the other there are processes to keep the business running with quality. In view of this concept, a new model of service sales was developed in a predictive way for customers, with solutions that anticipate equipment failures, thus increasing their availability and operation, adding value to the product and positioning the company as a business that delivers the complete solution and not just the sale of goods.

The RD&I department was established and to enhance its capacity, an open innovation community was created on the Reddit platform, with the objective of creating and designing solutions that anticipate equipment failures, with the participation of all electronic technicians who service the equipment.

With an emphasis on increasing operational efficiency and increasing revenue through new forms of business, the Research, Development and Innovation area will bring agility to processes and solution proposals, applying the methodologies of Design Thinking and Human Centered Design.

In addition to inserting itself in the digital transformation scenario, this solution will allow the company to change its business model, rethink products and generate revenue related to After-Sales.

It will consolidate a scalable operation, easy to adapt to new scenarios and new product lines, and train its technicians more quickly, contributing to multidisciplinarity in performance.

It will generate more value for the customer, increasing the willingness to pay at a lower cost.

It will serve as a bridge for new discoveries and new developments, as a remote calibration standard, by which it will guarantee customers certificates of quality *on time* and reproducibility of results.

The implicit sale of "greater availability of equipment" and greater security in the quality of the results of the analyses to customers, will optimize their production process, becoming a target project to be presented to the key people interested: directors, CFO and CEO.

The chapter demonstrates that efficient management, based on innovation, strategic planning and inspiring leadership, has as its ultimate goal to create and deliver value to the customer. All the work of accurate diagnosis, maintenance planning with clear metrics, development of new solutions and disciplined execution with technical support is directly reflected in the salesperson's speech. The process is not just about presenting services or products, but rather reinforcing how the company is able to solve customer problems and boost its performance. This discourse becomes more powerful and assertive when supported by concrete results that show the positive impact generated.

By understanding that every step – from data collection and analysis to planned interventions and continuous improvements – contributes to the efficiency and predictability of customer operations, the seller can shape their narrative with confidence and clarity. It goes on to demonstrate how the company not only solves failures, but avoids unnecessary downtime, reduces operating costs and anticipates needs, offering personalized support and better decisions. Based on the history of the actions taken and the improvements implemented, the salesperson no longer presents an isolated technical service to deliver a consistent account of how the company monitors and supports the customer's growth over time.

In this model, the salesperson reinforces that the true value lies in the continuity and reliability of the results. It shows how the management and leadership practices described in the chapter ensure a service that evolves in an intelligent and integrated way. By presenting concrete examples of reducing downtime, improving service times and increasing asset productivity, the discourse becomes anchored in real benefits, which connect technical performance to strategic results.

More than "selling maintenance", the salesperson positions the company as a partner that delivers operational efficiency with predictability and confidence, contributing directly to the customer's business objectives. It shows that the data-driven and planning approach allows you to deeply understand the context of each customer and offer proactive solutions, ranging from optimizing the use of resources to identifying opportunities for operational improvements. With this, the customer realizes that the value offered goes beyond the simple exchange of parts or service at critical moments; It is a complete and continuous solution, which makes your operation safer, more productive and profitable.

This discourse, therefore, is based on results that reinforce how much the customer gains by having more control over their assets, greater predictability in their costs and reliable support at all stages of the life cycle of their equipment. The trust generated throughout this journey transforms the customer's perception of the price paid, increasing their willingness to pay, as they are not only hiring a technical service, but ensuring that their operation remains competitive and sustainable.

Ultimately, the salesperson becomes a strategic consultant who connects the entire narrative of the chapter into a single objective: to deliver real and sustainable value to the customer, consolidating the company as an indispensable partner for its success.

Eyes on the Future

The evolution of the data-driven maintenance and remote support model opens up a promising range of opportunities to transform technical services into a powerful platform for generating continuous value for customers and the company. With the integration of digital technical reports, telemetry, and remote service, it is possible to create a new dynamic of self-service and strategic consulting, enhancing operational and commercial results.

With detailed and centralized technical reports, we know exactly how much maintenance has been performed, the parts replaced, the causes of failures and the total downtime of each piece of equipment. This level of control allows you to identify consumption and performance patterns, as well as accurately predict the optimal time for upcoming interventions, aligning them with each client's specific MTBFs (Mean Time Between Failures). Automated management of consumables and critical parts, based on recorded inventory and historical average consumption, enables proactive replenishment, ensuring that the customer is always prepared for any maintenance without unexpected interruptions.

In addition, telemetry allows you to monitor the performance of assets in real time and detect early signs of anomalies. With the support of the remote service hotline, which uses video resources and pre-programmed guides, it is possible to offer early and effective solutions, often even before the customer realizes the need for intervention. When telemetry identifies warning signs, explanatory videos can be automatically sent to the customer, accompanied by a call from the technician to confirm the diagnosis and guide the procedure, reinforcing preventive and consultative care.

Consolidating data on maintenance histories, equipment behavior, and operator interactions creates a complete and customized dossier for each customer. This structure not only increases the predictability and efficiency of services, but also positions the company as a strategic partner, capable of offering valuable insights. For example, it is possible to identify recurring operational failures caused by human errors and recommend targeted training to correct these deviations. In specific cases, personalized videos can be sent to operators, with clear and objective instructions on how to act in certain critical situations.

The potential of a self-service platform goes beyond immediate technical support. By transforming the data collected into intelligent reports and predictive consulting, the company can offer a premium service, allowing customers to make more assertive decisions about their assets. These reports can include efficiency analyses, recommendations for operational improvements, and cost projections, creating a competitive differentiator for the company and strengthening long-term relationships with customers.

In addition, the standardization of data allows a comparative view between different units and operators, highlighting points for improvement and best practices. With this information, it is possible to build a solid and replicable knowledge base, with tutorial videos, rapid response protocols, and automated reports that anticipate corrective and preventive actions. This level of service generates trust and loyalty, positioning the company as a reference in innovation and excellence in technical service.

In summary, the centralization of information, the use of telemetry, and intelligent remote service allow not only to meet customer demands proactively and efficiently, but also to open new business fronts with consulting services, performance analysis, and personalized training. The creation of a self-service platform with detailed equipment dossiers transforms the maintenance process into a strategic management solution, generating value at different levels and consolidating the company as an indispensable partner in its customers' journey of efficiency and innovation.

REFERENCES

Bennis, W. G. (2009). *On Becoming a Leader* - The leadership classic. 4th edition. Basic Books.

Brown, T. (2008). Design thinking. *Harvard Business Review, 86*(6), 84.

Cavalcanti, C. C., & Filatro, A. (2016). *Design Thinking in face-to-face, distance and corporate education* (1st ed.).

Salum, F. (2022). *Value Creation – PDE. 72 slides. PDE Lesson, Slide 38.* FDC - Fundação Dom Cabral.

Salum, F. (2022). *Value Creation – PDE. 72 slides. PDE Lesson, Slide 22.* FDC - Fundação Dom Cabral.

ADDITIONAL RESOURCES

Mobley, K. (2014). *Maintenance Engineering Handbook* (8th ed.). McGraw-Hill Professional Publishing.

Norm, SAE JA1011. (1999). Evaluation Criteria for Reliability-Centered Maintenance (RCM). *Processes (Basel, Switzerland).*

Silva, E. V. (2024). *Article: Maintenance Costs.* Uninove University.

Smith, D. J. (2005). *Reliability, Maintainability and Risk* - Practical Methods for Engineers. 7th edition. Butterworth-Heinemann Moubray, John (1997). Reliability-Centered Maintenance. Industrial Press.

ENDNOTES

1 SALUM, Fabian; Value Creation – PDE. 72 slides. PDE Lesson, Slide 38. FDC - Fundação Dom Cabral. 07/11/2022

2 SALUM, Fabian; Value Creation – PDE. 72 slides. PDE Lesson, Slide 22. FDC - Fundação Dom Cabral. 07/11/2022

3 Text extracted from the book: CAVALCANTI, Carolina Costa; FILATRO, Andrea. Design thinking in face-to-face, distance and corporate education, 1st edition.

Chapter 3
Six Steps to Implement Innovation in Companies:
Innovation Hubs, Internal Hackathons, Innovation Communities

Thierry Cintra Marcondes
https://orcid.org/0009-0007-4923-9556
Accenture, Brazil

ABSTRACT

Innovation is critical for business survival and growth. This article presents six steps to implement innovation: mindset shift, education, exposure to new ideas, maker culture, agile methodologies, and measuring ROI. Supported by success stories from Microsoft, L'Oréal Brazil, Dyson, and Ford, it highlights how fostering innovation drives competitiveness, cost reduction, and talent retention. By embracing digital transformation and AI, businesses can future-proof operations and achieve sustainable growth.

INTRODUCTION

"Innovate or die." What once sounded like a cliché has become an undeniable truth for businesses. Organizations that do not prioritize innovation face a high risk of losing competitiveness, efficiency, and relevance in the market. This is the worst-case scenario: not only do they fail to expand their market, but they also miss the opportunity to create competitive advantages. Without innovation, profit margins shrink, customer experience deteriorates, and the ability to retain talent and offer a safe and inspiring environment is severely compromised. Even the sense of belonging — essential for engaging teams — is at risk.

The data is undeniable. According to Deloitte, companies that foster a culture of continuous learning are 92% more likely to innovate successfully. McKinsey reinforces this point, revealing that organizations with an innovative mindset are 39% more likely to outperform their competitors financially. The Global Innovation Index also highlights that companies embracing digital transformation achieve 20% better performance in innovation.

DOI: 10.4018/979-8-3693-4030-1.ch003

This raises the question: how can we implement innovation effectively and practically? Innovation implementation has evolved from ad-hoc processes to structured frameworks. Early approaches focused on R&D and technology push, while more recent models emphasize user-centric design and open innovation (Chesbrough, 2003). Today, a holistic approach is required, integrating various methodologies and considering both internal and external factors (Teece, 2010).

In this article, I present 6 proven steps to foster innovation, create an innovative culture, and build strong connections with the business ecosystem. These same steps can be applied to implement artificial intelligence and digital transformation in any organization. These steps draw upon established theories of innovation, such as Rogers' Diffusion of Innovations (Rogers, 2003) and Christensen's Disruptive Innovation (Christensen, 1997), tailored for practical application.

Moreover, I bring real-life examples of companies that have achieved extraordinary results, establishing themselves as global benchmarks in their industries. These stories demonstrate how these strategies can transform businesses and generate lasting impact.

1. MINDSET SHIFT: THE FOUNDATION OF EVERYTHING

The first step to any successful transformation is cultural change. People need to see technology and innovation not as threats but as powerful allies to achieve efficiency, solve problems, and create new opportunities. These tools not only optimize processes but also deliver better experiences within the company and to customers. This requires careful change management to address resistance (Kotter, 1996).

This is, without a doubt, the most important step in the process. Fear — of losing jobs, of privacy being invaded, or of becoming obsolete or feeling useless — can be a significant barrier. For this reason, it is essential to approach this step with patience, clear communication, and transparency to build an environment of psychological safety (Edmondson, 1999). Additionally, it is necessary to foster a continuous learning mindset, promoting autonomy and engagement.

How to Implement It?

This step varies depending on the company's level of maturity and specific characteristics, requiring a tailored approach to meet business needs. Planning is essential: create a schedule aligned with the organization's goals and identify priorities.

For example: some companies urgently need to reduce costs and increase operational efficiency, while others should focus on improving customer experience.

The mindset shift must be aligned with managers, ensuring both moral and financial support, as well as with the individual goals of each employee. It's important to note that a lack of clear objectives can stall progress, but care must also be taken to avoid creating a climate of fear.

Common Best Practices Across Most Companies

1. Education and Training
    - *Inspiring Talks:* Bring in experts to share success stories and demonstrate the positive impact of technologies.
    - *Interactive Workshops:* Organize sessions with industry leaders to engage and inspire teams.

- *Digital Literacy:* Train teams in topics like artificial intelligence, blockchain, and automation, preparing them for the future.

2. Exposure to New Ideas
    - *Visits to Innovation Hubs:* Take teams to visit startups and innovative ecosystems.
    - *Internal Hackathons:* Host events to encourage creative solutions to real problems.
    - *Participation in Innovation Communities:*
    - C-Rebels: A community that connects leaders passionate about innovation, promoting the exchange of ideas, learnings, and best practices.
    - CRI Inovação: A committee led by Fundação Dom Cabral, which presents trends, promotes networking, and discusses both success stories and failures.
3. Psychological Safety
    - Emphasize that digital transformation is a complement, not a threat, to human work.
    - Value learning and experimentation over simply avoiding mistakes. Innovative companies adopt the mantra of "testing and learning", encouraging greater creative freedom.
4. Inspirational Success Stories
    - *Microsoft and Satya Nadella:*
    - Under Nadella's leadership, Microsoft transformed from a rigid company into one with a culture of continuous learning. Initiatives like Growth Mindset and artificial intelligence training helped Microsoft surpass competitors like Google and Amazon in areas such as cloud computing.
    - *L'Oréal Brazil:*
    - During its Innovation Week, L'Oréal invested R$ 50,000 in talks, augmented reality demonstrations, and 3D printers. The event fostered a maker culture and technological experimentation, resulting in R$ 15,000 in savings within just one week.
    - At its São Paulo office, the company created a Digital Transformation and Industry 4.0 area. The initiative demonstrated that digital transformation is an ally, showcasing technologies like virtual reality, blockchain, and 3D printers. In the first week alone, a R$ 10,000 investment in the 3D printer led to immediate savings and sparked discussions about new ways of working.

Cost and Return on Investment

The mindset shift requires continuous investment. The more mature the company, the greater the financial and strategic effort required, and often the return is not immediate. Measuring the success of a mindset shift can be done through employee surveys, tracking participation in innovation initiatives, and monitoring overall employee satisfaction (Kaplan & Norton, 1992).

- *Initial Phase:* The average monthly cost is around R$ 5,000, covering materials, small tests, and talks.
- *Advanced Phase:* Investment can reach R$ 30,000 per month, especially when incorporating Design Thinking workshops, hackathon methodologies, and Startup Weekend processes. Although more expensive, these methodologies broaden team perspectives and lay the groundwork for a more robust maker culture.

Key Benefits of a Mindset Shift

- *Development of Critical Thinking:* Encourages reflection and curiosity, resulting in process improvements, new ideas, and innovative projects.
- *Familiarity with New Technologies:* Reduces fear, increases the sense of belonging, and prepares the organization for future changes.
- *Attraction and Retention of Talent:* Environments that invest in training and development retain talent longer and attract diverse profiles.
- *Integration with AI:* By fostering a culture of inquiry and exploration, teams become better equipped to work with artificial intelligence and other emerging technologies.

Transitional Summary: A mindset shift creates the fundamental cultural and psychological conditions necessary for innovation to thrive. It sets the stage for hands-on experimentation and collaborative problem-solving, which are essential components of a Maker Culture.

2. MAKER CULTURE: BUILD, TEST, AND LEARN

Maker Culture promotes hands-on experimentation, allowing teams to test hypotheses quickly, learn from mistakes, and refine their solutions based on data and results. It is an environment that values prototyping and the mindset of "testing and learning" rather than striving for perfection from the start.

By adopting this approach, companies create a virtuous cycle of innovation where failures are not seen as setbacks but as opportunities to grow. Maker Culture is not just a philosophy; it is a powerful strategic tool for transforming ideas into tangible results. This approach aligns with the principles of Agile development, emphasizing iterative processes and customer feedback (Schwaber & Beedle, 2001).

How to Implement Maker Culture

1. Creating Internal Labs
    - Set up dedicated innovation spaces equipped with practical tools like 3D printers, robots, and simulation software. These labs act as environments where new ideas can be tested without fear of failure.
    - Provide specialized training, empowering employees to use these tools productively and creatively.
    - Avoid the misconception that these spaces are merely "fun zones" or relaxation areas. Labs should be focused, creative work environments with structured rituals, such as regular brainstorming and prototyping sessions.
2. Agile Methodologies
    - Adopt practices such as Lean Startup (Ries, 2011) and Design Thinking, which encourage rapid development and validation of ideas.
    - Foster a mindset of "failing fast and cheap", or more subtly, a mindset of "testing and learning." These methodologies help teams experiment, adjust, and iterate their solutions with agility.

Inspirational Success Stories

- *L'Oréal Brazil and 3D Printers*
- In just six months, L'Oréal Brazil achieved remarkable results by adopting Maker Culture:
- Saved R$ 560,000 by producing parts for production lines in-house, avoiding external purchases.
- Reduced downtime in production lines by 35%, addressing delays caused by broken parts and replacement lead times.
- Shortened Time to Market (T2M) by 2 months, enabling faster testing, adjustments, and earlier product launches, with significant indirect gains.
- Optimized the setup of a shampoo and conditioner production line using 3D printers and color adjustments, cutting product changeover time from 3h40 to 1h20, resulting in immeasurable gains due to weekly product switches.
- One team participating in the labs came up with an innovative idea to minimize waste and reduce losses from near-expired raw materials, preventing R$ 600,000 in losses. This solution became recurring, saving R$ 60,000 per month, which was enough to cover all training costs.
- *Dyson*
- Dyson is a global example of the practical application of Maker Culture. Before launching its famous bagless vacuum cleaner, the company tested 5,127 prototypes. This success was made possible by an iterative approach focused on hands-on learning.
- *Ford*
- The automaker uses digital prototyping to accelerate vehicle development. A model that used to take 5 years to produce is now completed in 3 years, thanks to tools like digital twins and 3D printing, which have become essential to reducing costs and streamlining innovation.

Cost and Return on Investment

3D Printer as an Example

- *Initial Investment:* R$ 10,000 for the printer and R$ 300 for filaments in different materials and colors.
- *Investment Recovery:* The initial cost was recovered in just the first week due to the savings generated.
- *Additional Resources:* An investment in a specialized professional was required to manage designs, manufacturing, and continuous improvements.

Overall ROI

The implementation of Maker Culture depends on the company's maturity level and type, but this stage typically delivers excellent short- and medium-term gains, with an average ROI of 6 months to 1 year. Key performance indicators (KPIs) beyond ROI can include the number of prototypes created, time to prototype, and employee engagement in lab activities (Davila et al., 2003).

What Are the Benefits of Maker Culture?

- *A Mindset of Building and Questioning*
- Encourages employees to take a practical approach to problems, formulate hypotheses, ask questions, and seek solutions.
- Develops critical skills for integrating artificial intelligence into the organization, such as data analysis and insight-driven decision-making.
- *Agile Culture and Squad Implementation*
- Hands-on, iterative work fosters new organizational structures, such as squads and cross-functional teams.
- These setups make the company more adaptable and innovative, enabling faster deliveries and collaborative solutions.
- *Internal Cost Reductions*
- Maker Culture is one of the best strategies to rethink processes, products, and operations, generating significant savings.
- Prototyping and rapid testing reduce errors and optimize resource utilization.
- *Increased Engagement and Creativity*
- Providing tools and space for innovation boosts employee satisfaction and their sense of belonging, helping to retain talent and attract creative profiles.
- Engaged and trained teams are more likely to contribute innovative ideas and collaborate toward the company's success.

Transitional Summary: Maker Culture enhances internal capabilities for innovation. To further accelerate and broaden innovation efforts, organizations can leverage external partnerships through Open Innovation.

3. OPEN INNOVATION: EXPANDING BOUNDARIES

Open Innovation promotes collaboration with startups, universities, and other companies to accelerate solutions (Chesbrough, 2003). This approach reduces costs and enhances results by integrating external knowledge into the internal process.

How to Implement

- *Strategic Partnerships (according to the company's maturity level):*
- Startups: Work with startups to solve specific problems.
- Universities: Establish collaborations with universities to develop joint research.
- Clients and Suppliers: Collaborate with clients and suppliers to create customer-centric products and processes. This stage requires a higher level of maturity but offers significant gains.
- Competitors: Work on creating products and processes with potential competitors. The definition of competitors is changing; they used to be from the same sector, but now they compete for the customer's attention.
- *Participation in Innovation Hubs:*

- Engage with hubs like Cubo or Inovabra in São Paulo, which connect companies to a vibrant innovation ecosystem.

Success Stories

- *Procter & Gamble:*
- With its Connect + Develop program, 50% of innovations come from external sources. An example is the collaboration with startups to create advanced personal care technologies.
- *L'Oréal and the Tank Problem:*
- L'Oréal faced significant issues with the loss of tanks (containers that carry sauces) in production, resulting in high costs and delays. The company collaborated with a tank supplier and a research center to solve these problems. L'Oréal provided valuable information, a testing environment, and specialists to monitor the process, resulting in a customized solution that reduced costs and increased competitiveness.
- *BASF and AgroStart:*
- The program connects agricultural startups to BASF, accelerating the creation of sustainable solutions for agribusiness.

Cost and Return on Investment

This process does not always bring immediate ROI, but it reduces errors, enhances continuous improvement, and fosters the creation of new products. Some interesting data:

- Companies that adopt open innovation are 60% more likely to increase their revenues, according to Accenture.
- A Stanford University study shows that companies that adopt open innovation are 3.5 times more innovative than those that maintain internal innovation.
- Open innovation can reduce R&D costs by up to 30%, according to Capgemini.
- Investing in innovation hubs like Cubo and Inovabra can be expensive, but it can result in strategic partnerships and collaborations that bring significant returns. Metrics for open innovation success include the number of external partnerships, the value of co-created products, and the speed of innovation cycles (Trott, 2008).

Benefits of this Stage

- *Creation of New Products and Processes:* Can bring relevant gains in the short, medium, and long term.
- *Acceleration of Knowledge and Mindset Change:* Fundamental for mindset change and maker culture.
- *New Business and Real Economy:* Example: two cosmetic industries negotiated to have their products come with RFID from the factory, benefiting both.
- *Talent Attraction:* Contact with many people who can become talents for the company or provide consulting services.

- *Cost and Risk Reduction:* Collaborating with external partners allows sharing costs and risks, resulting in faster and more economical innovation.

Transitional Summary: The synergy of a shifted mindset, maker culture, and open innovation allows for substantial gains. It is important to celebrate quick wins to maintain momentum, demonstrate value, and boost morale.

4. QUICK WINS: SMALL VICTORIES WITH BIG IMPACT

When it comes to innovation and organizational transformation, focusing on quick wins is a powerful strategy. This moment of pause to celebrate achievements ties in with the three steps mentioned above. These initial victories are crucial for gaining traction, quickly validating ideas (especially to keep the board and supporters engaged), and most importantly, increasing the engagement of the teams involved. Additionally, they provide concrete evidence for leaders and boards, justifying the continuation and expansion of investments.

Why Quick Wins Are Important

Small victories offer an effective way to demonstrate tangible results in a short period. Instead of waiting months or years for a major final launch, quick wins allow teams to celebrate progressive advances and build confidence. According to a Deloitte study, projects that deliver quick results are 36% more likely to succeed than those that follow a long schedule without intermediate validations.

Moreover, quick wins help test and validate concepts on a smaller scale before investing significant resources. They create a continuous learning cycle: mistakes can be quickly corrected, and successes can be scaled more safely.

How to Implement Quick Wins

To effectively implement this strategy, it is important to align small victories with the company's strategic priorities. This includes clearly defining the innovation horizons (H1, H2, and H3) (Nagji & Tuff, 2012):

- H1: Incremental improvements in existing processes or the core business.
- H2: Development of new products or services that expand the company's offerings.
- H3: Disruptive or transformative initiatives that can be explored in the future.

Practical Steps to Identify and Achieve Quick Wins

- *Map Internal Opportunities:* Identify processes that can be quickly optimized, such as improvements in manufacturing, quality control, or administrative flows. These initiatives are generally linked to H1.

- *Develop Simple Prototypes:* Test ideas on a small scale before implementing larger changes. For example, conduct a pilot of a new system in a single department before expanding it to the entire organization.
- *Align Expectations with Leadership:* Ensure that expected results are clearly defined and communicated to stakeholders. This increases the visibility and impact of small victories.
- *Celebrate and Amplify Results:* When a quick win is achieved, share the learnings and impact with teams and leaders. This increases motivation and strengthens the culture of innovation.

Practical Example: Quick Wins in Internal Processes

Imagine a company that wants to reduce the manufacturing time of its products. Instead of immediately investing in an expensive automation system, it can implement a pilot in a single production line. With simple adjustments, such as reorganizing the workflow or adopting a visual control tool, the company can reduce setup time from 3 hours to 1.5 hours, generating immediate gains.

This small advance not only increases productivity but also serves as a concrete example of the positive impact that innovation can bring, strengthening the case for new investments in technology or training.

Quick Wins in the Context of Open Innovation and Maker Culture

Many quick wins can be derived from open innovation or maker culture initiatives mentioned earlier in the book. For example:

- A company can collaborate with startups to develop a prototype that solves a specific problem.
- Internal teams can use 3D printers to create replacement parts in production processes, reducing costs and machine downtime.

These initial results show that a large budget or long-term plan is not necessary to start seeing concrete benefits.

Transitional Summary: After establishing quick wins, companies should work to create a culture that embraces and celebrates diversity and creates opportunities for all.

5. DIVERSITY, EQUITY, AND INCLUSION (DE&I): INNOVATING FOR EVERYONE

Diverse companies generate more creative solutions (Hong & Page, 2004). Inclusion broadens the impact of innovation, allowing more people to benefit from changes. Diverse teams generate more creative solutions, while accessibility broadens the impact and adoption of innovations by more people. This also involves considering ethical implications, such as data privacy and algorithmic bias (O'Neil, 2016).

At this stage, we have a high level of maturity. People feel safe and included in the environment, can develop, have potential, autonomy, and tools to express their opinions, participate in building solutions, make mistakes without being reprimanded, and collaborate with various groups, stimulating diversity of thought.

This stage permeates the entire process. Depending on the company's level, it may be alongside the maker culture. However, studies and visits to companies indicate that this stage is more advanced due to the required maturity level. According to McKinsey, companies with diverse teams are 36% more likely to outperform their competitors in financial performance. The global accessibility market moves around $1 trillion per year, according to Accenture.

Success Stories

- *Nike Go FlyEase:*
- Created based on feedback from a boy with cerebral palsy, the shoe also benefited obese people and the elderly, showing how inclusion generates innovation. Nike is an example of a successful journey that led to the creation of a diversity-based council. An American boy with cerebral palsy wrote to the brand saying he had difficulty tying his shoes. Instead of just sending a letter and a gift, Nike decided to solve the problem in practice. They called the boy, a design and engineering team, and created the Go FlyEase shoe. The shoe was a success not only for him but also for obese people, the elderly, and people with other disabilities. The product's reach was so great that other products in this line were developed. In 2021, the brand created an accessibility directorate, and in 2023, opened a board position for a person focused on the topic.
- *Apple VoiceOver:*
- An accessibility tool for blind people that increased the user base and consolidated Apple's reputation as a leader in inclusion.

Cost and Return

The returns from this stage are difficult to measure, such as well-being, improved mental health, increased productivity, and greater creativity. However, investments are constant, from literacy and training, closely resembling investments in mindset change. Metrics include employee diversity rates, employee satisfaction scores, and the impact of inclusive products on new markets (et al., 2020).

Transitional Summary: The natural evolution of innovation with diversity, equity, and inclusion will lead to venture building and venture capital.

6. VENTURE BUILDER AND VENTURE CAPITAL: BETTING ON THE FUTURE

Structures like Venture Builder create internal startups, while Venture Capital finances high-risk, high-impact ideas. These structures allow companies to create internal startups, while Venture Capital provides the financial resources needed to scale ideas and create impact. This level is quite advanced, as there may be more losses than gains in the process, but when successful, the gains can be very high.

Remember, we are not talking about M&A (acquisitions), which would be another company system. We are talking about investing in startups or risky projects to enter the company's strategic portfolio, usually new products for the market that can generate revenue - the focus is on strategic gains.

How to Start

The company's structure must have the maturity to take risks, know the processes, and the board and directors must be able to take risks that can bring a lot of revenue but can also result in losses. Therefore, it is impossible to measure the return on investment, and the investments are high, in the millions.

Success Stories

- *Zé Delivery:*
- Created by Ambev through its innovation hub, ZX Ventures, Zé Delivery is a beverage delivery app that started in 2016. During the pandemic, the demand for deliveries grew exponentially, and Zé Delivery managed to scale its operations impressively. In 2020, the service went from 1.6 million annual orders to over 27 million. This was possible thanks to a series of technology and product squads that worked in various areas, from integrating new sellers to optimizing payments and fraud prevention. Zé Delivery's success was not a stroke of luck but the result of years of planning and building a robust ecosystem and startup mindset. The company managed to maintain high customer satisfaction rates and good reviews in app stores, even with rapid growth.
- *Vem de Bolo:*
- Vem de Bolo is an interesting example of a venture builder developed by Nestlé. The idea emerged in 2017 within the company's innovation hub, aiming to create a marketplace for independent cake makers and bakeries. Nestlé, in partnership with the innovation consultancy Play Studio, invested R$ 4.5 million in the project. After six months of market research, Vem de Bolo was launched, initially operating only in São Paulo. The platform allows cake makers to sell their products more conveniently, using Nestlé's scale to facilitate the purchase of supplies and delivery logistics. Vem de Bolo registered over 150 cake makers in São Paulo and sold 25,000 cakes in its first year of operation. The platform not only offers a sales channel but also helps cake makers professionalize their businesses, providing training in areas such as product photography, content creation, and pricing. This case shows how Nestlé used open innovation to explore an underserved market and create value for both the company and local entrepreneurs. The initiative caught the attention of Nestlé's global teams, standing out as a successful example of venture building.

Return on Investment

As mentioned, it is impossible to measure the return on investment at this stage, but there are several indications: According to Startup Genome, venture-building ecosystems can increase the success rate of startups from 10% to 30%. According to the book "Startup Studio," the cost of creating a startup via Venture Builder is 1.5 times lower than the traditional method, with twice the speed and greater accuracy, as it has design specialists and programmers that early-stage startups often lack. Qualitative metrics, such as the number of new ventures created and their strategic alignment with the company's goals, are also important (Block et al., 2021).

CONCLUSION

Digital transformation and innovation are no longer luxuries; they are conditions for market survival. By implementing these six steps, companies not only adapt to changes but lead disruption in their sectors. Success is reserved for those who act with strategy, courage, and long-term vision.

REFERENCES

Block, J. H., Colombo, M. G., & Cumming, D. J. (2021). Venture capital, corporate venture capital, and angel investing: Updated review and research agenda. *Venture Capital*, *23*(1), 1–27.

Chesbrough, H. W. (2003). *Open innovation: The new imperative for creating and profiting from technology*. Harvard Business School Press.

Christensen, C. M. (1997). *The innovator's dilemma: When new technologies cause great firms to fail*. Harvard Business School Press.

Davila, T., Epstein, M., & Shelton, R. (2003). *Making innovation work: How to manage it, measure it, and profit from it*. Wharton School Publishing.

Edmondson, A. C. (1999). Psychological safety and learning behavior in work teams. *Administrative Science Quarterly*, *44*(2), 350–383. DOI: 10.2307/2666999

Hong, L., & Page, S. E. (2004). Groups of diverse problem solvers can outperform groups of high-ability problem solvers. *Proceedings of the National Academy of Sciences of the United States of America*, *101*(46), 16385–16389. DOI: 10.1073/pnas.0403723101 PMID: 15534225

Kaplan, R. S., & Norton, D. P. (1992). The balanced scorecard—

Chapter 4
Analyzing the Impact of Managerial Innovation Dimensions on the Financial Performance of Moroccan Listed Companies

Mohamed Alae Lamrhary
https://orcid.org/0009-0002-8691-7540
Université Sidi Mohamed Ben Abdallah, Morocco

Samira Slaoui
FST FES, Morocco

ABSTRACT

This review article examines the impact of various dimensions of managerial innovation—namely process, organizational, strategic, and technological innovation—on financial performance, with a specific focus on Moroccan companies listed on the stock exchange. Through a systematic analysis of recent literature, the study explores how these innovation types contribute to organizational competitiveness and financial success. The findings highlight that each dimension influences financial performance uniquely, mediated by factors such as firm size, industry sector, and market environment. Specifically, the research underscores that while process and technological innovations directly enhance operational efficiency, organizational and strategic innovations are crucial for long-term adaptability and market positioning. The study further discusses implications for Moroccan managers, offering targeted recommendations for innovation practices that could optimize financial outcomes in emerging markets.

1. INTRODUCTION

In an era of increasing globalization and rapid economic transformations, managerial innovation has emerged as a crucial strategic lever for ensuring business competitiveness and sustainability. Faced with market uncertainties and constant technological advancements, organizations are rethinking their

DOI: 10.4018/979-8-3693-4030-1.ch004

management practices to enhance operational efficiency and optimize financial performance. On a global scale, studies have highlighted the key role of managerial innovation in helping firms adapt to economic shifts and evolving stakeholder expectations (Crossan & Apaydin, 2010). This type of innovation encompasses changes in organizational processes, management practices, and strategic approaches that enhance a company's competitiveness (Birkinshaw et al., 2008). However, while the benefits of managerial innovation are well-documented in developed economies, its impact in emerging markets particularly in Morocco remains an underexplored area of research.

The concept of managerial innovation originated from early studies on organizational change and strategic management. Research by Damanpour & Aravind (2012) emphasized that innovation extends beyond products and technologies to include structural and procedural transformations that enable firms to adapt to their environments. Moreover, Kafouros et al. (2008) found that managerial innovation can significantly impact profitability and corporate growth, although its effect depends on institutional contexts and the maturity of financial markets. Consequently, firms operating in dynamic environments, such as emerging markets, must develop innovative management approaches to maximize their financial performance.

However, empirical research on the relationship between managerial innovation and financial performance remains fragmented, particularly in developing economies. While numerous studies in industrialized nations have established a positive link between these two variables, few have examined this relationship in Morocco, where economic, institutional, and regulatory specificities may alter the expected outcomes. Additionally, most existing research focuses on technological innovation, overlooking critical organizational and strategic dimensions that are essential for the transformation of publicly listed Moroccan firms.

From a practical perspective, the adoption of managerial innovation presents several challenges. Moroccan firms, especially those listed on the stock exchange, face increasing demands from investors and regulators for transparency and performance. In this context, they must integrate managerial innovations to comply with new governance standards and market expectations. However, these transformations require specific resources and competencies that are not always readily available. Furthermore, the lack of a clear framework for measuring the impact of managerial innovation on financial performance remains a barrier to its widespread adoption.

This study seeks to bridge these gaps by providing a comprehensive analysis of managerial innovation dimensions and their influence on the financial performance of Moroccan publicly listed firms. Through a systematic review of the literature, this chapter examines the mechanisms through which managerial innovation enhances financial indicators such as return on investment (ROI), return on equity (ROE), and revenue growth (Martínez-Conesa et al., 2017).

Thus, this research is structured around the following central question:

How do different dimensions of managerial innovation influence the financial performance of publicly listed companies in Morocco?

By addressing this question, this chapter offers both theoretical and practical contributions, providing insights for decision-makers and researchers on leveraging managerial innovation to improve financial performance. It also outlines future directions for the integration of managerial innovation in Moroccan firms, considering local specificities and institutional challenges.

2. CONTEXT OF THE STUDY

The implementation of managerial innovation is increasingly recognized as a strategic necessity for businesses seeking to enhance competitiveness and financial performance. (SA Zahra,2017)

However, its adoption is often challenged by organizational, cultural, and institutional constraints, particularly in emerging markets such as Morocco. (C Marquis, M Raynard, 2015)

While many global firms have successfully integrated managerial innovation to optimize operational efficiency, Moroccan publicly listed companies face specific barriers that hinder the effective application of these innovations.

2.1. The Economic and Financial Landscape in Morocco

Morocco has been making significant strides in economic reforms, with a focus on modernizing its financial sector and improving corporate governance. The Casablanca Stock Exchange (Bourse de Casablanca), one of Africa's leading financial markets, plays a pivotal role in attracting investments and promoting transparency in corporate management(K Elkouiri, A Kadouri,2022). However, despite these advancements, Moroccan companies still struggle with challenges such as:

-Rigid organizational structures that slow down the adoption of new management practices(EA Fang, Q Wu, C Miao, J Xia, D Chen,2013).
-Resistance to change from employees and executives who are accustomed to traditional management methods .
-Regulatory and bureaucratic complexities that create uncertainties in implementing innovative business models (C Henning Reschke, D Bogenhold,2010).

A study by Bouazza et al. (2015) highlights that Moroccan publicly listed firms face difficulties in aligning their management strategies with international best practices due to a lack of resources, expertise, and digital transformation capabilities.

2.2. Managerial Innovation Challenges: Case Studies from Moroccan Companies

To illustrate the challenges associated with managerial innovation implementation, we examine three Moroccan companies from different sectors that have faced difficulties in applying innovative management practices (Hoskisson et al., 2011).

- **Case Study 1: Attijariwafa Bank – Digital and Process Innovation**

Attijariwafa Bank, one of Morocco's leading financial institutions, attempted to implement a digital transformation strategy to enhance operational efficiency and customer experience (LF Battanta, G Lancioni, M Giolfo,2025).

- The initiative included:

-Automation of banking processes to reduce paperwork and waiting times (A Adewumi, SE Ewim, NJ Sam-Bulya,2024).
-Introduction of AI-powered customer service for better engagement(BN Kaluarachchi, D Sedera,2024).
-Expansion of mobile banking services to reach unbanked populations (A Lema – Inkany-iso,2017) .

- Challenges encountered:
  -Employee resistance due to fear of job loss with automation.
  -Cybersecurity concerns that required significant investment in IT infrastructure.
  -Regulatory constraints in adapting digital innovations to Moroccan banking laws (A Allioui, H Allioui,2023) .

Despite these challenges, the bank successfully integrated process innovations by implementing extensive employee training programs and collaborating with regulators to ensure compliance with evolving digital banking standards (O Moromoke, O Aro, A Adepetun, O Iwalehin – 2024).

- **Case Study 2: Maroc Telecom – Organizational and Strategic Innovation**

Maroc Telecom, the largest telecommunications provider in the country, sought to restructure its management hierarchy and introduce more agile decision-making processes to improve efficiency (C Abdelfatah, 2018).

- The company aimed to:
  -Decentralize decision-making to enable regional managers to respond to market demands faster (VM Papadakis, S Lioukas,1998).
  - Implement performance-based incentives to encourage innovation at all levels(K Sanders, F Jorgensen, H Shipton,2018) .
  - Foster a culture of continuous learning by investing in leadership training.
- Challenges encountered:
  -Bureaucratic resistance, as top management was reluctant to cede control (KJ O'brien, 1996).
  -Difficulties in measuring performance due to a lack of standardized KPIs (T Haponava, S Al-Jibouri, 2012).
  - Employee skepticism regarding the fairness of the incentive system (D Sliwka, 2007) .

Over time, Maroc Telecom overcame these obstacles by gradually phasing in changes, increasing internal communication efforts, and ensuring transparency in performance evaluations (C Abdelfattah, 2018).

- **Case Study 3: OCP Group – Technological and Strategic Innovation**

The OCP Group, a global leader in the phosphate industry, aimed to integrate sustainability-driven innovation into its operations.

- The company adopted:

-AI-driven logistics optimization to reduce transportation costs (N Neagu, K Dorer, D Greenwood, 2006)
-Smart supply chain management to minimize waste and improve efficiency (H Winkler, B Kaluza, 2006)
-Digital monitoring systems to enhance sustainability and compliance.

- Challenges encountered:
  -High initial investment costs in advanced digital technologies.
  -Lack of skilled workforce to operate AI-based logistics systems.
  -Global market volatility affecting innovation financing.

OCP Group addressed these issues by partnering with international tech firms and local universities to develop workforce training programs and secure strategic funding for long-term sustainability innovations.

2.3. Key Challenges of Managerial Innovation in Moroccan Firms

From these case studies, several overarching challenges in the implementation of managerial innovation in Morocco can be identified:

- Cultural Barriers and Resistance to Change
  -Many firms in Morocco operate under hierarchical management structures, making it difficult to implement decentralized or agile management approaches (H Hammouch, O Manta, M Palazzo,2024).
  -Employees often fear job displacement with the introduction of automation and new technologies (DA Spencer, 2018).
- Regulatory and Institutional Constraints
  -Government policies and labor laws sometimes slow down the adoption of new managerial practices (C Ichniowski, K Shaw, RW Crandall, 1995).
  -Compliance requirements add complexity to strategic restructuring initiatives (D Jerab, T Mabrouk, 2023)
- Resource Limitations and Financial Constraints
  -Many publicly listed companies lack the necessary financial and human resources to experiment with innovative business models (R Bolton, M Hannon, 2016).
  -Access to funding for large-scale managerial innovation projects remains limited, particularly for firms outside the banking and telecommunications sectors (AL Delbecq, PK Mills, 1985).
- Limited Digital Readiness
  -While some industries, such as finance and telecom, are making progress, many Moroccan firms still lack advanced IT infrastructure to support digital transformation(J Sundbo, 1997)
  -Cybersecurity and data protection concerns further complicate the adoption of technology-driven managerial innovations (C Windahl, 2015).

2.4. Addressing the Challenges: Research Contributions and Future Directions

Given the complexities surrounding managerial innovation in Morocco, this study seeks to:

- Analyze how different dimensions of managerial innovation (process, organizational, strategic, and technological) impact financial performance S Gopalakrishnan,2000
- Identify the key factors that influence the successful implementation of managerial innovation in Moroccan publicly listed companies. S Abdallah-Ou-Moussa, MG Wynn,2025
- Provide actionable insights for decision-makers and business leaders to navigate the challenges of innovation adoption. J Nedeljkovic, T Dim - 2024

By exploring these issues through a systematic literature review, this research will help fill the knowledge gap in understanding the link between managerial innovation and financial outcomes in Morocco. S Achli, V Ongoma, TE Epule, D Dhiba, W Salih., 2025 .Furthermore, it will offer recommendations for future researchers and policymakers, enabling them to formulate strategies that facilitate the adoption of effective and sustainable managerial innovations. S Wiltsey Stirman, J Kimberly, N Cook, A Calloway, 2012

The Moroccan business environment presents a unique set of challenges and opportunities for the adoption of managerial innovation. H Rachidi, M El Mohajir, 2021

While some firms have made progress in process, organizational, and strategic innovations, they often face barriers related to resistance to change, regulatory constraints, and resource limitations. A Hadjimanolis, 2003

By studying real-world cases and systematically reviewing existing literature, this research will contribute to a better understanding of the relationship between managerial innovation and financial performance in Morocco. F El Fallahi, A Ibenrissoul,2023

It will also provide practical guidance for companies looking to navigate the complexities of innovation adoption, ultimately strengthening Morocco's position as a competitive and innovative economy. I Boulaid, M Ennadi,2025

3. STATEMENT OF THE PROBLEM

3.1. Ideal State: Toward Innovative and High-Performance Management

In a globalized economy characterized by intensifying competition and rapid technological advancements, companies strive to optimize their financial performance while enhancing operational efficiency. ME Porter - 2008 The ideal state assumes that managerial innovation which includes organizational, process, strategic, and technological changes is systematically integrated into businesses to maximize competitiveness M Tushman, D Nadler, 1986. Ideally, publicly listed companies in Morocco should adopt these innovative management practices, adapt them to their local context, and achieve measurable benefits in profitability, growth, and investor attractiveness. Abetti, P., Ben Rayana, N., Durgee, J., El Abassi, M., & Hasan, I. (2004)

However, in reality, several gaps and challenges hinder this smooth transition to an innovative and high-performance management model. John Wiley & Sons, Inc. (2004)

3.2. Gaps in Conflicting Evidence

Existing research on the relationship between managerial innovation and financial performance remains contradictory and context-dependent. O Gök, S Peker, 2017

While studies conducted in developed economies (Damanpour & Aravind, 2012) suggest a positive correlation between innovative management practices and corporate financial growth, other research highlights that this impact heavily depends on institutional, cultural, and economic factors (Kafouros et al., 2008).

In Morocco, there is a lack of empirical evidence regarding the actual impact of managerial innovation on the profitability of publicly traded companies. Z Samlal,2020 Some firms that adopt innovative management practices do not necessarily experience immediate improvements in performance, raising questions about the specific conditions required for managerial innovation to yield tangible financial benefits(M Makri, PJ Lane, 2006)

3. 3. Inconclusive Findings

Several international studies present inconclusive results regarding the specific effects of each dimension of managerial innovation (process, organization, strategy, and technology) on financial performance. Specifically:

-Organizational innovation may sometimes lead to internal resistance and high adaptation costs, neutralizing its short-term benefits (Birkinshaw et al., 2008).
-Strategic innovation is often influenced by macroeconomic and competitive factors, making it difficult to establish a direct link to financial performance (Crossan & Apaydin, 2010).
-Technological innovation, while generally associated with productivity improvements, requires substantial investments that do not always translate immediately into revenue growth (Martínez-Conesa et al., 2017).

These findings emphasize the need for a more refined and context-specific analysis to determine under what conditions managerial innovation positively impacts the financial performance of publicly listed Moroccan firms(B Kawtar, B Khadija,2025).

3. 4. Theoretical Gaps

Although organizational innovation theory and the resource-based view (RBV) explain how firms can leverage managerial innovation, these frameworks remain generalized and poorly adapted to emerging market contexts (S Gupta, N Malhotra,2013).

Current theoretical models fail to adequately account for:

-The impact of local institutions and regulations on the implementation of managerial innovations. A Arslan - 2011
-The role of investors and shareholders in either promoting or obstructing the adoption of innovative management practices. SA Zahra,1996

-The cultural and organizational specificities of Moroccan companies, which influence change dynamics and resistance management. G Forster, J Fenwick, 2015

Thus, it is crucial to develop an integrated theoretical framework tailored to emerging markets, explaining the mechanisms of adoption and impact of managerial innovation in Morocco. H Hammouch, O Manta, M Palazzo, 2024

3.5. Gaps in Empirical Knowledge

The lack of local empirical studies is one of the primary gaps that hinder the practical application of managerial innovation in Morocco. D Dhiba, W Salih, L Ousayd, A Chehbouni, 2025 Unlike developed markets, where firms have access to historical databases and standardized financial performance indicators, research on Moroccan publicly listed companies remains limited in scope and depth. I Benhayoun, I Zejjari,2024

Key missing insights include:

-The extent to which Moroccan companies adopt innovative management practices. S Zaoui, SA Hamou-ou-Brahim, H Zhou, 2021
-The moderating factors influencing the relationship between innovation and performance (regulation, organizational culture and digitalization). MA Mollah, D Pal, MB Amin, MA Rahaman, 2024
-Sectoral comparisons, which could reveal whether certain industries benefit more from managerial innovation than others. H Forsman, 2011

These gaps call for a systematic study to identify the key success factors and determinants of financial performance linked to managerial innovation in Morocco. Y Sefiani - 2013

3.6. Contextual Gaps in the Moroccan Business Environment

Although Morocco has introduced policies to modernize its economy and financial sector, several structural and institutional constraints limit the adoption of managerial innovation. RB Ayed Mouelhi, M Ghazali, 2021

Some of these contextual challenges include:

-Corporate governance is still evolving, with varying levels of transparency and reporting across companies. A Vickneswaran, 2025
-A shifting regulatory framework, making it difficult to integrate managerial and strategic innovations. L Lescrauwaet, H Wagner, C Yoon,2022
-A corporate culture dominated by rigid hierarchical structures, slowing down the adoption of agile and collaborative management practices. H Koski - 2024

These factors highlight the need for a contextualized approach that considers local constraints while drawing insights from global best practices. MB Teagarden, MA Von Glinow, K Mellahi, 2018

3.7. Practical Challenges in Implementing Managerial Innovation

Publicly listed companies in Morocco face multiple practical obstacles when attempting to integrate innovative management practices:

-Resistance to change from employees and senior executives. PR Lawrence - 1968
-Lack of specialized expertise to implement organizational and strategic innovations. R Modranský, S Jakabová, A Oláh,2020
-Financial constraints, limiting investments in new management approaches. S Fazzari, RG Hubbard, BC Petersen - 1987
-Difficulty in measuring innovation outcomes, making its adoption riskier for companies. AKA Ghadim, DJ Pannell, MP Burton, 2005

Addressing these challenges requires a well-defined strategy and the right tools to facilitate a successful transition to innovative management practices. CM McDermott, GC O'connor, 2002

3.8. How This Study Fills These Gaps

This study aims to bridge these gaps by providing:

- A detailed analysis of different dimensions of managerial innovation and their impact on the financial performance of Moroccan publicly listed firms. H Rachidi, M El Mohajir, 2021
- An empirical approach based on a systematic literature review and real-world case studies. T Olsson, S Sentilles, E Papatheocharous, 2022
- A theoretical framework adapted to the Moroccan context, integrating institutional and cultural constraints. T Bouguerba, 2024
- Practical recommendations for businesses, policymakers, and researchers seeking to deepen their understanding of managerial innovation. M Bogers, H Chesbrough, 2018

By exploring these issues, this research will contribute to a better understanding of the key drivers of managerial innovation in Morocco and provide guidance for future researchers and decision-makers on how to effectively integrate these practices. T Laajini, H Tadjousti, 2025

4. RELATED THEORIES AND THEORETICAL FRAMEWORK

Managerial innovation, as a strategic tool for enhancing corporate performance, is grounded in several organizational and economic theories that explain its impact, mechanisms, and implementation challenges. This study examines the key related theories that support the concept of managerial innovation and justifies the choice of the primary theoretical framework used for analysis.

4.1. Organizational Innovation Theory

- Definition and Origin

The Organizational Innovation Theory asserts that companies must adopt new managerial practices, processes, and organizational structures to remain competitive in a rapidly changing environment. This approach originates from Schumpeter's (1934) work on innovation, which differentiates between product innovation, process innovation, and organizational innovation. Later, Damanpour & Aravind (2012) expanded this theory, emphasizing that organizational innovation extends beyond technology and includes structural and strategic transformations within firms.

- Relevance to This Study

This theory is crucial to the research as it explains how publicly listed companies in Morocco can enhance their financial performance by integrating innovative management practices. The study will analyze how different types of managerial innovation (process, organizational, strategic, and technological) influence financial performance, while considering the unique challenges faced by Moroccan firms.

4.2. Resource-Based View (RBV) and Dynamic Capabilities Theory

- Definition and Origin

The Resource-Based View (RBV), introduced by Barney (1991), argues that a company's performance depends on its internal resources and how effectively they are utilized. This concept evolved with the Dynamic Capabilities Theory, proposed by Teece et al. (1997), which highlights a firm's ability to adapt, integrate, and reconfigure its resources in response to market changes.

- Relevance to This Study

This theory is fundamental in understanding how Moroccan firms can develop and leverage their internal resources (human capital, organizational structures and technological tools) to innovate in management and enhance financial performance. Here, managerial innovation is viewed as a dynamic capability that enables firms to adapt to shifting market conditions and gain a sustainable competitive advantage.

4.3. Institutional Theory

- Definition and Origin

The Institutional Theory, developed by DiMaggio & Powell (1983), suggests that companies adopt organizational practices not only for efficiency but also in response to institutional pressures, regulatory requirements, and social norms. This theory identifies three mechanisms that drive innovation adoption:

-Coercive isomorphism: Adoption due to regulatory or legal pressures.
-Mimetic isomorphism: Adoption by imitating leading companies in the industry.
-Normative isomorphism: Adoption influenced by professionalization and stakeholder expectations.

- Relevance to This Study

This theory is particularly relevant in the Moroccan context, where publicly listed companies face governance regulations, investor expectations, and transparency requirements. It helps analyze how institutional constraints influence managerial innovation adoption and to what extent these practices contribute to financial performance.

4.4. Organizational Change Theory

- Definition and Origin

This theory suggests that organizational change is necessary for growth and performance improvement. Lewin (1951) proposed a three-step model for change:

-Unfreeze: Challenging existing practices.
-Change: Implementing new strategies and methods.
-Refreeze: Stabilizing and integrating innovations.

Later models, such as Kotter's (1996) eight-step change process, emphasize the importance of clear vision, strong leadership, and employee involvement in driving successful innovation.

- Relevance to This Study

The adoption of managerial innovation in Moroccan companies cannot succeed without a well-defined change management strategy. This theory helps understand organizational resistance to innovation and provides insights into how companies can effectively implement managerial changes.

4.5. Primary Theoretical Framework: Dynamic Capabilities Theory

Among these theories, Dynamic Capabilities Theory (Teece et al., 1997) serves as the main theoretical framework for this study.

- Justification for This Choice
  -It explains how firms can develop and leverage managerial resources to achieve a sustainable competitive advantage.
  -It highlights the ability of Moroccan companies to integrate, modify, and adapt their management practices in response to economic and technological challenges.
  -It provides a structured framework to analyze the internal and external factors influencing a firm's ability to innovate in management and achieve financial benefits.

This study will therefore rely on this theory to examine how managerial innovation can drive financial performance in an emerging market like Morocco.

4.6. Definition and Origin of Key Concepts

- Definition of Managerial Innovation

Managerial innovation refers to the adoption of new practices, processes, and organizational structures aimed at improving efficiency, flexibility, and overall business performance (Birkinshaw et al., 2008). Unlike technological innovation, which focuses on new products or services, managerial innovation concerns how companies are managed and structured.

- Origin of the Managerial Innovation Concept

The idea that management itself can be a field of innovation dates back to early studies on bureaucracy (Weber, 1922) and scientific management (Taylor, 1911). However, the modern concept of managerial innovation was formalized by Schumpeter (1934), who recognized that changes in management and organization could be as disruptive and beneficial as technological innovations.

- Definition of Financial Performance

Financial performance refers to a firm's ability to generate value for shareholders and stakeholders. It is typically measured using indicators such as:

-Return on Investment (ROI)
-Return on Equity (ROE)
-Revenue growth

- Origin of the Financial Performance Concept

Early studies on financial performance trace back to Fama & French (1992), who developed models to analyze the factors influencing corporate profitability.

This study is primarily based on Dynamic Capabilities Theory, while integrating perspectives from Organizational Innovation Theory, Institutional Theory, and Organizational Change Theory. This approach will help understand how Moroccan firms can leverage managerial innovation to enhance financial performance, while overcoming specific economic and institutional challenges.

5. EMPIRICAL LITERATURE REVIEW

The impact of managerial innovation on financial performance has been the subject of numerous empirical studies across management and economic disciplines. This literature review focuses on the main empirical studies that have examined the dimensions of managerial innovation and their influence on financial performance, while highlighting contradictory findings and gaps in the context of emerging markets such as Morocco.

5.1. Managerial Innovation and Financial Performance: A Generally Positive Relationship

Several empirical studies conducted in developed economies have highlighted a positive relationship between managerial innovation and financial performance. For example, Damanpour & Aravind (2012) conducted a meta-analysis of over 20 years of research and found that firms adopting innovative management practices experience faster revenue growth, improved profitability, and greater resilience to economic crises.

Similarly, Martínez-Conesa et al. (2017) analyzed a sample of European firms and demonstrated that organizational and strategic innovation enhances financial indicators such as return on investment (ROI) and return on equity (ROE). These findings suggest that firms investing in managerial innovation benefit from greater flexibility and process optimization, which ultimately leads to better financial performance.

However, these results are not uniform and depend on contextual factors such as organizational structure, regulations, and corporate culture.

5.2. Process Innovation and Corporate Productivity

Process innovation, which involves implementing new management methods, digital tools, and more efficient decision-making systems, has been widely studied. A study by Gunday et al. (2011) in Turkey found that improving internal processes, particularly through digitalization and automation, leads to higher corporate productivity and reduced operational costs.

Similarly, Hollen et al. (2013) observed that companies adopting lean management and continuous improvement practices report higher operational efficiency and better profit margins.

However, some studies suggest that process innovation can face internal resistance, particularly from employees who perceive these changes as a threat to job security (Birkinshaw et al., 2008). This underscores the importance of change management and employee training to ensure a smooth transition to innovative work methods.

5.3. Organizational Innovation and Business Agility

Organizational innovation, which includes changes in governance structures, decentralization of decision-making, and new leadership models, is a key driver of financial performance.

A study by Volberda et al. (2013) analyzed how organizational innovation affects firms' ability to adapt to economic crises. Their results indicate that firms with more flexible structures and participatory decision-making systems performed better during financial downturns and maintained profitability.

Conversely, research by Christensen & Overdorf (2000) found that large firms with rigid hierarchical structures tend to be less responsive to managerial innovation, limiting their ability to adapt to changing business environments. This is particularly relevant for Moroccan publicly listed companies, which often have highly centralized decision-making processes, making it difficult to adopt more flexible management models.

5.4. Strategic Innovation and Long-Term Performance

Strategic innovation involves adopting new business models, entering new markets, and restructuring competitive strategies. Teece et al. (2016) showed that companies embracing innovative strategies, such as revenue diversification and digital transformation, achieve sustainable growth and greater resilience to economic crises.

A concrete example is provided by Kesting & Ulhøi (2010), who studied Scandinavian companies and found that those investing in strategic innovation (such as customer co-creation and flexible offerings) achieved higher profitability and stronger customer loyalty.

However, these benefits are not immediate. Chesbrough (2010) highlighted that strategic transformation requires significant initial investments, which can impact short-term financial results. This raises a key issue for Moroccan publicly listed companies: how to balance investment in strategic innovation while maintaining financial stability?

5.5. Empirical Studies on Managerial Innovation in Emerging Markets

While empirical studies on managerial innovation are abundant in developed economies, research in emerging markets is more limited. However, some studies have examined the impact of these innovations in contexts similar to Morocco.

-Kafouros et al. (2008) studied Chinese firms and found that investments in managerial innovation are more profitable in companies operating in stable regulatory environments.
-Bouazza et al. (2015) analyzed Algerian firms and emphasized that lack of infrastructure and funding hinders the adoption of managerial innovations, limiting their impact on financial performance.
-Khanna & Palepu (2010) demonstrated that firms in emerging markets must adapt managerial innovations to local constraints, such as regulatory challenges and corporate culture, to maximize their effectiveness.

These findings suggest that conclusions drawn from developed economies cannot be directly applied to Moroccan companies. Thus, conducting specific studies on the Moroccan context is essential to better understand the links between managerial innovation and financial performance.

This empirical literature review highlights several key points:

- Managerial innovation generally has a positive impact on financial performance, but results vary based on institutional and economic contexts.
- Process innovation improves productivity, but may generate internal resistance.
- Organizational innovation enhances agility and adaptability, but adoption is challenging for centralized companies.
- Strategic innovation provides long-term benefits, but requires substantial investment.
- Few studies have examined the impact of managerial innovation in Morocco, leaving a gap in the academic literature.

This study aims to fill these gaps by providing a comprehensive analysis of the dimensions of managerial innovation and their impact on the financial performance of Moroccan publicly listed firms. It will also help identify best practices to maximize the impact of managerial innovations in the Moroccan business context.

6. CRITICAL ANALYSIS

Process innovation has been widely recognized for its potential to enhance operational efficiency and cost-effectiveness, directly impacting a company's profitability and cost management. Studies have shown that firms investing in process innovations, such as lean manufacturing or total quality management

(TQM), often report improved financial metrics, including lower operational costs, higher productivity, and improved profit margins (Chen et al., 2016). For instance, Demeter and Losonci (2015) found that lean manufacturing significantly improved financial performance by reducing production costs and cycle times, allowing firms to optimize their resources and increase output. Moreover, He and Wong (2004) argue that process innovation enables firms to better align their resource allocations with operational needs, thus boosting overall profitability.

Despite these positive impacts, some studies caution that the financial benefits of process innovation may be limited by high implementation costs and the need for substantial managerial oversight (Damanpour & Gopalakrishnan, 2001). Furthermore, the effectiveness of process innovation on financial outcomes can be highly context-dependent, requiring specific organizational conditions such as a strong culture of continuous improvement and alignment with corporate strategy (Crossan & Apaydin, 2010).

Organizational innovation, particularly structural and managerial reforms, has shown mixed yet promising results in improving financial outcomes. Research highlights that organizational innovations, such as flattening hierarchies or fostering cross-functional teams, can enhance financial performance by reducing bureaucratic inefficiencies and increasing employee engagement (Walker et al., 2011). For example, Camison and Villar-Lopez (2014) observed that firms that adopted more flexible organizational structures reported increased financial returns, attributing these gains to improved communication and faster decision-making processes.

However, the financial impact of organizational innovation may be indirect, as it relies heavily on fostering a culture of innovation and resilience (Birkinshaw et al., 2008). Some studies suggest that the success of organizational innovation in driving financial performance depends on its alignment with other innovation practices within the firm, such as technological and strategic innovations (Lam, 2004). Additionally, Damanpour and Schneider (2006) note that the relationship between organizational innovation and financial outcomes can be moderated by factors like leadership style and organizational size.

Strategic innovation often has the most direct impact on financial performance, as it involves redefining the firm's competitive position and exploring new revenue streams. Studies have found that strategic innovation, such as entering new markets or redefining business models, is associated with significant financial gains, particularly in terms of revenue growth and market share (Markides, 1997; Casadesus-Masanell & Ricart, 2010). Firms that prioritize strategic innovation are better able to anticipate and adapt to market changes, ensuring long-term financial viability. For instance, a study by Teece (2010) emphasized that companies with dynamic strategic capabilities could navigate complex market dynamics more effectively, translating to improved financial stability.

However, strategic innovation carries certain risks, including the possibility of misalignment with core competencies or overextension into unfamiliar markets (Govindarajan & Trimble, 2004). Some firms may experience short-term financial strain due to the initial investment required for strategic shifts, though these investments often pay off in the long term (Baden-Fuller & Haefliger, 2013).

Technological innovation in management practices—such as the adoption of big data, artificial intelligence, and digital platforms—has been found to significantly enhance financial performance by increasing operational efficiency and improving decision-making accuracy. According to Brynjolfsson and McAfee (2014), firms that integrate advanced technologies experience faster growth and higher profitability, as these technologies enable better resource management and customer engagement. Additionally, studies indicate that technological innovations allow firms to optimize supply chains and respond to customer needs more effectively, thereby enhancing revenue generation and reducing costs (Kane et al., 2015).

Nevertheless, some scholars argue that the impact of technological innovation on financial performance is subject to diminishing returns, particularly in saturated markets where competitive advantages can be quickly eroded by technological diffusion (Mol & Birkinshaw, 2014). Furthermore, technological innovation often requires substantial initial investments and a workforce skilled in digital tools, which may limit the extent to which financial benefits are realized immediately (Schilling, 2017).

7. KEY FINDINGS FROM INTERNATIONAL CONTEXTS AND APPLICABILITY TO THE MOROCCAN MARKET

In reviewing international studies on the relationship between managerial innovation dimensions and financial performance, several key insights emerge, many of which offer practical implications for Moroccan publicly listed companies:

a. **Process Innovation**: International studies consistently highlight that process innovation enhances financial performance by reducing costs and improving productivity, an insight that is particularly relevant to Moroccan firms seeking operational efficiency (Demeter & Losonci, 2015). Moroccan companies could benefit from lean manufacturing or agile process management practices, which are widely adopted in more mature markets to streamline production and reduce waste.
b. **Organizational Innovation**: Studies in diverse contexts, such as those conducted by Walker et al. (2011) in the U.S. and Camison and Villar-Lopez (2014) in Europe, show that organizational innovation fosters adaptability and resilience, leading to better financial outcomes. In Morocco, where firms must often adapt to regulatory changes and economic fluctuations, flexible organizational structures could enable companies to respond to market dynamics more efficiently, thereby securing financial stability.
c. **Strategic Innovation**: The adoption of strategic innovation has been shown to provide firms with competitive advantages and new revenue streams in various international markets, including the U.S. and Asia (Teece, 2010). For Moroccan companies, strategic innovation could be a powerful tool for growth, allowing them to differentiate in both domestic and international markets. By exploring new market segments and developing innovative business models, Moroccan firms can capture value and boost financial performance.
d. **Technological Innovation**: The implementation of advanced technologies has been widely recognized for its positive impact on financial performance, especially in developed economies where digital transformation is rapidly progressing (Brynjolfsson & McAfee, 2014). Moroccan companies could similarly leverage digital tools to enhance operational efficiency and customer interaction, although they may face challenges related to infrastructure and digital literacy. The gradual adoption of technologies like data analytics and ERP systems could provide Moroccan firms with a pathway to achieving the financial benefits observed in international contexts (Aubert & Hamadache, 2015).

8. SYNTHESIS OF LITERATURE FINDINGS

This section synthesizes the primary trends and contradictions identified within the literature regarding the impact of managerial innovation on financial performance. Furthermore, it addresses specific research gaps, particularly within the Moroccan context, where empirical studies are limited. Understanding these gaps will provide direction for future research and actionable insights for Moroccan firms aiming to leverage managerial innovation for improved financial outcomes.

8.1 Summary of trends and contradictions in the literature

A strong consensus within the literature underscores the positive correlation between managerial innovation across process, organizational, strategic, and technological dimensions and enhanced financial performance. Several studies affirm that firms that adopt innovative managerial practices generally experience better profitability, improved resource allocation, and heightened competitiveness (Crossan & Apaydin, 2010; Chen et al., 2016). For instance, empirical research demonstrates that process innovations are associated with cost reduction and efficiency improvements, which in turn positively impact profit margins (Demeter & Losonci, 2015). Similarly, strategic innovations are linked to revenue growth and market expansion, as firms that redefine their competitive strategies are better positioned to capture new market opportunities (Casadesus-Masanell & Ricart, 2010; Markides, 1997).

Organizational and technological innovations also consistently show a favorable impact on financial outcomes, albeit in more indirect ways. Organizational innovations enhance adaptability and employee productivity, while technological innovations streamline operations, reduce costs, and improve decision-making accuracy (Damanpour & Schneider, 2006; Brynjolfsson & McAfee, 2014). These studies suggest that the integration of various managerial innovation dimensions can foster a holistic transformation within organizations, enabling firms to achieve long-term financial stability and growth.

Despite these positive associations, several contradictions exist within the literature, particularly concerning the degree of impact each dimension of innovation has on financial performance. For example, while some studies argue that process innovation yields immediate financial benefits due to cost reductions (Damanpour et al., 2009), others caution that the high implementation costs of such innovations may delay financial gains, creating a temporary negative impact on short-term profitability (He & Wong, 2004). This divergence suggests that the financial impacts of process innovation may depend on the scale of change and the firm's capacity for cost absorption.

Moreover, the effectiveness of organizational innovation in enhancing financial performance remains debated. While some research asserts that flexible structures and collaborative cultures promote productivity and ultimately improve financial outcomes, other studies argue that organizational innovation may result in increased administrative costs without a guaranteed improvement in profitability (Mol & Birkinshaw, 2014). The impact of organizational innovation may also be contingent upon contextual factors such as industry type, company size, and regional differences in workforce culture (Damanpour & Aravind, 2012).

Another area of contradiction pertains to strategic innovation. While strategic shifts often lead to long-term financial success, some firms experience short-term financial strain due to the resource investment required to redefine competitive positioning (Govindarajan & Trimble, 2004). Strategic innovation carries inherent risks, particularly in volatile markets where rapid shifts in strategy may not yield anticipated returns, highlighting the importance of aligning innovation with market stability and firm capabilities.

8.2 Identification of Research Gaps Relevant to Moroccan Companies

While the positive impact of managerial innovation on financial performance is well-documented in international research, specific gaps exist in understanding how these innovations function within the Moroccan market. Below are the key research gaps that future studies could address to better inform Moroccan companies about the potential of managerial innovation to enhance their financial performance:

-Lack of empirical evidence specific to Moroccan firms: Few studies have investigated the impact of managerial innovation on financial performance specifically within Moroccan companies. While international studies offer valuable insights, the unique economic, cultural, and regulatory environment in Morocco requires localized research to validate these findings. The effectiveness of managerial innovations may differ significantly due to Morocco's distinct market characteristics, such as regulatory requirements, labor dynamics, and access to technology. Research focusing on Moroccan firms could provide contextually relevant insights into how managerial innovations can be adapted to local market conditions (Bouazza et al., 2015).

- Limited analysis of contextual moderators: Existing literature rarely explores the moderating effects of contextual factors—such as industry type, firm size, and market volatility—on the relationship between managerial innovation and financial performance in emerging markets. In Morocco, these factors likely influence the adoption and success of managerial innovations. For instance, larger firms may have more resources to invest in comprehensive process and technological innovations, whereas smaller firms may focus on incremental improvements (Elbanna & Child, 2007). Investigating how these contextual variables affect innovation outcomes could help Moroccan firms better align their innovation strategies with organizational capabilities and market demands.
- Short-term versus long-term financial impacts: Much of the literature emphasizes the long-term financial benefits of managerial innovation, often overlooking the short-term financial strain that can accompany such initiatives. Moroccan firms may face immediate financial pressures due to the upfront costs associated with implementing process, technological, or strategic innovations. Research that differentiates between short-term and long-term financial outcomes could provide a more nuanced understanding of the financial risks and benefits of managerial innovation in Morocco (Martínez-Conesa et al., 2017).
- Role of cultural and organizational factors in adoption of innovation: Cultural and organizational factors play a significant role in the adoption and effectiveness of managerial innovations, yet there is limited research exploring these dynamics within Moroccan companies. Given that organizational culture, leadership style, and employee attitudes towards change can influence the success of innovations, studies focusing on these factors could yield valuable insights for Moroccan firms. This line of inquiry would allow managers to tailor innovation strategies that are culturally sensitive and more likely to be accepted within the organization (Hofstede, 2001; Zahra & Covin, 1995).
- Impact of technological infrastructure limitations: While technological innovation is widely regarded as a key driver of financial performance, Moroccan companies may encounter specific challenges related to technological infrastructure, such as limited access to advanced technologies and skilled personnel. Research exploring how these limitations affect the financial impact of technological innovations could help Moroccan firms identify realis-

tic strategies for leveraging technology within existing constraints (Aubert & Hamadache, 2015). Studies could also investigate how Moroccan firms can adopt scalable, cost-effective technological innovations that align with local infrastructure capabilities.

9. METHODOLOGY

9.1 Systematic review approach

The methodology for this review follows a systematic approach designed to ensure a comprehensive and unbiased analysis of the literature on the impact of managerial innovation on financial performance. The systematic review methodology is grounded in established protocols, particularly the Preferred Reporting Items for Systematic Reviews and Meta-Analyses (PRISMA) framework. PRISMA is widely recognized for its structured approach to identifying, screening, and evaluating research articles, making it highly suitable for synthesizing complex literature in the field of management and organizational studies (Moher et al., 2009).

The systematic review process involved several key stages. First, a literature search was conducted across multiple academic databases, including Scopus, Web of Science, JSTOR, and Google Scholar. Search terms included combinations of keywords such as "managerial innovation," "financial performance," "process innovation," "organizational innovation," "strategic innovation," and "technological innovation," as well as region-specific terms like "Morocco" and "emerging markets" to ensure the capture of contextually relevant studies.

To refine the scope of articles selected, inclusion and exclusion criteria were established. Studies were included if they addressed at least one dimension of managerial innovation (process, organizational, strategic, or technological), provided empirical data on the relationship between managerial innovation and financial performance, and included analyses applicable to corporate contexts similar to Morocco, particularly emerging markets. Articles were excluded if they lacked empirical data, focused solely on non-managerial types of innovation (such as product or service innovation), or were purely theoretical without a clear link to financial outcomes. Non-English articles were also excluded unless translations were available to maintain consistency in interpretation.

Each selected article underwent a quality assessment to evaluate its methodological rigor, clarity of research design, and robustness of findings. Studies using recognized quantitative methods, including regression analysis and structural equation modeling, as well as qualitative approaches such as case studies and interviews, were prioritized to ensure that the review synthesized insights from methodologically sound research. This assessment facilitated a focus on studies that contributed valuable data relevant to the research question (Tranfield et al., 2003).

9.2 Analysis of sources and data

Once the articles were selected, a multi-step analysis process was employed to systematically organize and interpret the findings. First, data extracted from each article were coded and categorized according to the four main dimensions of managerial innovation: process, organizational, strategic, and technological. The coding process used a thematic analysis approach, which involves identifying, analyzing, and reporting patterns (themes) within data (Braun & Clarke, 2006). Each article was reviewed to extract findings relevant to how each innovation dimension impacts financial performance, with a focus on indicators such as profitability, cost reduction, revenue growth, and productivity. Process innovation was coded for findings related to operational efficiency, cost management, and resource optimization; organizational innovation for findings on employee productivity, structural changes, and cultural adaptability; strategic innovation for findings on competitive positioning, market expansion, and revenue diversification; and technological innovation for findings on technological infrastructure, data analytics, and decision-making efficiency. This coding process allowed for a structured comparison of results across studies, highlighting consistencies and discrepancies in how each innovation dimension affects financial performance.

For the majority of the studies included, qualitative content analysis was applied to explore recurring themes and patterns in the findings. The qualitative analysis focused on understanding the contextual factors influencing the success of each innovation dimension, such as organizational culture, leadership, and market conditions (Patton, 2002). This approach provided a nuanced understanding of the impact of managerial innovation on financial performance, particularly within the broader cultural and economic environment. Where applicable, quantitative data from the articles were synthesized through a meta-analysis to evaluate the effect size of managerial innovation dimensions on financial performance. Meta-analysis techniques, such as calculating standardized mean differences or pooled correlation coefficients, were employed to quantitatively assess the strength of associations across studies (Borenstein et al., 2011). Given the variability in methodologies, a random-effects model was preferred to account for differences in study populations, methods, and contexts, providing a more generalized understanding of the relationship between managerial innovation and financial outcomes.

Following data coding and analysis, findings were synthesized to form a cohesive understanding of the role of managerial innovation in driving financial performance. This synthesis involved aggregating results by innovation dimension, identifying key factors that moderate or mediate the impact of innovations on financial outcomes, and contextualizing these findings within the Moroccan market, where local conditions may influence the effectiveness of these innovations (Aguinis et al., 2011). The systematic and rigorous approach outlined ensures that the review presents a reliable synthesis of the literature on managerial innovation and financial performance, grounded in high-quality research and contextual relevance to Moroccan firms.

10. RESULTS OF THE REVIEW

10.1 Synthesis of the effects of managerial innovation on financial performance

The review highlights how different dimensions of managerial innovation process, organizational, strategic, and technological contribute uniquely to financial performance. Each dimension demonstrates specific impacts on financial metrics such as profitability, cost reduction, revenue growth, and operational efficiency, which vary based on organizational context and external market conditions. This synthesis organizes the findings by innovation dimension and identifies patterns and insights that are particularly relevant to Moroccan companies.

10.1.1 Results organized by innovation dimension

Process Innovation has been shown to positively impact financial performance primarily through cost reduction and enhanced efficiency. Studies indicate that firms adopting process innovations, such as lean manufacturing and Six Sigma practices, experience immediate gains in operational efficiency and reductions in production costs (Demeter & Losonci, 2015). These improvements stem from streamlining workflows, optimizing resource use, and minimizing waste, which contribute to higher profit margins and greater productivity. Process innovations are also associated with enhanced flexibility, allowing firms to adapt quickly to shifts in demand or production needs. In industries such as manufacturing, where efficiency is crucial, process innovations offer a pathway to sustainable profitability by reducing cost structures and enabling firms to scale production without proportionally increasing costs (Damanpour et al., 2009).

Organizational Innovation contributes to financial performance through structural changes that promote adaptability, communication, and employee engagement. Research suggests that companies implementing flatter hierarchies, cross-functional teams, or decentralized decision-making processes benefit from increased productivity and a stronger alignment with corporate goals (Walker et al., 2011). These structural reforms reduce bureaucratic bottlenecks, accelerate decision-making, and enhance collaborative work environments, which ultimately lead to improved financial outcomes. Organizational innovation is also linked to higher employee satisfaction and engagement, as more collaborative and empowered workplaces encourage employee investment in company success. However, the financial benefits of organizational innovation may take time to materialize, as these innovations often involve cultural shifts that require gradual implementation and strong leadership (Birkinshaw et al., 2008).

Strategic Innovation is identified as one of the most direct drivers of financial performance, as it involves reconfiguring a firm's competitive positioning, exploring new revenue streams, and expanding market reach. Strategic innovations—such as entering new markets, redefining business models, and adopting differentiation strategies—allow companies to capture new customer segments and create unique value propositions (Casadesus-Masanell & Ricart, 2010). Firms that successfully integrate strategic innovation often report higher revenue growth and increased market share, as they are better positioned to meet evolving customer demands and adapt to market dynamics. Teece (2010) found that strategic innovation drives financial performance by fostering dynamic capabilities that enable firms to respond effectively to competitive pressures and capitalize on emerging opportunities.

Technological Innovation improves financial performance by enhancing operational efficiency and decision-making accuracy. Studies show that firms leveraging advanced technologies, such as data analytics, artificial intelligence, and enterprise resource planning (ERP) systems, can achieve faster response times, optimized resource allocation, and reduced operational costs (Brynjolfsson & McAfee, 2014). Technological innovation also facilitates better customer relationship management (CRM), enabling firms to increase customer retention and satisfaction. This innovation dimension is particularly beneficial for companies operating in fast-paced industries, where staying competitive requires quick adaptation to technological advancements. Mol and Birkinshaw (2014) emphasize that technological innovation is essential for firms to remain resilient in volatile markets, as it provides a robust foundation for sustainable financial growth.

10.1.2 Identification of patterns and specific insights for Moroccan companies

The synthesis of results reveals several patterns that can guide Moroccan companies in leveraging managerial innovation for financial performance improvement. First, process innovation emerges as an accessible and immediate approach to cost management and operational efficiency, making it particularly relevant for Moroccan firms in industries like manufacturing, where cost pressures are high. By adopting lean and agile production practices, Moroccan companies can reduce operational costs and increase productivity, enabling them to compete effectively on both domestic and international markets.

Organizational innovation provides Moroccan firms with tools for building flexible and resilient structures. Given the variability in the Moroccan economic landscape, organizational innovation offers a pathway to adaptability, particularly in industries that face regulatory and economic shifts. However, the success of organizational innovation depends on fostering a culture that supports open communication, knowledge sharing, and cross-functional collaboration. Companies that invest in developing such a culture are likely to benefit from increased employee productivity and alignment with organizational goals, which are crucial for sustained financial success.

Strategic innovation appears to be highly impactful for Moroccan firms seeking growth through market expansion and diversification. By rethinking their business models and exploring new market segments, Moroccan companies can differentiate themselves in a competitive landscape and capitalize on new revenue sources. This is particularly relevant in Morocco's emerging sectors, where firms have the opportunity to pioneer innovative business strategies. Strategic innovation can position Moroccan firms as market leaders, allowing them to attract investment and establish competitive advantages that are sustainable over the long term.

10.2 Discussion of results

The results of this review highlight the varied and multidimensional impact of managerial innovation on financial performance across process, organizational, strategic, and technological dimensions. A comparison with theoretical expectations and initial hypotheses confirms that managerial innovation has a generally positive effect on financial performance, as anticipated. However, the strength and nature of this effect vary according to contextual factors, revealing the importance of moderating and mediating variables in shaping these outcomes.

10.2.1 Comparison with theoretical expectations and initial hypotheses

The review findings align closely with the theoretical frameworks of the Resource-Based View (RBV) and Organizational Innovation Theory, both of which posit that firms can achieve competitive advantages through unique internal resources and adaptive organizational practices. The RBV framework anticipates that innovations contributing to operational efficiency, strategic flexibility, and technological integration would have a positive impact on financial performance by leveraging a firm's unique resources and capabilities (Barney, 1991). This review supports this theoretical view, demonstrating that process and technological innovations, in particular, are effective in reducing costs and optimizing resource allocation, which directly enhance profitability.

Similarly, Organizational Innovation Theory posits that adaptive organizational structures promote innovation and financial resilience, aligning with the observed benefits of organizational innovation on employee engagement and productivity. The findings suggest that companies that invest in flexible and decentralized structures benefit from improved productivity and responsiveness to market changes, as predicted by Organizational Innovation Theory (Damanpour & Schneider, 2006). Strategic innovation also aligns with theoretical expectations by providing firms with a pathway to market differentiation and new revenue streams, supporting long-term financial growth and stability (Casadesus-Masanell & Ricart, 2010).

However, there are some nuanced deviations from theoretical expectations. While strategic and technological innovations are often seen as high-impact approaches for competitive positioning and market leadership, the review indicates that these dimensions may require substantial initial investments, potentially straining financial performance in the short term. This finding challenges the assumption that all dimensions of managerial innovation yield immediate financial benefits and suggests that certain innovations may require a longer time horizon to realize financial returns fully (Govindarajan & Trimble, 2004).

10.2.2 Identification of moderating and mediating variables

The impact of managerial innovation on financial performance appears to be influenced by several key moderating and mediating variables, which include company size, industry sector, market environment, and organizational culture. These variables shape how effectively innovations are implemented and determine the degree to which they contribute to financial outcomes.

Company Size is a significant moderating variable that affects the scale and type of managerial innovation adopted. Larger companies typically have greater resources to invest in comprehensive innovation initiatives, allowing them to adopt complex process and technological innovations that can lead to substantial efficiency gains and competitive advantages (Aguinis et al., 2011). In contrast, smaller firms may lack the financial and human capital to implement large-scale innovations and may focus instead on incremental improvements or organizational innovations that are less resource-intensive. The review findings indicate that the financial benefits of managerial innovation are often more pronounced in larger firms, as they can leverage economies of scale to maximize the impact of their innovation investments.

Industry Sector also plays a crucial moderating role, as different industries prioritize various dimensions of innovation according to operational needs and market pressures. For instance, process and technological innovations are more impactful in manufacturing and technology-intensive sectors, where efficiency and speed are critical for maintaining competitiveness. Meanwhile, strategic and organizational innovations may be more relevant in service-oriented industries, where adaptability and employee engagement directly influence customer satisfaction and retention (Damanpour & Aravind, 2012). This sectoral variation highlights the importance of aligning innovation strategies with industry-specific demands to optimize financial performance.

Market Environment is another moderating factor that affects how managerial innovations translate into financial outcomes. In stable markets, strategic innovations may offer substantial returns by allowing firms to explore new market segments or introduce differentiated offerings. Conversely, in highly volatile markets, such as those often experienced in emerging economies like Morocco, technological and process innovations may provide more immediate benefits by enhancing operational flexibility and cost-efficiency, helping firms to weather economic fluctuations more effectively (Elbanna & Child, 2007). This insight suggests that the choice of innovation dimension should consider the stability and predictability of the external environment.

Organizational Culture acts as a mediating variable that influences the success of innovation implementation. Companies with a culture that encourages risk-taking, learning, and continuous improvement are better positioned to realize the financial benefits of managerial innovation. For example, a culture that supports open communication and cross-functional collaboration can enhance the effectiveness of organizational innovations, as employees are more likely to adopt new practices and work collaboratively towards common goals (Zahra & Covin, 1995). Conversely, firms with rigid, hierarchical cultures may struggle to implement innovations effectively, as resistance to change can undermine adoption and limit financial gains.

11. DISCUSSION AND IMPLICATIONS

11.1 Theoretical contribution

This review contributes significantly to the theoretical understanding of managerial innovation and its impact on financial performance, particularly through the lens of the Resource-Based View (RBV) and Organizational Innovation Theory. By organizing managerial innovation into four dimensions—process, organizational, strategic, and technological—this review advances the understanding of how each innovation type uniquely influences financial outcomes. It reinforces the RBV's proposition that firms' internal resources, such as innovation capabilities, can be instrumental in creating competitive advantages (Barney, 1991). However, it also reveals that the financial benefits associated with managerial innovation are not uniform across dimensions or contexts, which adds depth to the theory by emphasizing the importance of situational factors, such as firm size and industry type. The review further builds on Organizational Innovation Theory by illustrating how adaptive organizational structures and cultures mediate the success of innovations, affirming that innovations are most effective when aligned with organizational culture and strategic objectives (Damanpour & Schneider, 2006). This synthesis thus highlights that managerial innovation must be viewed as a multifaceted construct that contributes to financial performance through complex interdependencies, rather than as a one-size-fits-all approach.

11.2 Practical implications for Moroccan companies

The findings of this review provide practical recommendations for managers in Moroccan companies, emphasizing the importance of adopting a strategic approach to managerial innovation. Moroccan managers should tailor their innovation strategies to the specific needs and resources of their companies. Process innovation, for instance, should be prioritized in industries with high operational costs, such as manufacturing, as it directly contributes to cost efficiency and productivity. Managers are encouraged to adopt lean production methods and streamlined workflows to maximize output while minimizing resource wastage. Implementing such innovations can make Moroccan companies more competitive locally and internationally by reducing their cost base and enhancing operational efficiency.

For organizational innovation, Moroccan companies should foster a culture of collaboration and adaptability. Given the economic fluctuations and regulatory shifts that characterize emerging markets like Morocco, flexible structures that support cross-functional collaboration and decentralized decision-making are essential for resilience. Managers should prioritize the development of a workplace culture that values innovation and employee empowerment, as this will improve productivity and facilitate the successful implementation of managerial changes. Organizational structures that are adaptable and employee-centered will likely result in sustained financial improvements, particularly in service industries where customer engagement and response times are critical.

Strategic innovation, while resource-intensive, offers Moroccan firms an avenue to differentiate themselves and expand market reach. For managers in companies with the resources to pursue ambitious growth objectives, strategic innovation can unlock new revenue streams and position the firm as a leader within its sector. Managers are advised to explore new business models, target previously underserved market segments, and leverage competitive positioning strategies to capture greater market share. This approach is particularly relevant in sectors such as renewable energy and technology, where Morocco has potential for growth on an international scale.

Technological innovation also holds substantial potential for Moroccan companies, especially in sectors like finance, telecommunications, and logistics. Although technology investments can be costly, incremental adoption of digital tools—such as data analytics, CRM systems, and ERP solutions—can significantly enhance decision-making efficiency and customer engagement. Given the infrastructural constraints in Morocco, managers should consider phased technology investments to minimize risk and manage financial costs effectively. By strategically adopting technological innovations, Moroccan companies can enhance both customer satisfaction and operational efficiency, thereby strengthening their competitive positions.

11.3 Limitations of the review

While this review provides valuable insights into the impact of managerial innovation on financial performance, certain limitations must be acknowledged. First, the availability of data poses a limitation, as the literature on managerial innovation in emerging markets, particularly in Morocco, is still relatively sparse. Many of the studies included in this review are based on international contexts that may not fully reflect the unique economic, regulatory, and cultural dynamics of Morocco. Additionally, the reliance on peer-reviewed studies limits the scope of findings, as industry reports, case studies, and practitioner insights were not included. Methodologically, the review's use of thematic analysis and meta-analysis introduces potential biases due to the variability in study designs and sample characteristics across the literature. Finally, the cross-sectional nature of many studies reviewed limits the ability to assess long-term financial impacts, as most data capture only short-term or immediate effects of managerial innovations.

11.4 Directions for future research

To build upon the findings of this review, future research should consider several avenues to deepen understanding and applicability for Moroccan firms. First, there is a need for more **empirical studies within the Moroccan context** to provide insights that are directly relevant to local companies. Such research could use longitudinal designs to assess the long-term financial impacts of managerial innovations and provide a more comprehensive understanding of how these innovations evolve over time.

Future research should also consider **larger and more diverse samples** within Morocco to capture differences across industries and firm sizes. Expanding the range of sample populations would help identify sector-specific factors and refine recommendations for different types of organizations. Additionally, **mixed-method approaches** that combine quantitative and qualitative data could provide a richer perspective on how managerial innovations are implemented and the challenges encountered. Case studies of successful Moroccan firms, for instance, could yield valuable insights into best practices and key factors that drive the success of innovations in similar contexts.

Finally, future research should explore **new moderating and mediating variables** that may affect the relationship between managerial innovation and financial performance. For example, factors such as regulatory changes, digital infrastructure, and workforce skills may play critical roles in determining the effectiveness of innovations in Morocco. By examining these additional variables, researchers can develop a more nuanced model that better reflects the complex interplay of factors influencing the success of managerial innovations. Such research would provide Moroccan managers with actionable insights that are closely aligned with the realities of the Moroccan market, ultimately contributing to enhanced financial performance and sustained competitive advantage.

12. CONCLUSION

This review synthesizes extensive literature on the impact of managerial innovation on financial performance, organizing insights across four key dimensions: process, organizational, strategic, and technological innovation. The findings reveal that each dimension contributes uniquely to financial outcomes, with process and technological innovations primarily enhancing operational efficiency, organizational innovation fostering adaptability and productivity, and strategic innovation supporting market positioning and revenue growth. Together, these innovations form a multifaceted approach that organizations can leverage to achieve sustained financial success.

One of the principal contributions of this review is the identification of critical factors that moderate and mediate the relationship between managerial innovation and financial performance. Company size, industry type, market environment, and organizational culture emerge as significant influences, underscoring the need for a contextualized approach to innovation. This nuanced understanding highlights that while managerial innovation offers substantial financial benefits, its effectiveness depends on alignment with organizational resources, market conditions, and strategic priorities.

The importance of managerial innovation is particularly pronounced in the context of Moroccan companies, where the challenges of a developing economy require firms to innovate in order to compete both domestically and internationally. Process innovation provides Moroccan firms with tools to reduce operational costs and improve productivity, a critical advantage in cost-sensitive industries. Organizational innovation promotes flexibility and resilience, essential qualities in a volatile economic landscape. Strategic innovation offers a pathway for Moroccan firms to differentiate themselves, exploring new markets and business models that can drive long-term growth. Finally, technological innovation, although resource-intensive, enhances decision-making and customer engagement, enabling Moroccan companies to maintain a competitive edge in an increasingly digitalized world.

Managerial innovation is a powerful driver of financial performance, with significant potential to enhance the competitiveness and resilience of Moroccan firms. By adopting tailored innovation strategies that reflect their unique organizational needs and market conditions, Moroccan companies can leverage managerial innovation to strengthen financial outcomes and position themselves for sustainable success in an evolving global economy. This review thus underscores the vital role of managerial innovation as a strategic asset, providing both theoretical insights and practical guidance for managers and researchers focused on maximizing financial performance through innovation in Morocco and beyond.

REFERENCES

Adewumi, A., Ewim, S. E., Sam-Bulya, N. J., & Ajani, O. B. (2024). Advancing business performance through data-driven process automation: A case study of digital transformation in the banking sector. *International Journal of Multidisciplinary Research Updates*, *8*(2), 12–22. DOI: 10.53430/ijmru.2024.8.2.0049

Adewumi, A., Ewim, S. E., Sam-Bulya, N. J., & Ajani, O. B. (2024). Advancing business performance through data-driven process automation: A case study of digital transformation in the banking sector. *International Journal of Multidisciplinary Research Updates*, *8*(2), 12–22. DOI: 10.53430/ijmru.2024.8.2.0049

Adewumi, A., Ewim, S. E., Sam-Bulya, N. J., & Ajani, O. B. (2024). *Améliorer la performance de l'entreprise grâce à l'automatisation des processus basée sur les données: une étude de cas de la transformation numérique dans le secteur bancaire*. Mises à jour de la Revue internationale de recherche multidisciplinaire, 08(02), 012–022. https://doi.org/DOI: 10.53430/ijmru.2024.8.2.0049

Aguinis, H., Boyd, B. K., Pierce, C. A., & Short, J. C. (2011). Walking new avenues in management research methods and theories: Bridging micro and macro domains. *Journal of Management*, *37*(2), 395–403. DOI: 10.1177/0149206310382456

Allioui, A., & Allioui, H. (2024). *Une forte croissance financière dynamique des PME familiales marocaines: quelle faisabilité de l'intelligence artificielle*. RAIS Journal for Social Sciences.

Allioui, H., & Mourdi, Y. (2023). Exploring the full potentials of IoT for better financial growth and stability: A comprehensive survey. *Sensors (Basel)*, *23*(19), 8015. DOI: 10.3390/s23198015 PMID: 37836845

Aubert, B. A., & Hamadache, K. (2015). Business Process Management: The Moroccan Experience. *Journal of Business Process Management*, *21*(4), 665–678.

Baden-Fuller, C., & Haefliger, S. (2013). Business models and technological innovation. *Long Range Planning*, *46*(6), 419–426. DOI: 10.1016/j.lrp.2013.08.023

Barney, J. B. (1991). Firm resources and sustained competitive advantage. *Journal of Management*, *17*(1), 99–120. DOI: 10.1177/014920639101700108

Barney, J. B., & Clark, D. N. (2007). *Resource-based theory: Creating and sustaining competitive advantage*. Oxford University Press. DOI: 10.1093/oso/9780199277681.001.0001

Battanta, L. F., Lancioni, G., Giolfo, M., & Magli, F. (2025). The Moroccan banking system and FinTech deployment: An overview of the literature. *European Journal of Islamic Finance*. https://doi.org/DOI: 10.13135/2421-2172/10550

Battanta, L. F., Lancioni, G., Giolfo, M., & Magli, F. (2025). The Moroccan banking system and FinTech deployment: An overview of the literature. *European Journal of Islamic Finance*, *12*(1), 52–73. DOI: 10.13135/2421-2172/10550

Battanta, L. F., Lancioni, G., Giolfo, M., & Magli, F. (2025). The Moroccan banking system and FinTech deployment: An overview of the literature. *Home / Archives, 12*(1). https://doi.org/DOI: 10.13135/2421-2172/10550

Birkinshaw, J., Hamel, G., & Mol, M. J. (2008). Management innovation. *Academy of Management Review, 33*(4), 825–845. DOI: 10.5465/amr.2008.34421969

Bolton, R., & Hannon, M. (2016). Gouverner les transitions durables par l'innovation des modèles d'affaires: Vers une compréhension systémique. *Research Policy, 45*(9), 1731–1742. DOI: 10.1016/j.respol.2016.05.003

Borenstein, M., Hedges, L. V., Higgins, J. P. T., & Rothstein, H. R. (2011). *Introduction to Meta-Analysis*. Wiley.

Bouamor, D., Semmar, N., & Zweigenbaum, P. (2012). Identifying bilingual multi-word expressions for statistical machine translation. *Proceedings of the Eighth International Conference on Language Resources and Evaluation (LREC 2012)*, 674–679. European Language Resources Association (ELRA). https://perso.limsi.fr/pz/FTPapiers/Bouamor_LREC2012.pdf

Bouazza, A. B., Ardjouman, D., & Abada, O. (2015). Establishing the factors affecting the growth of small and medium-sized enterprises in Algeria. *American International Journal of Social Science, 4*(2), 101–115.

Bouazza, A. B., Ardjouman, D., & Abada, O. (2015). Establishing the factors affecting the growth of small and medium-sized enterprises in Algeria. *American International Journal of Social Science, 4*(2), 101–115. https://www.aijssnet.com

Bouazza, A. B., Ardjouman, D., & Abada, O. (2015). Establishing the factors affecting the growth of small and medium-sized enterprises in Algeria. [Center for Promoting Ideas, USA.]. *American International Journal of Social Science, 4*(2), 101–115.

Braun, V., & Clarke, V. (2006). Using thematic analysis in psychology. *Qualitative Research in Psychology, 3*(2), 77–101. DOI: 10.1191/1478088706qp063oa

Brynjolfsson, E., & McAfee, A. (2014). *The Second Machine Age: Work, Progress, and Prosperity in a Time of Brilliant Technologies*. W.W. Norton & Company.

Burns, T., & Stalker, G. M. (1961). *The Management of Innovation*. Tavistock Publications.

Camison, C., & Villar-Lopez, A. (2014). Organizational innovation as an enabler of technological innovation capabilities and firm performance. *Journal of Business Research, 67*(2), 2891–2902. DOI: 10.1016/j.jbusres.2012.06.004

Casadesus-Masanell, R., & Ricart, J. E. (2010). From strategy to business models and onto tactics. *Long Range Planning, 43*(2-3), 195–215. DOI: 10.1016/j.lrp.2010.01.004

Chen, Y. S., Lin, M. J. J., & Chang, C. H. (2016). The influence of intellectual capital on new product development performance – The manufacturing companies of Taiwan as an empirical study. *Journal of Business Research, 69*(5), 1849–1853.

Crossan, M. M., & Apaydin, M. (2010). A multi-dimensional framework of organizational innovation: A systematic review of the literature. *Journal of Management Studies, 47*(6), 1154–1191. DOI: 10.1111/j.1467-6486.2009.00880.x

Cuervo-Cazurra, A., & Genc, M. (2008). Transforming disadvantages into advantages: Developing-country MNEs in the least developed countries. *Journal of International Business Studies*, *39*(6), 957–979. DOI: 10.1057/palgrave.jibs.8400390

Damanpour, F., & Aravind, D. (2012). Managerial innovation: Conceptions, processes, and antecedents. *Management and Organization Review*, *8*(2), 423–454. DOI: 10.1111/j.1740-8784.2011.00233.x

Damanpour, F., & Evan, W. M. (1984). Organizational innovation and performance: The problem of "organizational lag". *Administrative Science Quarterly*, *29*(3), 392–409. DOI: 10.2307/2393031

Damanpour, F., & Gopalakrishnan, S. (2001). The dynamics of the adoption of product and process innovations in organizations. *Journal of Management Studies*, *38*(1), 45–65. DOI: 10.1111/1467-6486.00227

Damanpour, F., & Schneider, M. (2006). Phases of the adoption of innovation in organizations: Effects of environment, organization, and top managers. *British Journal of Management*, *17*(3), 215–236. DOI: 10.1111/j.1467-8551.2006.00498.x

Davenport, T. H. (1993). *Process Innovation: Reengineering Work through Information Technology*. Harvard Business Press.

Delbecq, A. L., & Mills, P. K. (1985). Des pratiques managériales au service de l'innovation. *Dynamique Organisationnelle*, *14*(1), 24–34. DOI: 10.1016/0090-2616(85)90041-5

Demeter, K., & Losonci, D. (2015). Lean production and business performance: International empirical results. *Competitiveness Review*, *25*(2), 205–223.

Elbanna, S., & Child, J. (2007). Influences on strategic decision effectiveness: Development and test of an integrative model. *Strategic Management Journal*, *28*(4), 431–453. DOI: 10.1002/smj.597

Elkouiri, K., & Kadouri, A. (2022). Concentration de la propriété et performance des entreprises industrielles cotées à la Bourse de Casablanca. *International Journal of Accounting, Finance, Auditing, Management and Economics, 3*(3-2), 45–62. https://doi.org/DOI: 10.5281/zenodo.6581539

Elkouiri, K., & Kadouri, A. (2022). *Actionnariat, concentration et performance des sociétés industrielles cotées à la Bourse de Casablanca*. International Journal of Accounting, Finance, Auditing [IJAFAME]. *Management and Economics*, *3*, 3–2. DOI: 10.5281/zenodo.6581539

Elkouiri, K., & Kadouri, A. (2022). Actionnariat, concentration et performance des sociétés industrielles cotées à la Bourse de Casablanca. *Domicile/Archives, 3*(3-2). https://doi.org/DOI: 10.5281/zenodo.6581539

Fang, E. A., Wu, Q., Miao, C., Xia, J., & Chen, D. (2013). L'impact des nouvelles pratiques technologiques en matière de produits et d'opérations sur la structure de l'organisation. *Revue internationale d'économie de production, 145*(2), 733–742. https://doi.org/DOI: 10.1016/j.ijpe.2013.06.003

Govindarajan, V., & Trimble, C. (2004). *Strategic Innovation and the Science of Learning*. Harvard Business School Press. DOI: 10.1109/EMR.2004.25051

Haber, E., & Zarsky, T. (2018). Cybersecurity for infrastructure: A critical analysis. *Florida State University Law Review. Florida State University. College of Law*, *44*(2), 515–577. https://ir.law.fsu.edu/lr/vol44/iss2/3

Hammer, M., & Champy, J. (1993). *Reengineering the Corporation: A Manifesto for Business Revolution*. Harper Business.

Hammouch, H., Manta, O., & Palazzo, M. (2024). Adapter les systèmes de contrôle de gestion aux facteurs de contingence organisationnelle: Une étude sur les entreprises industrielles marocaines. *Entreprises, 4*(4), 883–898. DOI: 10.3390/businesses4040048

Haponava, T., & Al-Jibouri, S. A. (2011). Proposed system for measuring project performance using process-based key performance indicators. *Journal of Management Engineering, 28*(2), 140–149. Advance online publication. DOI: 10.1061/(ASCE)ME.1943-5479.0000078

Haponava, T., & Al-Jibouri, S. H. S. (2011). Système proposé pour mesurer le rendement du projet à l'aide d'indicateurs clés de rendement fondés sur les processus. *Journal of Management Engineering, 28*(2), 149. DOI: 10.1061/(ASCE)ME.1943-5479.0000078

Haponava, T., & Al-Jibouri, S. H. S. (2011). Système proposé pour mesurer le rendement du projet à l'aide d'indicateurs clés de rendement fondés sur les processus. *Journal of Management Engineering, 28*(2). Advance online publication. DOI: 10.1061/(ASCE)ME.1943-5479.0000078

He, Z. L., & Wong, P. K. (2004). Exploration vs. exploitation: An empirical test of the ambidexterity hypothesis. *Organization Science, 15*(4), 481–494. DOI: 10.1287/orsc.1040.0078

Hitt, M. A., Xu, K., & Carnes, C. M. (2016). Resource based theory in operations management research. *Journal of Operations Management, 41*(1), 77–94. DOI: 10.1016/j.jom.2015.11.002

Hofstede, G. (2001). *Culture's Consequences: Comparing Values, Behaviors, Institutions, and Organizations Across Nations*. SAGE Publications.

Hoskisson, R. E., Castleton, M. W., & Withers, M. C. (2009). Complémentarité dans le suivi et la cohésion: un suivi plus intense conduit à une rémunération plus élevée des dirigeants. *Perspectives de l'Académie de gestion, 23*(2), 57–74. https://doi.org/DOI: 10.5465/amp.2009.39985541

Hoskisson, R. E., Covin, J., Volberda, H. W., & Johnson, R. A. (2010). Revitaliser l'entrepreneuriat: À la recherche de nouvelles possibilités de recherche. *Journal of Management Studies, 47*(6), 1203–1221. DOI: 10.1111/j.1467-6486.2010.00997.x

Hoskisson, R. E., Covin, J., Volberda, H. W., & Johnson, R. A. (2010). Revitalizing entrepreneurship: Seeking new research opportunities. *Strategic Management Journal, 31*(12), 1051–1059. DOI: 10.1111/j.1467-6486.2010.00997.x

Ichniowski, C., Shaw, K., & Crandall, R. W. (1995). Vieux chiens et nouveaux tours: Déterminants de l'adoption de pratiques de travail améliorant la productivité. [Brookings Institution Press]. *Brookings Papers on Economic Activity: Microeconomics, 1995*, 1–65. DOI: 10.2307/2534771

Jérab, D., & Mabrouk, T. (2023). Le rôle du leadership dans le changement de culture organisationnelle. *Journal Name, 13*, 1–13.

Kafouros, M. I., Buckley, P. J., Sharp, J. A., & Wang, C. (2008). The role of internationalization in explaining innovation performance. *Technovation, 28*(1-2), 63–74. DOI: 10.1016/j.technovation.2007.07.009

Kaluarachchi, B. N., & Sedera, D. (2024). Améliorer l'efficacité grâce à l'engagement client alimenté par l'IA en fournissant des solutions personnalisées dans le secteur bancaire. Dans *Intégration des technologies basées sur l'IA dans le marketing de services* (chap. 16, 44 pages). DOI: 10.4018/979-8-3693-7122-0.ch016

Kane, G. C., Palmer, D., Phillips, A. N., Kiron, D., & Buckley, N. (2015). *Strategy, not technology, drives digital transformation. MIT Sloan Management Review and*. Deloitte University Press.

Kimberly, J. R., & Evanisko, M. J. (1981). Organizational innovation: The influence of individual, organizational, and contextual factors on hospital adoption of technological and administrative innovations. *Academy of Management Journal*, *24*(4), 689–713. DOI: 10.2307/256170 PMID: 10253688

Kowalkowski, C., Windahl, C., Kindström, D., & Gebauer, H. (2015). Quelle transition de service? Repenser les hypothèses établies sur les stratégies de croissance des fabricants axées sur les services. *Gestion du Marketing Industriel*, *45*, 59–69. DOI: 10.1016/j.indmarman.2015.02.016

Lam, A. (2004). Organizational innovation. In Fagerberg, J., Mowery, D. C., & Nelson, R. R. (Eds.), *The Oxford Handbook of Innovation* (pp. 115-147). Oxford University Press.

Le Breton-Miller, I., & Miller, D. (2018). Regard sur le passé et l'avenir: « Gouvernance familiale et performance de l'entreprise: agentivité, intendance et capacités ». *Family Business Review*, *31*(2). Advance online publication. DOI: 10.1177/0894486518773850

Lema, J. M., & Suarez Martinez, S. (Eds.). (2017). *Innovative wastewater treatment & resource recovery technologies: Impacts on the environment*. Elsevier., DOI: 10.1016/B978-0-12-811034-3.00001-0

Makri, M., Lane, P. J., & Gomez-Mejia, L. R. (2006). Incitatifs des PDG, innovation et performance dans les entreprises à forte intensité technologique: Un rapprochement des systèmes d'incitation basés sur les résultats et le comportement. *Strategic Management Journal*, *27*(9), 931–956. DOI: 10.1002/smj.560

Markides, C. (1997). Strategic Innovation. *Sloan Management Review*, *38*(3), 9–23.

Marquis, C., & Raynard, M. (2015). Stratégies institutionnelles dans les marchés émergents. *Annales de l'Académie de gestion, 9*(1), 291-335. https://doi.org/DOI: 10.5465/19416520.2015.1014661

Marquis, C., & Raynard, M. (2015). Stratégies institutionnelles dans les marchés émergents. *Annals of the Academy of Management*, *9*(1). Advance online publication. DOI: 10.5465/19416520.2015.1014661

Marquis, C., & Raynard, M. (2015). *Stratégies institutionnelles dans les marchés émergents.* Annales de l'Académie de gestion, 9(1). https://doi.org/DOI: 10.5465/19416520.2015.1014661

Martínez-Conesa, I., Soto-Acosta, P., & Palacios-Manzano, M. (2017). Corporate social responsibility and its effect on innovation and firm performance: An empirical research in SMEs. *Journal of Cleaner Production*, *142*, 2374–2383. DOI: 10.1016/j.jclepro.2016.11.038

McClure, P. K. (2017). "You're fired," says the robot: The rise of automation in the workplace, technophobes, and fears of unemployment. *Social Science Computer Review*, *36*(2), 139–156. Advance online publication. DOI: 10.1177/0894439317698637

Moher, D., Liberati, A., Tetzlaff, J., & Altman, D. G.PRISMA Group. (2009). Preferred reporting items for systematic reviews and meta-analyses: The PRISMA statement. *PLoS Medicine*, *6*(7), e1000097. DOI: 10.1371/journal.pmed.1000097 PMID: 19621072

Mol, M. J., & Birkinshaw, J. (2009). The sources of management innovation: When firms introduce new management practices. *Journal of Business Research*, *62*(12), 1269–1280. DOI: 10.1016/j.jbusres.2009.01.001

Moromoke, O. A., Aro, O. E., Adepetun, A. O., & Iwalehin, O. (2024). Navigating regulatory challenges in digital finance: A strategic approach. *International Research Journal of Modernization in Engineering Technology and Science*, *6*(10). Advance online publication. DOI: 10.56726/IRJMETS62892

Moromoke, O. A., Aro, O. E., Adepetun, A. O., & Iwalehin, O. (2024). Navigating regulatory challenges in digital finance: A strategic approach. *International Research Journal of Modernization in Engineering Technology and Science*, *6*(10). https://www.doi.org/10.56726/IRJMETS62892

Moromoke, O. A., Aro, O. E., Adepetun, A. O., & Iwalehin, O. (2024). Surmonter les défis réglementaires en finance numérique: Une approche stratégique. *International Research Journal of Modernization in Engineering Technology and Science*, *06*(10). Advance online publication. DOI: 10.56726/IRJMETS62892

Neagu, N., Dorer, K., Greenwood, D., & Calisti, M. (2006). *LS/ATN: Reporting on a successful agent-based solution for transport logistics optimization*.

Neagu, N., Dorer, K., Greenwood, D., & Calisti, M. (2006). *Rapport sur une solution efficace basée sur des agents pour l'optimisation de la logistique de transport*. IEEE Workshop on Distributed Intelligent Systems: Collective Intelligence and Its Applications (DIS'06), 15-16 juin 2006, Prague, République tchèque. IEEE. DOI: 10.1109/DIS.2006.46

O'Brien, K. J. (1996). Résistance légitime. *Politique mondiale, 49*(1), 31–55. https://doi.org/DOI: 10.1353/wp.1996.0022

O'Brien, K. J. (1996). Légitime résistance. *World Politics*, *49*(1), 31–55. DOI: 10.1353/wp.1996.0022

Papadakis, V. M., Lioukas, S., & Chambers, D. (1998). Strategic decision-making processes: The role of management and context. *Strategic Management Journal*, *19*(2), 115–147. DOI: 10.1002/(SICI)1097-0266(199802)19:2<115::AID-SMJ941>3.0.CO;2-5

Papadakis, V. M., Lioukas, S., & Chambers, D. (1998). Processus décisionnels stratégiques: Le rôle de la direction et le contexte. *Strategic Management Journal*, *19*(2), 115–147. DOI: 10.1002/(SICI)1097-0266(199802)19:2<115::AID-SMJ941>3.0.CO;2-5

Papadakis, V. M., Lioukas, S., & Chambers, D. (1998). Strategic decision-making processes: The role of management and context. *Strategic Management Journal*, *19*(2), 115–147. DOI: 10.1002/(SICI)1097-0266(199802)19:2<115::AID-SMJ941>3.0.CO;2-5

Patton, M. Q. (2002). *Qualitative Research and Evaluation Methods*. SAGE Publications.

Ramdani, B., Rothwell, B., & Boukrami, E. (2020). Open Banking: l'émergence de nouveaux modèles d'affaires numériques. *Revue internationale de l'innovation et de la gestion de la technologie, 17*(5), 2050033. https://doi.org/DOI: 10.1142/S0219877020500339

Reschke, C. H., Bogenhold, D., & Kraus, S. (2010). Comment l'innovation et l'entrepreneuriat peuvent vaincre l'incertitude et la complexité: Apprendre de l'inattendu. *International Journal of Complexity in Leadership and Management, 1*(1), 55–71. DOI: 10.1504/IJCLM.2010.035789

Reschke, C. H., Bogenhold, D., & Kraus, S. (2010). How innovation and entrepreneurship can overcome uncertainty and complexity: Learning from the unexpected. *International Journal of Complexity in Leadership and Management, 1*(1), 55–71. DOI: 10.1504/IJCLM.2010.035789

Ribeiro, D., Roig, S., & Cuervo, Á. (Eds.). (2007). *Entrepreneurship: Concepts, theory, and perspective.* Springer Berlin Heidelberg., DOI: 10.1007/978-3-540-48543-8

Sanders, K., Jorgensen, F., Shipton, H., Van Rossenberg, Y., Cunha, R., Li, X., Rodrigues, R., Wong, S. I., & Dysvik, A. (2018). Performance-based rewards and innovative behaviors. *Human Resource Management, 57*(6), 1339–1352. DOI: 10.1002/hrm.21918

Sanders, K., Jorgensen, F., Shipton, H., Van Rossenberg, Y., Cunha, R., Li, X., Rodrigues, R., Wong, S. I., & Dysvik, A. (2018). Récompenses basées sur la performance et comportements innovants. *Human Resource Management, 57*(3). Advance online publication. DOI: 10.1002/hrm.21918

Schilling, M. A. (2017). *Strategic Management of Technological Innovation.* McGraw-Hill Education.

Schumpeter, J. A. (1934). *The theory of economic development.* Harvard University Press.

Simons, R. (1994). *Levers of control: How managers use innovative control systems to drive strategic renewal.* Harvard Business School Press.

Sliwka, D. (2007). La confiance comme signal d'une norme sociale et les coûts cachés des systèmes d'incitation. *The American Economic Review, 97*(3), 999–1012. DOI: 10.1257/aer.97.3.999

Spencer, D. A. (2018). Fear and hope in the age of mass automation: Debating the future of work. *New Technology, Work and Employment, 33*(1), 1–12. Advance online publication. DOI: 10.1111/ntwe.12105

Sundbo, J. (1997). Management of innovation in services. *Service Industries Journal, 17*(3), 432–455. DOI: 10.1080/02642069700000028

Teece, D. J. (2010). Business models, business strategy and innovation. *Long Range Planning, 43*(2-3), 172–194. DOI: 10.1016/j.lrp.2009.07.003

Teece, D. J., Pisano, G., & Shuen, A. (1997). Dynamic capabilities and strategic management. *Strategic Management Journal, 18*(7), 509–533. DOI: 10.1002/(SICI)1097-0266(199708)18:7<509::AID-SMJ882>3.0.CO;2-Z

Tranfield, D., Denyer, D., & Smart, P. (2003). Towards a methodology for developing evidence-informed management knowledge by means of systematic review. *British Journal of Management, 14*(3), 207–222. DOI: 10.1111/1467-8551.00375

Volberda, H. W., Van Den Bosch, F. A., & Heij, C. V. (2014). Management innovation: Management as fertile ground for innovation. *European Management Review, 11*(1), 1–15. DOI: 10.1111/emre.12007

Wan, W. P., Hoskisson, R. E., & Yiu, D. W. (2011). Théorie des ressources et diversification des entreprises: Réalisations et opportunités. *Journal of Management*, *37*(5). Advance online publication. DOI: 10.1177/0149206310391804

Winkler, H., & Kaluza, B. (2008). Utilisation d'un calcul de création de valeur pour la détection et l'évaluation des innovations de produits et de processus dans les réseaux de valeur. In Specht, D. (Ed.), *Innovations de produits et de processus dans les chaînes de valeur* (pp. 1–31). Gabler., DOI: 10.1007/978-3-8349-9765-4_1

Winkler, H., Schemitsch, H. B., & Kaluza, B. (2008). Problèmes d'agence principal dans les réseaux de création de valeur orientés projet: identification des problèmes et approches de solution. In *Réseaux de création de valeur* (pp. 65–85). Springer., DOI: 10.1007/978-3-540-89936-5_4

Winkler, H., Schemitsch, H. B., & Kaluza, B. (2008). Problèmes Principal-Agent dans les réseaux de création de valeur orientés projet — Identification des problèmes et approches de solutions. In Becker, J., Knackstedt, R., & Pfeiffer, D. (Eds.), *Réseaux de création de valeur* (pp. 65–85). Physica-Verlag HD., DOI: 10.1007/978-3-7908-2056-0_4

Zahra, S. A. (2017). Gouvernement, propriété et entrepreneuriat d'entreprise: L'impact modérateur des opportunités technologiques de l'industrie. *Academy of Management Journal*, *39*(6). Advance online publication. DOI: 10.5465/257076

Chapter 5
Information Science as the Base for Multi-, Inter-, and Trans-Disciplinary Studies: A More Robust Way to Innovate Through Digital Transformation

George Leal Jamil
https://orcid.org/0000-0003-0989-6600
Informacoes em Rede Consultoria e Treinamento Ltda, Brazil

ABSTRACT

Nowadays, several topics are approached by the Business arena before the scientific community could think, study and publish their thoughts in order to provide a robust background to it. Among them are Digital Transformation, the new waves of Artificial Intelligence and Data Science. From several facts and perceived occurrences, it is possible to relate failures in some technological projects as originated by the lack of scientific background, which led to impulsive decisions to implement, resulting in precarious solutions. The main topic addressed is innovation produced by digital transformation implementation. For this purpose, three different methodological approaches were considered: Multidisciplinarity, Interdisciplinarity and Transdisciplinarity. For this chapter, Information Science, practicing multi, inter and transdisciplinarity, can be a productive and open scientific field where digital transformations can occur and be analyzed and studied, in a robust and coherent way, leading to innovations which can offer better chances to be developed and, finally, succeed.

INTRODUCTION - A DEMAND FOR INNOVATION THROUGH SCIENTIFIC METHODOLOGY

Innovation has been a frequent demand for organizations, becoming a strategic need for their survival and overall competitiveness. Management topics, especially those concerning the application of technologies quickly become a key focus of continuous approaches and discussions to innovate. In an

DOI: 10.4018/979-8-3693-4030-1.ch005

unstable, vulnerable and volatile context, it remains a strategic priority for organizations, aiming to design, manufacture and offer new forms of value to negotiate innovations in real competitive scenarios.

In this chapter, the objective is to analyze how scientific approaches for methodologies can be applied to allow organizations to align information and knowledge to innovate. A specific observation will be dedicated to information technology resources, and how they can be applied with scientific background. This way, Information Science (IS) is a field where, through an open application of multi, inter and transdisciplinary methods that have been used for many years, is selected as the study paradigm to be adopted.

The problem, to be revisited during the discussion of this chapter, is to apply such scientific methodologies to design, develop, test and deliver tools, resources and artifacts which will be supported and applied with fundamental scientific bases, resulting in more robust and durable market solutions. Conversely, without these fundamentals, it is possible to reason that technology can be applied for various reasons, not providing essentials to promote a solid solution, merely functional. These "functional" results will suffer from lack of stability and adaptability, leading to unstable applications, which will bring more impact to the organizational structure.

Faced with this scenario that foreshadows the precariousness of organizational solutions linked to information that did not consider the practices and experiences arising from a scientific field or background, as that of Information Science, the objective is to define also in the main focus for our study, the phenomenon called "Digital Transformation" (DT) seeking to restore that effective contribution of a scientific base to comprehend and design an organizational change towards digital resources applications.

This last event, DT, therefore, show problems commonly seen for fast and incompletely comprehended adoption of information technology solutions, such as: Ambitious visions for the results, lack of a basic foundation for project management principles, precarious evaluation of information resources and databases availability, failures of integration between systems, technology resources and processes, and, finally, excessive focus on technology and not on the organizational social and market demands. For a brief confirmation of this scenario, we can observe this analysis applied to the initial adoption of artificial intelligence resources, highlighting similar problems according to business analysts and end-users (Siddique, 2021).

Additionally, given the opportunity of the study in question, it is relevant to improve the understanding of the concept of digital transformation, with the support of the basis of Information Science. This presents an additional benefit, when it is built into a reasoning, with the contribution by evaluating through the methodological forms of multidisciplinarity, interdisciplinarity, and transdisciplinarity. The richness provided by multi and interdisciplinary research, as well as the potential of transdisciplinarity, in the context of Information Science (Rascão, Jamil, and Marques, 2021), a field genuinely interdisciplinary in its origins, bringing a superior degree of clarity that will enable a greater understanding of digital transformation, given the challenges that this dynamic entails.

Therefore, the objective of the discussion of this chapter is synthesized in perceiving digital transformation from the perspective of Information Science, addressing to further the discussion of methodological forms of interdisciplinarity, multidisciplinarity, and transdisciplinarity practices. The chapter begins with this statement of objectives, evolving to the discussion of the conceptual background, advancing to its own research exercise section where the concepts will be related to the objective of the study. Ahead, in the next section, some cases will be enunciated and discussed, ending with the conclusions and reports of the cited references.

It is believed that this text makes it possible to promote a better view of the conceptual interrelationship, its application, and an additional and perennial motivation, so that Information Science is appreciated as a contextual basis to analyze facts and organizational, market processes that involve information management, providing a greater understanding of the fundamentals, perspectives, and unfolding for the themes, allowing the reader to better avoid problems technical and business pitfalls such as those described above.

CONCEPTUAL BACKGROUND

This conceptual background will be developed analyzing some phenomena where some market propositions were launched in the market and attracted a lot of attention - investments, plans and projects - without the background which is proposed in the chapter. Later, the methodological approach for multi, inter and transdisciplinarity will be defined, followed by a discussion about Information Science as the scientific field which will host the studies about digital transformation. A final approach regarding DT will also be discussed, composing the needed background for the following discussion.

Basic Scenario - How To Innovate In A Potential Information Conundrum

From Schumpeter (1934) it is possible to learn about two sources of continuous innovation: Technology and Business models. Interestingly, some signals and facts from scenarios, where modified positionings were hastily introduced by competitors can also be recognized from these two origins.

It is possible to describe events and facts which suggest the fundamental problem of lack of scientific background or its misinterpretation. Some decades ago, for instance, we had the proposition of Reengineering as a revolutionary approach to modernize and give focus to organizations (Hammer, 1990). From Harrington (1998) and (Wang, 2008), it became evident that the expression of focus - reengineering - was misused as an argument for revising processes to optimize business productivity, and even the life cycle management for the new version of organizational processes, resulting from the application of reengineering, portraying its mismanagement. It was a clear misapplication of a strategic to operational context, which was precariously adopted by market organizations, leading to several signals of confusion, ending with the general rejection of the concept of reengineering, as companies struggle with cuts and forced, scientifically unsupported, reductions in their activities. The lack of scientific comprehension of a definition, led to an immediate, impulsive adoption of actions, resulting in unnecessary losses of jobs and structures.

Addressing the application of computing services in organizations, a first moment to be highlighted was the decentralization of processing, related to the massive event of implementation of microcomputers in workplaces (Jamil, 2001; The Economist, 2018). These proposals were radical in proposing the "disconnection of mainframes", phasing out large mainframe computer systems, expensive and complex to manage (a fact that did not occur as predicted).

This "wave" only considered the processing power of microcomputers and the consequent migration of small applications to these environments, potentially reducing the demand for large computers used in a centralized manner. Besides not having occurred in the intended and predicted proportion, the decentralization without basic scientific support to design plans and projects, ended up causing problems of information assets control and potential future security incidents, as noted later (Ghezzi, 2009). This phenomenon led to a cycle of alternating decentralization and centralization of computing systems.

A recent event was the adoption of enterprise information systems, such as ERPs, WMSs SCMs, and CRMs, among several others (O'Brien & Marakas, 2006; Stair and Reynolds, 2009). This fact can be considered a harbinger of the approach developed in this chapter, since such systems were generally offered to the market as technological solutions with rectilinear, unidirectional application (Jamil, 2001; Lacerda and Jamil, 2021). Failures in projects associated with these systems occur, mainly, because of the lack of comprehension of organizational problems, processes and projects to solve and compose, resulting in an uncertain implementation, which was designed to react to events, not to address information and its composition towards knowledge, a main topic studied by Information Science. The IS field study, as it will be seen in this chapter, could be applied to allow more robust and effective solutions.

It is to consider the general disregard for the principles defined for information systems design and implementation, such as the composition of elements for information management (consequently of data and knowledge), oriented to implement alignments, structuring of processes, and business models (Lacerda and Jamil, 2021). Given this "immediatism" approach, there were many cases of allegations of strategic, time, investment, and organizational synergy losses. It is to be verified that even in their most modern versions, such as those of implementation through cloud computing resources and services, it is possible to detect that such problems of lack of conceptual background persist (Ecosys, 2014).

Some other events and cases could be cited and commented on, resulting in a broader composition which always signals that information technology resources, as technological support for innovation through digital transformation, were adopted without paying attention to the scientific base. In the following, a work will be done to present an alternative which could comprise a scientific background for these projects.

Methodological Contexts - Multi, Inter And Transdisciplinarity

Multidisciplinarity, interdisciplinarity and transdisciplinarity can be regarded, in a simple view, as ways to correlate studies from different fields, for a possible wider and evolutionary process of potential relationship to enable a broader scientific contribution (NCSU, 2021). Their differences rely both on delimitations for each contributive scientific field and the outcome of their practices and results, as it will be studied in the following.

Initially, the multidisciplinary methodology is approached as the situation where several specialties, scientific fields, are exercised in simultaneous or sequential action, with a view to contribute to a co-operative behavior, observing a certain objective of analysis (Klein, 2010; Badecki, 2019). This action keeps each of the sources of methodological contribution in its place of origin, without necessarily having a permanent collaboration that results in "crossing borders" (Badecki, 2019), maintaining the individualities of those original fields when the final treatment of the objective of analysis is produced (Nicolescu, 2014; Loureiro and Guimarães, 2019).

In multidisciplinary studies, it is possible to have the participation and contribution with diverse points of view, conceptual bases, evaluation methodologies, diagnosis of realities and analyses, in simultaneous contribution, in an objective, productive and collaborative way, essentially without requiring the creation of a unified, permanent research framework (Alvargonzález, 2011). The contribution of several specialists and sources of analysis, within their own exercise, serves to elaborate complementary views on the object and associated phenomena, leading to a multiple set of complementary and distinct views on the fact under discussion. Without the demand for proof regarding their cooperation, as it is defined as temporary and with a limited analytical goal, no additional experiments are demanded to affirm a new methodological course.

A simpler form in the preparation, in its requirements, as stated above, multidisciplinarity is performed, for example, when clients from various backgrounds examine a value proposition by a commercial communication company. In these cases, their reactions are equivalent to analyses, based on culture and previous experiences, leading to diverse views, allowing their contributions in a brief and quick way (Jamil, 2001).

The challenge for multidisciplinary analysis, however, is to generate a consolidated view with the studies and productions coming from the various contributory fields. Since, according to the concept, after this individualized analysis, multidisciplinarity does not advocate carrying out the final consolidation, it is up to a delegating body - who promotes, who receives the research - to bring this consensus (Alvargonzález, 2011; Loureiro and Guimarães, 2019). The final production, as a message of consistent analysis of the object of study, should contemplate a balanced view between the various contributions and a message with fluidity and access, allowing the recipient of the research to understand it as integral and treated by diverse views (Jamil, 2021).

The study of interdisciplinarity, in turn, shows the first method or methodological context that advances towards some level of permanent relationship between scientific fields. In a typical interdisciplinary study, diverse fields contribute, initially, with the collaborative definition of a methodological basis that prescribes instruments, definitions of application and analysis of research objects (Zaman and Goschin, 2010). Faced with this proposition of a context to be immediately applied for data collection process design, as well as in the analyses to be done with this data, there is a proposal for mutual observation and potential understanding between the fields (Madsen, 2012; Rascão, Jamil and Marques, 2022).

Such a proposition, however, brings new charges and questions such as: Are the forms arising from the joint observation, once again, balanced in terms of the fundamentals and voice of each of the fields involved? Is there an expression of the same in the resulting analyses? Is there therefore a contribution from the fields involved or just something collateral from one of the fields in addition to another predominant one (would this be a desirable fact or not)?

The practice of interdisciplinarity, something that has been longed for and dealt with over a long period of research, brings with it additional responsibility in the definition and practice of methods (Loureiro and Guimarães, 2019). These requests, which must be addressed by researchers and practitioners, demand answers that are expensive, time consuming and of high level, to be produced to consider the scientific level for the outcome.

Interdisciplinarity, by maintaining the objectives in analysis of more than one scientific field and having demands delimited to one study or studies of the same object of analysis, is considered a fundamental stone even for scientific fields formation. A special highlight is that of Information Science, where information - whether considered "information object" or "information process" - is already an object of study of an interdisciplinary nature (Saracevic, 1995; Huang and Chang, 2011; Jamil, 2011).

In Jamil (2021) it has seen that in the context of interdisciplinarity, at the essential Bauhaus movement, about a century ago, various contexts have benefitted from this scientific exercise, as is the elaborated case of the arts, especially in the case of painting. For example, the works of Vasily Kandinsky, in his evolution to abstractionism - precisely when a deeper conceptual field exercise is intended - the artist developed, during several years of his career, an approach that added experiences, forms and feelings from various fields, reaching the human sciences. One characteristic phase was marked when Kandinsky approached Biology and Human Sciences in his studies, then proposing new forms and figures, with smoother painting, in his artworks.

The interdisciplinary approach is a very successful way, if we analyze the frequency of positive citations to it in several fields, expressed by scientific communications, and even the contributions referenced in the productions of practical applications such as the development of market solutions. For example, it is possible to perceive interdisciplinary postures in the studies for interfaces of computer programs and applications to be used on mobile platforms (Bresciane, 2015). In these contexts, generally, designers add technical, technological and formal contents from the end-use areas of the application are considered towards a composition of research instruments and objectives, as a result, without, essentially, modifying those native scientific fields which cooperated for it (IDEO, 2022).

Starting from interdisciplinarity, finally, the promise of transdisciplinarity is reached, where the construction of a new field of analysis, potentially of scientific foundation, arises from the fusion between antecedent or originating fields. Here, in transdisciplinarity, in addition to the technical construction, the analytical and critical context is developed, resulting in a new basis of paradigm, of scientific construction (Flores and Rocha Filho, 2016). The possibilities of conducting isolated research only, as it occurs in the cases stated for multidisciplinarity and interdisciplinarity, are far surpassed, since the goal is to create a whole new field of reflection, thought and, finally, methodological construction (Max-Neef, 2005; Rigolot, 2020).

In contrast, the challenges are much greater in transdisciplinary practice. In order to affirm that a new field has been reached, rigorous scientific standards must be achieved, the interrelationship with the original fields needs to be clear and the practical definitions of this new context also need robustness to be addressed (Lawrence and Despres, 2004; NCSU, 2020). A scientific field, therefore, can be generated by the transdisciplinary objective, now serving as a basis for new practices and exercises of philosophical reasoning, delimitations, reflections and application in studies to be developed.

A practical portrait of the search for the transdisciplinary is the generation of programs and courses that come from the combination of basic forms of thought, such as those of the confluence of Medicine and technology-based areas, such as Computing and Engineering. For example, when analyzing the proposals for emphases and formations in Biomedicine, Bioengineering and Biosafety, it is verified not only delimited fields in agglutination, but the birth of ways of thinking colligated from the previous bases, giving flow to innovative forms and contents.

Here, it is important to affirm that the complete exercise that will identify the resulting new scientific field in the future will only occur after time. Answers where the community will perceive the innovative contribution, first in the search for expository practice, then advancing to inductions and, finally, in obtaining answers that make the dialogue effective on behalf of the created field will be observed only after a thorough path, which composes from the basic conceptual background definitions and a solid test of methodological contexts (Zaman and Goscin, 2010; Rigolot, 2020).

A comment must be made to elaborate an observation of evolutionary relationship between these methodological contexts. Much is sought to associate something like a progression that begins in multidisciplinarity, advances towards interdisciplinarity, reaching transdisciplinarity (Alvargonzález, 2011; Loureiro and Guimarães, 2019). This evolution can be considered in aspects, such as the proficiency of researchers, the communications of studies, the emphasis on models and instruments (in interdisciplinarity) or even in the initial formation of the transdisciplinary field.

This evolution of methodological approaches will constitute a way to apply Information Science background, first as one of the basic observation fields in multidisciplinary studies, then advancing with IS own capabilities for an interdisciplinary observation, resulting in methods to be used in advance. Finally, as it happened with Information Systems, Information Technology, Internet literacy and several other scientific fields, Information Science can perform quite well as an original source to form an innovative context, which, in result, composes a new scientific paradigm. This is a potential background to allow robust, managed and coherent innovations, such as those pertaining to the digital transformation development.

Digital Transformation - The Recall For Continuous Innovation

One of the most cited expressions in terms of the occurrence of new technological waves and their application is "Digital Transformation" (DT). Under this denomination, an attempt was made to encompass all experiences of introducing digital resources for organizational use, extending it to business partners, customers and stakeholders, arguably identifying innovative projects and processes, in the sense of proposing modifications to organizational strategic positioning. (Tadeu *et al.,* 2019; Lacerda and Jamil, 2021).

In a contribution widely accepted as foundational for understanding DT, Rogers (2017) argues that this process is the effective result of relevant perspectives of modification in the organizational value-added chains due to the insertion of new resources derived from digital technology. The author promotes an observation of market cases, dividing it into the perspectives of data management, value proposition, decisions and customer journeys, organizational structuring and its preparation for competition in markets, and finally, the innovative potential. Associating these perspectives, a strategic alternative scenario, supported by digitalization, will potentially offer an increased level of value aggregation, with the possibility of agility and cost scalability, due to the application of digital technologies (Rogers, 2017; Ross, Beath and Moker, 2019).

An important way to understand digital transformation is to bring the main attraction to the transformation concept and subsequently assess how digital technologies enhance these changes. Here, citing the contributions of Kotter (2012) in the observation of organizational innovative positionings, there is a basis for assessing that the changes themselves and the proposals of management in their leadership and maintenance lead to a consistent level for the application of new technologies, which result in innovative processes and methods, enlightening the relationship with the "digital" resources to implement such organizational changes.

The choice offered by technology, however, may be effective, but it carries the risk of overemphasis on technological focus, deviating from the planned organizational strategic objectives. This fact eventually results in technologically admissible results but without connection to the changes of value proposition positioning sought by organizations (Jamil and Silva, 2021; Lacerda and Jamil, 2021). This description entails exactly those facts described earlier, as projects without the scientific foundation and its derived problems.

Therefore, the definition that Digital Transformation, as a concept, is the organizational effort to effectively change its structures, methods, and processes, orienting itself according to its way of planning the strategy, adopting digital technologies for this purpose. This definition does not exclude the supposedly inverse path, of having the availability of digital technology and from there building new value-aggregation positionings. This brings potential innovation through a digital project which encompasses scientific principles, as intended in our development.

Innovation - A Motivation For Strategic Digital Transformation

Innovation is a top-level topic both for academic and scientific studies and for organizational management. In the last decades, it has been frequently studied and aimed for application through plans, academic designs, prospective studies and investments, reaching the level of an intended and optimized goal for any change of traditional positionings and strategic decisions (Jamil, Briones and Lema, 2019).

As it was defined in the modern era by Joseph Schumpeter in Schumpeter (1934), innovation is a way to promote, define and offer a change in any practical market solution, process or organizational standard, also defining a completely new, radical or "disruptive" positioning by one organization (OECD, 2005; Christensen, 2024).

Along with an object, process specification and organizational design for a new structure, innovation studies pay attention also to the process which can originate innovations, encompassing the opportunities shown by the market, provocations from value aggregation chains and customers´ reactions, to plans and projects and, finally, reaching the innovative strategic design for one organization as to position itself in a different, changed way (Firestein, 2019; Christensen, 2024).

Finally, an essential topic is how to consider the aspects of managing an innovation-driven or "to innovation" organizational process, which is approached as strategic innovation management. As it can be perceived, this topic is discussed by strategy authors, such as it can be seen in Davenport (1998), along with studies from Michael Porter, Henry Mintzberg and several other fundamental researchers and practitioners. This top-level topic regards advances and ways to manage the process for innovation, both turning it into a radically innovation-oriented process and attempting to optimize, leverage costs, identify and attract skilled people, grant safety and social respect, among many other strategic topics, regarding one new innovation positioning.

Innovation was discussed in here just for the chapter purpose, as it is a deep and wide context, socially and economically critical for mankind existence, far beyond of an important, although strict view, of products and services positioning.

Concluding the references background, it is time to discuss Information Science capabilities, as a modern and developing scientific field to allow a better way to apply multi, inter and transdisciplinary studies to improve digital transformation comprehension as an innovation enabling process.

THE INHERENTLY INTERDISCIPLINARY NATURE OF INFORMATION SCIENCE

Interdisciplinary is an inherent characteristic of Information Science, given especially the nature of the central object of focus of the field—Information—as it is treated as a dynamic, integrated, and densely interrelated concept with others (Buckland, 1995). These interrelationships form the basis for interdisciplinary and potential transdisciplinary approaches within Information Science curricula. This also applies to its collaborations with other scientific contexts, such as Information Systems, Information Technology and Info-communication (Stair and Reynolds, 2009; Turban *et al.*, 2010).

In Saracevic's (1995) approach, it is possible to read the affirmation of this inherently interdisciplinary nature cf the field. This starts with the analysis of its main topic of study - information - which is configured as genuinely interdisciplinary. In the perspectives developed by the author, including a historical development of IS, from its own origins, an intense contributory relationship was produced that goes beyond the initially considered delimitations of Library Science and Information Retrieval. This thesis is also considered when we evaluate the reflections of Bergeron (1996), where an important beginning of IS relationship is evaluated in extension from the initial conceptions, greatly assisting in the observation of the development, at that time still initial, of systems based on the Internet for data and information management.

In 1968, Harold Borko highlighted the interdisciplinary character of Information Science, providing a comprehensive definition that detailed the processes of information creation, use, decision-making, equipment, methods, and resources. According to Borko (1968), IS involves the integrated study of these elements, as well as interdisciplinarity, as it relates to the various disciplines that govern the specific contexts of the sub-processes in question, such as Technology, Engineering, Mathematics, and Systems Thinking.

It is provocative to verify how these studies and analyses have developed over time, perceiving the treatment of large volumes of data and their immense repositories—Kimball and Ross (2013)—reaching the perception of large technological and market complexes. In these cases, it is perceived that when there is no interdisciplinary view, there is an absence of conditions for cooperative exercise with Information Science, resulting in a situation for a lack of perception of the treatment and processing of these data for the composition of information and knowledge (Jamil, 2001; Kearns and Lederer, 2003; Johnson, 2012), exactly reaching the potential usages for information systems or, as it can be defined, a digital transformation resource.

There is also to be noticed the broad discussion between two different aspects for the concept of information. In the discussions of the field's formation, authors generally approach the contexts of "information—object" and "information—process" (Jamil, 2001). In the first case, it is the object of analysis, taken as a particular item, an element to be analyzed in an individualized way (Allee, 1997; Davenport, 1997; Jamil, 2001; Jamil, 2021). Here, information is the result of a management process, possibly composed or created from data and enabling a larger definition and architecture, in the destination of knowledge production (Choo, 1998).

In this dynamic scenario, it is possible to check how information studies and information management, under the observation of Information Science, born with an interdisciplinary intention: as these two-way characteristics challenge researchers and practitioners, it is practically impossible to develop studies without adopting other fields´ tools, methods and, moreover, knowledge to advance analysis and conclusions. This way, a dynamic situation where interdisciplinarity plays a significant and basal role.

In the conception of information—process, the authors refer predominantly to the act of informing, involved and addressed by the field of Communication, where a context is transmitted to another (Davenport and Pruzak, 1997, Choo, 1998). This process, therefore, involves steps pertinent to the perception or development of the content to be transmitted, its forwarding through instances from the source to the recipient, culminating in the acceptance of the final reception. In this research, although perceiving greater observation to "information—object," it is considered that both approaches enable, notably, the perception of the interrelationship and connection of information, therefore, of its study, with other elements, factors, processes, methods, and resources. It is the interdisciplinary nature of Information Science.

Thus, the inherently interdisciplinary nature, of historical development, of Information Science is presented. It is important to highlight that, given the study previously done for the discussion of methods and methodological contexts, IS places itself as a potential contributor to the multidisciplinary and transdisciplinary exercise, as has been verified in practice, especially in the statements of those previously mentioned formations.

Nowadays it is possible to foresee the need of specific fields and contexts to be applied for a better understanding of artificial intelligence, data analytics and several other innovative resources, as to produce knowledge for strategic decision making in new organizations and business models. Information Science scientific field evolves to answer these needs.

There is, evidently, a greater risk in this evolution, as already discussed by Jamil and Silva (2021), which is the market pressure, mainly originating from the technology markets, in terms of products, resources, and services, when adopting a vague term of "information" that is, at the same time, valued and vague and imprecisely defined.

In the following section, a discussion around digital transformation will be conducted, as to explore more the multi, inter and transdisciplinary powerful relations over information, enabling the adoption of Information Science principles to comprehend and better estimate innovations arising from digital transformation.

THE CHALLENGE AROUND DIGITAL TRANSFORMATION THROUGH THE LENSES OF INFORMATION SCIENCE

It was stated earlier about the precariousness of certain decisions in the management of projects and implementations and without the needed scientific basis, not only when these solutions are introduced to the market, but also for the long-term use.

In the beginning of this chapter, the theme of Digital Transformation is also exposed to this risk of adoption without consistency, identifying any processes involving digital resources, highlighting the pressures on the actual use of social media, as one valid communication way, but positioned without the needed socio-technological moderation.

Digitization of processes, methods, systems and communications has undeniably been evolving rapidly in recent years. Even entire organizations have come to exist only in the digital realm, justifying the term "virtual companies". This way, the expression Digital Transformation has emerged as a promoter and justification for new digital projects, with broad and ill-defined objectives, incorporating both minimal solutions, ranging from minor modifications processes to proposals for breakthrough disruptive innovations that would deserve better studies, such as empirical bases for building knowledge from the proposed changes.

This section seeks the objective of the chapter, to present Digital Transformation through the interpretation of Information Science, mainly having the methodological contexts of multi, inter and transdisciplinarity as elements of this observation as the path which allows better understanding and development. This text will aim, therefore, to provide the opportunity for IS to host and cooperate on research that aims to discern digital transformation with a foundation for the solid construction of knowledge.

Reflections on Digital Transformation

It is reiterated, therefore, that the focus in this expression is *transformation*. From this point of view, it is considered the intended or provoked change, for example actions or reactions by one company when perceiving competition, being implemented by digital resources. Even when undertaken by the support coming from the availability of digital resources, these changes are the phenomena that generate the permanent consequences, in the internal and external relationships of organizations. Digital technology, always evolutionary and revolutionary, does not in itself carry them out, although it is essential in promoting the intended change if these are effectively the center of strategic formulation.

These changes are exemplified on a large scale, as already discussed, and several examples can be used, such as:

- The updating, optimization or complete revision of operational processes, being implemented gradually or completely using digital technology. This aspect includes everything from the simple usage of an application oriented to the end user, to a complete service automation of one *chatbot* composed using artificial intelligence resources, integrated with composed environments via communication connection interfaces, such as "APIs" (*Application Program Interfaces*).
- Definitions for new organizational communication services, such as those oriented for internal processes information and management, through a set of dashboards - including the strategic planning process itself - to a complete automation of integrated plants, even those which are composed of intercontinental platforms.
- The scope of an entire tactical management environment, from the proposition of detailing strategic goals and their description in the business units, departments or areas of a company, remodeling delegations, communications and promoting continuity between the processes. In this way, genuinely, an integrated management platform is built, as can be seen in integrated management systems such as ERPs (*Enterprise Resource Planning*) and SCM (*Supply Chain Management*), among several others.
- Large scale adoption of software robots which will allow researchers and economic agents to communicate and interact over a new product or service put available on the market, such as the design, development, production and offering of a new medicine drug.
- Complete integration between the strategic environments, involving pre-planning and formulation, the design and composition of the plans and, finally, the delegated management of their execution, relating to the tactical and operation control instances mentioned above. Here, special attention should be paid to the fact that there are better prospects for alignment and communication throughout the strategic environment and its relationship to the effective organizational operation, promoted by digitization, bringing organizational managers to a level of better allocation of time and resources.
- Automation of harvesting, storage and distribution for food and food products, reaching the level of complete robotization, with the application, for example, of drones which are guided by artificial intelligence - powered machinery and software layers towards an optimal arrangement to provide these critical products to the market.

In these and other cases, digitization, taken as the insertion of computerized or digital integrated resources and platforms, enables the optimization of information management, bringing here, intentionally, the focus of the study to the target of the discussion in this chapter. A digital perspective for information management is approached; the scientific field of Information Science offers perspectives for its understanding, improving the overall chances of management as discussed earlier in this text.

The Methodological Contexts In Information Science

Methodological diversity has been a constant feature in the field of Information Science. The native interdisciplinary character, explored earlier in this study, fundamentally contributes to this diversity. Additionally, as it is a field that is always directly or indirectly involved in topics closely related to innovation, to business information systems and to technological application, IS becomes open to various participations and relationships in the field of reflecting on questions and proposing ways to answer them within the scope of scientific research.

Reducing isolation in knowledge silos makes Information Science more necessary and receptive to interactions with other fields and forms, further dynamizing the exchange of experiences and related productions, additionally leading to new contributory results for the field itself and for others, related temporarily during the exercise of a single research, or permanently, resulting in a transdisciplinary perspective, as discussed below. One example are those studies, sponsored by Information Science research, which approach the usage of information. These studies treat the situations and methods adopted to manage - collect, compose, store, share, evolve and develop (to knowledge, for example) - information in organizational contexts. It is an undeniable contribution of Information Science for a better perception in these scenarios.

Historically, within a multidisciplinary context, IS has received contributions from areas assessing technological repercussions for processes optimization, standardization and modernization. This mark was the source of the analysis by information resource management (IRM), a body of knowledge predecessor of topics such as information management and knowledge management.

The contributions of IRM and its developments are clearly related to what is known as "analytics" studies and the area called Data Science, a field still evolving and expressively related in multi and interdisciplinary forms with technological resources and ways of interacting with informational users, mainly regarding commercial and market applications. These resources are, nowadays, considered as the basis for data science applications and several artificial intelligence implementations, such as machine learning-oriented solutions.

As an example of this multi- and interdisciplinary context that characterize the process described above, it is possible to verify the citations of the "user journeys" or "customer journeys" currently addressed with prominence as an evolution of digital marketing. In these analysis, there is a trend of interpretation, which can be automated, of the tracking and action of customers in use or potential use of interaction, approximation and engagement resources on commercial platforms, aiming to understand and possibly advancing their reactions and intentions as to promote a better engaging process, improving their decision interaction.

Customer experiences, or CX, refer, mainly, to what the customer absorbs about the entire profile of the organization broadly offers and negotiates value. Here, the current highlight is on brand perception studies, where relationships of proximity and affection are sought by practitioners in general, expanding in new directions the contexts discussed for several years under the aegis of *branding*. It is an area for carrying out multidisciplinary studies, mainly, with analysis by interdisciplinarity still fitting at a second level. A great opportunity for applying the principles of Information Science lies in the development of multi and interdisciplinary studies, seeking research with greater longevity and application of results in real cases.

This field has the potential to become a transdisciplinary area in the coming years, receiving contributions from Organizational Management - Marketing, Strategy, Strategic planning among others - Statistics, Computing - in here the real opportunity towards a digital transformation - Commercial and Branding sources. This future “blended” scientific field will likely focus on the boundary of organizational / products and services value perceptions. Its interactions, promoted through digital platforms, enabling the existence of virtual companies, such as some actual from Education, Tourism, Healthcare, Insurance and Banking sectors, can be better explained with the observation of users' interactions, preferences, reactions and developments towards a retention of the organizational message.

User Experience, or UX, refers to a potential customer´s perception of a specific product or service, within the organization´s broader offerings. Therefore, UX becomes a specific subset of CX, each with its objective development, but with a broad relationship. Here, transdisciplinarity can also be sought, even aware that the path is long and demanding tests to ensure maturity and the possibility of development, receiving potential contributions also from fields such as Engineering, Automation and Operations management, among many others.

Clearly, the various fields involved, such as Marketing, Technology, Systems and Information Science, Process Management, Design, Brand Management, among others, do not immediately contribute to the formation of unified contexts for the practice of interactions (as characterized by an interdisciplinary contribution) or to the formation of a new field of theoretical-practical connection (the transdisciplinary). The multidisciplinary scope, mainly if supported by IS, can answer the questions with the effectiveness and precision possible for emerging demands, defining itself as a starting point for a potential evolution to inter and transdisciplinary plans and projects.

It is important to highlight the practical aspect of multidisciplinarity in the approach by Information Science as a basis for studies of associations such as the one described above. Within the scope of IS, there is an open space for all these contributions to develop in the design of interfaces, methods of interaction and integration of users, search and retrieval of information, studies focused on the forms and desires of end customers, among other topics. This observation, without a doubt, brings expressive justification for the adoption of multidisciplinarity in current applied studies and research.

The context of interdisciplinarity, given that it is affirmed as native to Information Science, has already been discussed at the desired level earlier, only reinforcing here the perspective as being a path practically aligned with the fundamentals of research projects carried out in the field. For example, AI-based projects, which are composed mainly from an extensive combination of different, shared and displaced datasets, consolidated in one environment, usually supported by a main structure of a neural network, could be considered, for study purposes, as a typical interdisciplinary asset. This way, IS could make possible a better comprehension, due to its extensive scientific production, about how the logics and modelling produced can evolve further to new capabilities and applications, helping designers, engineers and implementation professionals to succeed in their plans and actions.

Finally, the expectation of transdisciplinarity is affirmed for some time when it comes to the contribution of IS as a collaborative field in terms of methodology. These contributions of Information Science allow, given the methodological, structural and philosophical openness of the field, to build new complete contexts of research and knowledge development. Constant practices are verified when entire fields are defined based on information, such as those of "Information Technology", "Information Systems" and even "Data Science", the latter still maturing as its own scientific context, for now generally receiving the fundamentals from IS for its consistent and robust proposition.

It is a point suggested in this chapter that both areas - Data Sciences and Artificial Intelligence - could deserve a thorough observation, to provoke origins for new scientific fields, with an organization of likely sub-fields, associated to one main focus of thought, improving their robustness connecting to the immense chain of results already obtained by the application of resources and methods under these commercial identifications.

The decades-long emergence and evolution of these fields demonstrate the openness of Information Science as a source for research contexts and knowledge building. These are the fundamentals that define the bases and propositions for defining research objects, the context of applying methodologies and practical models and, finally, in the interpretation of evidence and results. The transdisciplinary contribution of CI results in potential robustness

EVALUATING DIGITAL TRANSFORMATION WITH THE CONTRIBUTIONS OF INFORMATION SCIENCE

To conclude this development, it is appropriate to insert some practical evocations of the application of Information Science in the interpretation of Digital Transformation, recognizing the phenomenon as one huge opportunity as to be treated in the contexts of multi, inter and transdisciplinarity under Information Science observation.

Initially, there is the consideration of multidisciplinarity for informational studies and related concepts of data and knowledge. Evidences of this motivation is when it is verified that systems oriented to data processing, involve techniques known from IS, as well as from other fields, producing the due results, without, however, making any progress in the edification of joint methods (interdisciplinarity) or proposing something totally new from a base of scientific context (transdisciplinarity).

Among these techniques, one can cite those of identification and classification of information, the searches to relate information to other concepts (data and knowledge, for example) making relevant the composition of a solid conceptual base and the analyses of informational user behavior. These approaches could be useful, for example, to widen and combine studies from social media interactions - maybe in commercial services - to other sources and scientific fields, resulting in a more detailed analysis, resulting in a better way for UX and CX applications.

Although considering the analysis as satisfactory, these data collection and processing environments, which currently receive the additional incentive of "data science" models and techniques and technologies such as artificial intelligence could advance in the sense of building methods and processes of greater interrelationship both in the methodologies and techniques adopted and in the generation of a new scientific field.

As previously discussed, there are significant perspectives for, in the face of a more frequent production of results for these studies, reaching the later levels of interdisciplinarity and transdisciplinarity to be structured according to constant learnings within the multidisciplinary scope.

Considering how to deal with the possibilities for these studies, it would be possible to think about composing instruments to measure the behavior of the informational user as a function of the implementation of artificial intelligence resources and tools. Advancing to the interdisciplinary level would lead to a greater understanding and involvement of end-users, in mass service environments, public sectors and of high-engagement contexts (such as those oriented to improving *user* and *customer experience*), bringing up the combination of knowledge scientifically produced by other scientific backgrounds.

It is worth noting that client engagement can be obtained via sharing of databases and cooperation in storage, processing and classification processes and analysis, thereby enhancing the repercussions of success for multidisciplinarity. Among those fields, it is possible to cite Geography, Demography, Marketing, Strategy, etc.

Therefore, multidisciplinarity is well-suited within Information Science, supported by the longstanding and frequent dialogue of the scientific field with the other sciences, such as Engineering, Natural Sciences, Applied Social Sciences, among others. Here, given the constancy of productions, there is a culture of receptivity of contributions from diverse fields, even if this does not result in a permanent edification of any context, being acted to obtain a result of focus of certain research.

Development of interfaces presents a simple and crucial evocation to understand the perspectives of interdisciplinarity, by IS, in the development of systems and environments oriented to Digital Transformation. This is a critical aspect when proposing and launching innovations in the market, ranging from ergonomics of products and services usage to its functionalities and eventual "navigability". Innovations can be perceived or proposed through the proposition of new ways for a user to interact with a product or service.

Classic methods of Technology Acceptance studies, identified by the Technology Acceptance Model - TAM, and the Unified Theory of Acceptance and Use of Technology - UTAUT, both with intense production by Information Science in the composition of methods of collecting evidence and processing, define new visions in the design and implementation of systemic interfaces, being an important practical example of interdisciplinarity in DT with observation of Information Science.

The interdisciplinary context emerges in the design, development and implementation of interfaces, ranging from the appreciation of cultural factors in the treatment of interactions with potential clients, passing through system-to-system interfaces (existing, for example, in fully automated environments) and, finally, in the composition of *dashboards*, as integrated panels intended to provide interpretation of results for end users, such as those of monitoring organizational planning work (for example, the results of strategic indicators).

Here, the recommendation of IS culture in "user-centered studies" (of information), a fundamental methodology of the field and with expressive practice already carried out over a vast period and covering various perceptions, such as those of technological acceptance, rejection of insertions, evaluation of interface performance, affection for data interchange elements, among others.

In the case of Interdisciplinarity, therefore, there is in IS a strong focus and cultural base, as it is a foundation of the field, it is of its proposed origin. The interpretation of Digital Transformation only must benefit from this fundamental alignment of IS, incorporating techniques that have already become classic and effective in the practice of scientific research.

Finally, transdisciplinarity is provoked with the bases of IS, as stated in the previous theoretical-practical study. The exercise, for example, of the field of Information Systems, has provided researchers and practitioners with a complete view of processes, integration with people, organizations and methods around information and all its potential treatment.

It must be affirmed, as a conclusion of this development, that the very emergence and affirmation of the field "Digital Transformation" attests to a potential transdisciplinary perspective. And, undeniably, Information Science participates in this evolution, by contributing with its perspectives for the most adequate study and with generation of foundations for analysis of a field undefined in terms of its borders, still in internal formation and submitted to an intense dynamic. If we evaluate the recent facts, DT was already discussed, mainly in the aspect studied earlier in this chapter - seeing Transformation as a priority element to Digital - when the current wave of commercial availability of data sciences and artificial intelligence appeared.

Recognizing in an unquestionable way that we are in a moment of evolution before the first offers of this type in artificial intelligence, as well as data sciences - collection, processing, storage, distribution, valuation and final use of data in various forms - the formation of new fields derived from the appraisals and contexts about the processing of "data" becomes timely and expected. In here, innovation can be regarded as the occurrence of a new scientific field of even for its innovative productions, such as new business models propositions, based on automation, robotization or new ways to deal with customer and user experiences.

A topic which could reemerge for this analysis is Multimedia, corresponding to the composition of scientific methods applied to treat the fusion of mediums, such as image, audio, video, texts and messages (clearly, text, but with a different goal, to interact over a chatting platform). This possible new field could receive a good contribution of Information Science, in several ways already discussed in this chapter.

Transdisciplinarity may become a hallmark methodological basis of the current moment ahead in the foundation of specific and comprehensive methods of scientific research in the contexts of AI application. The awareness of the contribution of the various original fields, as always done in the context of Information Science, will be decisive even to promote an eventual replacement in currently treated bases by others, as occurred previously with "Data Processing" and "Informatics" by Computer Engineering and the very and fundamental definition of Computer Science.

Finalizing this development, it is appropriate to mention that Information Science has history and bases, even constituting a scientific posture, in the multi, inter and transdisciplinary treatment of elements such as data, information and knowledge and in the amplitude of the various methods, processes and resources that connect for their processing. The study of the management of informational resources, of information systems, of the identification and classification of contents, which comes from the elementary formations of Library Science and reached the necessary support for the creation and construction of machine and deep learning models are well carried out and discussed in IS and by IS in contribution to the other fields.

This way, digital transformation can become a more comprehensive, understood and managed fact, addressed by Information Science, encompassing several contributions by other fields, through a receptive context for combined methodologies, tools and dialogs which will allow this academic and scientific intended result. Information Science has become a highly recommended scientific field to be adopted and practiced allowing a better application of different experiences and sources, even contributing to form new fields, in a transdisciplinary approach.

CONCLUSIONS

Innovations are demanded nowadays from various stakeholders. Educational institutions, Airlines, Banks, Food processing companies, among many others seek to be different from its competitors, aiming to develop a competitive advantage from specific, well-addressed ways to aggregate value to their products and services. Companies, everywhere, discovered the power of leveraging data, electing this content as a precious asset. No doubt data is valuable. But not only *per se*, as an isolated asset.

Data is a source for information and knowledge. Over the years, resources for this essential and strategic production have evolved from logical methods practiced through manual procedures to modern solutions involving software layers, advanced algorithms, data structures, and sophisticated AI platforms.

But these contributions do not reach IS in an unplanned way. They integrate this rich context through multidisciplinary, interdisciplinary and transdisciplinary dialogues, which allow several different degrees of contributions and productions. IS has been decisive to build a new scientific context, such as the reviews of Information Technology or the formation of Information Systems. IS capabilities and resources being available to develop a new way to improve scientific discussion in outcomes.

In this text, an observation was held regarding innovation from digital transformation, observed from the Information Science field, applying multi-, inter- and transdisciplinary methodological postures. Fortunately, it is an ongoing process, identifying dynamics which will provide more and more results in the future, as it is desired and demanded by researchers and practitioners from new organizations, to enhance their plans and projects with a solid and receptive scientific background.

REFERENCES

Allee, V. (1997). *The Knowledge Evolution: Expanding Organizational Intelligence*. Butterworth-Heinemann.

Alvargonzález, D. (2011). Multidisciplinarity, Interdisciplinarity, Transdisciplinarity, and the Sciences. *International Studies in the Philosophy of Science*, *25*(4), 387–403. DOI: 10.1080/02698595.2011.623366

Bardecki, M. J. (2019). Multi - disciplinarity. In Leal Filho, W. (Ed.), *Encyclopedia of Sustainability in Higher Education*. Springer., DOI: 10.1007/978-3-030-11352-0_90

Bergeron, P. (1996). Information resources management. (1996) [ARIST]. *Annual Review of Information Science & Technology*, *31*, 263–300.

Borko, H. (1968, January). Information Science: What is it? (1968). *American Documentation*, *19*(1), 3–5. DOI: 10.1002/asi.5090190103

Bresciani, M. S. M. (2015). Interdisciplinaridade – transdisciplinaridade nos estudos urbanos. URBANA: Revista Eletrônica do Centro Interdisciplinar de Estudos sobre a Cidade, Campinas, SP, v. 7, n. 1, p. 10–62, 2015. DOI: . disponível em https://periodicos.sbu.unicamp.br/ojs/index.php/urbana/article/view/8642547. com acesso em March, 2021.DOI: 10.20396/urbana.v7i1.8642547

Buckland, M. (1995). *Information and Information Systems*. Praeger.

Choo, C. W. (1998). *The Knowing Organization: How organizations use information to construct meaning, create knowledge and make decisions*. Oxford University Press.

Christensen, C. (2024). The Innovator's Dilemma, with a New Foreword: When New Technologies Cause Great Firms to Fail. Harvard Business Review.

Davenport, T. (1998). *Ecologia da Informação*. Futura.

Davenport, T., & Prusak, L. (1998). *Working Knowledge*. Harvard Business School Press.

Ecosys (2014). Falta de conhecimento de empresas reduz a adoção de ERPs em nuvem. Disponível em https://eccosys.com.br/guia/falta-de-conhecimento-de-empresas-reduz-adocao-de-erp-na-nuvem/ com acesso em setembro de 2024.

Firestein, R. (2019). *Create in a Flash: A Leader's Recipe for Breakthrough Innovation*. Green Tractor Publishing.

Flores, J. F., & da Rocha Filho, J. B. (2016). Transdisciplinaridade e Educação. *RevistAleph*, (26). Advance online publication. DOI: 10.22409/revistaleph.v0i26.25103

Ghezzi, C. (2009). Decentralized Software Development: Pitfalls and Challenges. In Gotel, O., Joseph, M., & Meyer, B. (Eds.), *Software Engineering Approaches for Offshore and Outsourced Development*. DOI: 10.1007/978-3-642-02987-5_1

Hammer, M. (1990). Reengineering work: Do not automate, obliterate. In Harvard Business Review. Available at https://hbr.org/1990/07/reengineering-work-dont-automate-obliterate, accessed on January, 2025.

Harrington, H. J. (1998). Performance improvement: The rise and fall of reengineering. *The TQM Magazine*, *10*(2), 69–71. DOI: 10.1108/09544789810211353

Huang, M.-H., & Chang, Y.-W. (2011). A study of interdisciplinarity in information science: Using direct citation and co-authorship analysis. *Journal of Information Science*, *37*(4), 369–378. DOI: 10.1177/0165551511407141

IDEO. (2022). Human Centered Design: Design Thinking Toolkit. Disponível em https://www.ideo.com/post/design-kit com acesso em fevereiro de 2022.

Jamil, G. L. (2001). Repensando a TI na empresa moderna, Axcel Books do Brasil, 2001, 645 páginas. Disponível em https://eccosys.com.br/guia/falta-de-conhecimento-de-empresas-reduz-adocao-de-erp-na-nuvem/, com acesso em Setembro de 2024.

Jamil, G. L. (2021). Can new technologies define a “Bauhaus” for information science? An exam of interdisciplinarity in the face of innovations. In Rascão, J. P., Marques, M. B., & Jamil, G. L. (Eds.), *Information and Technology in Business, the Digital society*. Lambert.

Jamil, G. L., Briones-Peñalver, A. J , & Lema, D. G. P. (2019). Reflecting on Industrial Business Models: A History of Tradition, Challenges, and Potential Innovations. In Alcaraz, J. L. G. *et al.*, Best Practices in Manufacturing Process: Experiences from Latin America. Chapter #10, pp. 211-237.

Jamil, G. L., & Silva, A. R. (2021). Emerging technologies in a modern competitive scenario: Understanding the panorama for security and privacy requirements. In Anunciação, P. F. (Ed.), *Pessoa, C. R. M. and Jamil, G. L. (2021) Digital Transformation and challenges for data security and privacy* (pp. 1–16). IGI Global. DOI: 10.4018/978-1-7998-4201-9.ch001

Johnson, J. E. (2012). Big data + Big Analytics + Big opportunity. Financial e Executive, July/August 2012, pp. 51-53

Kearns, G. S. e Lederer, A. (2003). A resource-based view of strategic IT alignment: How knowledge sharing creates competitive advantage. Journal of decision sciences, v.34, n.1 winter 2003.

Kimball, R., & Ross, M. (2013). *The Datawarehouse Toolkit: the definitive guide to data modelling*. John Wiley and Sons.

Klein, J. T. (2010). A Taxonomy of Interdisciplinarity. In Frodeman, R., Klein, J. T., & Mitcham, C. (Eds.), *The Oxford Handbook of Interdisciplinarity* (pp. 15–30). Oxford University Press.

Lacerda, B., & Jamil, G. L. (2021). Digital Transformation for Businesses: Adapt or Die! Reflections on How to Rethink Your Business in the Digital Transformation Context. (2021). In *Anunciação, P. A., Pessoa, C. R. M. and Jamil, G. L. Digital Transformation and Challenges for Data security and privacy* (pp. 252–268). IGI Global Publishers.

Lawrence, R., & Despres, C. (2004). Futures of Transdisciplinarity. *Futures*, *36*(4), 397–405. DOI: 10.1016/j.futures.2003.10.005

Loureiro, F., & Guimarães, F. S. (2019). *Multidisciplinarity and Interdisciplinarity in International Relations: teaching History and Theory of International Relations* (Vol. 10). Mural Internacional., DOI: 10.12957/rmi.2019.37636

Madsen, D. (2012) Interdisciplinarity in information field. in Proceedings of the American society for information science and technology. Disponível em https://asistdl.onlinelibrary.wiley.com/doi/10.1002/meet.14504901016, com acesso em julho de 2024.

Max-Neef, M. A. (2005). Foundations of Transdisciplinarity. Ecological Economics. V. 53, N.1, April 2005, pp. 5-16. Available in https://www-sciencedirect.ez93.periodicos.capes.gov.br/science/article/pii/S0921800905000273?via%3Dihub, com acesso em March, 2021.

NCSU - North Carolina State University. Research Development Office - The difference between multidisciplinary, interdisciplinarity and convergent research (2020). August, 31st, 2020. In https://research.ncsu.edu/rdo/2020/08/the-difference-between-multidisciplinary-interdisciplinary-and-convergence-research/, acessed in January, 2021.

O'Brien, J. A., & Marakas, G. M. (2006). *Management information systems* (Vol. 6). McGraw-Hill Irwin.

OECD. (2005) Oslo Manual. Available at http://www.finep.gov.br/images/a-finep/biblioteca/manual_de_oslo.pdf, accessed on February, 2025.

Rascão, J. P., Jamil, G. L., & Marques, M. B. (2021). *Debate on Inter-Transdisciplinarity of Information Science: From theory to practice*. Lambert.

Rigolot, C. (2020). Transdisciplinarity as a discipline and a *way of being*: Complementarities and creative tensions. *Humanities & Social Sciences Communications*, *7*(1), 100. DOI: 10.1057/s41599-020-00598-5

Rogers, D. L. (2017). *The digital transformation playbook: rethink your business for the digital age*. Columbia University Press.

Ross, J. W., Beath, C. M., & Moker, M. (2019). Designed for Digital: How to Architect Your Business for Sustained Success (Management on the Cutting Edge). MIT Press, 2019.

Saracevic, T. (1995). Interdisciplinary nature of Information Science. In Ciência da Informação - Vol 24, número 1, 1995.

Schumpeter, J. (1934) The theory of Economic Development. Reprinted by Routledge Publishing, 1980.

Siddique, I. (2021). Harnessing Artificial Intelligence for Systems Engineering: Promises and Pitfalls (September 30, 2022). European Journal of Advances in Engineering and Technology, 2022,9(9):67-72, Available at *SSRN*: https://ssrn.com/abstract=4885910 or http://dx.doi.org/DOI: 10.2139/ssrn.4885910

Stair, R., & Reynolds, G. (2009). *Principles of information systems*. Course Technology.

Tadeu, H.; Duarte, A. L.; Taurion, C. e Jamil, G. L. (2019). Digital Transformation: Digital Maturity Applied to Study Brazilian Perspective for Industry 4.0: Experiences from Latin America. DOI: .DOI: 10.1007/978-3-319-99190-0_1

The Economist (2018), "Blockchain technology may offer a way to re-decentralise the internet". Available at https://www.economist.com/special-report/2018/06/28/blockchain-technology-may-offer-a-way-to-re-decentralise-the-internet accessed on June, 2024.

Turban, E., Leidner, D., McLean, E., & Wetherbee, J. (2010). *Tecnologia da informação para a gestão*. Ed. Bookman.

Wang, P. (2008), "Whatever Happened to Business Process Reengineering? The Rise, Fall, and Possible Revival of Business Process Reengineering from the Organizing Vision Perspective". Published in "Business Process Transformation" by Routledge Ed., 2008, pp. 1 - 18.

Zaman, G. e Goschin, Z. (2010). Multidisciplinarity, Interdisciplinarity and Transdisciplinarity: Theoretical Approaches and Implications for the Strategy of Post-Crisis Sustainable Development. Theoretical and Applied Economics, Asociatia Generala a Economistilor din Romania - AGER, vol. 0(12(553)), pages 5-20.

Chapter 6

Impact of Innovation Dimension Mediation on the Performance of Ethiopian SMEs With an Entrepreneurship Orientation

Gada Gizachew Wakjira
https://orcid.org/0000-0002-7725-1161
Bule Hora University, Ethiopia

Shashi Kant
https://orcid.org/0000-0003-4722-5736
Bule Hora University, Ethiopia

Kenenisa Lemi Debela
Jimma University, Ethiopia

ABSTRACT

Innovation and entrepreneurship go hand in hand since applying creative thinking to SMEs manufacturing firms gives the sector a competitive advantage. Thus, the purpose of this study was to investigate how entrepreneurship orientation affects the performance of small and medium-sized firms (SMEs) in Ethiopia. The study used an explanatory research design and a quantitative methodology. To gather data from a sample of 451, the researchers used a probability sampling technique, more precisely a multi-stage clustering sampling. Based on information from a pilot study with 95 respondents, an exploratory factor analysis (EFA) was carried out to evaluate the internal consistency and reliability of the study variables. Confirmatory factor analysis (CFA) and structural equation modeling (SEM) were used to analyse the gathered data. The findings show that entrepreneurship orientation and innovation mediation as a combination have a big impact on organisational success.

DOI: 10.4018/979-8-3693-4030-1.ch006

INTRODUCTION

Entrepreneurial orientation (EO), according to Lee et al. (2023) and Amari (2023), is a crucial element of progress in society and the economy, especially when it comes to using knowledge and tangible assets and looking ahead. Kirchhoff and Phillips (1988) defined entrepreneurial orientation (EO) as the process, practice, and decision-making activities that lead to the creation of new businesses and entities. Entrepreneurial orientation has gained more attention globally, especially in developing economies, since it is believed to play a crucial role in the creation of jobs, the reduction of poverty, and economic development in the functioning of SMEs (Ganguly et al., 2023; Khan et al., 2020; Lumpkin and Dess, 1996). There are several EO metrics, including as risk-taking, aggressiveness, proactivity, autonomy, and invention, all of which are investigated from theoretical foundations (Karimi et al., 2021; Vallaster and Angerer, 2018). The combination of innovativeness, pro-activeness, and risk-taking constitutes a uni-dimensional tactical orientation, according to Ajzen's theory of planned behaviour (TPB), which was presented in 1985 (Al-Swidi & Mahmood, 2012; Ferreira and Coelho, 2020). The additional elements of aggressive rivalry and autonomy that Lumpkin & Dess presented in 1996 are pertinent to this (Abrehet Mehari & Prof. Rovshan Guliev, 2020; Karaev, 2023).

Because SMEs augmented both domestically and internationally, which incentivized enterprises to innovate in order to achieve a lasting competitive edge, Perdana and Prasasti (2023) looked at innovation in the context of SMEs' success. Irawan et al. (2023) assert that corporates need to enhance their skills to adjust to the swiftly evolving and suddenly altering situations (Peng et al., 2021). This has the ability to bring about transformation and opportunity for all business sectors and can also play a big role in economic development (Arslan, 2020; Becerra-Vicario et al., 2023). Innovation provides small, middle, and micro-sized firms (SMEs) with opportunities and a supportive environment in terms of knowledge and assets, according to Maaodhah et al. (2021). Utilising innovative technologies to boost SMEs' performance is common in Eastern and Horn of Africa. Additionally, small and medium-sized businesses are being forced to find creative and enhanced methods to stay in business due to competition in today's business environment (Wahyuda et al., 2023; Perdana & Prasasti, 2023; Šlogar et al., 2023).

STATEMENT OF PROBLEM

There are still a number of inconsistencies in the evidence, as well as theoretical, methodological, and knowledge gaps that require attention, despite the augmented interest in comprehending the connection among innovation and performance among Small and Medium Enterprises (SMEs) with an entrepreneurial orientation. By examining the influence of innovation dimension mediation on the performance of Ethiopian SMEs with an entrepreneurial orientation, this study seeks to close these gaps. The outcomes of previous studies on the connection among innovation and SME performance have been incongruous. According to some research, there is a good correlation among innovation and SME performance (Michael Nnaemeka Ajemba & Ebube Chinwe Arene, 2022; Perdana & Prasasti, 2023 Mustafa, Rehman, Zaidi, & Iqbal, 2015). Other research, also, has discovered ambiguous or even detrimental associations (Mohamed et al., 2023; Rauch, Wiklund, Lumpkin, & Frese, 2009; Mata & Aliyu, 2014).. More research is necessary to fully comprehend the link among innovation and performance, particularly in the context of Ethiopian SMEs, in light of these contradicting data gaps.

By examining the mediating effect of innovation characteristics in the context of Ethiopian SMEs with an entrepreneurial orientation, this study aims to close this theoretical gap. From a methodological standpoint, earlier research has mostly used cross-sectional designs, which makes it difficult to determine causal linkages or look at how the innovation-performance relationship changes over time (Yeni, 2015; Zehir et al., 2015; Fatoki, 2012; Mahmood & Hanafi, 2013). Also, few studies use longitudinal data to evaluate how innovation aspects affect the success of SMEs (Tesfa, 2023; Farooq et al., 2021). In order to close these methodological limitations and provide a more thorough analysis, this study uses a longitudinal research approach. Regarding the particular innovation features that mediate the link among innovation and performance in Ethiopian SMEs with an entrepreneurial orientation, there is a lack of thorough understanding. By performing empirical research to determine and evaluate the mediating influence of several innovation aspects on the performance of Ethiopian SMEs, this study aims to close this knowledge gap. This study created a conceptual framework that specifically includes the mediating role of innovation dimensions in order to close the theoretical gap. The framework will offer a basis for comprehending how various aspects of innovation mediate the link among innovation and performance in Ethiopian SMEs with an entrepreneurial orientation by drawing on pertinent theoretical views.

OBJECTIVES

1. To examine the effect of Entrepreneurial orientation on firm performance.
2. To look at the effect of Entrepreneurial Orientation with Innovation.
3. To inspect the effect of Innovation t on firm performance
4. To analyze the effect of Innovation mediated among EO and firm performance.

THEORETICAL LITERATURE

Concept of Entrepreneurial Orientation

Miller (1985) first conceptualised EO as initiative, risk-taking, and inventiveness. Two additional dimensions were later added: autonomy and competitive aggressiveness, by Lumpkin and Dess (1996). According to Kanaan-jebna et al. (2022), each of the five dimensions has a different level depending on business and environmental factors, but they are all crucial for EO performance associations in SMEs' performance. According to Siranesh & Mebratu (2020) & Etim, Adabu, and Ogar (2017), entrepreneurial orientation is a collection of standards, guidelines, and decision-making processes that a company employs to boost its propensity for innovation, proactivity, and willingness to take risks.

Entrepreneurial Orientation Determinants

Autonomy

Lumpkin and Dess (1996) define autonomy as the capacity to make decisions on one's own and go forth in pursuit of fresh chances. According to Šlogar et al. (2023), efficiency and a company's independent mindset are related. The ability and willingness of an individual to pursue market possibilities independently is defined as autonomy in the literature on economic corporates. With the help of creative goods and process innovations, an corporate can swiftly and independently make decisions that open up new markets (Aroyeun et al., 2019; Bruhn et al., 2018; Adeniyi et al., 2020).

Competitive Aggressiveness

Competitive aggressiveness is a measure of a company's openness to its rivals (Lumpkin and Dess 1996; Runyan et al., 2006; Wahyuda et al., 2023). The ability of a business to overcome obstacles quickly and surpass rivals in a competitive market is known as aggressiveness. Proactive companies are likely to be proactive in difficult situations, as manifest by (Covin and Covin, 1990 & Kiyabo & Isaga, 2020). Moreover, an individual's capacity to manage fierce competition in concurrence is considered by the EO dimension of competitive aggressiveness (Miller 1983). Perdana & Prasasti (2023) define competitive aggressiveness as using creativity to outsmart competitors (Ledi & Ameza–Xemalordzo, 2023).

Innovativeness

It has been manifest that SMEs' receptivity to innovation influences their performance by guiding their association with the theory of contingencies (Wakjira & Ibrahim, 2023). Following Schumpeter's initial introduction of the concept of innovativeness, there was a long-standing idea that innovation was restricted to internal corporations or R&D department activities. Because of this, invention and creativity became important strategic assets that were safeguarded by laws and stringent oversight (Miller 1983; Susanto, 2019).

Pro-Active

According to Aloulou (2023) & Adesoga et al. (2018), being proactive manifests that a company is anticipating market demands and acting accordingly to gain a competitive edge over its rivals. This is followed by opportunity scanning, as suggested by Miller (1983) and Wales et al. (2016). In terms of pro-activeness, EO has been widely hailed as a key component for improving firm performance. Proactive business firms are able to capitalize on first mover advantage and dominate over market distribution channel, as demonstrated by Zahra and Covin (1995); Celikyay et al., (2022) and Beck et al., (2021).

Risk-Taking

In the words of Siam et al. (2022), risk is the degree of unpredictability surrounding a decision's outcome (Abbaszadeh et al., 2023). When creating a novel service or good, a manager or business owner must take into account both the monetary and advertising ramifications, including the cost associated with launching the new good or service, how it ought to be marketed, and whether there currently is a demand for it (Yaskun et al., 2023). Risks are involved in the decisions that must be made regarding the new good or service (Almacena Platform, 2021).

Concept of Innovation

Modern research on innovation in the marketing discipline will address the term "innovation," which has different meanings in different business sectors (Hunt and Morgan, 1995; Lavoie & Main, 2022). Innovation is defined as a process that starts with an idea, develops it further, and ends with implementation (Brief, 2022). According to Aloulou (2023), innovative products, services, marketing prowess, and business inventiveness are all factors in SMEs' success.

As a result, innovation happens when a company possesses the capacity to innovate, which is defined as the capacity to create, develop, and apply new concepts, products or services, strategies, and procedures that benefit those involved in the business (Hult et al. 2004; Sufa et al., 2022). In this regard, innovation capabilities of firms have been defined by (Tesfa, 2023; Farooq et al., 2021) as a crucial element that permits innovation culture within an corporate, the presence of enabling activities, and the capacity to comprehend and react suitably to the outside influences of SMEs (Calantone et al. 2002; Hussain et al., 2023). Sufa et al. (2022) contended that innovation is a decision that corporations make in response to marketplace competition and effective leadership (Manickam et al., 2023) and refers to an organization's endeavour to create, manufacture, and sell novel goods for the industry through the use of information and technology (Al-Shourah, 2021).

Determinants of Innovation

Product Innovation

Perdana and Prasasti (2023) define product innovation as the introduction of a new product or one that has undergone substantial improvement or modification in terms of features, functionalities, and ease of use (Issau et al., 2022). It also includes improvements made by a business to its technological specifications, its line of products, or other helpful characteristics. Product innovation remains one of the company's main sources of competitive advantage (Yesuf et al., 2023). This is because new product innovation can increase efficacy, which can increase efficiency and give SMEs a competitive advantage (Alhamami et al., 2023). Several studies have confirmed the correlation among a company's success and its ability to innovate new products (Alhamami et al., 2023; Patel & Patel, 2023).

Process Innovation

In accordance with Wahyuda et al. (2023), Process innovation is defined as "improving the process of manufacture, shipment technique, or advancing operations" (Abbaszadeh et al., 2023). This can involve making substantial machinery and approach modifications as well as introducing major improvements in the software, technological advances, and instruments used in a manufacturing or delivery method business. As per Issau et al. (2022), Numerous studies, including those by Morone and Testa (2008), Abraham and Viswanadham (2021), and Essa and Ph (2021), have discovered a beneficial association among process innovation and company performance. Ersoda AniyanK (2021) came to the conclusion in the investigation they conducted that a company's functioning output, client loyalty, and economic growth could all be improved through the introduction of process innovation.

Marketing Innovation

According to Wahyuda et al. (2023), marketing innovation is the application of a novel marketing strategy that entails material modifications to a product's positioning, upward mobility, design, or cost (Abraham and Viswanadham, 2021; Essa and Ph, 2021). To put it another way, it's the application of a novel marketing idea or approach that is untested and drastically different from the company's current marketing strategies (Hota, 2021).

Organizational Innovation

Business innovation, in accordance with Dessie (2021), is the use of novel business techniques in business operations that weren't used by the company before, such as the management of knowledge, novel leadership approaches, operations restructuring, workplace corporate, or interactions with the outside world (Borah, Iqbal & Akhtar, 2022).

In line with Abraham & Viswanadham (2021), an extremely substantial factor for overall sales is business innovation instead of process or product innovation. In agreement, (Abbaszadeh et al., 2023) revealed that business innovation could support and foster strong processes for business learning and skills.

Concept of Organizational Performance

According to Abdul Adis and Jublee (2020), firm performance has grown to be an essential issue for contemporary unique executives, auditors, and entrepreneurs across numerous nations (Chelliah et al., 2023). In order to address the limitations raised by Johnson and Kaplan (1987), the Balance Score Card (BSC) incorporates multiple perspectives, including learning and growth, internal processes, and customer satisfaction. Previously, firm performance was determined using the conventional cost or financial accounting process, which was created in the early 1900s. Financial metrics were the primary means of evaluating a company's performance before the BSC was introduced (Rafiq et al., 2020). Thus, nonfinancial factors were ignored, such as learning and development (Shi and Trinidad, 2023).

Organizational Performance Indicators

Financial Performance

According to Thuong & Singh (2023), one component of the BSC is the company's financial performance. When evaluating an enterprise's performance of SMEs in the production sector, financial performance points of view reveal a substantial beneficial association with achievement of SMEs with their approach execution by means of the application of the BSC and it incorporates money and non-financial indicators for SMEs (Yuliana and Kristiana, 2021).

Business Process

According to Thuong & Singh (2023), the Business Process additionally serves as a component of the BSC. The benefits of implementing internal business processes like the BSC consist of improvements in efficiency in operation and the effectiveness of business operations viewpoints (Alhamami et al., 2023). According to a number of examinations, a more thorough operational framework of small and medium-sized enterprises can support an organization's long-term success by incorporating and covering the entire the internal business process (Yaskun et al., 2023).

Learning and Growth Perspectives

HR professionals and entrepreneurs have started to use more contemporary approaches that center on the cultivation of human capital in order to gain a competitive edge in the worldwide corporate financial system (Adis and Jublee, 2020). Kaplan & Norton (1996) indicated that learning and growth perspectives within SMEs is an additional component of the BSC business benefits that are related to the learning & development outcomes achieved by the BSC in business industries, including business research and business success, SMEs boost the spirits of team members creative productivity contribution (Yuliana and Kristiana, 2021).

Customer Perspectives

The views of customers are a further aspect of the business supply chain (BSC), according to Thuong & Singh (2023). Rafiq et al. (2020) proposed that the viewpoint of SMEs in the BSC of customer view-points that reveals beneficial thoughts about their fulfilment of SMEs and the reinforces maintaining clients and encourages new client acquisition. In accordance with Osuagwu, et al., (2019), the emphasis is on how long-term product and service shelf life targets relate to SMEs' capacity to provide superior goods and services activities that aligned to SMEs' requires and preferences (Girma et al., 2022).

EMPIRICAL LITERATURE REVIEW

Entrepreneurial Orientation and Firm Performance

In accordance with Barney (2001), the outcomes of previous investigations indicate that the success of small businesses is influenced by their entrepreneurial orientation (Mohamed et al., 2023; Rauch, Wiklund, Lumpkin, & Frese, 2009; Yeni, 2015; Zehir et al., 2015; Fatoki, 2012; Mahmood & Hanafi, 2013; Mata & Aliyu, 2014). These outcomes are consistent with the resource-based theory, which postulates that a firm's unique resources and capabilities are the source of its edge over others and out-standing accomplishment (Barney, 1991). Alhamami et al., (2023) pointed out that previous research has not focused much on examining competitive advantage's mediating function on the entrepreneurial mindset and company performance association (Shaher & Ali, 2020), despite the resource-based view

(Maaodhah et al., 2021) suggesting the substantial function of advantage over rivals for boosting the success of a company.

Nonetheless, some research has found that entrepreneurial orientation substantially and favorably affects a company's competitive edge (Michael Nnaemeka Ajemba & Ebube Chinwe Arene, 2022; Perdana & Prasasti, 2023 Mustafa, Rehman, Zaidi, & Iqbal, 2015). Mahmood and Hanafi (2013) discovered in a separate investigation that there is a partial mediation among the achievements of SMEs and entrepreneurial orientation (Karaev, 2023). Also, prior research has manifest that advantage over rivals has a positive and substantial impact on company efficiency (Sufa et al., 2022). There is a wealth of empirical data demonstrating the association among EO and firm performance (Bibi et al., 2021). According to Xie & Hanafiah (2023), there is a substantial association among the expansion of market share and the EO dimensions of pro-activeness, risk-taking, and innovativeness as well as overall EO (Alhamami et al., 2023; Karimi et al., 2021; Regan, 2021). Based on 51 empirical studies, Rauch et al. (2009) performed a statistical meta-analysis and discovered a positive association among EO and corporate expansion (Wakjira, 2023; Negeri et al., 2023; Moeuf et al., 2016).

H1:*Entrepreneurial orientation has a statistical substantial effect on firm performance.*

Entrepreneurial Orientation and Innovation

In the opinion of Crossan & Apaydin (2010), entrepreneurship and innovation are inextricably linked because the implementation of creative thinking in SMEs manufacturing companies increases the industry's edge over competitors (Ayalu et al., 2022). Additionally, there is an ongoing connection among entrepreneurship and approaches to innovation (Alegre & Chiva, 2013) and entrepreneurship and groundbreaking discoveries (Lumpkin & Dess, 1996), which encompass the procedures of opportunity learning, assessment, and exploitation (Kinyanjui, 2022). Since the primary goal of EO is to launch an emerging field, enter an innovative concept, establish marketplaces, or introduce new products into already-existing markets, innovation and EO go hand in hand. This is demonstrated by the fact that EO and innovation work together to achieve the accomplishments of small business units (Zhou et al., 2005; Lumpkin & Dess, 1996). Accordingly, the more innovative products SMEs use, the more competitive advantage their firm performance has (Yaskun et al., 2023) and the more challenges their rivals will have in creating competitive solutions.

H2:*Entrepreneurial Orientation has substantial association with Innovation.*

Innovation Influence and Firm Performance

The association among innovation and firm performance has been extensively researched in a number of distinct nations and industries, according to Mairesse and Robin (2009). Their statistical outcomes are covered in the section that follows. The current literature on innovation has yielded a number of studies that have generally found a positive association among innovation and various company performance metrics (Ledi & Ameza–Xemalordzo, 2023). These outcomes repeatedly highlight how crucial innovation is to a company's ability to maintain and grow revenues, which in turn leads to better performance. Zhu et al. (2019) assert that MSEs require inventiveness to enhance the performance of their SMEs (Shaher & Ali, 2020). The authors' claim was based on the outcomes of academic research regarding the association among innovation and performance. For example, Moshi and Matotola (2023) used longitudinal data from an assortment of 378 SMEs.

Innovation has a major impact on SMEs' performance, as confirmed by Kinyanjui (2022). Accordingly, studies on the relative importance of innovation by Yousif Ali (2023); Ayinaddis, 2023; and Shaher & Ali (2020) discovered an immediate and beneficial connection among innovation and various performance levels. Likewise, it was found that method and end product innovation have a positive impact on firm performance in studies conducted by Yaskun et al. (2023) and Rosli and Sidek (2013) on the impact of innovation on the earnings of manufacturing SMEs.

H3:*Innovation has statistical substantial effect on firm performance*

Innovation Mediation in Association Among EO and SMEs Performance

In accordance with Damanpour (1991), EO and innovation are typically associated with the firm performance of SMEs as a driver of the manufacturing industry (Alhamami et al., 2023) and the most substantial industry determinant is innovation as mediated by EO and the firm performance of SMEs (Birbirsa & Kero, 2022). Rubera & Kirca (2012) posit that the reason for the positive association among innovation achievement is that the company's efficiency can achieve normal rents through revenue the mining in a short-term quasi-monopoly establish (Khairuddin et al., 2020).

Being receptive to shifts in the market and seeing new trends are essential components of an entrepreneurial mindset. By creating new goods or services that meet changing consumer demands, innovation aids SMEs in adjusting to these changes (Ayalu et al., 2022). SME performance may be enhanced by efficiently satisfying consumer requests, expanding into new market categories, and remaining relevant by adopting innovation (Alhamami et al., 2023; Patel & Patel, 2023). Businesses with an entrepreneurial mindset are driven to look for ways to increase efficiency and enhance operations. Streamlining workflows, cutting expenses, and improving operational procedures are all possible with innovation (Ledi & Ameza– Xemalordzo, 2023). SME's can improve their financial performance, increase efficiency, and streamline their operations by implementing cutting-edge technology, processes, or business models (Alhamami et al., 2023; Karimi et al., 2021; Regan, 2021).

H4:*Innovation has a mediated among EO and firm performance.*

Figure 1. Empirical construct

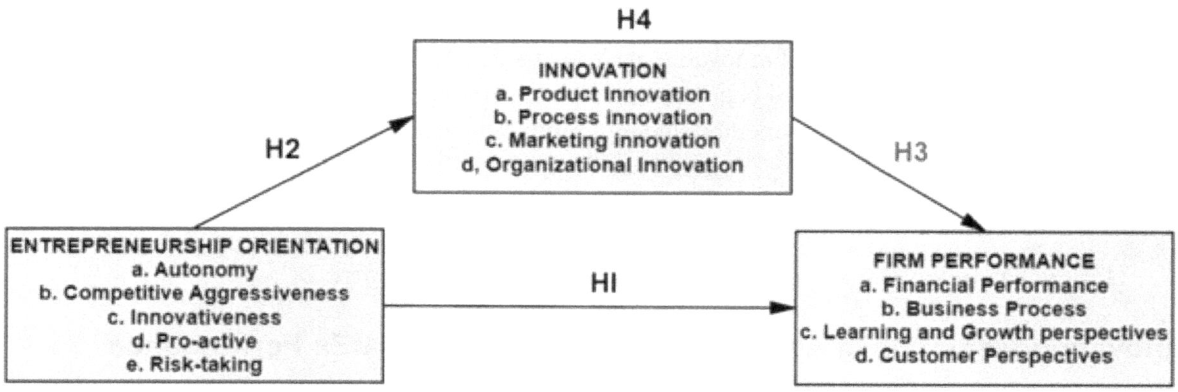

Source: Empirical Reviewed Literature (2023)

RESEARCH METHODOLOGY

The current study employs a quantitative research approach that involves assessing data on variables (EO, innovation, and firm performance) using quantitative scales and quantitative data analysis. Quantitative research approach and explanatory research design was used by the researchers. Before beginning field duties, researchers conduct an exploratory factor exploration (EFA) based on the pilot study's 90 responses (20% of sample size) to verify the scope of the proxies and check the internal reliability of study variables with help of KMO test of reliability and data adequacy. Confirmatory factor analysis (CFA) and structural equation modelling (SEM) are used to analyze data from a sample of 451 employees working in the textile industrial parks in Ethiopia in order to test the framed hypothesis.

Description of the Study Area and Total Population

In variable of the consequence of Entrepreneurship Orientation on SMEs performance the Mediating function of Innovation practices in SMEs Textile, Garment and Apparel manufacturing industry of targeted population covers textile, Garment and apparel industries located in Oromia regional state located in Finfine zuria and East Shewa, Ethiopia. According to Ethiopian Textile Industry Development Institute, out of the total 139 textiles and apparel manufacturers in Ethiopia, textile, garment and apparel manufacturers located within the vicinity of Finfine zuria are 7 and in East Shewa Adama Zuria are 3 industry, totally 10 textile, garment and apparel manufacturing industry . Also, from the aforementioned manufacturers which are located in Ethiopia. Because due to distance and budget constraints, during multi stage sampling from Ethiopia, researchers selected 10 textiles from Oromia region and East Shewa zone. From Oromia region they surveyed the Finfine area (capital city), Garment and apparel industries were selected on base of their cluster consists of Bole Lemi Indusial Park and Adama textile Indusial Park, Garment and Apparel manufacturing industrial area as a target population.

Sampling Techniques and Sample Size

In quantitative data research the probability sampling methods and through the major types, from general to specific, to use a multi-stage clustering sampling will be used to collect data a sample from large population. For this study, 392 respondents from manufacturing Textile, Garment and Apparel industry of SMEs was suggested as sample size that was calculated by using the Yemane (1967) method. Thus, to maximise the outcomes' generalizability and counteract an expected inadequate reply or answered rate from among 10 and 20 percent (Lai Van Voi 2023), respondents will be selected proportionally from textile, garment and apparel manufacturing sectors (Le Danih Vinih 2023).Therefore:-

***n*=392, 392+15% of 392 of =451**

DATA ANALYSIS

Evaluation for Sampling Adequacy

To show the ordinary allocation for each variable product, investigators used the KMO test. The KMO metric evaluates whether the sample size is sufficient for factor analysis. It has a value range of 0 to 1, where values nearer 1 are more appropriate for factor analysis. The KMO value in your instance is 0.839, indicating that the sample is generally suitable for factor analysis. To ascertain whether the variables in the dataset are correlated and, thus, appropriate for factor analysis, Bartlett's test is employed. It examines the null hypothesis that the association matrix is a unique matrix by proving that there are no associations among the variables. An approximate chi-square value serves as the test statistic.

Table 1. Sampling adequacy test

Sampling Adequacy by Kaiser-Meyer-Olkin appraise		**.839**
Test of Sphericity by Bartlett's	Likelihood	1985.056
	freedoms Degree	114
	p-unit	.000

Source: SPSS Output, 2023

The test statistic value is 1985.056 on base of table 1 outcomes. The degree of disagreement among the identity matrix and the observed association matrix is indicated by this value. Greater discrepancy is suggested by higher values, which prove that the variables are correlated. The significance level, represented by the letter "Sig.," manifests the likelihood that the observed chi-square value happened by accident. Since the chi-square value in this instance is statistically substantial at a very low p-value (less than 0.001), the significance level is reported as.000. This implies that there is evidence to support the conclusion that the variables are correlated and to reject the null hypothesis. Overall, based on the KMO measure and Bartlett's test, the data appears suitable for factor analysis, indicating that there are associations among the variables in the dataset.

Confirmatory Examination

To evaluate the model's fitness using indices, investigators employed confirmatory factor analysis (CFA). The index values demonstrated the covariance among the model's measuring items for transformational leadership, public sector performance, and employee commitment. Following the confirmation of the CFA description and the satisfaction of all reliability and validity thresholds, analysts found that the measurement model specified for each and every dormant (latent) variable included in the measurement model had been authenticated. Researchers then incorporate the variables into the structural equation model to perform structural equation modeling, or SEM. Researchers in figure 2, begin with external variables at the extreme right, move to the middle for mediator variables, and endogenous variables at the far left.

Figure 2. Confirmatory examinations of determinants

Source: Output of AMOS, 2023

In Figure 2, investigators use a single headed arrow on the assumption's route, followed by substrate, to connect the exogenous variable to its particular endogenous variable. The weights for both identical and non-standardized regression models were determined by researchers using the structural model based on AMOS 26, as depicted in figure 2. After using fit indices to test the model's fitness, the model is statistically fitted under the desired parameters. The factor loading of all residual items exceeded the required value of 0.4, as can be observed in Figure 2. This demonstrated to the investigators that the measurement model's uni-dimensionality had been reached. According to the researcher's outcomes from the confirmatory factor investigation (CFA) and the examined index fitness, it was the modus operandi that was partied at this point rather than the earlier structural modelling. The following table 4 manifests the average variance extracted (AVE) and achieved Composite reliability (CR) requirements for factoring all components to check discriminate validity.

Table 2. Discriminate validity check

	CR	AVE	MSV	MaxR(H)	Organization Performance	Entrepreneurship Orientation	Innovation
Organization Performance	0.793	0.529	0.139	0.796	**0.586**		
Entrepreneurship Orientation	0.752	0.517	0.156	0.781	0.302	**0.604**	
Innovation	0.785	0.522	0.152	0.803	0.253	0.503	**0.698**

Source: AMOS 2023

The outcomes of a validity check for the three constructs of innovation, entrepreneurship orientation, and business performance are manifest in Table 2. To evaluate the validity of these constructs, the table offers a range of statistical values. Let's review each metric. The CR (Composite Reliability) metric gauges the dependability of internal consistency. The composite reliability for each construct is represented by the values in the table. As an illustration, the composite reliability for innovation is 0.785, entrepreneurship orientation is 0.752, and corporate performance is 0.793. Greater dependability is indicated by higher values. The measure of convergent validity known as AVE (Average Variance Extracted) manifests how much variance the construct is able to capture. The AVE for each construct is manifest by the values in the table. The AVE, for instance, is 0.529 for corporate performance, 0.517 for entrepreneurial orientation, and 0.522 for innovation. An acceptable AVE value is generally defined as 0.5 or higher. A measure of discriminant validity called maximum shared variance, or MSV, manifests how much a construct shares more variance with other constructs than with itself. The MSV for each construct is represented by the values in the table. The MSV for corporate performance, for instance, is 0.139; the MSV for innovation is 0.152; and the MSV for entrepreneurial orientation is 0.156. Better discriminant validity is indicated by lower values.

The highest association among a construct and any other construct in the study is represented by MaxR(H) (Maximum Association with Any Other Construct). The MaxR(H) for each construct is represented by the values in the table. The MaxR(H) for innovation is 0.803, the MaxR(H) for entrepreneurship orientation is 0.781, and the MaxR(H) for corporate performance is 0.796. Better discriminant validity is indicated by lower values. The constructs appear to have good convergent validity (AVE values above 0.5) and reliability (CR values above 0.7) overall based on the values in the table. The MSV values for each construct are less than the AVE values, indicating reasonably good discriminant validity. The relatively high association (0.604) among innovation and entrepreneurial orientation, also, is noteworthy and may point to a possible problem with the discriminant validity of each of these constructs.

Figure 3. Structure equation model

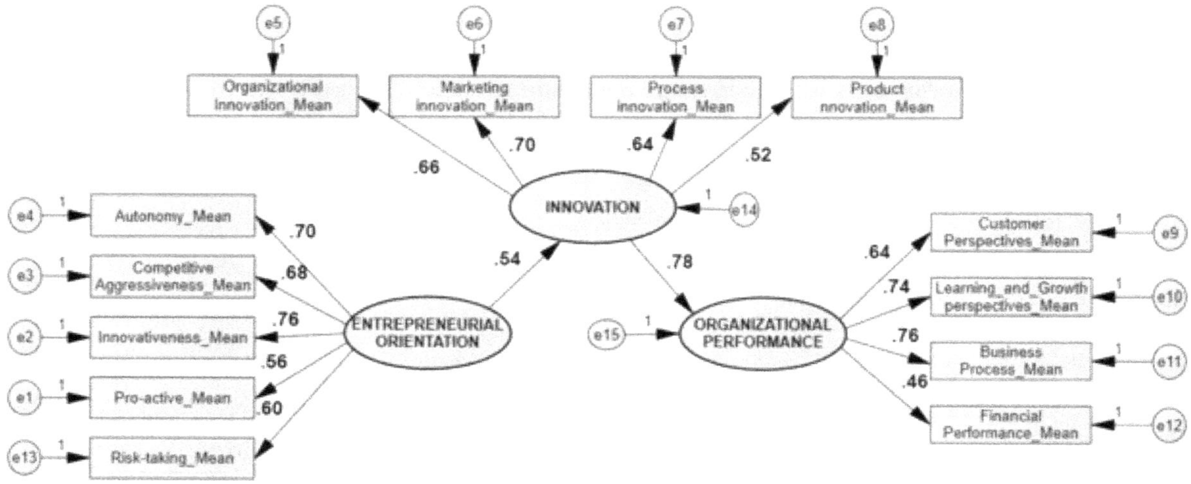

Source: AMOS Output, 2023

Table 3. Hypothesis test

			Estimate	S.E.	C.R.	P	H	Hypothesis Test
Innovation	<---	Entrepreneurship Orientation	1.583	.139	11.380	***	H2	Acknowledged
Organization Performance	<---	Innovation	.748	.172	4.353	***	H3	Acknowledged
Organization Performance	<---	Entrepreneurship Orientation	.158	.084	1.889	.084	H1	Rejected

Source: AMOS Output, 2023

The outcomes for a structural equation model are manifest in figure 3 and Table 3 and include estimates, standard errors (S.E.), critical ratios (C.R.), p-values, hypotheses, and the outcomes of hypothesis tests. The estimate of 1.583 points manifested a favorable co-association among innovation and entrepreneurship orientation. The estimate's accuracy is indicated by the standard error (0.139). A measure of the association ship's strength, the critical ratio (11.380) manifests that the association ship is statistically substantial. Three asterisks (***) usually indicate a very low p-value (e.g., $p < 0.001$), suggesting a highly substantial association ship; also, the significance level is not stated. Consequently, it can be said that Hypothesis H2, according to which Entrepreneurship Orientation influences Innovation, is true. The estimate of 0.748 points manifested a favorable co association among business performance and innovation. The estimate's accuracy is manifest by the standard error (0.172). The statistical significance of the association ship is indicated by the critical ratio of 4.353. Once more, the significance level is not stated; also, the three asterisks (***) point to an extremely low p-value. Consequently, it is acknowledged that Hypothesis H3, which claims that Innovation affects Corporate Performance, is true.

The estimate of 0.158 points manifested a favorable co association among business performance and entrepreneurial orientation. In contrast to the other association ships, the standard error (0.084) manifests a comparatively higher degree of uncertainty surrounding the estimate. The statistical significance threshold is typically exceeded by the critical ratio (1.889). The reported p-value is 0.084, indicating a higher level of significance than the standard 0.05 threshold. Thus, it can be concluded that Hypothesis H1, which posits that Entrepreneurship Orientation influences firm Performance, is not supported. In conclusion, the outcomes imply that innovation and an entrepreneurial orientation both substantially improve business performance. In contrast to the other associations, the one among Corporate Performance and Entrepreneurship Orientation is comparatively weaker.

Model Fitness

Table 4. Fitness indices

Indices	Manifested Value	Output
IFI	0.924	*Customary fit*
CMIN/DF	2.06	*Moderate fit*
CFI	0.923	*Customary fit*
RMSEA	0.063	*likelihood Considered*
GFI	0.959	*Customary fit*
TLI	0.929	*Customary fit*
PNFI	0.652	*Customary fit*
PCFI	0.671	*Customary fit*

Source: AMOS Output, 2023

The outcomes of the fitness indices for a structural equation model are manifest in Table 4. The model's fit to the observed data is evaluated using these indices. In Table 4, IFI (Incremental Fit Index) value is 0.924. IFI provides an overall measure of fit by comparing the hypothesized model's fit with an independence model. In this instance, 0.924 indicates a reasonably good fit; a value closer to 1 indicates a better fit. The value of CMIN/DF (degrees of freedom/Chi-square) is 2.06. A measure of the model's fit to the data that accounts for model complexity is called CMIN/DF. A fit of approximately 2 or less is generally regarded as satisfactory. A moderate fit is suggested in this instance by 2.06. 0.923 is the CFI (Comparative Fit Index) value. The CFI compares the fit of the proposed model with an independence model and is comparable to the IFI. In this instance, 0.923 indicates a reasonably good fit; a value closer to 1 indicates a better fit. The value of 0.063 is the root mean square error of approximation, or RMSEA. The difference among the population covariance matrix and the proposed model is measured by RMSEA. A better fit is indicated by smaller values; generally, values less than 0.08 are regarded as acceptable. 0.063 in this instance indicates a likelihood of fit.

GFI (Goodness-of-Fit Index) value is 0.959. The model's percentage of the observed covariance that is explained is measured by the GFI. In this instance, 0.959 indicates a reasonably good fit; a value closer to 1 indicates a better fit. 0.929 is the Tucker-Lewis Index value. An additional metric called TLI contrasts the fit of the proposed model with an independence model. In this instance, 0.929 indicates a reasonably good fit; a value closer to 1 indicates a better fit.

The value of the Parsimony Normed Fit Index (PNFI) is 0.652. Models with more parameters are penalised by the PNFI measure. In this instance, 0.652 indicates a reasonably good fit; a value closer to 1 indicates a better fit. The value of the PCFI (Parsimony Comparative Fit Index) is 0.671. An additional metric that penalizes models with more parameters is PCFI. In this instance, 0.671 indicates a reasonably good fit; a value closer to 1 indicates a better fit. In general, the model manifests a reasonable fit to the observed data based on the values in table 6. A typical or fairly good fit is indicated by the values of the IFI, CFI, GFI, TLI, PNFI, and PCFI. There is some opportunity for improvement in terms of model fit, as indicated by the CMIN/DF value, which indicates a moderate fit. There is a chance of fit because the RMSEA value is less than the cutoff of 0.08. All things considered, the model seems to give a fair representation of the data, but more analysis and thought may be needed to guarantee a good fit.

Mediation Analysis

The outcomes of a mediation analysis examining the connection among innovation, entrepreneurship orientation, and SMEs firm performance are manifest in figure 4 and Table 5. The direction, approximate (identical) coefficients, and p-values for the determinants are manifest in table 5. The direction of the association ship among the determinants is indicated by the arrows in the "Direction" column. The association ship indicated by the arrow "<---" is one in which the determinant on the left, SMEs performance, acts as an independent variable or predictor, and the determinant on the right, entrepreneurial orientation and innovation, acts as an intermediary or mediator.

Figure 4. Full mediation model

Source: AMOS, 2023

Table 5. Mediation of innovation

	Entrepreneurship Orientation	Innovation	Organization Performance
Innovation	0.540	.000	0.000
Organization Performance	0.156	0.779	0.000

Source: AMOS result, 2023

The outcomes of a mediation analysis that looked at the connections among innovation, organizational performance, and entrepreneurial orientation are manifest in Table 5 and figure 4. The path coefficients among the variables are manifest in the table. The co-association coefficient among innovation and entrepreneurial orientation is 0.540. This manifests that there is a substantial and positive co association among these two variables. The co-association coefficient among Corporate Performance and Innovation is 0.779. This manifests that there is a substantial and positive co association among these variables. The co-association coefficient among Corporate Performance and Entrepreneurship Orientation is 0.156. This manifests that these variables have a positive but not statistically substantial association ship. The association ships are statistically substantial at a high degree of confidence ($p < 0.001$) when the p-values in the table are represented as 0.000. It's crucial to remember that the table doesn't include the precise p-values, making it impossible to pinpoint the precise degree of significance.

The table's outcomes suggest that innovation positively impacts SMEs firm performance and that entrepreneurship orientation positively influences innovation. But there isn't a meaningful co-association found among corporate performance and entrepreneurship orientation. Rather, the outcomes point to innovation as a mediating factor in the association ship among corporate performance and entrepreneurial orientation.

DISCUSSION

The results of this study in Table 5 provide valuable insights into the mediation of innovation within Ethiopia, especially in terms of entrepreneurial orientation as well as organizational performance. The substantial mediation impact caused by innovation, represented by an internationally accepted estimate of 0.540 as well as a p-value of.000, implies that innovation is an important driver in improving the link between entrepreneurial orientation as well as organizational performance among Ethiopian firms. This is especially important in a country because entrepreneurial ventures are developing as critical components of economic expansion and advancement. In Ethiopia, where many enterprises are still in their early phases and face specific constraints such as restricted access to funding, infrastructural shortfalls, and regulatory barriers, cultivating a creative culture can give a competitive advantage. The outcomes demonstrate that merely having an entrepreneurial mindset is inadequate to achieve high organizational performance. Instead, Ethiopian firms have to engage in creative methods, like as technology adoption, process changes, or product creation, in order to transform their mindset of entrepreneurship into actual performance results. This conclusion emphasizes the importance of organizations prioritizing innovation to be a strategic priority, particularly in a quickly changing market where adaptation is critical. Furthermore, the relatively low generalized estimate of 0.156 for the direct association between entrepreneurial orientation as well as organizational performance, along with a p-value of 0.779, emphasizes the importance of innovation being a mediating factor.

Considering the Ethiopian context, this demonstrates that businesses with a strong entrepreneurial mindset may fail to achieve large performance benefits until they include innovation into their administrative structures. This is especially important in industries like agriculture, clothing, overall technology, where new techniques may boost efficiency, production, and competitiveness in the markets. Furthermore, the outcomes demonstrate the potential for Ethiopian governments and corporate leaders to foster innovation-friendly settings. Strategy might include improving access to funding for creative initiatives, providing entrepreneurs with expertise and tools, and encouraging cooperation involving educational organizations and enterprises to promote R&D efforts. Ethiopian can unleash the complete potential of its entrepreneurial sector by cultivating an innovation-friendly ecosystem, resulting in long-term economic development with increased organizational performance. To summarize, the intermediary role of innovation is an important factor for Ethiopian firms looking to improve their performance through an entrepreneurial mindset. The emphasis on innovation not only coincides alongside worldwide trends, but it also tackles the specific issues encountered by Ethiopian businesses. Future study could concentrate on practical techniques for applying innovative practices in diverse industries, and guaranteeing the drive to innovate leads to measurable results for organizational success.

CONCLUSION

The study's findings highlight the critical role of innovation being a mediator between entrepreneurial approach and business accomplishment, especially in Ethiopia. The significant mediating influence of innovation demonstrates that simply having an entrepreneurial mentality is insufficient to achieve high performance. To fully realize their potential, firms must actively foster new processes. This requirement is especially important for Ethiopian businesses, which confront unique problems such as resource limits and infrastructural issues. As a result, developing a creative culture is vital to increasing competence and guaranteeing long-term success Likewise, the results show a modest direct association between entrepreneurial attitude and organizational success. This emphasizes the value of taking a quantitative approach to innovation. To effectively encourage innovation, lawmakers and business leaders must work together to provide entrepreneurs with the tools, skills, and support networks they require. By boosting innovation, Ethiopia may leverage on its labor force's entrepreneurial spirit, resulting in better workplace results that are beneficial to the country's broader growth in business. Future study ought to concentrate on developing practical ways for combining innovation across industries. This ensures that business ventures result in considerable performance gains. As Ethiopia's economy evolves, the collaboration between innovation and entrepreneurship will come to be critical to developing a strong and dynamic business environment. Table 5 illustrates this link by demonstrating the intermediary nature of innovation in the environment of Ethiopian SMEs exhibiting an entrepreneurial orientation. The measures show that innovation has a large influence on entrepreneurial attitude, as well as a significant direct effect on organizational success. This emphasizes the importance for firms to prioritize innovation as a major driver of success in a market that is increasingly competitive.

REFERENCES

Abawa, A. (2020). *Micro, Small and Medium Enterprises(MSMEs)Development Strategies in Ethiopia: Retrospective and Prospective Analysis. January 2017.*

Abbaszadeh, M., Abdavi, F., & Golmohammadi, H. (2023)... *Improving the Validity of Mixed and Multi-Methods Through Triangulation in Sports Management Research.*, *4*(2), 16–27.

Abraham, A., & Viswanadham, P. (2021). *Innovations The effect of risk management practice on the financial performance of Ethiopian commercial banks. 6*, 1294–1311.

Abraham, A. Ph. B., & Viswanadham, P. (2021). Innovations Tax Assessment and Collecti on Problems of Category " A. *Tax Payers : The Case of Gulelle Sub City Revenues and Customs Authority Branch Office in Addis Ababa City Administration*, *6*, 869–878.

Abrehet Mehari, Abrehet Mehari. (2020). Practice of Economic Integration of Small and Medium Enterprise in Manufacturing Sector: The case of Tigray Regional State, Ethiopia. *International Journal of Social. Political and Economic Research*, *7*(1), 77–91. DOI: 10.46291/IJOSPERvol7iss1pp77-91

Adam, N. A. (2021). *Innovation practices for survival of small and medium enterprises (SMEs) in the COVID-19 times : the role of external support. 6.*

Adesoga, D., Asikhia, P., Olalekan, U., & Taiwo, F. (2018)... *The Effect of Pro-Activeness on Growth of Selected Small and Medium Scale Enterprises in Ogun State Nigeria*, *20*(12), 14–21. DOI: 10.9790/487X-2012051421

Alhamami, A. A., Hashim, N. A., Hamid, R. A., & Hamid, S. N. A. (2023). The Mediating Role of Marketing Innovation Among Internal Social Media Utilization and Business Performance of Smes in Saudi Arabia. *International Journal of Professional Business Review*, *8*(5), e01811. DOI: 10.26668/businessreview/2023.v8i5.1811

Almacena Platform. (2021). *Challenges and Opportunities for Smallholders. October.* https://www.almacenaplatform.com/blog/challenges-and-opportunities-to-improve-the-livelihoods-of-smallholder-farmers/

Aloulou, W. J. (2023). Be Innovative and Resilient: Empirical Evidence from Saudi Firms on How to Translate Entrepreneurial Orientation into Firm Performance. *Administrative Sciences*, *13*(7), 168. Advance online publication. DOI: 10.3390/admsci13070168

Alshammari, S. A., Rehman, A. U., Alreshoodi, S. A., & Rab, M. A. (2023). How entrepreneurial competencies influence the leadership style : A study of Saudi female entrepreneurs. *Cogent Business & Management*, *10*(2), 2202025. Advance online publication. DOI: 10.1080/23311975.2023.2202025

Amari, A. (2023). Which kind of perceived business support fosters self-initiated expatriate engineers ' creativity in high-adversity contexts? The moderating role of cross-cultural resilience. *Cogent Business & Management*, *10*(3). Advance online publication. DOI: 10.1080/23311975.2023.2259150

Arslan, I. K. (2020). the Importance of Creating Customer Loyalty in Achieving Sustainable Competitive Advantage. *Eurasian Journal of Business and Management*, *8*(1), 11–20. DOI: 10.15604/ejbm.2020.08.01.002

Ayalu, G., Abbay, A. G., & Azadi, H. (2022). The role of micro- and small-scale enterprises in enhancing sustainable community livelihood: Tigray, Ethiopia. *Environment, Development and Sustainability*, *25*(8), 7561–7584. DOI: 10.1007/s10668-022-02359-7 PMID: 35645605

Ayinaddis, S. G. (2023). The effect of innovation orientation on firm performance: Evidence from micro and small manufacturing firms in selected towns of Awi Zone, Ethiopia. *Journal of Innovation and Entrepreneurship*, *12*(1), 26. Advance online publication. DOI: 10.1186/s13731-023-00290-3

Beck, S., LaFlamme, M., Bergenholtz, C., Bogers, M., Brasseur, T. M., Conradsen, M. L., Crowston, K., Di Marco, D., Effert, A., Filiou, D., Frederiksen, L., Gillier, T., Gruber, M., Haeussler, C., Hoisl, K., Kokshagina, O., Norn, M. T., Poetz, M., Pruschak, G., & Xu, S. M. (2021). Examining Open Innovation in Science (OIS): What Open Innovation can and cannot offer the science of science. *Innovation (North Sydney, N.S.W.)*, *25*(3), 221–235. DOI: 10.1080/14479338.2021.1999248

Bibi, P., Ahmad, A., & Bilal, H. (2021). Disintegrated Market Orientation and Firm Performance. *Moderating Effect of Entrepreneurial Orientation.*, *7*(1), 155–168.

Celikyay, M., Sonmez Cakir, F., & Adıguzel, Z. (2022). Effects of supplier orientation and competitive strategies on ICT firms' performance. *Journal of Asia Business Studies*, (July). Advance online publication. DOI: 10.1108/JABS-09-2021-0384

Chelliah, M. K., Aravindan, K. L., & Muthaiyah, S. (2023). Entrepreneurial Orientation and Open Innovation Promote the Performance of Services SMEs: The Mediating Role of Cost Leadership. *Administrative Sciences*, *13*(1), 1–19. DOI: 10.3390/admsci13010001

Dele-ijagbulu, O. (2020).. . *Dimensions of Firm-Level Entrepreneurial Orientation as Antecedents to Employment Growth in SMMEs Chukuakadibia Eresia-Eke.*, *15*(3), 89–111.

Dessie, M. M. (2021). *Innovations The Effect of Innovation Orientation on Employees Creativity and Organizational Culture (A Study on Textile Factories in Amhara. 6*, 1639–1647.

Faiz, G. A., Ahmed, F., & Swidi, A. K. A. (2015). The Mediating Effect of Market Orientation on the Relationship among Entrepreneurial Orientation Dimensions and Organizational Performance: A Study on Banks in Libya. *Games Review*, *1*(2), 40–51. DOI: 10.18488/journal.100/2015.1.2/100.2.40.51

Fentie, Y. A., & Devi, P. M. U. (2021). Innovations Investigation on Macroeconomic Determinants of Non-Performing Loans. *The Case of Ethiopian Commercial Banks.*, *6*, 1582–1595.

Ganguly, B., Dey, B. K., Pareek, S., & Sarkar, B. (2023). Cost-Effective Imperfect Production-Inventory System under Variable Production Rate and Remanufacturing. *Mathematics*, *11*(15), 1–25. DOI: 10.3390/math11153417

Girma, B., Lakew, D. M., & Nurebo, B. Y. (2022). *Horn of Africa Journal of Business and Economics (HAJBE), 2022, 5 (2), PP : 15-29 ISSN : 2617-0078 (Print), 2617-0086 (Online) Determinants of Commercial Banks Liquidity in Ethiopia Horn of Africa Journal of Business and Economics (HAJBE), 2022. 5*(2), 15–29.

Herlinawati, E., Suryana, Ahman, E., & Machmud, A. (2019). The effect of entrepreneurial orientation on smes business performance in Indonesia. *Journal of Entrepreneurship Education*, *22*(5), 1–15.

Hota, E. T. (2021). Innovations Integration of Labour Rights Issues into Bilateral Investment Treaties. *Particular Emphasis to Bilateral Investment Treaties.*, *6*, 1715–1725.

Irawan, T. T., Mansor, M. N. B., ... (2023). Entrepreneurial Orientation and Firm Performance: A Systematic Review. In *European Business & ...* (Vol. 9, Issue 3, pp. 51–62). DOI: 10.11648/j.ebm.20230903.12

Issau, K., Acquah, I. S. K., Gnankob, R. I., & Hamidu, Z. (2022). Innovation orientation and performance of small and medium-sized enterprises (SMES) in Ghana: Evidence from manufacturing sector. *Innovation & Management Review*, *19*(4), 290–305. DOI: 10.1108/INMR-07-2020-0092

Jemal, S., Ayenew, Z., & Lemmi, K. (2022). *Horn of Africa Journal of Business and Economics (HAJBE), 2022, 5 (2), PP : 180-197 ISSN : 2617-0078 (Print), 2617-0086 (Online) Systematic Literature Review on Corporate Social Responsibility Practice of Organization Horn of Africa Journal of Bu. 5*(2), 180–197.

Kanaan-jebna, A., Baharudin, A. S., Tawfeeq, T., & Alabdullah, Y. (2022). Entrepreneurial Orientation, Market Orientation. *Managerial Accounting and Manufacturing SMEs Satisfaction.*, *6*(1), 1–14. DOI: 10.21070/jas.v6i1.1590

Karimi, A., Daryani, M. A., & Rahmani, S. (2021). The influence of entrepreneurial orientation on firm growth among Iranian agricultural SMEs: The mediation role of entrepreneurial leadership and market orientation. *Journal of Global Entrepreneurship Research*, *11*(1), 519–531. DOI: 10.1007/s40497-021-00282-1

Khan, W. A., Hassan, R. A., Arshad, M. Z., Arshad, M. A., Kashif, U., Aslam, F., & Wafa, S. A. (2020). The effect of entrepreneurial orientation and business culture on firm performance: The mediating role of innovation. *International Journal of Innovation. Creativity and Change*, *13*(3), 652–677.

Kinyanjui, J. (2022). *Influence of innovation orientation on organizational performance of suppliers in the telecommunication industry in Kenya* .

Kiyabo, K., & Isaga, N. (2020). Entrepreneurial orientation, competitive advantage, and SMEs' performance: Application of firm growth and personal wealth measures. *Journal of Innovation and Entrepreneurship*, *9*(1), 12. Advance online publication. DOI: 10.1186/s13731-020-00123-7

Ledi, K. K., & Ameza–Xemalordzo, E. (2023). Rippling effect of corporate governance and corporate social responsibility synergy on firm performance: The mediating role of corporate image. *Cogent Business & Management*, *10*(2), 2210353. Advance online publication. DOI: 10.1080/23311975.2023.2210353

Lee, H., Zhang, Y., Zhang, H., & Liu, X. (2023). Challenges and opportunities of utilizing design thinking in the industry 4. 0 era : Based on the case study of industry wastewater treatment system development project process. *Journal of Innovation and Entrepreneurship*, *12*(1), 56. Advance online publication. DOI: 10.1186/s13731-023-00315-x

Long, B. (2022). Factors Affecting Organizational Performance : A Study on Four Factors. *Motivation, Ability, Roles, and Organizational Support.*, *1*, 1–15.

Lustrilanang, P., Rizki, L. T., & Said, J. (2023). *The Role of Control of Corruption and Quality of Governance in ASEAN : Evidence from DOLS and FMOLS Test The Role of Control of Corruption and Quality of Governance in ASEAN : Evidence from DOLS and FMOLS Test*. DOI: 10.1080/23311975.2022.2154060

Maaodhah, A. S. A., Singh, H., Al-Juboori, Z. M. A., Pitchy, A. L. B. A., & Ekene, I. (2021). The Impact of Market Orientation and Entrepreneurial Orientation on Firm Performance of Wholesale and Retailer SMES in Malaysia. *International Journal of Academic Research in Business & Social Sciences*, *11*(6), 729–743. DOI: 10.6007/IJARBSS/v11-i6/10019

Makhloufi, L., Laghouag, A. A., Sahli, A. A., & Belaid, F. (2021). Impact of entrepreneurial orientation on innovation capability: The mediating role of absorptive capability and organizational learning capabilities. *Sustainability (Basel)*, *13*(10), 5399. Advance online publication. DOI: 10.3390/su13105399

Manickam, S., Pachamuthu, S., & Chavan, S. (2023). Effect of Thermal Barrier Coatings On the Stability, Performance, and Emission Characteristics of Pongamia Water Emulsion Biodiesel in Ci Engines Case Studies in Thermal Engineering The effect of thermal barrier coatings and neural networks on the stabi. *Case Studies in Thermal Engineering*, *49*(September), 103079. DOI: 10.1016/j.csite.2023.103079

Maruta, I. G. N. A., Sularso, R. A., & Susanti, N. (2017). The Effect of Market Orientation, Entrepreneurship Orientation, and Imitation Strategy on Competitive Advantage (Study on SME of Leather Bag and Suitcase in East Java). *International Journal of Business and Management Invention ISSN*, *6*(7), 24–35. www.ijbmi.org

Moshi, R., & Matotola, S. (2023). Entrepreneurial Competence and Small and medium enterprises' Profitability: A Case of sunflower oil processing enterprises in Tanzania. *Cogent Business & Management*, *10*(2), 2208431. Advance online publication. DOI: 10.1080/23311975.2023.2208431

Osuagwu, C. (2019). Market Orientation Conceptualizations, Components and Performance-Impacts: A Literature Review and Conceptual Framework. *International Journal of Marketing Studies*, *11*(2), 102. DOI: 10.5539/ijms.v11n2p102

Peng, J., Qin, Q., & Tang, T. (2021). The influence of marketing innovations on firm performance under different market environments: Evidence from china. *Sustainability (Basel)*, *13*(18), 1–15. DOI: 10.3390/su131810049

Perdana, R., & Prasasti, A. (2023). Entrepreneurial orientation, company performance, and competitive advantage in Indonesian culinary SMEs. *Small Business International Review*, *7*(1), e547. DOI: 10.26784/sbir.v7i1.547

Regan, D. B. (2021). The Effect of Small and Medium Enterprises in Employment Creation and Income Generation in Somali Regional State : A Case of Kebridahar Town. *Munich Personal RePEc Archive, 105879*, 32. https://mpra.ub.uni-muenchen.de/105879/

Schulze, A., Townsend, J. D., & Talay, M. B. (2022). Completing the market orientation matrix: The impact of proactive competitor orientation on innovation and firm performance. *Industrial Marketing Management, 103*(April), 198–214. DOI: 10.1016/j.indmarman.2022.03.013

Shaher, A. T. H. Q., & Ali, K. A. M. (2020). The effect of entrepreneurial orientation and knowledge management on innovation performance: The mediation role of market orientation. *Management Science Letters, 10*(15), 3723–3734. DOI: 10.5267/j.msl.2020.6.020

Siam, S. T., Shaari, J. A. N., & Heriyadi, H. (2022). The Relationship of Organizational Competition Orientation, Marketing Capability and Customer Networks in Achieving Seller Achievement. *European Journal of Business and Management Research, 7*(2), 168–175. DOI: 10.24018/ejbmr.2022.7.2.1219

Šlogar, H., Mori, B., & Hrvatin, S. (2023). *administrative sciences Does the Relationship among Entrepreneurial Orientation and Subjective Financial Firm Performance Have an Inverted U-Shape ? Evidence from Southeast European SMEs.*

Susanto, Y. B. (2019). The Impact of Market Orientation and Dynamic Marketing Capability on the Marketing Performance of "Make-To-Order" SMEs. *GATR Journal of Management and Marketing Review, 4*(3), 157–170. DOI: 10.35609/jmmr.2019.4.3(1)

Tesfa, S. Y. (2023). The Effect of Corporate Social Responsibility on Consumer Loyalty. *The Mediating Role of Corporate Image., 73*, 1536–1553.

Van Thuong, C., & Singh, H. (2023).. . *The Impact of a Balanced Scorecard on Enterprise Performance in Ho Chi Minh., 12*, 198–215.

Wahyuda, O., Syukur, A., Bagus, M.-H., Purmono, B., Pebrianti, W., & Setiawan, H. (2023). Empowerment of Indonesian MSME Business Performance: An Empirical Analysis of Knowledge Sharing and Innovation. *International Journal of Scientific Research and Management, 11*(01), 4458–4467. DOI: 10.18535/ijsrm/v11i01.em06

Wakjira, G. G., & Ibrahim, M. I. (2023). *Assessments of Structural Equation Modeling (SEM) of Internal Marketing Practice with Sustainable Marketing Performance : The Case of Small and Medium Enterprise (SMEs) in Ethiopia. 1*(2), 48–56.

Wani, T. A., & Ali, S. W. (2015). Innovation Diffusion Theory (IDT). *Journal of General Management Research, 2*(2), 98–115.

Wedajo, T. D. (2021). *Innovations Assessing the internal factors that hinder the performance of Small and Micro Enterprises in Ethiopia : the case of Ficthe town. 6*, 1453–1464.

Wolfenden, S. F. (2020). *Entrepreneurial Marketing Orientation : The Influence on the Adoption and Use of Digital Marketing Technology in Small Tourism Businesses. October.*

Yaskun, M., Sudarmiatin, S., Hermawan, A., & Rahayu, W. P. (2023). The effect of market orientation, entrepreneurial orientation, innovation and competitive advantage on business performance of Indonesian MSMEs. International Journal of Professional Business Review: Int. *J. Prof. Bus. Rev.*, *8*(4), 39.

Yesuf, Y. M., Getahun, D. A., & Debas, A. T. (2023). Factors affecting " employees ' creativity ": The mediating role of intrinsic motivation. *Journal of Innovation and Entrepreneurship*, *12*(1), 31. Advance online publication. DOI: 10.1186/s13731-023-00299-8

Yousif Ali, M. S. (2023). The effect of market information on export performance: The mediating role of notional export support and encouragement. *Uncertain Supply Chain Management*, *11*(3), 1305–1318. DOI: 10.5267/j.uscm.2023.3.017

Chapter 7 Internal Control Advisory to Support Innovation in the Corporate Governance Diplomacy: A Case Study of a Brazilian Science and Technology Institution (Oswaldo Cruz Foundation - Fiocruz)

Regina Lúcia de Carvalho Januário Dias
https://orcid.org/0009-0004-3185-0865
Institute of Technology in Drugs (Farmanguinhos), Fiocruz, Brazil

Vitor Coutinho
University of Aveiro, Portugal

Jorge Lima de Magalhães
https://orcid.org/0000-0003-2219-5446
Institute of Technology in Drugs (Farmanguinhos), Fiocruz, Brazil

ABSTRACT

This work aims to present a specific case, in internal control, and to report the actions taken in the planning, creation and implementation of the Internal Control Advisory of the Institute of Technology in Pharmaceuticals - Farmanguinhos. For the development of this article, firstly, the organizational charts of Oswaldo Cruz Foundation (Fundação Oswaldo Cruz – Brazilian name) / Fiocruz and Farmanguinhos were brought into context, to briefly situate the reader in the complex environment of the activities developed in both institutions. That done, the steps adopted by the Institute's Executive Board were described step by step to structure the Internal Control Advisory - ICA, in view of the need to meet the demands arising during the internal audits carried out by Audin (Fiocruz's auditing body) / Fiocruz. Finally, the performance of ICA presents the interaction with the several sectors inside and outside the Institute of Drugs Technology (Farmanguinhos/Fiocruz)

DOI: 10.4018/979-8-3693-4030-1.ch007

1. INTRODUCTION

Corporate management is part is an integral part of organizations and has been increasingly discussed in the business world as a tool for the development and growth of companies. Therefore, to achieve planned results, there is a set of best practices aimed at aligning the entire corporate body. These actions bring to management a new form and a new concept in the model of managing the corporate organization. This ideological current provides the business group with people with common interests and activities. Given the above, these groups have more strength to pursue their goals than if individuals were acting in isolation (Arantes, 2020).

However, one cannot think about strategic management without mentioning Mintzbert (1987). He first wrote about the 5 "P"s of strategy. Each "P" is a different approach to strategy, namely: Plan; Ploy; Pattern; Position and, Perspective. Thus, by understanding each P, a robust business strategy can be developed to make the most of each organization's strengths and resources. (Mintzberg, 1987).

For the purposes of this study, the strategic and corporate management of the Institute of Pharmaceutical Technology (Farmanguinhos) of the Oswaldo Cruz Foundation (Fiocruz) was analyzed. This Institute is a technical-scientific unit of Fiocruz-Fiocruz, created in 1976, which operates in the areas of education, research, innovation, technological development and production of medicines. It is characterized as the largest Official Pharmaceutical Laboratory (OPL) linked to the Ministry of Health. This OPL also stands out for its pursuit of reducing the costs of medicines, where this action enables better access by society to medicines and Brazilian public health programs (Figueiredo et al., 2017). Thus, Farmanguinhos, due to its complex logistical activity in the production of medicines and scientific knowledge, became a plausible corporate scenario to be studied together with the Federal Public Administration since it implemented the Internal Control Advisory in its organizational structure.

In this sense, this chapter aims to present an overview of corporate and strategic management through the administrative experience of OPL Farmanguinhos/Fiocruz in the creation and implementation of an Internal Control Advisory.

The concept of management can be better understood from the perspective of the administration area. Mintzberg (1987) states that management means influencing action. Therefore, management is thought of in organizations, where their corporate units do what needs to be done, that is, it means the action of each unit of the business organization. The term corporatism comes from the word body (from the Latin, corpus), and is associated with the idea that each group performs its specific function, just like the parts of the body. Therefore, this would be the most efficient configuration for the corporate management society (Mintzberg, 1987).

In this business management, it is complemented by a series of actions that add value to its processes, which transforms into an economy. These values are added through information, ideas and business intelligence, which also involves people management, material growth, environmental resources, etc. Therefore, management tools must be enhanced with non-material values (Handy, 1995).

Corporate strategic management arose from the need to anticipate potential surprises in the business world. This process involves formulating strategies and skills to add value to the business by optimizing resources (people, time, financial resources and other material resources). The strategy is to determine objectives and how to achieve them. In order to have a successful company, it is essential to implement an action plan with clear goals and objectives, since the greater the complexity and variation of the corporate environment, the greater its impact on the company (Camargos & Dias, 2003). But despite the benefits of strategic implementation within the corporate environment, its insertion can present several difficulties, such as people management, carrying out inadequate diagnoses, stipulating complex schedules, resistance to changes in the work system, lack of team engagement or lack of employee commitment (CERTO & PETER, 2010; Paladini, 2011).

People management processes and practices impact not only the performance of professionals, but also the organizational climate, which is directly related to the satisfaction and motivation of professionals and, consequently, to their commitment. When the environment is positive and constructive, there is satisfaction, responsibility, motivation of professionals and better performance. Strategic people management can add value to the business and people's lives when it combines the expectations of both parties and manages to align them in unique objectives that generate an advantage for everyone involved (Cavalcanti et al., 2018).

It is emphasized that the era of the digital economy creates prerequisites for improving the efficiency of enterprises. Digital innovations generate new means of social interaction between employees; they transform economic relations between the enterprise and its business partners. More rational and cost-effective methods of cooperation and coordination are emerging to jointly solve business problems. However, at the same time, digital innovations also generate negative effects that are informational in nature (Krap et al, 2024).

2. BACKGROUND

2.1. Decision-Making

The term "decision making" was coined in the middle of the last century by Chester Barnard. Executives inserted this term into the business world. Since then, this expression has replaced more limited descriptions, such as "resource allocation" and "policy definition". The definition of policies can be something interminable and there will always be resources to allocate, whereas the decision implies the end of the deliberations and the beginning of the action (BUCHANAN; O'CONNEL, 2006).

Other theoretical thinkers have laid the foundations of the study for decision-making in management, such as James March, Herbert Simon and Henry Mintzberg. The decision-making process in companies is only a small wave in a current of thought born in a time when man, faced with uncertainty, sought guidance in the stars. Knowing who makes decisions, and in what way, is what shaped systems of government, justice and social order worldwide (Albert Camus, 1996 apud BUCHANAN, O'CONNEL, 2006).

Buchanan and O'Connel (2006) argue that the study of decision making is a mixture of various disciplines of knowledge, such as mathematics, sociology, psychology, economics and political science. Philosophy reflects what a decision reveals about ourselves and our values. History has said that the decision is made by leaders at critical moments. Already the study of risk and organizational behavior is born of a more practical desire: to help the administrator obtain better results.

Although a good decision does not guarantee a good result, the prerogative generally compensates. The growing sophistication of risk management, understanding the variations in human behavior, and the technological advancement that supports and simulates cognitive processes have, in many situations, improved decision making (BUCHANAN & O'CONNEL, 2006; DIAS et al, 2012).

Authors like Silva and Martins (2003), Buchanan and O'Connel (2006), Cordeiro (2009), Guimarães (2009), Franganito (2010), Gomes and Mota (2016), affirm the history of decision making for a long time. It is rich and diverse, since it is only a small sample of individuals, facts, research and thoughts that contribute to the notion of what we have today on the subject.

2.2. Strategic Planning

Strategic planning is a business management tool widely used by organizations. According to Oliveira (1991), it is a tool to sustain the development and implementation of business strategies. However, strategic planning, as it has been used, results from an evolution from changes in the management of organizations, due to the constant modifications in the environment and the increase of competitiveness. Ferreira, Reis and Pereira (1997) affirm that one of the greatest tasks of the administrator of that time is to anticipate the changes. The enterprise value maximization strategy is executed with a focus on risk and uncertainty (MICHALSKI, 2013).

According to Terence (2002), individually analyzing each part of the term strategic planning, it is observed that literature treats planning as one of the most important tasks of an administrator, which presupposes: choosing a destination, evaluating alternative paths and deciding the specific course to reach the chosen destination.

Planning means developing a program for achieving organizational objectives and goals, involving choosing a course of action, anticipating what needs to be done, determining when and how action should be taken (MEGGINSON, MOSLEY, PIETRI JUNIOR, 1986; DAVID, 2011).

In this way, planning provides the basis for effective action that results from management's ability to anticipate and prepare for changes that could affect organizational goals (MEGGINSON; MOSLEY; PIETRI JUNIOR, 1986).

Planning has two basic aspects that are vital to the organization: determining the organization's objectives and choosing the means to achieve these goals. Such basic aspects of planning rely on methods, plans or logic, therefore, are not intuitive. In this way, the organization sets goals and the best procedure to achieve them.

On strategic, although the term originated from military art, it assumed a new connotation in the field of administration, which denotes an appropriation of the term without the corresponding appropriation of the concept. The meaning of strategy in organizations is relatively complex, since the literature shows innumerable definitions, addressing different aspects (TERENCE, 2002). The company strategy consists of a set of competitive changes and business approaches that managers execute to achieve the best performance, seeking to strengthen the organization's position in the market, promoting customer satisfaction and achieving their goals.

According to Mintzberg, Raisinghani and Théoret (1976), there may not be a simple interpretation for strategy because it is understood to concern both the organization and the environment and can affect the organization. In addition, it implies content and process issues involving several conceptual and analytical aspects. Considering that the changes refer to new circumstances for the organization, the essence of the strategy should remain unchanged.

The elaboration of the business strategy is carried out at several organizational levels and the activities involved in this process are analysis, planning and selection of strategies that increase the chances that the objectives of the organization will be achieved. The processes of definition, implementation and monitoring of business strategies occur in constantly changing circumstances (Mintzberg, RAISINGHANI and THÉORET, 1976).

According to Gaj (1987), among several approaches to strategy, he highlighted one which associated strategy with planning, observing that the final product, obtained from strategy, is usually expressed through a written plan for a period and with periodic adjustment mechanisms.

In this way, the strategy should be considered as a guiding scheme within the flow of business decisions (OLIVEIRA, 1991).

2.3. Decision Models

According to Freitas and Kladis (1995), to better understand and lead the decision-making process, theoretical and empirical researchers, as well as administrators, devote much of their efforts to organizations, whether public or private, to this question. Quade (1989), referring to decision-making in the public sector, noted that, this way, in the past, events were happening more slowly, where empiricism and experience were far more efficient than they are today.

Through the process of trial and error, and the policy of "give and take" it was possible for the public official machine to develop its goals, estimates and values in society, or in the part which its actions had influence. However, these procedures are no longer possible today. This is no longer the case: technology and events move much more rapidly and the natural attempt, error, the politics of giving and receiving become very catastrophic, not only in a war situation, but in pressures on the population, scarcity environmental degradation (FREITAS and KLADIS, 1995; GOVINDAN, K. et al., 2015).

Managers and people involved in the various decision-making processes of organizations need support (even scientific) to happen in a more satisfactory way which, in the model of Freitas (1993), is drawn as values, beliefs and resources. This process needs to be well understood, and tools, methods and models need to be available now for decision making (Freitas, KLADIS, 1995).

Due to the complexity and increasing challenges of the external environment, the organizational decision-making process has been the subject of research and studies. The work "Administrative Behavior", by Simon (1947), triggered a growing interest in the decision-making process in organizations, where an administrative theory based on limited rationality and fragmented knowledge of people was presented, as well as the impossibility of indefinite processing of information, thus, causing an interference in the dynamics of organizations (OLIVEIRA, 2009).

According to Oliveira (2009) *apud* Simon (1965), in the development of the theory about the decision process, a difference was established between programmed decisions (those more routine) and non-programmed (the most unpredictable and complex) decisions. Some decisions are repetitive, occurring even in each cycle of time, while others happen unexpectedly. Simon (1977), as creator of the terms "programmed decisions and unscheduled decisions," describes that decisions are not distinct types, but a continuous whole, with highly programmed decisions at one end, and decisions of all the nuances in this continuum. The decisions are:

- **Programmed**: these are explained by a set of rules and pre-established procedures. They are taken in an environment of certainty or/of low uncertainty because almost all results are already known and, therefore, are easily delegated.
- **Non-programmed**: in turn, have no rules to follow and do not have a specific scheme to be used, may be known or unpublished. In unpublished decisions, the decision maker is faced with a

completely new situation and cannot count on any pre-established rules to assist him, hardly all variables are available, or there is much difficulty to be gathered and organized in time skillfully.

The exemplification of the types of decision is shown in Figure 1, in which they are positioned according to the degree of structure of the problem. Problems already known will be more quickly solved because of highly programmed decisions, as opposed to the occurrence of new problems, where being unaware of their impact resulted in unscheduled decisions.

Figure 1. Types of decision making

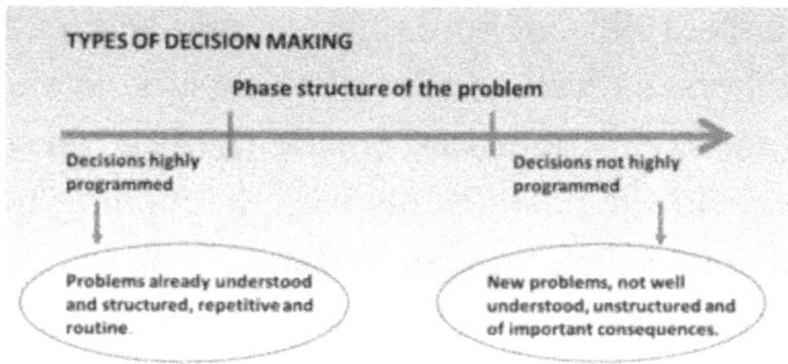

Source: Created by the authors adapted Oliveira (2009) apud Simon (1965).

On the other hand, Mintzberg, Raisinghani and Théoret (1976) believe that when the decision maker finds himself faced with an unscheduled situation, he seeks to reduce his decision to sub-decisions. Thus, to apply already known procedures, thus, elaborating his own model.

According to Oliveira (2009), the decisions can be classified in administrative activity in three levels:

- Operational Level: focuses on the company's operations and involves the efficient and effective use of existing facilities and all resources to execute them. At this level, the decision is a process by which it ensures that the operational activities are well developed, where control uses pre-established procedures and rules of decisions. A large part of these decisions is scheduled and the procedures to be followed are generally very stable. Operational decisions and their actions often result in an immediate response.
- Tactical level: there is concern about the generic acquisition of resources and tactics for the acquisition, location of projects and new products. Decisions are usually related to administrative control, and used to decide on control operations, formulate new decision rules that will be applied by the operations personnel and resource assignment.

This level, the decision about the planned operation (standards, expectations, assumptions) and changes from a planned operation are required. The explanation of these changes or variations and the analysis of the possibilities of decision during actions are considered, in Freitas's model (1993), as criteria of rationality, effectiveness and efficiency.

- Strategic level: considers the definition of objectives, policies and general criteria to plan the way to be followed by the organization. The purpose of the decisions is to develop strategies, so the organization can achieve its macro-objectives. At this level, the activities do not have a period with standardized cycle. They can be irregular, although some strategic plans are made within annual plans or in pre-established periods.

The predictability of the alternatives available in each situation within the organization can be considered in the model of Freitas (1993), interpreting how the actions will take place after each decision. Figure 2 graphically depicts the relationship between decisions and administrative levels (FREITAS; KLADIS, 1995).

Figure 2. Types of decision by administrative level

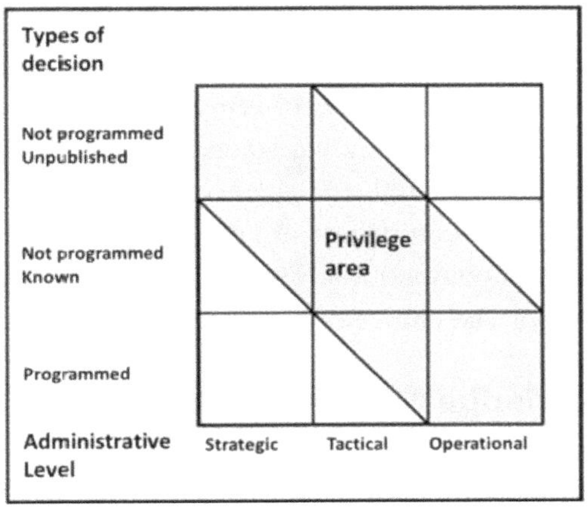

Source: Freitas e Kladis (1995).

The model proposed by Simon (1965) *apud* Carvalho (2006) admits the existence of a limited rationality and works with a simplified model of reality, considering that many aspects of this reality are substantially irrelevant at a given moment. So, it makes its choice of an alternative based on the satisfactory standard of the actual situation, considering only some of the most relevant and crucial factors.

Even if the simplicity imposed by the limited rationality is adopted, the models of decision methods must be classified and analyzed in detail. From the point of view of decision making, problems can be classified into three categories (Turban, Aronson, 1998):

- **Structured**: or also known as well defined, if its definition and phases of operation to achieve the desired results are very clear and it is always possible to repeat the execution. As, for example, payroll, accounting postings and general data processing operations.
- **Semi-structured**: problems with operations are well known but contain some variable factors or criterion that can influence the outcome, as with the problem of forecasting sales or purchases.
- **Unstructured**: when the problems are not fixed, or the scenarios are known, as well as the decision criterion. As, for example, the operation of choosing the cover of a weekly magazine (or first page of a newspaper of daily circulation), in which several alternatives are foreseen, but all can be replaced in the last hour, if some important fact occurs.

Organizational decision-making is also described, in which more teamwork and participation are required, where dialogue procedures are emphasized based on the understanding that, in organizations, communication should be stimulated to establish common thinking (ANGELONI, 2003).

According to Oliveira (2009), the decision can be made under the following conditions:

- **Conditions of certainty**: when the decision is made with full knowledge of all states of nature. There is certainty of what will occur during the period in which the decision is made. It is possible to assign 100% probability to a state of nature. The probability indicates the degree of certainty in which 0% will be the complete uncertainty, and 100% or 1 indicates the complete certainty.
- **Risk conditions**: when the probabilities associated with each of the states of nature are known. The total number of states of nature is known. Contrary to the condition of certainty, which used 100% certainty in the result, this certainty varies from 0 to 100%.
- **Conditions of uncertainty or in conditions of ignorance**: when the total state of nature has not been obtained, or even the part of the known states of nature has data obtained with uncertain probability, or the probability associated with the events is unknown.
- **Conditions of competition or in conditions of conflict**: when the strategies and states of nature are determined by the action of competitors. There are, necessarily, two or more decision makers involved; the result depends on the choice of each of the decision makers (Gomes et al., 2006).

2.4. Information and Knowledge Management

Information is a data or set of data articulated to construct a meaningful message. The way to organize the data is an intention. It is therefore partially subjective. The information implies a transmitter and a receiver but also a media whose nature is far from neutral (Kira TARAPANOFF (org.), 2015).

These poles suppose the existence of an aptitude, in the form of selective understanding, to extract the meaning of the information from the surrounding noise. Information is therefore a set of data placed in a context, mainly organizational for what interests us here, and bearing a meaning. Again, the concept has given rise to considerable developments in the information system and the development of protocols to make information systems as efficient as possible. Three notions are generally associated: that of information system (with the question of its economic performance), that of quantity of information (with the tension that appears between the amount of information made available and the super information) and ambiguity (which opens the question of relevance, a question linked to the imprecise notion of "quality" of information) (Pesqueux, 2005).

Information is a condition for survival, given that it expands the context of communication, rescuing and preserving social memory. Its value is intangible and resists all mechanisms of forgetfulness and destruction, since the collection of information reconstruction allows the cognitive evaluation and the knowledge of a certain reality in question (Gavirneni et al., 1999).

In the 21st century, thanks to the evolution of the Web, the speed of information generation is unprecedented and unprecedented even in the world. The world's per capita technological capacity to store information has almost doubled every 40 months since the 1980s. The Internet evolved from a purely static platform and assumed to be dynamic and an interactive role (Minelli, M. et al., 2013), allowing users to exchange large amounts of information instantly.

Once the proliferation of data and information through the internet and the flow of information is established and there are no restrictions on distance and availability, the big question that arises is the ability to screen, interpret and convert this volume of information into knowledge. The identification and analysis of the amount of information of a given area, whether scientific, technological or business; identify their state of the art and their respective correlations have become hard work (ALEIXO & DUARTE, 2015; Magalhães & Quoniam, 2013).

Therefore, a management analysis is necessary to integrate actions and to provide the effective actions to generate technological innovation and, consequently, wealth of the country. Thus, a better management for information architecture, also known as business architecture, encourage a new understanding (Ross et al., 2006).

Knowledge Management requires the involvement and support of all the company's stakeholders to preserve, transmit and develop knowledge. Indeed, it is the individuals who are at the center of the creation of value and who hold the keys to the success of such a project. The management of the knowledge and know-how of the company is therefore not universal, it depends strongly on the culture of the country in which it is practiced (Balmisse, 2006).

To motivate individuals to share their knowledge, it is important to consider cultural and human factors, and not just tools and procedures define "culture" as the set of values, norms, habits, and customs specific to a society that will influence the personality of the individual (Pesqueux & Ferrary, 2011).

According to Morin (1982), it is necessary to elevate the concept of "system" of this theoretical level to the paradigmatic level. A paradigm, for the author, is the set of fundamental relationships of association and/or opposition between a small number of "master concepts" that command/control all existing knowledge in all the speeches and theories. Thus, to sustain the action cycle (application) of the knowledge to the action progress, it is necessary to try to identify/define what are these notions, featuring the new paradigm and to resume as the (re)definition of the system where the logic of complexity in modeling the actual performance interact with a posture of the researcher and the actors involved in evaluative process ((Hartz, 1997; PIAGET & GARCIA, 1983).

3. THE CASE OF THE ORGANIZATIONAL STRATEGY OF FIOCRUZ AND FARMANGUINHOS

The Oswaldo Cruz Foundation's mission is to promote health and social development, generate and disseminate scientific and technological knowledge, and be an agent of citizenship. It is also a Foundation linked to the Ministry of Health, acting as a science and technology institution in health. Fiocruz plays a relevant role on the national and international scene, particularly in Latin America. To this end, Fiocruz has approximately 47 (forty-seven) departments distributed throughout its organization, namely: Presidency, Collegiate Departments, Vice-Presidency, Departments of Direct and Immediate Assistance to the President, Decentralized Units, Sectoral Departments, Regional Offices, and Specific Singular Departments. In other words, Fiocruz is a reference in health, scientific knowledge, and innovation, of which the Institute of Pharmaceutical Technology (Farmanguinhos) is one of its units (FIOCRUZ, 2024).

3.1. Corporate and Strategic Management of the Institute of Technology in Drugs (Farmanguinhos)

The Institute of Pharmaceutical Technology is located at the base of the Foundation's organizational structure, comprising the set of Specific Singular Bodies. Farmanguinhos has a strategic organizational structure, common to public institutions, thus following the same design as Fiocruz's organizational chart. It is worth noting that the Institute's mission and vision are, respectively: To promote public health by offering integrated and sustainable solutions, generating and disseminating knowledge, innovating and providing medicines; To be a protagonist and innovator in providing integrated and sustainable solutions for the Unified Health System (SUS – Brazilian term), until 2031. The Internal Control Advisory is included in the strategic level of the Institute's organizational structure, linked to the executive management, its advisors and vice-directors. The organizational structure of the Institute includes a General Assembly, Deliberative Council, 18 (eighteen) advisory bodies, including the Internal Control Advisory Board, and 4 (four) Vice-Directors.

3.2. Regarding Corporate Governance

Corporate governance is a system that establishes that organizations are managed, monitored and encouraged through relationships between the board of directors, directorates, oversight and control bodies and the entire organizational structure. Considering the need for constant alignment in improving knowledge management and innovation in organizations, the management of the organization. According to Cadbury (apud Andrade; Rossetti, 2011, p. 26), corporate governance is expressed by a system of values that governs organizations in their network of internal and external relationships. In the case of Farmanguinhos, the performance of corporate governance was briefly detailed, where the Executive Director is chosen through an electoral process, after the presentation of candidacies, validated by the Deliberative Council, and the debate of the candidates' management program with the unit's employees.

3.3. Regarding the General Assembly

Collegiate body that represents the Institute's community, composed of public servants in office, who, meeting in assemblies, must approve proposals and deliberate, ultimately, on issues relevant to the Farmanguinhos community.

3.4. Regarding the Deliberative Council

The Deliberative Council (CD) is a collegiate body of the institution. It is responsible for strategic issues of a technical-scientific or administrative nature affecting the Institute, and is chaired by the unit's Executive Director, composed of the board of directors and other members, elected for a two-year term.

3.5. Regarding the Executive Board

Farmanguinhos' management team is composed of the following members: the director, the five deputy directors (deputy director of institutional management, deputy director of education, research and innovation, deputy director of labor management, deputy director of operations and production and deputy director of quality management), coordinators and the chief of staff of the Institute.

3.6. Regarding Internal Control Advisory as Support for Innovation in Corporate Governance

After illustrating the organizational structure, it is possible to describe the case study for the implementation of the Internal Control Advisory (ICA) of OPL Farmanguinhos.

In 2019, there was still no ICA in the organizational structure of OPL. In 2018, after an Internal Audit by Fiocruz, the Executive Board began planning for the creation and implementation of the ICA in the organizational chart of OPL Farmanguinhos, in order to strengthen internal management controls, as well as comply with Decree No. 9,203, of November 22, 2017. This Decree created the Integrity Program of the Comptroller General of the Union. The Program created functioned as a governance tool capable of promoting the adoption of institutional measures and actions aimed at preventing, detecting and punishing fraud, acts of corruption, irregularities and misconduct. These actions aim at better management and transparency. For better understanding, please see articles 1 to 3 of the Decree, which provide:

> Art. 10 This Decree establishes the governance policy of the direct, autonomous and foundational federal public administration.
> Art. 20 For the purposes of the provisions of this Decree, the following are considered:
> I - public governance - a set of leadership, strategy and control mechanisms put into practice to evaluate, direct and monitor management, with a view to implementing public policies and providing services of interest to society;
> II - public value - products and results generated, preserved or delivered by the activities of an organization that represent effective and useful responses to the needs or demands of public interest and modify aspects of society as a whole or of some specific groups recognized as legitimate recipients of public goods and services;
> III - senior management - 'ministers of State, holders of special positions, holders of level 6 positions of the Senior Management and Advisory Group - DAS and presidents and directors of autonomous agencies, including special ones, and of public foundations or authorities of equivalent hierarchy; and
> IV - risk management - a permanent process, established, directed and monitored by senior management, which includes the activities of identifying, evaluating and
> V managing potential events that may affect the organization, aimed at providing reasonable assurance regarding the achievement of its objectives.
> Art. 30 The principles of public governance are: I - responsiveness; II - integrity; III - reliability; IV - regulatory improvement; V - accountability and responsibility; and VI - transparency. (BRASIL, 2017).

3.7. Regarding Planning

The institution's strategic planning allowed the executive board of OPL Farmanguinhos to present the new structure and decision flow with internal controls for managing quality tools, based on Ordinance No. 787/PR, of June 26, 2018, from the presidency of Fiocruz (FUNDAÇÃO OSVALDO CRUZ, 2018) and the Joint Normative Instruction MPOG/CGU, No. 1, of May 10, 2016, which provides for internal controls, risk management and governance within the scope of the federal Executive Branch, in which are established, among others, namely:

> Art. 10 The bodies and entities of the federal Executive Branch must adopt measures to systematize practices related to risk management, internal controls, and governance.
> Art. 30 The bodies and entities of the federal Executive Branch must implement, maintain, monitor, and review internal management controls, based on the identification, assessment, and management of risks that may impact the achievement of the objectives established by the Public Authority. Internal management controls constitute the first line (or layer) of defense of public organizations to enable the achievement of their objectives. These controls are operated by all public agents responsible for conducting activities and tasks, within the scope of the final and support macro processes of the bodies and entities of the federal Executive Branch. The definition and operation of internal controls must take into account the risks that are intended to be mitigated, in view of the objectives of public organizations. Thus, in view of the objectives established by the public administration bodies and entities, and the risks arising from internal or external events that may hinder the achievement of these objectives, the most appropriate internal controls must be in place to mitigate the likelihood of tax incidents occurring, or their impact on organizational objectives.
> Art. 60 In addition to internal management controls, the bodies and entities of the federal Executive Branch may establish second line (or layer) defense bodies to supervise and monitor these internal controls. Thus, committees, boards or specific advisory bodies to deal with risks, internal controls, integrity and compliance, for example, may constitute internal control oversight bodies (BRASIL, 2016, p. 14-15).

In this sense, the OPL Deliberative Council followed the assessment in the first quarter of 2019, where, previously, benchmarking studies were carried out at the Afonsos Support Group and the Afonsos Air Force City Hall, two Executive Management Units of the Air Force Command, located in the city of Rio de Janeiro/RJ, as well as at the Internal Control of the Attorney General's Office of the Municipality of Rio de Janeiro.

In this process, a successful ICA was identified in the Military Commands, and in the Air Force Command (COMAER), which is consolidated in the Air Force Command Instruction No. 174-1 (Internal Control in Management Units). This fact can be verified in Ordinance No. 19/SEFA, of April 9, 2007, which provides for internal control actions in the Management Units of the Air Force Command, and which aims to guide and standardize the execution of procedures related to internal control actions within the scope of COMAER (BRASIL, 2007).

In view of the above, based on the scope of the activities carried out in the agencies visited, a proposal for a corporate management project was developed, with the proposal for the creation and implementation of ICA/Farmanguinhos. It is worth noting that, in this action alone, it was possible to estimate the comparison of the financial movement of the institutions and the number of Administrative Management

Processes (AMP) opened annually. This process is essential for OPL Farmanguinhos to have a global view of its physical structure and personnel allocated.

3.8. Regarding Creation and implementation

Internal management controls can be considered tools to ensure operation within the estimated standards. Gomes (1999) states that a control system comprises the structure and the control process. In the same sense, De Souza (2009) states that the organization of the internal control system and its efficient operation are the sole responsibility of the Administrator. Thus, the strategic senior management must properly manage and report on the management of assets and the human resources entrusted to it. The entire corporate management process must seek to be free of waste and deviations. It is worth highlighting that the practice of good administration allows external control bodies to verify that in these management processes, the administrator acted correctly and competently. To this end, it must be kept under permanent surveillance and evaluation, so that its results are positive. In this sense, the internal controls of corporate and strategic management permeate the entire organization. For this reason, the senior management leader seeks mechanisms to monitor efficiency, effectiveness and efficacy in constant evaluation of its performance.

Thus, the Internal Control Advisory Board of the Pharmaceutical Technology Institute/Farmanguinhos was created and implemented in 2019. To this end, the characteristics, competencies and attributions of the ICA were established, namely:

a) **Characteristics:** the ICA is provided for in art. 30 of the Joint Normative Instruction MPOG/CGU No. 1, of May 10, 2016, and is directly subordinate to the organization's top manager. It is intended to support the development of internal management controls and to carry out supervision and monitoring actions for the activities developed in the various sectors of the OPL, which includes risk management, compliance, quality assurance, financial control, guidance and training of managers. It must also be organized with the necessary human and material resources, with the objective of strengthening management and streamlining internal control actions, as well as providing the necessary support to the various sectors of the OPL.
b) **Competencies**: The ICA is part of the Unit Committee for Integrity Management, Risks and Internal Management Controls, as specified in Clause III of art. 15 of Ordinance 787/2018-PR, of June 26, 2018, of Fiocruz, and will analyze the various AMP, considering the results obtained through the application of the risk matrix.
c) **Responsibilities:** The ICA must act positively in the institutional environment so that the practices recommended by the Governance, Risks and Controls Committee are disseminated in the OPL. The purpose is to advise the organization's top manager on the adoption of effective management measures and strategies, as well as on the continuous improvement of organizational processes, decision-making and standards, providing the reduction or elimination of rework. It is also responsible for issuing technical guidelines on improprieties and/or irregularities found in the AMP and in administrative acts or facts. This step aims to minimize the probability of failures and, to this end, monitoring of accounting, financial and operational facts must be established.

It is important to note that balance sheets, trial balances, inventories or other relevant documents, as well as administrative management processes, must be checked for legality, legitimacy and cost-effectiveness.

Finally, the ICA must monitor the organization's external audits, namely: Audin from Fiocruz or others that may arise.

3.9. Performance of the Internal Control Advisory of Farmanguinhos

The ICA became a specialized sector with a role to play in the interrelationship of the organization's organizational structure.

Thus, the first initiatives were organizational awareness-raising, based on interviews with members of the Executive Board (executive director, deputy directors, coordinators and the chief of staff). The step-by-step implementation stages, through awareness-raising, were important, since they enabled demystification and provided ICA with a safe performance based on dialogue, knowledge, experience and trust.

The ICA identified the needs of the internal public and, in this way, enabled training managers in the Fundamental Aspects of Internal Control, with reflections on Weaknesses and Risks in bids and contracts; Cost and price formation spreadsheet; and SIAFI, Concepts, foundations and good practices. In the same sense, the contract inspection teams were also trained with an exchange of experience between the teams. This action brought the institution's contract inspection teams closer together.

After the implementation of the ICA was consolidated, all due diligences forwarded by the external and internal control bodies (TCU, CGU and AUDIN/FIOCRUZ) were transferred to the ICA, which, in turn, is the body for analysis, technical support and preparation of reports for the audit bodies.

It is worth highlighting the preparation of a risk matrix for administrative management processes, as a way of preventing possible errors. This action made the analysis of processes an ongoing activity, since the matrix made it possible to identify those processes that require a closer look by the Administration. In this sense, the Risk Management Committee was created, in alignment with Fiocruz's Risk Management Policy. This action aimed to prepare the Risk Management and Internal Controls Plan, which was submitted for approval by the OPL Farmanguinhos Deliberative Council and the Fiocruz Governance, Risks and Control Committee.

4. CONCLUSIONS

Strategic and corporate management are essential for the success of business organizations, whether public or private. This chapter sought to present a case of a successful public institution that implemented the concepts of strategy and control mentioned by Mintzber (1987).

The path taken by public agents in the planning, creation and implementation of the Internal Control Advisory at OPL Farmanguinhos provided mechanisms to develop, assess and supervise the internal controls of corporate management.

It is worth noting that the process of awareness, transparency and dialogue are fundamental for the success of a new organizational structure, such as that of an Internal Control Advisory. It was possible to observe a better structure, confidence of strategic and operational managers and also of the risk matrix tool implemented.

Another fundamental point was the technical guidelines and the search for interconnection between the ICA and other sectors of the OPL and external control bodies, such as the Federal Attorney's Office at Fiocruz, the General Coordination of Administration (COGEAD/Fiocruz) and Internal Audit — AUDIN/Fiocruz.

Nevertheless, detailed future studies are needed to assess continuity, as well as replication in other public and/or private entities, to validate or improve the performance of strategic and corporate management processes.

REFERENCES

Andrade, A., & Rossetti, J. P. (2004). *Corporate governance: fundamentals, development and trends*. Atlas.

Arantes, J. R. (2020). The challenges of corporate management. https://repositorio.pgsscogna.com.br//handle/123456789/42724

Araújo, L. C. G. D. (2010). Organization, Systems and Methods: and organizational management technologies. In Organization, Systems and Methods: and organizational management technologiesl (pp. 360-360).

Balmisse, G. (2006). GKnowledge Management Tools Guide: Overview, Choice and Implementation. https://www.leslivresblancs.fr/informatique/applications-pro/knowledge-management/livre-blanc/outils-du-km-panorama-choix-et-mise-en-oeuvre-62.html

Camargos, M., & Dias, A. (2003). STRATEGY, STRATEGIC MANAGEMENT AND CORPORATE STRATEGY: A THEORETICAL SYNTHESIS.

Cavalcanti, M., Farah, O. E., & Marcondes, L. P. (2018). *Entrepreneurship: SURVIVAL STRATEGY FOR SMALL BUSINESSES*. SARAIVA EDITORA.

de Souza, A. A. S. (2009). Internal control in public administration. *Control Magazine: Doctrines and Articles*, *7*(1), 225–234.

Dias, J. A., & Duarte, P. (2015). Big data opportunities in healthcare. How can medical affairs contribute. *Rev Port Farmacoter*, *7*(4), 230–236.

Figueiredo, T. A., Schramm, J. M. de A., & Pepe, V. L. E. (2017). The public production of medicines compared to the National Policy of Medicines and the burden of disease in Brazil. Cadernos de Saúde Pública, 33. https://doi.org/DOI: 10.1590/0102-311X00179815

FIOCRUZ. O. C. F. (2024). FIOCRUZ. Fiocruz. https://portal.fiocruz.br/en

Gavirneni, S., Kapuscinski, R., & Tayur, S. (1999). Value of Information in Capacitated Supply Chains. *Management Science*, *45*(1), 16–24. DOI: 10.1287/mnsc.45.1.16

Gomes, J. S. (1999). *Management control: a contextual and organizational approach: texts and cases*. Atlas.

Handy, C. (1995, maio 1). Trust and the Virtual Organization. Harvard Business Review. https://hbr.org/1995/05/trust-and-the-virtual-organization

Hartz, Z. M. de A. (1997). Health evaluation: From conceptual models to practice in the analysis of program implementation. Editora FIOCRUZ. https://books.scielo.org/id/3zcft

Kira, T. A. R. A. P. A. N. O. F. F. (org.). (2015). Analysis of Information for Decision Making Challenges and Solutions (Vol. 1). Editora Intersaberes. http://www.estantevirtual.com.br/b/kira-tarapanoff/analise-da-informacao-para-tomada-de-decisao-desafios-e-solucoes/158550962

Krap, A., Bataiev, S., Bobro, N., Kozub, V., & Hlevatska, N. (2024). Examination of digital advancements: Their influence on contemporary corporate management methods and approaches. Multidisciplinary Reviews, 7, 2024spe026. .DOI: 10.31893/multirev.2024spe026

Magalhães, J. L., & Quoniam, L. (2013). Perception of the Information Value for Public Health: A Case Study for Neglected Diseases: Library and Information Science Book Chapter | IGI Global. Em Rethinkin the Conceptual Base for New Practical Applications in Information Value and Quality (p. 345). IGI Global. https://www.igi-global.com/chapter/perception-information-value-public-health/84218

Minelli, M., Chambers, M., & Dhiraj, A. (2013). Big Data, Big Analytics. John Wiey & Sons, Inc. https://books.google.com.br/books/about/Big_Data_Big_Analytics.html?hl=pt-BR&id=Mg3WvT8uHV4C

Mintzberg, H. (1987). The Strategy Concept I: Five Ps for Strategy. *California Management Review*, *30*(1), 11–24. DOI: 10.2307/41165263

Paladini, E. P. (2011). *Strategic quality assessment (2a edição)*. Atlas.

Pesqueux, Y. (2005). Knowledge Management: An Organizational Model? KNOWLEDGE MANAGEMENT: AN ORGANIZATIONAL MODEL? https://halshs.archives-ouvertes.fr/halshs-00004005/document

Pesqueux, Y., & Ferrary, M. (2011). Knowledge management (2o ed). Economica. /Entreprise/Livre/management-de-la-connaissance-9782717860153

Piaget, J., & Garcia, R. (1989). *Psychogenesis and the history of science*. Columbia University Press.

Ross, J. W., Weill, P., & Robertson, D. (2006). *Enterprise Architecture As Strategy: Creating a Foundation for Business Execution*. Harvard Business Review Press.

Chapter 8
Performance Management Program as a Disruption in Public Management:
The Case of the Administrative Council for Economic Defense (CADE)

Marcos Paulo de Oliveira Corrêa
https://orcid.org/0000-0002-7186-6624
SKEMA Business School, Conselho Administrativo de Defesa Econômica, Brazil

Karine Lustosa Panerai
Conselho Administrativo de Defesa Econômica, Brazil

Luisa Souza Lima Macedo
Conselho Administrativo de Defesa Econômica, Brazil

ABSTRACT

The increasing demand for greater efficiency in public administration has driven the adoption of innovative human resource management approaches, such as the Management and Performance Program (PGD). The PGD modernizes public administration by replacing traditional attendance-based control with a productivity-driven model focused on results. Its implementation at the Brazilian Administrative Council for Economic Defense (CADE) has fostered greater flexibility in task execution without compromising service quality. This chapter employs a case study methodology, enabling an in-depth analysis of the PGD's implementation and its organizational impact. In 2025, the primary challenge will be to consolidate this new model while ensuring its sustainability and securing leadership engagement. The establishment of a dedicated team to oversee the PGD at CADE strengthens structured governance, which is crucial for the modernization of the federal public administration.

DOI: 10.4018/979-8-3693-4030-1.ch008

1. INTRODUCTION

The increasing demand for a more effective and transparent public administration committed to efficiency has become a contemporary imperative, particularly in light of the need to achieve goals and results that generate positive, measurable, and socially relevant impacts (Sousa, 2024). This context has driven the adoption of innovative approaches and the development of new paradigms in human resource management, emphasizing the implementation of strategic practices aimed at maximizing organizational performance. These practices seek not only to deliver concrete and sustainable outcomes but also to continuously strengthen public governance, enhance accountability, and align institutional actions with collective interests, thereby reinforcing the commitment to service quality in the public sector.

To support the advancement and modernization of human resource management in public administration, the Brazilian federal government has, over the years, introduced regulations establishing guidelines, criteria, and procedures for implementing the Management and Performance Program (PGD) across public institutions. As analyzed by Santiago and Castro (2022), early regulations primarily sought alternatives to rigid attendance control, including the possibility of waiving time tracking in specific circumstances. Over time, the focus shifted toward regulating and structuring teleworking practices, reflecting an adaptation to evolving labor dynamics and the pursuit of greater flexibility and efficiency in the public sector.

The PGD has emerged as an innovative and disruptive instrument in public administration, with potentially broader and more significant impacts than those brought about by traditional administrative reforms (Aguiar et al., 2024). As analyzed by Elvira and Bezerra (2023), one of the PGD's primary functions is to replace the conventional attendance control system for federal civil servants with an evaluation model centered on productivity and the achievement of concrete results. This shift necessitates a restructuring of work arrangements, particularly regarding the location where activities are performed, allowing tasks to be carried out either in person or via teleworking, whether partially or fully. Additionally, working hours are made more flexible, as deliverables can be submitted remotely, granting civil servants greater autonomy in managing their time.

A key advantage of the Management and Performance Program (PGD) is its emphasis on results-oriented management. Individual and institutional performance is assessed based on concrete deliverables and predefined targets, reinforcing a culture of meritocracy and accountability within the public sector. This approach prioritizes the quality and impact of work performed, fostering continuous improvement in both individual and collective performance while also promoting skill development and alignment with organizational objectives.

Furthermore, the PGD plays a crucial role in enhancing civil servants' quality of life by providing flexibility in work schedules and locations, facilitating a better balance between professional and personal life. This increased flexibility not only helps individuals reconcile these two aspects but also enhances the attractiveness of public sector careers, contributing to talent retention and engagement in public service (Sousa, 2024). Another significant benefit is the promotion of digital transformation and the modernization of work processes. The PGD drives the adoption of advanced technologies and management tools, fostering innovation and automating repetitive tasks, which, in turn, leads to more agile, efficient, and error-resistant processes. Additionally, the program encourages a more structured institutional planning culture, providing greater clarity in defining objectives and allocating resources, thereby enhancing efficiency and optimizing the use of public funds.

Following extensive discussions on new regulations and a revised model for the PGD, which will be addressed later in this chapter, Decree No. 11.072, of May 17, 2022 (Brasil, 2022), was enacted, introducing several innovations to the PGD, including the types of activities that can be performed under work plans; the possibility of participation for individuals residing abroad; and the number of available positions, among other aspects. The decree establishes that the management program applies to civil servants working both in person and remotely, emphasizing the need for an agreement between the civil servant and the administration, the alignment of the chosen work arrangement with the assigned tasks, and the expected productivity levels for teleworking civil servants.

Since then, multiple federal public administration agencies have adopted the PGD to modernize human resource management practices and enhance institutional efficiency. In this context, on December 4, 2020, the Administrative Council for Economic Defense (CADE) issued Ordinance No. 395, which introduced and regulated the PGD within the agency, initially on a pilot basis. With this initiative, CADE became one of the pioneering institutions in implementing the program within the federal public service. This development marked the beginning of a paradigm shift in work management, fostering a results-oriented culture aimed at improving the efficiency and effectiveness of public services provided to society.

After an initial six-month implementation period, the program was formally established with the issuance of Ordinance No. 317/2021, which took effect on July 1, 2021. It is worth noting that CADE had already been advancing its workflow modernization efforts, introducing teleworking on an experimental basis as early as December 2017, through Ordinance No. 438/2017, in accordance with paragraph 6 of article 6 of Decree No. 1,590, of August 10, 1995 (Brazil, 1995). The implementation of the PGD and its subsequent developments will be explored in greater detail in the following sections.

This chapter aims to critically examine the challenges encountered in implementing the PGD at CADE, outlining the phases of the implementation process, the methods and tools employed, and the results achieved thus far. It also explores the transition to a new management system, emphasizing the implications of this shift, the progress made, and future prospects for enhancing performance management within the agency.

To establish a solid conceptual foundation for understanding the program at CADE, the theoretical framework presents key concepts related to New Public Management, organizational performance, results-oriented strategic human resource management, and the PGD as implemented by the federal government. The objective is to equip readers with the essential theoretical underpinnings that inform strategic decisions in adopting more flexible, results-driven human resource management models, where civil servants can perform their duties and meet their objectives regardless of their physical work location.

2. THE NEW PUBLIC MANAGEMENT: IMPACTS ON ORGANIZATIONAL PERFORMANCE

Public management has undergone multiple reforms over the years, driven by changes in the role of the state and evolving organizational and administrative structures. Some scholars argue that these reforms occurred in waves, as they were implemented simultaneously across various countries (Sousa, 2024). However, it is important to note that these waves did not signify abrupt breaks from previous models but rather gradual transformations, resulting in hybrid structures that integrate elements of traditional management with newer approaches (Chowdhury & Shil, 2021).

According to Kettl (2005), from the 1980s onward, a wave of reforms reshaped the public sector with the goal of modernizing and streamlining state administration. These changes aimed to reduce the size of the state and enhance efficiency in service delivery at lower costs. However, two central dilemmas emerged: identifying which management models optimize efficiency while minimizing costs and determining the appropriate functions for government intervention. Kettl (2005, p. 76) observes that "public administration reformers have been trying to create a cheaper and more efficient state" but underscores the necessity of mechanisms that balance short-term political interests with long-term objectives to ensure meaningful and sustainable reforms. Additionally, Secchi (2009) highlights several key drivers of public management reforms, including the state's fiscal crisis, intensified competition for private investment and skilled labor, advancements in organizational and technological knowledge, the rise of pluralist and neoliberal values, and the increasing complexity of contemporary societies.

Indeed, scholars have identified three major waves of reform in public management:

1. Progressive Public Administration, rooted in Weberian bureaucracy, which dominated from the 19th to the 20th century.
2. New Public Management (NPM), introduced in the 1980s and 1990s, with a strong emphasis on efficiency and results.
3. The Post-NPM Era, emerging in the early 21st century, incorporating new governance models.

New Public Management marked a paradigm shift in public administration, redirecting the focus from process management to service delivery. According to Yilmaz (2022), this approach prioritized efficiency, transparency, and adaptability in public service provision. Bureaucracy, once regarded as essential for control and standardization, has, in some contexts, come to be seen as an impediment to organizational performance—particularly when applied excessively or rigidly. This inflexible model can constrain the creativity and autonomy of civil servants while also limiting innovation and responsiveness to societal demands.

The negative impacts of bureaucracy are most pronounced when they arise internally within organizations rather than being imposed by external forces. This phenomenon underscores how excessively rigid administrative practices can undermine both organizational efficiency and the quality of services provided. In this context, New Public Management (NPM) has emerged as an approach to modernizing public administration, emphasizing results, efficiency, and institutional performance, with the primary objective of enhancing the quality of public services delivered to society (Ugyel, 2021). The adoption of this paradigm has significantly transformed the focus of public organizations, shifting from a purely process-oriented approach to strategic management centered on achieving tangible outcomes and improving citizen services.

Within the New Public Management framework, the Results-Oriented Management approach is particularly noteworthy, as it seeks to enhance organizational performance and is central to discussions on public sector reforms (Velasco, Pantoja & Oliveira, 2023). These reforms aim to modernize state administration and improve its efficiency by implementing structural and procedural changes. Key aspects of this approach include human resource management and the pursuit of competitiveness, efficiency, efficacy, and effectiveness in organizational practices (Montenegro et al., 2021).

By fostering a results-driven culture, New Public Management enhances organizational capacity by optimizing the strategic use of available resources, promoting innovation, and developing new service delivery models. Consequently, this approach leads to substantial improvements in both organizational performance and the quality of public services (Alosani, Al-Dhaafri & Awadh Abdulla, 2021).

Despite the transformations introduced by New Public Management, adapting civil servants to new practices remains a significant challenge. The shift toward a more flexible and results-oriented model can be perceived as a departure from traditional bureaucratic practices, often triggering resistance among civil servants. This resistance impedes the effective implementation of changes and exposes weaknesses in change management strategies and competency development aligned with the new organizational model. Given this scenario, it is crucial to reformulate Human Resource Planning processes in public administration, ensuring that people management practices align with the principles of efficiency and effectiveness while fostering an adaptive and engaging organizational environment. Such measures help minimize internal conflicts and enhance employees' commitment to institutional objectives (Yilmaz, 2022).

Human resource management strategies have a direct impact on individual employee performance, which in turn influences the overall effectiveness of the organization. In this context, a win-win approach is adopted, in which human resource management is regarded not only as a tool for achieving institutional goals but also as a critical factor in fostering employee well-being and engagement. This perspective highlights that the implementation of consistent and effective human resource management practices benefits both individuals and the organization by enhancing skill development, increasing employee motivation, and ensuring alignment with institutional objectives. As a result, these improvements drive organizational performance, regardless of whether the institution operates in the public or private sector (García-Juan, Escrig-Tena & Roca-Puig, 2023).

2.1 Strategic People Management: A Results-Oriented Approach

According to Aguiar et al. (2024), performance management plays a crucial role in public administration and is a fundamental subsystem of human resource management. This process encompasses the planning, monitoring, and evaluation of employees' activities. In this context, the ability of civil servants to remain flexible and adaptable amid constant technological and organizational changes is essential, necessitating a transformation in public sector workforce management. This perspective underscores the importance of innovative practices, such as the PGD, which aim to align institutional objectives with individual development, fostering a results-driven culture and continuous improvement in public administration.

2.2 The Performance and Management Program (PGD)

The PGD was formally introduced into the Brazilian legal framework in 1995 through Decree No. 1,590, of August 10, which allowed the exemption of attendance records in cases where work results could be objectively measured (Brazil, 1995). However, as noted by Pena (2024), for nearly two decades, little progress was made in regulating this mechanism, leading to a period of regulatory stagnation and limited application. This situation began to shift in subsequent years as certain public institutions initiated efforts to regulate teleworking, driven by a growing demand for greater flexibility and adaptation to evolving labor dynamics.

On August 31, 2018, the Secretariat for Personnel Management published the first formal regulation of the PGD at the federal level through Normative Instruction No. 1, which established general guidelines, criteria, and procedures for agencies and entities within the Federal Administration's Civilian Personnel System (SIPEC). This milestone marked a significant conceptual advancement, setting the groundwork for implementing results-based management practices across the federal public administration.

The regulatory landscape evolved considerably in 2020, propelled by the global health emergency caused by the COVID-19 pandemic. The urgent need to transition to remote work accelerated the expansion and consolidation of the PGD, leading to the revision of Normative Instruction No. 1/2018 and the publication of Normative Instruction No. 44, of June 10, 2020, which further enhanced the flexibility of remote management practices, enabling a more dynamic and adaptable approach to new working conditions.

Shortly thereafter, on July 30, 2020, both regulations were repealed by Normative Instruction No. 65, which introduced updated guidelines to facilitate, enhance flexibility, and personalize adherence to the PGD within public institutions. This new regulation formally established the modalities of remote work, allowing employees to perform their duties either partially or fully outside the physical premises of government agencies, leveraging technological resources to ensure the efficient execution of activities (Brasil, 2020).

On May 17, 2022, Decree No. 11.072 was enacted, establishing a new regulatory framework for the PGD within the direct, autarchic, and foundational federal public administration. This decree consolidated previous regulations, reinforcing the emphasis on results-based management and productivity monitoring for civil servants, replacing traditional attendance control. It also set guidelines for the standardized implementation of the PGD across different agencies, outlining clear criteria for target setting, performance indicators, and monitoring mechanisms, with the objective of modernizing public administration and enhancing service delivery to society.

Regulatory developments continued with the publication of Normative Instruction No. 89, of December 13, 2022, which repealed Normative Instruction No. 65/2020. However, its validity was extremely brief, as it was revoked just 28 days later by Normative Instruction No. 2, of January 10, 2023. This new directive required central agencies to issue specific regulations aligned with Decree No. 11.072/2022 within a maximum of ninety days from its publication, reaffirming the commitment to a more consistent and standardized approach to performance management in the public sector.

The PGD continued to evolve in 2023, when the Secretariat for Management and Innovation issued Joint Normative Instruction No. 24, on July 28, providing detailed guidelines on its implementation and execution. Subsequently, on December 21, 2023, Joint Normative Instruction No. 52 was published, reinforcing regulations on human resource management within the PGD framework. Both directives aim to solidify a results-driven management culture, clearly defining commitments, targets, and indicators, while also promoting strategic planning, continuous process improvement, and greater efficiency and transparency in the federal public administration. Finally, certain provisions of Normative Instructions No. 24/2023 and No. 52/2023 were amended or repealed by Joint Normative Instruction No. 21/2024, which introduced significant changes, including an extension of the deadline for agencies to fully comply with the new regulations.

The Implementation Panel of the Ministry of Management and Innovation (2025) reports that approximately 86% of federal agencies and entities have already adopted the Management and Performance Program (PGD), underscoring the widespread implementation of new guidelines aimed at modernizing public administration. These guidelines emphasize strategic planning, effective results management, and enhanced transparency in service delivery.

In 2025, efforts to consolidate and continuously improve management processes are expected to intensify, particularly in program data administration and the reinforcement of transparency measures. The government has made significant strides in developing the Transparency Panel, a tool designed to strengthen accountability mechanisms and enhance public oversight of government performance.

3. METHODOLOGY

The methodological approach adopted in this study is the case study, as outlined by Marconi and Lakatos (2021). In a case study, the objective is not to establish a rigid structure in advance but rather to enable the broad and detailed collection of information, employing various research techniques to explore the complexity of the phenomenon under investigation. The primary goal is to achieve an in-depth understanding of a specific situation, capturing the intricacies of the facts and contexts involved. This approach allows for a comprehensive analysis of particular organizational issues, such as the procedures for implementing the PGD.

This research focuses on the implementation of the PGD and its regulatory framework in Brazil, with particular emphasis on the case of CADE. Within this federal agency, the adoption of the PGD seeks to enhance people management efficiency and optimize institutional outcomes in accordance with the guidelines established by current legislation. The study follows a mixed-methods approach, integrating qualitative and quantitative methodologies. It combines an exploratory literature review with a documentary analysis of key regulations related to the PGD, such as Decree No. 11.072/2022, alongside other relevant directives. This methodological combination ensures a comprehensive understanding of the processes and challenges associated with implementing the program.

Additionally, a systematic literature review was conducted using Scopus, Web of Science, and Google Scholar to identify relevant studies on the implementation of management and performance programs in public organizations. The search was carried out using the keywords: "Management and Performance Program," "PGD public administration," and "management and performance in public administration." The most relevant studies were selected to form the theoretical foundation of this chapter, providing a robust basis for a critical and contextualized analysis of the challenges and impacts of implementing the PGD at CADE.

The selection of articles was based on thematic relevance, timeliness, and alignment with the research objective, prioritizing publications from the last ten years. This selection criterion broadened the theoretical framework and provided a solid foundation for the case study, enabling a critical analysis of best practices, challenges, and impacts associated with the implementation of the PGD within CADE.

Finally, after establishing the theoretical framework, the research aimed to systematize CADE's experience in implementing the PGD. This process involved the collection and analysis of institutional documents, internal regulations, and program monitoring records, with the goal of providing a detailed account of the stages, challenges, and outcomes observed. The study also examined the results of the PGD's implementation, offering a comprehensive understanding of its impact on human resource management and organizational efficiency. This analysis helped identify the best practices and areas for improvement, providing valuable insights for the future adoption of the program in other public institutions.

4. THE IMPLEMENTATION OF THE MANAGEMENT PROGRAM AT THE ADMINISTRATIVE COUNCIL FOR ECONOMIC DEFENSE

In October 2020, following the publication of Normative Instruction No. 65/2020, CADE formally submitted a request to the Ministry of Justice and Public Security (MJSP) to implement the Management Program (PGD). The authorization was officially published in the Federal Official Gazette (DOU) on November 25, 2020, allowing the program to begin its pilot phase, which was formalized on December 7, 2020, and implemented on January 4, 2021, with completion scheduled for June of the same year.

The pilot phase of CADE's PGD (internally referred to in Portuguese as "PG CADE") lasted six months and was characterized by continuous monitoring and rigorous evaluation, enabling ongoing refinements. During this period, the activity tables established with the units were systematically reviewed to ensure greater alignment with the institution's work processes. To support implementation, a range of informational materials—including manuals, instructional videos, and frequently asked questions—was developed to address CADE's specific needs. Oversight was carried out by a management committee composed of CADE employees, responsible for monitoring and assessing unit performance and recommending continuous improvements. This committee also prepared management reports, analyzed non-compliance cases, and proposed adjustments to enhance program effectiveness.

At the conclusion of the pilot phase, definitive regulations were established, culminating in the publication of Ordinance No. 317/2021, which formalized the permanent implementation of the PGD within the agency. This milestone institutionalized PG CADE, solidifying a paradigm shift toward results-based management. As no standardized system had been developed for use across the federal government, CADE partnered with the Comptroller General of the Union (CGU) to adopt CGU's system, incorporating adaptations to accommodate the agency's specific requirements.

In 2022, the need arose for a qualitative deepening of the PGD, particularly regarding its perception as a strategic tool for people and performance management. In this context, investments were made in the development of business intelligence panels to ensure efficient management of program data and results. Simultaneously, awareness initiatives were launched for both leaders and participants, emphasizing the importance of performance analysis, identifying activities of strategic value, and driving continuous improvement. These efforts also included the implementation of structured evaluation and feedback routines, which are essential for establishing the PGD as a strategic pillar within the organization.

In 2023, PG CADE underwent substantial changes in response to regulatory updates following a change in government, including the revocation of Normative Instruction No. 89 and the publication of new normative instructions. During the first half of the year, the focus was on regulatory compliance, involving the review of internal processes such as the registration of legal leave and the procedures of the General Coordination of Strategic People Management (CGESP). Additionally, automated weekly reports were sent to leaders to enhance engagement in performance monitoring and continuous improvement. In the second half of the year, following the publication of Joint Normative Instruction SEGES-SGPRT/MGI No. 24, CADE formed a multidisciplinary working group to align the program with new regulatory requirements and drafted an updated ordinance proposal. Concurrently, a task force was established to resolve pending work plan registrations from the past three years, ensuring the accuracy and regularization of program data. Efforts to align with Joint Normative Instruction No. 52, of December 2023, further reinforced compliance with the evolving regulatory landscape.

The year 2024 marked a significant transformation in the PGD, driven by regulatory changes and the adoption of a new operational system. The updates fundamentally shifted the program's focus, transitioning from an individual-centered, time-allocation approach to a model emphasizing institutional performance. The PGD now aligns participants' work with unit-level deliverables and broader organizational strategies, adopting a more strategic, results-driven framework compared to the previously operational approach. This evolution strengthens the integration of individual and institutional objectives, enhancing the program's overall effectiveness and impact.

One of the most significant innovations in the new model was the replacement of the previous activity table with a delivery plan, which has become the core operational element of the PGD. This change fosters a stronger emphasis on collective performance and unit-level outcomes, rather than prioritizing individual tasks. Additionally, the new legislation introduced a consequence policy, establishing clear mechanisms to address deficiencies in work plan execution. This policy aims to reinforce accountability, drive continuous improvement, and ensure alignment with institutional objectives, applying a structured and equitable approach to performance management.

Another significant milestone was CADE's adoption of the PETRVS system. Developed by the Secretariat of Management and Innovation of the Ministry of Management and Innovation in Public Services (SEGES/MGI), PETRVS was designed to support the execution, control, and monitoring of the PGD cycle across the federal public administration. The system complies with regulatory requirements, integrating data via an Application Programming Interface (API) to enhance transparency and oversight. It also enables deliverables to be linked to up to three levels of strategic planning or the value chain, ensuring that unit-level activities are aligned with institutional objectives.

The PETRVS System is a platform developed by the central body of SIORG (the Secretariat of Management and Innovation at SEGES/MGI) to operationalize the PGD cycle. Its primary function is to manage, monitor, and enhance the transparency of deliverables at the execution unit level, as well as to track participants' work plans, ensuring alignment with strategic, tactical, and operational planning in the federal public administration. The system is structured according to the guidelines set forth in Normative Instruction No. 24/23 and is built upon a governance model designed to integrate various hierarchical levels within institutions, ensuring that unit-level activities remain consistent with institutional objectives and responsive to public demands.

At the strategic level, the PETRVS System plays a critical role in formulating, monitoring, and tracking institutional guidelines, ensuring that the execution units' actions are fully aligned with strategic planning, organizational priorities, and high-level public administration directives. This alignment is essential to guarantee that each deliverable produced by the execution units not only meets immediate operational demands but also contributes directly to achieving broader institutional objectives, fostering efficient, results-driven management.

The PETRVS structure allows execution units to record their delivery plans systematically and uniformly, ensuring that each deliverable maintains a clear connection to the organization's strategic guidelines. This capability provides greater visibility into how individual and collective efforts contribute to institutional macro-objectives, reinforcing the institution's value chain and promoting a continuous cycle of evaluation and improvement. Through this integration, the implementation of public policies and government programs becomes more structured, enabling detailed monitoring of strategic execution and the timely adoption of corrective measures when necessary.

At the tactical level, the PETRVS System plays a crucial role in facilitating the alignment of individual work plans, ensuring that participants' activities correspond to the deliverables previously defined by the execution unit. This alignment is essential for strategically distributing tasks and considering individual competencies, institutional priorities, and operational needs. The system supports an integrated workforce management model, allowing each civil servant to allocate their workload proportionally to the unit's demands. This approach promotes balance across different operational fronts and ensures that deliverables are met within established deadlines.

In addition to ensuring more efficient workforce planning, PETRVS helps prevent both the overloading and underutilization of human resources, promoting a fair and balanced distribution of responsibilities among PGD participants. This is particularly important for optimizing team productivity, as it enables managers to track staff allocation across various projects and processes, identify bottlenecks, and make necessary adjustments. The system also allows managers to renegotiate work plans in response to shifting institutional priorities or emergency needs that require the realignment of previously planned activities. In this way, PETRVS enhances the units' adaptive capacity, providing greater flexibility in human resource management and enabling more agile and effective responses to public service demands.

At the operational level, PETRVS is an essential tool for executing agreed-upon activities, facilitating efficient delivery management, and providing a structured framework for comprehensive oversight of civil servants' work. The system offers functionalities that allow tracking the progress of deliverables, updating status reports, and documenting any challenges encountered during execution. This continuous data capture fosters dynamic and transparent monitoring, equipping management with a clear overview of task progression and enabling proactive intervention to ensure the successful completion of objectives.

The real-time monitoring capabilities of PETRVS empower managers to oversee activity execution continuously, allowing for timely adjustments whenever necessary. If challenges arise that jeopardize the timely completion of deliverables, the system facilitates redistribution of efforts, either by readjusting tasks within the team or reallocating personnel to higher-priority projects. Moreover, in cases requiring substantial modifications to the scope of work or team composition, PETRVS enables the renegotiation of work plans, ensuring that execution remains aligned with the unit's strategic goals and institutional commitments.

By functioning at both the tactical and operational levels, PETRVS provides a structured environment for intelligent workforce management in the public sector, enhancing predictability, control, and transparency over day-to-day activities. The system's ability to continuously adapt and adjust work plans and deliverables strengthens not only the internal efficiency of organizational units but also the quality of public services provided to society. This ensures that the PGD continues to serve as a key instrument for modernizing and enhancing public administration.

Additionally, PETRVS plays a crucial role in transparency and accountability, as it enables data export via APIs to populate the PGD Transparency Panel, allowing citizens and oversight bodies to monitor the progress of delivery plans and the results achieved. The system also facilitates integration across multiple data sources, generating key performance indicators (KPIs) that support data-driven decision-making and foster a more transparent and evidence-based management environment. However, as of January 2025, these indicator functionalities have not yet been implemented and are planned for future system updates. This integration reinforces the principle of accountability, ensuring that public administration operates under a governance model driven by data, thereby enhancing efficiency, reliability, and overall effectiveness in workforce and performance management.

In addition to execution and monitoring, PETRVS incorporates a structured work plan evaluation model, enabling managers to assess compliance with work plans based on pre-established parameters. Participant evaluations may consider criteria such as productivity, application of appropriate methods and techniques, teamwork, commitment to unit objectives, and adaptability to organizational changes. Performance is classified into various levels, ranging from exceptional to unsatisfactory, ensuring a transparent and results-driven evaluation process. This aspect of the system reinforces the governance-by-results framework, in which public administration shifts its focus from attendance control to the effective delivery of value to society.

The implementation of PETRVS also underscores the need to foster a results-oriented organizational culture. Traditionally rooted in rigid process control, public administration now finds in this system a tool to promote a new mindset among managers and civil servants, where the primary focus shifts to delivering high-quality products and services to society. This paradigm shift is essential for modernizing public management and developing more efficient institutions aligned with citizens' needs.

As a result, PETRVS has become an indispensable tool for results-based management in the federal public administration. By integrating strategic, tactical, and operational levels, the system enables more efficient, transparent, and data-driven management. Its structure not only facilitates effective delivery planning but also ensures continuous monitoring and precise evaluation of both unit and individual performance. While some functionalities remain under development, the platform already represents a significant advancement in the modernization of public administration, contributing to a more dynamic, responsive, and excellence-driven service model for society.

It is important to highlight that PETRVS was developed in PHP, ensuring flexibility and potential integration with other government systems in the future. This allows its structure to be expanded to accommodate emerging demands and challenges within public administration, further strengthening its role in results-based management and public sector innovation. Through this advancement, the federal administration takes a significant step toward greater efficiency, transparency, and excellence in service delivery to society.

The migration to the PETRVS System was a necessary initiative to modernize and enhance the management of deliverables and work plans within the PGD framework. The transition to this new platform aimed to consolidate a more efficient governance model, ensuring greater transparency, predictability, and strategic alignment of institutional activities. However, as with any organizational change, the implementation of the new system required careful planning and a well-structured communication strategy to secure leadership engagement and participant buy-in.

To facilitate the transition, structured meetings were held with agency leadership to present the new methodology and clarify the guidelines for registering the Delivery Plan. These meetings were designed to raise awareness among managers about the importance of the new system and to equip them with the knowledge needed to guide their teams through the implementation process. Discussions covered technical aspects of the system, including the workflow for registering deliverables, the linking of deliverables to the organization's strategic planning, and the short-, medium-, and long-term monitoring mechanisms. Additionally, the critical role of managers as knowledge multipliers was reinforced, emphasizing their responsibility in assisting their teams' adaptation to the new model.

Following this initial leadership alignment, informational sessions were conducted with all agency participants to ensure that each civil servant understood how to structure their Individual Work Plan within the system. These sessions detailed the criteria for distributing workload percentages across different deliverables, along with the monitoring guidelines for activities throughout the management cycle. The focus was to provide technical support and clarify doubts, ensuring a smooth and transparent transition. Additionally, the importance of participant autonomy in defining their work plans—while maintaining alignment with institutional guidelines and securing validation from their immediate supervisors—was strongly emphasized.

Although CADE already has a well-established Management Program culture, resistance to change is natural, particularly regarding the adoption of new technological platforms and work methodologies. Addressing these challenges required a continuous communication approach, consistently reinforcing the benefits of the new system and ensuring that all stages of the migration were carried out with full transparency. Moreover, actively listening to the concerns of civil servants and managers was essential in facilitating a smoother transition and engaging participants in adopting PETRVS as a central tool for planning and monitoring deliverables.

Against this backdrop, the availability and engagement of the Strategic General Coordination of People team played a pivotal role in the successful migration. To ensure that questions were promptly addressed and challenges effectively resolved, direct support channels were established, allowing participants to receive close guidance and assistance throughout the adaptation period. The technical team also monitored system usage during the initial weeks, identifying potential difficulties and proposing enhancements to optimize usability. This ongoing support not only facilitated user adoption but also helped ensure that the new management model was understood and seamlessly integrated, strengthening a more structured and results-driven public administration.

The data recorded in PETRVS is aggregated to populate the PGD Transparency Panel, enabling accurate tracking and public disclosure of annual results through official channels. This system advances digitalization and standardization in performance management, streamlining results-based governance and promoting greater efficiency in public administration. The year 2024 marked a milestone in PGD innovation, consolidating a more dynamic and effective model, tailored to contemporary demands within the federal public administration. This evolution strengthens transparency and accountability while enhancing data-driven decision-making.

Looking ahead to 2025, the primary challenge will be to sustain and enhance the new model, ensuring its long-term effectiveness and stability. For CADE, it will be essential to foster a collaborative environment where leaders and teams are aligned with the program's strategic objectives and committed to its success. Developing an organizational culture centered on results-oriented management and continuous learning will be fundamental in solidifying this new phase. A key challenge will be to fully meet the transparency requirements set by the PGD's Monitoring and Public Reporting Panel, which will necessitate highly structured, reliable, and easily accessible data to ensure that information accurately reflects program progress and challenges. To achieve this level of maturity, it will be critical to establish robust mechanisms for continuous monitoring of deliverables, ensuring the quality and consistency of records. The adoption of best practices in data collection and analysis, combined with the strategic use of technology, will be instrumental in consolidating the model, enabling an agile, transparent, and continuously improving management cycle.

With this objective in mind, CADE has established a dedicated team to oversee the PGD, further reinforcing its commitment to efficiency, transparency, and institutional performance.

Table 1. PGD implementation timeline at CADE

2020	**Pilot implementation of PG CADE**
2021	Permanent implementation of PG CADE
2022	Qualitative leap in deliverables and leader-employee management
2023	Compliance and regulatory changes
2024	Strategic change management with a focus on planning and delivery efficiency
2025	Advancements in data management and transparency

Source: Prepared by the author.

Migrating between different systems presents a substantial challenge for the responsible team, particularly in ensuring the integrity and continuity of processes. The transition requires adapting to new interfaces, ensuring data compatibility, and mitigating operational impacts. Despite these challenges, the results achieved thus far demonstrate notable progress, including greater efficiency in information processing and enhanced data accessibility. However, the constraints of a limited team impose restrictions on the pace of implementation and the ability to address emerging demands, necessitating careful management of priorities and available resources. In response to this scenario, a structured training initiative has been essential in equipping staff with the necessary skills to facilitate an effective transition, ensuring that all users fully understand the changes and can operate the new systems confidently and autonomously.

Currently, the primary challenges center on strengthening data control tools to enhance reliability and security in information management. The success of these initiatives depends on ongoing awareness efforts among managers and participants, emphasizing their critical role in the responsible use of these systems. This educational process, combined with targeted training to support civil servants' adaptation, aims not only to promote engagement but also to mitigate resistance and uncertainty, enabling a smoother and more sustainable transition to a more robust and efficient digital environment.

5. CONCLUSION

The normative evolution of the Management and Performance Program (PGD) in Brazil reflects a trajectory of gradual advancements, from its formal introduction in 1995 to its consolidation as a strategic public management model. Initially designed to increase flexibility in public sector attendance policies, the program has undergone multiple reformulations over the decades, driven by emerging needs and shifts in the administrative landscape. The formal regulation in 2018 marked a turning point, establishing uniform guidelines for public institutions and promoting greater clarity in the adoption of results-oriented practices.

The COVID-19 pandemic significantly accelerated this transformation, necessitating the urgent adoption of mechanisms to ensure the continuity of government operations in a remote work environment. The normative instructions issued in 2020 reflected this need, broadening the flexibility of the PGD and enabling it to adapt more rapidly to the evolving nature of work. This period underscored the importance of a more dynamic, performance-driven management model that not only addressed crisis scenarios but also enhanced efficiency and transparency in public administration.

By 2022, the institutionalization of the PGD was further solidified with Decree No. 11.072/2022, which reinforced results-based management and the modernization of administrative processes. The introduction of clear metrics, performance indicators, and monitoring mechanisms reaffirmed the federal government's commitment to a more strategic and efficient public sector. However, the rapid succession of regulatory changes between 2022 and 2024 presented challenges in establishing a standardized implementation model, requiring ongoing adaptation efforts from public agencies.

The case of the Administrative Council for Economic Defense (CADE) exemplifies how PGD implementation is a dynamic process of continuous improvement. Since the pilot phase in 2021, CADE has invested in the systematic review of activities, the development of support materials, and the creation of a management committee to ensure the program's effectiveness. As new regulations were introduced, the agency responded proactively, integrating business intelligence tools and enhancing people management strategies to improve performance monitoring.

The adoption of the PETRVS system in 2024 marked a significant milestone in PGD management, facilitating the digitalization and standardization of processes while ensuring greater transparency in institutional performance tracking. This evolution reinforced a collective performance model and strengthened strategic integration, consolidating a more structured approach aligned with public administration objectives. The implementation of consequence policies and the restructuring of evaluation mechanisms reflect the maturity of the program, with an increased focus on accountability and continuous improvement.

Looking ahead to 2025, the primary challenge will be to solidify this new model, ensuring that these changes are embedded sustainably and effectively. Maintaining a commitment to transparency and efficient management will require continuous monitoring of deliverables, quality assurance in record-keeping, and strong leadership engagement in managing the PGD. Within CADE, the establishment of a dedicated team for program administration underscores the importance of structured governance, which is essential for sustaining progress and ensuring the definitive consolidation of the PGD as a key instrument in the modernization of federal public administration.

REFERENCES

Alosani, M. S., Al-Dhaafri, H. S., & Awadh Abdulla, A. (2021). Investigating the role of HRM practices on service innovation: Empirical evidence from UAE government agencies. *Management Research Review*, *44*(1), 1–24. DOI: 10.1108/MRR-03-2020-0141

Brasil. (1995). *Decree No. 1,590, of August 10, 1995*. Provides for the working hours of employees of the direct Federal Public Administration, autarchies and federal public foundations, and provides for other provisions. https://www.planalto.gov.br/ccivil_03/decreto/D1590.htm

Brasil. (2017). *Ordinance No. 438, of December 21, 2017*. Regulates telework, as a pilot experience, within the scope of CADE's organizational units and provides other provisions. Ministry of Justice and Public Security. https://cdn.cade.gov.br/Relatoriorios%20de%20gestao/2020/Cap.%204/Portaria%20Cade%20438_2017.pdf

Brasil. (2018). *Normative Instruction No. 1, of August 31, 2018*. Establishes guidelines, criteria and general procedures to be observed by the bodies and entities that are part of the Civil Personnel System of the Federal Administration – SIPEC regarding the implementation of the Management Program. Ministry of Planning, Development and Management. https://www.in.gov.br/materia/-/asset_publisher/Kujrw0TZC2Mb/content/id/39382838/do1-2018-09-03-instrucao-normativa-n-1-de-31-de-agosto-de-2018-39382704

Brasil. (2020). *CADE Ordinance No. 395, of December 4, 2020*. Establishes the Management Program within the scope of the Administrative Council for Economic Defense – CADE. Ministry of Justice and Public Security. https://cdn.cade.gov.br/Relatoriorios%20de%20gestao/2020/Cap.%204/Portaria%20Cade%20395_2020.pdf

Brasil. (2020). *Normative Instruction No. 44, of June 10, 2020*. Amends Normative Instruction No. 1, of August 31, 2018, which establishes guidance, criteria and general procedures to be observed by the bodies and entities that are part of the Civil Personnel System of the Federal Administration – SIPEC. Ministry of Economy. https://www.in.gov.br/en/web/dou/-/instrucao-normativa-n-44-de-10-de-junho-de-2020-261279454

Brasil. (2020). *Normative Instruction No. 65, of July 30, 2020*. Establishes guidelines, criteria and general procedures to be observed by the bodies and entities that are part of the Civil Personnel System of the Federal Administration – SIPEC regarding the implementation of the Management Program. Ministry of Economy.

Brasil. (2021). *CADE Ordinance No. 317, of June 18, 2021*. Establishes the Management Program (PG. CADE) within the scope of the Administrative Council for Economic Defense (CADE). Ministry of Justice and Public Security. https://cdn.cade.gov.br/Relatoriorios%20de%20gestao/2021/Cap%204/Portaria%20Cade%20317.pdf

Brasil. (2022). *Decree No. 11,072, of May 17, 2022*. Provides for the Management and Performance Program (PGD) of the direct, autarchic and foundational federal public administration. https://www.planalto.gov.br/ccivil_03/_ato2019-2022/2022/decreto/d11072.htm

Brasil. (2022). *Normative Instruction SGP-SEGES/SEDGG/ME no. 89, of December 13, 2022*. Establishes guidelines to be observed by the bodies and entities that are part of the Civil Personnel System of the Federal Administration – SIPEC. Ministry of Economy. https://www.in.gov.br/en/web/dou/-/instrucao-normativa-sgp-seges/sedgg/me-n-89-de-13-de-dezembro-de-2022-451152923

Brasil. (2023). *Joint Normative Instruction SGP-SRT-SEGES/MGI No. 52, of December 21, 2023*. Establishes guidelines regarding the rules for people management within the scope of the Management and Performance Program – PGD. Ministry of Management and Innovation in Public Services. https://www.in.gov.br/en/web/dou/-/instrucao-normativa-conjunta-sgp-srt-seges/mgi-n-52-de-21-de-dezembro-de-2023-532726070

Brasil. (2023). *Normative Instruction SGP-SEGES /ME no. 2, of January 10, 2023*. Repeals Normative Instruction SGP-SEGES/SEDGG/ME no. 89. Ministry of Economy. https://www.in.gov.br/en/web/dou/-/instrucao-normativa-sgp-seges-/me-n-2-de-10-de-janeiro-de-2023-457679698

Brasil. (2023). *SEGES-SGPRT/MGI Joint Normative Instruction No. 24, of July 28, 2023*. Establishes guidelines to be observed by the bodies and entities that are part of the Civil Personnel System of the Federal Administration – SIPEC. Ministry of Management and Innovation in Public Services. https://www.in.gov.br/en/web/dou/-/instrucao-normativa-conjunta-seges-sgprt-/mgi-n-24-de-28-de-julho-de-2023-499593248

Brasil. (2024). *Joint Normative Instruction SEGES-SGP-SRT/MGI no. 21, of July 16, 2024*. Amends Joint Normative Instruction SEGES-SGPRT/MGI No. 24. Ministry of Management and Innovation in Public Services. https://www.in.gov.br/en/web/dou/-/instrucao-normativa-conjunta-seges-sgp-srt/mgi-n-21-de-16-de-julho-de-2024-572617003

Brasil. (2025). *Management and Performance Program*. Ministry of Management and Innovation. https://www.gov.br/servidor/pt-br/assuntos/programa-de-gestao

Chowdhury, A., & Shil, N. C. (2021). Performance management systems in public sector under new public management regime: an australian case. *Journal of Public Administration, Finance and Law*. https://doi.org/10.47743/jopafl-2021-20-03

da Silva Aguiar, N. A., Barbosa, F. L. S., de Almeida Bizarria, F. P., & Barros, R. O. (2024). Management and Performance Program and Organizational Changes in the Brazilian Federal Public Administration: An integrative literature analysis. *Iberian Journal of Information Systems and Technologies*, (E69), 451–464.

Elvira, E. J. S., & Bezerra, L. F. (2023). *The management program: On the path to post-new performance management?* Enap. http://repositorio.enap.gov.br/handle/1/7641

Kettl, D. F. (2005). The Global Revolution: Public Sector Administration Reform. In Bresser Pereira, L. C., & Spink, P. (Eds.), *Reforma do estado e administração pública gerencial* (pp. 75–122). Editora FGV.

Marconi, M. de A., & Lakatos, E. M. (2021). *Fundamentals of scientific methodology* (9th ed.). Atlas.

Pena, K. W. (2024). *Teleworking in public administration* [Master's thesis, Universidade Federal de Goiás].

Santiago, V. B., & Castro, M. C. D. E. (2023). Management and Performance Program in the bodies of the Federal Public Administration. In *9th National Meeting of Public Administration and Governance – EMPRAD, São Paulo*. ANPAD.

Chapter 9
Innovation Management at the Fuzzy Front End:
AI, Business Cases, and Product Portfolios

Xiaomei He
University of Strasbourg, France

Marion Neukam
University of Strasbourg, France

Thierry Burger-Helmchen
https://orcid.org/0000-0001-8866-1919
University of Strasbourg, France

ABSTRACT

This chapter examines the strategic management of the Fuzzy Front End (FFE) in innovation, emphasizing its role in new product development for mature firms with extensive product portfolios. We show how FFE shapes competitive advantage and how artificial intelligence (AI) redefines early-stage decision-making, business case development, and business model innovation. AI reveals emerging opportunities, enhances decision-making under uncertainty, and strengthens risk assessment. We underline its impact on resource allocation, balancing disruptive innovation with operational stability. AI accelerates business model evolution, sharpening project selection, portfolio management, and strategic alignment.

1. INTRODUCTION

In the rapidly evolving market landscape, mature companies with extensive product portfolios face the dual challenge of sustaining competitiveness while fostering continuous innovation. The Fuzzy Front End (FFE) of innovation, often regarded as a critical yet ambiguous phase in the new product development (NPD) process, plays a pivotal role in addressing this challenge. The FFE, frequently referred to as "Phase 0" or "Pre-Project Activities," encompasses the earliest stages of the innovation process, including opportunity identification, concept development, and preliminary evaluation before formal product development begins (Gassmann and Schweitzer, 2013; Koen *et al.*, 2001; Tian *et al.*, 2024).

DOI: 10.4018/979-8-3693-4030-1.ch009

This phase is characterized by high uncertainty and complexity, requiring firms to balance creativity and control to successfully translate nascent ideas into actionable innovation pathways (Neukam, 2017).

The importance of the FFE lies in its potential to shape the trajectory of innovation within an organization. Research indicates that effective management of this phase can significantly influence the time-to-market, resource allocation efficiency, and overall success of new product introductions (Khosravi *et al.*, 2019). In the case of mature firms with extensive and diversified product portfolios, managing the FFE becomes particularly complex due to organizational inertia, competing strategic priorities, and entrenched decision-making processes (Globocnik *et al.*, 2020). These challenges underscore the need for structured methodologies and advanced analytical tools to navigate the ambiguity of early-stage innovation effectively.

In the recent period FFE has undergone several evolutions. One of a technological nature is the employment of AI in all steps of R&D processes, which consequently impacts FFE as the first stage of this process. Additionally, the development of multiple business models within large corporations has introduced specific management techniques in the FFE, especially considering that it can influence not only a single product line but an entire portfolio.

1.1 The Role of Artificial Intelligence in Enhancing the FFE and Business Case Development

In recent years, Artificial Intelligence (AI) has emerged as a transformative enabler in innovation management. Its effects on the FFE have yet to be properly investigated, both from a practical and a theoretical perspective, due to the numerous causal links it shares with the final outcome and business model construction. AI technologies can process vast amounts of structured and unstructured data, uncovering latent market trends, evolving customer preferences, and emerging technological opportunities that might otherwise go unnoticed (Bahoo *et al.*, 2023; Mariani and Dwivedi, 2024). This capability is particularly valuable in business case development, where AI-driven predictive analytics and machine learning models can enhance opportunity assessment, risk evaluation, and financial modelling, ultimately leading to more informed decision-making in the FFE.

For example, AI-powered natural language processing (NLP) and sentiment analysis can mine consumer feedback, social media data, and industry reports to identify unmet needs and emerging demand patterns (Pietronudo *et al.*, 2022). Similarly, AI-based simulations can evaluate multiple business case scenarios, providing firms with quantitative insights on potential market adoption, revenue projections, and competitive positioning. By leveraging generative AI tools, firms can automate ideation processes, augmenting creative problem-solving and accelerating the early-stage conceptualization of new products and services (Mariani and Dwivedi, 2024).

1.2 The Complexity of Managing the FFE in Firms with Extensive Product Portfolios

For mature companies with extensive product portfolios, effectively managing the FFE requires strategic coherence between multiple innovation initiatives, ensuring alignment with the existing business model portfolio (Joyce and Paquin, 2016). The challenge is twofold: firms must simultaneously explore disruptive innovations while maintaining the operational stability of established product lines (Gambardella and McGahan, 2010). In this context, AI can serve as a cognitive assistant, helping decision-makers navigate complex portfolio interdependencies, resource allocation trade-offs, and cross-functional collaboration challenges.

Structured portfolio management frameworks advocate for integrating AI-driven insights into FFE decision-making, enabling firms to optimize their innovation pipelines, minimize redundant efforts, and prioritize high-potential opportunities. By embedding AI-driven scenario planning and real-time market intelligence, firms can reduce the risk of late-stage failures, improve cross-functional coordination, and create more resilient business cases that support scalable and sustainable growth strategies (Eling and Herstatt, 2017).

1.3 Chapter Structure

In this chapter we examine the strategic challenges and opportunities associated with FFE management, emphasizing the role of AI in enhancing business case formulation and business model integration. The discussion begins with a review of FFE within innovation processes, followed by an exploration of how firms can operationalize FFE outcomes into structured business cases and business model innovations. We then analyze the impact of FFE-driven initiatives on business model portfolios, highlighting strategic alignment, portfolio management frameworks, and control mechanisms. Finally, we assess the transformative role of AI in optimizing innovation drivers, accelerating FFE activities, and dynamically managing business model portfolios. The chapter concludes by synthesizing key insights and proposing future research directions on AI-enhanced innovation management.

2. STRATEGIC MANAGEMENT OF THE FUZZY FRONT END AND THE STANDARD DRIVERS OF INNOVATION

This section explores the strategic management of the FFE and its role in innovation processes. We begin by reviewing key definitions and conceptualizations of the FFE, examining its challenges and strategic importance. Then, we analyze how companies can operationalize the FFE to transition from idea generation to structured business case development. Finally, we investigate how FFE-driven innovations influence business model transformation, particularly in the context of mature firms managing extensive product portfolios.

2.1 The FFE in Innovation Management

The FFE is characterized by its ambiguity and uncertainty, necessitating a structured approach to innovation management (Khosravi *et al.*, 2019). The process begins with preliminary opportunity identification, market, and technology analysis (Pre-phase zero), progressing to product concept and definition (Phase zero), and culminating in product definition and project planning (Phase one), as highlighted in Figure 1 depicting the three phases front end model (Khurana and Rosenthal, 1998). Therefore, a strategic approach to the FFE requires an understanding of the foundational elements of new product development execution, including specification and design, prototype test and validation, volume manufacturing, and market launch (Khosravi *et al.*, 2019).

Figure 1. The three phase front end model

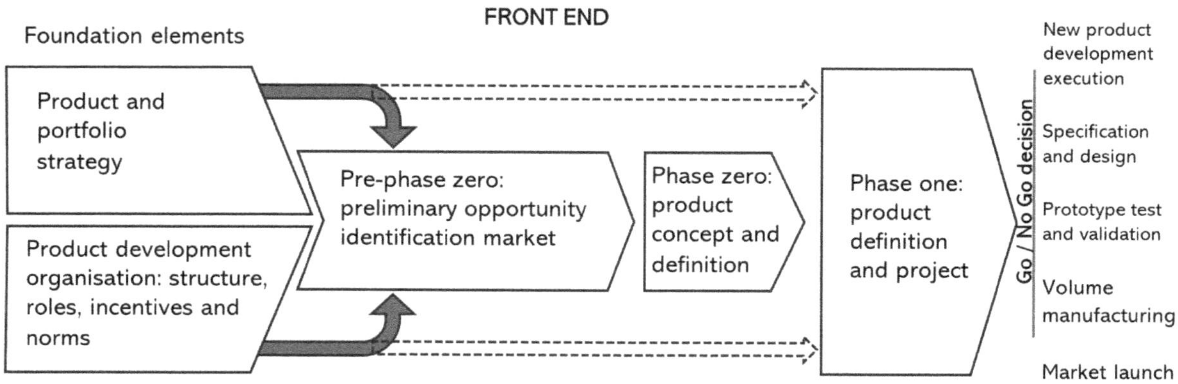

(Khurana and Rosenthal, 1998)

The FFE of innovation represents a critical, albeit complex, phase of the innovation management process, particularly for mature companies with extensive product portfolios. As a precursor to structured product development, the FFE is often marked by uncertainty and ambiguity. This phase is characterized by the initial identification of opportunities and the formulation of ideas before formal project development begins. Despite its unstructured nature, the FFE is widely recognized as a strategic leverage point for enhancing the overall effectiveness and efficiency of innovation processes (Bhatia *et al.*, 2017).

The importance of the FFE lies in its potential to shape the trajectory of innovation by addressing two critical challenges: managing uncertainty and accelerating time to market. Mature companies, which often operate in highly competitive environments, face a dual imperative during this phase. On the one hand, they must identify and explore disruptive opportunities; on the other, they must ensure that these opportunities align with their existing business model portfolio.

Historical examples highlight the consequences of poorly managed FFEs. For instance, Airbus's A380 program faced significant delays and cost overruns due to late-stage changes in customer specifications. Similarly, Boeing's Dreamliner project encountered severe integration issues stemming from a lack of supplier involvement in the early phases of development (Gassmann and Schweitzer, 2014). These cases underscore the necessity of "failing early" to succeed sooner, a principle that advocates for robust experimentation and decision-making during the FFE.

Despite these challenges, the FFE offers a high degree of strategic leverage. Decisions made during this phase influence the entire innovation process, from concept development to market launch. As Gassmann and Schweitzer (2014) emphasize, the ability to identify and address critical uncertainties early in the process is a hallmark of successful innovation management.

Effective management of the FFE requires a delicate balance between structure and flexibility. While structured processes can reduce uncertainty, excessive rigidity can stifle creativity and hinder the identification of disruptive opportunities. Mature companies must adopt an approach that integrates interdisciplinary perspectives, fosters collaboration, and encourages a tolerance for ambiguity. Hence, the model proposed by Koen et al. (2001, see Figure 2) highlights the dynamic interplay of front-end elements and influencing factors. A key characteristic of the New Concept Development model (NCD) is its circular, iterative structure, where the five front-end elements are not bound to a fixed sequence. Instead, they can occur in any order, repeatedly, in parallel, or sequentially. During opportunity identification, potential innovations emerge by aligning company objectives with various tools and problem-solving techniques, such as brainstorming and causal analysis. At the opportunity analysis stage, technological and market-related criteria help assess whether an opportunity is worth pursuing. The phase of idea genesis follows, where concepts evolve through an iterative and evolutionary process. The most promising ideas are then filtered in the idea selection stage. The front-end engine consists of controllable company-driven factors, such as leadership, culture, and business strategy, which foster an environment conducive to innovation. Additionally, both internal elements (e.g., organizational capabilities, technologies, and strategic direction) and external influences (e.g., distribution channels, customer expectations, and competitive forces) shape the front end of innovation.

The NCD model can also be seen as a form of new business case development, fostering the need for new business models. As a result, it serves as a tool that, while potentially less operationalizable than others, offers a smoother integration between the FFE, business models, and corporate-level strategy.

Figure 2. New concept development model

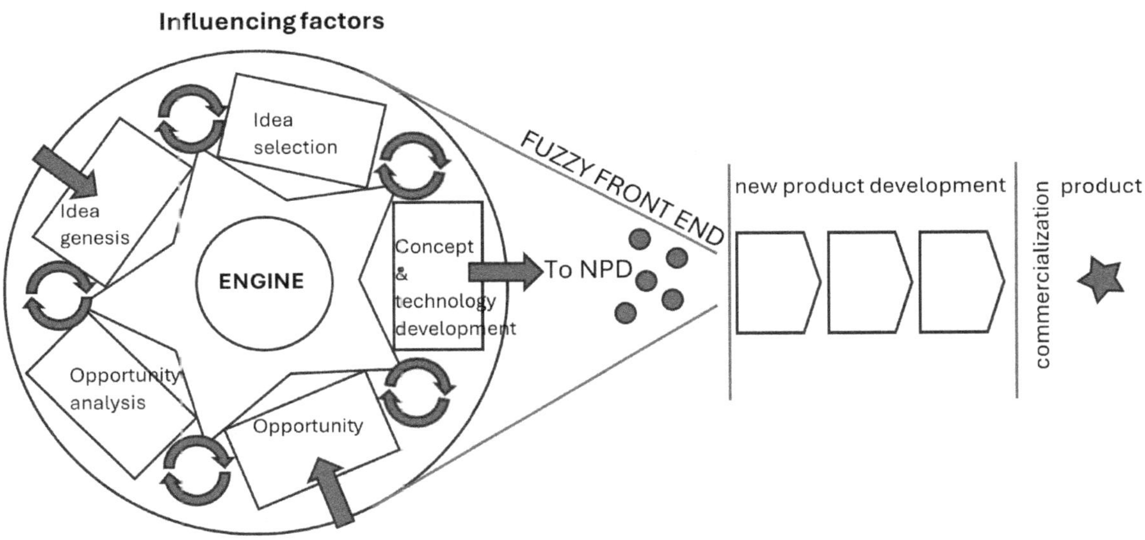

(source Koen et al.)

Strategic alignment is another critical component of FFE management. Mature companies must ensure that their innovation activities are consistent with their broader strategic goals. However, as illustrated by the case of Sulzer (Gassmann and Schweitzer, 2013), a rigid adherence to strategy can sometimes hinder the exploration of unconventional opportunities. In dynamic environments, companies may need to adopt a more flexible approach, allowing for the emergence of bottom-up initiatives that challenge existing paradigms.

The successful management of the FFE is supported by a range of tools and methods designed to address its inherent complexities. Techniques such as trend scanning, customer integration, and crowd-sourcing enable companies to identify latent needs and emerging opportunities. For instance, BMW's use of empathic design and netnography in the development of adaptive headlights illustrates how the FFE can be leveraged to anticipate future customer requirements (Stephan-Korus, 2021).

Moreover, the integration of interdisciplinary teams with various educational or cultural backgrounds during the FFE fosters a holistic approach to problem-solving (Neukam, 2017). By incorporating diverse perspectives, companies can enhance the quality of decision-making and reduce the risk of late-stage failures. The art of managing the FFE lies in striking a balance between creativity and discipline, flexibility and structure, and market pull and technology push.

2.2 Operationalizing Innovation Through the FFE: From Business Case To Business Model

Operationalizing the Fuzzy Front End (FFE) requires a structured yet flexible approach to transform early-stage ideas into viable business opportunities. This process is particularly complex for mature companies with extensive product portfolios, where innovation must align with both existing business models and broader strategic objectives. The FFE serves as the crucial bridge between creativity and structured development, enabling firms to evaluate, refine, and select ideas before committing to full-scale product development (Cohendet *et al.*, 2024). However, managing this transition effectively involves multiple challenges, including organizational inertia, risk assessment, and resource allocation constraints.

One of the most significant steps in the operationalization of the FFE is the creation of a business case, which serves as the first structured articulation of an innovation's market potential. A well-developed business case not only defines the technological feasibility and competitive advantage of an idea but also plays a decisive role in securing internal approval and funding (Bohnsack *et al.*, 2014; Sosna *et al.*, 2010). In firms with extensive product portfolios, this stage is even more critical, as multiple projects compete for limited resources, and decision-makers must evaluate whether a new idea complements or disrupts existing product lines.

Academic research on innovation management highlights several structured approaches to transforming ideas into robust business cases. As mentioned, Koen et al. (2001) propose the New Concept Development Model (NCD) as a framework to guide firms through the preliminary phases of innovation, emphasizing the need for cross-functional collaboration, iterative refinement, and early market validation (Cooper, 2008; Cooper and Sommer, 2016; Hollen *et al.*, 2013). Further studies, such as those by Hollen et al. (2013), stress the importance of incorporating external knowledge and interdisciplinary insights in FFE activities. By analyzing case studies from industries such as automotive, pharmaceuticals, and consumer electronics, they demonstrate how firms that integrate customer feedback, competitor intelligence, and technological foresight into their business case development achieve higher innovation success rates. This approach aligns with the principles of open innovation, where firms collaborate with external partners, research institutions, and even competitors to co-develop new concepts and mitigate early-stage risks (Burger-Helmchen and Siegel, 2019). Figure 3 depicts such an approach where innovation strategy leads, after a routine gate (RG), towards concept development, product development, but also, depending on the technology involved (new or old), strong change/adaptation in the business model. Therefore, FFE, business case-business model, and strategy are closely linked.

Figure 3. Holistic framework for the front end of innovation

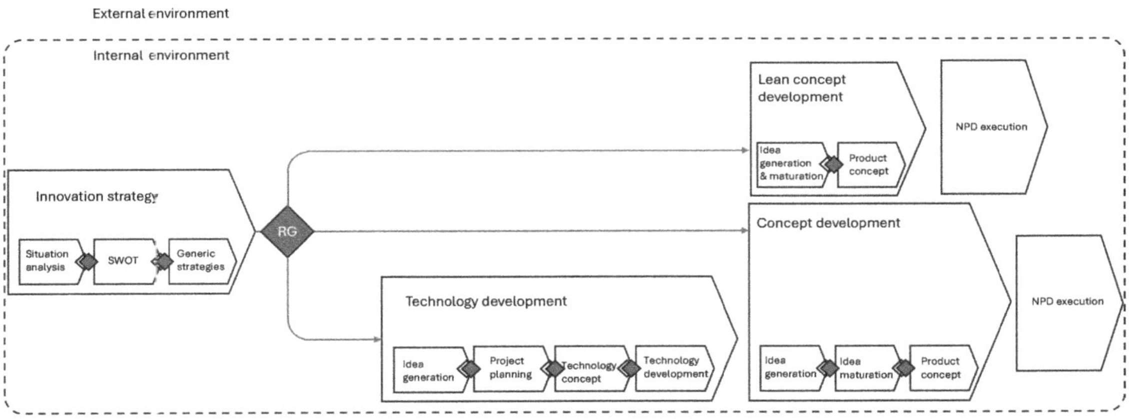

(sources: Gassman and Schweitzer, 2014; Cooper and Sommer, 2016)

Therefore, the business case serves a dual function in the innovation process. Internally, it provides a structured rationale for decision-makers to allocate financial, technical, and human resources to a project. Externally, it acts as a market-oriented validation tool, helping firms assess whether an idea has commercial viability and strategic alignment with existing or emerging business models (Sosna et al., 2010). A key challenge in this phase is (again) managing uncertainty, as many innovations face unpredictable market dynamics and technological risks (De Brentani and Reid, 2012). To address this, firms increasingly rely on AI-powered analytics and machine learning models to enhance opportunity evaluation and scenario planning (Bahoo, Cucculelli, & Qamar, 2023). Moreover, Joyce and Paquin (2016) argue that traditional financial models for business case assessment may not fully capture the long-term strategic impact of an innovation. This is particularly relevant for industries undergoing regulatory shifts and sustainability-driven transformation, such as energy, automotive, and technology sectors (Afuah, 2014).

Once a business case is approved, the challenge shifts to integrating new innovations into the company's business model portfolio. Business model innovation (BMI) can be defined as "designed, novel, and nontrivial changes to the key elements of a firm's BM and/or the architecture linking these elements" (Foss and Saebi, 2017, p.216). BMI requires firms to rethink value creation, delivery, and capture mechanisms (Amit and Zott, 2012; Baden-Fuller and Haefliger, 2013). Their study highlights that many large firms struggle to transition from validated business cases to scalable business models due to structural inertia and misalignment with existing corporate strategies.

A critical factor in business model innovation success is the ability to adapt organizational structures and processes to accommodate disruptive changes. For example, research by Gambardella and McGahan (2010) illustrates how firms like Netflix, Tesla, and Amazon leveraged incremental business model adaptation to ensure continuous innovation scalability. These companies iteratively refined their business models based on real-world market feedback, allowing them to pivot and scale efficiently.

Firms that adopt agile innovation methodologies—where iterative development cycles allow for continuous refinement of business models—tend to outperform competitors in highly volatile industries (Neukam, 2017). This approach contrasts with traditional stage-gate models, where rigid sequential processes often hinder the rapid adaptation of business cases into scalable business models. This is why the NCD model of Koen et al. (2001) proposes an iterative and cyclic model to overcome the pitfalls of a traditional sequential thinking.

For mature companies managing diverse product portfolios, the transition from business case to business model innovation presents unique challenges. These firms must carefully evaluate how new innovations align with their existing product ecosystems, ensuring that new business models enhance rather than cannibalize existing revenue streams (Globocnik et al., 2020). One effective strategy, as highlighted in empirical studies of multinational corporations, is the adoption of dual business models, where firms simultaneously manage legacy operations while scaling new ventures (Casadesus-Masanell and Tarziján, 2012). This allows for greater flexibility in innovation execution, enabling firms to gradually transition from traditional to emerging business models without disrupting core revenue streams. Moreover, firms that implement structured portfolio management frameworks can better balance risk and reward when integrating FFE-driven innovations into their business model portfolio. Their research suggests that organizations that establish clear governance structures and cross-functional collaboration mechanisms achieve higher success rates in business model transformation efforts.

2.3 Innovative Business Models as a Result of the FFE

During the FFE, firms should foster an environment conducive to experimentation and creative exploration, enabling them to identify unmet customer needs, assess emerging technologies, and conceptualize new market opportunities that may not align with their current business models. This phase allows for iterative learning and adaptation, facilitating the development of new revenue streams, customer engagement strategies, and value propositions without necessitating immediate large-scale investments. For example, the adoption of sustainable business models has gained traction where companies are pushed to align economic objectives with environmental and social responsibilities. Align FFE activities with sustainability requirements not solely depend on external requirements, but needs to be integrated throughout all steps of the FFE process (Eling, 2020). Some works provide already frameworks for integrating these dimensions during the FFE, enabling organizations to design business models that meet stakeholder expectations and comply with regulatory standards while remaining competitive (Neukam and Bollinger, 2022). From this perspective, the FFE obtains a crucial role in Business Model transformation as it is during this phase that sustainability is not exclusively considered as a constraint for companies representing increasing costs for customers. Integrating such sustainability issues to the FFE can also be a stimulating factor for team creativity and ultimately team performance (Bollinger *et al.*, 2023).

Despite the potential benefits, mature companies often face significant barriers to business model innovation, including organizational inertia, rigid decision-making structures, and a general aversion to risk. These challenges are particularly pronounced in industries with well-established value chains and tightly coupled processes, where the costs and risks associated with deviating from standard practices can be substantial (Randhawa *et al.*, 2021). To overcome these obstacles, some companies adopt a dual approach, maintaining the stability of their core operations while experimenting with innovative models in separate or semi-autonomous units. This strategy allows organizations to test new ideas without jeopardizing existing business activities, thereby mitigating the risks associated with radical innovation (Cohendet and Llerena, 2005).

Innovations emerging from the FFE can vary from incremental enhancements to existing business models to entirely new approaches that redefine industry standards. Disruptive innovations often arise when companies adopt a customer-centric approach during the FFE, focusing on understanding and addressing latent customer needs (Guittard *et al.*, 2015). For instance, the use of empathic design and real-time customer feedback during product development can inform novel value propositions and revenue models. Additionally, open innovation and crowdsourcing techniques enable companies to co-create with external stakeholders, generating business models that emphasize collaboration and shared value creation. Platforms that utilize crowdsourcing to develop designs demonstrate how the FFE can support the development of decentralized, community-driven models that challenge traditional production paradigms (Afuah and Tucci, 2013).

In addition to disruptive innovations, the FFE also supports the development of hybrid models that integrate elements of both traditional and novel approaches. These models often arise in response to shifting consumer preferences, technological advancements, or regulatory pressures. For instance, the integration of subscription-based services into traditionally product-oriented industries represents a hybridization of business models that has been catalyzed by insights gained during the FFE.

For mature companies, the innovations generated during the FFE must be strategically integrated into their business model portfolios to maximize their impact. This requires a nuanced understanding of how new models interact with existing ones, as well as the ability to identify and exploit synergies across the portfolio. Globocnik et al. (2020) propose a formal control framework for managing business model portfolios, emphasizing the importance of alignment between strategic goals and operational capabilities. The integration process often involves iterating on initial FFE outputs to ensure their feasibility and scalability. This iterative approach enables companies to refine their business models based on real-world data and stakeholder feedback, thereby increasing the likelihood of successful implementation. Moreover, organizations must remain agile and responsive to emerging trends, as the dynamic nature of the FFE often results in the need for continuous adaptation and evolution of business models.

By embedding FFE-driven innovations into their business model portfolios, mature companies can enhance their resilience and adaptability in competitive markets. This integration not only supports incremental growth but also positions organizations to capitalize on disruptive opportunities that may reshape their industries.

3. IMPACT ON BUSINESS MODEL PORTFOLIO

In this section, we go deeper into the synergistic relationship between the Fuzzy Front End (FFE) in mature companies and their business model portfolios that we started to explore in section 2.3. For this, we rely on the formal control framework proposed by Globocnik et al. (2020), which offers a structured approach for companies to leverage their internal environments and strategic capabilities to innovate effectively (Figure 4). Building upon the discussions in Section 2, we adapt this framework to address the unique challenges faced by mature organizations, focusing on elements such as Internal Environment, Strategy, Business Model Portfolio, Tactics, and Operations. This extended framework serves as a blueprint for managing the complexities inherent in the FFE, ensuring alignment between strategic objectives and operational execution.

As organizations navigate increasingly complex and fast-paced markets, the challenge of managing multiple interrelated business models while ensuring alignment with strategic goals has become more pressing. A formal control framework provides mechanisms to maintain consistency between strategic objectives and the operationalization of business models, emphasizing the dynamics of business model portfolios and their responsiveness to environmental changes. This approach ensures that strategic planning and business model management are not siloed activities but are integrated processes that co-evolve in response to market dynamics.

3.1 Integrating Strategy and Business Model Portfolios

Traditionally, strategic planning and business model management have been treated as separate domains in both theory and practice. However, recent research underscores the importance of aligning these activities to ensure that strategy not only guides but also evolves alongside business models in response to environmental dynamics. A business model serves as an operative definition of strategy, translating high-level strategic objectives into specific activities related to value creation, delivery, and capture. Several authors highlight that a well-defined business model acts as a reflection of the firm's realized strategy, bridging the gap between strategic intent and operational execution (Casadesus-Masanell and Zhu, 2013).

The formal control framework proposed by Globocnik et al. (2020) (Figure 4) decomposes this alignment challenge into four interconnected levels: strategy, business model portfolio, tactics, and operations. At the strategic level, organizations determine their positioning within an industry by selecting target markets, bases of competition, and resource configurations.

At the business model portfolio level, these strategic choices are translated into a set of interrelated business models that collectively realize the organization's goals. This involves managing a diversified portfolio of business models to leverage existing resources and capabilities while exploring new opportunities. The concept of a diversified business model portfolio enables firms to increase long-term value for their shareholders and customers by leveraging existing resources. While companies need to build new capabilities for new business activities, these benefit the entire business portfolio, especially when comprised of

The tactics level involves the specific actions and initiatives undertaken to implement the business models, while the operations level focuses on the day-to-day activities that deliver value to customers. Each level is governed by distinct goals and decision-making processes but is interlinked through feedback and control loops. This interconnectedness ensures that adjustments at one level can propagate through the others, maintaining alignment and coherence across the organization.

By integrating strategy and business model portfolios through a formal control framework, mature companies can enhance their agility and responsiveness to environmental changes. This alignment ensures that strategic objectives are effectively operationalized, and that business models remain dynamic and adaptable in the face of evolving market conditions. Such an approach not only facilitates effective strategy execution but also fosters a culture of continuous innovation and improvement.

3.2 Managing Business Model Portfolios

The management of business model portfolios is essential for firms operating in diversified and dynamic environments. In the context of mature companies, a portfolio perspective allows for a strategic balance between exploiting existing business models and exploring new opportunities (Globocnik *et al.*, 2020). Managing multiple interrelated business models necessitates a structured approach that ensures resource optimization, risk mitigation, and strategic coherence. For instance, Amazon's diversified portfolio—including Amazon Prime, AWS, and Amazon Marketplace—demonstrates how an integrated approach to business model management can enable firms to maximize synergies while adapting to emerging opportunities and competitive pressures (Teece *et al.*, 2016).

Figure 4. Business model (BM) decision making levels

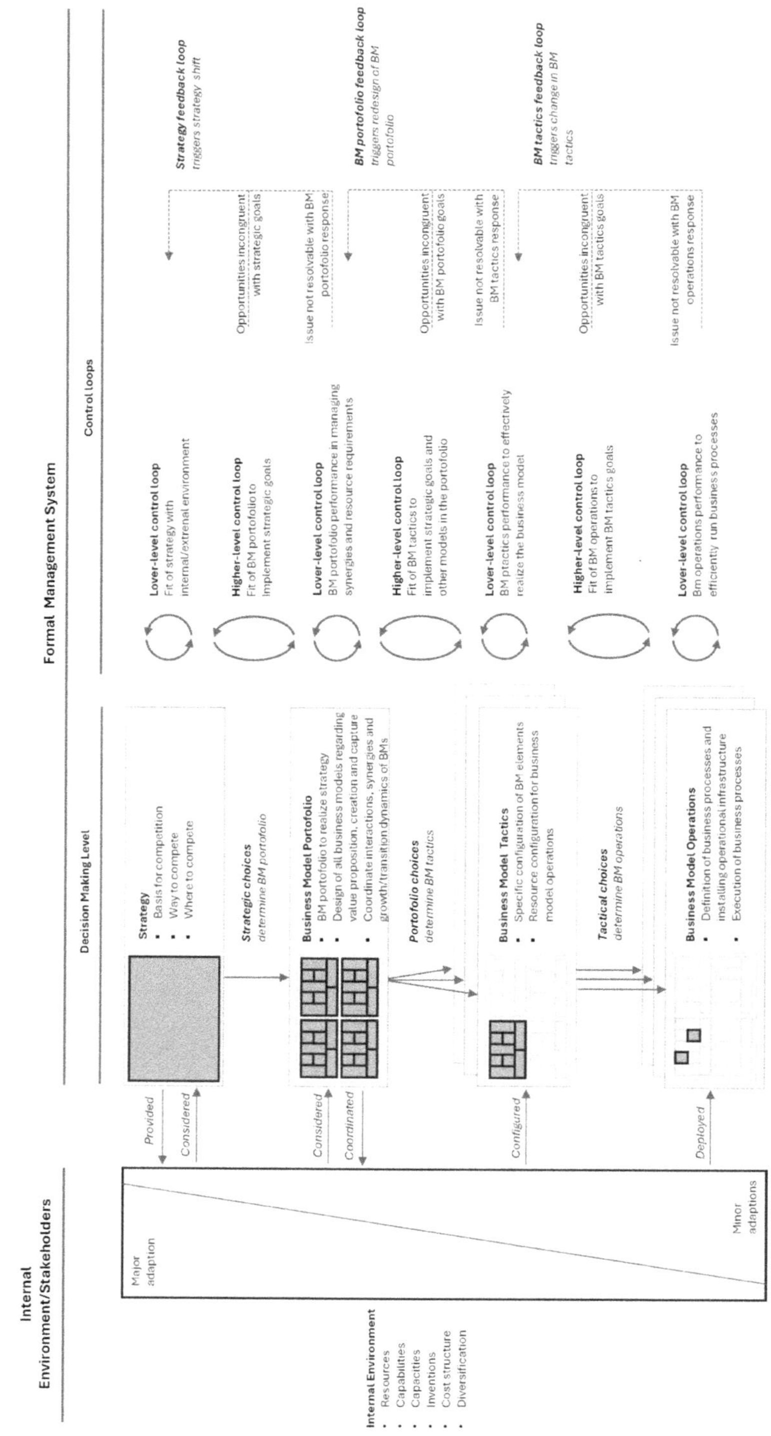

(source: Globocnik et al., 2020).

Effective portfolio management involves evaluating alternative models, ensuring strategic alignment, and dynamically reallocating resources across business units (Gassmann et al., 2016). This requires formal mechanisms to monitor external environmental changes and internal business model interactions. For example, Daimler's shift towards mobility-as-a-service through its car2go initiative highlights the need for coordinated resource management across traditional manufacturing and service-oriented business models (Globocnik et al., 2020). Such transitions illustrate how established firms can leverage existing capabilities while simultaneously incubating new models.

A key challenge in business model portfolio management is the balance between innovation and operational stability (Bohnsack *et al.*, 2014). Traditional firms often exhibit inertia, as business model innovation requires overcoming entrenched decision-making structures and legacy systems. A bifurcated strategy—where firms operate both their core business and an experimental innovation unit—can mitigate these challenges (Gassmann *et al.*, 2016). This dual approach enables companies to test new models without disrupting their existing operations, ensuring gradual adaptation and risk containment.

Business model portfolio management is also deeply linked to the Fuzzy Front End (FFE), as emerging ideas and business cases often dictate shifts in existing models or the development of new ones (Kim and Wilemon, 2002). The FFE acts as a strategic experimentation phase where organizations assess potential new business models before committing substantial resources (Neukam, 2017). Firms that excel in business model innovation leverage FFE insights to refine value propositions, revenue mechanisms, and operational structures before transitioning to full-scale implementation.

One prominent example of business model transformation informed by the FFE is Netflix's evolution from DVD rental services to a streaming platform and subsequently to a content producer (Parker *et al.*, 2016). This transition required Netflix to rethink its business model portfolio, reallocating resources from physical distribution to digital content creation and platform management. The FFE insights derived from user engagement metrics and emerging content consumption patterns enabled the company to adapt its strategic direction effectively.

Effective business model portfolio management ensures that mature firms remain competitive in dynamic markets. By linking the FFE to strategic decision-making, companies can navigate the complexity of business model transitions, leveraging AI-driven insights and structured frameworks to optimize resource allocation and enhance innovation outcomes. As business environments continue to evolve, firms must continuously reassess and refine their business model portfolios to maintain resilience and strategic agility.

3.3 Dynamics of Strategy and Business Models

The relationship between strategy and business models is not static but rather dynamic, shaped by external market forces, technological advancements, regulatory shifts, and evolving competitive landscapes. Strategic management frameworks traditionally emphasize long-term positioning and resource allocation, while business model theory focuses on the mechanisms through which firms create, deliver, and capture value. The interaction between these two domains becomes particularly complex in diversified firms managing multiple business models simultaneously (Johanson and Vahlne, 2009).

Business model dynamics refer to the continuous evolution, transformation, and adaptation of business models in response to internal and external pressures. Companies must balance the exploitation of existing business models with the exploration of new opportunities—a challenge that has become increasingly pronounced in fast-moving industries (Demil and Lecocq, 2010). Firms that fail to recognize the need for business model evolution risk obsolescence, while those that successfully integrate strategic foresight into their business model management gain a competitive edge (Fjeldstad and Snow, 2018).

Strategic adaptation plays a crucial role in ensuring that business models remain aligned with long-term corporate objectives. Companies frequently face inflection points that require them to reassess and adjust their business models. These strategic adjustments can take multiple forms:

- **Incremental Adaptation**: Small, continuous refinements that improve efficiency and enhance existing models without significant structural changes (e.g., Apple's iterative updates to its iPhone business model while maintaining its high-margin strategy).
- **Business Model Extension**: Expanding an existing model to serve adjacent markets or introduce complementary offerings (e.g., Tesla's expansion from electric vehicles into energy storage and solar solutions).
- **Radical Transformation**: Overhauling a business model to respond to disruptive forces or shifts in market demand (e.g., IBM's shift from hardware manufacturing to a service-oriented model centered on cloud computing and AI).
- **Business Model Diversification**: Developing entirely new business models alongside existing ones to hedge against industry uncertainties (e.g., Google's investment in both search engine advertising and cloud computing services).

These different adaptation pathways demonstrate the interplay between strategic intent and business model flexibility. A firm's ability to navigate these transitions depends on its capacity to anticipate industry shifts, allocate resources effectively, and engage in proactive learning processes (Burger-Helmchen and Llerena, 2012).

Business model evolution is not solely dictated by internal strategic choices but is also heavily influenced by external pressures and constraints. Some of the most significant factors shaping business model dynamics include:

- **Technological Innovation**: Advances in digitalization, artificial intelligence, and automation create new opportunities for business model innovation. Companies that fail to integrate emerging technologies into their business models risk being outcompeted by more agile players (Mariani & Dwivedi, 2024).
- **Regulatory and Policy Changes**: Government regulations, data privacy laws, and environmental policies often necessitate business model adjustments. For example, the shift toward sustainable practices has led automotive companies to develop electric and hybrid vehicle models to comply with emission regulations.
- **Market Demand and Consumer Behavior**: Shifts in customer preferences, driven by social trends or economic conditions, require firms to rethink their value propositions (Eling, 2020). The rise of the subscription economy, exemplified by businesses like Spotify and Netflix, reflects a fundamental shift in how consumers access and pay for services.
- **Competitive Pressures**: As industries become increasingly competitive, firms must continuously refine their business models to maintain differentiation. This often involves adopting new monetization strategies, streamlining operations, or leveraging partnerships to strengthen market positioning (Gassmann et al., 2016).

Internally, corporate culture, leadership vision, and organizational capabilities play a significant role in determining how effectively a company adapts its business models. Companies with rigid hierarchies and risk-averse cultures may struggle to pivot, whereas those that embrace experimentation and agility can more readily align their business models with evolving strategic priorities (Fjeldstad & Snow, 2018).

The following case studies illustrate how leading companies have successfully managed the dynamics between strategy and business models:

- ***Adobe's Transition to a Subscription-Based Model***: Adobe's shift from selling packaged software (Photoshop, Illustrator) to a cloud-based subscription service (Adobe Creative Cloud) exemplifies how strategic foresight can drive business model transformation. The transition enabled Adobe to generate recurring revenue, increase customer lifetime value, and improve software accessibility. This move also aligned with broader industry trends favouring software-as-a-service (SaaS) over one-time licensing (Mariani and Dwivedi, 2024).
- ***Spotify's Continuous Model Refinement***: Spotify has continuously adapted its business model by experimenting with freemium tiers, exclusive content, and podcast monetization strategies. These iterative changes reflect the company's commitment to aligning its revenue streams with evolving consumer preferences and competitive dynamics.
- ***Unilever's Portfolio Approach to Sustainability***: Unilever has integrated sustainability-focused business models into its corporate strategy, introducing purpose-driven brands like Dove's self-esteem initiatives and Ben & Jerry's activism-based campaigns. These models help Unilever differentiate its products while addressing regulatory and consumer expectations around corporate social responsibility.

For firms operating in dynamic industries, effectively managing business model evolution requires a structured approach that balances strategic intent with flexibility. Key strategic implications include:

- **Scenario Planning**: Organizations should leverage predictive analytics and scenario modeling to assess potential disruptions and preemptively adjust business models.
- **Modular Business Model Design**: Designing business models that can be easily adapted or reconfigured in response to market shifts increases resilience and scalability.
- **Organizational Learning and Agility**: Embedding a culture of continuous learning, rapid experimentation, and cross-functional collaboration enables firms to pivot when necessary without destabilizing existing operations.

By embracing business model evolution as an ongoing strategic process rather than a one-time adjustment, firms can sustain long-term competitive advantage and remain responsive to industry changes. As the intersection between strategy and business model management continues to evolve, organizations must cultivate dynamic capabilities that allow them to anticipate, navigate, and capitalize on emerging opportunities.

3.4 Formal Control Mechanisms

The successful management of business model portfolios necessitates robust control mechanisms to ensure alignment between innovation initiatives and overarching strategic objectives. Control mechanisms serve to mitigate risks, optimize resource allocation, and enhance adaptability, all of which are crucial for companies operating in dynamic environments. The integration of formal control frameworks with creative innovation processes remains a complex but necessary endeavour, requiring firms to balance structured oversight with entrepreneurial flexibility (Bollinger and Burger-Helmchen, 2021).

Control mechanisms in business model innovation function across multiple dimensions, including strategic alignment, financial oversight, performance measurement, and organizational adaptation. Traditional management control systems focus on financial metrics, yet these alone are insufficient in the context of innovation-driven business models. A more holistic approach incorporates diagnostic, interactive, and cultural controls to guide business model evolution while maintaining strategic coherence (Simons, 1995).

- ***Diagnostic Control Systems***: These focus on tracking performance indicators and ensuring that business models meet predefined financial and operational targets. Organizations like Amazon employ data-driven analytics to continuously monitor the effectiveness of their evolving business models, adjusting pricing strategies and service offerings based on real-time insights (Globocnik et al., 2020).
- ***Interactive Control Systems***: These emphasize collaboration and adaptability, fostering an iterative approach to business model refinement. Spotify, for example, engages in frequent strategy reviews that incorporate user feedback, competitor analysis, and market experimentation to fine-tune its revenue model (Mariani & Dwivedi, 2024).
- ***Cultural Controls***: These leverage corporate values, leadership vision, and organizational norms to encourage innovation while maintaining cohesion. Google's 20% time policy, which allows employees to dedicate a portion of their work hours to creative projects, exemplifies how cultural control mechanisms can facilitate business model innovation (Bollinger and Burger-Helmchen, 2021).

A significant challenge in managing business model innovation is determining the optimal balance between formal and informal control mechanisms (Bollinger and Burger-Helmchen, 2021). Formal control mechanisms, such as budgetary constraints, performance metrics, and governance structures, provide accountability and risk mitigation but can also stifle creativity if applied too rigidly. Informal controls, such as peer evaluations, leadership guidance, and cross-functional collaboration, offer greater flexibility but require strong organizational trust to function effectively (Chesbrough, 2006).

- ***Formal Control Mechanisms***: These are particularly relevant in highly regulated industries, such as pharmaceuticals and finance, where compliance with external regulations necessitates rigorous oversight (Gassmann et al., 2016). Companies like Novartis employ structured innovation funnels to assess the feasibility of emerging business models while ensuring regulatory adherence.
- ***Informal Control Mechanisms***: Tech firms and creative industries, such as Tesla and Pixar, rely heavily on informal control mechanisms to nurture disruptive business models. Tesla's approach

to continuous iteration, for instance, involves decentralized decision-making where engineers and designers operate with significant autonomy within broad strategic guidelines.

The transition between business models presents a critical inflection point for companies, requiring a structured yet flexible control framework. Business model transitions typically involve significant resource reallocation, stakeholder engagement, and strategic recalibration. Three core control mechanisms facilitate effective transitions:

- ***Resource Allocation and Investment Controls***. Firms must ensure that emerging business models receive adequate funding without jeopardizing existing operations. Adobe's move from a software licensing model to a cloud-based subscription service required gradual investment shifts, ensuring financial stability during the transition (Mariani & Dwivedi, 2024).
- ***Stakeholder Management and Strategic Communication***. Engaging internal and external stakeholders is critical when shifting business models. When Netflix transitioned from DVD rentals to streaming and later into content production, it had to redefine its partnerships with content providers, technology firms, and customers while managing resistance from legacy stakeholders (Demil & Lecocq, 2010).
- ***Continuous Performance Monitoring and Iterative Refinement***. Business model innovation is rarely a one-time event; it requires continuous refinement. Amazon's multi-stage expansion—from online retail to AWS, Prime memberships, and physical stores—demonstrates how firms can use feedback loops to iteratively adjust and optimize their business models (Bucherer *et al.*, 2012)

Control mechanisms play a crucial role in helping firms achieve organizational ambidexterity, the ability to balance exploration of new business models with exploitation of existing ones (Tushman and O'Reilly, 1996). Successful business model management hinges on the simultaneous pursuit of efficiency and adaptability, ensuring that firms do not become overly reliant on either incremental improvements or radical shifts (Gibson and Birkinshaw, 2004).

- ***Exploratory Controls***: Encourage experimentation, flexibility, and risk-taking. Companies such as SpaceX operate with high degrees of uncertainty, necessitating loose control mechanisms that allow for rapid iteration and bold innovation experiments.
- ***Exploitative Controls***: Focus on efficiency, standardization, and scalability. McDonald's, for example, employs highly standardized control mechanisms to ensure the consistency of its business model across global markets while introducing localized adaptations where necessary (Gassmann et al., 2016).

The successful implementation of control mechanisms requires a nuanced understanding of industry dynamics, organizational structure, and corporate strategy. Several best practices emerge from the literature:

- ***Scenario Planning and Predictive Analytics***: Leveraging data-driven forecasting tools enables firms to anticipate potential disruptions and proactively adjust their business models.
- ***Decentralized Decision-Making with Centralized Oversight***: Allowing autonomy at operational levels while maintaining centralized strategic control fosters innovation without compromising alignment.

- ***Modular Business Model Design***: Creating adaptable business models that can be easily reconfigured in response to market shifts enhances resilience and scalability (Gassmann et al., 2016).

By embedding effective control mechanisms within their innovation processes, companies can navigate complex business model transitions, sustain competitive advantage, and enhance long-term resilience. The ability to dynamically adjust control frameworks in response to evolving market conditions will remain a critical competency for firms striving to manage business model innovation successfully.

3.5 Implications for Mature Companies

For mature companies with established business model portfolios, maintaining competitiveness requires a structured approach that integrates strategic planning with business model management. This integration enables firms to navigate complexities related to resource allocation, stakeholder coordination, and environmental uncertainty. By aligning their business models with long-term objectives, mature firms can simultaneously refine existing operations and explore new market opportunities, ensuring they remain adaptable in dynamic industries.

Mature firms often operate in industries characterized by regulatory scrutiny, operational standardization, and entrenched competition. While these factors contribute to stability, they can also limit the flexibility required for business model transformation. Managing a business model portfolio helps firms align innovation efforts with existing operational capabilities, allowing them to introduce new models without disrupting core revenue streams. Companies like Philips and Schneider Electric have successfully transitioned from hardware-centric businesses to service-oriented models by leveraging their technological expertise and developing complementary digital solutions (Sosna *et al.*, 2010).

One of the significant challenges for mature firms is overcoming internal resistance to change. Large organizations often develop rigid structures and a preference for operational efficiency over experimentation, which can hinder the adoption of new business models. Governance frameworks that encourage structured experimentation, combined with cross-functional collaboration, facilitate the integration of innovation while preserving operational stability (Neukam and Guittard, 2018). The ability to balance tradition with innovation was exemplified by Bosch, which integrated smart technology into its traditional manufacturing business, allowing it to compete in the emerging IoT market while maintaining its core engineering focus (Winterhalter *et al.*, 2017).

Resource allocation plays a central role in managing business model portfolios, particularly in environments where financial and human capital are constrained. Decision-making processes must align investment priorities with long-term strategic goals while minimizing risks. Companies must continuously evaluate which business models to expand, which to maintain, and which to phase out based on market trends and organizational capabilities. Cisco's pivot from networking hardware to a cloud and cybersecurity-focused business model underscores the importance of directing resources toward emerging growth areas while gradually reducing reliance on legacy revenue streams (Amit and Zott, 2012).

Managing interdependencies between business models also strengthens a firm's resilience against external shocks. Diversification within a business model portfolio helps firms hedge against market volatility and technological disruption, a strategy particularly critical in industries with rapid regulatory and technological shifts. In the financial services sector, firms like JPMorgan Chase have expanded their digital banking services while maintaining traditional banking operations, ensuring they remain competitive against fintech challengers (Stefani *et al.*, 2019).

Strategic flexibility is essential for long-term growth, allowing firms to test, scale, and retire business models as market conditions evolve. The ability to experiment with new offerings while maintaining stability in core business areas enables mature firms to remain competitive in uncertain environments. L'Oréal's expansion into direct-to-consumer digital channels while sustaining its established retail partnerships illustrates how firms can dynamically manage business models to capture new opportunities without disrupting existing operations (Markides and Charitou, 2004).

By integrating formal control mechanisms into their business model portfolios, mature companies can systematically navigate transformation, ensuring innovation efforts are both effective and sustainable. This structured approach not only safeguards long-term market positioning but also enables firms to continuously adapt to emerging opportunities in a rapidly changing economic landscape.

4. THE TRANSFORMATIVE IMPACT OF AI ON INNOVATION DRIVERS, THE FUZZY FRONT END, AND BUSINESS MODEL PORTFOLIO MANAGEMENT

The integration of artificial intelligence into business model innovation and strategic management has become an essential factor in maintaining a competitive edge in rapidly evolving markets. AI is increasingly being recognized as a key enabler in optimizing FFE of innovation, enhancing decision-making processes, and managing business model portfolios more effectively. As organizations face mounting pressure to adapt to digital transformation, AI-driven solutions offer the potential to refine market analysis, forecast industry trends, and streamline innovation pathways (Madsen and Brix, 2024).

A growing number of companies are adopting AI technology, while managers seek to identify the most promising business models (Burström *et al.*, 2021; Coskun-Setirek and Tanrikulu, 2021). As underlined by Jorzik *et al.* (2024) BMI often emerges as an evolutionary by product of AI implementation.

AI contributes significantly to improving the efficiency of innovation ecosystems by reducing uncertainty in the early phases of new product development. AI-based predictive analytics allow firms to harness extensive data sources to evaluate market potential and technological feasibility at an unprecedented scale (Chattopadhyay *et al.*, 2024). Additionally, machine learning algorithms can accelerate the ideation process by identifying emerging trends and patterns that may not be immediately apparent to human analysts. The role of AI extends beyond data processing to include strategic applications, such as optimizing business model transitions, automating decision-making frameworks, and enhancing portfolio-level synergies (Csaszar et al., 2024).

As organizations increasingly rely on AI-powered innovation processes, it is essential to examine the mechanisms through which AI transforms the FFE and how it aligns with business model portfolio management. AI-enabled strategic planning tools can help firms navigate complex market environments, mitigate risks, and uncover opportunities for sustainable growth. This section explores these implications, focusing on how AI enhances innovation drivers, strengthens business case development, and facilitates the strategic alignment of business model portfolios.

This study adopts the widely accepted definition by Kaplan and Haenlein (2019, p. 17), which describes AI as a system capable of correctly interpreting external data, learning from it, and using these learnings to achieve specific goals and tasks through flexible adaptation. In addition to definitional variations, researchers classify AI according to different criteria. Some distinguish AI based on its level of intelligence, such as strong versus weak AI (Lee *et al.*, 2019), while others focus on its applications in business, differentiating between assisted, augmented, and autonomous intelligence (Garbuio and Lin, 2019).

The number of studies examining the relationship between AI and business model innovation has grown significantly in recent years (Burström *et al.*, 2021; Mishra and Tripathi, 2021). These contributions adopt diverse analytical perspectives, ranging from narrowly focused process improvements to large-scale industry disruptions. In our work, we specifically consider the role of generative AI in business model innovation, acknowledging its potential to transform value creation, delivery, and capture mechanisms. Given the evolving nature of AI, different perspectives within the literature may overlap, and transformative innovations driven by generative AI could lead to disruptive industry changes, albeit with a potential time lag (Warner and Wäger, 2019).

4.1 AI and Innovation Drivers

Artificial intelligence has emerged as a transformative force in reshaping innovation processes, particularly within the early-stage activities of the Fuzzy Front End. AI-driven tools enhance an organization's ability to identify, evaluate, and pursue new opportunities by leveraging vast amounts of real-time data. Advanced machine learning models improve the accuracy of market forecasts, optimize decision-making processes, and help firms anticipate shifts in consumer preferences, regulatory landscapes, and technological advancements (Felin and Holweg, 2024).

One of the most significant contributions of AI to innovation is its capacity to process and synthesize large volumes of unstructured data, which traditional analytical methods struggle to manage. AI-driven natural language processing (NLP) algorithms extract actionable insights from patents, scientific publications, social media discourse, and industry reports, providing firms with a competitive advantage in knowledge acquisition and trend detection (Csaszar *et al.*, 2024). This capability allows companies to stay ahead of competitors by identifying new business opportunities at an accelerated pace.

AI also enhances the creative dimension of innovation. While traditional models of business innovation rely heavily on human intuition and experience, AI-assisted generative design tools introduce new perspectives by exploring potential product configurations, business strategies, and process optimizations beyond human cognitive limitations. AI-facilitated creativity is particularly valuable in industries where disruptive innovations emerge rapidly, requiring organizations to explore unconventional solutions and optimize existing business models.

Moreover, AI significantly influences strategic decision-making by improving risk assessment models. Predictive analytics and simulation-based approaches allow firms to evaluate multiple business scenarios and assess potential risks before committing to large-scale investments. AI's ability to enhance risk management strategies enables firms to allocate resources more efficiently, reducing the likelihood of costly failures in the innovation process.

Another key function of AI in innovation management is its role in enhancing collaboration within ecosystems. AI-powered platforms facilitate knowledge-sharing between firms, universities, research institutions, and startups, fostering co-innovation and reducing redundancy in R&D efforts (Chattopadhyay *et al.*, 2024). By integrating AI-enabled collaboration tools, organizations can accelerate the time-to-market for new products while minimizing development costs.

Finally, AI contributes to optimizing the alignment between business models and market dynamics. As AI continues to evolve, its integration into strategic innovation processes will play an increasingly central role in determining how firms adapt to rapidly changing industries. By incorporating AI-driven insights into the FFE, businesses can refine their innovation strategies, improve the efficiency of decision-making, and enhance their overall competitive positioning. These advancements highlight the transformative impact of AI on innovation ecosystems, reinforcing its status as a critical driver of business model evolution.

4.2 Revolutionizing the Fuzzy Front End and Business Case Formulation Phase

Artificial intelligence has revolutionized the Fuzzy Front End (FFE) by significantly enhancing the processes of ideation, concept evaluation, and business case formulation. AI technologies enable firms to conduct real-time assessments of market opportunities, consumer preferences, and technological feasibility with unprecedented precision. These capabilities allow organizations to make data-driven decisions, reducing uncertainties and improving the strategic alignment of innovation initiatives (He and Burger-Helmchen, 2024).

One of the fundamental ways AI contributes to the FFE is through predictive analytics. By leveraging machine learning models trained on historical market data, AI can forecast demand trends, detect early signals of emerging technologies, and identify gaps in existing product offerings. These insights enable firms to refine their initial concepts, ensuring that innovation efforts align with market needs and strategic objectives (Burger-Helmchen, 2024). Moreover, AI-powered tools help firms assess potential barriers to market entry, providing early warnings about regulatory compliance issues, intellectual property risks, and competitive threats. These insights allow companies to modify their innovation strategies proactively, avoiding costly missteps in product and business model development (He and Burger-Helmchen, 2024).

AI also enhances the efficiency of business case formulation by automating key analytical processes. Traditionally, developing a business case required extensive market research, financial modeling, and stakeholder analysis—processes that were often time-consuming and prone to biases. AI-driven platforms, however, can rapidly synthesize data from diverse sources, offering comprehensive business case assessments that integrate real-time economic indicators, competitor strategies, and regulatory trends (He and Burger-Helmchen, 2024). Additionally, AI-powered scenario modeling enables firms to simulate various business conditions, helping decision-makers evaluate the robustness of different innovation pathways under fluctuating market conditions.

Beyond predictive capabilities, AI plays a role in generating novel ideas and exploring creative solutions. Generative AI tools use deep learning models to simulate potential product configurations, alternative business models, and consumer engagement strategies. This enhances the creative process by providing a broader range of viable options, thereby expanding the innovation horizon for mature firms (Burger-Helmchen, 2024). AI's ability to recognize hidden connections between disparate market signals allows for the identification of novel value propositions, helping firms develop innovative products and services that address unmet consumer needs.

AI-driven simulations allow firms to test the feasibility of new concepts before committing to development. Digital twins, for example, create virtual representations of products and business models, enabling real-time experimentation and performance analysis under different market conditions. This iterative approach significantly improves decision-making accuracy while minimizing financial risks. Through digital prototyping, AI facilitates a faster and cost-effective validation of new business ideas, reducing the time required to bring innovations from conceptualization to commercialization.

AI also strengthens collaboration in the FFE by facilitating knowledge integration across different departments and external partners. AI-powered platforms enable organizations to tap into global innovation ecosystems, connecting R&D teams with academic institutions, startups, and industry experts. By streamlining access to collective intelligence, AI enhances the robustness of business case development and increases the likelihood of successful innovation outcomes even in difficult environments (Irissarry and Burger-Helmchen, 2024). Additionally, AI enhances cross-functional teamwork by automating knowledge-sharing processes, enabling seamless communication between different innovation stakeholders.

The role of AI in shaping business model innovation extends to optimizing resource allocation for early-stage projects. By leveraging AI-driven decision-support systems, firms can allocate resources more efficiently, prioritizing projects with the highest potential for success. AI enhances capital allocation decisions by identifying synergies between new innovations and existing business models, ensuring that firms maximize the value of their investments while minimizing operational redundancies. AI-assisted funding models also optimize R&D expenditure by dynamically adjusting budgets based on real-time project performance indicators.

As AI continues to evolve, its role in the FFE will become even more prominent. AI's integration into business case development will further refine opportunity assessment methodologies, reducing the cognitive biases that often affect human-driven innovation decisions. The increasing sophistication of AI algorithms will lead to improved predictive accuracy, allowing firms to anticipate market shifts with greater confidence. Additionally, AI-driven optimization frameworks will enhance the strategic alignment between innovation initiatives and overarching corporate objectives, ensuring that early-stage projects contribute to long-term business success.

The broader implications of AI's role in the FFE suggest that firms that leverage AI-driven tools will gain a significant competitive advantage in managing early-stage innovation. The ability to process vast amounts of data, identify emerging trends, optimize resource allocation, and test business models before full-scale deployment positions AI as a transformative force in business model innovation. As organizations continue to refine their AI capabilities, the synergy between AI and FFE will drive more efficient, data-driven, and strategically aligned innovation processes. This transformative impact underscores the importance of AI as a core enabler of business model adaptation and sustainable competitive advantage.

Studying AI as an innovation driver, Jorzik *et al.* (2024) proposed a matrix illustrating the impact of AI on BMI. Reworking this matrix, we propose a version where AI influences business cases, attracting attention during the FFE.

Table 1. AI to shape new business cases (adapte from Jorzik et al., 2024)

Fuzzy Front End Business Case →	**Evolutionary (AI as an enabler of adaptation of Business case)**	**Deliberate (AI actively shaping new Business Cases)**
Peripheral Role of AI in Value Creation	**Technology-Supported Business Cases** AI enhances efficiency in existing business cases without fundamentally altering value creation. AI assists in evaluating potential business opportunities but does not autonomously generate new cases.	**AI-Assisted Business Case Refinement** AI provides structured support in refining and testing business cases, leveraging data-driven insights to assess feasibility and market fit.
Core Role of AI in Value Creation	**Emergent AI-Driven Business Cases** AI-enabled operations gradually shape new business cases as firms explore AI applications. Business opportunities emerge naturally as AI-driven efficiencies reveal new strategic possibilities.	**AI-Generated Business Case Formation** Generative AI plays a central role in ideation and evaluation, autonomously proposing and validating new business cases based on market trends, customer needs, and operational data.

AI-driven business case development can be classified along two primary dimensions: the role of AI in value creation (peripheral used in addition to other tools by humans vs. core, as an almost autonomous simulation engine) and the extent to which AI deliberately shapes business cases (evolutionary vs. deliberate). These dimensions result in four AI-enabled business case perspectives: technology-supported business cases, AI-assisted business case refinement, emergent AI-driven business cases, and AI-generated business case formation.

Technology-supported business cases emerge in contexts where AI primarily functions as an analytical tool to enhance efficiency but does not fundamentally alter the strategic approach to business case development. In this perspective, AI assists managers by processing large volumes of market and operational data, enabling a more informed assessment of potential business opportunities. However, AI remains a supportive mechanism rather than a generative force. For example, AI-driven analytics platforms can evaluate emerging market trends, customer behaviours, and financial feasibility, helping decision-makers refine business case parameters. Despite its utility, AI in this context does not autonomously propose new business models; rather, it aids in optimizing pre-existing managerial hypotheses.

AI-assisted business case refinement represents a more interactive role for AI in the fuzzy front end of innovation. Here, AI is leveraged not only for analytical support but also for structured decision-making, allowing organizations to refine and test business case scenarios dynamically. This perspective recognizes AI's potential in simulating market conditions, forecasting demand, and running scenario analyses to evaluate business model feasibility before commitment. By systematically assessing different strategic alternatives, AI enables firms to adjust and optimize business cases based on real-time insights. This process significantly reduces the risk of uncertainty associated with early-stage decision-making, providing organizations with a more structured foundation for investment and resource allocation.

Emergent AI-driven business cases occur when firms implement AI primarily for operational or efficiency-related improvements but, in doing so, uncover entirely new business opportunities. Unlike the previous perspectives, where AI is used as a deliberate tool in decision-making, emergent AI-driven business cases arise organically as a byproduct of AI implementation. Organizations may initially integrate AI to enhance customer service, optimize supply chains, or streamline internal processes, only to realize that AI-generated efficiencies create new market possibilities, revenue streams, or strategic advantages. This perspective aligns with the concept of serendipitous discovery, where business case evolution is not intentionally planned but rather emerges from AI's ongoing interaction with operational environments.

AI-generated business case formation represents the most transformative role of AI in the Fuzzy Front End of innovation. In this perspective, Generative AI actively participates in ideation, evaluation, and validation of entirely new business models. AI autonomously generates potential business cases by identifying gaps in the market, assessing technological feasibility, and predicting competitive advantages. Through advanced simulations, AI can test multiple business case variations, adjusting key parameters such as pricing strategies, customer segmentation, and partnership models. This data-driven approach significantly enhances decision-making speed and accuracy, allowing firms to explore a broader range of opportunities with reduced uncertainty.

We believe that the impact of AI-driven business case development extends beyond individual projects and into broader product portfolio management. Traditional portfolio decisions rely on static evaluations of potential projects, often constrained by historical performance data and incremental improvements. AI, however, introduces a more dynamic approach to portfolio management by continuously analyzing market shifts, customer preferences, and technological advancements. By integrating AI-generated business cases into product portfolio decision-making, organizations can achieve greater flexibility and responsiveness to emerging opportunities.

4.3 AI as a Bridge Between the Fuzzy Front End and New Product Development

FFE and new product development (NPD) serve distinct roles in the innovation process. At the beginning of FFE, business case generation is crucial, as firms explore opportunities, define value propositions, and assess feasibility. This phase is highly exploratory, relying on creativity and strategic foresight. In contrast, NPD follows a more structured approach, using traditional marketing assessment techniques such as consumer research, competitive analysis, and financial forecasting to refine and commercialize innovations. As the transition from one phase to the other represents important managerial challenges (Markham *et al.*, 2010), both are subject to AI enhancement.

Mariani and Dwivedi (2024) explore the implications of GenAI for NPD, emphasizing its potential to reshape existing frameworks. They highlight that traditional NPD models aim to balance three key objectives: minimizing development cycle time, maximizing alignment with customer needs, and compressing development costs. According to their findings, GenAI has the potential to simultaneously achieve these goals without major trade-offs by accelerating research and development, enabling real-time product testing, and reducing costs through digital prototyping.

They further discuss how GenAI can enhance established NPD methodologies such as stage-gate processes or quality function deployment, and computer-aided design and manufacturing. GenAI can augment these processes by refining creative brainstorming, optimizing prototyping through digital rendering and 3D printing, and validating business models in real time. These applications build on industry research that highlights the relevance of GenAI in design and development (Verganti *et al.*, 2020). Verganti et al. (2020) argued that as algorithms play an increasingly prominent role in creative problem-solving, human design shifts toward an act of sense making, where individuals focus on identifying and defining which problems should or could be addressed.

Mariani and Dwivedi (2024) also examine how GenAI can transform product testing by making it more responsive to dynamic consumer preferences. They cite previous studies that emphasize the role of AI in improving market responsiveness (Mariani and Fosso Wamba, 2020). Additionally, they suggest that GenAI can be used to simulate customer personas and predict product adoption, contributing to the emerging field of innovation analytics (Mariani and Nambisan, 2021).

Those works on how the integration of AI into innovation processes transforms the transition from FFE ideation to structured new product development (NPD) leads us to table 2. The FFE is characterized by high uncertainty, weakly defined opportunities, and unstructured exploration, whereas NPD focuses on refining, testing, and launching concrete innovations. AI's progressive information processing capabilities—ranging from exploiting existing knowledge to exploring new frontiers—enable a structured yet flexible approach to innovation management.

Table 2. AI's Role in Bridging the Fuzzy Front End (FFE) and New Product Development (NPD), inspired by Haefner et al. (2021).

AI Information Processing Capability	Role in the Innovation Process	Application in the Fuzzy Front End (FFE)	Application in New Product Development (NPD)
Level 3: Exploring	AI autonomously identifies novel patterns, reframes problems, and generates unconventional ideas.	Overcomes human cognitive biases by identifying unexpected insights beyond conventional knowledge domains, fostering disruptive innovation.	Supports radical innovation by exploring alternative design approaches, reconfiguring product concepts, and proposing novel market opportunities.
Level 2: Expanding	AI expands the innovation space by refining ideas, identifying connections across domains, and optimizing search processes.	Enhances idea selection by filtering and clustering diverse ideas, increasing the likelihood of identifying viable innovation pathways.	Improves feasibility analysis by synthesizing diverse data sources, supporting decision-making in early-stage product development.
Level 1: Exploiting	AI optimizes and accelerates structured innovation processes by processing large datasets efficiently	Increases efficiency in evaluating early-stage concepts by analyzing market trends and historical data.	Enhances refinement, testing, and optimization of product prototypes, improving cost-efficiency and perfo

- At the highest capability level (**Exploring**), AI extends ideation beyond human cognitive limitations, identifying disruptive opportunities that challenge conventional industry assumptions. This is particularly valuable in the early FFE stage, where ambiguity is high, and breakthrough innovations require radical shifts in perspective.
- At the **Expanding** level, AI bridges FFE and NPD by structuring idea selection and refining concept development. By clustering, ranking, and filtering ideas, AI reduces uncertainty and supports decision-making, allowing firms to prioritize the most promising innovation pathways.
- Finally, at the **Exploiting** level, AI enhances later-stage NPD by optimizing prototype testing, refining market fit, and streamlining commercialization strategies. This ensures that innovations generated in the FFE are effectively transformed into market-ready products.

By systematically overcoming information processing constraints and local search limitations, AI could create a seamless transition from open-ended exploration to structured product development, enhancing both creativity in ideation and efficiency in execution.

4.4 Business Model Portfolio Management

Finally, we discuss how AI improves BM portfolio management. In the realm of business model portfolio management, AI introduces sophisticated analytical tools that enable companies to simulate the impact of different strategic choices and market scenarios on their business models. This predictive capability allows for dynamic portfolio management, where business models can be adapted in real-time based on emerging data and strategic insights. Furthermore, AI can optimize resource allocation across the portfolio, ensuring that investments are directed towards the most promising innovations with the highest potential for value creation (Pietronudo *et al.*, 2022).

Table 3. FFE with AI (source: authors)

Element	FFE with AI (New Product Development)	FFE with AI (Extensive Product Portfolio)
Time-to-Market	AI accelerates ideation and prototyping, reducing the time needed to move from concept to market.	AI optimizes product lifecycle management, reducing inefficiencies in launching iterative product updates.
Business Case Formulation	AI-driven analytics assess market feasibility and predict customer adoption rates with greater accuracy.	AI evaluates the viability of integrating new products into existing lines, improving strategic fit.
Resource Allocation	AI optimizes investment in R&D, ensuring that only high-potential concepts receive funding.	AI dynamically reallocates resources between existing products and potential innovations.
Market Forecasting	Machine learning models predict market trends and emerging demands, enhancing strategic decisions.	AI continuously monitors competitor movements and consumer trends, guiding portfolio adjustments.
Risk Assessment	AI simulations assess potential risks and regulatory compliance issues early in the development phase.	AI enhances risk analysis for managing multiple products, ensuring alignment with evolving industry standards.
Product Differentiation	Generative AI enhances creativity by suggesting novel configurations and features for new products.	AI suggests variations and improvements for existing products based on consumer feedback and data analytics.
Portfolio Management	AI assists in determining whether a new product fits within the company's business model strategy.	AI helps balance product introductions and discontinuations to maintain an optimized portfolio.
Customer Insights	AI tools analyze consumer behavior to identify unmet needs and tailor product features accordingly.	AI-driven sentiment analysis informs adjustments to product marketing and development strategies.

Table 3 illustrates the transformative role of artificial intelligence in optimizing the FFE for both new product development and companies managing an extensive product portfolio. AI accelerates the time-to-market by streamlining ideation and prototyping for entirely new products, while in firms with existing product portfolios, AI enhances product lifecycle management and reduces inefficiencies in iterative updates. The business case formulation process benefits from AI's advanced predictive analytics, which evaluate market feasibility and anticipate customer adoption trends with greater accuracy. Additionally, AI optimizes resource allocation, ensuring that high-potential innovations receive priority while maintaining an equilibrium between sustaining existing products and introducing new ones.

AI also plays a crucial role in market forecasting by continuously monitoring competitor movements and detecting emerging trends, enabling firms to refine their strategic direction. Risk assessment is significantly improved through AI simulations, which evaluate potential pitfalls in regulatory compliance and market dynamics before committing to large-scale product development. In the context of product differentiation, AI suggests novel design configurations and business models that maximize competitive advantage. For companies with extensive product portfolios, AI contributes to portfolio management by ensuring an optimal balance between introducing new offerings and discontinuing underperforming ones. Furthermore, AI-driven customer insights provide a data-backed foundation for refining product positioning and marketing strategies.

5. CONCLUSION AND FUTURE DIRECTIONS

This chapter has synthesized insights from various scholarly contributions to examine the intersection of business model innovation and the Fuzzy Front End (FFE) in mature companies. The analysis has highlighted the central challenge these firms face in linking structured business model portfolios with the uncertain and exploratory nature of early-stage innovation. While no new empirical data was introduced, this synthesis of existing research has provided a comprehensive understanding of how firms navigate innovation management in dynamic environments. The discussion has underscored that while AI-driven tools and formal control frameworks offer significant advantages in optimizing business model adaptation, challenges remain in aligning disruptive innovation with existing corporate structures.

One of the key insights from this chapter is the transformative role of AI in enhancing decision-making and risk management within the FFE. AI-powered predictive analytics, generative design tools, and automated decision-support systems enable firms to accelerate the ideation process, improve business case formulation, and enhance strategic alignment. AI has also proven to be particularly valuable for companies managing extensive product portfolios, where it facilitates real-time portfolio optimization, resource allocation, and risk assessment. These technological advancements indicate that firms leveraging AI effectively can reduce market uncertainty, improve the success rate of new product development, and ensure better synchronization between innovation processes and business model evolution.

Despite these advancements, our analysis has also revealed persistent gaps in the way mature firms integrate disruptive innovation into their existing business frameworks. Many established organizations exhibit structural inertia, where rigid operational models and ingrained decision-making hierarchies hinder the adoption of radical business model transformations. While AI and digitalization can provide predictive insights, organizational resistance to change often limits the impact of these tools. This resistance is not only linked to the introduction of a new technology such as AI into the team functioning. Using AI as a creative and idea generating tool also risks to be perceived as a competing factor to the innovation teams, strengthening the Not-invented-here-Syndrome for idea adoption (Hussinger and Wastyn, 2016).

Furthermore, although AI contributes to data-driven strategic planning, it cannot fully replace the human creativity and intuition that drive breakthrough innovations. This interplay between automation and human ingenuity remains a critical area for further investigation.

Future research should explore how mature companies can cultivate greater organizational flexibility to better absorb and integrate disruptive innovations. One promising avenue is the study of ambidextrous organizations—firms that successfully balance exploitation of existing business models with exploration of new ones. Understanding how these companies structure decision-making, allocate resources, and foster a culture of continuous experimentation can offer valuable insights for firms struggling to break free from structural rigidity.

Empirical studies are needed to assess the long-term impact of AI on business model adaptation. While theoretical frameworks suggest that AI enhances business model innovation, real-world case studies analyzing the effectiveness of AI-driven decision-making in different industry contexts would provide valuable practical insights. Comparative analyses across industries could reveal sector-specific challenges and best practices for integrating AI into FFE processes.

Another critical area for future exploration is the relationship between sustainability-driven business models and AI-enhanced innovation management. Since its emergence and widespread application, AI solutions are confronted to increasing criticism about related energy consumption – which represents a counter evolution regarding worldwide CO2 emission reduction goals. Similar as in other energy intensive sectors (Bollinger and Schaeffer, 2024; Henderson, 2020), current observations seem to suggest that the global costs of AI including the potential negative environmental impact are not transparent and therefore not sufficiently considered by strategic management. Coupled with an increasing dependence on AI technologies throughout entire business processes, these costs risk being transferred to the consumers as most important end-consumers, therefore negatively impacting their financial strategy. We suggest future research to focus on potential trade-offs between increasing energy consumption due to AI technologies and efficiency gains for firms through AI-generated solutions. From this point of view, while AI has proven effective in optimizing efficiency and resource allocation, its potential to support sustainable innovation remains underexplored. Future studies could examine how AI can assist firms in aligning economic, social, and environmental objectives within their business model portfolios, ensuring that innovation contributes to long-term societal benefits. Here, it would be necessary to analyse the value of human capital compared to AI's pure technology dimension as well as overall potential critical substitutions through AI of human intuition on a firm's processes. Also, from a societal point of view, it will be necessary to take into account AI's impacts on labour market evolutions which might be counterintuitive to the common good of societies.

Finally, while AI provides new tools for business model innovation, it remains a facilitator rather than a substitute for strategic leadership. Companies must cultivate innovation-driven leadership that embraces AI not just as a technological enabler but as part of a broader transformation strategy. Research should further investigate how executives and managers can effectively integrate AI insights into strategic decision-making while fostering a culture that supports both incremental improvements and radical transformations.

REFERENCES

Afuah, A. (2014). *Business Model Innovation: Concepts, Analysis, and Cases*. Routledge. DOI: 10.4324/9780203584583

Afuah, A., & Tucci, C. (2013). Value capture and crowdsourcing. *Academy of Management Review*, *38*(3), 457–460. DOI: 10.5465/amr.2012.0423

Amit, R., & Zott, C. (2012). Creating Value Through Business Model Innovation. *MIT Sloan Management Review*, *53*(3), 41–49.

Baden-Fuller, C., & Haefliger, S. (2013). Business Models and Technological Innovation. *Long Range Planning*, *46*(6), 419–426. DOI: 10.1016/j.lrp.2013.08.023

Bahoo, S., Cucculelli, M., & Qamar, D. (2023). Artificial intelligence and corporate innovation: A review and research agenda. *Technological Forecasting and Social Change*, *188*, 122264. DOI: 10.1016/j.techfore.2022.122264

Bhatia, A., Cheng, J., Salek, S., Chokshi, V., & Jetter, A. (2017), "Improving the Effectiveness of Fuzzy Front End Management: Expanding Stage-Gate Methodologies through Agile", 2017 Portland International Conference on Management of Engineering and Technology (PICMET), presented at the 2017 Portland International Conference on Management of Engineering and Technology (PICMET), pp. 1–8, DOI: 10.23919/PICMET.2017.8125390

Bohnsack, R., Pinkse, J., & Kolk, A. (2014). Business models for sustainable technologies: Exploring business model evolution in the case of electric vehicles. *Research Policy*, *43*(2), 284–300. DOI: 10.1016/j.respol.2013.10.014

Bollinger, S., & Burger-Helmchen, T. (2021). "From the control of innovation to creativity: Towards an integrative framework", Revue d'économie industrielle. *De Boeck Supérieur, Louvain-la-Neuve*, *174*(2), 223–247. DOI: 10.4000/rei.10365

Bollinger, S., Neukam, M., & Guittard, C. (2023). Towards strategic CSR: A constraint stimulating innovation. *Innovations*, *72*(3), 65–102. DOI: 10.3917/inno.pr2.0150

Bollinger, S., & Schaeffer, V. (Eds.). (2024). *Strategic Management. Creativity and Transition: Howto Reinvent Strategic Management?* EMS Éditions.

Bucherer, E., Eisert, U., & Gassmann, O. (2012). Towards Systematic Business Model Innovation: Lessons from Product Innovation Management. *Creativity and Innovation Management*, *21*(2), 183–198. DOI: 10.1111/j.1467-8691.2012.00637.x

Burger-Helmchen, T. (2024). Crowdsourcing of Inventive Activities, AI, and the NIH Syndrome. *Administrative Sciences*, *14*(11), 300. Advance online publication. DOI: 10.3390/admsci14110300

Burger-Helmchen, T., & Llerena, P. (2012). Creativity, human resources and organizational learning. In Buenstorf, G. (Ed.), *Evolution, Organization and Economic Behavior* (pp. 155–184). Edward Elgar Publishing., DOI: 10.4337/9780857930897.00016

Burger-Helmchen, T., & Siegel, E. (2019). From open design to open strategy: Managing the evolution of crowdsourcing-based business models. In Barlatier, P.-J., & Mention, A.-L. (Eds.), *Managing Digital Open Innovation*. WorldScientific Publishing.

Burström, T., Parida, V., Lahti, T., & Wincent, J. (2021). AI-enabled business-model innovation and transformation in industrial ecosystems: A framework, model and outline for further research. *Journal of Business Research*, *127*, 85–95. DOI: 10.1016/j.jbusres.2021.01.016

Casadesus-Masanell, R., & Tarziján, J. (2012). When One Business Model Isn't Enough. *Harvard Business Review*, *90*(1/2), 132–137.

Casadesus-Masanell, R., & Zhu, F. (2013). Business model innovation and competitive imitation: The case of sponsor-based business models. *Strategic Management Journal*, *34*(4), 464–482. DOI: 10.1002/smj.2022

Chattopadhyay, S., Honoré, F., & Won, S. (2024), "Free range startups? Market scope, academic founders, and the role of general knowledge in AI", Strategic Management Journal, John Wiley & Sons, Ltd, Vol. n/a No. n/a, DOI: 10.1002/smj.3685

Chesbrough, H. W. (2006). *Open Business Models: How to Thrive in the New Innovation Landscape* (1st ed.). Harvard Business Press.

Cohendet, P., Dupouët, O., Naggar, R., & Rampa, R. (2024). "Orchestrating orphan ideas in the fuzzy front end of a large firm's R&D department", R&D Management, John Wiley & Sons. *R & D Management*, *54*(3), 558–573. DOI: 10.1111/radm.12661

Cohendet, P., & Llerena, P. (2005). A Dual Theory of the Firm between Transactions and Competences: Conceptual Analysis and Empirical Considerations[[doi:]. *Revue d'Economie Industrielle*, *110*(1), 175–198. http://www.editecom.com/index_rei.htm. DOI: 10.3406/rei.2005.3078

Cooper, R. G. (2008). Perspective: The Stage-Gate® Idea-to-Launch Process—Update, What's New, and NexGen Systems. *Journal of Product Innovation Management*, *25*(3), 213–232. DOI: 10.1111/j.1540-5885.2008.00296.x

Cooper, R. G., & Sommer, A. F. (2016). The Agile–Stage-Gate Hybrid Model: A Promising New Approach and a New Research Opportunity. *Journal of Product Innovation Management*, *33*(5), 513–526. DOI: 10.1111/jpim.12314

Coskun-Setirek, A., & Tanrikulu, Z. (2021). Digital innovations-driven business model regeneration: A process model. *Technology in Society*, *64*, 101461. DOI: 10.1016/j.techsoc.2020.101461

Csaszar, F. A., Ketkar, H., & Kim, H. (2024). Artificial Intelligence and Strategic Decision-Making: Evidence from Entrepreneurs and Investors. *Strategy Science*, *9*(4), 322–345. DOI: 10.1287/stsc.2024.0190

De Brentani, U., & Reid, S. E. (2012). The Fuzzy Front-End of Discontinuous Innovations: Insights for Research and Management. *Journal of Product Innovation Management*, *29*(1), 70–87. DOI: 10.1111/j.1540-5885.2011.00879.x

Demil, B., & Lecocq, X. (2010). Business Model Evolution: In Search of Dynamic Consistency. *Long Range Planning*, *43*(2/3), 227–246. DOI: 10.1016/j.lrp.2010.02.004

Eling, K. (2020). Green, Social and Profitable-The Role of Front End of Innovation Decision Making in Achieving More Sustainable New Products. In Tiwari, R., & Buse, S. (Eds.), *Managing Innovation in a Global and Digital World* (pp. 305–319). Springer., DOI: 10.1007/978-3-658-27241-8_19

Eling, K., & Herstatt, C. (2017). Managing the Front End of Innovation—Less Fuzzy, Yet Still Not Fully Understood. *Journal of Product Innovation Management*, *34*(6), 864–874. DOI: 10.1111/jpim.12415

Felin, T., & Holweg, M. (2024). Theory Is All You Need: AI, Human Cognition, and Causal Reasoning. *Strategy Science*, *9*(4), 346–371. DOI: 10.1287/stsc.2024.0189

Fjeldstad, Ø. D., & Snow, C. C. (2018). Business models and organization design. *Long Range Planning*, *51*(1), 32–39. DOI: 10.1016/j.lrp.2017.07.008

Foss, N. J., & Saebi, T. (2017). Fifteen Years of Research on Business Model Innovation: How Far Have We Come, and Where Should We Go? *Journal of Management*, *43*(1), 200–227. DOI: 10.1177/0149206316675927

Gambardella, A., & McGahan, A. M. (2010). Business-Model Innovation: General Purpose Technologies and their Implications for Industry Structure. *Long Range Planning*, *43*(2/3), 262–271. DOI: 10.1016/j.lrp.2009.07.009

Garbuio, M., & Lin, N. (2019). Artificial Intelligence as a Growth Engine for Health Care Startups: Emerging Business Models. *California Management Review*, *61*(2), 59–83. DOI: 10.1177/0008125618811931

Gassmann, O., Frankenberger, K., & Sauer, R. (2016). *Exploring the Field of Business Model Innovation*. Springer International Publishing., DOI: 10.1007/978-3-319-41144-6

Gassmann, O., & Schweitzer, F. (2013). *Management of the Fuzzy Front End of Innovation*. Springer International Publishing AG.

Gibson, C., & Birkinshaw, J. (2004). Building Ambidexterity Into an Organization. *MIT Sloan Management Review*, *45*(4), 47–55.

Globocnik, D., Faullant, R., & Parastuty, Z. (2020). Bridging strategic planning and business model management – A formal control framework to manage business model portfolios and dynamics. *European Management Journal*, *38*(2), 231–243. DOI: 10.1016/j.emj.2019.08.005

Guittard, C., Schenk, E., & Burger-Helmchen, T. (2015). Crowdsourcing and the Evolution of a Business Ecosystem. In Garrigos-Simon, F., Gil-Pechuan, I., & Estelles-Miguel, S. (Eds.), *Advances in Crowdsourcing* (pp. 49–62). Springer., Springer. DOI: 10.1007/978-3-319-18341-1_4

Haefner, N., Wincent, J., Parida, V., & Gassmann, O. (2021). Artificial intelligence and innovation management: A review, framework, and research agenda . *Technological Forecasting and Social Change*, *162*, 120392. DOI: 10.1016/j.techfore.2020.120392

He, X., & Burger-Helmchen, T. (2024). Evolving Knowledge Management: Artificial Intelligence and the Dynamics of Social Interactions. *IEEE Engineering Management Review*, •••, 1–30. DOI: 10.1109/EMR.2024.3519342

Henderson, R. (2020). *Reimagining Capitalism in a World on Fire*. PublicAffairs.

Hollen, R. M. A., Van Den Bosch, F. A. J., & Volberda, H. W. (2013). The Role of Management Innovation in Enabling Technological Process Innovation: An Inter-Organizational Perspective. *European Management Review*, *10*(1), 35–50. DOI: 10.1111/emre.12003

Hussinger, K., & Wastyn, A. (2016). In Search for the Not-Invented-Here Syndrome: The Role of Knowledge Sources and Firm Success. *R & D Management*, *46*(S3), 945–957. DOI: 10.1111/radm.12136

Irissarry, C., & Burger-Helmchen, T. (2024). Using Artificial Intelligence to Advance the Research and Development of Orphan Drugs. *Businesses*, *4*(3), 453–472. DOI: 10.3390/businesses4030028

Johanson, J., & Vahlne, J.-E. (2009). The Uppsala internationalization process model revisited: From liability of foreignness to liability of outsidership. *Journal of International Business Studies*, *40*(9), 1411–1431. DOI: 10.1057/jibs.2009.24

Jorzik, P., Klein, S. P., Kanbach, D. K., & Kraus, S. (2024). AI-driven business model innovation: A systematic review and research agenda. *Journal of Business Research*, *182*, 114764. DOI: 10.1016/j.jbusres.2024.114764

Joyce, A., & Paquin, R. L. (2016). The triple layered business model canvas: A tool to design more sustainable business models. *Journal of Cleaner Production*, *135*, 1474–1486. DOI: 10.1016/j.jclepro.2016.06.067

Kaplan, A., & Haenlein, M. (2019). Siri, Siri, in my hand: Who's the fairest in the land? On the interpretations, illustrations, and implications of artificial intelligence. *Business Horizons*, *62*(1), 15–25. DOI: 10.1016/j.bushor.2018.08.004

Khosravi, P., Newton, C., & Rezvani, A. (2019). Management innovation: A systematic review and meta-analysis of past decades of research. *European Management Journal*, *37*(6), 694–707. DOI: 10.1016/j.emj.2019.03.003

Khurana, A., & Rosenthal, S. R. (1998). Towards Holistic 'Front Ends' In New Product Development. *Journal of Product Innovation Management*, *15*(1), 57–74. DOI: 10.1111/1540-5885.1510057

Kim, J., & Wilemon, D. (2002). "Focusing the fuzzy front–end in new product development", R&D Management, John Wiley & Sons. *R & D Management*, *32*(4), 269–279. DOI: 10.1111/1467-9310.00259

Koen, P., Ajamian, G., Burkart, R., Clamen, A., Davidson, J., D'Amore, R., Elkins, C., Herald, K., Incorvia, M., Johnson, A., Karol, R., Seibert, R., Slavejkov, A., & Wagner, K. (2001). Providing clarity and a common language to the fuzzy front end. *Research Technology Management*, *44*(2), 46–55. DOI: 10.1080/08956308.2001.11671418

Lee, J., Suh, T., Roy, D., & Baucus, M. (2019). Emerging Technology and Business Model Innovation: The Case of Artificial Intelligence. *Journal of Open Innovation*, *5*(3), 44. DOI: 10.3390/joitmc5030044

Madsen, K. M., & Brix, J. (2024), "Building Capacity for Digital Innovation—A Game-Study", Creativity and Innovation Management, John Wiley & Sons, Ltd, Vol. n/a No. n/a, DOI: 10.1111/caim.12650

Mariani, M., & Dwivedi, Y. K. (2024). Generative artificial intelligence in innovation management: A preview of future research developments. *Journal of Business Research*, *175*, 114542. DOI: 10.1016/j.jbusres.2024.114542

Mariani, M. M., & Fosso Wamba, S. (2020). Exploring how consumer goods companies innovate in the digital age: The role of big data analytics companies. *Journal of Business Research*, *121*, 338–352. DOI: 10.1016/j.jbusres.2020.09.012

Mariani, M. M., & Nambisan, S. (2021). Innovation Analytics and Digital Innovation Experimentation: The Rise of Research-driven Online Review Platforms. *Technological Forecasting and Social Change*, *172*, 121009. DOI: 10.1016/j.techfore.2021.121009

Markham, S. K., Ward, S. J., Aiman-Smith, L., & Kingon, A. I. (2010). The Valley of Death as Context for Role Theory in Product Innovation. *Journal of Product Innovation Management*, *27*(3), 402–417. DOI: 10.1111/j.1540-5885.2010.00724.x

Markides, C., & Charitou, C. D. (2004). Competing with dual business models: A contingency approach. *The Academy of Management Perspectives*, *18*(3), 22–36. DOI: 10.5465/ame.2004.14776164

Mishra, S., & Tripathi, A. R. (2021). AI business model: An integrative business approach. *Journal of Innovation and Entrepreneurship*, *10*(1), 18. DOI: 10.1186/s13731-021-00157-5

Neukam, M., & Bollinger, S. (2022). Encouraging creative teams to integrate a sustainable approach to technology. *Journal of Business Research*, *150*, 354–364. DOI: 10.1016/j.jbusres.2022.05.083

Neukam, M., & Guittard, C. (2018). Reach for the stars: Knowledge sharing in international organizations. *Journal of Innovation Economics Management*, *27*(27), 9–35. DOI: 10.3917/jie.027.0009

Neukam, M. N. (2017). Managing the fuzzy front-end in multicultural teams. *European Journal of Innovation Management*, *20*(4), 578–598. DOI: 10.1108/EJIM-11-2016-0112

Parker, G.G., Alstyne, M.W.V. and Choudary, S.P. (2016), Platform Revolution: How Networked Markets Are Transforming the Economy - and How to Make Them Work for You, WW Norton & Co, Erscheinungsort nicht ermittelbar.

Pietronudo, M. C., Croidieu, G., & Schiavone, F. (2022). A solution looking for problems? A systematic literature review of the rationalizing influence of artificial intelligence on decision-making in innovation management. *Technological Forecasting and Social Change*, *182*, 121828. DOI: 10.1016/j.techfore.2022.121828

Randhawa, K., West, J., Skellern, K., & Josserand, E. (2021). Evolving a Value Chain to an Open Innovation Ecosystem: Cognitive Engagement of Stakeholders in Customizing Medical Implants. *California Management Review*, *63*(2), 101–134. DOI: 10.1177/0008125620974435

Simons, R. L. (1995). *Levers of Control: How Managers Use Innovative Control Systems to Drive Strategic Renewal, Harvard Business School Press*. Massachusettes.

Sosna, M., Trevinyo-Rodríguez, R. N., & Velamuri, S. R. (2010). Business Model Innovation through Trial-and-Error Learning: The Naturhouse Case. *Long Range Planning*, *43*(2/3), 383–407. DOI: 10.1016/j.lrp.2010.02.003

Stefani, U., Schiavone, F., Laperche, B., & Burger-Helmchen, T. (2019). New tools and practices for financing novelty: A research agenda. *European Journal of Innovation Management*, *23*(2), 314–328. DOI: 10.1108/EJIM-08-2019-0228

Stephan-Korus, A. (2021), "Bootlegging in a Technology-Driven Organization: Process, Challenges, and Opportunities", Corporate Underground, Vol. Volume 39, WORLD SCIENTIFIC (EUROPE), pp. 103–132, .DOI: 10.1142/9781800612266_0004

Teece, D., Peteraf, M., & Leih, S. (2016). Dynamic Capabilities and Organizational Agility: Risk, Uncertainty, and Strategy in the Innovation Economy. *California Management Review*, *58*(4), 13–35. DOI: 10.1525/cmr.2016.58.4.13

Tian, B., Fu, J., Li, C., & Wang, Z. (2024). "Determinants of competitive advantage: The roles of innovation orientation, fuzzy front end, and internal competition", R&D Management, John Wiley & Sons. *R & D Management*, *54*(1), 21–38. DOI: 10.1111/radm.12633

Tushman, M. L., & O'Reilly, C. A.III. (1996). Ambidextrous organizations: Managing evolutionary and revolutionary change. *California Management Review*, *38*(4), 8–30. DOI: 10.2307/41165852

Verganti, R., Vendraminelli, L., & Iansiti, M. (2020). Innovation and Design in the Age of Artificial Intelligence. *Journal of Product Innovation Management*, *37*(3), 212–227. DOI: 10.1111/jpim.12523

Warner, K. S. R., & Wäger, M. (2019). Building dynamic capabilities for digital transformation: An ongoing process of strategic renewal. *Long Range Planning*, *52*(3), 326–349. DOI: 10.1016/j.lrp.2018.12.001

Winterhalter, S., Zeschky, M.B., Neumann, L. and Gassmann, O. (2017), "Business Models for Frugal Innovation in Emerging Markets: The Case of the Medical Device and Laboratory Equipment Industry", Technovation, Vol. 66–67 No. Supplement C, pp. 3–13, .DOI: 10.1016/j.technovation.2017.07.002

Chapter 10
Disruptive Maturity Through the Incorporation of Strategic Innovation in Ethiopian Higher Education

Shashi Kant
https://orcid.org/0000-0003-4722-5736
Bule Hora University, Ethiopia

Metasebia Adula
https://orcid.org/0000-0001-5732-2850
Bule Hora University, Ethiopia

Gada Gizachew Wakjira
https://orcid.org/0000-0002-7725-1161
Bule Hora University, Ethiopia

ABSTRACT

The links between the predictor variables—Strategic Alignment, Human Capital, Customer Focus, Technological Capabilities, and Institutional Revitalization in Ethiopia—are examined in this regression study. The direction and intensity of these associations are shown by the standardized regression weights. The results demonstrate that while human capital in Ethiopia has a somewhat favorable association with institutional revitalization, technological capabilities and customer focus have a large beneficial impact. On the other hand, it seems that institutional revitalization and strategic alignment in Ethiopia have little in common. These results have consequences for how managers make decisions about how to allocate resources for technology capabilities, customer-centric initiatives, and human capital development. Institutions in Ethiopia should also make investments in people management, evaluate strategy alignment, and keep an eye on renewal. Further investigation into other aspects, longitudinal investigations, and comparative analysis are among the avenues for future study.

DOI: 10.4018/979-8-3693-4030-1.ch010

INTRODUCTION

Ethiopian higher education system's capacity for strategic innovation is directly related to its health and long-term viability (Kant et al., 2025; Alhakimi & Mahmoud, 2020). Educational institutions in Ethiopia must be able to change and adapt to meet the changing needs and demands of learners, businesses, and society at large in a world that is becoming more competitive and changing quickly (Celikyay et al., 2022). In order for colleges and universities establishments to remain influential and relevant, they must create innovative academic programs, delivery strategies, research projects, and institutional structures (Kamandi et al., 2021). Higher education in Ethiopia can better position itself to draw and keep top talent, adapt to changing labor market demands, use novel teaching methods and technologies, optimize educational outcomes and the student experience, and diversify revenue sources to maintain its fiscal viability by adopting an innovative culture (Maruta et al., 2017). In the end, strategic innovation ensures that the system of higher education continues to be a crucial driver of economic growth, knowledge production, and social mobility (Lumpkin & Pidduck, 2021).

Universities in Ethiopia have to deal with a number of important strategic innovation challenges. Development of curricula and programs is essential (Karabulut, 2015), as educational institutions have to keep reviewing and updating their offerings to meet the changing demands of both employers and students (Srivastava et al., 2001). Particularly in the aftermath of the COVID-19 pandemic, delivery model innovation—such as the introduction of on the internet, mixed, which and adaptable educational modalities—has gained substantial importance (Wakjira, 2023b). In higher education, integrating cutting-edge technology like learning analytics and artificial intelligence can improve education, investigation, and management responsibilities (Adula et al., 2025; Siam et al., 2022; Kithinji & Misuko, 2022; Nduati, 2019).

Incorporating strategic innovation throughout courses can help Ethiopia's higher education system attain disruptive maturity. For example, the university in Addis Ababa has worked with industry leaders to provide hands-on projects that allow students to apply academic knowledge in real-world scenarios (Lisdiono et al., 2022). Similarly, Bahir Dar University prioritizes outreach to the community, allowing students to address local issues via service-learning projects (Karaev, 2023). These examples demonstrate how integrating theory and practice increases critical thinking, employability, especially entrepreneurial abilities. Ethiopian universities can create an additional dynamic and responsive learning ecosystem by staying innovative and engage with other sectors (Yousaf et al., 2021).

In the rapidly evolving world of higher education, technology skills have emerged up a crucial engine of innovation, notably in Ethiopian institutions (Chege & Kinyua, 2023). The epidemic boosted the widespread use of digital technologies, emphasizing the importance of strong technology infrastructure. To improve teaching and learning experiences, institutions should prioritize investments in modern technology like educational administration systems and interactive platforms (Salisu & Abu Bakar, 2020). Universities can create more interesting and successful educational environments by providing professors and students with the resources they need. Also, incorporating technology throughout the curriculum is critical for encouraging creativity. This involves using online materials, virtual exercises, and collaborative technologies to encourage active learning (Kithinji & Misuko, 2022).

These technological improvements not only increase accessibility to data, however they also educate students for a digital workforce by matching their talents to business expectations (Joneidi Jafari & NiliPourTabataba'i, 2017). While institutional leadership and culture play important roles in innovation, the critical requirement for improved technology capabilities cannot be emphasized. Universities are required to concentrate on building a solid technical basis that allows for fast adaptability to evolving trends (Bonney, 2017). Ethiopian higher education can considerably enhance educational outcomes while remaining competitive during a increasingly globalized educational context, eventually promoting a creative atmosphere driven by technology growth (Arif Anjum, 2018).

STATEMENT OF PROBLEM

Although strategic innovation is widely acknowledged as being important in higher education, further study is required to determine how specifically it affects the revival of these institutions in Ethiopia (Science & Papers, 2021). Despite the growing acceptance of strategic innovation, research on its efficacy and the degree to which it impacts organizational rejuvenation in the Ethiopian setting is lacking (Yousaf et al., 2021). Closing this gap is essential to lead the creation of successful plans for the nation's educational institutions and to influence policy and decision-making processes (Raymond et al., 2013). Researchers may obtain thorough understanding of how strategic innovation affects these institutions' overall revival by investigating these dimensions (Kamandi et al., 2021; Karabulut, 2015; Djampagau et al., 2018). Also, studies might look at the factors that facilitate the effective adoption of strategic innovation in higher education institutions in Ethiopia (Farooq et al., 2021). Moreover, comparative studies that examine the experiences of universities in other nations might offer Ethiopia institutions insightful comparisons and useful standards (Nur & Salim, 2014). Educational institutions in Ethiopia can improve their efficacy, relevance, and overall contribution to the growth of the nation by comprehending the relationship among organisational rejuvenation and strategic innovation (Khairuddin et al., 2020).

Any organisation, including educational establishments, frequently faces resistance to change (Kar-aev, 2023). Staff, administrators, and faculty members might be wary of novel techniques, procedures, or technological advancements connected to strategic innovation. To overcome resistance, one must employ efficient change management strategies, maintain open lines of communication, and foster an atmosphere that values innovation and experimentation (Suliyanto & Rahab, 2012). Institutional cultures can have strong roots and be difficult to modify. The adoption of strategic innovation may be hampered by the bureaucratic structures, long-standing customs, and hierarchical decision-making processes seen in Ethiopian educational institutions (Farooq et al., 2021). Successful implementation requires changing the institutional culture to value creativity and taking calculated risks (Michael Nnaemeka Ajemba & Ebube Chinwe Arene, 2022).

LITERATURE REVIEW

Strategic Innovation

Strategic innovation is the ability to act creatively, take calculated risks, collaborate effectively with others, recognise and seize new opportunities, and do all of the things that are required to accomplish strategic objectives (Drnevich and Croson 2013; Lichtenthaler and Lichtenthaler, 2009). It also ensures that day-to-day operations are aligned with the fulfillment of the company's strategic goal and that a skilled workforce with strong leadership is developed. According to Regan and Ghobadian (2017), it is recognized as a purposeful and systematic endeavor to recognize novel prospects and challenges, as well as to create novel models of business, propositions of value, and organisational capabilities in university to tackle them. Strategic innovation, according to Chesbrough and Rosenblum (2002), is the capacity to recognize opportunities when others do not, mould those opportunities to create value, and take a piece of that value (Salisu & Abu Bakar, 2020). Henderson and Clark (1990) defined strategic innovation as changing the parameters of industry competition, usually via the introduction of new goods, services, or technological advancements. Hamel (2006) asserts that rather than only utilizing pre-existing resources, strategic innovation entails creating new strategic alternatives and skilled human capital for the business (Almujaini et al., 2021).

Strategic Innovation Determinants

Four proxies are used and identified in literature, that are customer focus, strategic alignment, technological capabilities and human capital (Moki, Ndung'u & Kinyua, 2019; Salisu & Abu Bakar, 2020, Muithya, Muathe & Kinyua, 2021).

Customer Focus: In the context of strategic innovation, customer focus refers to the methodology and frame of mind that centres the innovation process on the needs and wants of the customer (Lichtenthaler and Lichtenthaler, 2009). To effectively meet and exceed customers' expectations, it entails comprehending and predicting their wants, preferences, and behaviours in order to produce goods, services, and strategies (Drnevich and Croson 2013). Conversely, strategic innovation is the methodical and purposeful investigation and use of novel concepts, technology, and business structures in order to gain a competitive edge and spur expansion (Salisu & Abu Bakar, 2020; Kamandi et al., 2021).

Strategic Alignment: The synchronization and integration of innovation activities with the overarching strategic direction and objectives of an organisation is referred to as strategic alignment in the context of strategic innovation (Lichtenthaler and Lichtenthaler, 2009). It entails making certain that innovation initiatives support the organization's long-term performance and are consistent with its larger strategic objectives and priorities (Abdi & Kinyua, 2018).

To gain a competitive edge and spur expansion, strategic innovation entails investigating and using novel concepts, technology, and business models (Kamandi et al., 2021; West et al., 2014).

Technological Capabilities: In the context of strategic innovation, technological capabilities pertain to an organization's capacity to comprehend, use, and harness technology in an efficient manner to propel innovation and secure a competitive edge (Kamandi et al., 2021). It includes all of the information, abilities, tools, infrastructure, and procedures needed to create, implement, and oversee technological advances (Salisu & Abu Bakar, 2020). Technology is a key enabler and shaper of innovation in the ever changing corporate world. Establishing and maintaining technology skills is essential for organisations to remain competitive, take advantage of new possibilities, and adapt to changing consumer demands (Almujaini et al., 2021; Kagermann, Wahlster, & Helbig, 2013).

Human Capital: When discussing strategic innovation, human capital refers to the abilities, know-how, inventiveness, and capacities of the people who work for an organization and who support that organization's capacity for strategic innovation (Salisu & Abu Bakar, 2020). It includes workers of all levels and in a variety of roles, as well as their combined skill and intellectual capital (Abdi & Kinyua, 2018; Moki, Ndung'u & Kinyua, 2019).

Institutional Disruptive maturity: In the context of strategic innovation, institutional disruptive maturity is the process of rethinking and changing an organizations or institution's structures, procedures, and practices in order to promote an innovative culture and support strategic innovation projects

(Drnevich and Croson 2013). It entails bringing the institutional structure back to life and bringing it into line with the strategic objectives and aspirations of the company (Lichtenthaler and Lichtenthaler, 2009). It may be difficult for institutions or organisations that have been using outdated or inflexible structures and procedures to adjust to the quickly evolving business environment and seize new possibilities (Almujaini et al., 2021).

INSTITUTIONAL DISRUPTIVE MATURITY

The process of revitalizing institutions to improve their efficacy, effectiveness, and relevancy in accomplishing their goals is known as institutional disruptive maturity (Yelmi et al., 2021). It entails implementing major adjustments to an institution's structures, procedures, and policies in order to solve current issues, adjust to evolving conditions, and enhance overall performance (Rafiq et al.; 2020). The idea behind institutional rejuvenation is the recognition that institutions may age or become stagnant over time (Tesfa, 2023; Farooq et al., 2021), which reduces their ability to serve the reasons for which they were created Shiferaw et al. (2023). According to Jublee and Abdul Adis (2020) through the introduction of novel concepts, the modernization of procedures, and the alignment with new demands and trends, disruptive maturity seeks to overcome institutional stagnation and revitalise the organization.

Determinants of Institutional Disruptive Maturity

Strategic Planning: One of the most important factors in determining the success of disruptive maturity is creating a well-defined strategic plan that describes the institution's objectives, strategies, and priorities (Almacena Platform, 2021). This entails evaluating the institution's existing situation, spotting possibilities and weaknesses, and creating a change management plan (Chelliah et al., 2023).

Stakeholder Engagement: According to Bahiaa et al. (2019), gaining support, coming up with fresh ideas, and creating a feeling of ownership all depend on including important stakeholders in the disruptive maturity process. Involving stakeholders—staff, clients, partners, and the community—can yield insightful information and guarantee that disruptive maturity initiatives meet their needs and expectations.

Capacity Building: Effective regeneration depends on enhancing institutional capacity via education, training, and information sharing. Developing the organisational capacity allows personnel to adjust to new procedures and technology (Tesfa, 2023; Farooq et al., 2021).

***Change Management*:** During regeneration, effective change management is essential (Perdana & Prasasti, 2023). To promote seamless transitions and minimise disruptions, this entails resolving opposition, cultivating a culture of innovation and continuous improvement, and putting suitable change management tactics into practice (Tesfa, 2023).

Monitoring and Evaluation: Setting up strong processes for monitoring and evaluating disruptive maturity activities is helpful in determining their impact, assessing progress, and locating bottlenecks (Sufa et al., 2022). Frequent monitoring guarantees accountability at every stage of the process and permits prompt modifications (Rafiq et al., 2020).

THEORETICAL FOUNDATION

The integration of strategic innovation has resulted in a disruptive maturity process for Ethiopia higher education system (Samsudin & Ismail, 2019; Menberu, 2020; Alsaad et al., 2022). The Knowledge-Based View (KBV) and the Dynamic Capabilities Theory (DCT), two well-known theories, have been linked to this field of study. According to the Knowledge-Based View (KBV), an organization's knowledge is its most valuable asset, and gaining a competitive edge depends on its capacity to produce, transfer, and apply knowledge (Gant, 1996; Roos, 1998; Hoskisson et al., 1999; Bontis, 2002, De Caloos, 2002; Huizing & Bouman, 2002 and Balogun Kenin, 2003). The capacity of an entity to adjust and react to changes in its environment is emphasized by the Dynamic Capabilities Theory (DCT) (Torraco et al., 2022; Barinta, 2022). The DCT theory highlights the necessity for Ethiopian higher education institutions to cultivate dynamic capacities that enable them to perceive and react to new trends, technologies, and market demands in the quickly changing environment of higher education. This could entail encouraging collaborative partnerships with industry and foreign counterparts, investing in faculty development, and implementing agile curriculum development (Gant, 1996, Roos, 1998; Bonts, 2002; De Carlos, 2002; Hoskisson et al., 1999; Huizenga & Bouman, 2002 and Kenin, 2003; Kamandi et al., 2021).

The Knowledge-Based View (KBV) with Dynamic Capabilities Theory (DCT) is important for understanding strategic innovation in Ethiopian higher education. KBV asserts that knowledge is a crucial resource over enterprises, highlighting the need of exploiting intellectual capital to create a competitive advantage (.Kamandi et al., 2021). This approach emphasizes the vital importance of harnessing local knowledge and skills in Ethiopian colleges, while additionally creating an atmosphere conducive to the growth of new ideas. Universities may better match their curriculum with economic along with societal requirements through fostering knowledge development as well as dissemination (Alsaad et al., 2022).

In contrast, DCT concentrates on an organization's ability to adapt, integrate, and restructure both inside and outside skills in response to quickly changing circumstances (Menberu, 2020). Regarding Ethiopian higher education, this entails creating flexible institutional frameworks that can adapt to technology improvements and the needs of the job market. By developing dynamic capacities, colleges may innovate more effectively (Menberu, 2020; Alsaad et al., 2022), guaranteeing that graduates have relevant skills. Together, KBV and DCT present a complete framework for this research, demonstrating how Ethiopian institutions may improve their strategic methodologies for innovation. By integrating knowledge management with adaptive skills, establishments may better educate students for the complexity that constitutes the modern workforce while also contributing to national development goals (Torraco et al., 2022; Barinta, 2022).

EMPIRICAL LITERATURE REVIEW

Strategic Innovation and Disruptive Maturity

According to Colander (1962), Porter (1980), Barney (1991), Grant (1991), Kumar & Saha (2008), Bergh & Lim (2008), and Barreto and Karia (2020), there is a relationship among strategic innovations and the disruptive maturity of university. These strategic partnerships involve the development of new products to enhance an existing portfolio of goods and services and meet the ever-evolving needs of both present and potential customers (Kamandi et al., 2021). Also, Barnes (2001) states that the fundamental dynamic changes throughout both the internal and outside domains of the corporate environment in the revival of university are the core focus of strategic alignment.

Empirical evidence across Ethiopia's higher education institutions demonstrates substantial obstacles as well as possibilities in the pursuit of innovation and excellence (Powell, 2016; Islamgaleyev et al., 2020). A recent survey found that technology adoption within universities is still in its infancy, with many institutions missing basic digital infrastructure (Happy Ananda et al., 2023). For example, just 30% of institutions said they had full learning management systems in place, impeding the move to successful online learning settings (Malkomson, 2015; Arif Anjum, 2018). This study emphasizes the importance of continuing to invest in technology to improve educational delivery as well as accessibility (Kelly & Palaniappan, 2023). Also, polls of students and teachers reveal a growing need for more participatory and technology-driven educational environments. Over 70% of the pupils preferred blended learning approaches that included conventional and digital techniques (Almujaini et al., 2021).

This attitude corresponds with worldwide educational trends, which highlight the necessity of incorporating technology into the curriculum (Timothy, 2022). Also, data indicates that colleges that have invested in technology capabilities have witnessed an increase in student involvement and academic achievement. Institutions like as Addis Abeba University have begun to use new technological solutions, resulting in increased student satisfaction ratings (Ciemleja and Lāce, 2008; (OECD, 2011; Ismail et al., 2019). These outcomes show that, while problems exist, there is a clear route for Ethiopian higher education to use technology to stimulate creativity, enhance educational outcomes, and effectively prepare students for the needs of the modern workforce.

Strategic Alignments and Disruptive Maturity

Strategic alignment, which is primarily concerned with the changing relationships among both the internal and external domains of company life in firm disruptive maturity in university, is primarily concerned with the content and process of business strategy (Wernfelt, 1984; Beck, 1991; Kenedy & Schndeir, 2000; Barnes, 2001 & Wand, 2012). In this case, information strategy is seen to encourage and assist business strategy (Salisu & Abu Bakar, 2020). A large body of conceptual literature (Sirah & Woldetensay, 2021) and institution infrastructure identify the following as major perspectives of strategic alignment determinants: plan operation, scientific potential, comparable possibility, quality of service, institution technological capabilities, and information information technology strategy (Happy Ananda et al., 2023). Angulo-Ruiz, Pergelova, and Wei (2019) carried out an empirical test of strategy alignment utilising organisational rejuvenation measures in university disruptive maturity. A multifaceted notion, strategic alignment may be defined as being market-, quality-, or product-oriented (Kamandi et al., 2021). According to Al Khalifa (2016), strategic alignment is a method used to maintain staff members' adherence to the institution's work ethics and principles. Another name for it is employee participation.

A rigid interaction among information science and business is often required for an institution to be incorporated as a strategic innovation into a service, product, or business process (Tarofder & Gordon, 2007). According to Martinez-Simarro et al. (2015), companies that continuously look for strategic innovation in information science are believed to generate and profit from informational science alone, giving them a possible competitive advantage. Businesses may be able to save expenses by improving production and efficiency (Mithas and Rust, 2016). The objective of quality and productivity orientation is a greater level of service quality and internal efficiency of university disruptive maturity (Marinova et al., 2008; Joneidi Jafari & Nili Pour Tabataba'i, 2017).

H1: Strategic alignment has a considerable association with in university disruptive maturity.

Customer Focus and Institution al Disruptive Maturity

Customer-focus is the second factor that influences strategic innovation. It involves satisfying the needs and expectations of current and potential customers by understanding their needs thoroughly and then offering them valuable services (Narver & Slater, 1990; Barney, 2001; Estensoro et al., 2002; Bartley et al., 2007; Islamgaleyev et al., 2020). A customer-focus strategy should lead to business institution success and the creation of value for the customer, which encourages customer loyalty (Aziz et al., 2021; Thuong & Singh, 2023). Customer-focused cultures and practices enhance business companies' overall rejuvenation in university (customer satisfaction, sales, market share, and profitability) and ensure the expansion of competitive advantage (Abrokwah-Larbi, 2023). It seems that this finding provides an insight into the fact that customer focus will only lead to business success when customers evaluate suppliers' customer focus orientation. However, Bartley et al. (2007) provided helpful guidance on how businesses may apply a framework that evaluates the customer focus culture and level of customer attention inside a firm to raise their customer focus (Timothy, 2022).

However, most of their customer-oriented programmes concentrate on areas such as relationship management, customer service, customer forecasting, planning for induction and training, customer surveys, and customer data understanding (Wakjira, 2023a). An institution that places a high priority on meeting the needs of its current and potential customers has developed its skills and partnered with customer insights to improve customer service and product development. Artificial intelligence is enabled by network ties and operates an organisational network application that facilitates relationships among third parties and businesses. It implies a simulation of human intelligence of business processes and other machines, particularly in computer systems in to specific application systems in Institutions (Yau *et al.*, 2021; Madhani *et al.*, 2020; Hamidi *et al.*, 2020; Neneh, 2018).

H2: Customer Focus has a considerable association with university disruptive maturity

HUMAN CAPITAL AND DISRUPTIVE MATURITY

According to Cefos and Marsi (2006), human capital is essential to the continued existence and disruptive maturity in small and medium-sized enterprises (university). It is often understood as the commercial application of skilled knowledge that is gained through training, experience, and education and that enables an individual to perform specific tasks and activities in university; technological advances usage, software, and tools to address a particular business challenge and it is required to stream and grow an industry market trend, as well as a growing awareness of policy. rising policy consciousness and proficiency in the workplace (Grant 197; Becker, 1993; Malkomson, 2015; Arif Anjum, 2018). Human capital, or the expertise, competencies, and capabilities, is the cornerstone of business institutions (Alqershi, 2021). Moreover, employees who generate intellectual capital due to their aptitude, personality, and quick thinking (Subramaniam & Youndt, 2005; Alqershi et al., 2019). Competence, according to Khalique, Shaari, Abdul, and Isa (2011), covers the behavioural components of an employee's employment and includes knowledge, training, and attitude (Science & Papers, 2021). It is also recognised that human capital has a major role in the performance of university and is essential to their renewal in a competitive market (Djampagau et al., 2018).

The main constituents of the production-based economy were labour, capital, and the physical assets of university (Features et al., 2019). When it comes to creating value to university, human capital has become more important than physical assets (Hejazi et al., 2016). Additionally, it has been recognised as the primary driver of organisational disruptive maturity across a range of companies, hence enhancing strategic innovation competence, organisational disruptive maturity, and national economic growth (Sivalogathasan & Wu, 2015). This is due to the fact that human capital today adds to the success of institutions by utilising the knowledge-based approach and the Dynamic Capabilities theory.

H3: Human Capital has a substantial relationship with Institution al disruptive maturity

Technological Capabilities and Institutional Disruptive Maturity

Intellectual talents have become recognized as critical drivers of innovation, especially within higher education. The term disruptive maturity describes how institutions progress in their embrace and integration of new technology. In the framework of Ethiopian higher education, its path to maturity entails not just implementing new tools, but also reforming pedagogical paradigms to successfully exploit this technology (Kelly & Palaniappan, 2023). At the heart of this transition is the requirement for a robust digital infrastructure. Investing in modern technology capabilities, such as the use of cloud computing, AI, and data analytics, enables establishments to provide more customized and adaptable learn experiences. Learning management tools, for example, can provide personalized educational routes, allowing students to participate at their own speed and in accordance with their preferences for teach (Kelly & Palaniappan, 2023). The shift to individualized learning is an indication of disruptive maturity since it substantially alters the relationship between educators and students. Furthermore, the use of electronic devices promotes collaborative learning situations. Virtual learning environments and collaborative platforms allow students from many geographical places to participate in collaborative educational experiences. This not only increases accessibility, but also extends the educational experience by introducing students to a variety of viewpoints (Kelly & Palaniappan, 2023). . As Ethiopian colleges adopt these technologies, they may foster an improved inclusive educational environment that breaks down conventional barriers. As institutions grow through the disruptive maturity continuum, they must also prioritize the development and an innovative culture. This includes preparing staff to effectively use new technology and encouraging participants to experiment with novel teaching methods (Tees, Risano & Shun, 1997; Luftman &Barier, 1999)

Universities may encourage continual improvement in teaching and learning by creating an atmosphere in which educators feel empowered to experiment with and use technology solutions. In essence, technological skills are a critical driver of innovation in Ethiopian colleges and universities (Kim, Lee, 2011; Wang, 2012; Amdo & Kou, 2015). Institutions that embrace disruptive maturity can not only improve their educational offerings, but also improve the readiness of students for a quickly changing labor market. Ethiopian universities may position themselves as global educational leaders by making prudent investments regarding technology and committing to cultivating a creative culture, thereby altering the learning procedure for subsequent generations (Chen et al., 2015).

H4: Technological Capabilities has a substantial relationship with Institutional disruptive maturity

Figure 1. Strategic innovations dimensions and institutional disruptive maturity

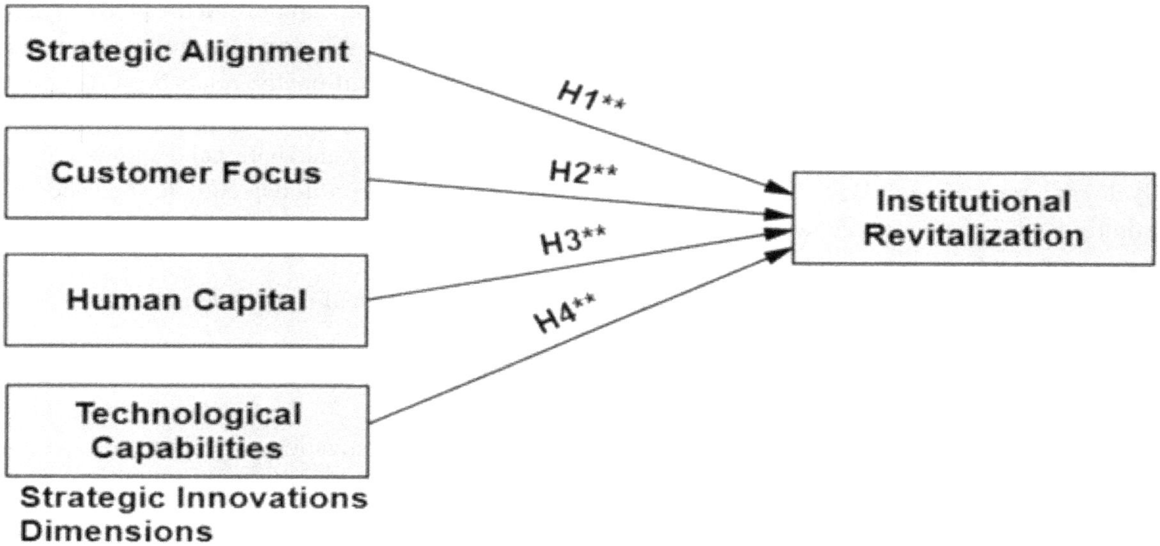

Source: Own adapted by researcher from literature (2024)

RESEARCH METHODOLOGY

This study is cross-sectional and quantitative in nature, providing a full examination and exploration of the strategic innovation dimensions across small and medium firms in Ethiopia. To ensure the validity, reliability, and accuracy of the scale, the study was pre-tested by six practitioners and five academics in Ethiopia before the final data collection. The questionnaire was quickly modified pursuant to the outcomes in order to improve its validity and suitability for use in Ethiopia. With the SEM data processing method used in this study, a minimum sample size of 425 may be used to analyses the data. A 5-point Likert scale was used in the design and collection of all samples and data, and the study's data collection was completed in 2023. Before beginning the survey, each respondent was informed of the general purpose and content of the study's questionnaire, with the aim of safeguarding their right to data and choice, as well as the security and privacy of their personal information. This action was taken with ethical considerations in mind. Because this research was specific and specialized, and because it required respondents to have knowledge of corporate internal business strategies and procedures, purposeful sampling was used in this study. The researchers were able to concentrate on a specific group that possessed unique characteristics and may provide informative data that would assist them address their study objectives through the use of intentional sampling. The intended audience was managers of medium- and large-scale Ethiopian industrial businesses. The target companies were found using the list of small and medium-sized enterprises that were registered with the Industrial Park Development Authority. This research then made contact with the relevant businesses to get information from managers at the medium and upper levels of management.

DATA ANALYSIS

To assess the validity and reliability of table 1, researchers look at the following metrics: Cronbach's alpha, average variance extracted (AVE), composite reliability (rho_a and rho_c), and variance inflation factors (VIF). These metrics are widely used in psychometrics to assess a measuring tool's reliability and validity.

Table 1. Reliability and validity

Variables	No. items	Mean	Standard deviation	Cronbachs alpha	Composite reliability (rho_a)	Composite reliability (rho_c)	Average variance extracted	Variance inflation factors
Customer Focus	7	3.02	0.985	0.803	0.806	0.803	0.459	1.275
Strategic Alignment	6	3.35	0.963	0.789	0.872	0.898	0.386	1.261
Technological Capabilities	5	3.05	0.913	0.783	0.968	0.871	0.540	1.063
Human Capital	6	3.16	0.951	0.768	0.890	0.863	0.376	1.042
Institution Disruptive maturity	5	3.43	0.964	0.784	0.947	0.975	0.537	1.052

Source: SPSS, 2024

The reliability and validity study outcomes for the five variables are manifests in table 1. The analysis yields data on the number of items, mean, standard deviation, variance inflation factors, Cronbach's alpha, composite reliability (rho_a, rho_c), average variance extracted, and variance. There are seven items in the Customer Focus variable. With a standard deviation of 0.985 and an average score of 3.02, the respondents demonstrated a modest level of customer concern. With an internal consistency dependability of 0.803, the scale's components are deemed to be consistently measuring the same construct, according to Cronbach's alpha. The scale's dependability is further supported by the composite reliability (rho_a and rho_c) values of 0.806 and 0.803, respectively. 45.9% of the variation in the observable variables is explained by the underlying construct, according to the average variance retrieved (0.459). It appears from the variance inflation factor (1.275) that multicollinearity is not a major problem for this variable.

Six items make up the Strategic Alignment variable. There is a considerable degree of strategic alignment manifests by the mean score of 3.35. A somewhat narrow distribution of responses is manifests by the standard deviation of 0.963. With a Cronbach's alpha of 0.789, the internal consistency dependability is deemed acceptable. Good dependability is suggested by the composite reliability ratings (rho_a = 0.872, rho_c = 0.898). 38.6% of the variation in the observable variables is explained by the underlying construct, according to the average variance retrieved (0.386). Variance inflation factor (1.261) indicates that there are no noteworthy problems with multicollinearity. According to Cronbach's alpha and composite reliability values, the variables generally show acceptable to high levels of dependability based on the data that have been supplied. Different levels and distributions of the constructs are suggested by the means and standard deviations. Higher values imply more convergent validity. The average variance extracted values show the percentage of variation explained by the underlying component. The values of the variance inflation factor indicate that multicollinearity is not very problematic.

Table 2. Assessment of normality

Variable	min	max	skew	c.r.	kurtosis	c.r.
Technological Capabilities	1.800	5.000	-.470	-4.230	.168	.754
Customer Focus	1.600	5.000	-.278	-2.499	.410	1.847
Human Capital	1.400	5.000	-.375	-3.371	.423	1.904
Strategic Alignment	1.800	5.000	-.569	-5.125	1.024	4.607
Institutional Disruptive maturity	1.200	5.000	-.809	-7.284	.926	4.169
Multivariate					8.146	10.732

Source: AMOS, 2024

The outcomes of the normalcy evaluation for the following five factors are manifests in Table 2. The evaluation provides details on skewness, kurtosis, and the lowest and highest values. The range of values for the Technological Capabilities variable is 1.800 to 5.000. The data may have a left-sided tail, as manifests by the somewhat negatively skewed distribution suggested by the skewness score of -0.470. A somewhat peaky, comparatively normal distribution is indicated by the kurtosis value of 0.168. The range of the Strategic Alignment variable is 1.800 to 5.000. A somewhat negatively skewed distribution with a potential left-side tail is indicated by the skewness rating of -0.569. Given the 1.024 kurtosis score, the distribution appears to have a substantial peakness. Overall, the variables show varied degrees of deviations from normalcy based on the data that have been supplied. While kurtosis values above or below zero imply different degrees of peachiness or flatness in the distribution, respectively, skewness values closer to zero indicate less skewness.

Table 3. Regression weights

			Estimate	S.E.	C.R.	P
Institution Disruptive maturity	<---	Strategic Alignment	1.376	.239	5.757	***

Source: AMOS, 2024

The regression weights are manifests in table 3. The estimated impact of the independent variable (strategic alignment) on the dependent variable (institutional disruptive maturity) is represented by the regression weight. The estimate, standard error (S.E.), critical ratio (C.R.), and p-value are all manifests in table 3. According to the regression weight estimate of 1.376, institutional disruptive maturity and strategic alignment are affirmatively correlated. This indicates that there is a correlation among higher degrees of institutional disruptive maturity and higher levels of strategic alignment.

Table 4. Identical regression weights

			Estimate
Institution Disruptive maturity	<---	Strategic Alignment	0.489

Source: AMOS, 2024

The identical regression weight is manifests in table 4. After accounting for the standardization of both variables, the identical regression weight estimate of 0.489 shows the direction and intensity of the association among institutional disruptive maturity and strategic alignment. A affirmative correlation among the two variables is indicated by the affirmative value. More specifically, there is a 0.489 standard deviation rise in institutional disruptive maturity for every one standard deviation increase in strategic alignment.

Figure 2. SEM, 2024

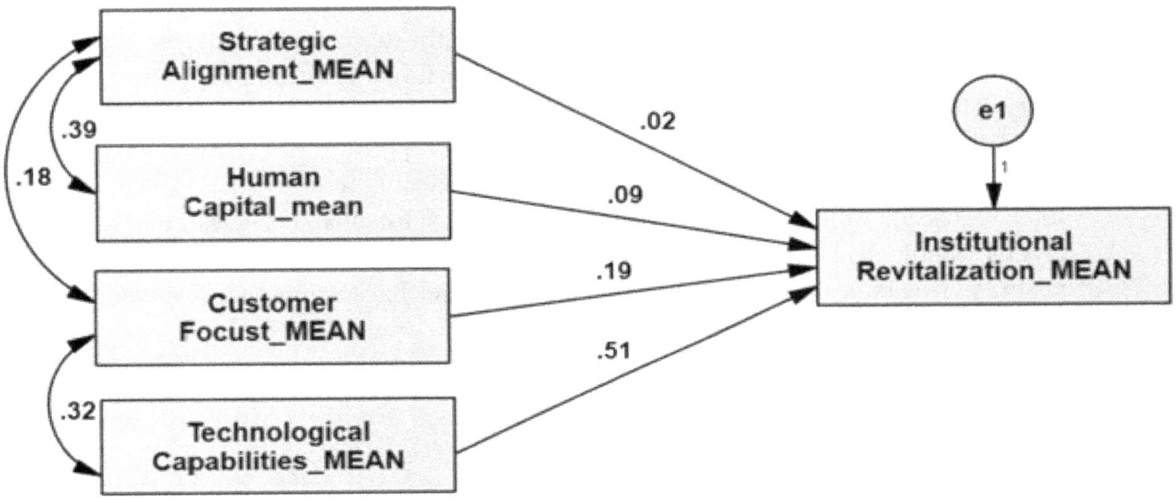

Source: AMOS, 2024

Table 5. Regression weights and hypothesis testing

Direction of Path			Estimate	S.E.	C.R.	P	Label
Institutional Disruptive maturity	**<---**	**Strategic Alignment**	**.023**	**.054**	**.461**	**.645**	**H1**
Institutional Disruptive maturity	<---	Human Capital	.090	.054	1.773	.076	H3
Institutional Disruptive maturity	<---	Customer Focus	.192	.050	3.783	***	H2
Institutional Disruptive maturity	<---	Technological Capabilities	.510	.045	11.393	***	H4

Source: AMOS, 2024

To interpret the direction of the paths among Institutional Disruptive maturity and other variables (Strategic Alignment, Human Capital, Customer Focus, and Technological Capabilities), we need to analyze the regression weights or path coefficients and conduct hypothesis testing.

H1. Institutional Disruptive maturity <--- Strategic Alignment:

- If the regression weight is affirmative and statistically substantial, it suggests that higher levels of Strategic Alignment are associated with higher levels of Institutional Disruptive maturity. This indicates that Strategic Alignment has a direct affirmative impact on Institutional Disruptive maturity.

H2. Institutional Disruptive maturity <--- Human Capital:

- If the regression weight is affirmative and statistically substantial, it suggests that higher levels of Human Capital are associated with higher levels of Institutional Disruptive maturity. This indicates that Human Capital has a direct affirmative impact on Institutional Disruptive maturity.

H3. Institutional Disruptive maturity <--- Customer Focus:

- If the regression weight is affirmative and statistically substantial, it suggests that higher levels of Customer Focus are associated with higher levels of Institutional Disruptive maturity. This indicates that Customer Focus has a direct affirmative impact on Institutional Disruptive maturity.

H4. Institutional Disruptive maturity <--- Technological Capabilities:

- If the regression weight is affirmative and statistically substantial, it suggests that higher levels of Technological Capabilities are associated with higher levels of Institutional Disruptive maturity. This indicates that Technological Capabilities have a direct affirmative impact on Institutional Disruptive maturity.

Table 6. Identical regression weights

			Estimate
Institutional Disruptive maturity	**<---**	**Strategic Alignment**	**.019**
Institutional Disruptive maturity	<---	Human Capital	.079
Institutional Disruptive maturity	<---	Customer Focus	.168
Institutional Disruptive maturity	<---	Technological Capabilities	.499

Source: AMOS, 2024

The identical regression weights among the four variables—Strategic Alignment, Human Capital, Customer Focus, and Technological Capabilities—and Institutional Disruptive maturity are manifests in Table 6. There is a very modest affirmative correlation among institutional disruptive maturity and strategic alignment, as indicated by the identical regression weight of 0.019. This implies that Institutional Disruptive maturity is not directly impacted by Strategic Alignment. There is a weakly affirmative correlation among institutional disruptive maturity and human capital, as indicated by the identical regression weight of 0.079. This implies that the relationship among human capital and institutional disruptive maturity is not very strong. Customer Focus and Institutional Disruptive maturity have a somewhat affirmative association, as indicated by the identical regression weight of 0.168. This implies that Institutional Disruptive maturity is moderately impacted directly by Customer Focus. Technological capabilities and institutional disruptive maturity have a high affirmative association, as indicated by the identical regression weight of 0.499. This implies that Institutional Disruptive maturity is directly impacted by Technological Capabilities in a substantial way.

DISCUSSION

The regression outcomes manifests in Table 5 sheds light on the variables impacting institutional disruptive maturity within Ethiopian higher education, especially in regards to digital transformation. The outcomes suggest that technical skills having the greatest influence on institutional maturity, with a path estimate of 0.510. This argues that colleges should emphasize the advancement of their technology infrastructure in order to stimulate innovation. Institutions such as Addis Abeba University are starting to incorporate cutting-edge technologies into the curriculum and administrative procedures, exhibiting a proactive attitude to digital transformation. Continuous investment in digital resources and tools is required to stay up with contemporary educational standards. Customer attention also plays an important role in fostering disruptive maturity, as indicated by a path projection of 0.192. It also indicates that institutions that connect with their students and the community are more able to innovate. For example, Bahir Dar University introduced initiatives to address local socioeconomic issues, ensuring its educational programs are relevant and helpful. Universities may improve educational achievements and societal impact by emphasizing student and stakeholder needs and input. Human capital, increasing a path measurement of 0.090, has a affirmative but less substantial association with institutional maturity. This implies that, while producing qualified instructors and educators is crucial, it must be supplemented by other strategic activities. Institutions should invest in broad courses that provide instructors with the required abilities to understand and successfully use digital technology. The coupling of faculty abilities with technology progress is critical for developing an innovative culture. On the other hand, the route estimate that strategic alignment is non-substantial (0.023), implying that simply matching institutional strategies is inadequate to promote disruptive maturity. This emphasizes the importance of a comprehensive approach which involves technology capabilities, a customer-centric emphasis, and human capital development. To help with these developments, numerous policy proposals arise. First, the Ethiopian government should expand financing for higher education's technology infrastructure, to guarantee institutions have access to cutting-edge digital technologies. Second, institutions' curricula must be revised to include crucial digital skills that satisfy labor-market demands. Third, encouraging collaboration among institutions and communities at large can improve the relevance of educational programs. In the end, faculty members' continual professional development should be stressed so that they can successfully incorporate new technology into their research and classroom instruction. By applying these ideas, Ethiopian educational institutions may improve potential disruptive maturity, contributing to the country's socioeconomic growth as well as positioning themselves as innovation leaders.

CONCLUSION

In a nutshell the regression weights analysis provides critical insights into the determinants driving institutional disruptive maturity in Ethiopian educational institutions, particularly in the setting of digital transformation. The most important discovery is the importance of technical skills, which appeared as the biggest predictor of organizational maturity. The following emphasizes the need for colleges to invest in digital infrastructure and new technology to improve teaching and administrative procedures. Addis Abeba University exemplifies the capability advantages of such investments by demonstrating how a strong technology infrastructure may dramatically enhance educational achievements and operational efficiency. The strong influence of consumer focus emphasizes the necessity of engaging students as well as local communities. Higher education institutions that actively gather input and tailor their curricula to social requirements might provide more meaningful educational experiences. The instance of Bahir Dar University demonstrates how community participation may result in mutually beneficial collaborations that improve both academic relevance and social effect. While human capital improvement is important, the decreased correlation shows that colleges should take a more complete strategy that includes technology, client focus, and faculty training. This comprehensive plan serves as crucial for creating a flexible and creative educational environment. The fact that strategic alignment had no meaningful impact suggests that coordinating institutional strategies solely is insufficient. On the contrary, a more integrated approach that incorporates a variety of aspects is required for successful digital transformation.

MANAGERIAL AND PRACTICAL IMPLICATIONS

The previously post-pandemic landscape has dramatically altered the direction of higher education, resulting in developments with administrative and practical ramifications for Ethiopian institutions. One of the most noticeable trends is the increased use of digital learning technology. During the pandemic, schools throughout the world moved to online education, and Ethiopian colleges increasingly understand the importance of using hybrid learning approaches. To enable effective online and blended learning environments, managers must invest in solid digital infrastructure as well as faculty training. This investment would improve not just educational delivery but also accessibility for students living in rural places. Another developing trend is a greater emphasis towards student-centered learning. The epidemic underlined the significance of teaching practices that are flexible and adaptable. As a consequence, institutions are driven to create courses that are more tailored to learner requirements and interests. This transformation necessitates a strategic emphasis on including students in the curriculum design procedure and instilling a feeling of ownership and relevance in their educational experiences. Managers should offer workshops as well as forums that allow students to provide comments and collaborate on course development. Also, the global trend of multidisciplinary study is gaining traction. As complex global concerns develop, it is clearer that education must overcome conventional departmental boundaries. Ethiopian organizations should promote cross-departmental collaboration and establish programs that combine many subjects of study. This strategy in addition improves the learning experience, but also equips graduates to deal with complex situations in the workplace. In the end, the value of mental health and well-being in education has been highlighted. Managers have to prioritize the development of support structures and resources towards student mental health, realizing that a comprehensive approach to education is critical for instilling resilience as well as success in children. By adopting these trends, Ethiopian institution of higher learning can adapt to the needs of a fast changing global context, increasing its global significance and effectiveness.

RECOMMENDATIONS AND FUTURE DIRECTIONS

Ethiopian educational institutions confront a number of difficulties that might be solved by strategic suggestions, especially in light of worldwide standards of excellence. One major concern is insufficient investment in infrastructure. Ethiopian universities sometimes function with antiquated infrastructures and minimal technical resources. To address this, spending for infrastructure development must be prioritized, as in the Finnish model, which places substantial expenditures in educational institutions. Partnerships with overseas universities might enhance knowledge transfer and resource exchange, therefore improving the total number of species learning environment. Another substantial difficulty is the mismatch among academic curricula and labour market demands. Graduates frequently find their capacity incompatible with company expectations.

To overcome this, its Ethiopian institutions should embrace an approach that collaborates with industry partners, similar to Germany's dual educational system, which combines academic learning involving practical practice. This alignment can guarantee that graduates were more prepared for the workforce, hence increasing their employability. Academic growth is also an important area that requires improvement. Many Ethiopian educators lack access to continued professional growth, affecting teaching quality. Regular faculty development programs, influenced by methods in the United States, can improve instructional techniques and guarantee that educators are ready to change to new technology and teaching trends. Governance and administrative mistakes impede advancement in Ethiopian university systems. Ethiopian authorities should simplify administrative processes to strengthen accountability and institutional autonomy, drawing on effective management models from countries such as Australia. Finally, cultivating a culture for research and innovation is critical to the progress of higher education. Ethiopian institutions may boost economic growth and globally competitiveness by supporting research collaborations similar to those found in Canada, where collaboration among industry as well as academia encourages innovation. By embracing these ideas, Ethiopian higher education will be better positioned to face future difficulties in achieving success.

REFERENCES

Adula, M., Kant, S., Jabo, D., & Ahmed, E. (2025). Marketing Strategy Effect on Over-Tourism Management by Mediation of Public-Private Partnerships in Ethiopia. In *Solutions for Managing Overtourism in Popular Destinations* (pp. 37–60). IGI Global Scientific Publishing. DOI: 10.4018/979-8-3693-8347-6.ch003

Ajemba, M. N., & Arene, E. C.Michael Nnaemeka AjembaEbube Chinwe Arene. (2022). Research gaps for future research and their identification. *World Journal of Advanced Research and Reviews*, *16*(1), 575–579. DOI: 10.30574/wjarr.2022.16.1.1062

Almujaini, H., Hilmi, M. F., Abudaqa, A., & Alzahmi, R. (2021). Corporate foresight institution al learning and disruptive maturity: The moderating role of digital transformation and mediating role of innovativeness in SME. *International Journal of Data and Network Science*, *5*(4), 703–712. DOI: 10.5267/j.ijdns.2021.7.011

Alqershi, N. (2021). *Strategic thinking, strategic planning, strategic innovation and the disruptive maturity of SME : The mediating role of human capital. 11*, 1003–1012. DOI: 10.5267/j.msl.2020.9.042

Alqershi, N., Abas, Z., Sanuri, S., & Mokhtar, M. (2019). *The Mediating Effect of Human Capital on the Relationship Among Strategic the mediating effect of human capital on the relationship among strategic innovation and performance of manufacturing in. July.*

Alshura, M. S. (2018). Customer Relationship Management (CRM) Capabilities & Building a Sustainable Competitive Advantage in Mobile Phone Operators in Jordan. *International Journal of Business and Management*, *13*(3), 262. DOI: 10.5539/ijbm.v13n3p262

Arif Anjum, D. (2018). Impact of Technology Adoption on the Disruptive maturity of Small and Medium Enterprises in Ethiopia. *The Journal of Social Sciences Research*, *5*(Special Issue), 857–867. DOI: 10.32861/jssr.spi5.857.867

Arslan, I. K. (2020). the Importance of Creating Customer Loyalty in Achieving Sustainable Competitive Advantage. *EurAfican Journal of Business and Management*, *8*(1), 11–20. DOI: 10.15604/ejbm.2020.08.01.002

Auka, D. O. (2016). Strategic Management Practices and Disruptive maturity of Public Secondary Schools in Kenya: A Case of Nakuru District. *International Journal of Developmental Research*, *6*(6), 8216–8226. http://www.journalijdr.com/sites/default/files/issue- pdf/5389.pdf

Ayalu, G., Abbay, A. G., & Azadi, H. (2022). The role of micro- and small-scale enterprises in Barinta, D. D. (2022). *Entrepreneurial Orientation and Market Orientation as Drivers of Micro, Small, and Medium Enterprises Disruptive maturity. 3*(8), 548–562. DOI: 10.47153/jbmr38.4172022

Ayalu, G., Abbay, A. G., & Azadi, H. (2023, August). enhancing sustainable community livelihood: Tigray, Ethiopia. *Environment, Development and Sustainability*, *25*(8), 7561–7584. DOI: 10.1007/s10668-022-02359-7

Bonney, L. (2017). Journal of Innovation in emerging economies. *Suma de Negocios*, *3*, 1–12. DOI: 10.1016/j.jik.2017.03.008

Celikyay, M., Sonmez Cakir, F., & Adıguzel, Z. (2022). Effects of supplier orientation and competitive strategies on ICT firms' disruptive maturity. *Journal of Asia Business Studies*, (July). Advance online publication. DOI: 10.1108/JABS-09-2021-0384

Chege, M. M., & Kinyua, G. (2023). Strategic Innovation as an Antecedent of Firm Disruptive maturity. *Critical Review*, *11*(9), 1–20.

Chelliah, M. K., Aravindan, K. L., & Muthaiyah, S. (2023). Entrepreneurial Orientation and Open Innovation Promote the Disruptive maturity of Services

Chelliah, M. K., Aravindan, K. L., & Muthaiyah, S.SME. (2022, December 20). The Mediating Role of Cost Leadership. *Administrative Sciences*, *13*(1), 1–19. DOI: 10.3390/admsci13010001

Dadi, T., & Bacha, D. (2021).. . *Innovations The Mediating Role of Perceived Usefulness in Mobile Payment Services Adoption in Ethiopia*, *6*, 857–868.

Djampagau, H. R. D., Salim, U., & Wijayanti, R. (2018). *The Relationship of Human Capital, Innovation, and Corporate Disruptive maturity (A Study of Small and Medium Rattan Businesses in Palu City Central Sulawesi Province). 2018*, 453–463. DOI: 10.18502/kss.v3i10.3394

Elangovan, N., & Sundaravel, E. (2021). Method of preparing a document for survey instrument validation by experts. *MethodsX*, *8*, 101326.

Farooq, R., Vij, S., & Kaur, J. (2021). *Innovation orientation and its relationship with business disruptive maturity : moderating role of fi rm size. June*. DOI: 10.1108/MBE-08-2020-0117

Features, S., Investing, O. F., Human, I. N., The, I. N., & Of, E. R. A. (2019). *Social & Behavioural Sciences CIEDR 2018 The International Scientific and Practical Conference " Contemporary Issues of Economic Development of Russia : Challenges and Opportunities " STRATEGIC FEATURES OF INVESTING IN HUMAN CAPITAL.*

Howell, A. (2018). Innovation and Firm Disruptive maturity in the People's Republic of Ethiopia: A Structural Approach with Spillovers. In SSRN *Electronic Journal* (Issue 805). https://doi.org/DOI: 10.2139/ssrn.3191735

Joneidi Jafari, M., & NiliPourTabataba'i, S. A. (. (2017). Corporate foresight and its effect on innovation, strategic decision making and institution al disruptive maturity (case study: Iranian banking industry). *Foresight*, *19*(6), 559–576. DOI: 10.1108/FS-07-2017-0035

Jung, S. U., & Shegai, V. (2023). The Impact of Digital Marketing Innovation on Firm Disruptive maturity: Mediation by Marketing Capability and Moderation by Firm Size. *Sustainability (Basel)*, *15*(7), 5711. Advance online publication. DOI: 10.3390/su15075711

Jyoti, J., & Sharma, J. (2012). Impact of Market Orientation on Business Disruptive maturity: Role of Employee Satisfaction and Customer Satisfaction. *Vision: The Journal of Business Perspective*, *16*(4), 297–313. DOI: 10.1177/0972262912460188

Kamandi, F. G., Kinyua PhD, G., & MuchemiPhD, A. (2021). Strategic Alignment as an Antecedent of Customer Satisfaction: A Review of Literature. *International Journal of Managerial Studies and Research*, *9*(12), 38–57. DOI: 10.20431/2349-0349.0912005

Kant, S., Adula, M., Kumari, M., & Nimbrain, N. (2025). Role of SMEs in Greening the Economy by Mediation of Eco-Innovation in the Horn of Africa. In *Examining Green Human Resources Management and Nascent Entrepreneurship* (pp. 173–200). IGI Global., DOI: 10.4018/979-8-3693-7046-9.ch007

Kant, S., Niguse, T., Terefe, A., & Adula, M. (2025). Emerging Trends and Innovations: Effect on Social Sustainability in Manufacturing Supply Chains by Shifts in Consumer Behavior in the Horn of Africa. *Enhancing Social Sustainability in Manufacturing Supply Chains*, 135-162. http://dx.doi.org/ DOI: 10.4018/979-8-3693-9740-4.ch005

Karabulut, A. T. (2015). Effects of Innovation Strategy on Firm Disruptive maturity: A Study Conducted on Manufacturing Firms in Turkey. *Procedia: Social and Behavioral Sciences*, *195*, 1338–1347. DOI: 10.1016/j.sbspro.2015.06.314

Karaev, F. (2023). The Impact of Competitive Strategies on Firm Disruptive maturity: The Mediating Role of Market Orientation and Innovation: An Empirical Study of the Georgian Beverage Sector. *Technology and Investment*, *14*(02), 119–135. DOI: 10.4236/ti.2023.142007

Kelly, A. E., & Palaniappan, S. (2023). Using a technology acceptance model to determine factors influencing continued usage of mobile money service transactions in Ghana. *Journal of Innovation and Entrepreneurship*, *12*(1), 34. Advance online publication. DOI: 10.1186/s13731-023-00301-3

Khairuddin, N. H., Kamarulzaman, N. H., Hashim, H., & Hussin, S. R. (2020). The relationship among marketing strategies and innovative market orientation on disruptive maturity of halal- certified agro-food SME. *Food Research*, *4*, 124–132. DOI: 10.26656/fr.2017.4(S1).S22

Khan, W. A., Sciences, O., Arshad, M. Z., & Kashif, U. (2020). *The Effect of Entrepreneurial Orientation and Organisational Culture on Firm Disruptive maturity : The Mediating Role of Innovation. June.*

Kithinji, E. T. B., & Misuko, D. N. (2022). Effect of Innovative Strategies on the Disruptive maturity of Small and Medium Enterprises in Nairobi County, Kenya. *Journal of Strategy and Management*, *6*(6), 127–139. DOI: 10.53819/81018102t2112

Li, J., Pan, Q., Peng, Y., Feng, T., Liu, S., Cai, X., Zhong, C., Yin, Y., & Lai, W. (2020). Perceived quality of urban wetland parks: A second-order factor structure equation modeling. *Sustainability (Basel)*, *12*(17), 1–15. DOI: 10.3390/su12177204

Lisdiono, P., Said, J., Yusoff, H., & Hermawan, A. A. (2022). Examining Leadership Capabilities, Risk Management Practices, and Institution al Resilience: The Case of State- Owned Enterprises in Indonesia. *Sustainability (Basel)*, *14*(10), 6268. DOI: 10.3390/su14106268

Lumpkin, G. T., & Pidduck, R. J. (2021). Global entrepreneurial orientation (Geo): An updated, multidimensional view of eo. *Advances in Entrepreneurship, Firm, Emergence, and Growth*, *22*, 17–68. DOI: 10.1108/S1074-754020210000022002

Maruta, I. G. N. A., Sularso, R. A., & Susanti, N. (2017). The Effect of Market Orientation, Entrepreneurship Orientation, and Imitation Strategy on Competitive Advantage (Study on SME of Leather Bag and Suitcase in East Java). *International Journal of Business and Management Invention ISSN*, *6*(7), 24–35. www.ijbmi.org

Ngo, Q. H. (2023). The effectiveness of market orientation in the logistic industry: A focus on SME in an emerging country. *Heliyon*, *9*(7), e17666. DOI: 10.1016/j.heliyon.2023.e17666 PMID: 37539317

Nur, N., & Salim, U. (2014). Entrepreneurship Orientation, Market Orientation, Business Strategy, Management Capabilities On Business Disruptive maturity; Study At Small And Medium Enterprise Printing In Kendari " *International Journal of Business and Management Invention ISSN (Online*, *3*(12), 2319–8028.

Nurwahdah, A. (2022). *Research in Business & Social Science The influence of green transformational leadership and green attitude on green organisational citizenship behaviour mediated by emotional intelligence*. *11*(3), 99–111.

Nyuur, R. B., Brecic, R., & Sobiesuo, P. (2015). Foresight capabilities and SME product/service adaptiveness: The moderating effect of industry dynamism. *International Journal of Foresight and Innovation Policy*, *10*(2–4), 145–164. DOI: 10.1504/IJFIP.2015.074395

Online) Determinants of Commercial Banks Liquidity in Ethiopia Horn of Afica Journal of Business and Economics (HAJBE),2022. *5*(2), 15–29.

Osuagwu, C. (2019). Market Orientation Conceptualizations, Components and Disruptive maturity-Impacts: A Literature Review and Conceptual Framework. *International Journal of Marketing Studies*, *11*(2), 102. DOI: 10.5539/ijms.v11n2p102

Peng, J., Qin, Q., & Tang, T. (2021). The influence of marketing innovations on firm disruptive maturity under different market environments: Evidence from Ethiopia. *Sustainability (Switzerland)*, *13*(18), 1–15. DOI: 10.3390/su131810049

Perdana, R., & Prasasti, A. (2023). Entrepreneurial orientation, company disruptive maturity, and competitive advantage in Indonesian culinary SME. *Small Business International Review*, *7*(1), e547. DOI: 10.26784/sbir.v7i1.547

Pescaroli, G., Velazquez, O., Alcántara-Ayala, I., Galasso, C., Kostkova, P., & Alexander, D. (2020). A Likert Scale-Based Model for Benchmarking Operational Capacity, Institution al Resilience, and Disaster Risk Reduction. *International Journal of Disaster Risk Science*, *11*(3), 404–409. DOI: 10.1007/s13753-020-00276-9

Poerwanto, I. (2022). *Research in Business & Social Science Market orientation and innovation on corporate disruptive maturity : A study of food and beverage sector in Surabaya*. *11*(3), 27–32.

Raymond, L., Bergeron, F., & Croteau, A. M. (2013). Innovation Capability and Disruptive maturity Of Manufacturing SME: The Paradoxical Effect of IT Integration. *Journal of Institution al Computing and Electronic Commerce*, *23*(3), 249–272. DOI: 10.1080/10919392.2013.807714

Rianto, M. R., Woestho, C., & Noor, A. W. (2022). *The Role of Mediating Innovation and Social Media : Market Orientation and Entrepreneurial Orientation on the Disruptive maturity of SME ' s Processed by Sea Products in Labuan Village, Banten*. *1*(8), 1703–1714.

Ryan, G. (2018). Introduction to positivism, interpretivism and critical theory. *Nurse Researcher*, *25*(4), 14–20. DOI: 10.7748/nr.2018.e1466 PMID: 29546962

Salisu, Y., & Abu Bakar, L. J. (2020). Technological capability, relational capability and firms' disruptive maturity: The role of learning capability. *Revista de Gestao*, *27*(1), 79–99. DOI: 10.1108/REGE-03-2019-0040

Schulze, A., Townsend, J. D., & Talay, M. B. (2022). Completing the market orientation matrix: The impact of proactive competitor orientation on innovation and firm disruptive maturity. *Industrial Marketing Management*, *103*(April), 198–214. DOI: 10.1016/j.indmarman.2022.03.013

Shaher, A. T. H. Q., & Ali, K. A. M. (2020). The effect of entrepreneurial orientation and knowledge management on innovation disruptive maturity: The mediation role of market orientation. *Management Science Letters*, *10*(15), 3723–3734. DOI: 10.5267/j.msl.2020.6.020

Shi, C., & Trinidad, M. E. (2023). *A Quantitative Analysis on the Satisfaction of Residents on the Integration and Development of the Chengdu-Chongqing Economic Circle*. *73*, 1672–1681.

Shiferaw, R. M., Birbirsa, Z. A., & Werke, S. Z. (2023). *Entrepreneurial leadership, learning institution and institution al culture relationship : a systematic literature review*. DOI: 10.1186/s13731-023-00305-z

Siam, S. T., Shaari, J. A. N., & Heriyadi, H. (2022). The Relationship of Institution al Competition Orientation, Marketing Capability and Customer Networks in Achieving Seller Achievement. *European Journal of Business and Management Research*, *7*(2), 168–175. DOI: 10.24018/ejbmr.2022.7.2.1219

Siddiqui, S. (2019). Research Paradigms: Their Assumptions and Relevance. *International Journal of Research in Social Sciences*, *9*(5), 254–265.

Sirah, S., & Woldetensay, A. (2021). *Innovations An Empirical Study of the Long and Short Run Causation among Being without a job and Inflation in Ethiopia* . *6*, 1177–1190.

Srivastava, R. K., Fahey, L., & Christensen, H. K. (2001). The resource-based view and marketing: The role of market-based assets in gaining competitive advantage. *Journal of Management*, *27*(6), 777–802. DOI: 10.1177/014920630102700610

Sufa, W. G., Birbirsa, Z. A., & Mogose, T. E. (2022). *Practices of Marketing strategy and Supply chain management practices of the textile Manufacturing industry in the case of selected Industrial Parks, Ethiopia* . *70*, 798–827.

Suliyanto, & Rahab. (2012). The role of market orientation and learning orientation in improving innovativeness and disruptive maturity of small and medium enterprises. *Afican Social Science, 8*(1), 134–145. DOI: 10.5539/ass.v8n1p134

Taherdoost, H. (2017). Validity and Reliability of the Research Instrument ; How to Test the Validation of a Questionnaire / Survey in a Research. *International Journal of Academic Research in Management (IJARM), 5*(2296–1747), 28–36. DOI: 10.2139/ssrn.3205040

Taherdoost, H. (2022). What are Different Research Approaches? Comprehensive Review of Qualitative, Quantitative, and Mixed Method Research, Their Applications, Types, and Limitations. *Journal of Management Science & Engineering Research*, *5*(1), 53–63. DOI: 10.30564/jmser.v5i1.4538

Tesfa, S. Y. (2023). The Effect of Corporate Social Responsibility on Consumer Loyalty. *The Mediating Role of Corporate Image.*, *73*, 1536–1553.

Thuong, C. Van, & Singh, H. (2023). *The Impact of a Balanced Scorecard on Enterprise Disruptive maturity in Ho Chi Minh. 12*, 198–215.

Timothy, V. L. (2022). The effect of top managers' human capital on SME productivity: The mediating role of innovation. *Heliyon*, *8*(4), e09330. DOI: 10.1016/j.heliyon.2022.e09330 PMID: 35520617

Tranmer, M., Murphy, J., Elliot, M., & Pampaka, M. (2020). Multiple Linear Regression (2 nd Edition) Mark Tranmer Maria Pampaka January 2020 License and attribution. *Cathie Marsh Institute Working Paper 2020-01.*, *1*, 1–59. https://hummedia.manchester.ac.uk/institutes/cmist/a rchive-publications/working- papers/2020/2020-1- multiple-linear-regression.pdf

Ugwu, C. I., Ekere, J. N., & Onoh, C. (2021). Research Paradigms and Methodological Choices in the Research Process. *Information Science and Technology*, *14*(2), 116–124.

Wakjira, G. G. (2023a). Assessments of CFA measurement model of digital marketing success with business performance: the mediating role of customer loyalty: the case of commercial banks of Ethiopia (CBE), Ethiopia. *Partners Universal International Innovation Journal*, *1*(4), 205–218.

Wakjira, G. G. (2023b). Influencing Factors of Micro Finance Institute on Innovative Entrepreneurship Growth: In Case of West Guji Zone bule Hora Woreda. Partners Universal International Innovation Journal, 1(1), 60-74.

Wakjira, G. G., Ibrahim, M. I., Gizachew, G., Mohammad, W., & Ibrahim, I. (2023). *Effect of Market Orientation on Institution al Disruptive maturity : The Mediating role of Entrepreneurial and Innovation Capability Multiple Regression Model : In Oromia Regional state West Guji Zone Coffee trading and Factory Enterprise ' s PLC. 2022.*

Yaskun, M., Sudarmiatin, S., Hermawan, A., & Rahayu, W. P. (2023). The effect of market orientation, entrepreneurial orientation, innovation and competitive advantage on business performance of Indonesian MSMEs. International Journal of Professional Business Review: Int. *J. Prof. Bus. Rev.*, *8*(4), 39.

Yousaf, Z., Radulescu, M., Sinisi, C. I., Serbanescu, L., & Maria, L. (2021). *Towards Sustainable Digital Innovation of SME from the Developing Countries in the Context of the Digital Economy and Frugal Environment.*

Chapter 11

Challenges to the Adoption of the Balanced Scorecard as a Managerial Innovation in Moroccan SMEs

Azzouz Elhamma
https://orcid.org/0000-0003-3382-0278
Ibn Tofaîl University, Morocco

Nabila Snoussi
https://orcid.org/0009-0003-6816-0289
Ibn Tofaîl University, Morocco

ABSTRACT

Small and Medium-Sized Enterprises (SMEs) represent over 95% of the Moroccan economy, making it crucial to focus on their management practices to enhance performance and ensure their sustainability. This paper aims to explore the reasons for the non-applicability of the Balanced Scorecard (BSC) in Moroccan SMEs. The BSC, a comprehensive performance measurement tool that simultaneously considers both financial and non-financial data, is highly relevant for improving SME performance. The study seeks to address the question: what are the barriers to the implementation of the BSC within Moroccan SMEs? A quantitative methodology was employed, using an online survey distributed to 31 SMEs across various sectors. The findings show that the primary challenges are the long time required for implementation (37%), lack of financial resources (23%), and the complexity of the tool (21%). This research has several important contributions, both theoretical and managerial.

INTRODUCTION

In the contemporary, volatile, and intricate business landscape, organizations must differentiate themselves to thrive, leading to heightened competition. In this context, innovation, particularly managerial innovation, has become a critical driver of competitive advantage. According to Dereli (2015, p. 1365), "in the global competitive environment, organizational and managerial innovations are the keys

DOI: 10.4018/979-8-3693-4030-1.ch011

to success for companies". Certainly, the Balanced Scorecard (BSC) is regarded as one of the most significant managerial innovations developed over the past three decades (Elbanna et al., 2022; Tawse and Tabesh, 2023). Since its introduction in 1992 by Kapan and Norton, the BSC has been used by various organizations around the world to improve performance measurement systems and to better develop, communicate, and implement their overall strategy. According to Kaplan & Norton (1992, p. 71), "The BSC is like the dials in an airplane cockpit: it gives managers complex information at a glance". The BSC was introduced as a strategic tool for performance measurement and management, designed to address not only financial performance and shareholder perceptions of the organization, but also customer satisfaction, internal business processes, and innovation and learning, with the objective of fostering continuous improvement and value creation (Kaplan & Norton, 1992).

During the 2000s, Rigby and Bilodeau (2009) reported that 53% of companies worldwide used the BSC. However, the adoption rate of the BSC had decreased to 29% (Rigby & Bilodeau, 2018). Similarly, Kumar et al. (2024) reviewed existing research and suggested potential avenues for advancing the BSC. To accomplish this, the study employed a bibliometric and systematic approach, analyzing a corpus of 1,294 BSC-related studies. The results indicate that BSC research has followed a bell-curve trajectory, with a notable and concerning decline in both productivity and popularity. On the other hand, in the early 2020s, 2GC (2021) reported a significant rise in BSC usage at the executive level, from 44% in 2019 to 88% in 2020. Additionally, 80% of surveyed organizations across 21 countries found the BSC as useful, or more useful, than before the COVID-19 pandemic, highlighting its value as a strategic tool for managing and responding to change. Given this situation, the BSC has continued to be a fundamental performance measurement tool across various contexts (Fatima & Elbanna, 2020) and remains the focus of numerous research studies in both developed and developing countries. In this context, Tawse and Tabesh (2023) note that a comparison of Google Scholar search results reveals that, over the past five years, the term "balanced scorecard" has appeared in 4,660 article titles, while during the same period, the Performance Prism—the second most researched PMS—has appeared in only 127 article titles.

Research studies conducted over the past three decades have shown that the traditional focus on financial performance is no longer sufficient, as it is often perceived as being short-term oriented. Companies are now encouraged to adopt a broader concept of organizational performance, which includes new non-financial dimensions and emphasizes the importance of innovation in both strategy and management (Elhamma, 2014). Based on a comparative analysis of performance tracking systems conducted by Garengo et al. (2005), the BSC has emerged as the most prominent and effective managerial innovation for assessing organizational performance across various dimensions (Dudic et al., 2020). The BSC was developed in the 1990s to address the needs of large companies, but it can also be beneficial for SMEs. Based on a survey of 201 SMEs in Spain, Malagueño et al. (2018, p. 221) demonstrated that "firms using BSC for feedforward control obtained better financial performance and presented higher levels of exploitative innovation. We also found that the positive effect of BSC on perceived and attained financial performance is stronger in more established SMEs". Despite these advantages, numerous studies have demonstrated that its implementation remains limited in SMEs, which encounter significant obstacles to its adoption.

Certainly, SMEs are essential to the economy, especially in developing countries. The World Bank reports that formal SMEs represent about 90% of businesses and provide over 50% of global employment, contributing up to 40% of GDP in emerging economies (www.worldbank.org). In Morocco, SMEs are crucial, making up 95% of the economy, according to the SME Confederation. They operate in various sectors, including tourism, communications, industry, transport, agriculture, and financial services.

SMEs are key to Morocco's economy, accounting for 46% of employment, 31% of exports, and 40% of production (Ehamma, 2025).

The objective of this study is to examine the challenges faced by small and medium-sized enterprises (SMEs) in Morocco when implementing the Balanced Scorecard (BSC). This focus is justified by the high failure rate of BSC implementation, which, according to a survey conducted by the management consulting firm KPMG, stands at 70% (DeBusk & Crabtree, 2006). In this context, we will explore the main question research: what are the primary barriers to the successful implementation of the BSC in Moroccan SMEs?

To address our research question, a questionnaire survey was conducted with 31 Moroccan SMEs. The primary findings indicate that the dashboards of the surveyed companies are unbalanced and, as a result, do not align with the Balanced Scorecard (BSC). The key barriers to the adoption of the BSC include, primarily, the lengthy implementation time, insufficient financial resources, and the complexity of the tool. The results of the study have significant contributions both theoretically and managerially.

This article will be divided into three sections. The first section is dedicated to the literature review related to our research problem. The second section addresses the presentation of the methodological framework. The third section is reserved for the presentation of the study's results.

1. THEORETICAL FRAMEWORK AND LITERATURE REVIEW

1.1. The Balanced Scorecard: Genesis and Evolution

The Balanced Scorecard was introduced in the early 1990s, the result of extensive research conducted by two Harvard University professors, David Norton and Robert Kaplan. In 1992, following their studies on performance measurement, the two researchers focused on developing a comprehensive tool based on empirical studies conducted between 1984 and 1992 (Méric, 2003). Their aim was to move beyond the traditional financial perspective of performance measurement and adopt a more holistic view of the organization (Kaplan & Norton, 1992). The BSC integrates non-financial aspects, considering them to be of equal importance as financial metrics (Elhamma, 2023). This tool, on one hand, enables the assessment of the impact of actions taken through financial measures, and on the other, allows the anticipation of future performance based on operational criteria (Méric, 2003). The researchers drew an analogy to an aircraft's dashboard, "where the information must constantly be interconnected so that the pilot can successfully carry out his mission" (p. 71), thus positioning the BSC as a strategic management tool.

Norton and Kaplan present the BSC in a relatively straightforward manner as a tool that enables managers to obtain a comprehensive view of the organization through four distinct perspectives. Each of these perspectives corresponds to an axis designed to address a specific question. The financial perspective focuses on how investors perceive the company, while the customer perspective aims to consolidate all elements necessary to retain customers and meet their needs. The organizational learning perspective is responsible for identifying innovative solutions to improve and create value. Finally, the internal processes perspective defines the actions to be taken within the company to achieve the objectives set by the other perspectives (Kaplan & Norton, 1992).

The BSC continues to evolve, incorporating new elements into its framework. Indeed, the Balanced Scorecard must now enable the organization to develop a strategy that aligns with its goals. To achieve this, four novel processes are integrated into the framework (Kaplan & Norton, 1996):

- **Goal cascading (Translating the vision):** This process involves expressing the company's vision in the form of clear and measurable objectives.
- **Communication (Communicating and Linking):** Managers are required to inform the entire organization about the adopted strategy and link it to both individual and departmental objectives.
- **Operational planning (Business Planning):** By balancing its operational and financial needs, the organization defines its priorities and allocates the necessary resources, thereby bringing itself closer to achieving its long-term strategic goals.
- **Feedback and Learning (Strategic Learning):** By positioning the Balanced Scorecard at the core of its management system, the organization can monitor its short-term results using the four perspectives of the Balanced Scorecard. This allows the company to assess its strategy in light of recent performance, make adjustments as needed, and implement a more effective long-term strategy.

In order for organizations to better visualize the appropriate strategy, Kaplan and Norton introduced a new element to the BSC: the strategy map. This approach considers each of the four perspectives as a domain in which actions are undertaken, with effects that must be interconnected (Kaplan & Norton, 2000; Méric, 2003). The strategy map, therefore, serves two key purposes: first, it explicitly and clearly defines the causal relationships between the assumptions made by the organization to reach its desired future position in an uncertain environment; second, it allows for continuous testing of these relationships to adapt the results as necessary (Kaplan & Norton, 2000; Cifalinò & Lisi, 2019).

The BSC has continued to evolve since its inception in 1992. Initially regarded as a simple diagnostic tool primarily for large commercial enterprises, it now enables all types of organizations, including those in the public and non-profit sectors, to establish an appropriate strategy and support its implementation (Kaplan, 2009). Today, there is even a redefinition of three of the four perspectives that comprise the Balanced Scorecard. The financial perspective has transformed into the "Results" perspective to reflect the triple bottom line of financial, environmental, and societal performance. The customer perspective has been redefined as the "Stakeholder" perspective to capture the interests of the various participants in the ecosystem. Finally, the organizational learning perspective has been renamed the "Driver" or "Catalyst" perspective to ensure that all stakeholders in the new ecosystem are well-coordinated, fostering the sustainability of their coalition (Kaplan & McMillan, 2020).

1.2. The Balanced Scorecard: Advocates and Critics

The BSC is now considered the quintessential performance measurement tool. Indeed, within a decade of its creation, the Balanced Scorecard had already been adopted by the majority of the *Fortune 1000* companies (Hendricks et al., 2004), and its use has since expanded to thousands of private, public, and non-profit organizations (Kaplan & Grossman, 2010). This success can be attributed to several factors. In fact, Quesado et al. (2018) have highlighted and categorized the numerous advantages of this tool through a study that examined both theoretical literature and empirical research on the subject. Among the most significant contributions, it is noteworthy that the Balanced Scorecard, with its four perspectives, represents a balance between external measures related to shareholders and customers, and internal measures related to critical processes such as innovation, learning, and growth (Kaplan & Norton, 2000).

Malagueño et al. (2018), in their study conducted with 201 Spanish SMEs, demonstrated that when properly implemented as a proactive control tool, the BSC not only enabled better performance outcomes (primarily financial), but also facilitated the consideration of operational innovation at higher levels.

Other authors, such as Lueg (2015) and Hoque (2014), describe the BSC as the "cornerstone" of a company's management system. It clarifies the organization's strategy, translates objectives into concrete and achievable actions, facilitates internal communication of the strategy, improves the use of available resources while promoting continuous learning, and aligns individual and departmental objectives with the company's overall strategy. Furthermore, the BSC has transformed managerial practices within organizations, requiring managers to develop and articulate strategies with greater care and foresight, considering the organization as an integrated and coordinated set of activities (Atkinson & Epstein, 2000). Thus, it embodies the essence of the organization and adapts to each level of responsibility, communicating strategic priorities relevant to each organizational level (Zazueta Salido et al., 2019), establishing a flexible and formal mechanism between the company's strategy and operational control (Quesado et al., 2018).

Due to its nature as an interactive performance measurement system (Cardoni et al., 2020), the BSC is, by default, self-immunized from criticism regarding the failure of its implementation. Such failure can only be attributed to the poor selection of indicators or ineffective communication within the organization (Trébucq, 2011; Benková et al., 2020). In fact, a review of the literature reveals that instances discussing the failure of the BSC are rare. However, criticisms of the tool are more prevalent. For example, Méric (2003) considers the BSC as a complementary tool to other methods, such as Activity-Based Costing (ABC) or Economic Value Added (EVA), rather than as a central management tool alternative to the budgeting system.

This observation is shared by several researchers who propose integrating and combining certain practices with the BSC for greater efficiency and effectiveness, notably Total Quality Management (TQM) (Mehralian et al., 2017) or the Analytical Network Process (Hu et al., 2015). Its comprehensiveness has also been questioned by Tsai et al. (2009), who argue that the Balanced Scorecard should not only be used internally but should also take into account the organization's external environment, which is now highly unstable and uncertain. They point out that the tool fails to address risk and exogenous factors (Trébucq, 2011). Nielsen et al. (2017) are more critical of the utility of the Balanced Scorecard, labeling it a "dead tool" that should be replaced by new methods that place greater emphasis on the importance of intangible assets.

Other researchers, such as Pessanha and Prochnik (2006), criticize the BSC for focusing solely on the interests of shareholders and overlooking the interests of other stakeholders, such as suppliers. The authors also highlight the lack of involvement and engagement of the BSC with employees in terms of objectives and measures, which prevents employees from becoming more familiar with the tool and using it optimally. BizShifts (2010) challenges the top-down design of the Balanced Scorecard, as this one-way approach places the entire responsibility for success or failure on management. In this sense, the BSC is somewhat associated with a tool that fosters a bureaucratic, highly hierarchical management style, which contrasts with its intended role as an interactive system that promotes communication and cross-functional collaboration within the organization (Ates et al., 2013).

1.3. The Balanced Scorecard in SMEs

The BSC was initially presented by its creators as a universal tool, applicable to any organization, and thus, its use in small enterprises was considered feasible (Kaplan & Norton, 2001). As Julien (2005) suggests, small businesses are simply "large enterprises in a smaller scale." However, this view is contested by Robinson and Pearce (1984) and Choi et al. (2018), who identify several distinguishing characteristics between the two types of organizations. These authors primarily highlight that SMEs are limited in terms of market opportunities and clientele, that their leaders adopt a personalized management style (Julien & Carrier, 2005), and are often reluctant to disclose the company's strategy. Furthermore, SMEs tend to adopt a reactive rather than proactive approach (Garengo & Sharma, 2012), and operate in environments that change more frequently than those of large firms, which have greater experience and are better prepared for change (Benková et al., 2020). Garengo et al. (2005) summarize these divergences in three key aspects: innovation, evolution, and uncertainty. From these points, it can be understood that the Balanced Scorecard has been more successful in large enterprises than in small and medium-sized organizations.

Indeed, the only case of a small business where the implementation of the BSC was deemed successful, according to Kaplan and Norton, is that of Southern Garden Citrus. However, this case is considered biased (Oriot & Misiaszek, 2012) as the company in question was a subsidiary of a large agri-food group.

A review of the literature reveals few cases addressing the implementation and success of the BSC in SMEs. Nonetheless, several authors advocate for its adoption within these businesses, suggesting that it would help formalize their strategy, assess it more effectively, and better manage uncertainties (Tennant & Tanoren, 2005; Cooper, 2017). Additionally, it could increase the rigor and alignment between strategy and operational processes (McAdam, 2000).

It is important to note that the implementation of the BSC is a continuous process that translates the company's strategy into operational actions, making it a daily concern for everyone (Rompho, 2011). According to Kaplan and Norton (2001), the failure to implement the BSC can result from either inadequate construction (e.g., having too few or too many indicators for each perspective or selecting measures that do not align with the chosen strategy) or limited use of the tool (e.g., treating it as a one-time management tool rather than a permanent system, or lacking commitment from employees and management) (Benková et al., 2020).

Furthermore, SMEs tend to focus more on financial and operational performance than on other areas (Tennant & Tanoren, 2005; Takoudjou Nimpa & Teulon, 2018). Moreover, such businesses face several barriers to implementing the Balanced Scorecard, as they often lack a clear strategy and possess limited resources (Takoudjou Nimpa & Teulon, 2018). Other factors contributing to the failure of the BSC in SMEs include frequent changes in strategy to adapt to environmental shifts. In such cases, regular revisions of the tool require the acquisition of new data, which can create confusion among employees and result in a loss of time, money, and effort (Frederico et al., 2020).

In light of the previous discussion, it would be more appropriate to align with Mintzberg's (1982) contingent view, which contextualizes the use of management tools and challenges the universality of the BSC. This explanation seems to be the most plausible in accounting for the difficulties faced by SMEs in implementing the Balanced Scorecard, which does not necessarily align with their specific needs (Takoudjou & Teulon, 2018).

2. METHODOLOGICAL FRAMEWORK

2.1. Analytical Approach and Epistemological Positioning

In management sciences, researchers have the option of choosing between quantitative and qualitative methods (Elhamma, 2024). In this study, we have chosen to adopt a deductive approach. In fact, the use of a quantitative method facilitates this type of reasoning, as we rely on the results obtained from the sample under analysis to consolidate and corroborate a previously established finding from the reviewed literature. Furthermore, the choice of the quantitative method is justified by its ability to provide an overall view of the studied sample, to visualize it systematically, and to easily exclude outlier cases that could distort the general profile of the studied population. This study also adopts a positivist epistemological stance, as our analysis is based on empirical and objective data derived from observable and measurable facts, in contrast to constructivism, which places a greater emphasis on the personal involvement of the researcher and social interaction (Dehbi & Angade, 2019).

We have also chosen a non-probabilistic sampling method based on purposive selection. Our sample was selected according to specific distribution criteria, namely, SME managers who are best positioned to understand the challenges their companies face regarding performance monitoring and management. To analyze the obtained results, we used two software tools: Excel and SPSS. On the one hand, we conducted an in-depth analysis using graphical representations of the responses. On the other hand, we aimed to demonstrate the relationship between the nature of the obstacles encountered (the dependent variable) in implementing the Balanced Scorecard and the characteristics of the SMEs (size as the independent variable), in order to address our previously defined research question.

2.2. Presentation of the Sample

This study aims to examine the main barriers to the implementation of the BSC in Moroccan SMEs. Morocco is located in North Africa, south of the Mediterranean. Historically, the country was under Spanish and French protectorates, which is why many of the management tools used by Moroccan companies are derived from the French system (Elhamma and Moalla, 2015). However, in recent years, Moroccan companies have begun to adopt modern tools, such as international accounting standards (IFRS) (Elhamma, 2023, 2024, 2025), contemporary management control tools (Elhamma, 2012, 2014, 2015), etc.

In management sciences, researchers have the option to choose between quantitative or qualitative methods (Elhamma, 2024). Given the nature of our research problem, we will adopt quantitative methods, specifically using questionnaires as the data collection tool. The questionnaire distributed over a period of 20 days through two modes: via professional social networks targeting the relevant businesses, specifically Moroccan SMEs, and directly sent to managers of small and medium-sized enterprises. Out of the 34 responses received, 31 were retained (2 responses were from large enterprises, which were outside the intended target, and one questionnaire was incomplete). The distribution of the responses is as follows: Six medium-sized enterprises, nineteen small enterprises, and six very small enterprises.

Table 1. Presentation of the sample by company size

Type of Companies	Number of Companies by Type
Medium-sized company (50 to 200 employees and/or revenue between 10 and 175 million MAD)	**6**
Trade	5
Industry	1
Small company (10 to 50 employees and/or revenue between 3 and 10 million MAD)	**19**
Trade	10
Industry	5
Information technology	1
Services	3
Microenterprise (< 10 employees and/or revenue < 3 million MAD)	**6**
Trade	2
Services	4
Total	**31**

Source: Authors

The sample of 31 SMEs, though relatively small, is justified for several reasons. First, only SMEs that have already implemented dashboards were included in the study. However, the use of such tools is still relatively rare in Morocco, which naturally limits the available sample size. Second, while the sample size is modest, it exceeds the minimum threshold of 30 units, which is generally considered sufficient for ensuring representativeness and conducting reliable statistical analyses.

2.3. Variable Measurement and Data Collection

The first variable examined in this study is the extent of BSC implementation by the surveyed SMEs. To measure this variable, respondents were asked to indicate the degree to which the components associated with the four perspectives of the BSC are integrated into their dashboards. These perspectives include the "Financial Perspective", "Customer Perspective", "Learning and Growth Perspective", and "Internal Process Perspective". Respondents were given three response options: "unimportant", "moderately important", and "Very important". For each perspective, examples of indicators were provided to identify which indicators are most commonly used for each perspective.

The second variable examined in this study pertains to the barriers to BSC adoption by Moroccan SMEs. To measure this variable, five potential barriers were presented to the respondents: formalized and standardized tools, lack of human resources, the time required for implementation, the complexity of the tool, and lack of financial resources. Each respondent was given the option to indicate the barriers they encountered in achieving full implementation of the BSC.

3. RESULTS AND DISCUSSION

3.1. The Applicability of the Balanced Scorecard in SMEs

The table below (Table 2) presents the results regarding the level of BSC adoption among the surveyed SMEs.

Table 2. Degree of importance of the indicators integrated into the dashboards of the companies surveyed

Types of indicators	Degree of importance	%
Financial Perspective	Very important	80.65%
	Moderately important	12.90%
	Unimportant	6.45%
Customer Perspective	Very important	77.42%
	Moderately important	12.90%
	Unimportant	9.68%
Organizational Learning Perspective	Very important	22.58%
	Moderately important	38.71%
	Unimportant	38.71%
Internal Processes Perspective	Very important	19.35%
	Moderately important	45.16%
	Unimportant	35.49%

Source: Authors

Figure 1. The level of importance of the indicators included in the dashboards of the surveyed companies

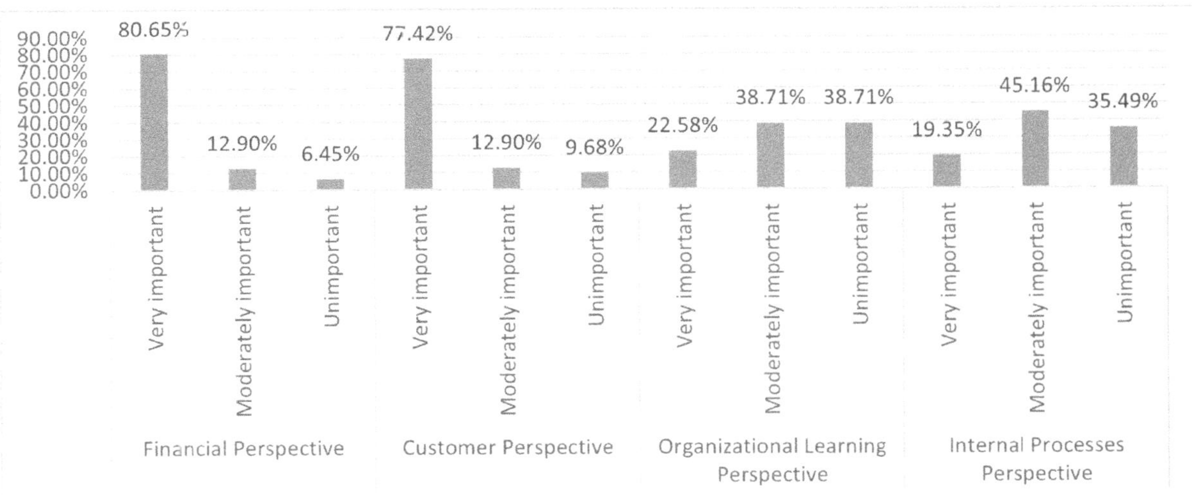

Source: Authors

The study's results show that the financial and customer perspectives are considered highly important by the majority of companies, with 80.65% and 77.42% of responses, respectively. This reflects the emphasis placed on profitability and customer satisfaction in the strategic management of companies. In contrast, the perspectives of organizational learning and internal processes are viewed as less essential.

3.1.1. The Financial Perspective

For each perspective, four key indicators were selected. For the financial perspective, we chose sales, return on investment, net profit, and cash flow. The results are summarized in the following table.

Table 3. Degree of importance of integrating financial indicators into the dashboards of the surveyed SMEs.

Types of indicators	Degree of importance	%
Sales	Very important	87.10%
	Moderately important	9.70%
	Unimportant	3.2%
Return on investment	Very important	54.80%
	Moderately important	22.60%
	Unimportant	22.6%
Net income	Very important	41.90%
	Moderately important	29%
	Unimportant	29.1%
cash-flow	Very important	38.70%
	Moderately important	29%
	Unimportant	32.3%

Source: Authors

Figure 2. Importance of integrating financial indicators into the dashboards of the surveyed SMEs.

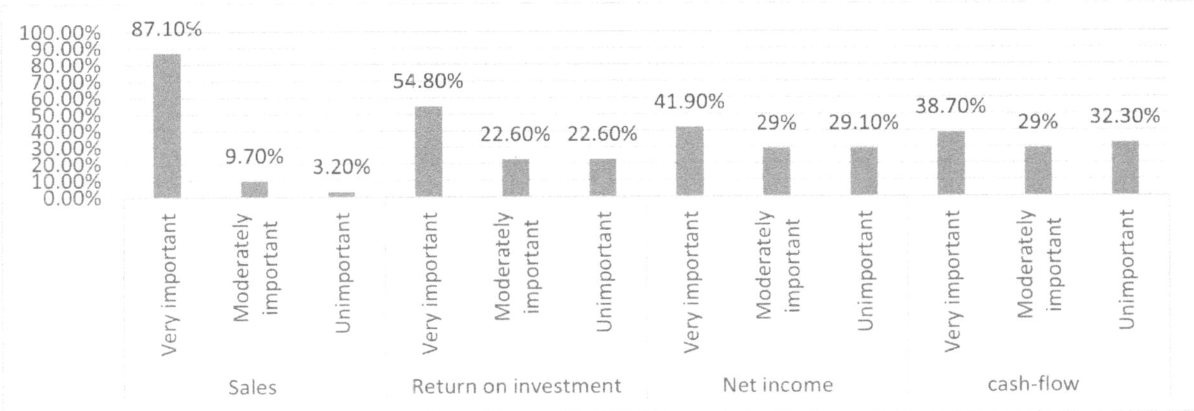

Source: Authors

Sales (87.1%) is by far the most important indicator for the surveyed SMEs, while cash flow (38.7%) is the least valued. This suggests that SMEs tend to prioritize indicators reflecting short-term performance over long-term objectives. Consequently, it can be inferred that SMEs may be constrained by time, which limits their ability to focus on longer-term strategic goals.

3.1.2. The Customer Perspective

Similar to the financial perspective, four indicators were selected for the customer perspective: customer loyalty rate, customer satisfaction, new customer rate and market share. The following table summarizes the study's results.

Table 4. Degree of importance of integrating customer-related indicators into the dashboards of the surveyed SMEs.

Types of indicators	Degree of importance	%
Customer loyalty rate	Very important	80.6%
	Moderately important	9.7%
	Unimportant	9.7%
Customer satisfaction	Very important	67.7%
	Moderately important	19.4%
	Unimportant	12.9%
New customer rate	Very important	45.2%
	Moderately important	19.4%
	Unimportant	35.4%

continued on following page

Table 4. Continued

Types of indicators	Degree of importance	%
Market share	Very important	38.7%
	Moderately important	32.3%
	Unimportant	29%

Source: Authors

Figure 3. Importance of integrating customer-related indicators into the dashboards of the surveyed SMEs.

Source: Authors

The SMEs surveyed primarily focused on customer loyalty rate (80.6%), followed by customer satisfaction (67.7%), new customer rate (45.2%), and finally, market share (38.7%), which ranked last. The two key indicators, in turn, enhance financial performance, as loyal, satisfied customers save the SME money, time, and effort.

3.1.3. The Organizational Learning Perspective

As for the organizational learning perspective, the selected indicators are the improvement of the information system, product/service innovation, recruitment of experienced personnel, and enhancement of employee skills. The following table summarizes the survey results.

Table 5. Degree of importance of integrating organizational learning-related indicators into the dashboards of the surveyed SMEs

Types of indicators	Degree of importance	%
Improving employee skills	Very important	41.9%
	Moderately important	35.5%
	Unimportant	22.6%
Improving the information system	Very important	41.9%
	Moderately important	25.8%
	Unimportant	32.3%
product/service innovation	Very important	32.3%
	Moderately important	45.2%
	Unimportant	22.5%
recruitment of experienced staff	Very important	29%
	Moderately important	29%
	Unimportant	42%

Source: Authors

Figure 4. Importance of integrating organizational learning-related indicators into the dashboards of the surveyed SMEs.

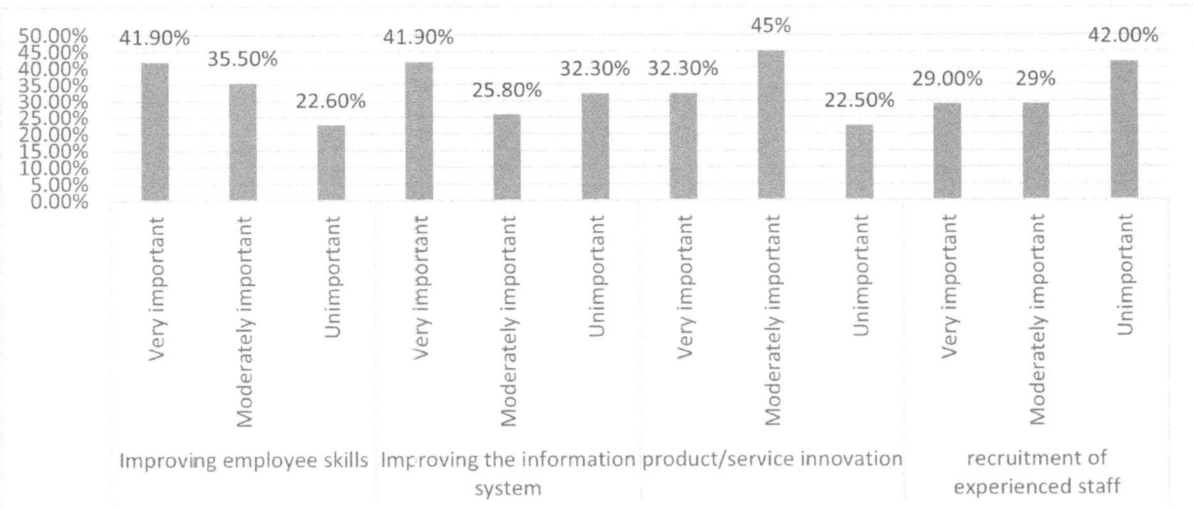

Source: Authors

Based on our results, the improvement of skills and the enhancement of the information system are considered equally important (41.9%), followed by innovation, and, lastly, the recruitment of experienced personnel. This may be explained by the fact that employees in these types of companies often hold multiple roles simultaneously to perform daily tasks, reducing the need to recruit new specialized and experienced staff.

3.1.4. The Internal Processes Perspective

Finally, regarding the internal processes perspective, the four selected indicators are the implementation of a procedures manual, the development of partnerships, the establishment of a performance-based bonus system, and the improvement of financial statement synthesis. The table below summarizes the key findings of our survey.

Table 6. Degree of importance of integrating internal processes-related indicators into the dashboards of the surveyed SMEs.

Types of indicators	**Degree of importance**	%
Improvement of financial statement synthesis	Very important	48.8%
	Moderately important	29%
	Unimportant	22.2%
Implementation of a procedures manual	Very important	41.9%
	Moderately important	32.3%
	Unimportant	25.8%
Development of partnerships	Very important	41.9%
	Moderately important	22.6%
	Unimportant	35.5%
Establishmet of a performance-based bonus system	Very important	29%
	Moderately important	16.1%
	Unimportant	54.9%

Source: Authors

Figure 5. Importance of integrating internal processes-related indicators into the dashboards of the surveyed SMEs.

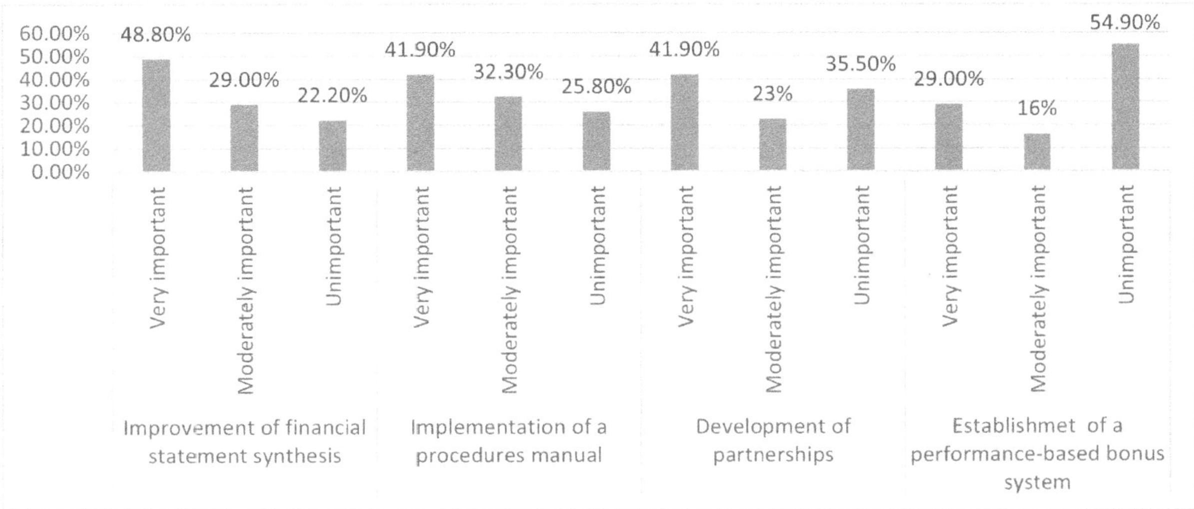

Source: Authors

The top-ranked indicator is financial, specifically the synthesis of financial statements. The implementation of a procedures manual ranks second, reflecting SMEs' awareness of the need to formalize their managerial practices, although this is not a top priority for them.

3.2. Barriers to the Implementation of the Balanced Scorecard in SMEs

The results of the survey on the main barriers to the successful implementation of the BSC are summarized in the Table 7.

Table 7. The main obstacles to the implementation of the Balanced Scorecard (BSC) in the companies studied.

Types of obstacles	%
Formalized and standardized tool	15%
Lack of human resources	2%
Long time required to implement the tool	37%
The complexity of the tool	21%
Lack of financial resources	23%
No obstacles	2%

Source: Authors

Figure 6. The key obstacles to implementing the Balanced Scorecard (BSC) in the studied companies.

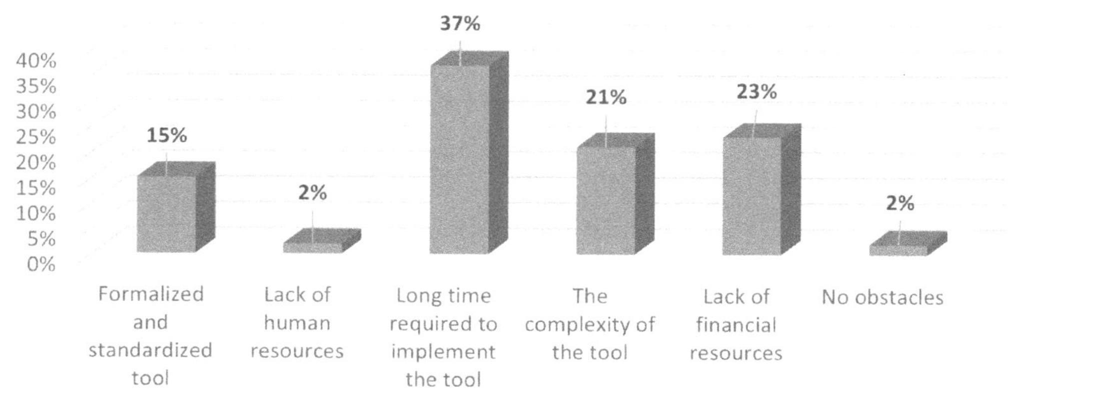

Source: Authors

Among the responses gathered, the most frequently cited obstacle to the implementation of the tool is the long time required for its adoption, with 37% of respondents highlighting this issue. This is followed by a lack of financial resources, mentioned by 23% of respondents. The complexity of the tool is also identified by 21% of participants as a significant barrier. Additionally, the Balanced Scorecard is criticized for being a formalized and standardized tool, as noted by 15% of respondents. Finally, although less frequently cited, the lack of qualified human resources is recognized by 2% of respondents as an obstacle. Notably, 2% of respondents reported no obstacles to implementing the tool.

Furthermore, we explored the potential dependence between the various obstacles identified in relation to the implementation of the BSC and the characteristics of SMEs, with a particular emphasis on the factor of size.

Table 8. The main obstacles to the implementation of the Balanced Scorecard (BSC) in SMEs based on their size.

Types of obstacles	Medium-sized company	Small company	Micro-enterprises
Formalized and standardized tool	30%	12%	11%
Lack of human resources	10%	0	0
Long time required to implement the tool	30%	46%	11%
The complexity of the tool	30%	15%	33%
Lack of financial resources	0	24%	45%
No obstacles	0	3%	0
Total	100%	100%	100%

Source: Authors

Figure 7. The key obstacles to implementing the Balanced Scorecard (BSC) in SMEs, categorized by their size.

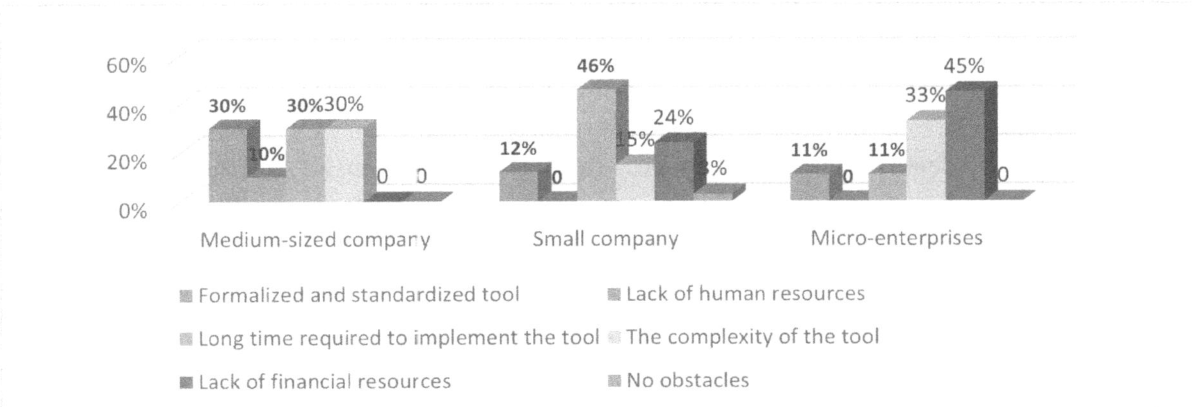

Source: Authors

For medium-sized companies, 30% of respondents express concerns about the formalized and standardized nature of the BSC, which may not align well with their established processes. They also report challenges related to the time required for implementation (30%) and the tool's complexity (30%), reflecting the need for substantial resources and adaptation. A minor concern is the lack of human resources (10%). In contrast, small companies face the greatest challenge with the time required for implementation (46%), due to limited staff and resources. Financial constraints (24%) are also significant, although they are less concerned with the tool's complexity (15%) or its formal structure (12%). Notably, no small companies reported issues related to human resources. Finally, micro-enterprises are most hindered by financial limitations (45%) and the complexity of the tool (33%). Despite their smaller size, they are less concerned with time constraints (11%) and human resource issues. To test the dependence between the obstacles to BSC implementation and firm size, we employed the Fisher-Freeman-Halton exact test, which yielded a value of 27.931 with a p-value of 0.018 ($p < 0.05$). This result confirms a significant dependency at the $\alpha = 5\%$ level between the obstacles and the firm size of the surveyed companies.

3.3. Reflection and Discussion

As demonstrated in the existing literature in the field, the findings of the study indicate that SMEs place greater emphasis on operational and financial performance compared to other internal aspects. These finding align with studies conducted on SMEs in sub-Saharan Africa (Takoudjou Nimpa & Teulon, 2018), which highlight the predominant role of accounting data within companies, as well as with the study by Tennant and Tanoren (2005), who state that "SMEs are weak in organizational learning and innovation" (p. 134). Recently, Elhamma (2023) found that most surveyed SMEs incorporate financial metrics (e.g., sales, results, cash flow) into their performance management systems (PMS). However, metrics related to internal business processes, as well as learning and growth metrics, are largely absent. Overall, the four dimensions of the BSC are unevenly integrated into the PMSs of these companies. In general, SMEs often prioritize financial and operational performance over other areas (Tennant &

Tanoren, 2005; Takoudjou Nimpa & Teulon, 2018). These results can be explained by several obstacles and barriers faced by SMEs in adopting the BSC.

The results of our study identify the key challenges faced by companies in implementing the BSC. In general, the most significant obstacles are the time required for implementation (37%) and the complexity of the tool (21%). These findings suggest that companies view the BSC as a complex system that demands a substantial time investment. The shortage of human and financial resources is also a notable challenge, although its impact is less significant (2% and 23%, respectively). These findings align with common challenges identified in previous studies (Takoudjou and Teulon, 2018; Benková et al., 2020). According to Gowindasamy and Jantan (2018), the lack of human resources, the absence of specialist support, and the lack of IT support or an adequate information system can all make implementing the BSC system more challenging. In the Sudanese context, Abdalla et al. (2022) investigated the barriers to implementing the BSC in the Sudanese banking sector. Using a qualitative approach, they conducted 12 semi-structured interviews with managers and senior executives from 10 banks operating in Sudan. The results revealed that the lack of training for human resources and the high cost of BSC implementation are key barriers to its adoption.

When analyzing the results by company size, several interesting trends emerge. Mid-sized companies seem to face greater challenges with the complexity of the tool and the time required for its implementation, suggesting they have more resources but may lack the necessary structures to manage such a change. In contrast, small companies are more impacted by the time needed for implementation, which may reflect a limited capacity to dedicate time to strategic projects. Micro-enterprises, however, are strongly affected by a lack of financial resources and the complexity of the BSC. This indicates that these companies, with their smaller structures and budgets, struggle to integrate the BSC into their operations. Finally, it is worth noting that some companies did not identify any major obstacles, which could suggest either easier adoption or an underestimation of the challenges involved in implementation. These results highlight the crucial role of company size in shaping how obstacles to BSC implementation are perceived and addressed, with challenges varying depending on available resources and organizational structures. Several researchers (Rherib et al., 2021) have demonstrated that firm size is a key contingency factor that impedes the applicability of the BSC in such organizations, as it is positively correlated with the effectiveness of the scorecard. Elhamma (2011), based on the findings of an empirical study in Morocco, examined the impact of firm size on the differentiation of indicators integrated into the dashboards of 62 companies. The results of this research indicated that dashboards tend to become more 'balanced' and align more closely with the BSC as the size of the company increases.

CONCLUSION

The objective of this study was to examine the degree of adoption of the BSC and the main barriers to its implementation in Moroccan SMEs. Using a sample of 31 SMEs, two key findings emerged. First, the dashboards used by the studied SMEs are not balanced like the BSC and primarily focus on financial indicators (sales, results, etc.). Second, the main barriers to the implementation of the BSC are primarily related to the long time required for implementation (37%), lack of financial resources (23%), and the complexity of the tool (21%).

This research offers several theoretical and managerial contributions. From a theoretical perspective, the study provides new insights into the adoption of the BSC in SMEs and the key barriers to its implementation in such companies. From a managerial standpoint, this research helps SME leaders understand this performance measurement tool and, more importantly, gain insight into the barriers to its successful implementation, enabling them to overcome these challenges and benefit from this modern management control tool.

However, the results obtained should be interpreted with caution for several reasons. First, the sample size may not accurately reflect the current situation of all Moroccan SMEs. Second, the diversity of the sample is limited, as it is predominantly composed of companies from the commercial and industrial sectors, with only one company representing each of the information technology and service sectors. As a result, it is difficult to clearly determine the specific needs of each sector or make recommendations that could be universally applicable.

To overcome the limitations mentioned, the ideal approach would be to extend this analysis to a larger sample over a longer period, using a combination of both quantitative and qualitative methods. This would allow for a deeper interpretation of the results through both exploratory and confirmatory processes. Another avenue would be to employ a random probability sampling method to obtain more relevant and less biased results, possibly by utilizing survey databases obtained from official organizations, such as the Moroccan Confederation of SMEs.

REFERENCES

Abdalla, Y. A., Ibrahim, A. M., Lasyoud, A. A., & Warsame, M. H. (2022). Barriers of implementing the balanced scorecard: Evidence from the banking sector in the developing market. *Journal of Governance & Regulation*, *11*(2), 173–180. DOI: 10.22495/jgrv11i2art15

Ates, A., Garengo, P., Cocca, P., & Bititci, U. S. (2013). The development of SME managerial practice for effective performance management. *International Journal of Production Research*, *51*(28), 8549–8567.

Atkinson, A., & Epstein, M. (2000). Measure for measure: Realizing the power of the balanced scorecard. *CMA Management*, *74*(7), 23–28.

Benková, E., Gallo, P., Balogová, B., & Nemec, J. (2020). Factors affecting the use of balanced scorecard in measuring company performance. *Sustainability (Basel)*, *12*(3), 1178. DOI: 10.3390/su12031178

Cantele, S., & Zardini, A. (2018). Is sustainability a competitive advantage for small businesses? An empirical analysis of possible mediators in the sustainability–financial performance relationship. *Journal of Cleaner Production*, *182*, 166–176. DOI: 10.1016/j.jclepro.2018.02.016

Cardoni, A., Zanin, F., Corazza, G., & Paradisi, A. (2020). Knowledge management and performance measurement systems for SMEs' economic sustainability. *Sustainability (Basel)*, *12*(7), 2594. DOI: 10.3390/su12072594

Choi, J., Kim, S., & Yang, D.-H. (2018). Small and medium enterprises and the relation between social performance and financial performance: Empirical evidence from Korea. *Sustainability (Basel)*, *10*(6), 1816. DOI: 10.3390/su10061816

Cifalinò, A., & Lisi, I. E. (2019). Managing multiple forms of strategic training fit through the balanced scorecard. *International Journal of Training and Development*, *23*(3), 240–252. DOI: 10.1111/ijtd.12160

Cooper, D. J., Ezzamel, M., & Qu, S. Q. (2017). Popularizing a management accounting idea: The case of the balanced scorecard. *Contemporary Accounting Research*, *34*(2), 991–1025. DOI: 10.1111/1911-3846.12299

DeBusk, G. K., & Crabtree, A. D. (2006). Does the balanced scorecard improve performance? *Management Accounting Quarterly*, *8*(1), 44–48.

Dehbi, S., & Angade, K. (2019). What is the approach for research in management science? *Journal Name*, 16.

Dereli, D. D. (2015). Innovation management in global competition and competitive advantage. *Procedia: Social and Behavioral Sciences*, *195*, 1365–1370. DOI: 10.1016/j.sbspro.2015.06.323

Dudic, Z., Dudic, B., Gregus, M., Novackova, D., & Djakovic, I. (2020). The innovativeness and usage of the balanced scorecard model in SMEs. *Sustainability (Basel)*, *12*(8), 3221. DOI: 10.3390/su12083221

Elbanna, S., Kamel, H., Fatima, T., & Eid, R. (2022). An investigation of the causality links in the balanced scorecard: The case of the Gulf Cooperation Council hospitality industry. *Tourism Management Perspectives*, *41*, 100934. DOI: 10.1016/j.tmp.2021.100934

Elhamma, A. (2011). Impact of size on the content of dashboards in companies in Morocco: Results of an empirical study. *Moroccan Journal of Management and Marketing Research, 317.*

Elhamma, A. (2012). The activity-based costing in Morocco: Adoption and diffusion. *Arabian Journal of Business and Management Review*, *1*(6), 33–45. DOI: 10.12816/0002111

Elhamma, A. (2014). Performance du Balanced Scorecard: Perception of company managers. *International Journal of Management and Strategy*, *5*(2), 1–9.

Elhamma, A. (2015). The relationship between budgetary evaluation, firm size, and performance. *Journal of Management Development*, *34*(8), 973–986. DOI: 10.1108/JMD-06-2014-0053

Elhamma, A. (2023). Impact of mandatory IFRS adoption on foreign direct investment: The moderating role of conflict of interest regulation. *Journal of Financial Reporting and Accounting, Vol. ahead-of-print*, No. ahead-of-print.

Elhamma, A. (2023). The relationship between organisational decentralisation, balanced scorecard and its perceived benefits in Moroccan SMEs. *International Journal of Accounting. Auditing and Performance Evaluation*, *19*(2), 185–202. DOI: 10.1504/IJAAPE.2023.132384

Elhamma, A. (2024). Determinants of national IFRS adoption: Evidence from the Middle East and North Africa region. *International Journal of Accounting. Auditing and Performance Evaluation*, *20*(1-2), 69–90. DOI: 10.1504/IJAAPE.2024.135535

Elhamma, A. (2024). Participant observation as a qualitative research method in management sciences: Interests, conditions, and limitations. In Bentalha, B., & Alla, L. (Eds.), *Applying qualitative research methods to management science* (pp. 75–91). IGI Global., DOI: 10.4018/979-8-3693-5543-5.ch004

Elhamma, A. (2025). Lack of resilience and failure of small and medium-sized enterprises (SMEs) in Morocco: Current situation and key determinants. In Dias Rouco, J. C., & Nunes Figueiredo, C. (Eds.), *Enhancing resilience in business continuity management* (pp. 245–266). IGI Global., DOI: 10.4018/979-8-3693-8809-9.ch008

Elhamma, A. (2025). Perceived benefits of IFRS adoption in Morocco and French-speaking Sub-Saharan African countries. In F. Albuquerque & P.G. dos Santos (Eds.), *Contemporary innovations in reporting and analysis* (pp. 183–210). IGI Global.

Elhamma, A., & Moalla, M. (2015). Impact of uncertainty and decentralization on activity-based costing use. *International Journal of Accounting and Economics Studies*, *3*(2), 148–155. DOI: 10.14419/ijaes.v3i2.4817

Fatima, T., & Elbanna, S. (2020). Balanced scorecard in the hospitality and tourism industry: Past, present, and future. *International Journal of Hospitality Management*, *91*, 102656. DOI: 10.1016/j.ijhm.2020.102656 PMID: 32904633

Frederico, G. F., Garza-Reyes, J. A., Kumar, A., & Kumar, V. (2020). Performance measurement for supply chains in the Industry 4.0 era: A balanced scorecard approach. *International Journal of Productivity and Performance Management*, *70*(4), 789–807. DOI: 10.1108/IJPPM-08-2019-0400

Garengo, P., Biazzo, S., & Bititci, U. S. (2005). Performance measurement systems in SMEs: A review for a research agenda. *International Journal of Management Reviews*, *7*(1), 25–47. DOI: 10.1111/j.1468-2370.2005.00105.x

Garengo, P., & Sharma, K. M. (2012). Performance measurement system contingency factors: A cross analysis of Italian and Indian SMEs. 1–21.

2GC. (2021). *Balanced scorecard usage survey 2020*. 2GC. Available at https://2gc.eu/resources/survey-reports/2020-survey

Gowindasamy, M., & Jantan, A. H. (2018). Barriers to the successful implementation of balanced scorecard in a few selected non-profit organizations in Malaysia. *European Journal of Business and Management*, *10*(15), 35–50.

Hendricks, K. B., Wiedman, C. I., Menor, L., & MenorIssues, L. (2004). *The balanced scorecard*. 7.

Hoque, Z. (2014). 20 years of studies on the balanced scorecard: Trends, accomplishments, gaps and opportunities for future research. *The British Accounting Review*, *46*(1), 33–59. DOI: 10.1016/j.bar.2013.10.003

Hu, Y., Wen, J., & Yan, Y. (2015). Measuring the performance of knowledge resources using a value perspective: Integrating BSC and ANP. *Journal of Knowledge Management*, *19*(6), 1250–1272. DOI: 10.1108/JKM-10-2014-0431

Julien, P.-A., & Carrier, C. (2005). *Les PME bilan et perspectives* (3rd ed.).

Kaplan, R. S. (2009). Conceptual foundations of the balanced scorecard. 37.

Kaplan, R. S., & Grossman, A. S. (2010). How market mechanisms from the private sector could energize the nonprofit world.

Kaplan, R. S., & McMillan, D. (2020). Updating the balanced scorecard for triple bottom line strategies. SSRN *Electronic Journal*. DOI: 10.2139/ssrn.3682788

Kaplan, R. S., & Norton, D. P. (1992). Measures that drive performance. *Harvard Business Review*. PMID: 10119714

Kaplan, R. S., & Norton, D. P. (1996). Using the balanced scorecard as a strategic management system. *Harvard Business Review*.

Kaplan, R. S., & Norton, D. P. (2000). Having trouble with your strategy? Then map it. *Harvard Business Review*. PMID: 11143152

Kaplan, R. S., & Norton, D. P. (2001). Transforming the balanced scorecard from performance measurement to strategic management: Part I. *Accounting Horizons*, *15*(1), 87–104. DOI: 10.2308/acch.2001.15.1.87

Klovienė, L., & Speziale, M.-T. (2015). Is performance measurement system going towards sustainability in SMEs? *Procedia: Social and Behavioral Sciences*, *213*, 328–333. DOI: 10.1016/j.sbspro.2015.11.546

Kumar, S., Lim, W. M., Sureka, R., Jabbour, C. J. C., & Bamel, U. (2024). Balanced scorecard: Trends, developments, and future directions. *Review of Managerial Science*, *18*(8), 2397–2439. DOI: 10.1007/s11846-023-00700-6

Lueg, R., & Vu, L. (2015). Success factors in balanced scorecard implementations. [PDF].

Madsen, D. O., & Stenheim, T. (2014). Perceived problems associated with the implementation of the balanced scorecard: Evidence from Scandinavia. *Problems and Perspectives in Management*, *12*(1), 121–131.

Malagueño, R., Lopez-Valeiras, E., & Gomez-Conde, J. (2018). Balanced scorecard in SMEs: Effects on innovation and financial performance. *Small Business Economics*, *51*(1), 221–244. DOI: 10.1007/s11187-017-9921-3

Manzaneque-Lizano, A., Alfaro-Cortés, L., & De La Priego Cruz, M. (2019). Stakeholders and long-term sustainability of SMEs: Who really matters in crisis contexts, and when? *Sustainability (Basel)*, *11*(23), 6551. DOI: 10.3390/su11236551

McAdam, R. (2000). Quality models in an SME context. *International Journal of Quality & Reliability Management*, *17*(3), 305–323. DOI: 10.1108/02656710010306166

Mehralian, G., Nazari, J. A., Nooriparto, G., & Rasekh, H. R. (2017). TQM and organizational performance using the balanced scorecard approach. *International Journal of Productivity and Performance Management*, *66*(1), 111–125. DOI: 10.1108/IJPPM-08-2015-0114

Méric, J. (2003). The emergence of a discourse on managerial innovation? The case of the Balanced Scorecard. *Accounting - Control - Auditing, 9*(3), 129.

Meyssonnier, F. (2015). What management control for start-ups? *Accounting - Control - Audit, 21*(2), 33.

Mintzberg, H. (1982). *Structure and dynamics of organizations* (1st ed.).

Nielsen, C., Lund, M., & Thomsen, P. (2017). Killing the balanced scorecard to improve internal disclosure. *Journal of Intellectual Capital*, *18*(1), 45–62. DOI: 10.1108/JIC-02-2016-0027

Oriot, F., & Misiaszek, E. (2012). The Balanced Scorecard through the filter of a French SME. Or why SMEs prefer “tailor-made”. *Revue française de gestion, 38*(225), 27–43.

Papulová, Z., Gažová, A., Šlenker, M., & Papula, J. (2021). Performance measurement system: Implementation process in SMEs. *Sustainability (Basel)*, *13*(9), 4794. DOI: 10.3390/su13094794

Pessanha, D. S. da S., & Prochnik, V. (2006). Practitioners’ opinions on academics’ critics on the balanced scorecard. SSRN *Electronic Journal*. DOI: 10.2139/ssrn.1094308

Quesado, P., Aibar Guzmán, B., & Lima Rodrigues, L. (2018). Advantages and contributions in the balanced scorecard implementation. *Intangible Capital*, *14*(1), 186. DOI: 10.3926/ic.1110

Rherib, N., El Amili, O., & Baba, E. K. (2021). The Impact of Company Size and Age on the Adoption of the Balanced Scorecard in Morocco: Results of an Empirical Study. *Revue Internationale des Sciences de Gestion*, *4*(2), 671–685.

Rigby, D., & Bilodeau, B. (2009). *Management tools and trends*. Bain & Company.

Rigby, D., & Bilodeau, B. (2018). *Bain, Brief Management tools and trends.*

Rigby, D., & Bilodeau, B. (2018). *Management tools and trends*. Bain & Company.

Robinson, R. B.Jr, & Pearce, J. A.II. (1984). Research thrusts in small firm strategic planning. *Academy of Management Review*, *9*(1), 128–137. DOI: 10.5465/amr.1984.4278109

Rompho, N. (2011). Why the balanced scorecard fails in SMEs: A case study. *International Journal of Business and Management*, *6*(11), 39. DOI: 10.5539/ijbm.v6n11p39

Sardi, A., Sorano, E., Ferraris, A., & Garengo, P. (2020). Evolutionary paths of performance measurement and management system: The longitudinal case study of a leading SME. *Measuring Business Excellence*, *24*(4), 495–510. DOI: 10.1108/MBE-01-2020-0016

Takoudjou Nimpa, A., & Teulon, F. (2018). Viewpoint: An exploratory study of the applicability of the balanced scorecard in SMEs in Francophone sub-Saharan Africa. *Gestion 2000, 35*(3), 129.

Tawse, A., & Tabesh, P. (2023). Thirty years with the balanced scorecard: What we have learned. *Business Horizons*, *66*(1), 123–132. DOI: 10.1016/j.bushor.2022.03.005

Tennant, C., & Tanoren, M. (2005). Performance management in SMEs: A balanced scorecard perspective. *International Journal of Business Performance Management*, *7*(2), 123. DOI: 10.1504/IJBPM.2005.006486

Trébucq, S. (2011). The balanced scorecard in France. A communication tool that is still misunderstood. *French Journal of Management*, *37*(211), 131–143.

Tsai, W.-H., Chou, W.-C., & Hsu, W. (2009). The sustainability balanced scorecard as a framework for selecting socially responsible investment: An effective MCDM model. *The Journal of the Operational Research Society*, *60*(10), 1396–1410. DOI: 10.1057/jors.2008.91

Zazueta Salido, R. A., Lagarda-Leyva, E. A., & Lozoya Diaz, D. G. (2019). Strategic plan for a regional innovation center and business accelerator of southern Sonora using megaplanning and balanced scorecard. *287–323.*

Chapter 12

Business Model Effect on Innovation With Mediation of Open Innovation Analysis in Africa

Fikeralem Toma
https://orcid.org/0009-0006-5078-7763
Bule Hora University, Ethiopia

Shashi Kant
https://orcid.org/0000-0003-4722-5736
Bule Hora University, Ethiopia

Tafese Niguse
https://orcid.org/0009-0008-6536-6392
Bule Hora University, Ethiopia

ABSTRACT

This chapter investigates how models of business influence innovation and the mediation part innovation in open sources plays in this process especially for African industries. Our investigation draws data from different industry employees to test theories around the Business Model Canvas and Innovation in open sources Theory. Our initial Kaiser-Meyer-Olkin evaluation demonstrated that the investigation sample meets suitable requirements for statistical investigation. We used Exploratory Factor Investigation to detect the essential elements behind the data. CFA validated our stated measurement tool. The investigation utilized AMOS software to test how models of business connect with innovation while innovation in open sources functions as an intermediary. Reducing business model design flaws improves innovative results and innovation in open sources helps make this influence stronger.

DOI: 10.4018/979-8-3693-4030-1.ch012

INTRODUCTION

In the swiftly changing international ecosystem, the association among models of business and innovation has gained substantial focus, especially regarding developing nations like those in Africa (Kant et al., 2025). As nations endeavor to improve their competitive advantage, comprehending how models of business influence innovation particularly through the framework of innovation in open sources has become essential (Spieth et al., 2025). Innovation in open sources, a notion brought to prominence by Henry Chesbrough in the early 2000s, highlights the implication of external open innovations and implicit knowledge exchange in the innovation process. This viewpoint is especially pertinent in Africa, where limited resources and varied market demands require inventive methods for business development and sustainability (Ciampi, et al., 2021). Investigation literature concerning models of business and innovation has largely concentrated on developed countries, frequently neglecting the distinctive challenges and opportunities found in African markets (Tetteh et al., 2025). Investigations have indicated that innovative models of business can spur fiscal development, yet numerous African companies face difficulties in influence executing such models due to infrastructural limitations, restricted access to financial resources, and a shortage of technical advanced workers (Li & Zhang, 2025). Also, while innovation in open sources has been a implicit knowledge as a viable means to navigate these constraints, empirical investigates exploring its mediating function in the African context are limited (Kant, et al., 2025).

Many African countries are experiencing rapid fiscal development driven by a youthful population (Spieth et al., 2025) and augmenting digital networking (Kant et al., 2025). Also, realizing this potential is often hindered by structural issues, such as inadequate infrastructure, limited access to seed capital, and a lack of technical advanced workers (Li & Zhang, 2025). This case exemplifies how even innovative models of business can struggle to thrive in ecosystems lacking supportive infrastructure (Solaja et al., 2025). Another compelling case study is found in South Africa, where the government has made strides in nurturing innovation through policies aimed at fostering entrepreneurship (Alsaadi et al., 2025). Also, many small and medium-sized enterprises (SMEs) still grapple with bureaucratic red tape and regulatory barriers that stifle creativity and slow down the implementation of innovative practices (Bennett et al., 2025). An investigation conducted by the World Bank highlighted that (Jenkinson, et al., 2024), while South Africa has a robust framework for innovation, the actual uptake among SMEs is limited due to these practical challenges (Njinyah, et al., 2024). This case investigates underscore the complexity of implementing innovative models of business and the critical need for innovation in open sources practices that can facilitate and implicit knowledge sharing across sectors.

In spite of the augmenting interest in innovation and models of business, numerous practical obstacles remain (Bennett et al., 2025). These obstacles comprise insufficient infrastructure, regulatory challenges, and a lack of inclusive policies that promote open innovation among businesses and external organizations such as universities and investigation institutions (Kant & Adula, 2025). Also, cultural elements can influence the adoption of innovation in open sources methods, influencing how businesses interact with external open innovation (Alsaadi et al., 2025). This chapter intends to fill these investigation voids by examining the influence of models of business on innovation, particularly through the mediation of innovation in open sources in Africa. The central question of the survey for this survey is: ***"How corporate models influence innovation in African companies, and what role does innovation in open sources play as a mediation factor in this association?"***

STATEMENT OF PROBLEM

Despite the a implicit knowledge implication of models of business in fostering innovation, especially through innovation in open sources, considerable gaps remain in the comprehension and execution of these concepts within the African setting (Yanada et al., 2025). The desired outcome would involve African enterprises utilizing innovative models of business and innovation in open sources to improve their competitive advantage, resulting in sustainable fiscal development. Also, contradictory evidence is present, with some investigation suggesting that although innovative models of business can stimulate development, numerous firms still struggle due to influence implementation strategies (Tetteh, et al., 2024). The ambiguous outcomes from current investigate reveal theoretical deficiencies, particularly in grasping how innovation in open sources influences the connection among models of business and innovation results. Implicit knowledge deficiencies also arise from a scarcity of empirical investigation tailored to the African setting, complicating the process of applying outcomes from developed regions (Solaja et al., 2025). As well, contextual deficiencies are apparent in the limited investigation of how local cultural, fiscal, and political elements influence the adoption of innovative practices (Bennett et al., 2025. Practical obstacles are numerous, including insufficient infrastructure, regulatory limitations, and restricted access to financing, which also obstruct the influence function of innovative models of business (Akhigbe, et al., 2024).

This investigation intends to address these deficiencies by delivering a thorough examination of how models of business influence innovation in African firms, with a specific focus on the mediating function of innovation in open sources. By doing so, it aspires to provide insights that can guide future inquiries and assist policymakers in developing ecosystems that promote innovation and fiscal advancement.

THEORITICAL FOUNDATION

The investigation relies on multiple theories to structure its examination of the connection among models of business, innovation, and innovation in open sources. Central to this are the ***"Resource-Based View (RBV)",*** which highlights the implication of internal resources and capabilities in gaining competitive advantage, and the Innovation in *"Open sources Theory",* which asserts that companies can gain from both internal and external innovation sources (Bashir, et al., 2023). The RBV emphasizes how the distinctive resources available to African firms can be utilized to develop innovative models of business (Ciampi, et al., 2021). Conversely, Innovation in open sources Theory illustrates the importance of open innovation and implicit knowledge sharing in fostering innovation, especially in ecosystems with limited resources (Tetteh et al., 2025).

The primary theoretical perspective for this investigation is the Innovation in ***"Open sources Theory".*** This framework is especially apt for exploring the African setting, where open innovation across different sectors such as partnerships among businesses, universities, and government bodies can aid in the creation and execution of innovative models of business (Li & Zhang, 2025). By concentrating on this theory, the investigation seeks to reveal how innovation in open sources can influence the connection among models of business and innovation results, laying a strong groundwork for comprehending the dynamics involved (Yanada et al., 2025).

The "***Dynamic Capabilities Theory***" states that business entities must create and change their skills in accordance with changing surroundings. In terms of operating models then innovation, this idea implies that African enterprises must develop distinctive characteristics that enable opportunities to detect opportunities, exploit them, and alter the way they operate accordingly. Companies that integrate their business models alongside innovative practices can better handle external issues and use open innovation to boost resilience and agility (Spieth et al., 2025).

"Innovation Diffusion Theory" describes the reasons for, how, including how quickly new ideas and technologies spread within a culture. In Africa, knowing the variables that influence the adoption of new business models is critical (Alsaadi et al., 2025). This framework can assist identify the qualities of innovations which promote or impede their spread among organizations, as well as recognize the role of open innovation in encouraging the sharing of innovations and procedures across organizational borders. Businesses may use open innovation to expedite the process of dissemination and enhance their inventive outcomes (Bennett et al., 2025). The philosophical foundation for studying the influence of business models upon innovation, with the mediation of open innovation in Africa, consists of many frameworks. Dynamic Capabilities Theory stresses adaptation, whereas the concept of innovation diffusion is concerned with the dissemination of new methods.

DEFINITION AND ORIGIN OF CONCEPTS

Business Models: The framework that describes how an organization generates, provides, and acquires value is known as a business model. It includes the target market, income streams, cost structure, and value proposition of the business. Over time, the idea has changed, with several models appearing to take into account shifts in consumer behavior, market dynamics, and technology (Yanada et al., 2025).

Innovation: The process of converting concepts or innovations into products and services that add value or enhance already-existing commodities or procedures is known as innovation. This phrase refers to a broad spectrum of endeavors, from small-scale enhancements to game-changing innovations (Yanada et al., 2025).

***Open Innovation*:** A paradigm known as "innovation in open sources" pushes businesses to boost creativity by combining their own resources with outside concepts and technology. This idea first surfaced in the early 2000s and went against the conventional wisdom that innovation should only take place inside businesses (Spieth et al., 2025). Since they emphasize the interaction among internal resources and external partnerships in nurturing sustainable development and development, these words are essential to comprehending the dynamics of innovation in the context of African economies (Li & Zhang, 2025).

Empirical Literature Review

Models of Business and Innovation

Organizations' innovation ecosystem is greatly shaped by their models of business, which have a big influence on their capacity to develop and use creative ideas (Alsaadi et al., 2025). Businesses' understanding and response to client demands are directly influence by the value proposition outlined in a business model, which motivates them to customize goods and services that improve customer pleasure and experience (Bennett et al., 2025). Also, a business model's structure dictates how resources are distributed; models that place a high priority on innovation frequently devote substantial financial, human, and technological resources to investigation and development (R&D), creating an atmosphere that encourages experimentation and the creation of new products (Snihur & Bocken, 2022). Also, partnerships and cooperation are often emphasized in creative models of business, which are critical to innovation in open sources. Companies can access a larger pool of ideas and resources by interacting with external stakeholders like universities, investigation institutes, and other enterprises. This can result in the development of solutions that may not be possible through internal efforts alone. A business model's flexibility and adaptability are also essential because they enable companies to react quickly to shifts in the market and customer needs, nurturing ongoing innovation. In the software sector, for case study, subscription-based models of business allow for quick product changes based on customer input, which promotes innovation and advancement (Clauss, et al., 2022).

A business model's revenue model has an influence on innovation tactics as well; organizations with a variety of income sources are frequently better able to seed capital creative endeavors. Freemium models of business are a prime case study of this; they draw sizable user bases while making money off of a portion of them, enabling ongoing investment in innovation. Also, successful models of business assist companies in controlling the risks related to innovation (Solaja et al., 2025). Agile approaches that facilitate quick prototyping and market testing are fostered by models that encourage experimentation and iterative development, which foster a culture where failure is viewed as a teaching moment (Clauss et al., 2022). Lastly, a clearly defined business model propels innovation by improving a company's competitive edge and market placement. Businesses may stand out in the marketplace by knowing their distinct value proposition, which encourages them to keep innovating to maintain their competitive edge. In summary, the interplay among models of business and innovation is complex and multifaceted, with innovative models serving as critical catalysts for sustainable development and competitiveness.

H1: Business model has substantial influence on innovation

Business Model On Innovation And Open Investigation

For businesses looking to improve their capacity for innovation and gain a competitive edge, the association among innovation in open sources and models of business is essential. Models of business have a substantial influence on how organizations participate in innovation in open sources, influencing their methods for value creation, resource allocation, and open innovation tactics (Alsaadi et al., 2025). A business model that prioritizes a open innovation framework, for case study, enables firms to specify the functions of different stakeholders, including suppliers, customers, and investigation institutions, in the innovation process. Platform-based businesses may aggressively solicit user contributions from outside sources, creating a co-creation atmosphere that improves innovation results (Moradi, et al., 2022). Also, the distribution of resources for innovation in open sources projects is determined by commercial models. Businesses that place a high priority on innovation in their models of business are more likely to make investments in building capacities that enable productive cooperation with outside partners (Spieth et al., 2025). These investments may include educating staff members on open innovation techniques and purchasing equipment that promotes candid communication. The leveraging of innovation in open sources is also influence by the unique value creation processes created by various models of business.

In a similar way, a subscription-based model promotes ongoing user input, enabling companies to iteratively enhance their products in response to outside insights (Costa & Moreira, 2022).

Also, risk management tools that allow companies to split the costs and risks of new endeavors are frequently included in models of business that support innovation in open sources (Bennett et al., 2025). By fostering a culture of experimentation, this cooperative method encourages businesses to investigate new ideas and technologies without fully assuming the risk of failure. An organization's capacity to react quickly to shifts in the market and consumer needs is improved when its business model is in line with innovation in open sources methods, guaranteeing competitiveness in ever-changing contexts (Valdez-Juárez, et al., 2021). In the end, a company's competitive advantage is greatly increased by the synergy among models of business and innovation in open sources, which allows businesses to stand out from the competition with distinctive value propositions derived from outside partnerships. In an augmenting competitive ecosystem, businesses may improve their inventive skills, promote cooperation, and achieve sustainable success by incorporating innovation in open sources ideas into their models of business (Solaja et al., 2025).

H2: Business model has substantial influence on innovation in open sources investigation

Open Innovation and Innovation

Organizations' innovation processes are greatly influence by innovation in open sources investigation, which changes the way they think through, create, and apply new concepts. By utilizing outside expertise, resources, and alliances, this strategy pushes businesses to expand their innovation activities beyond internal borders. Innovation in open sources expands the breadth of idea creation by utilizing a variety of sources, including suppliers, consumers, and investigation institutions (Alsaadi et al., 2025). This gives firms access to a multitude of viewpoints and insights that help shape the innovation process. New product offers and more inventive solutions may result from this flood of ideas. As well, integrating outside concepts and technology may quicken development cycles, allowing businesses to leverage pre-existing solutions or work with outside partners to launch innovations faster (Palumbo, et al., 2022). As well, by enabling businesses to split the expenses and risks involved in creating new goods or services, innovation in open sources promotes better risk management (Spieth et al., 2025). Because the danger of failure is lessened via cooperation, this risk-sharing approach incentivizes businesses to take on more ambitious initiatives. Interacting with external stakeholders also improves an organization's comprehension and responsiveness to market demands, resulting in innovations that are more consumer-friendly and relevant (Bennett et al., 2025). This market-driven strategy aids in avoiding the mistakes made while creating solutions that don't appeal to the intended market. As well, by allowing businesses to stand out from the competition with distinctive products and services, innovation in open sources may boost consumer loyalty and brand reputation (Srisathan, et al., 2022).

Last but not least, adopting innovation in open sources encourages an innovative culture within businesses, encouraging cooperation and information exchange that gives staff members the confidence to experiment with new concepts and open innovation with outside partners (Solaja et al., 2025). In conclusion, innovation in open sources investigation has a substantial influence on the innovation process through improving idea generation, speeding up development cycles, enhancing risk management, improving market relevance, bolstering competitive advantage, and fostering an innovative culture. In the end, this leads to more influence and successful innovation strategies that result in long-term development and success in a competitive ecosystem (Kant et al., 2025).

H3: innovation in open sources investigation has substantial influence on innovation

Mediation of Innovation, Business Models And Open Innovation

Implicit knowledge how businesses may successfully convert their strategic frameworks into observable inventive results requires a implicit knowledge of the mediation influence of innovation in open sources investigation among models of business and innovation (Alsaadi et al., 2025). By facilitating the incorporation of outside resources and implicit knowledge into current models of business, innovation in open sources enhances the innovation process by bringing in a variety of ideas, skills, and technological advancements (Bennett et al., 2025). Open innovation-focused models of business support co-creation and information sharing by establishing an atmosphere that is conducive to open innovations with external stakeholders, including suppliers, customers, and investigation institutions. This partnership guarantees that new innovations closely match market demands while also accelerating innovation (Ciampi, et al., 2021).

As well, by nurturing a culture of carrying out tests and risk-sharing, innovation in open sources investigation enables businesses to undertake large-scale initiatives that would not be possible with just conventional internal procedures (Bennett et al., 2025). Businesses may improve their ability to react to changes in the market and produce more user-centric innovations by enabling real-time feedback from external sources (Spieth et al., 2025). In the end, innovation in open sources enhances the competitive advantages provided by creative models of business, allowing companies to stand out in the market and continually adjust to changing circumstances. In order to match business model goals with influence innovation techniques, it is imperative to an implicit knowledge and capitalizes on the mediation influence of innovation in open sources (Kant, et al., 2025).

H4: *The linkage among models of business and innovation has substantially influence by innovation in open sources investigation*

CONCEPTUAL FRAMEWORK

A conceptual framework directs investigation and real-world functions by providing an organized method for comprehending the connections among different components within a given ecosystem. The framework outlines the ways in which models of business, innovation in open sources, and innovation results interact and influence one other. It asserts that models of business, which describes how value is produced and distributed, serve as the cornerstone of an organization's strategy. By serving as a mediating tool, innovation in open sources makes it easier to incorporate outside resources and expertise into the innovation process.

Figure 1. Conceptual framework

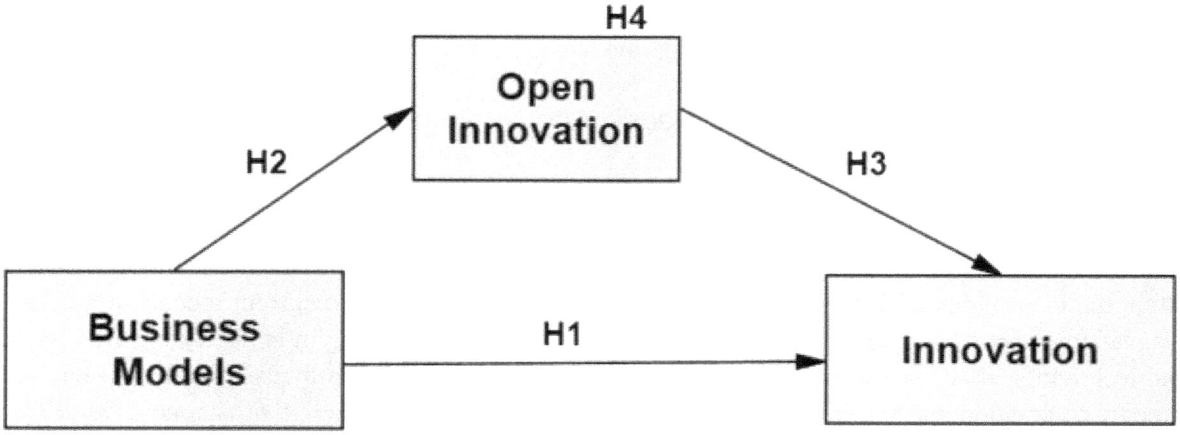

INVESTIGATION METHODOLOGY

This investigation used a quantitative investigation technique with a sample size of 390 individuals to collect empirical data on the connections among innovation in open sources, models of business, and innovation results. The Kaiser-Meyer-Olkin (KMO) measure was computed to make sure the data was suitable for factor investigation, and the result showed sufficient sampling for the investigation (Solaja et al., 2025). After the scale was complete, it was added to the survey that was given to the 390 respondents. The hypotheses pertaining to innovation in open sources and innovation outcomes were then validated and underlying dimensions were found using exploratory factor investigation (EFA). This investigation made it possible to identify substantial aspects influencing innovation processes and to reduce the complexity of the data Following EFA, the proposed correlations among the variables in the conceptual framework were examined using AMOS software and Structural Equation Modeling (SEM). SEM made it possible to examine both direct and indirect influences, giving investigations a thorough grasp of how models of business influence creativity through innovation in open sources techniques (Alsaadi et al., 2025). To make sure the structural model was accurate and dependable; the model fit indices were evaluated (Bennett et al., 2025)

Scale Development

The construction of the scale for the constructs was based on a careful investigation of pertinent literature, which guided the development of measuring questions that faithfully encapsulated each construct. In order to ensure content validity and relevance to the investigation setting, the items were drawn from validated scales utilized in prior investigations. The literature investigation emphasized substantial aspects of innovation in open sources that were converted into quantifiable metrics, including cooperation, information exchange, and integration of external resources (Solaja et al., 2025, Alsaadi et al.,

2025, Spieth et al., 2025, Bennett et al., 2025). *Every* item was thoughtfully crafted to capture the unique features of the structures it was intended to assess A smaller group pre-tested the scale in order to make necessary adjustments to the items based on comments about their relevance and intelligibility. After the scale was complete, it was added to the survey that was given to the 390 respondents. Through the EFA process, the gathered data was subsequently examined to verify the validity and reliability of the scale, guaranteeing that the measuring tools successfully captured the relevant structures. The thorough scale construction procedure was essential in providing a strong basis for the subsequent quantitative investigation, which in turn strengthened the validity of the investigation's conclusions.

DATA INVESTIGATION

Table 1. KMO and Bartlett's test

Kaiser-Meyer-Olkin Measure of Sampling Adequacy.		**.850**
Bartlett's Test of Sphericity	Approx. Chi-Square	1419.738
	df	91
	Sig.	.000

Source: Authors

The outcomes of the Bartlett's Test of Sphericity and Kaiser-Meyer-Olkin (KMO) Measure of Sampling Adequacy, which are essential for determining if the data is appropriate for factor investigation, are manifests in Table 1. A high degree of sampling adequacy is manifests by the KMO value of 0.850, which implies that there is enough sample size to do factor investigation. KMO levels near 1.0 indicate high sample adequacy, while values over 0.70 are often regarded as satisfactory. The idea that the variables in the dataset have sufficient association to merit more investigation is manifested by this high KMO score. The data's suitability for component investigation is also manifested by Bartlett's Test of Sphericity, which has an estimated Chi-square value of 1419.738 and a implication level (p-value) of 0.000. The null hypothesis that the correlation matrix is an identity matrix may be rejected since the substantial result shows that there are enough correlations among the variables. This finding is substantial because it suggests that the variables are connected and that factor investigation might provide valuable information about the data's underlying structure. When taken as a whole, these outcomes support the validity of the choice to use exploratory factor investigation (EFA) and strengthen the soundness of the investigation design used in this investigation.

Table 2. Total variance explained

Component	Initial Eigenvalues			Extraction Sums of Squared Loadings			Rotation Sums of Squared Loadings		
	Total	**% of Variance**	**Cumulative %**	**Total**	**% of Variance**	**Cumulative %**	**Total**	**% of Variance**	**Cumulative %**
1	4.51	32.235	32.235	4.51	32.235	32.235	3.00	21.451	21.451
2	1.37	9.787	42.022	1.37	9.787	42.022	2.16	15.491	36.942
3	1.27	9.087	51.110	1.27	9.087	51.110	1.98	14.168	51.110
4	.949	6.782	57.892						
5	.871	6.223	64.115						
6	.773	5.519	69.634						

Extraction Method: Principal Component Investigation.
Source: Authors

Total Variance Explained Investigation

The outcomes of the Total Variance Explained, which was manifested from the Principal Component Investigation (PCA) carried out as a component of the Exploratory Factor Investigation (EFA), are compiled in Table 2. For the components that were detected, the table displays the rotation sums of squared loadings, extraction sums of squared loadings, and starting eigenvalues. According to the initial eigenvalues, the first component is responsible for 32.235% of the cumulative variance, or 4.51 of the total variance. This implies that the initial element plays a crucial role in encapsulating the data's fundamental structure. With eigenvalues of 1.37 (9.787% of variance) and 1.27 (9.087% of variance), respectively, the second and third components also make a substantial contribution. These three factors collectively explain more than half of the variation (51.10%). The first component continues to explain 32.235% of the variation after extraction, while the first three components are still substantial. The rotation sums of squared loadings show that the first component contributes 21.451% of the variance after rotation, whilst the second and third components each contribute 15.491% and 14.168%. Their importance in the overall structure is confirmed by the tight alignment of the cumulative variance described by these three components with the earlier outcomes.

Despite being over 0.7, the eigenvalues for the fourth, fifth, and sixth components point to decreasing returns in terms of explained variance, suggesting that they might not make a substantial contribution to the model. The inclusion of the first three components for additional investigation in the investigation is validated by this investigation, which shows that they are all strong and together account for a substantial amount of the variation. This result bolsters the idea that the underlying components found by EFA are substantial and deserving of more inquiry within the parameters of the investigation design.

Confirmatory Principle Component Examination

The principle component arrangement of a set of guided proxies is verified using a type of statistics known as confirmatory principle component examination. Based on the underlying frame job and correlatives among observed proxies, investigations can evaluate hypotheses utilizing CFA. To evaluate the hypothesis that there is a association among the proxies that are being steered and the latent conceptions that underpin them, the inquiry manifested CFA under table 3.

Table 3. Covariances

Covariance			Approximation	S.E.	C.R.	P	Hy.
Open Innovation	<-->	**Business model**	**.242**	**.021**	**8.231**	***	**H2**
innovation	<-->	Open Innovation	.303	.032	7.692	***	**H3**
Business model	<-->	innovation	.254	.24	7.831	***	**H1**

Source: Authors

The correlation results among the three main constructs innovation, business model, and innovation in open sources investigation are manifests in Table 3. The approximation, standard error (S.E.), critical ratio (C.R.), and implication level (P) of each pairwise covariance are provided. With a critical ratio of 8.231 and a implication level of *** ($p < 0.001$), the correlation among innovation in open sources investigation and business model is 0.242. This large positive correlation suggests a considerable association among the efficacy and flexibility of the company model and the improvement of innovation in open sources techniques. This result demonstrates that innovation in open sources investigation has a favorable influence on business model influences, supporting Hypothesis 2 (H2).

Likewise, the correlation among innovation in open sources investigation and invention is 0.303, with a implication level of *** and a C.R. of 7.692. This finding supports Hypothesis 3 (H3) by indicating a strong positive correlation among improved innovation in open sources methods and improved innovation results. Finally, the correlation among innovation and business model is 0.254, with a implication level of *** and a C.R. of 7.831. This high correlation supports Hypothesis 1 (H1) by highlighting how crucial a clear fiscal strategy is to supporting creative endeavors.

Table 4. Validity concern

	CR	AVE	MSV	MaxR(H)	BM	ONA	INN
BM	0.732	0.643	0.219	0.852	**0.654**		
OIA	0.758	0.657	0.132	0.763	0.171	**0.665**	
INN	0.793	0.603	0.274	0.797	0.383	0.193	**0.684**

Note: BM= Business Model; OIA= open sources investigation; INN= Innovation
Source: Authors

The validity metrics for the constructs of innovation (INN), innovation in open sources investigation (OIA), and business model (BM) are manifests in Table 4. Maximum Shared Variance (MSV), Average Variance Extracted (AVE), Composite Reliability (CR), and Maximum Reliability (MaxR(H)) are among the measurements. Each construct's Composite Reliability (CR) values BM at 0.732, OIA at 0.758, and INN at 0.793 are over the permissible cutoff of 0.7. The reliability of the scales manifested in the investigation is also manifested by these results, which show a high degree of internal consistency among the items assessing each construct. The manifested criterion of 0.5 is also met by the Average variation Extracted (AVE) values, which quantify the proportion of variation collected by a construct compared to the variance caused by measurement error. The AVEs for BM, OIA, and INN are 0.643, 0.657, and 0.603, respectively. The validity of the scales is also manifested by these numbers, which demonstrate that each construct successfully captures a substantial percentage of its variance.

All constructions have low Maximum Shared Variance (MSV) values: BM is 0.219, OIA is 0.132, and INN is 0.274. These low MSV values imply that the constructs measure different ideas because there is little overlap among them. Finally, with BM at 0.852, OIA at 0.763, and INN at 0.797, the Maximum dependability (MaxR(H)) values are likewise good, suggesting strong dependability in the measurement model.

Mediating Role Examination

Figure 2. Structure equation model

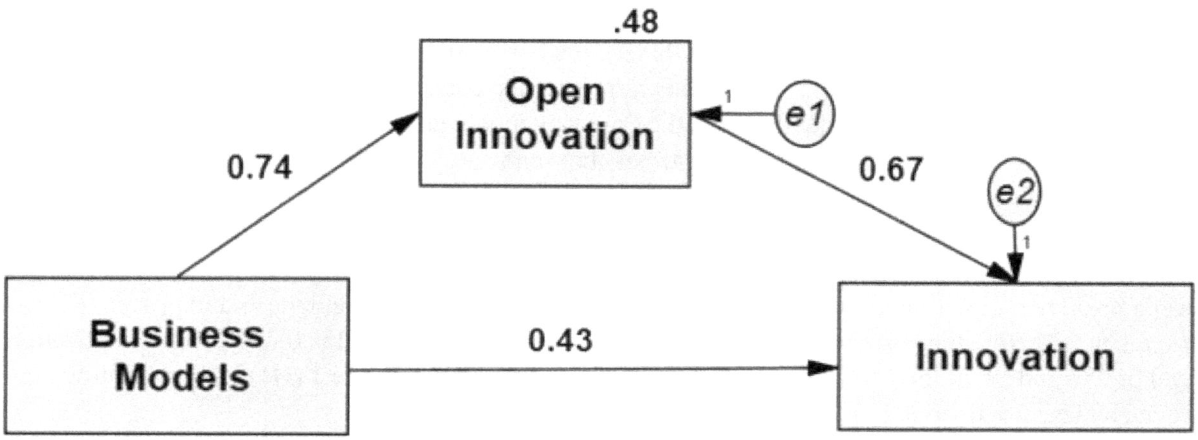

Model fit indices

Table 5. Indices for model fit

Sig.	Chi-Sq	RMR	Fitness Goodness	Fitness Confirmatory	TLI	RMSEA
0.004	1.954	.036	0.930	0.912	0.911	.028

Source: Authors

The model fit indices are manifests in Table 5 manifests substantial information about how well the manifested structural model captures the connections among the components. The outcomes include the Root Mean Square Residual (RMR), Chi-Square (Chi-Sq), substantial level (Sig.), and other goodness-of-fit indices, including RMSEA (Root Mean Square Error of Approximation) and TLI (Tucker-Lewis Index). The model fit is statistically substantial, as indicated by the implication level of 0.004. Although the model is statistically substantial, more interpretation is required to evaluate practical fit, according to a substantial Chi-Sq value of 1.954 A good match among the actual and projected covariance matrices is manifests by the low RMR value of 0.036. This implies that the model does well in reducing the differences among the predictions made by the model and the actual data. According to the goodness-of-fit indices, the Fitness Confirmatory value is 0.912 and the Fitness Goodness value is 0.930. Both numbers are higher than the generally recognized cutoff point of 0.90, indicating that the model fits the data well. The robustness of the model is also manifested by the TLI value of 0.911, which is likewise over the acceptable cutoff of 0.90. Lastly, a great match is manifests by the RMSEA value of 0.028, which is far below the ideal limit of 0.05. This low RMSEA result indicates that the population covariance matrix is sufficiently approximated by the model.

Table 6. Regression examination

Relative			**Approx.**	**S.E.**	**C.R.**	**P**	**Ass.**
Open Innovation	**<---**	**Business model**	**.864**	**.143**	**7.583**	***	**H2**
innovation	<---	Open Innovation	.328	.131	2.853	***	H3
innovation	<---	Business model	.957	.192	7.958	***	H1

Source: Authors

The outcomes of the regression investigation, which show the connections among the constructs of innovation, innovation in open sources investigation, and business model, are manifests in Table 6. The relevant coefficient (Approx.), standard error (S.E.), critical ratio (C.R.), implication level (P), and hypothesis (Ass.) for each connection are described in detail.

The correlation among Business Model and Innovation in open sources Investigation has a regression coefficient of 0.864, a critical ratio of 7.583, and a implication level of *** ($p < 0.001$). Hypothesis 2 (H2) is manifested by this substantial positive coefficient, which shows that a clearly defined business model has a considerable influence on how influence innovation in open sources approaches are.

The correlation among Innovation in open sources Investigation and Innovation has a coefficient of 0.328, a implication level of ***, and a C.R. of 2.853. This finding supports Hypothesis 3 (H3) by indicating a positive correlation among improved innovation in open sources methods and improved innovation results. Lastly, the correlation among business model and innovation has a regression coefficient of 0.957, a implication level of ***, and a C.R. of 7.958. This supports Hypothesis 1 (H1) by showing that the business model has a very substantial beneficial influence on innovation results. Overall, Table 6's regression results confirm the investigation's theoretical underpinnings by highlighting the noteworthy and constructive correlations among the constructs. The outcomes give practitioners substantial insights for improving organizational performance through successful innovation strategies by highlighting the crucial roles that models of business and innovation in open sources play in generating innovative outcomes.

Table 7. Mediating role influence

	Influence	worth	Path Influence
Business model→ Innovation	Direct Influence	.43	Direct influence stated
Business model→ Open Innovation → Innovation	Indirect Influence	.74*.67=.48	Indirect Influence Ensued
	Whole influence	.91	Partial mediation

Source: Authors

A thorough examination of Innovation in open sources Investigation's mediating function in the link among Business Model and Innovation is given in Table 7. Influence business strategies greatly improve innovation results, as evidenced by the business model's direct influence on innovation, which is measured at 0.43. Also, an indirect influence of 0.48 is manifested by multiplying the influence of Innovation in open sources Investigation on Innovation (0.67) by the influence of the Business Model on Innovation in open sources Investigation (0.74). This implies that the business model influences innovation both directly and indirectly by enabling innovation in open sources investigation. The combined influence of direct and indirect channels is demonstrated by the total or entire influence of 0.91, underscoring the fact that the business model substantially fosters innovation in a number of ways. Since Innovation in open sources investigation strengthens the link among Business Model and Innovation, the investigation comes to the conclusion that there is a partial mediation influence. This emphasizes how crucial it is to include innovation in open sources techniques into company plans in order to improve innovation potential. It also offers insightful information for companies looking to streamline their innovation procedures.

DISCUSSION

The results from this investigation illuminate the complex connections among Business Model, Innovation in open sources Investigation, and Innovation. The notable direct influence of the Business Model on Innovation highlights the importance for organizations to implement adaptive and efficient business strategies that can promote innovative results. As well, the mediating function of Innovation in open sources Investigation implies that organizations can improve their innovation abilities by incorporating innovation in open sources practices into their models of business. This interaction emphasizes the necessity for a comprehensive approach to innovation, in which businesses not only concentrate on internal processes but also actively open innovation with external ideas, technologies, and partnerships. The outcomes also suggest that the positive association among Business Model and Innovation is not simply linear; on the contrary, it is enriched through the processes of innovation in open sources, indicating that firms should focus on implicit knowledge sharing and teamwork with external parties. The research in Table 7 helps us understand how influence works within the connection between business models and innovation results. The data delineates two distinct types of influence: direct and indirect. The direct relationship between how a company operates its business and innovation stands at 0.43 showing strong evidence of impact. When organizations develop effective business models they naturally become better at executing innovative solutions.

Analysis shows two results with direct influence at 0.43 and indirect influence totaling 0.48. The study results show that business models influence innovation through additional elements that modify this basic relationship. Our outcomes show that business models drive innovation through open source examination in collaborative environments that help companies build better innovation capabilities. Our research proves external collaboration and open-source methods drive better innovation results for businesses who involve broader communities in their development programs. Both direct and indirect factors combined to powerfully impact innovation according to the overall influence score of 0.91. Partial mediation shows direct impacts remain considerable even though open source investigation now contributes to shaping innovation outcomes. Companies need to enhance both their business models and open-source collaboration methods to achieve better innovation results. Our results from Table 7 show how a business model's relationship with innovation involves multiple interlinked factors. Companies should combine traditional business model strategies with collaborative and open-source practices based on their different types of impact. This will help them reach better innovative outcomes. Our outcomes help explain how innovation works in modern interconnected business practice.

CONCLUSION

This research reveals a complete way to link Business Model to Open Source Innovation and General Innovation. Our outcomes show that a strong Business Model helps spark new ideas both right away and indirectly when applied to open-source study techniques. Your business needs to stay creative by accepting outside open innovations and working together with others to achieve success. The direct impact of organizational models on innovation shows how clear frameworks make better process flow and helps businesses use resources well toward new ideas. Strong business models make it easier for our companies to find new opportunities because they include built-in approaches to adaptation and flexibility that support fast action on emerging market trends. Open-source investigation connections help innovation thrive by showing us that teamwork and knowledge exchange drive the innovation process. Since today's world connects many businesses operate better when they tap into external methods. When organizations work closely with their customers suppliers and market competitors they discover more ideas that lead to innovation prospects. Through open-source frameworks companies share knowledge to build a creative environment where they test new ideas together and produce better innovative results.

These results create fundamental meaning for everyone involved. Organizations succeed when they won't their business model to relate with open-source innovation frameworks. By aligning business models with open-source innovation frameworks companies tap into external knowledge plus save money and get new ideas to market faster. By using both approaches companies establish better unique qualities that help them compete effectively in their industry segment. Leadership drives successful experiments and risks while getting suitable support for joint team work. Companies achieve better results when they allow employees to partner with external experts because their study proves that both strong business models and open-source innovation work together effectively. Combining both methods helps organizations advance innovation while getting ready to handle present business changes. Organizations that synchronize their business strategies effectively will achieve peak performance and future development amid market competition.

MANAGERIAL IMPLICATIONS

This research delivers essential outcomes managers need to help organizations better support innovation efforts. Top leaders need to design adaptable company plans that use their strength while working with outside partners. In current business trends organizations need to modify their operations by adding external expertise to earn long-term competitive advantages. Company success in developing new ideas relies heavily on managers who build open-source innovation platforms. When leaders let open innovation work they bring together customers suppliers and research partners to build successful projects together. Companies gain better results and transmit knowledge when multiple parties team up to bring fresh ideas and assets to life. Networking with partners helps organizations achieve better innovation results by accessing the creative intelligence of everyone involved. When managers partner with customers they gain useful product improvement and customer experience data and when they connect with research groups they find new ways to do things and discover advanced scientific methods. Managers should make employee learning an organizational priority by training staff in open-source innovation practices. Helping staff get better at open innovation requires managers to develop their employees' skill sets in using external sources. Our training should teach people better digital and collaborative skills while helping them think creatively about future plans and strategies. When organizations teach their workers to work successfully with outside partners and adopt open innovation approaches they build an agile team that generates better ideas.

Organizations need both employee technical abilities and mental attitudes embracing creative exploration. Managers should support employees who test unexpected ideas while seeing problems as chances to improve existing methods. Building innovation spaces and joint work teams help employees develop new ways of thinking by sharing ideas from outside partners. Managers must understand how to evaluate if their open innovation efforts produce desired results. Measuring performance results from partnerships helps managers determine which partnerships succeed and which ones fail. Our data-based system helps companies modify their strategies regularly to achieve better results. Our study shows that effective managers control how organizations create new ideas. Managers can build a successful innovation environment when they adapt their business models and partner selection while training employees. Organizations that take action first improve innovation results while building better ways to succeed in today's challenging business world. Organizations that use these methods achieve better business growth while staying strong during competitive challenges.

PRACTICAL IMPLICATIONS

Business models should make open-source innovation initiatives a central functional priority. A true commitment to open innovation goes beyond simply implementing tools with practices; it entails actively engaging with external sources of hidden information and new ideas. Businesses that interact and pass on ideas are better positioned to manage today's quickly changing workplace than those that remain closed. To make the process simpler, businesses must create organized procedures that encourage cooperation and information exchange. Creating specialized venues for invention, such as innovation centers, enables businesses to stimulate cooperation among their employees and external visitors. These specialized surroundings promote both the creation of project strategies and the discovery of novel insights. Corporations may boost creativity and link new ideas by cultivating both physical and digital innovation pathways. Strategic collaboration with startups and institutions is critical towards strengthening innovation skills. Startups provide innovative ways to consider and agile development methodologies, allowing enterprises to improve the efficiency of their present day operations. Collaborating with these fast-moving entities gives businesses access to new technology and market insights. Furthermore, collaboration with universities introduces experts and new ideas, allowing corporations to investigate burgeoning areas of science. For successful management of open-source innovation initiatives, firms must form organized teams that define how cooperation occurs. Clear mechanisms should be built to facilitate relationships with outside suppliers, such as managing ownership of intellectual property as well as sharing digital data.

Defining responsibilities and responsibilities before collaborating helps to reduce risks as well as assure positive outcomes. It is essential to cultivate a work atmosphere that encourages information sharing and expertise exchange. Whenever employees interact with other sources, they frequently uncover novel responses to challenges. To support this strategy, firms need to establish incentive systems that recognize effective cooperation on inclusive innovation efforts. Investing in tools and technologies that improve communication between employees is also important. Digital communication tools may dramatically increase cooperation by enabling teams as well as other stakeholders to share documents and manage undertakings far better. Based on the data in Table 7, business models have a considerable immediate impact on innovation, with a value of 0.43. Furthermore, commercial models have an indirect impact on innovation through openness to innovation, which is calculated as 0.48 (0.74 multiplied by 0.67). The aggregate effect is 0.91, suggesting partial mediation. This highlights the need of incorporating free and open-source innovation tactics into corporate designs, since the competitive landscape requires such approaches. Organizations may improve their ability to adapt quickly to market disruptions by establishing innovation centers, cultivating partner connections, having clear governance frameworks, and encouraging team cooperation and technological investments. Innovative management can help businesses preserve a competitive advantage and make ongoing progress via innovation.

THEORETICAL IMPLICATIONS

Our research adds value to business model and innovation research studies through its finding about Innovation acting as a mediator between open source investigations. In past studies business models and innovation outcomes have been studied directly without examining their underlying connections. Our research shows that these connections operate through many interacting forces that produce better innovation. The study reveals the subtle ways innovation interacts with these elements to help us better see how these relationships really function. The study results show that innovation depends heavily on external partnerships and openness as well as internal capabilities. Our analysis expands business model theory through open innovation principles by emphasizing their overlooked significance in this field. Our research analysis demonstrates a path for further exploring how partners and external resources support innovation.

By adding open-source innovation to business model discussions our theoretical methods now show innovation dynamics from multiple perspectives. By adding open-source innovation ideas into organizational analysis models researchers can now study how businesses use external and internal tools to improve their innovation methods. Research becomes stronger when scientists examine how multiple business connections impact the innovation process and results. Research in this area establishes a fundamental starting point for further research projects that study innovation dynamics. The research community can use this study's results as a foundation to analyze why open-source innovation performs better in unique business settings. Future research should analyze exactly how connecting with outside sources affects innovation results while revealing the steps that make these collaborations successful. As digital technologies emerge new business forms demand research about their role in nurturing open innovation. Our research should explore how modern technology systems like artificial intelligence and block chain help teams collaborate and share knowledge to direct progress in open innovation. Our study has improved our understanding of how Innovation connects with open sources investigation and advanced our understanding of business models and innovation theory. Examining these relationships between business models innovation and surrounding fields produces integrated models for studying how innovations operate. This new theory framework lets researchers better understand how businesses innovate in a fast-moving industry that connects global markets.

RECOMMENDATIONS

Organizations need to assess their business model frequently to spot clashes between their approved methods and open-source innovation systems. Organizations need to take positive actions to keep up with market collaboration trends and outperform their competitors. Periodic evaluations show businesses their practice weaknesses versus new innovation trends to help them improve their performance. Organizations should transform their current business plans into joint work approaches to match open-source methods. Our organization needs to create a space that lets staff share their thoughts and connects us beyond our existing networks. Organizations achieve better innovation results when they build spaces where teams can work together and learn from one another. Organizations can create successful new ideas through open source projects by setting up cross-strength teams and innovation task forces made of many skill types. Organizations must look for business connections that stretch outside their normal range of partners. Working with startup partners, academic universities and other companies brings important knowledge and resources that power innovation. These partnerships use different models – joint business ventures, research linkups, and innovative challenge programs – to develop value together in a shared knowledge network. Through teamwork organizations discover brand-new methods of innovation that surpass their own resources.

Funding initiatives that promote open-source development forms an important part of our strategy. Businesses need to use tools and platforms that make it easier for people to share and exchange knowledge without direct explanation. Projects run better through shared software platforms that let team members talk live and plan their work together. Organizations that use open innovation platforms get immediate access to more new ideas from all sources which they can effectively blend into their projects. Organizations must regularly check their innovations' results and strategy performance to build this culture successfully. Setting important innovation performance measures helps companies check their advancement and find places to improve. By checking these performance markers often teams and companies can track their open-source progress toward set goals. The regular assessment and improvement process leads to both a responsible work environment and changes necessary for market survival. To succeed with open-source practices organizations must continuously track their business strategies and stay adaptable in their assessments. Organizations can boost their innovation capacity through both strategic changes and performance reviews when they partner with others, adopt technology support, explore new connections and evaluate their innovation results. Ultimately this dedication to get better will give them a market advantage by letting them deal properly with new market chances and difficulties.

FUTURE DIRECTIONS

Research in the future should examine the specific ways Innovation in open sources investigation connects with Business Model and Innovation. Our research needs to discover what makes open-source methods drive innovation across different business settings. Researchers might explore questions such as: What steps take place to transfer knowledge from outside sources into company innovation efforts? How do organizations achieve better results with startup collaborations compared to university partnerships in their innovation efforts? Future research needs to examine details about the different ways that open-source innovation helps companies be successful. Research through time could explore how relations to innovation work. Analyzing open-source innovation and business model performance at various phases of adoption helps researchers find recurring patterns that cross-sectional work misses. Research about innovation processes needs to examine how organizations modify their strategy in reaction to marketplace changes and new technology development. When researchers examine the same open-source innovation practices at different organizational stages over time they show us how these practices advance and evolve into strong organizational practices. Our research needs to explore how cultural and organizational aspects affect open-source innovation to benefit both theoretical knowledge and business practices. Organizational culture determines how employees see collaboration risk-taking and knowledge sharing behaviors. Research now needs to test how various national cultural settings affect open-source innovation methods during their rollout and achievement period. By studying cultural norms organizations can design open-source strategies that achieve better results.

Examining how leadership methods and reward systems work together with communication systems helps us create usable outcomes for open-source success. Our research here looks at how transformational leaders create spaces for open innovation to thrive while examining which employee rewards motivate knowledge exchange. The analysis reveals better ways for leaders to build successful innovation strategies and organizational systems. Future studies will greatly improve our knowledge about how Business Models affect Innovation and open-source practice relationships. Researchers will learn more by looking at how influence works along with long-term study results and organizational cultural analysis. Our complete study design produces both academic insight and clear business strategies for companies to use open-source innovation for success.

REFERENCES

Akhigbe, E. O., Bowale, E. I., Urhie, E., & Amonu, O. (2024). Technological innovation, prosperity and corruption outlook in Africa: A mediation investigation. *Journal of Money Laundering Control*, *27*(4), 658–673. DOI: 10.1108/JMLC-08-2023-0133

Alsaadi, O., Alpkan, L., & Yildiz, B. (2025). Business model innovation as a mediator between construction 4.0 and firm performance: Evidence from Turkish construction companies. *International Journal of Innovation Studies*, *9*(1), 77–89. DOI: 10.1016/j.ijis.2024.12.002

Bashir, M., & Pradhan, S. (2023). The influence of implicit knowledge-oriented leadership on business model innovation and innovation in open sources: Mediating role of absorptive capacity. *Management Decision*. Advance online publication. DOI: 10.1108/MD-10-2022-1481

Bennett, R., Kottasz, R., & Yuan, P. Y. (2025). The mediating role of digital innovation capability on the relationship between organisational agility and performance: The case of the UK arts and culture sector. *Cogent Business & Management*, *12*(1), 2465893. DOI: 10.1080/23311975.2025.2465893

Ciampi, F., Demi, S., Magrini, A., Marzi, G., & Papa, A. (2021). Exploring the influence of big data analytics capabilities on business model innovation: The mediating role of entrepreneurial orientation. *Journal of Business Investigation*, *123*, 1–13.

Clauss, T., Breier, M., Kraus, S., Durst, S., & Mahto, R. V. (2022). Temporary business model innovation–SMEs' innovation response to the Covid-19 crisis. *Research Management*, *52*(2), 294–312.

Costa, J., & Moreira, A. C. (2022). Public policies, innovation in open sources ecosystems and innovation performance. Investigation of the influence of funding and regulations. *Journal of Innovation in open sources. Technology, Market, and Complexity*, *8*(4), 210.

Jenkinson, N., Chiba, M. D., Mthombeni, M., & Verachia, A. H. (2024). Big data analytics influence on competitive performance: Mediating role of business model innovation. *South African Journal of Business Management*, *55*(1), 4261. DOI: 10.4102/sajbm.v55i1.4261

Kant, S., & Adula, M. (2025). Integrating Gender and Indigenous Knowledge in Digital Innovations for Ethiopia's Renewable Energy Transition. In Digital Innovations for Renewable Energy and Conservation (pp. 167-194). IGI Global.

Kant, S., Adula, M., Kumari, M., & Nimbrain, N. (2025). Role of SMEs in Greening the Economy by Mediation of Eco-Innovation in the Horn of Africa. In *Examining Green Human Resources Management and Nascent Entrepreneurship* (pp. 173–200). IGI Global., DOI: 10.4018/979-8-3693-7046-9.ch007

Kant, S., Niguse, T., & Adula, M. (2025). Family Business Entrepreneurial Orientation Influence on Improving Entrepreneurial Processes by Mediation of Innovation in open sources in the Horn of Africa. In *AI, Corporate Social Responsibility, and Marketing in Modern Organizations* (pp. 1-30). IGI International Scientific Publishing.

Kant, S., Niguse, T., Terefe, A., & Adula, M. (2025). Emerging Trends and Innovations: Effect on Social Sustainability in Manufacturing Supply Chains by Shifts in Consumer Behavior in the Horn of Africa. *Enhancing Social Sustainability in Manufacturing Supply Chains*, 135-162. http://dx.doi.org/ DOI: 10.4018/979-8-3693-9740-4.ch005

Li, S., & Zhang, K. (2025). The moderating effect of environmental dynamism on entrepreneurship and open innovation. *Suid-Afrikaanse Tydskrif vir Ekonomiese en Bestuurswetenskappe*, *28*(1), 14. DOI: 10.4102/sajems.v28i1.5767

Moradi, E., Jafari, S. M., Doorbash, Z. M., & Mirzaei, A. (2021). Influence of organizational inertia on business model innovation, innovation in open sources and corporate performance. *Asia Pacific Management Review*, *26*(4), 171–179. DOI: 10.1016/j.apmrv.2021.01.003

Palumbo, R., Manesh, M. F., Pellegrini, M. M., & Flamini, G. (2022). Setting the conditions for innovation in open sources in the food industry: Unravelling the human dimension of innovation in open sources. *British Food Journal*, *124*(6), 1786–1809. DOI: 10.1108/BFJ-02-2021-0162

Snihur, Y., & Bocken, N. (2022). A call for action: The influence of business model innovation on business ecosystems, society and planet. *Long Range Planning*, *55*(6), 102182. DOI: 10.1016/j.lrp.2022.102182

Solaja, O. A., Oyedele, O. O., Olajugba, O. J., Abiodun, A. J., Edewor, O. J., Solaja, O. O., & Akerele, F. O. (2025). SME Performance And Job Creation In Sub-Saharan Africa: The Role Of Digital Innovation. *Revista de Administração de Empresas*, *65*(2), e2023–e0553. DOI: 10.1590/s0034-759020250204x

Spieth, P., Breitenmoser, P., & Röth, T. (2025). Business model innovation: Integrative review, framework, and agenda for future innovation management research. *Journal of Product Innovation Management*, *42*(1), 166–193. DOI: 10.1111/jpim.12704

Srisathan, W. A., Ketkaew, C., Jitjak, W., Ngiwphrom, S., & Naruetharadhol, P. (2022). Innovation in open sources as a strategy for open innovation-based business model innovation: The moderating influence among multigenerational entrepreneurs. *PLoS One*, *17*(6), e0265025. DOI: 10.1371/journal.pone.0265025 PMID: 35679277

Tetteh, F. K., Nyantakyi, B., Owusu Kwateng, K., & Osei, H. V. (2025). The mediation role of innovation in the relationship between total quality management and performance of small and medium scale enterprises. *International Journal of Quality & Reliability Management*, *42*(2), 676–705. DOI: 10.1108/IJQRM-12-2021-0427

Yanda, L., Mwange, A., Moose, J., & Mwaanga, C. (2025). The Mediating Role of Innovation in Strategic Management Practices and SME Performance: Insights from Emerging Markets in Zambia. *East African Finance Journal*, *4*(1), 123–140. DOI: 10.59413/eafj/v4.i1.7

KEY TERMS AND DEFINITIONS

Business Model:: A structure that defines how a company creates, delivers, and acquires value, including its products, services, and overall market strategy.

Innovation in open sources investigation:: The method of systematically incorporating external ideas, technologies, and open innovations into the innovation process to improve organizational capabilities.

Innovation:: The procedure of transforming ideas or inventions into products and services that generate value or satisfy market demands, frequently involving new technologies or methods.

Direct Influence: The immediate influence one variable has on another without the mediation of additional variables.

Indirect Influence: The influence of one variable on another that is mediated by one or more intermediate variables.

Partial Mediation:: A situation in which a mediating variable accounts for part of the association among an independent variable and a dependent variable, but not the entirety of it.

Chapter 13
Artificial Intelligence and the Application of Augmented Reality Techniques as Disruptive Innovation:
A Thinking for Process Management in the Pharmaceutical Industry

Caio Mergh
https://orcid.org/0009-0009-2656-3905
Oswaldo Cruz Foundation, Brazil

Fábio Luiz Daudt Morais
https://orcid.org/0000-0002-0079-0933
Oswaldo Cruz Foundation, Brazil

Jorge Lima de Magalhães
https://orcid.org/0000-0003-2219-5446
Oswaldo Cruz Foundation, Brazil

ABSTRACT

Artificial Intelligence (AI) has been increasingly explored in various industries, and the application of the Retrieval-Augmented Generation (RAG) technique emerges as a promising solution for creating AI assistants that optimize complex processes, such as regulatory compliance in the pharmaceutical industry. The RAG technique combines information retrieval with text generation. This approach becomes especially relevant in a sector that faces growing challenges in interpreting and adhering to a large volume of frequently updated regulations. Regulatory compliance in the pharmaceutical industry demands not only information management but also precision and efficiency in interpreting standards, something that can be compromised by manual processes susceptible to errors. These challenges pose risks of non-compliance, legal sanctions, and damage to companies' reputations. The creation of the AI assistant using the RAG technique provides a significant advantage by allowing employees to obtain information quickly and accurately, without the need for manual analysis of databases

DOI: 10.4018/979-8-3693-4030-1.ch013

1. BACKGROUND

1.1. The Pharmaceutical Industry

The global pharmaceutical sector is one of the most influential and economically significant in the international scenario. In 2024, the sector continues to expand, with projections indicating that the global drug market will reach $1.9 trillion by 2027. This growth is driven by advances in innovative drugs, especially in areas such as oncology and biotechnology, which are expected to account for 35% of global spending in the same period. Cancer drugs, for example, are expected to reach an investment of $370 billion, almost double the current level, due to the increasing prevalence of chronic diseases and the search for more effective therapies (IQVIA, 2023).

The COVID-19 pandemic also played a significant role, accelerating the development and distribution of vaccines, which still influence global pharmaceutical markets. Despite this, the sector faces challenges such as the need to control costs, especially in more mature markets, and the increasing implementation of price controls, such as those in the US with the Inflation Reduction Act. Even so, the sector remains a pillar of innovation, with the growing use of technologies such as artificial intelligence to accelerate drug development and improve operational efficiency.

Additionally, regions in Asia, Africa, and Latin America lead in volume growth, driven by population growth and expanded access to medicines. This scenario reaffirms the central role of the pharmaceutical industry in global public health and technological progress, consolidating itself as one of the world's main economic forces (IQVIA, 2023; LEVY, 2024).

In this sense, Augmented reality (AR) and virtual reality (VR) emerged as a highly significant affordable and efficient approach in various sectors of health care and medicine such as surgery, diagnostic imaging, medical education, medical training, rehabilitation, nursing, telemedicine, palliative care, therapeutics, and patient care management (SING et al, 2022).

1.2. Regulation In The Pharmaceutical Industry

Globally, the pharmaceutical industry is a highly regulated sector, being one of the sectors with the highest regulatory rigor. The sector is impacted by national regulations and can be affected by national, international, and legislation from numerous federative entities at different levels (PONTIN et al., 2020; SENSEN et al., 2010; ULLAGADDI, 2024).

According to Pontin (2020), national regulations can originate from federal, state, and municipal legislators, and national regulatory bodies such as ANVISA – National Health Surveillance Agency, Ministry of Health, and bodies such as the National Commission for the Incorporation of Technologies in the Unified Health System – CONITEC, the National Health Council through its commissions such as the National Research Ethics Commission – CONEP, the Drug Market Regulation Chamber (CMED), ANS – National Supplementary Health Agency, professional councils such as the Federal Council of Medicine (CFM), Federal Council of Pharmacy (CFF).

As a significant part of the national market is composed of transnational industries (Pontin et al., 2020), international regulations also have a great impact on the Brazilian market. These are established by international regulatory agencies such as the US Food and Drug Administration (FDA), the European Medicines Agency (EMA), and the World Health Organization (WHO), and more recently, the guidelines of the International Council for Harmonization (ICH) (ULLAGADDI, 2024).

In addition to being highly regulated, this sector operates in a highly dynamic environment. The need for review and updating regulations by the bodies makes drug manufacturers constantly review their practices to adapt to changes, further increasing the importance of regulatory compliance, which ultimately aims to protect the health of the user/patient and the popular economy (LALLAS et al., 2022; PONTIN et al., 2020; SESEN et al., 2010).

1.3. Regulatory Compliance

Regulatory compliance is of fundamental importance in different segments. Mcintyre, S.R., (2008) defined compliance as "a desired outcome, with respect to laws and regulations, internal policies and procedures, and commitment to stakeholders that can be consistently achieved through managed investment of time and resources. Compliance management includes legal and tactical activities in daily business processes."

According to the definition of Lallas et al. (2022), "in the manufacture of pharmaceutical products, regulatory compliance refers to the ability of the management system of a pharmaceutical manufacturing plant to define and comply with a set of regulations, originating from pharmaceutical production processes."

Regulatory authorities inspect the development and manufacturing facilities of pharmaceutical products for compliance with good practices, which can be laboratory (GLP), clinical laboratory (GCLP), good clinical practices (GCP), and good manufacturing practices (GMP). Therefore, the large volume of legislation and consequently, the large volume of guidelines to be met, combined with their technical and specific details, creates a challenge for companies to quickly find and apply relevant information. This generally results in higher costs, as teams spend valuable time navigating the vast repository of guidelines (KIM; MIN, 2024; KUMAR, 2017).

Inadequate practices that lead to violations and regulatory sanctions are not uncommon. The number of notices of concern, declarations of non-compliance, and warning letters issued by WHO, EMA, and FDA has shown an alarming increase in recent years (PÉREZ, 2017).

In Brazil, only in 2023, ANVISA conducted nine (9) on-site inspections of Good Clinical Practices (GCP) in clinical trial centers in Brazil, where 251 findings were identified, 33 critical, 172 majors, and 46 minor (ANVISA, 2023). Regarding Good Manufacturing Practices, only in 2023, 19 inspections were carried out in national pharmaceutical chemical plants producing active pharmaceutical ingredients (API), with most establishments classified as "Establishment in AVI" (Voluntary Action Indicated), that is, establishments where conditions or practices were found to be non-compliant with current sanitary standards/GMP requirements during the inspection, which resulted in the immediate and voluntary adoption by the company of standardized sanitary actions to the affected batch, product, or production line (ANVISA, 2024).

Regulatory compliance management is generally a manual task that requires a lot of labor and creates excessive overhead (YIP; RAY; PARAMESH, 2006). In a pharmaceutical company that operates in different segments such as clinical studies and the manufacture of drugs of different categories and pharmaceutical forms, this management can be even more complex and costly.

A recent study highlighted the financial impact of compliance with regulatory guidelines. It revealed that compliance efforts can consume up to 25% of the operational budget of a medium or large pharmaceutical manufacturing site (CRUDELI, 2020 APUD KIM, 2024).

1.3.1. Documentation in Compliance

The documentation of a company has a significant impact and relationship with regulatory compliance, making adherence to Good Documentation Practices (GDP) a common requirement by global regulatory agencies. According to RDC 658/2022 (ANVISA, 2022), legislation that regulates Good Manufacturing Practices for medicines in Brazil, "the main objective of the documentation system used must be to establish, control, monitor, and record all activities that, directly or indirectly, affect all aspects of the quality of medicines." The RDC also states that "the documentation that constitutes the Pharmaceutical Quality Management System must include sufficient instructional details to facilitate common understanding of the requirements, as well as allow satisfactory recording of various processes and evaluation of any observations, so that the continuous application of the requirements can be demonstrated."

RDC 09/2015 (ANVISA, 2015), which regulates the conduct of clinical trials with medicines in Brazil, also requires documentation of its activities, with audits being a "systematic and independent analysis of the activities and documents related to the study."

Compliance with Good Distribution Practices (GDP) is required by regulatory authorities in the US and Europe, i.e., FDA and EMA (KUMAR, 2017).

Thus, it is possible to characterize that the verification of adherence of activities, operations, and strategies of a pharmaceutical industry to the requirements of Good Practices is largely evidenced through its documentation.

1.4. Innovation And New Technologies In The Pharmaceutical Industry

The pharmaceutical industry is constantly transforming, especially when it comes to technology (ULLAGADDI, 2024; MIOZZA; BRUNETTA; APPIO, 2024). It is common to find the term "Industry 4.0" in publications, studies, and even in service offerings by companies.

This term is recent and is related to the evolution of Information and Communication Technologies, referring to the integration of these technologies in production processes. It also refers to the fourth in a series of technological revolutions, which, in addition to significant impacts on economic growth, brought through cyber technologies, the recognition of the value of information and data along with IT technology (PEREIRA; BARRETO; AMARAL, 2017). According to Pereira et al (2017), "the value proposition of the Industry 4.0 concept is closely associated with end-to-end digitization of all physical assets and integration into digital ecosystems of all partners in the value chain."

Some works that proposed solutions for the regulatory compliance process, such as Sesen et al. (2010), proposed OntoReg, the prototype of an ontological knowledge base infrastructure applied to the domains of pharmaceutical process and regulation, to facilitate the management of validation requirements. The proposed solution platform was integrated with an OWL reasoner and a rule engine through the use of a Java integrated development environment for automated validation and non-compliance reporting. However, since its proposal until today, technology has evolved exponentially.

According to Miozza (2024), the pharmaceutical industry is at a moment when digital transformation offers numerous opportunities to improve the value chain of medicines. Technologies and themes such as the Internet of Things (IoT), Augmented Reality (AR), Artificial Intelligence (AI), Blockchain, Cybersecurity, Cloud Computing, 3D Printing, Digital Twins, Virtual Reality (VR), Robotics, and Big Data are ready to drive transformative changes.

Even with the use of software that automates compliance activities and searches in databases, they require great effort or manual work, in addition to the human capacity to interpret this information.

1.4.1. Artificial Intelligence (AI)

When we use the term AI, it is directly linked to Large Language Models (LLM), which are, in fact, a type of generative AI specifically designed to generate text-based content, more popularly known as chatbots. According to Kerner (2024), "a Large Language Model (LLM) is a type of artificial intelligence algorithm that uses deep learning techniques and extremely large datasets to understand, summarize, generate, and predict new content." These were designed to understand and generate human-like language.

These models are trained on vast datasets of text and can generate text that is often indistinguishable from that written by humans. They can answer questions, perform analyses, write code, and even generate creative content based on the patterns they learned from the data they were trained on.

LLMs have revolutionized the way people search for information online, from research to directly requesting answers from chatbots (XIONG et al., 2024).

With the increasing use of LLMs, their capabilities continue to expand, making them a transformative, promising, and crucial technology in the field of AI (KERNER, 2024; MIAO et al., 2024).

The use of generative AI is transforming numerous aspects of the pharmaceutical industry, reshaping the way companies operate and potentially unlocking billions of dollars in value. It is estimated that technology could generate $60 billion to $110 billion per year in economic value for the pharmaceutical and medical products industries (SHAH et al., 2024). This can be leveraged to ensure compliance with regulatory guidelines. Navigating the complex and extensive guidelines provided by agencies such as ANVISA, FDA, and EMA is often a daunting and time-consuming task for industry participants (Kim & Min, 2024).

Advanced LLM models have evolved from simple text processors to tools capable of deep understanding of technical language, offering innovative ways to handle data related to various topics, such as health, processes, and industrial practices, among others (MIAO et al., 2024).

However, LLMs are also subject to failures. Traditional text generation models, typically based on encoder/decoder architectures, can translate languages, respond in different styles, and answer simple questions. But since these models rely on statistical patterns found in their training data, they sometimes provide incorrect or irrelevant information, called hallucinations. In these cases, they can generate responses that seem plausible but are actually incorrect, meaning an LLM can provide an inaccurate response that is not based on trained data, or even fictitious (KERNET, 2024; MIAO et al., 2024; XIONG et al., 2024).

These hallucinations can be a major problem in the search for technical information, such as specific good practice legislation, making the information provided directly by LLM unreliable.

An important observation is that even public data used by an LLM may be outdated, depending on the update date of its database by its developer. Additionally, LLMs do not have access to private data stored in an organization (LEWIS et al., 2020).

2. RETRIEVAL-AUGMENTED GENERATION (RAG) TECHNIQUE

2.1. What is RAG

This technique was proposed by Lewis et al. (2020) as a way to improve generation performance in knowledge-intensive tasks, such as text generation, by integrating relevant retrieved information. It represents an innovative approach in the field of artificial intelligence, combining information retrieval with text generation. RAG was initially developed to improve the accuracy and relevance of responses in question-and-answer systems, but its potential goes beyond, extending to areas such as regulatory compliance (LEWIS et al., 2020).

This mechanism allows the model to extract the most relevant and updated information, essentially "augmenting" the model's knowledge base in real-time (LEWIS et al., 2020).

In other words, it is a technique to improve the accuracy and reliability of generative AI models with facts obtained from external sources (MERRIT, 2023).

To explain the application of RAG in practice, it is necessary to know that LLMs are trained on a wide range of online textual data and, in some cases, are trained for a specific industry or domain expertise in mind.

Thus, companies can create applications or assistants using RAG to enhance the capabilities of LLMs, for example, creative writing assistants for marketing, document summarization for legal teams, code writing for software development (MERRIT, 2023).

The pharmaceutical industry is a field characterized by its highly specialized and detailed nature, so it is likely that all required knowledge is not fully included in the existing LLM databases and training. As a result, they will not be prepared to accurately answer questions in this field, making RAG an excellent choice and solution for the issue, allowing the analysis of private company documentation and regulatory compliance with applicable norms and legislation (KIM; MIN, 2024).

2.1.1. How the Technique is Structured

Lewis et al. (2020) called RAG "a general-purpose fine-tuning recipe" because the technique can be used by almost any LLM to connect with virtually any external resource, regardless of its content. Additionally, it not only uses the innate knowledge of these models but also seeks additional information from external sources to generate responses (MERRIT, 2023).

Its operation is as follows: When users ask a question to an LLM, the AI model sends the query to a Retriever that searches for similarity and semantics in the available external database. When it finds a match or several matches, it retrieves the related data, converts it into human-readable words, and passes it back to the LLM, which generates a natural language response leveraging all the power of human language it learned from its generic training data, but also with the specific data provided in the retrieval stage (Alejandro Alija, 2024).

Figure 1. Functioning and framework of the RAG technique

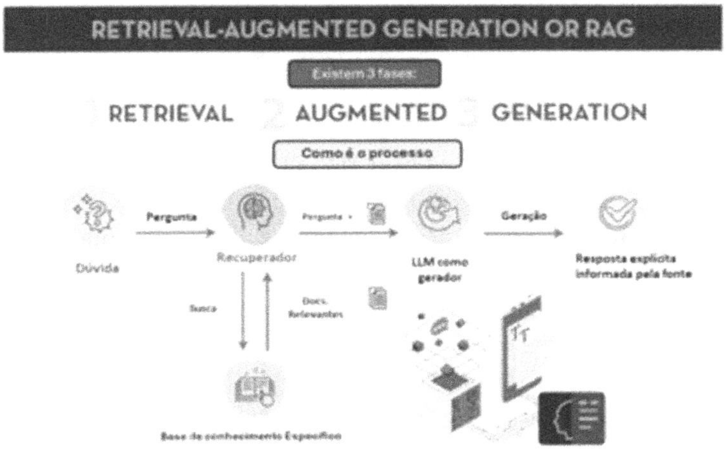

Source: Adapted from https://www.ml6.eu/blogspot/leveraging-llms-on-your-domain-specific-knowledge-base

2.1.2. Benefits of the Technique

The technique brings great benefits, both for the necessary activities in the pharmaceutical industry and for the problems presented by LLMs with their standard training. Access to an extensive private database effectively increases the knowledge of LLMs, improving their performance especially in generating responses that require specific factual information or updated knowledge (LI et al., 2022; MIAO et al., 2024). Recent studies indicate that RAG can automate significant parts of this process, improving both the efficiency and accuracy of responses in various health areas (LI et al., 2022).

Kim; Min, 2024, presented positive results in the application of a RAG model called Question and Answer RAG (QA-RAG) applied to regulatory compliance in the pharmaceutical industry, ensuring superior performance and accuracy driven by the use of responses, compared to the direct use of LLMs without the technique. Additionally, a reduction in incorrect or fabricated responses, known as hallucinations, was also observed.

Finally, due to the need to create a database with specific content, the technique allows databases to be updated quickly and constantly, as needed by companies, without relying on LLM updates, making it an extremely useful tool for regulated environments (MERRIT, 2023). For example, we can use the review of an ANVISA RDC. The replacement of the revoked RDC with a new one can be quickly carried out in the database that composes the RAG framework, allowing LLMs to bring responses based on the new content.

These recent studies, published between 2022 and 2024, highlight the ability of RAG to transform regulatory compliance in the pharmaceutical industry, offering tangible benefits such as greater accuracy, agility, and security.

2.1.3. Applications in Other Health Areas

RAG has stood out as an innovative approach in the application of artificial intelligence (AI) in highly complex contexts, including the health sector. Recent studies reinforce its growing relevance and present empirical evidence of its applicability in different subfields of health, highlighting its effectiveness in data integration, interpretation of clinical guidelines, and optimization of decision-making processes.

According to Yu et al., 2023, the application of RAG in health integration systems is shaping a new paradigm in clinical information management. In their study, the authors propose a roadmap for the integration of health systems, highlighting how the technique allows consolidating data from multiple sources and adapting them to the specific needs of professionals and patients. The RAG approach proved particularly efficient in reducing the time needed to access critical information, resulting in better patient outcomes and greater resource efficiency.

In the field of hepatology, Kresevic et al., 2024, explore the use of RAG to interpret specific clinical guidelines, a challenge traditionally marked by the complexity and volume of information involved. The technique was used to create models that help doctors navigate detailed medical recommendations, ensuring adherence to best practices and promoting personalized care. This study demonstrates how RAG can be applied in highly specialized areas, facilitating the practical application of scientific knowledge.

Additionally, Xiong et al., 2024, presented a detailed analysis of the use of RAG in the development of models for medication prescription support. In their work, the technique is employed to retrieve updated information on drug interactions and therapeutic guidelines, integrating them into clinical decision support systems. The research highlights the ability of RAG to handle large volumes of data and provide real-time responses, improving safety and effectiveness in medical practice.

Ke et al., 2024, developed an LLM-RAG model using 35 preoperative guidelines that were tested against human-generated responses, with a total of 1,260 responses evaluated (336 generated by humans, 336 generated by LLM, and 588 generated by LLM-RAG). As a result, the LLM-RAG model generated responses in an average time of 15 to 20 seconds, significantly faster than the 10 minutes normally required by humans, in addition to achieving 91.4% accuracy, compared to human-generated instructions, which had 86.3% accuracy, confirming the effectiveness of the technique.

In a recent study, Unlu et al. (2024) investigated the application of a specialized RAG framework, named RECTIFIER (RAG-Enabled Clinical Trial Infrastructure for Inclusion Exclusion Review), in the practice of clinical trial screening, compared to screening performed by a study team. Both RECTIFIER responses and the study team's responses were evaluated against the responses of a clinical specialist. The results showed that RECTIFIER performed better in determining the inclusion criteria for "symptomatic heart failure," and overall, greater sensitivity and specificity than the study team.

These studies, published between 2023 and 2024, highlight the innovative nature and broad applicability of RAG in different health contexts. By allowing AI systems to combine precision in information retrieval with the generation of contextualized responses, the technique consolidates itself as an essential tool to address the growing challenges in the area. This robust application base provides a promising perspective for the expansion of RAG use, including its potential for regulatory compliance in the pharmaceutical industry, the central theme of this work.

2.2. Application Of RAG In Regulatory Compliance

The application of the RAG technique in regulatory compliance in the pharmaceutical industry is an emerging field of study with great potential. Traditionally, compliance management relies on specialized teams that interpret and apply complex regulations. However, this approach can be inefficient and prone to errors, especially when dealing with a large volume of data.

The extensive set of guidelines, coupled with the complexity of their technical details, can make it difficult for companies to quickly locate and apply the most relevant information. Currently, regulatory compliance implies high costs, as it depends on specialized teams that invest time analyzing a vast collection of standards. Given this scenario, the pharmaceutical industry needs a more effective approach to interpret and navigate regulations, as well as reduce unidentified failures and gaps.

Large language models (LLMs) emerge as a possible solution to this challenge. In this context, the RAG technique stands out by acting as a link between the knowledge of these models and external sources of information, allowing for more precise and well-founded responses (Kim & Min, 2024).

Additionally, by enabling the integration of vast regulatory databases, RAG can help pharmaceutical companies stay updated with regulatory changes and respond quickly na accurately to audits and regulatory investigations.

3. REFLECTIONS FOR IMPLEMENTING FRAMEWORK

Figure 2. Framework proposal

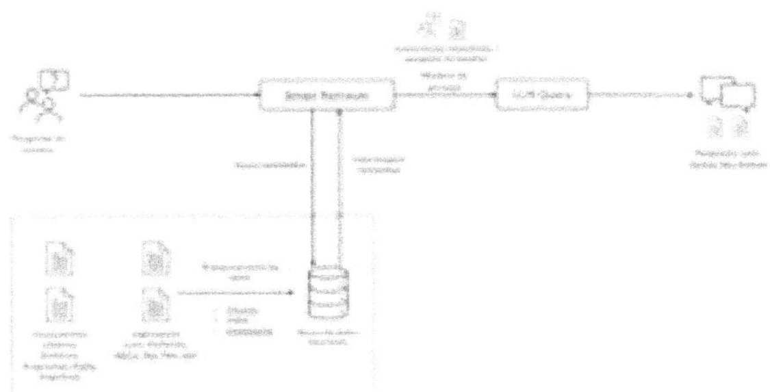

Considering the background of the observed theory, one can reflect on the possibility of the hypothesis that the application of Artificial Intelligence, using the RAG technique, can significantly improve the regulatory compliance process in the pharmaceutical industry, making it more precise, efficient, and safe. More specifically, it is believed that the use of large language models (LLMs) associated with RAG can overcome critical challenges faced by companies in the sector, such as precision and reliability. The combination of deep learning with active information retrieval can reduce ambiguous interpretations of regulations, ensuring more consistent and well-founded responses.

In this sense, hope so that thinking could represent a significant advancement for regulatory compliance management in the pharmaceutical industry, making the process more agile, economical, and reliable, while reducing risks arising from human errors and inaccurate interpretations.

3.1. Concerning the Analysis of Regulatory Documents

The pharmaceutical industry operates under a stringent set of rules and regulations imposed by global and local regulatory bodies such as the FDA (Food and Drug Administration), EMA (European Medicines Agency) and ANVISA (National Health Surveillance Agency). To ensure compliance with these requirements, companies adopt a hierarchical document structure, often represented by the so-called document pyramid (Chingale et al., 2024).

Unlike external regulatory documents, which are mandatory and serve as a reference, the document pyramid organizes internal documents that guarantee a consistent flow of information and processes, ensuring compliance at all stages of the production chain, e.g. like a pyramid with level 1, level 2, level 3 and level 4. Such as:

- Level 1: At the top of the document pyramid are documents that establish overarching principles and guidelines for developing, documenting, and implementing a compliant quality system applicable to all departments. The main document at this level is the Quality Manual, which defines the company's structure and commitments to quality and regulatory compliance.
- Level 2: At the second level, there are more detailed documents that regulate specific topics, ensuring consistency between departments by establishing guidelines for lower-level procedures. These documents provide general intent and guidelines for critical programs or systems, without providing specific directives or forms. They are universally applicable within the organization and include Programs, Policies, and Master Plans.
- Level 3: At the third level, there are documents that provide detailed step-by-step instructions for operational tasks. These documents are specific to each area or activity and ensure that employees follow established standards to ensure compliance and quality. Examples include Procedures, Work Instructions, and Protocols. • Level 4: The base of the document pyramid is made up of records of activities performed, which serve as evidence of the completion of a given action. These documents are essential to demonstrate adherence to processes established at higher levels and to facilitate regulatory audits. Examples include Records and Reports.

In this way, document analysis allows us to identify and classify the main internal and external regulations applicable to the pharmaceutical industry. Such as, for example, documents issued by regulatory agencies (ANVISA, FDA, EMA, etc.), in addition to internal guidelines of the companies participating in the study.

3.2. Concerning the Characterization of Technical Components of the RAG

The main challenges presented in the literature are:

- Quality of retrieval: the need to select the most relevant documents without losing context.
- Scalability: the system's ability to process large volumes of documents efficiently;
- Accuracy of the responses generated: the effective integration between the retrieval module and the generating model, ensuring coherent and well-founded responses.
- Confidentiality of information: the protection of private industry information contained in the documents processed by the RAG, ensuring that this data remains restricted and is not incorporat-

ed or exposed in the LLMs' knowledge bases. This includes mechanisms to prevent the leakage of sensitive data and to ensure that the information retrieved is used only in the authorized context, without compromising privacy and business security.

3.2.1. About the Definition of Functional and Non-Functional Requirements

A RAG-based framework for regulatory compliance in the pharmaceutical industry requires a robust and adaptable structure. Therefore, based on the literature review and analysis of available technologies, the following technical and functional requirements were defined to ensure efficient, safe and scalable performance:.

3.2.1.1. About Functional Requirements

In addition to technical requirements, the system must meet specific functional needs to ensure efficient and accessible experience for users, e.g.:

- **Automated collection of regulatory and internal documents**: The system must be able to collect and load documents from multiple sources and formats in an automated manner, reducing the need for manual processing.
- **Efficient segmentation and indexing**: To ensure agile and accurate retrieval of information, documents must be segmented into smaller units (chunks) and indexed in a vector database, allowing optimized and contextualized searches.
- **RAG-based QA engine**: The assistant must answer user questions using a language model (LLM) combined with semantic retrieval. Answers must always be based on valid regulatory documents, reducing the risk of misinformation.
- **Automatic update mechanism**: To ensure continuous compliance with new regulations, the system must have an automated update process. Whenever new guidelines are published by regulatory bodies or internal documents are changed, the database must be updated to reflect the changes.

3.2.1.2. About Non-Functional Requirements

Non-functional requirements establish the basis for the development and proper functioning of the system, ensuring that it meets the needs of the pharmaceutical industry in the regulatory context.:

- **Scalability**: The system must be able to process large volumes of regulatory and internal documents, maintaining efficient performance without degradation in response speed or accuracy. The architecture must support the addition of new documents and simultaneous queries from multiple users.
- **Accuracy and relevance**: The retrieval mechanism must ensure that the selected documents are highly relevant to the query performed. To this end, advanced semantic search techniques and vector embeddings will be used, minimizing the incidence of irrelevant or inaccurate information.
- **Security**: Since the system deals with internal regulatory documents and sensitive information from the pharmaceutical industry, it is essential to guarantee the integrity and confidentiality of the data. To this end, access controls, encryption and audit mechanisms will be implemented to track interactions with the system.

- **Interoperability**: The framework must be able to integrate with different data sources and APIs, allowing pharmaceutical companies to incorporate the assistant into their existing systems. The architecture must allow the ingestion of documents from public databases (e.g.: FDA, ANVISA) and private databases (internal industry documents). • Intuitive user interface: Interaction with the system must be simple and efficient, allowing regulatory experts and other users to make queries without the need for advanced technical training. To achieve this, a user-friendly interface will be developed, accessible via a browser.
- **Use of free software**: To ensure accessibility and reduce costs in the concept validation phase, the system must be developed exclusively with free software and free tools. This approach allows for greater transparency and flexibility, and avoids restrictions imposed by paid features.

With these requirements well defined, the framework will be able to offer robust support to companies in the pharmaceutical industry, optimizing regulatory compliance processes and ensuring greater efficiency in the analysis and interpretation of standards and internal documents.

3.2.2. About Study and Analysis of Existing Technologies

To ensure efficient implementation of the Retrieval Augmented Generation (RAG) technique in the regulatory context of the pharmaceutical industry, a careful analysis of the available technologies was conducted. The objective was to select tools and frameworks that met the defined requirements.

For this chapter, the study exclusively considered free software, that is, free tools and frameworks that met the previously defined requirements.

The first component analyzed was the data collection and loading module, responsible for ingesting regulatory and internal documents in different formats, such as PDF, CSV, structured texts and web pages. For this task, LangChain was selected, which offers specific modules, such as PyPDFLoader, CSVLoader, WebBaseLoader and TextLoader. This choice was made due to the versatility of the tool, which allows the integration of multiple data sources without the need for manual conversions, ensuring that the documents are accessible for subsequent processing.

In the text preprocessing and segmentation stage, it was essential to adopt an approach that preserved the context of the documents by dividing them into smaller segments. For this purpose, LangChain (RecursiveCharacterTextSplitter) was used, which fragments texts while maintaining overlap between segments. This technique ensures that critical information is not lost during retrieval, increasing the accuracy of the system.

Document retrieval was implemented using a vector database, allowing documents to be stored and retrieved based on their semantic similarity. For this task, FAISS (Facebook AI Similarity Search), one of the most efficient frameworks for vector search, was used, combined with the use of OpenAI Embeddings to convert documents into high-dimensional mathematical representations. The choice of FAISS was motivated by its ability to perform fast and accurate research even in large volumes of data, ensuring scalability to the system.

The implementation and monitoring of the system were planned to ensure scalability and reliability. To this end, architecture based on Google Cloud and Docker was defined, which allows the intelligent assistant to run in cloud environments in a safe and efficient manner. The use of containerization with Docker allows for easy maintenance and updating of the system, while cloud infrastructures ensure continuous availability and support for large volumes of data and requests.

With this set of technologies meeting the challenges of the RAG technique in the regulatory context, it is possible to carefully choose the components that allow the construction of an intelligent and robust assistant, capable of assisting pharmaceutical companies in the interpretation and updating of regulatory standards, ensuring compliance in an agile and efficient manner.

4. FINAL CONSIDERATIONS

The application of Artificial Intelligence (AI), specifically the Retrieval-Augmented Generation (RAG) technique, in regulatory compliance within the pharmaceutical industry represents significant advancement. The theoretical framework consolidated throughout this chapter highlights several key points:

a) **Enhanced Precision and Reliability**: The integration of deep learning with active information retrieval significantly reduces ambiguous interpretations of regulations, ensuring more consistent and well-founded responses. This addresses one of the critical challenges faced by the pharmaceutical industry in maintaining compliance with complex and frequently updated regulations.
b) **Operational Efficiency**: Intelligent automation minimizes the time required to locate, interpret, and apply regulatory guidelines, reducing the dependence on intensive manual analysis and operational costs. This efficiency is crucial for pharmaceutical companies to remain competitive and compliant in a dynamic regulatory environment.
c) **Information Security**: By operating in controlled environments, AI solutions ensure that sensitive data is treated with confidentiality, mitigating risks of leakage or misuse of regulatory information. This is particularly important given the stringent data protection requirements in the pharmaceutical sector.
d) **Overcoming Language Barriers**: The ability of large language models (LLMs) to interpret and translate standards in multiple languages facilitates global compliance, especially for companies operating in markets regulated by different authorities. This capability supports the industry's need for a unified approach to regulatory adherence across diverse regions.
e) **Continuous Updating**: The RAG technique enables the dynamic incorporation of new guidelines without the need for complete model reconfiguration, ensuring that the information used is always aligned with the latest regulations. This adaptability is essential for maintaining compliance in an ever-evolving regulatory landscape.

The trends identified in recent studies underscore the transformative potential of RAG in various health contexts, including clinical decision support, medication prescription, and clinical trial screening. These applications demonstrate the broad applicability and effectiveness of RAG in improving accuracy, agility, and security in regulatory compliance processes.

Future research should focus on customizing these AI models to cater to specific regulatory requirements and exploring the potential of supercomputers in training more sophisticated models. The promising future for AI-driven compliance solutions ensures better adherence to regulations and fosters innovation in the pharmaceutical sector.

REFERENCES

Alejandro Alija. (2024, janeiro 24). *RAG - Retrieval Augmented Generation: The key that unlocks the door to precision language models* | datos.gob.es. https://datos.gob.es/en/blog/rag-retrieval-augmented-generation-key-unlocks-door-precision-language-models

ANVISA. (2023). *Relatório de Métricas de Inspeções de BPC 2023*. ANVISA.

ANVISA. (2024). *Revisão Anual das Inspeções em Farmoquímicas Nacionais Coins (RINIFA_2023)*. ANVISA.

Arcanjo, J. (2023, maio 1). Gradio—Crie e compartilhe seus machine learning apps. *Data Hackers*. https://medium.com/data-hackers/gradio-crie-e-compartilhe-seus-machine-learning-apps-88d8b3c5cca4

Chingale, N. D., Walunj, S. D., Kikale, M. S., Devade, O. A., & Redasani, V. (2024). *DOCUMENTATION ESSENTIALS IN THE PHARMACEUTICAL INDUSTRY: A REGULATORY PERSPECTIVE.*

IQVIA. (2023, janeiro). *Global Market for Medicines to Rise to $1.9 Trillion by 2027, says Report from IQVIA Institute*. https://www.iqvia.com/newsroom/2023/01/global-market-for-medicines-to-rise-to-19-trillion-by-2027-says-report-from-iqvia-institute

Ke, Y. H., Jin, L., Elangovan, K., Abdullah, H. R., Liu, N., Sia, A. T. H., Soh, C. R., Tung, J. Y. M., Ong, J. C. L., & Ting, D. S. W. (2024). *Development and Testing of Retrieval Augmented Generation in Large Language Models—A Case Study Report*.

Kerner, S. M. (2024, maio). *What are Large Language Models (LLMs)? | Definition from TechTarget*. S. Large Language Models (LLMs). https://www.techtarget.com/whatis/definition/large-language-model-LLM

Kim, J., & Min, M. (2024). *From RAG to QA-RAG: Integrating Generative AI for Pharmaceutical Regulatory Compliance Process* (No. arXiv:2402.01717). arXiv. http://arxiv.org/abs/2402.01717

Kresevic, S., Giuffrè, M., Ajcevic, M., Accardo, A., Crocè, L. S., & Shung, D. L. (2024). Optimization of hepatological clinical guidelines interpretation by large language models: A retrieval augmented generation-based framework. *NPJ Digital Medicine*, *7*(1), 102. DOI: 10.1038/s41746-024-01091-y PMID: 38654102

Kumar, K. (2017). Good Documentation Practices (GDPs) in Pharmaceutical Industry. *Journal of Analytical & Pharmaceutical Research*, *4*(2). Advance online publication. DOI: 10.15406/japlr.2017.04.00100

Lallas, E. N., Santouridis, I., Mountzouris, G., Gerogiannis, V. C., & Karageorgos, A. (2022). An SQWRL-Based Method for Assessing Regulatory Compliance in the Pharmaceutical Industry. *Applied Sciences (Basel, Switzerland)*, *12*(21), 10923. DOI: 10.3390/app122110923

Levy, V. (2024). *2024 Global Life Sciences Sector Outlook*. Delloite.

Lewis, P., Perez, E., Piktus, A., Petroni, F., Karpukhin, V., Goyal, N., Küttler, H., Lewis, M., Yih, W., Rocktäschel, T., Riedel, S., & Kiela, D. (2020). *Retrieval-Augmented Generation for Knowledge-Intensive NLP Tasks* (No. arXiv:2005.11401). arXiv. http://arxiv.org/abs/2005.11401

Lewis, P., Wu, L., Karpukhin, V., Edunov, S., Chen, D., & Yih, W. (2020). *Dense Passage Retrieval for Open-Domain Question Answering* (No. arXiv:2004.04906). arXiv. http://arxiv.org/abs/2004.04906

Li, H., Su, Y., Cai, D., Wang, Y., & Liu, L. (2022). *A Survey on Retrieval-Augmented Text Generation* (No. arXiv:2202.01110). arXiv. https://doi.org//arXiv.2202.01110DOI: 10.48550

Merritt, R. (2023, novembro 15). *What Is Retrieval-Augmented Generation aka RAG?* NVIDIA Blog. https://blogs.nvidia.com/blog/what-is-retrieval-augmented-generation/

Miao, J., Thongprayoon, C., Suppadungsuk, S., Garcia Valencia, O. A., & Cheungpasitporn, W. (2024). Integrating Retrieval-Augmented Generation with Large Language Models in Nephrology: Advancing Practical Applications. *Medicina, 60*(3). *Medicina (Kaunas, Lithuania)*, *3*(3), 445. Advance online publication. DOI: 10.3390/medicina60030445 PMID: 38541171

Miozza, M., Brunetta, F., & Appio, F. P. (2024). Digital transformation of the Pharmaceutical Industry: A future research agenda for management studies. *Technological Forecasting and Social Change*, *207*, 123580. DOI: 10.1016/j.techfore.2024.123580

Pereira, T., Barreto, L., & Amaral, A. (2017). Network and information security challenges within Industry 4.0 paradigm. *Procedia Manufacturing*, *13*, 1253–1260. DOI: 10.1016/j.promfg.2017.09.047

Pérez, J. R. (2017). Maintaining Data Integrity. *Quality Progress*, *50*, 14–16.

Pontin, A. L., Angélica, C., Miragem, B., & Santorsula, C. B. (2020). *Compliance na Área da Saúde*. Editora Foco.

Sesen, M. B., Suresh, P., Banares-Alcantara, R., & Venkatasubramanian, V. (2010). An ontological framework for automated regulatory compliance in pharmaceutical manufacturing. *Computers & Chemical Engineering*, *34*(7), 1155–1169. DOI: 10.1016/j.compchemeng.2009.09.004

Shah, B., Viswa, C. A., Zurkiya, D., Leydon, E., & Bleys, J. (2024). *Generative AI in the pharmaceutical industry: Moving from hype to reality*. Mckinsey & Company.

Singh, S., Kumar, P., Chaudharya, H., Khurana, R., & Shokeen, A. (2022). Augmented Reality and Virtual Reality Creating Ripple in Medical and Pharmaceutical World. In Verma, J. K., & Paul, S. (Eds.), *Advances in Augmented Reality and Virtual Reality. Studies in Computational Intelligence* (Vol. 998). Springer., DOI: 10.1007/978-981-16-7220-0_8

Unlu, O., Shin, J., Mailly, C. J., Oates, M. F., Tucci, M. R., Varugheese, M., Wagholikar, K., Wang, F., Scirica, B. M., Blood, A. J., & Aronson, S. J. (2024). Retrieval Augmented Generation Enabled Generative Pre-Trained Transformer 4 (GPT-4) Performance for Clinical Trial Screening. *medRxiv*, 2024.02.08.24302376. DOI: 10.1101/2024.02.08.24302376

Xiong, G., Jin, Q., Lu, Z., & Zhang, A. (2024). *Benchmarking Retrieval-Augmented Generation for Medicine* (No. arXiv:2402.13178). arXiv. http://arxiv.org/abs/2402.13178 DOI: 10.18653/v1/2024.findings-acl.372

Yip, F., Ray, P., & Paramesh, N. (2006). Enforcing Business Rules and Information Security Policies through Compliance Audits; XISSF - A Compliance Specification Mechanism. *2006 IEEE/IFIP Business Driven IT Management*, 81–90. DOI: 10.1109/BDIM.2006.1649214

Yu, P., Xu, H., Hu, X., & Deng, C. (2023). Leveraging Generative AI and Large Language Models: A Comprehensive Roadmap for Healthcare Integration. *Healthcare, 11*(20). *Healthcare (Basel)*, *20*(20), 2776. Advance online publication. DOI: 10.3390/healthcare11202776 PMID: 37893850

Chapter 14
Digital Risk Management in Digital Transformation Contexts

Heloisa Guimarães Menezes
https://orcid.org/0009-0008-3397-211X
Fundação Dom Cabral, Brazil

Oscar Miguel Rete
https://orcid.org/0009-0005-7213-6710
Universidad Tecnológica Nacional, Argentina

ABSTRACT

Digital transformation drives innovation and efficiency but introduces complex risks that challenge traditional risk management approaches. Based on a qualitative literature review, this paper explores the intersection of digital transformation, innovation, and risk management, emphasizing the need to adjust traditional risk management strategies to meet digital risk challenges, including holistic strategies that integrate data protection, information security, and stakeholder considerations. While existing literature often focuses on risks like cybersecurity, broader implications for governance, innovation, and entrepreneurial ecosystems remain underexplored. Additionally, philosophical, sociological, and political analyses of digital threats are rarely integrated into organizational strategies. This study proposes a comprehensive framework to guide future research on digital risks, ensuring secure, sustainable, and resilient digital initiatives. Understanding the complexity of digital risks helps organizations to balance opportunities and challenges in the evolving digital economy.

Digital transformation has become a key driver of economic and social change, fostering new business models, increasing efficiency, and stimulating innovation. Advances in information and communication technologies, digitization, and connectivity have created significant opportunities and risks for businesses and their ecosystems. While extensive literature highlights digital transformation's benefits, including competitive advantages and enhanced customer relationships, fewer studies address its constraints and associated risks. Empirical research largely focuses on strategy implementation, cultural shifts, and process adaptation to digital transformation, with major consulting firms and academic institutions

DOI: 10.4018/979-8-3693-4030-1.ch014

such as Deloitte, PwC, McKinsey, Massachusetts Institute of Technology (MIT), Harvard, and Cornell contributing practical insights.

Despite its advantages, digital transformation introduces complex risks that challenge traditional approaches to innovation and risk management. Extensive research on this intersection examines cybersecurity, data protection, and information security risks. The growing exposure of business ecosystems to cyber threats expands attack surfaces, intensifying the need for robust technological infrastructure and security mechanisms[1]. Meanwhile, broader digital risks remain underexplored. These issues are particularly relevant as digital economies increasingly rely on data as their most valuable asset.

From a competitive perspective, studies document the rapid transformation of production structures and the shift of economic power to major technology firms. Digital businesses, particularly those anchored in platform-based models, challenge traditional industrial conglomerates, disrupting workforce structures and reshaping strategic considerations related to innovation portfolios and risk management. In this evolving landscape, corporate leaders must balance innovation opportunities with emerging threats., including algorithmic biases, workforce displacement, misinformation, and evolving regulatory requirements. Additionally, increased supply chain interdependencies and reliance on monopolistic digital platforms introduce new commercial and technological vulnerabilities, further undermining trust in the digital economy.

Recent studies explore digitalization's broader consequences on individuals, society, and politics. However, these analyses primarily adopt philosophical, psychological, sociological, or economic lenses rather than business management perspectives. While academically and socially relevant, such studies seldom influence corporate decision-making, despite the direct impact of digital risks on human capital and business ethics. As a result, corporate executives, boards, and innovation leaders often overlook the broader implications of digital risks, focusing predominantly on cybersecurity rather than considering digital transformation's multifaceted risks and opportunities.

This paper examines digital risks or threats from multiple perspectives, emphasizing existing knowledge gaps and their relevance to organizational strategy. It argues that companies must extend their risk management frameworks beyond cybersecurity threats to encompass algorithmic biases, skills shortages in the technology sector, misinformation, regulatory compliance, and digital supply chain dependencies. Traditional risk and innovation management approaches may be inadequate for navigating this evolving and complex landscape, necessitating the development of more comprehensive frameworks integrating innovation and risk management.

Emerging frameworks increasingly prioritize data as a strategic asset, influencing business decision-making at multiple levels. This paper conducts a qualitative analysis of studies and frameworks addressing digital risks within organizations, contextualizing them within the growing prevalence of digital transformation. It explores the intersection of innovation and risk management, underscoring the theoretical and empirical challenges organizations face. Additionally, it highlights the necessity for business leaders to develop a holistic understanding of digital risks, ensuring that both digital transformation initiatives, and innovation and risk management strategies remain innovative and sustainable.

The following section examines how digital transformation reshapes innovation management, influencing risk assessment and mitigation strategies. It contextualizes digital risks within the evolving business environment, emphasizing the importance of digital ecosystems in shaping competitive dynamics. Additionally, it discusses shifts in analytical perspectives on digital technologies, outlining key developments in the field. The next section addresses whether traditional risk management methodologies adequately capture and mitigate digital risks, presenting insights from current debates. Finally, the paper offers recommendations for companies seeking to enhance their digital risk strategies, balancing resilience and performance within digital transformation initiatives. The study concludes by identifying future research directions that could further inform organizational strategies in this domain.

DIGITAL TRANSFORMATION AS A STRATEGIC OPPORTUNITY. AND RISKS

Digital transformation is a crucial process for modern businesses. As Rete (2020) states, "digital transformation refers to the conversion of processes, activities, and models to meet the demands of the digital economy until the company is a fully connected digital organization" (p. 8). Furthermore, Rete (2020) explains that "digital transformation is the integration of digital technology into all areas of an organization, fundamentally changing the way it operates and delivers value to customers. It also represents a cultural shift that requires organizations to continually challenge the status quo, experiment, and become comfortable with failure" (p. 19).

Digital transformation extends beyond just knowledge and technology usage. It necessitates a digital strategy, skills, a digital mindset, and collaboration in digital ecosystems (Balmaekers, 2023). It involves managing innovation by considering the wide range of possibilities enabled by disruptive technologies and new business models (Frenkel, 2021). It calls for a greater tolerance for failure and a willingness to embrace risks in innovation processes (Ballesta, 2021), as well as engaging in risk prediction practices (Gartner, 2022).

However, the types of risks considered in the management processes of innovation or digital transformation are those inherent to the innovation process, where mistakes are part of the journey. The portfolio of innovative projects should seek the right balance between projects with higher or lower risks, according to the company's digital transformation strategy. Nevertheless, in the digital age, both the value created and the risks stem from data. Collecting personal data is becoming essential to understanding customer behavior and, consequently, to creating or redesigning solutions that better address their needs. Business success hinges on mastering, continuously updating, and protecting valuable information (Finne, 2000).

By increasing the volume, intensity of data, digital transformation makes this flow more complex. New business models intensify the integration of data into ecosystems, platforms (Parker, Alstyne & Choudary, 2019), and supply chains. The growing involvement of third parties in the innovation process, through various open innovation mechanisms, along with the significance of data management, necessitates new approaches to managing supply chain and innovation processes.

With the rising significance of data for businesses, greater attention to potential vulnerabilities beyond the enterprise is also needed, as cautioned by Singapore's Minister for Communications and Information, Josephine Teo (WEF, 2022), and Oliveira & Menezes (2021). Consequently, there is a demand for more rigorous information security risk management (Gartner, 2022), as well as the need to consider other vulnerabilities that may not be so visible in the risk matrix.

In this context, another area of interest is the process of system integration and software development by third parties. This practice, which has been prevalent for decades, gains importance in digital transformation environments due to the creation of ecosystems and business platforms. July 2024 cyber blackout exemplifies the increasing vulnerability of digital ecosystems. A failure at CrowdStrike, a cybersecurity provider, disrupted Microsoft and its supply chain, revealing weaknesses in centralized security models (Gartner, 2022). Such incidents emphasize the need for robust risk management strategies that account for external dependencies and emerging threats. Consequently, organizations must adopt more rigorous information security protocols while also considering less visible risks that traditional risk matrices may overlook (Gartner, 2022).

The need for innovation and digital transformation might seem to conflict with data risk management, given that in the digital era, strategic business opportunities often rely on open data, community building, decentralization of decisions, flexibility, tolerance for errors, and decentralized organizational models (Finne, 2000; WEF, 2022). The foundation of the Internet emphasizes openness, broad participation, collaboration, and freedom for the ecosystem.

However, as the Internet, originally from the academic realm, became integrated into the lives of people, businesses, and governments (Oliveira & Menezes, 2021), freedom of use became crucial to everyone due to its ethical, social, psychological, political, concurrent, and inclusive impacts. For the business world, the ability to extract maximum value from data relies on appropriate business strategies, innovation management, and sound and responsible governance. Thus, companies must adopt self-regulatory ethical and responsible mechanisms in their innovation and digital transformation plans, managing risks even independent of public regulation.

INNOVATION MANAGEMENT: WHAT CHANGES WITH DIGITAL TRANSFORMATION?

Innovation has long been a critical driver of business development and competitiveness. However, digital transformation has fundamentally reshaped how innovation is managed, creating both new opportunities and challenges for firms striving to maintain relevance in an increasingly dynamic market. Digital transformation integrates digital technologies across all organizational functions, fundamentally altering how products and services are created and delivered (Westerman, Bonnet & McAfee, 2014).

Over the past decades, innovation management has evolved, driven by technological advancements and shifts in the business landscape. While digital transformation has facilitated the development of digital products, process automation, and data-driven decision-making, it has also accelerated market disruptions. Firms must now anticipate technological changes, and evolving customer demands at a much faster pace. As a result, collaboration and openness to external innovation ecosystems have become essential, as companies can no longer rely solely on internal resources for sustaining competitive advantage (Balmaekers, 2023).

Historically, innovation management followed a traditional research and development (R&D) model characterized by a linear, internally focused process (Rothwell, 1994; Chesbrough, 2003). This rigid approach limited adaptability to market shifts, leading to the emergence of more flexible and collaborative models, such as open innovation (Chesbrough, 2003). Open innovation promotes collaboration with suppliers, customers, competitors, and other stakeholders, fostering value co-creation by integrating external knowledge and resources.

Innovation management has since transitioned to a more holistic model, extending beyond product and service innovation to include process improvements, business model transformation, and strategic adaptation (Tidd & Bessant, 2018). Digital transformation has accelerated this evolution, allowing firms to leverage digital ecosystems for greater collaboration. Digital platforms facilitate seamless interactions among customers, suppliers, and partners, fostering co-creation and faster innovation cycles (Gawer & Cusumano, 2014). Moreover, emerging technologies such as artificial intelligence (AI), the Internet of

Things (IoT), and blockchain are increasingly embedded into innovation processes, enabling disruptive solutions (Wirtz et al., 2019).

One of the most significant advantages of digital transformation is the ability to analyze large volumes of real-time data, opening new possibilities for data-driven innovation (Nambisan et al., 2017). AI and big data analytics allow companies to identify market trends and optimize decision-making, leading to enhanced product development and improved customer experiences (Davenport & Ronanki, 2018). This shift towards data-centric innovation has revolutionized how firms create and refine business models, driving efficiencies and competitive differentiation.

However, digital transformation also presents challenges. Market dynamics are evolving at an unprecedented pace, shortening product life cycles[2] and intensifying competition. To address these challenges, firms are increasingly adopting agile and iterative innovation methodologies such as lean innovation (Ries, 2011). Agile approaches—such as design thinking, lean startup, and open innovation—prioritize rapid prototyping, experimentation, and continuous learning (Khosravi et al., 2019). Unlike traditional sequential models, these methodologies allow firms to quickly adapt to market shifts while minimizing risks and inefficiencies.

Additionally, digital transformation has enabled new forms of stakeholder engagement. Firms now leverage digital platforms and social networks to co-create products and services with users, incorporating real-time feedback into innovation processes (Prahalad & Ramaswamy, 2004; Bogers et al., 2018). This shift towards participatory innovation has become essential for firms seeking to develop customer-centric solutions and maintain a competitive edge.

Beyond technological integration, digital transformation necessitates a profound cultural shift within organizations. Successful innovation in the digital era requires fostering a culture of creativity, experimentation, and calculated risk-taking. Firms must embrace failure as part of the learning process and encourage cross-functional collaboration (Bughin et al., 2018). Resistance to change remains a major barrier, highlighting the need for companies to build adaptability and openness into their organizational cultures (Vial, 2019).

In summary, innovation management has evolved from a linear, internally driven model to a more open, collaborative, and dynamic framework, shaped by digital transformation and emerging technologies. This transition has enabled firms to leverage external knowledge, engage diverse stakeholders, and accelerate innovation cycles, thereby strengthening long-term competitiveness.

While digital transformation introduces new tools and methodologies, it also presents cultural and organizational challenges. Firms that embrace digital opportunities while fostering adaptability will be better positioned to sustain innovation and competitive advantage. Given this evolving landscape, it is critical to assess how firms manage digital risks within their innovation strategies. The next stage of analysis will explore risk concepts, digital threats, and the challenges associated with risk management in digital transformation. By doing so, this study aims to advance the understanding of the intersection between innovation and risk management in the digital age.

DIGITAL RISKS: CONCEPTS, CHARACTERISTICS, AND ANALYTICAL CHALLENGES

Risk, often associated with negative outcomes, has undergone significant conceptual evolution. Historically, risk was considered neutral, encompassing both positive and negative effects (Gerber & Solms, 2005). The Project Management Institute (PMI, 2017) similarly defines risk as an uncertain event with both beneficial and detrimental impacts on project scope, cost, and quality. Over time, however, the predominant focus has shifted toward mitigating adverse consequences. Gerber & Solms (2005) illustrate this evolution by analyzing definitions from the Royal Society (1983–1991), which increasingly emphasized probability, severity, and consequences.

Academic discussions on risk management are shaped by two distinct paradigms: the natural sciences and the social sciences. The former treats risk as an objective, quantifiable phenomenon tied to technology and systems performance, while the latter considers risk as subjective and socially constructed, influenced by values, beliefs, historical context, and regulatory environments. These differing perspectives affect digital risk analysis and management approaches.

Digital risks stem from data-driven innovations and can affect not only firms, but also individuals, governments, and even the entire society. They are dynamic, often exponential, and involve multiple stakeholders. The approaches to risks have evolved. Early digital risk discussions focused predominantly on IT-related risks, associating risk management with technological assets such as hardware, software, and networks. The Information Systems Audit and Control Association (ISACA, 1998) defined IT risk as the business risk arising from the use, ownership, operation, participation, influence, and adoption within a company, as outlined in its COBIT Framework. Therefore, risk is defined as the potential for IT-related events and conditions to impact the business.

However, this traditional IT-centric view is limited, as it prioritizes technical security while neglecting broader organizational and societal risks. Finne (2000) expanded the scope by defining information security as encompassing both technical and economic measures to mitigate risks. It involves a series of measures to minimise the risks associated with information, including actions, equipment, methods, people, and even economic resources.

Although these two last concepts are broader, when considering measures not limited to technology, conventional risk management still prioritizes the protection of tangible assets, instead of adopting a more holistic and complex perspective to innovation and risks in its broad concept[3]. No tangible results or risks such as misinformation, surveillance, and monopolistic platform dependencies, are privileged.

One defining characteristic of digital risk is the exponential growth of data - and its consequences, widely regarded as the most valuable asset in the digital age. The total volume of global data is expected to reach 175 zettabytes by 2025 (Fan, 2022). While techno-optimists (Diamandis & Kotler, 2019, 2020; Ismail, Malone & Van Geest,, 2015; Balmaekers, 2023) view data abundance as a driver of innovation, critics warn against excessive data collection. Castells (1996) and Dowbor (2025) examines how digitalization exacerbates inequalities, while Harari (2018, 2024) warns of a potential future where AI renders human labor obsolete. Sadin (2018, 2020, 2024) and Han (2017a, 2017b, 2018, 2022) explore the sociopolitical consequences of hyper-connectivity, digital surveillance, and misinformation. Gendler & Mallamaci (2024) address the economic and technological dependencies resulting from monopolistic digital platforms, and the political implications. Wark (2022), Dowbor (2025) and, (2025) Varoufakis go beyond, indicating the advent of a new paradigm, replacing the capitalist paradigm.

While those who are optimistic about the direction of the digital economy address the opportunities for companies and the gains for society, such as the democratization and dematerialization of products, critics address the problems or threats of digital from multiple perspectives, as can be seen above. However, the critical approaches are not directly related to risk analysis, providing few contributions to the debate on risk management. They are critical approaches to the new model of economic development based on the generation of value through data, and warn about its economic, social and even political consequences. Such approaches do not adopt risk analysis methods and do not propose the use of risk management methods to avoid, minimize, or mitigate the impacts. Thus, it is possible to say that there is, in literature, an analytical limitation that allows the treatment of such threats or negative consequences of the advance of digital technology as a risk.

Literature on digital risk remains concentrated on cybersecurity threats, overlooking broader risks associated with digital transformation. Fernandez (2022) highlights the necessity of broadening risk assessments to include economic, political, and societal dimensions. While these scholars focus on systemic risks, organizations such as the World Economic Forum (WEF) and consulting firms like Deloitte (2018) offer structured frameworks for classifying digital risks. They categorize threats into several domains, including individual privacy risks, national security risks, disruptions to democratic institutions, and attacks on critical infrastructure. Prazeres (2021) and Gupta & Wang (2021) emphasize the geopolitical implications of digital risks, advocating for global governance and regulatory measures.

This study states that digital risk is not limited to cybersecurity vulnerabilities. As Fernandez (2022) notes, digital risk encompasses "all transformations derived from digitalization that may threaten basic aspects of our current way of life in economic, political, or social terms" (p. 490). This definition underscores the need for a broader and more integrated approach to digital risk management.

Traditional risk management frameworks have focused on securing IT infrastructure and networks. However, digital risks require a more holistic strategy that includes data privacy, misinformation, algorithmic bias, and the economic impact of digital monopolies. Managing these risks requires continuous monitoring and strategic adaptation, given the fast-evolving nature of digital threats.

As digital risks become more complex, interdisciplinary collaboration is essential. Risk management should no longer be the sole responsibility of IT departments; rather, it should involve economists, sociologists, ethicists, and policymakers. Morin's (2009, 2015) theory of complexity supports this approach, highlighting the interconnectivity of technological, economic, and sociopolitical factors. Berger & Solms (2005) similarly advocate for an interdisciplinary perspective on information security that incorporates legal, political, and economic considerations.

In conclusion, digital risks have evolved beyond IT security concerns, encompassing a broad spectrum of threats to data integrity, societal stability, and governance structures. While cybersecurity remains a priority, organizations must adopt a comprehensive risk management approach that includes economic, political, and social risks, integrating multiple stakeholders into the analyses.

As digital transformation accelerates, companies must shift from traditional IT-centric risk models to integrated frameworks that incorporate systemic risks, interdisciplinary insights, and governance considerations. The ability to anticipate and adapt to emerging threats is crucial for building resilient digital ecosystems. Expanding digital risk analysis will help organizations align their innovation strategies with responsible governance, ensuring technological advancements contribute to economic growth and societal well-being while mitigating potential threats.

RISK AND INNOVATION MANAGEMENT IN THE DIGITAL AGE

The following section discusses risk management, starting with an overview of basic concepts and models, and then exploring new approaches to risk in the digital age. Key objectives highlighted by researchers like Finne (2020) include identifying, measuring, and controlling uncertain events to minimize losses and optimize the return on security investments. An essential reference for risk management is the ISO 31000 standard, which outlines risk management as a process with sequential, complementary stages that are monitored, recorded, and communicated.

Innovation inherently involves a certain degree of risk, as it requires the exploration of new ideas and approaches whose success in the market is not guaranteed. Therefore, effective risk management is essential in the innovation process, as explored in this paper. The main risks in innovation management are the market, the technological, and the financial risk (Teece, 2010; Byers et al., 2011; Hjalager, 2010).

Market risk refers to the possibility that the innovative product or service will not be accepted by customers or will fail to generate the expected income. This may be due to a lack of understanding of consumer needs and preferences, inadequate market segmentation, or a lack of differentiation from the competition (Teece, 2010). According to Mu et al. (2020), effective market risk management is crucial for innovation success and requires deep customer knowledge and a strong differentiation strategy.

In contrast, technological risk implies the possibility that the technology on which the innovation is based does not function as expected or is surpassed by a more advanced technology (Byers et al., 2011). This risk can be mitigated through careful technical evaluation and constant technological monitoring. Tian et al. (2021) further suggest that technological risk management involves a continuous analysis of technological trends and an intellectual property protection strategy.

Finally, financial risk refers to the necessity of making a significant investment of financial resources without a guarantee of return (Hjalager, 2010). It is incumbent upon companies to meticulously evaluate the potential costs and benefits of innovation and to establish appropriate financing and risk management mechanisms. Coad et al. (2021) posit that financial risk management in innovation necessitates meticulous planning and the allocation of resources, in addition to a rigorous evaluation of expected returns.

In the digital age, risk management in innovation has undergone a transformation. While some risks, such as technological risks, have been amplified due to the rapid evolution of digital technologies, other emerging tools, such as predictive analytics and simulation, have made it possible to mitigate and anticipate possible failures (Kohli & Melville, 2019; March & Shapira, 1987). Advanced analytical tools, such as Machine Learning, AI and Natural Language Processing, "allow (…) the ability to probabilistically predict the future. Data science techniques can provide probabilistic estimates of what can go wrong, which, when combined with the intuition and judgement of the risk manager, can be a potent weapon to better manage risk" (Pareek, M., CISA, ACA, ACMA, PRM (2019, Use Cases Within IT Risk).

All risk management focuses, in some form or another, on avoiding adverse risk events. These events run the gamut from data leakage to server compromise, from failed IT changes to misused privileges and from intrusion detection to flagging spam.

Furthermore, the integration of artificial intelligence and machine learning has enhanced the capacity of companies to identify and assess risks with greater precision, enabling the design of effective mitigation strategies (Wamba-Taguimdje et al., 2020). Market risks gain a new perspective with digital transformation, with the agility of innovation and the construction of Minimum Value Products (MVP), reducing the time and costs involved in innovation and market testing. The insertion of new technological solutions into the market via sales through large product and service platforms also allows market testing to be accelerated and eventual adjustments or pivots to be made quickly.

NEW APPROACHES: WHY ARE TRADITIONAL RISK ANALYSIS AND MANAGEMENT NO LONGER APPLICABLE?

Traditional Information Technology (IT) risk management frameworks encompass a collection of processes, policies, and procedures organizations employ to mitigate IT-related risks. Historically, these risks were primarily associated with failures in technological infrastructure, systems, or software. However, in the digital transformation era, new perspectives must be incorporated, with a particular emphasis on personal data protection. This shift raises a fundamental question: Are conventional IT risk analysis and management methods sufficient to address the complexities and variety of digital risks? Contemporary risk management methodologies are influenced by several emerging factors, suggesting that for digital environments to function effectively, risk frameworks must extend beyond traditional parameters (Deloitte, 2018, p. 5).

Digital risks are not confined to threats affecting IT infrastructure or systems in isolation, nor are they solely about cybersecurity attacks. Instead, this study asserts that digital risks encompass a much broader and more complex spectrum of threats.

New areas must be incorporated into the discussion on risk management and innovation. There are "invisible" risks that might not be identified or monitored through traditional risk management cycles. A new generation of market risks has emerged, primarily driven by the value generated through data. As extensively analyzed in digital transformation literature, organizations derive significant wealth from collecting and analyzing personal data, enabling a deeper understanding of consumer behavior. This facilitates hyper-personalization, redesigning solutions to meet consumer needs, and anticipation of market trends. However, how data is utilized represents a major source of risks and threats to both companies and consumers. This study integrates elements from the social sciences paradigm, where risks are examined not from an objective or evaluative perspective but rather as subjective and perceived phenomena.

For instance, reputational risks arise from the misuse of personal data for political purposes, as evidenced by the Cambridge Analytica and Facebook scandal. Finne (2000) had already warned that business success depends on mastering, continuously updating, and protecting valuable information. Other risks include disinformation, which poses threats to individuals and society and is often mishandled by digital platforms under the pretext of "freedom" on the internet. Algorithmic biases in security systems in major cities, excessive surveillance by digital platforms, psychological distress caused by hyper-exposure, and performance pressures in digital environments are additional risks that must be accounted for. The last edition of The Global Risks Report (WEF, 2025) points out that misinformation will be the main risk in 2027.

Risk management strategies must evolve to identify and monitor these emerging risks effectively. Digital transformation accelerates the need for adjustments in risk management methodologies, particularly in innovation projects. The speed of change, exponential data growth, real-time decision-making requirements, multi-channel business models, complex digital ecosystems, open innovation, and multi-stakeholder influences have all reshaped risk management. These trends have broadened the scope of risks, increased their frequency, exposed companies to new vulnerabilities, and altered stakeholder relationships across value chains.

The intersection of digital risk and innovation management presents a significant challenge. Digital transformation fosters open data ecosystems, decentralized decision-making, and rapid experimentation, which may conflict with risk management's emphasis on control and stability. Companies must balance the need for innovation with the necessity of mitigating digital risks.

One major area of concern is third-party system integration and software development. Increasing reliance on external platforms introduces new vulnerabilities. The cyber blackout of July 2024, which disrupted Microsoft and its entire value chain due to a failure at CrowdStrike, illustrates how a single failure can trigger widespread consequences (Gartner, 2022).

Another critical issue is the governance of digital platforms. As monopolistic companies dominate global digital infrastructure, businesses and governments become increasingly dependent on their ecosystems. Gendler & Mallamaci (2024) warn of the economic risks associated with such dependencies, emphasizing the need for regulatory oversight to prevent anti-competitive behavior and data exploitation.

Open innovation initiatives aim to share the risks of the innovation process among participants, typically involving companies, startups, and research institutions. Corporate Venture Capital projects represent high-risk investments in startup-led innovation, allowing major corporations to benefit from innovation outcomes without directly engaging in the innovation process. Additionally, the life cycle of digital solutions has shortened, while research and development processes have accelerated due to agile methodologies. As a result, metrics for performance, risk, and failure are continuously reassessed.

Risk management in the digital era also necessitates expanding the scope of stakeholders beyond internal organizational boundaries. Finne (2000) emphasizes that companies cannot eliminate risks associated with information—at best, they can control them. He highlights that "some risks are beyond the reach of the company" (p. 237). This challenge is particularly evident in the role of IT system providers, platform users, business partners, and customers, all of whom generate and use data that influence business reputation and success.

A critical area of interest in this context is third-party system integration and software development. This long-standing practice has become increasingly significant in digital transformation environments, particularly with the emergence of business ecosystems and digital platforms. A notable global incident occurred in July 2024, when a cyber blackout impacted companies such as Microsoft and its entire value chain due to a failure at CrowdStrike, a cybersecurity provider. Gartner (2022) suggests that centralized cybersecurity controls may be becoming obsolete; however, the development of new control and risk management mechanisms remains a challenge.

The interconnectedness of IT systems, participation in business and innovation ecosystems, and collaboration with multiple stakeholders necessitate a more systemic and holistic approach to risk management. Companies should no longer be viewed as isolated entities but rather as complex, interdependent components of broader ecosystems where economic, political, and sociological factors interact dynamically (Morin, 2009, 2015).

In the digital economy, businesses operate within increasingly intricate ecosystems where the roles of various actors have become more deeply intertwined. Consumers do not merely consume; they generate vast amounts of data that fuel digital platforms. Labor relations have evolved into more flexible and dynamic arrangements, and the informal economy has taken on new characteristics. Meanwhile, algorithmic decision-making raises new ethical challenges. Information and data have surpassed physical products in economic value, fundamentally altering power dynamics between technology firms and traditional businesses. Organizations both influence and are influenced by an increasingly volatile, uncertain, and rapidly evolving environment.

The advances in digital technology and the growing interconnectedness of business relationships have further complicated corporate ecosystems. Gartner (2022) advises companies to extend their risk monitoring efforts beyond IT and corporate risk management, incorporating their entire digital ecosystem into governance strategies. This includes security leaders, risk managers, and other departments, with a priority on digital supply chain risk. Additionally, suppliers should be engaged in adopting more robust security practices.

Within organizations, there are already noticeable shifts in approaches to risk management. Recent reports from the World Economic Forum (WEF, 2023) on cybersecurity highlight a transformation in the perception of business leaders, board members, and cybersecurity professionals. The gap in perspectives identified in 2022 regarding the likelihood of cyberattacks significantly impacting organizations within two years has narrowed. Furthermore, interactions between cybersecurity executives and business leaders have increased in frequency and structure. This marks a critical step toward fostering effective risk management strategies by bridging the gap between IT, compliance, and executive leadership.

Several fundamental questions remain unanswered in the study of digital enterprise risk management, particularly regarding the evolution of methodologies to address contemporary challenges and establish digital resilience. One of the key contributions of this paper is examining how digital transformation and risk management intersect and how organizations can integrate these concepts into their business strategies.

While it is impossible to address comprehensively all aspects of this complex issue, this study critically evaluates various risk frameworks in digital transformation. One pressing challenge is determining what assets require protection. Information security scholars emphasize the importance of safeguarding data as a critical asset. However, traditional IT security approaches have focused primarily on protecting physical infrastructure, with risks measured through quantitative or qualitative methodologies.

Lee (2021) argues that digital risk management must incorporate both technical and human aspects, particularly in cybersecurity frameworks. He critiques existing cybersecurity models for relying too heavily on qualitative guidelines, lacking a balanced approach to cyber risk management, and failing to integrate quantitative assessments for financial impact. Additionally, he highlights the emphasis on technological risks while neglecting human factors.

In an era of digital globalization and e-commerce, risk management strategies must evolve. Digital transformation has interconnected business processes, making information security an indispensable element of corporate governance. Given its centrality to business operations, information security should not be viewed in isolation but as an interdisciplinary field spanning law, politics, economics, history, and cultural studies (Berger & Solms, 2005, p. 25).

Organizations must address fundamental questions: What is the true value of an organization's digital ecosystem, which generates vast amounts of data every minute? How can companies protect platform users from surveillance risks or algorithmic biases? These issues must be incorporated into corporate risk management frameworks, innovation processes, and digital transformation strategies.

Building on the perspectives of Morin (2009, 2015) and Berger & Solms (2005), this study underscores the complexity of digital risks and their reciprocal influence across various domains. Effective risk management in the digital era requires a paradigm shift—one that extends beyond technological and organizational boundaries to include broader social, political, and ethical considerations.

Deloitte's (2018) framework, illustrated in Figure 1, provides a comprehensive approach to digital risk management. It encompasses multiple risk categories—including strategic, technological, third-party, regulatory, and digital ecosystem risks—while proposing tailored control measures based on the level of digitalization in business operations.

Figure 1. The digital risk framework

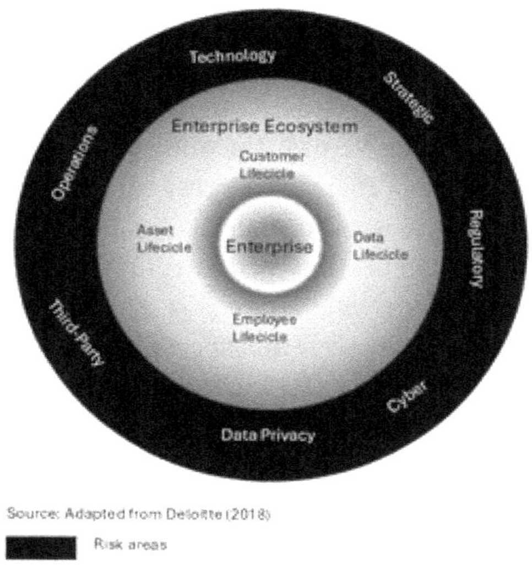

Source: Adapted from Deloitte (2018)

These approaches recognize the contribution of risk analysis while advocating for a broader perspective that incorporates insights from various social science disciplines. When designing a digital risk strategy, it is crucial to consider robust cybersecurity measures that extend beyond conventional information security. However, it is essential to question whether this is sufficient. Is cybersecurity the only risk for a digitally enabled organization? To create an effective digital environment that meets its intended objectives, it is imperative to consider risk areas beyond those traditionally addressed (Deloitte, 2018).

CONCLUSION

The rise of digital risks, driven by digital transformation and the pervasiveness of digital in life and companies, generates mistrust in the digital economy. Nevertheless, the inevitability of digital technology advancement and its ubiquity suggests that the increasing level of digital risks may require the imposition of limitations on the development of innovations, which would constitute a setback for current and future advances.

To ensure the continued advancement of digital innovation and digital transformation, expanding business opportunities and benefits for the entire society, it is recommended that corporate strategy management steers the way toward both innovation management and risk management. This will ensure that the relevant knowledge is available to deal with the multidimensional nature of phenomena and that decisions are made in a way that is agile and resilient in the face of complex problems. The safe, conscious, transparent, ethical and regulated use of technologies by companies should contribute to constant innovation and problem-solving, with the human being at the center.

Digital risk management is a vast and continuous domain, which recognises the growing and varied nature of these risks. To effectively address risks, especially those of a cyber nature, organisations must first understand their economic impact, align them with business needs and then design appropriate organisational structures. Systemic resilience is of paramount importance, as it highlights the interconnectivity of cyber risk and the necessity for unified responses. Entrepreneurs play a pivotal role in fostering trust, adequately preparing to meet the challenges of the modern age and considering business risks more comprehensively. An effective digital risk management strategy necessitates a strategic vision, an integrated and collaborative multi-stakeholders ecosystemic, and the participation of strategic levels within the company. This entails grasping the company as a complex entity, a component of interconnected and interdependent systems and ecosystems. Companies are not alone. Risk management can no longer consider the company in its internal universe. Digital risk management amplifies this requirement, because the variety, impact and speed of digital threats makes the importance of preventing and managing them even more acute and accelerated.

In the context of digital risk management during digital transformation, several questions arise that can guide future research. There is a need to deeply explore non-traditional risk management frameworks and understand why firms, even in the era of "stakeholder capitalism", have a lack of interest in digital risks not directly related to cybersecurity. How can organisations effectively balance new technologies adoption with managing the associated risks? What role do cybersecurity policies play in mitigating risks in specific sectors, such as healthcare or finance? It is necessary to determine how culture change strategies can be optimised to minimise resistance to change among employees. Furthermore, it is important to investigate how emerging technologies, such as artificial intelligence and machine learning, can be more efficiently integrated into existing systems to improve risk prediction and management. These questions not only expand the field of study but are also crucial to developing a robust theoretical and practical framework that guides organisations in their transition towards secure and efficient digital environments.

This paper utilized the empirical literature review method to analyze the state of art on the themes of risk management, digital risk management, and innovation management in the digital age, highlighting opportunities to fill knowledge gaps from a complex, holistic and human perspective on digital risks and innovation management in the digital era. The article aimed to present risk analysis elements that support companies in their innovation and digital transformation processes, shedding light on aspects still absent or scarcely considered by companies when managing innovation.

The results demonstrate that, despite literature addressing various facets of digital risks, the approaches are disconnected, especially when considering risks in the business environment. Advances have been made in expanding the perspective on risks, moving beyond an exclusive focus on technological infrastructure assets to include data protection and, further, personal or social risks such as misinformation, surveillance and algorithmisation of everything. However, these analyses do not explore the potential intersection or complementarity, and no analyses with philosophical approaches were observed, for example, bringing elements that contribute to improving business management. Thus, it would be pertinent to consider digital risk management as part of the strategic issues of organisations' innovation management.

Similarly, approaches to innovation management and risk management have made little progress on social or personal issues, even considering the relevance of the ecosystem as a user and producer of data, a relevant asset for digital transformation. Thus, although advances have been observed in methodologies from international entities and consultancies, recognising the relevance of a holistic perspective on risks, beyond cyber risks, gaps remain. While the World Economic Forum's global risk analyses do not delve into the business environment, Deloitte's methodology signals to companies the need to broaden their risk management perspective yet still does not incorporate human and social aspects into its analysis. Is it possible to humanise risk management and innovation management in the digital era, placing humans at the center of the issue?

REFERENCES

ABNT (2009). *Risk Management – Principles and guidelines. NBR ISO 31000.*

Balestta (2021). The reconfiguration of the market: how to build dynamic and adaptable ecosystems for the BANI world? Oliveira, C. & Menezes, H. (Eds.). *Digital: business and digital transformation.* Dom Cabral Foundation. Cap. 19, p. 176-177. ISBN: 978-65-994597-3-3

Balmaekers, H. (2023). *The innovator's Handbook 2023. Innov8rs.*https://innov8rs.co/dl.file.php?file=2019/02/The-Innovators-Handbook-Digital-Book.pdf

Bogers, M., Chesbrough, H., & Moedas, C. (2018). Open innovation: Research, practices, and policies. *California Management Review, 60*(2), 5–16. DOI: 10.1177/0008125617745086

Bughin, J., Catlin, T., Hirt, M., & Willmott, P. (2018). Why digital strategies fail. *The McKinsey Quarterly.*

Byers, T. H., Dorf, R. C., & Nelson, A. J. (2011). *Technology ventures: From idea to enterprise.* McGraw-Hill Education.

Castells, M. (1999). *The network society.* Peace and Land.

Chesbrough, H. W. (2003). *Open innovation: The new imperative for creating and profiting from technology.* Harvard Business Press.

Coad, A., Segarra, A., & Teruel, M. (2021). Innovation and risk: What it means for productivity and profitability. *Journal of Business Research, 125,* 119–128.

Davenport, T. H., & Ronanki, R. (2018). Artificial intelligence for the real world. *Harvard Business Review, 96*(1), 108–116.

Deloitte. (2018). *Managing risk in digital transformation.*https://www2.deloitte.com/content/dam/Deloitte/in/Documents/risk/in-ra-managing-risk-in-digital-transformation-1-noexp.pdf

Diamandis, P., & Kotler, S. (2019). *The future is better than you think.* HarperCollins Brasil.

Diamandis, P., & Kotler, S. (2020). *The future is faster than you think.* Alta Books Editora.

Dowbor, L. (2025). *The challenges of the digital revolution: Unlocking knowledge for the common good.* Elefante.

Fan, S. *(07 de Juno de2022). Quantum chip takes microseconds to do a task a supercomputer would spend 9,000 years on.*https://singularityhub.com/2022/06/07/a-photonic-quantum-device-took-microseconds-to-do-a-task-a-conventional-computer-would-spend-9000-years-on/

Fernandez, J. (2022). The Governance of Digital Risks: Challenges and Advances in the Regulation of Artificial Intelligence. *Transnational Law Notebooks (Marzo 2022), Vol 14, Nº ¡,pp 489-503.*https://Doi.org/10.20318/cdt.2022.6695

Finne, T. (2000). Information Systems Risk Management: Key Concepts and Business Processes. *Computers & Security, 19*(3), 234–242. DOI: 10.1016/S0167-4048(00)88612-5

Frenkel, J. (2021). How to manage innovation in the digital age? Oliveira, C.; Menezes, H. (Eds.). *Digital: business and digital transformation*. Cap. 17, p. 146-158. DOI: 10.52959/2021235310

Gartner (13 de abril de 2022). *7 top trends in cybersecurity for 2022*. https://www.gartner.com/en/articles/7-top-trends-in-cybersecurity-for-2022

Gawer, A., & Cusumano, M. A. (2014). Industry platforms and ecosystem innovation. *Journal of Product Innovation Management*, *31*(3), 417–433. DOI: 10.1111/jpim.12105

Gendler, M. A., & Mallamaci, M. (2024). Bits (not only) are bits. Genealogy, dimensions and actuality of studies on digital platforms. *Hipertextos Magazine*, *12*(21), e079.

Gerber, M. y Solms, R. (2005). Management of risk in the information age. *Computers & Security*, 24, 16e30. https://doi:DOI: <ALIGNMENT.qj></ALIGNMENT>10.1016/j.cose

Gupta, A., & Wang, H. Por que a nova era da globalização é digital? Oliveira, C.; Menezes, H. (Eds.). *Digital: a economia que move as empresas e o mundo: onde estamos e aonde podemos chegar*. Cap. 7, p. 59-64. DOI: 10.52959/2021135284

Han, B. (2017a). *Society of tiredness*. Voices.

Han, B. (2017b). *Society of transparency*. Voices.

Han, B. (2018). *In the swarm: digital perspectives*. Voices.

Han, B. (2022). *Infocracy: digitalization and the crisis of democracy*. Voices.

Harari, Y. (2018). *21 Lessons for the 21st Century*. Companhia das Letras.

Harati, Y. (2024). *Nexus: A brief history of information networks, from the Stone Age to artificial intelligence*. Companhia das Letras.

Hjalager, A. (2010). A review of innovation research in tourism. *Tourism Management*, *31*(1), 1–12. DOI: 10.1016/j.tourman.2009.08.012

Isaca Information Systems Audit and Control Association (1998). *COBIT Framework*.

Ismail, S., Malone, M. & Van Geest, Y. (2018). *Exponential Organizations: Why They're 10 Times Better, Faster, and Cheaper Than Yours (and What to Do About It*. HSM Editora.

Khosravi, P., Newton, C., & Rezvani, A. (2019). Management innovation: A systematic review and meta-analysis of past decades of research. *European Management Journal*, *37*(6), 694–707. DOI: 10.1016/j.emj.2019.03.003

Kohli, R., & Melville, N. P. (2019). Digital innovation: A review and synthesis. *Information Systems Journal*, *29*(1), 200–223. DOI: 10.1111/isj.12193

Lee, I. (2021). Cybersecurity: Risk management framework and investment cost analysis. ScienceDirect, Business Horizons (2021) 64, 659e671.

March, J. G., & Shapira, Z. (1987). Managerial perspectives on risk and risk taking. *Management Science*, *33*(11), 1404–1418. DOI: 10.1287/mnsc.33.11.1404

Morin, E. (2009). *Introduction to Complex Thinking*. Gedisa Editorial.

Morin, E. (2015). *The head well set: Rethinking reform. Reform thinking*. New Vision.

Mu, J., Thomas, E., Tao, L., & Tan, J. (2020). Market risk and firm innovation: Evidence from China. *European Journal of Innovation Management*, *23*(5), 975–998.

Nambisan, S., Lyytinen, K., Majchrzak, A., & Song, M. (2017). Digital innovation management: Reinventing innovation management research in a digital world. *Management Information Systems Quarterly*, *41*(1), 223–238. DOI: 10.25300/MISQ/2017/41:1.03

OECD. (2005): *Proposed Guidelines for Collecting and Interpreting Technological Innovation Data, Oslo Manual*, third edition, OECD/EU/Eurostat.

Oliveira, C., & Menezes, H. (2021). Why is it so Critical to Have Trust in the Digital Economy? Oliveira, C., Menezes, H. (Eds.). *Digital: the Challenge of Trust and Security in the Digital Economy.* Nova Lima https://www.fdc.org.br/conhecimento/publicacoes/livro-eletronico-e-book-35383

Parker, G., Alstyne, M., & Choudary, S. (2019). *Platform: The Strategy Revolution*. Portfolio-Penguin.

Prahalad, C. K., & Ramaswamy, V. (2004). Co-creation experiences: The next practice in value creation. *Journal of Interactive Marketing*, *18*(3), 5–14. DOI: 10.1002/dir.20015

Project Management Institute. (2017). *PMBOK® Guide - A Guide to Project Management Knowledge*. Project Management Institute.

Rete, O. (2020). *Digital Transformation for UTN and UIA* [Presentation]. National Technological University and Argentine Industrial Union.

Ries, E. (2011). *The lean startup: How today's entrepreneurs use continuous innovation to create radically successful businesses*. Crown Business.

Rothwell, R. (1994). Towards the fifth-generation innovation process. *International Marketing Review*, *11*(1), 7–31. DOI: 10.1108/02651339410057491

Sadin, E. (2018). *The siliconization of the world: the irresistible expansion of digital liberalism*. Caja Negra.

Sadin, E. (2020). *Artificial Intelligence or the Challenge of the Century*. Caja Negra.

Sadin, E. (2024). *The era of the tyrannical individual and the power of social networks*. Youtube https://www.youtube.com/watch?v=7IPS1sWfyVs

Teece, D. J. (2010). Business models, business strategy and innovation. *Long Range Planning*, *43*(2-3), 172–194. DOI: 10.1016/j.lrp.2009.07.003

Tidd, J., & Bessant, J. (2018). *Managing innovation: Integrating technological, market and organizational change*. John Wiley & Sons.

Toffler, A. (1981). *The Third Wave*. Editora Record.

Varoufakis, Yanis (2025). *Techno-feudalism: What killed capitalism.* Crítica. ISBN-13 : 978-8542233193

Vial, G. (2019). Understanding digital transformation: A review and a research agenda. *The Journal of Strategic Information Systems*, *28*(2), 118–144. DOI: 10.1016/j.jsis.2019.01.003

Wamba-Taguimdje, S. L., Wamba, S. F., Kamdjoug, J. R. K., & Wanko, C. E. T. (2020). Influence of artificial intelligence (AI) on firm performance: The business value of AI-based transformation projects. *Business Process Management Journal*, *26*(7), 1893–1924. DOI: 10.1108/BPMJ-10-2019-0411

Wark, M. (2022). *Capitalism is dead: The rise of the vectorialist class*. Holobionte Editions.

Westerman, G., Bonnet, D., & McAfee, A. (2014). *Leading digital: Turning technology into business transformation*. Harvard Business Review Press.

Wirtz, B. W., Pistoia, A., Ullrich, S., & Göttel, V. (2019). Business models: Origin, development and future research perspectives. *Long Range Planning*, *51*(1), 44–54.

World Economic Forum. (2021a). *The Global Risks Report 2021*.

World Economic Forum. (2021b). *Principles for Board Governance of Cyber Risk*.https://www.weforum.org/reports/136f100d-381a-4b09-a891-de2299144992

World Economic Forum. (2022). *Global cybersecurity outlook 2022*.https://www.weforum.org/agenda/2022/07/global-cybersecurity-outlook-davos-2022/

World Economic Forum. (2023). *Global cybersecurity outlook 2023* https://www.weforum.org/publications/global-cybersecurity-outlook-2023/

World Economic Forum. (2025). The Global Risks Report 2025.

World Intellectual Property Organization (WIPO). (2023). *Global Innovation Index 2023: Innovation in the face of uncertainty*. Geneva: WIPO. https://DOI:DOI: 10.34667/tind.48220

KEY TERMS AND DEFINITIONS

Complexity: The concept used in this article is inspired by Edgar Morin and his Theory of Complexity, which critiques linear causality and reductionist simplifications. His theory promotes a relational and multidimensional view of phenomena. The Theory of Complexity is based on three fundamental principles: dialogic, recursion, and hologrammatic. The dialogic principle allows the coexistence of opposing concepts, such as order and disorder, which sometimes collaborate to generate organization and complexity. Recursion describes how products and their effects are simultaneously causes and producers of what generates them, challenging the linear cause-and-effect view. Finally, the hologrammatic principle recognizes that both the whole is in the parts and the parts are in the whole, transcending both reductionism and holism.

Digital economy: The concept of digital economy used in this paper was developed by the author Menezes in an unpublished research report prepared while she was at Cornell University as a visiting scholar in 2019. According to the report, entitled "Brazil in the Digital Economy: A Contribution to Rank Brazil Towards the Future," the author conceptualizes the digital economy as the set of values created by a global ecosystem composed of people, companies, and machines enabled and hyperconnected under digital platforms. Its components include an infrastructure (which enables connectivity) and a global and local environment that allows people to conduct digital transactions under digitally oriented businesses.

Digital risks: Digital risks are the technological risks associated with the digital economy. They stem from data-driven innovations and solutions and can affect not only firms, but also individuals, governments, and even the entire society. They are dynamic, often exponential, and involve multiple stakeholders. Digital risks go beyond information security risks or cyberattacks; they encompass all threats related to the increasing use of digital technologies by companies and technology users. Examples of digital risks include hacker attacks on critical digital infrastructures, personal data breaches, misinformation, loss of subjectivity of social media users, algorithm biases, the diminishing relevance of humans in the workforce, among others.

Digital transformation: Digital transformation is a comprehensive process involving the integration and adoption of digital technologies to reconfigure business processes, models and organisational dynamics to create value in a hyper-connected context. It has a strategic nature and goes beyond technology, linking it to the reinvention of products and services through tools such as AI and big data. Digital transformation is also associated with the creation of collaborative ecosystems and the management of emerging risks, such as platform dependency or accelerated obsolescence. Its success depends on balancing innovation with ethical governance, ensuring that initiatives are sustainable and human-centred.

Innovation management: Innovation management is a multifaceted discipline that involves the systematic approach to fostering, managing, and implementing new ideas within organizations to drive growth, competitiveness, and value creation. Innovation management can be defined as the process of managing an organization's innovation procedure, starting from the initial stage of ideation, through to the execution, and ultimately to the successful implementation and commercialization of new products, services, or processes. It aims to create a structured environment where creativity and innovative thinking are encouraged, nurtured, and systematically channeled towards tangible outcomes. Effective innovation management is crucial for organizations seeking to maintain relevance in rapidly changing markets. It enables firms to adapt to emerging trends and technologies, meet evolving customer needs, enhance efficiency and productivity, and drive sustainable growth and profitability.

Risk management: Risk management is a comprehensive and systematic process critical to the functioning of organizations in uncertain environments. It involves the identification, assessment, and prioritization of risks followed by the coordinated and economical application of resources to minimize, monitor, and control the probability or impact of unfortunate events, or to maximize the realization of opportunities. Academically, risk management is studied through various lenses, including quantitative methods such as statistical analysis and modeling, qualitative approaches like scenario planning and expert judgment, and hybrid models that integrate both. It is multidisciplinary in nature, drawing from fields such as finance, engineering, healthcare, environmental science, and information technology. Effective risk management ensures that organizations can navigate the complexities and uncertainties of their operational landscape, safeguarding their assets, reputation, and long-term viability while fostering resilience and strategic growth.

Stakeholders: The PMI (2017) defines stakeholders as interested parties, which include individuals, groups, or organizations that can influence, be influenced by, or feel affected by a decision, activity, or project outcome. This concept can also extend beyond project management.

ENDNOTES

1 The World Economic Forum's Global Cybersecurity Outlook (2022) report indicates that cyber-attacks increased by 125% globally in 2021 and suggests a continued rebound through 2022. The report estimates that global losses resulting from cybercrime will reach more than 10 billion dollars in 2025.

2 Life cycle management becomes increasingly relevant, not only from the customer's perspective, requiring the rapid and efficient management of the product and service portfolio. It becomes essential in digital transformation and risk management to analyze not only the company itself but also the *extended enterprise*, incorporating the life cycles of assets, data, and employees (Deloitte, 2018), all of which are subject to rapid obsolescence.

3 The third edition of the OECD Oslo Manual (2005) is taken, which, reviewing the models and methods of analysis and measurement of innovation, considers not only the innovative aspects in product and process technologies, but also includes everything that concerns to non-technological innovation, such as in marketing and business models.

Chapter 15
The Impact of Technological and Non-Technological Innovation on Export Performance for SMEs in Marrakech Safi Region in Morocco:
A Structural Equation Analysis

Ali Bouzerouata
https://orcid.org/0000-0003-3737-2919
National School of Business and Management, El Jadida, Morocco

Younesse El Menyari
Faculty of Economics and Social Sciences Ain Chock, Hassan II University, Morocco

Abdelhak Sahib Eddine
National School of Business and Management, El Jadida, Morocco

ABSTRACT

This study examines the impact of technological and non-technological innovation on the export performance of SMEs in the Marrakech-Safi region of Morocco. Using a structural equation model and data from 92 SMEs, the study finds that organizational and technological innovations significantly enhance export performance, while marketing innovation has a limited direct impact. The research highlights the importance of diversifying innovation strategies and the mediating role of innovation extensiveness. The findings offer practical implications for business managers and policymakers, suggesting that innovation-driven strategies can boost international competitiveness for SMEs

DOI: 10.4018/979-8-3693-4030-1.ch015

INTRODUCTION

The importance of innovation in driving business success and enhancing global competitiveness has long been recognized as a crucial factor in economic growth. Small and medium-sized enterprises (SMEs), which form the backbone of most emerging economies, rely on innovation not only to survive in competitive markets but also to improve their export performance. As globalization accelerates and markets become increasingly interconnected, the need for SMEs to innovate grows significantly. While technological innovations, particularly in product and process development, have received considerable attention, there remains a significant gap in understanding the role of non-technological innovations, such as organizational and marketing innovations, in driving export success for SMEs, especially in developing countries.

The concept of innovation dates back to its pioneer, Schumpeter (1934), who considered innovation a key driver of economic development. Furthermore, innovation is recognized as one of the main determinants of international competitiveness (Egger & Becker, 2013). Developing countries, characterized by an SME structure with low technological intensity, could particularly benefit from non-technological innovations (Teece, 2010). For instance, organizational innovation has been shown to strengthen the overall performance of exporting companies (Damanpour & Aravind, 2011). While empirical evidence on the impact of innovation on export performance is abundant in developed countries, studies examining the role of non-technological innovation in emerging economies remain scarce (Mol & Birkinshaw, 2009).

Empirical research on the link between innovation and export performance has primarily been conducted in developed economies, with a strong focus on product and process innovations. Most studies assume that technological innovations are the main drivers of export success, leaving non-technological innovations relatively underexplored. Moreover, much of this research overlooks the constraints faced by SMEs, such as financial limitations, restricted access to advanced technologies, and a lack of specialized human capital, which pose significant barriers to innovation. Consequently, the relationship between non-technological innovations and export performance in emerging economies remains insufficiently studied.

This chapter aims to fill this critical gap by examining the impact of different forms of innovation on the export performance of SMEs in an emerging economy such as Morocco, with a specific focus on the Marrakech-Safi region. Additionally, this study provides valuable insights by emphasizing the importance of non-technological innovations in helping Moroccan SMEs overcome challenges and enhance their competitiveness.

The study's central research question is: **What are the effects of technological and non-technological innovations on the export performance of SMEs in Marrakech, and what strategies can these SMEs adopt to become more competitive in global markets?**

Addressing this question will help policymakers and business leaders understand the different forms of innovation and their impact on export performance. By bridging these gaps, this chapter will serve as a roadmap for future research and policy development, offering valuable insights for both academia and business practice.

Background of the Study

In the context of increasing globalization and international competition, SMEs are compelled to enhance their export competitiveness. Historically, the literature has primarily focused on technological innovation, such as the adoption of new technologies and the digitalization of processes, while largely neglecting non-technological innovations, which encompass design, marketing, cultural adaptation, distribution strategies, and managerial practices. However, in regions such as Marrakech-Safi, where SMEs predominantly operate in tourism, crafts, agriculture, and industry, non-technological innovations play a crucial role in business success. These forms of innovation contribute to filling the empirical literature gap by providing insights into their impact on export performance in such contexts.

For instance, an artisanal SME in Marrakech specializing in traditional carpet production revised its design models to better align with foreign consumer preferences. Additionally, it implemented a digital marketing strategy to increase its international visibility. However, despite these efforts, the company faced multiple challenges, including the lack of reliable logistics partners in target markets, language and cultural barriers, and a limited expertise in e-commerce. This practical case illustrates how non-technological innovation, although essential, can encounter operational and structural obstacles during its implementation.

Another relevant example is that of a tourism SME in the Marrakech-Safi region that aims to enhance customer relationship management. This company introduced innovative practices based on digital tools to improve service quality and optimize customer interactions. This scenario underscores the importance of integrating different forms of innovation. Ideally, strong export performance requires the combined and strategic utilization of both technological (product and process) and non-technological (organizational and marketing) innovations.

Despite the growing body of empirical studies on innovation and export performance, research focusing on Moroccan SMEs remains scarce. Moreover, existing studies often yield contradictory findings. While many highlight technological innovation as a key driver of international success, emphasizing its role in enhancing competitiveness through product differentiation and process optimization, other studies suggest that non-technological innovations play an equally important role. This ongoing debate underscores the need for further empirical research to better understand the interplay between different types of innovation and export performance in emerging economies.

STATEMENT OF THE PROBLEM

Despite the growing number of studies highlighting the crucial role of innovation in export performance, several limitations remain that hinder the overall understanding of how different forms of innovation impact export performance, particularly in developing countries. An ideal framework should be developed to understand the effects of both technological and non-technological innovations on the export success of SMEs. Moreover, empirical evidence is often contradictory, with some studies suggesting that technological innovations are the primary determinants of export success, while others find that non-technological innovations also play a significant role.

Therefore, the existing literature reveals several key gaps:

Theoretical Gaps

Theoretical literature often examines the impact of technological and non-technological innovations on export performance separately, without considering the intermediary mechanisms that influence these relationships. Additionally, there is a lack of analysis regarding regional and sectoral specificities, which are essential for understanding how SMEs leverage innovation to enhance their international competitiveness.

Knowledge Gaps

Most studies have focused on developed countries, where SMEs benefit from a strong innovation system and favorable conditions for accessing financing. In contrast, research on emerging economies, such as Morocco, remains limited. Consequently, there is a lack of evidence on how Moroccan companies, particularly SMEs, leverage different forms of innovation to improve their competitiveness in international markets. Moreover, few studies have employed rigorous methodologies, such as Structural Equation Modeling (SEM), to analyze these relationships in depth.

Contextual Gaps and Practical Challenges

The Marrakech-Safi region has specific economic characteristics that distinguish it from other industrial regions in Morocco and influence the nature of the innovation adopted by exporting SMEs. This region is dominated by sectors such as crafts, tourism, agriculture, and certain industrial activities, where non-technological innovations, particularly in terms of design, cultural adaptation of products, distribution strategies, and management practices, play a crucial role. However, the current literature fails to consider these sectoral specificities. Furthermore, SMEs in the Marrakech-Safi region face several challenges related to innovation, particularly in terms of financial resources and low technological absorption capacity. Thus, identifying effective innovation levers to strengthen their international presence is critical.

How This Study Fills These Gaps

This research seeks to address these gaps by adopting an integrative approach that considers both technological and non-technological innovations in analyzing the export performance of SMEs in the Marrakech-Safi region. By administering a questionnaire to 92 exporting SMEs and employing SEM, this study aims to identify the precise mechanisms through which these innovations influence competitiveness in international markets.

From a theoretical perspective, this study will contribute to the literature by proposing a robust conceptual framework that incorporates the interactions between technological and non-technological innovations in an emerging country context. Empirically, the research will provide new insights into the nuanced relationship between various forms of innovation and export performance. Additionally, this study will offer practical recommendations for policymakers and managers, helping them to strengthen the competitiveness of exporting SMEs and identify the most effective innovation strategies for SMEs in the region.

THEORETICAL BACKGROUND AND HYPOTHESES DEVELOPMENT

Several theories have explored the impact of technological and non-technological innovation on export performance, each offering unique perspectives and underlying mechanisms. This study presents the key theories commonly employed in this context.

Schumpeter's Theory of Innovation

The theory proposed by Joseph Schumpeter (1934) is considered the foundational framework for innovation. Schumpeter identified five types of innovation: product, process, organizational, marketing innovation, and the opening of new markets. This theory remains pertinent to our study, as it highlights the critical role of both technological innovations—such as product and process innovations—and non-technological innovations, including organizational and marketing innovations, in the export success of companies

Resource-Based View (RBV) Theory

The Resource-Based View (Barney, 1991) asserts that a company's competitiveness is derived from the exploitation of resources that are unique, rare, and difficult to imitate. From this perspective, a company's innovative capabilities, whether technological or non-technological, represent a sustainable competitive advantage that can support international expansion. The adoption of organizational and marketing innovations helps optimize internal resources and facilitates adaptation to export strategies. This theory is particularly relevant to our study, as it enables an analysis of how SMEs in the Marrakech region, with limited resources, can leverage organizational and marketing innovations to enhance their export performance.

Porter's Theory of International Trade and Competitive Advantage

Porter's theory (1990) emphasizes the importance of competitive advantage in international business and explains how companies can enhance their global presence through innovation. According to Porter, innovation plays a critical role in reducing costs and increasing product differentiation. The theory also highlights the significance of non-technological innovations, such as marketing and distribution, in strengthening export competitiveness. This theory is particularly relevant to our study as it underscores how SMEs can leverage their regional and sectoral specificities to differentiate themselves in export markets.

The Uppsala Model of SME Internationalization

The Uppsala Model (Johanson & Vahlne, 1977) emphasizes the importance of a gradual approach to internationalization. SMEs adopt a progressive strategy in their export activities as they gain experience and reduce uncertainties in foreign markets. This theory is particularly relevant to our study as it highlights the role of non-technological innovations, such as cultural adaptation of products and managerial practices, in facilitating SMEs' entry into foreign markets.

Main Theoretical Optics

Among these theories, Schumpeter's theory of innovation provides the main perspective for this study, as it offers a holistic framework for analyzing different forms of innovation and their impact on export performance. However, this study also draws on the Resource-Based View (RBV) theory to explain how SMEs can transform innovation into a sustainable competitive advantage and Porter's theory of competitive advantage to contextualize internationalization strategies based on innovation. Exporting is the most common mode of internationalization for SMEs (Majocchi et al., 2005) because it is generally less risky compared to other entry modes (Laufs & Schwens, 2014). Additionally, exporting has become a key driver of business growth in competitive markets (Gassman & Golovko, 2011). However, SMEs face several challenges, particularly in terms of limited financial resources compared to larger firms. Consequently, SMEs must adopt strategies to keep pace with larger companies in a competitive context (Zou & Cavusgil, 1994). One of the most effective strategies employed by firms is innovation, which serves as a fundamental tool for competing in international markets (Azar & Ciabuschi, 2017). Innovation is a multi-faceted concept, encompassing various typologies. In this regard, the Oslo Manual (2005) defines innovation in two main categories: technological innovations (such as product and process innovations) and non-technological innovations (such as organizational and marketing innovations)

Technological Innovation

The Oslo Manual (2005) defines technological innovation as the introduction of technological novelty or improvement in products or production processes. Consequently, technological innovation is divided into two categories: product innovation and process innovation. Product innovation is defined as the market launch of new or significantly improved products or services. This can involve either the creation of an entirely new product that did not previously exist on the market or the significant improvement of an existing product in terms of design, performance, quality, or other intrinsic characteristics. On the other hand, process innovation is defined as the introduction of new production or distribution methods. This can include the adoption of new production technologies, such as automation and artificial intelligence, improvements in supply chain management, and resource optimization, among others.

Non -Technological Innovation

Non-technological innovations are not directly based on technological or scientific advances. According to the OECD and Eurostat (2005), non-technological innovation encompasses two main types: marketing innovation and organizational innovation.

Marketing innovation involves introducing new changes or improvements in the 4 Ps of marketing: product, packaging, promotion, placement, and pricing.

On the other hand, organizational innovation refers to changes affecting the structure or management of an organization. This can include the adoption of more flexible working methods, changes in organizational structure, and improvements in administration and management. Damanpour and Aravind (2011) define organizational innovation as changes in strategy, structure, administration, and management (pp. 429-432).

Technological Innovation and Export Performance

Technological innovation is considered a key source of competitive advantage in international markets (Pla-Barber & Alegre, 2007). The relationship between innovation and export performance, particularly in the context of SMEs, has garnered significant attention from researchers due to the essential role SMEs play in economies. This relationship is often described as a two-way causal link (Love & Roper, 2015), drawing from two traditional theories: the “self-selection” hypothesis and the “learning by exporting” hypothesis.

The self-selection theory suggests that innovation grants firms a competitive advantage, making it easier for them to enter international markets. As a result, firms experience improved performance due to the productivity gains stemming from the positive externalities of innovation. Conversely, the “learning by exporting” hypothesis proposes that exporting strengthens innovation, as companies gain experience in foreign markets, thereby fostering further innovation.

Both theories have been explored through empirical studies. The self-selection hypothesis seems more applicable to SMEs, asserting that innovation is a necessary condition for entering export markets. The results vary depending on the type of innovation, the sector, the timing, and the countries studied. Generally, technological innovation positively influences exports for SMEs. For instance, Tavassoli (2018) found that product innovation positively affects export intensity. Additionally, product innovations significantly improve export turnover, demonstrating that the extensiveness of technological innovation positively impacts export performance, as confirmed by Cohen & Levinthal (1990) and Hurley & Hult (1998). The rationale for this is aligned with the progressive internationalization model proposed by Johanson & Vahlne (1977), which supports incremental entry into foreign markets.

Cassiman & Golovko (2010) highlighted the importance of SMEs introducing improved, high-quality products and focusing on process innovation to succeed in export markets. Caldera (2010) also showed that both process and product innovations positively influence export performance in manufacturing firms. Other studies, such as those by Van Beveren & Vandenbussche (2009), indicate that the simultaneous and complementary effects of product and process innovation positively affect export propensity in Belgium. Roman et al. (2019) argued that product innovation drives business expansion and promotes export activity. Similarly, Chaudhry (2018) observed a positive link between innovation and export performance for manufacturing companies in Pakistan, noting that these companies are more inclined to innovate when engaging in export activities. This positive relationship between technological innovation and export performance can be attributed to the dual comparative advantage created by innovation: increased profitability and a differentiated market position (Grant, 1991), especially as technological innovation boosts productivity (Vernon, 1966).

Technological innovation is multi-dimensional, and this study focuses on two key dimensions: radicalness and extensiveness. Radical innovation refers to the extent to which innovation brings about revolutionary changes in the market, while extensiveness pertains to the number of technological and non-technological innovations within an organization. Studying these two dimensions provides a more nuanced analysis of technological innovation, offering insights into the scope and nature of the changes introduced by innovation.

Empirical studies have emphasized the importance of both radicalness and extensiveness for a company’s success. Yusof et al. (2023) confirmed a positive relationship between the radicalness of innovation and firm performance, with radical innovation playing a crucial role in export performance by introducing new products to the market, thus providing a competitive edge and attracting market share (Zhou & Li, 2012). While radical innovation is critical, the extensiveness of innovation, or the total number of innovations a company introduces, also contributes to enhanced competitiveness (Damanpour & Gopalakrishnan, 1998). Studies such as Azar et al. (2017) and Yazdani & Kazemi (2022) also found that both the radicalness and extensiveness of innovation positively impact export performance.

Based on these previous studies, we can hypothesize:

- **H1a**: The radicalness of technological innovation has a positive impact on export performance.
- **H1b**: The extensiveness of technological innovation has a positive impact on export performance

Non Technological Innovation and Export Performance

Organizational Innovation

Non-technological innovations refer to innovations that do not rely on technological advancements. These innovations primarily include organizational innovations (such as advanced techniques in work organization), strategic innovations aimed at improving a company's strategic orientations, and marketing innovations that focus on enhancing the 4Ps of the marketing mix (product, price, place, and promotion).

Empirical studies have shown that these types of innovations have positive impacts on export performance (Rodil et al., 2016; Selma et al., 2018; Heredia et al., 2019). Damanpour and Aravind (2011) confirm that relying solely on one form of innovation does not lead to optimal performance. This highlights the need for companies to introduce other non-technological innovations, such as organizational innovations, to enhance their performance. While the relationship between technological innovation and export performance is well-documented, technological innovation tends to be easier to implement and less complex than non-technological innovations (Damanpour & Evan, 1984).

Organizational innovation, such as changes in strategy, administrative procedures, or structure, can foster a climate that supports export performance (Gunday et al., 2011). In a meta-analysis, Muhammad and Chelliah (2024) found that organizations that embraced organizational innovations performed exceptionally well in export markets.

It is also important to recognize the complementarity between technological and non-technological innovations, which often results in stronger performance (Hervas-Oliver et al., 2014). The Oslo Manual (OECD, 2005) identifies non-technological innovations, particularly organizational and marketing innovations, as key components of business innovation activities. Although SMEs face resource limitations, they are known for their effective internal communication and adaptability to challenging situations (Rothwell & Dodgson, 1991). As a result, SMEs often leverage non-technological innovations, such as marketing innovations or changes in organizational practices, to remain competitive (Hervas-Oliver et al., 2016).

Azar and Ciabuschi (2017) emphasize the complementarity between technological and non-technological innovations, suggesting that product and process innovations frequently require reorganization of business routines and workflow. As a result, organizational innovations positively influence export performance in competitive markets (Azar & Drogendijk, 2016). In a study of SMEs in Central and Eastern Europe, Bodlaj et al. (2020) found that organizational innovations positively impacted export growth. Additionally, marketing innovations were found to enhance export intensity in a study of 213 firms in Spain (Rodil et al., 2016). Another study of Spanish manufacturing SMEs revealed that the adoption of organizational innovations, such as new supply chain management strategies, significantly boosted their export competitiveness (Raymond & St-Pierre, 2010). However, some studies suggest that the impact of organizational innovation may be indirect, primarily acting through increased innovation extensiveness (Lewandowska, 2015).

Based on these studies and the literature review, we propose the following hypothesis:

- **H2a**: Organizational innovation has a positive impact on export performance.

Marketing Innovation

According to the OECD (2005), marketing innovation is a form of non-technological innovation that involves significant improvements in product design, pricing strategies, communication, and distribution channels. Numerous empirical studies have demonstrated the crucial role of marketing innovation in enhancing export performance, as this type of innovation allows companies to differentiate their products and services in foreign markets.

For example, Zou et al. (2002) conducted a study on Chinese SMEs and found that the export sales of these companies improved due to enhanced design and brand image in foreign markets. Similarly, Freeman and Cavusgil (2007) affirmed that export performance improved through investments in marketing differentiation strategies, such as advertising and new distribution channels. These findings were further confirmed by Hultman et al. (2011), who showed that improving international brand perception contributes to increased export performance.

Leonidou et al. (2002) emphasized that success in export markets often hinges on adapting distribution strategies to meet the regulatory and cultural expectations of foreign countries. Naidoo (2010) examined manufacturing SMEs in South Africa and found that companies employing innovative marketing strategies, such as product customization for local markets, outperformed their competitors in the export market by increasing their attractiveness abroad.

In line with these findings, Boso et al. (2013) investigated the impact of proactive marketing strategies on the export performance of Ghanaian companies. Their study showed that firms introducing innovations in their communication and pricing strategies, such as using digital media and dynamic pricing according to market conditions, enjoyed a greater share of the international market. These results highlight that marketing innovation enables companies to better navigate the uncertainties of foreign markets and improve their competitiveness.

Based on the literature and prior studies, we propose the following hypothesis:

- **H2b**: Marketing innovation has a positive impact on export performance.

Non Technological Innovation and Technological Innovation

Damanpour and Aravind (2011) define organizational innovation as changes in the strategy, structure, administration, and management of an organization (pp. 429-432). Damanpour (1991) proposes two measures of innovation: innovation radicalness and innovation extensiveness. Innovation radicalness refers to the extent to which an innovation triggers revolutionary changes in the market, while innovation extensiveness refers to the number of technological and non-technological innovations within an organization.

Organizational innovation involves improvements and changes in communication, policies, and strategies, and these changes are crucial for the successful adoption of technological innovations (Damanpour & Evan, 1984). Furthermore, organizational innovation, especially in terms of administrative changes, significantly contributes to the success of technological innovation (Damanpour et al., 1989).

Technological innovations, such as the launch of new products or processes, are often implemented alongside organizational innovations (Zahra et al., 1995; Hall & Mairesse, 1995). Gunday et al. (2011) found that organizational innovations—such as changes in structure or human resource management, create a conducive environment for adopting process innovations. Azar et al. (2017), in a study of 218

Swedish companies, confirmed that organizational innovation enhances the extensiveness of technological innovation. Similarly, Yazdani and Kazemi (2022) found a positive impact of organizational innovation on the extensiveness of technological innovation in their study of 311 companies in Pakistan. Teece (2010) also affirms the interaction between technological and organizational innovations, noting that the introduction of new business models facilitates technological innovation.

Damanpour et al. (1984) suggest that adopting organizational innovation as part of company routines results in an increased extent of technological innovation. Most empirical studies support this, finding that organizational innovations, such as the introduction of new structures, positively impact the number of technological innovations (Love & Ashcroft, 1999; Wakasugi, 1992). Based on these findings, we propose the following hypothesis:

- **H3a**: Organizational innovation positively impacts the extensiveness of technological innovation.

On the other hand, marketing innovation can also positively influence the radicalness of technological innovation. Marketing innovations, such as data analysis, design thinking, and consumer trend studies, help identify market gaps and opportunities, thereby encouraging the development of radical technologies. An innovative marketing strategy can transform market niches into viable segments, further motivating the development of radical technologies. A prime example is Tesla, which not only popularized electric cars but also revolutionized the global automotive market. Similarly, Airbnb transformed the travel industry by offering a radically different service.

Empirical studies support this perspective. Tellis et al. (2009) demonstrate that firms with a proactive marketing approach are more successful in commercializing radical innovations. Dupuis (1998) also highlights the role of marketing in facilitating the transformation of radical innovations, particularly in the distribution sector. Consequently, we propose the following hypothesis:

- **H3b**: Marketing innovation has a positive impact on the radicalness of technological innovation.

The Mediating Role of Technological Innovation

According to Pla Barber and Alegre (2007) and Sousa et al. (2008), the radicalness and extensiveness of technological innovations play a significant role in the relationship between non-technological innovations and export performance. Radicalness, which focuses on the introduction of competitive products, can enhance the impact of non-technological innovations on export markets. On the other hand, the extensiveness of technological innovation enables the full benefits of non-technological innovations to be realized by integrating them across various aspects of the firm. This, in turn, improves overall firm performance, particularly in export markets, by allowing the firm to adapt innovations effectively to different contexts and international markets.

Empirical studies have shown that technological innovations are key factors in the relationship between organizational innovations and export performance. Favre-Bonté (2015) found that the radicalness and extensiveness of technological innovations are critical levers for improving export performance. His research suggests that integrating technological innovations with organizational and marketing innovations enhances competitiveness in foreign markets.

Organizational innovations, such as changes in management practices, strategy, and administrative procedures, enhance the innovative capacity of firms, enabling them to better meet international expectations (Damanpour & Evan, 1984; Gunday, 2005). The launch of new products or processes is the result of technological innovations, which are often implemented alongside organizational innovations (Zahra et al., 1995; Hall & Mairesse, 1995). In their study, Azar et al. (2017) found that the radicalness and extensiveness of technological innovation mediate the relationship between organizational innovation and export performance. Similarly, Yazdani and Kazemi (2022) observed the same result.

Thus, it can be concluded that technological innovation plays a catalytic role in mediating the relationship between organizational innovation and export performance. Therefore, we propose the following hypotheses:

- **H4a**: The extensiveness of technological innovation mediates the relationship between organizational innovation and export performance.
- **H4b**: The radicalness of technological innovation mediates the relationship between marketing innovation and export performance.

Conceptual Framework

The conceptual framework of this study centers on the interaction between various forms of innovation and their influence on the export performance of SMEs. Specifically, it examines the impact of non-technological innovations, including organizational and marketing innovations, on the extensiveness and radicalness of technological innovation. These two dimensions: extensiveness and radicalness, serve as mediating variables between the different types of innovation (organizational and marketing) and export performance.

This framework facilitates an exploration of how non-technological innovations, through their degree of novelty and extensiveness, can enhance the international competitiveness of SMEs in the Marrakech-Safi region. The region's key sectors, such as crafts, tourism, and agriculture, provide a unique context for studying the role of innovation in boosting export performance.

Building on the literature and prior empirical studies, we propose the following conceptual model for our study, as illustrated in Figure 1 below:

Figure 1. Conceptual model

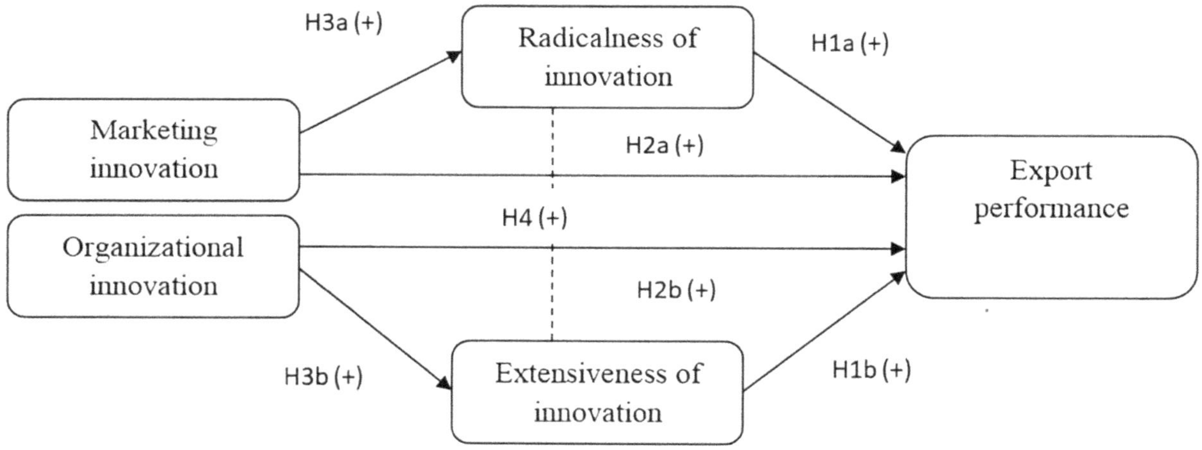

RESEARCH METHODOLOGY

This study adopts a quantitative approach through data collection using a questionnaire administered to 92 exporting SMEs operating in various sectors such as agriculture, crafts, tourism, and industry. The objective is to analyze the impact of different types of technological and non-technological innovations on export performance. The analysis method is based on structural equation modeling, a relevant and widely used technique for this type of research, allowing for a comprehensive examination of both direct and indirect effects of innovation on export performance.

Development of the Measurement Scale

The measurement scales used in this study were inspired by existing literature to ensure their conceptual validity. Marketing innovation was measured based on the works of Hurley and Hult (1998), Heredia et al. (2018), and Wadho and Chaudhry (2018), who define it as the introduction of new distribution, communication, and pricing strategies. Organizational innovation was assessed following the studies of Armbruster et al. (2008), Cieslik et al. (2015), and Angelo (2012), which define it as the adoption of new work methods and management practices.

The radicalness of innovation was measured according to Azar and Ciabuschi (2017), who define it as the degree of novelty of introduced innovations. The extensiveness of innovation was evaluated based on a quantitative approach measuring the number of innovations introduced, following Laursen and Salter (2006) and Weerawardena (2003).

Export performance was measured using widely recognized indicators in the literature, notably those from Zou and Stan (1998), Katsikeas et al. (2000), Azar and Ciabuschi (2017), and Vans et al. (2008). These indicators include export turnover, export rate, the number of export markets, and the degree of export success, providing a comprehensive assessment of SMEs' performance in international markets.

To ensure the relevance of these scales to the specific context of Moroccan SMEs, they were carefully adapted. A pre-test of the questionnaire was conducted with a small sample of business managers to assess the clarity of the items and the reliability of the measurements.

Data Analysis

Data analysis was performed using structural equation modeling (SEM), a suitable method for evaluating relationships between latent variables and their links to the conceptual model. First, a descriptive analysis was conducted to characterize exporting SMEs in the Marrakech-Safi region in terms of innovation levels and export performance. Next, validity and reliability analyses of the different constructs were performed to ensure that the structural equation method accurately captured the relationships between forms of innovation and export performance.

The results highlighted differences in the impact of technological and non-technological innovations, emphasizing the importance of adopting a holistic approach to enhance the competitiveness of exporting SMEs in an emerging economy.

Sample Selection Process and Collect Data Process

As part of this study, data collection was conducted anonymously to ensure the confidentiality of participants. Additionally, respondents were informed that the data collected from their responses would be used exclusively for academic and scientific purposes. All precautions were taken to ensure compliance with ethical research principles, particularly adherence to current regulations.

A non-random convenience sampling method was employed due to the reluctance of respondents, leading to the selection of 92 participants. The target population consists of SMEs in the Marrakech region that have exported at least once in the last three years. It is important to note that the official definition of SMEs, according to Law No. 53-00, establishing the SME charter promulgated in 2002, was used. This law defines an SME as a firm managed directly by natural persons, with a workforce not exceeding 200 employees and an annual turnover, excluding taxes, of less than or equal to 7.5 million dollars.

The total population size, before applying the exclusion criteria, was 254 firms, as recorded in the Regional Investment Center database. After excluding large firms and those that had not exported at least once in the last three years, the final population size was **N = 119 SMEs**. The sample size (**N = 92 SMEs**) was determined based on a margin of error of 5% and a confidence interval of 95%.

To collect data, a questionnaire was distributed to SME managers and owners. The questionnaire was administered online via Google Forms between April 2024 and November 2024, allowing for broad participation and ensuring rapid, anonymous data collection. Invitations were also distributed through site visits, LinkedIn, and telephone outreach. A total of 199 people were contacted, with 92 SMEs responding, resulting in a response rate of **46%**. This rate was achieved through multiple email and telephone reminders aimed at maximizing participation.

The questionnaire was designed based on existing literature (Evans & Mavondo, 2002; Evans, Mavondo, & Bridson, 2008; Weerawardena, 2003; Azar et al., 2017; Yazdani & Kazemi, 2022). Additionally, a professor-researcher reviewed the questionnaire to assess the relevance of the questions. A pre-test was conducted with a sample of 20 SME owners to identify potential misunderstandings, and feedback was incorporated to refine the questionnaire accordingly.

Table 1 presents the descriptive statistics of the sample, while Table 2 lists all constructs and measurement items.

Table 1. Descriptive statistics: mean and standard deviation (in brackets). N denotes the number of observations

Variables	Variable description	SMEs (N = 92)
NBR INNO PDT	Number of product innovations introduced by the company over the last three years: (Scale: 1 = 0 innovation and 4 = more than 6 innovations)	2.12 (0.78)
NBR NEW STRAT	Number of new strategies introduced by the company over the last three years: (Scale: 1 = 0 innovation and 4 = more than 6 innovations)	1.81 (0.58)
NBR NEW MANG	Number of new management styles introduced by the company over the last three years: (Scale: 1 = 0 innovation and 4 = more than 6 innovations)	1.89 (0.70)
NBR NEW WORK	Number of new work practices introduced by the company over the last three years: (Scale: 1 = 0 innovation and 4 = more than 6 innovations)	1.09 (0.61)
NBR NEW TRAI	Number of new training employee programs introduced by the company over the last three years: (Scale: 1 = 0 innovation and 4 = more than 6 innovations	1.94 (0.82)
NBR NEW COMM	Number of new communication strategies introduced by the company over the last three years: (Scale: 1 = 0 innovation and 4 = more than6innovations)	2.09 (0.57)
NBR INNO DIST	Number of new distribution strategies introduced by the company over the last three years: (Scale: 1 = 0 innovation and 4 = more than 6 innovations)	1.81 (0.57)
DEG INNO ORG	Degree of novelty of organizational innovations introduced by the company over the last three years: (Scale: 1 = 0 highly incremental and 4 = highly radical)	2.09 (1.03)
DEG INNO MARK	Degree of novelty of marketing innovations introduced by the company over the last three years: (Scale: 1 = 0 highly incremental and 4 = highly radical)	1 .92 (0.81)
CA EXPO	Average export sales over the last three years (Export turnover in billions dollars)	2.17 (1,14)
TAUX EXP	Average export rate reached over the last three years	2,59 (1.21)
NB MARCH EXP	Number of countries to which firm exported over the past three years)	2.32 (0.95)
DEG REUSS EXP	Degree of export success over the last three years. (Scale: 1= Very unsuccessful and 4 = Very successful)	2.88 (1.09)

Table 2. Constructs and items

Items et constructs	Scale
Technological innovation (based on Weerawardena .2003) **NBR INNO PDT** Number of product innovations introduced by the company over the last three years	1 = 0 innovations, 2 = 1 to 2 innovations 3= 3 to 5 innovations 4= more than 6 innovations
Technological innovation (based on Weerawardena .2003) **NBR INNO PROC** Number of process innovations introduced by the company over the last three years	1 = 0 innovations, 2 = 1 to 2 innovations 3= 3 to 5 innovations 4= more than 6 innovations
Non Technological innovation (based on Cieslik et al., 2015;Angelo, 2012) **NBR NEW STRAT** Number of new strategies introduced by the company over the last three years	1 = 0 innovations, 2 = 1 to 2 innovations 3= 3 to 5 innovations 4= more than 6 innovations
Non Technological innovation (based on Cieslik et al., 2015, Angelo, 2012) **NBR INNO MANG** Number of new management styles introduced by the company over the last three years	1 = 0 innovations, 2 = 1 to 2 innovations 3= 3 to 5 innovations 4= more than 6 innovations
Non Technological innovation (Cieslik et al., 2015;Angelo, 2012) **NBR NEW WORK**:Number of new work practices introduced by the company over the last three years:	1 = 0 innovations, 2 = 1 to 2 innovations 3= 3 to 5 innovations 4= more than 6 innovations
Non Technological innovation (Heredia et al., 2018;Wadho & Chaudhry, 2018) **NBR NEW COMM** Number of new communication strategies introduced by the company over the last three years	1 = 0 innovations, 2 = 1 to 2 innovations 3= 3 to 5 innovations 4= more than 6 innovations
Non Technological innovation (Heredia et al., 2018;Wadho & Chaudhry, 2018) **NBR NEW DIST** Number of new distribution strategies introduced by the company over the last three years	1 = 0 innovations 2 = 1 to 2 innovations 3= 3 to 5 innovations 4= more than 6 innovations
Technological innovation (based on Azar and Ciabuschi 2017) **DEG INNO PDT** The degree of novelty of product innovations introduced by the company over the last three years	1 = 0 highly incremental 2 = incremental 3= radical 4 = highly radical
Technological innovation (based on Azar and Ciabuschi 2017) **DEG INNO PROC** The degree of novelty of process innovations introduced by the company over the last three years	1 = 0 highly incremental 2 = incremental 3= radical 4 = highly radical
Non Technological innovation (Azar and Ciabuschi 2017) **DEG INNO ORG** The degree of novelty of organizational innovations introduced by the company over the last three years	1 = 0 highly incremental 2 = incremental 3= radical, 4 = highly radical

continued on following page

Table 2. Continued

Items et constructs	Scale
Non Technological innovation (Azar and Ciabuschi 2017)) **DEG INNO MARK**: The degree of novelty of marketing innovations introduced by the company over the last three years	1 = 0 highly incremental 2 = incremental 3= radical 4 = highly radical
NB_MARCH_EXP: Number of countries to which firm exported over the past three years	1 = 1 to 3 countries 2 = 4 to 6 countries 3= 7 to 10 countries 4 = Over 10 countries
Export performance (based on Evans & Mavondo 2002; Evans et al. 2008) **CA EXPO**: average export sales in billion dollars over the last three years?	1 = 500 to 999 2 = 1000 - 3999 3 = 4000 - 5999 4 = 6000- 7500
Export performance (based on Evans & Mavondo 2002; Evans et al. 2008) **TAUX EXP**: average export rate reached over total turnover in the last three years	1 = 10% - 30% 2 = 30% - 50% 3 = 50% - 70% 4 = 70% - 100%
DEG REUSS EXP (based on Azar,Ciabuschi, F., 2017) degree to which export activities have contributed to the company's overall performance over the past three years	1= Very unsuccessful 2 = unsuccessful 3= Moderately successful 4 =Very successful

Method

In our analysis, we employed the structural equation modeling (SEM) approach using Partial Least Squares (PLS) estimation, given the limited sample size. The analysis was conducted with Smart PLS 3 software, an advanced tool effective for examining complex relationships between latent and observed variables and for testing causal links within theoretical frameworks. The choice of SEM is justified for two main reasons: first, it enables the examination of intricate relationships between the latent dependent variable: export performance, and the explanatory variables related to different types of innovation. Second, given the complexity of our study's conceptual model, SEM provides an appropriate analytical framework to capture these interdependencies effectively.

RESULTS

Correlation Analysis

Following data collection, we conducted a validity and reliability analysis using Principal Component Analysis (PCA). The Kaiser-Meyer-Olkin (KMO) measure of sampling adequacy was 0.812, and Bartlett's test of sphericity was significant ($p < 0.005$), confirming that the correlations between variables were suitable for factor analysis. Principal Component Analysis extracted five factors, which were subsequently analyzed using structural equation modeling (SEM) with SmartPLS software to assess their validity and reliability. Table 3 presents the composite reliability (CR), Cronbach's alpha coefficient, and the average variance extracted (AVE) for the different constructs.

Table 3. Composite reliability (CR) and Cranach's Alpha coefficient and average variance extracted of different items

Latent Variable	Cronbach's Alpha	CR	AVE
Innovation Radicalness	0.717	0.838	0.633
Innovation Extensiveness	0.773	0.865	0.682
Marketing Innovation	0.617	0.796	0.566
Organizational Innovation	0.737	0.830	0.552
Export Performance	0.816	0.879	0.649

Source: Own elaboration (data from own survey)

All dimensions demonstrated acceptable reliability, with Cronbach's alpha values exceeding the 0.7 threshold, except for marketing innovation. However, its internal consistency remains satisfactory. Similarly, the convergent validity of all items was above 0.5, indicating that each construct adequately captures the explained variance.

Analyzing the relationship between innovation patterns and export performance is complex and requires a detailed examination. To test our hypotheses, we first evaluated the statistical relationships between marketing innovation, radicalness, and export performance (Table 4), as well as between organizational innovation, extensiveness, and export performance (Table 5). The constructs of marketing innovation (DEG INNO MARK, NBR NEW COMM, NBR NEW DIST) did not exhibit significant correlations with the constructs of innovation radicalness (DEG INNO PROC, DEG INNO PDT, NBR NOUV PDT). However, all constructs of innovation radicalness were positively and significantly correlated with export performance. In contrast, the correlations between organizational innovation, extensiveness, and export performance were predominantly positive and statistically significant.

Table 4. Correlations among marketing innovation, radicalness and export performance

	TAUX EXP	CA EXPO	DEG REUS EXP	NB NEW COMM	NBR NEW DIST	DEG NEW MARK	DEG INNO PDT	DEG INNO PROC	NBR NOUV PDT
NBR NEW COMM	0.074	0.258*	0.088	1	0.622**	0.438**	0.036	0.091	0.067
NBR NEW DIST	0.047	0.150	0.069	0.622**	1	0.323**	0.058	0.013	0.020
DEG INNO MARK	-0.054	0.273**	0.088	0.438**	0.323**	1	0.122	0.068	0.142
DEG INNO PDT	0.451**	0.441**	0.223*	0.036	0.058	0.122	1	0.404**	0.313**

continued on following page

Table 4. Continued

	TAUX EXP	CA EXPO	DEG REUS EXP	NB NEW COMM	NBR NEW DIST	DEG NEW MARK	DEG INNO PDT	DEG INNO PROC	NBR NOUV PDT
DEG INNO PROC	0.221*	0.153	0.511	0.091	0.013	0.068	0.404**	1	0.656**
NBR NOUV PDT	0.219*	0.345**	0.243*	0.067	0.020	0.142	0.313**	0.656**	1

*p < 0.10, **p < 0.05, ***p < 0.01

Table 5. Correlations among organizational innovation, extensiveness and export performance

	TAUX EXP	CA EXPO	DEG REUS EXP	DEG INNO ORG	NBR NEW WORK	NBR NEW STRA	NBR NEW MANG	NBR INNO PROC	NBR INNO PDT	NBR NEW TRAI
DEG INNO ORG	0.381**	0.707**	0.550**	1	0.451**	0.523**	0.376**	0.425**	0.400**	0.401**
NBR NEW WORK	0.173	0.399**	0.279**	0.451**	1	0.341**	0.408**	0.134	0.195	0.119
NBR NEW STRA	0.338**	0.641**	0.416**	0.523**	0.341**	1	0.372**	0.452**	0.404**	0.350**
NBR NEW MANG	0.149	0.584**	0.439**	0.376**	0.408**	0.372**	1	0.213*	0.146	0.254*
NBR INNO PROC	0.359**	0.474**	0.389**	0.425**	0.134	0.452**	0.213*	1	0.483**	0.584**
NBR INNO PDT	0.451**	0.441**	0.223*	0.400**	0.195	0.404**	0.146	0.483**	1	0.323**
NBR NEW TRAI	0.304**	0.461**	0.403**	0.401**	0.119	0.350**	0.254*	0.584**	0.323**	1

*p < 0.10, **p < 0.05, ***p < 0.01
Source: Own elaboration (data from own survey)

Assessment of Measurement Model

To validate the measurement model in structural equation modeling, outer loadings play a crucial role in assessing the quality of the measurement model. Outer loadings represent the correlation between observed indicators and their respective latent variables. Table 6 presents the results of the outer loadings for the observable variables related to the constructs of innovation radicalness, innovation extensiveness, marketing innovation, organizational innovation, and export performance.

All outer loadings exceed the 0.7 threshold, indicating that the indicators adequately measure their respective latent variables and exhibit strong convergent validity. However, certain variables, such as the average export rate (0.612), number of new management styles (0.678), number of new work practices (0.640), and degree of marketing innovation (0.683), have loadings below 0.7. Despite this, they remain relevant as they are theoretically grounded in the literature. According to Hair et al. (2017), loadings between 0.4 and 0.7 are acceptable, particularly when the average variance extracted (AVE) exceeds 0.5 and composite reliability is greater than 0.7, as demonstrated in Table 3.

Table 6. Outer loadings of observable variables

Latent variables	Observable Variables	Outer loading	T-sta (|O/ STDEV|)	P-Values
Innovation Radicalness	Degree of innovation products (DEG INNO PDT)	0.768	9.087	0.000
	Degree of innovation process (DEG INNO PROC)	0.810	10.143	0.000
	Number of novel products (NBR NOUV PDT)	0.808	11.086	0.000
Innovation Extensiveness	Number of innovation products (NBR INNO PDT)	0.860	33.864	0.000
	Number of innovation process (NBR INNO PROC)	0.841	16.235	0.000
	Number of new training programs(NBR NEW TRAI)	0.775	11.816	0.000
Marketing Innovation	Degree of marketing innovation (DEG INNO MAR)	0.683	3.386	0.001
	Number of new communication strategies (NBR NEW COMM)	0.805	5.462	0.000
	Number of new distribution strategies (NBR NEW DIST)	0.764	4.748	0.000
Organizational Innovation	Degree of organizational innovations (DEG INNO ORG)	0.841	22.490	0.000
	Number of new management styles (NBR NEW MANG)	0.678	9.248	0.000
	Number of new strategies (NBR NEW STAT)	0.795	23.380	0.000
	Number of new work practices (NBR NEW WORK)	0.640	8.297	0.000
Export Performance	Export turnover (CA EXPO)	0.905	60.237	0.000
	Degree of export success (DEG REUS EXP)	0.848	26.701	0.000
	Number of export market (NBR MARCH EXP)	0.828	17.400	0.000
	Average export rate (TX EXP)	0.612	6.572	0.000

Source: Own elaboration (data from own survey)

To assess discriminant validity, we employed the Heterotrait-Monotrait (HTMT) ratio, which is recognized as a more reliable method for determining the distinctiveness of constructs. According to Hair et al. (2017), HTMT values should remain below the threshold of 0.85 to confirm that constructs are unique and not redundant. As shown in Table 7, all constructs meet this criterion, confirming discriminant validity.

To evaluate model fit, we calculated the R-square (R^2) values for the constructs, as reported in Table 8. The model accounts for 64.3% of the variance in export performance through the exogenous variables. Additionally, 39% of the variance in innovation extensiveness is explained by the exogenous variables, while only 1.6% of the variance in innovation radicalness is accounted for. The low R^2 value for innovation radicalness may be attributed to the limited adoption of radical innovations among Moroccan SMEs, likely due to resource constraints, which in turn reduce the variance in innovation radicalness.

Table 7. HTMT ratio of constructs

Constructs	Export performance	Innovation extensiveness	Marketing innovation	Organizational innovation	Innovation radicalness
Export performance	1	0.819	0.406	0.840	0.588
Innovation extensiveness	0.819	1	0.314	0.733	0.692
Marketing innovation	0.406	0.314	1	0.455	0.218
Organizational innovation	0.840	0.733	0.455	1	0.493
Innovation radicalness	0.588	0.692	0.218	0.493	1

Source: Own elaboration (data from own survey)

Table 8. R-square values of constructs

Dependent variables	R^2	T-statistics (|O/STDEV|)	P-Values
Export Performance	0.643	3.799	0.000
Innovation Extensiveness	0.391	12.889	0.000
Innovation Radicalness	0.016	1.124	0.261

Source: Own elaboration (data from own survey)

Research Findings

To test the different hypotheses, we employed structural equation modeling using Smart PLS. The results are reported in Table 9 and are conceptualized in Figure 2.

Table 9. Structural equation model results for innovation types and export performance with mediating effect of extensiveness and radicalness innovation

Relation	Direct effect	Indirect effect	Total effect
(1) Radicalness innovation Export performance	0.091	-	0.091
(2) Extensiveness innovation Export performance	0.317***	-	0.317***
(3) Organizational innovation Export performance	0.489***	0.198***	0.688***
(4) Marketing innovation Export performance	0.062	0.012	0.073
(5) Marketing innovation Radicalness	0.128	-	0.128
(6) Organizational innovation Extensiveness	0.625***	-	0.625***

*p < 0.10, **p < 0.05, ***p < 0.01
Source: Own elaboration (data from own survey)

Relation 1: The radicalness of innovation does not have a significant direct effect on export performance in this sample. Contrary to expectations, radical innovation does not notably impact export performance, so **H1a** is not supported.

Relation 2: The extensiveness of innovation has a positive and significant direct effect on export performance. This suggests that companies adopting more varied or diversified innovations achieve better export results, confirming **H1b**.

Relation 3: Organizational innovation has a positive and significant direct effect on export performance. This indicates that the adoption of organizational innovations directly enhances the export performance of SMEs, confirming **H2a**. Additionally, part of the effect of organizational innovation on export performance is mediated by the extensiveness of innovation. This indirect effect is also significant, supporting **H4a**.

Relation 4: The direct effect of marketing innovation on export performance is not significant, meaning **H2b** is not supported. The indirect effect, mediated by radicalness innovation, is also not significant, so **H4b** is not supported.

Relation 5: Marketing innovation does not have a significant effect on innovation radicalness in this sample, so **H3b** is not supported.

Relation 6: This relationship shows a strong positive and significant impact of organizational innovation on innovation extensiveness. This suggests that organizational innovations facilitate the implementation of extensive innovations, confirming **H3a**.

Figure 2. Structural equation model results for innovation types and export performance with mediating effect of extensiveness and radicalness innovation

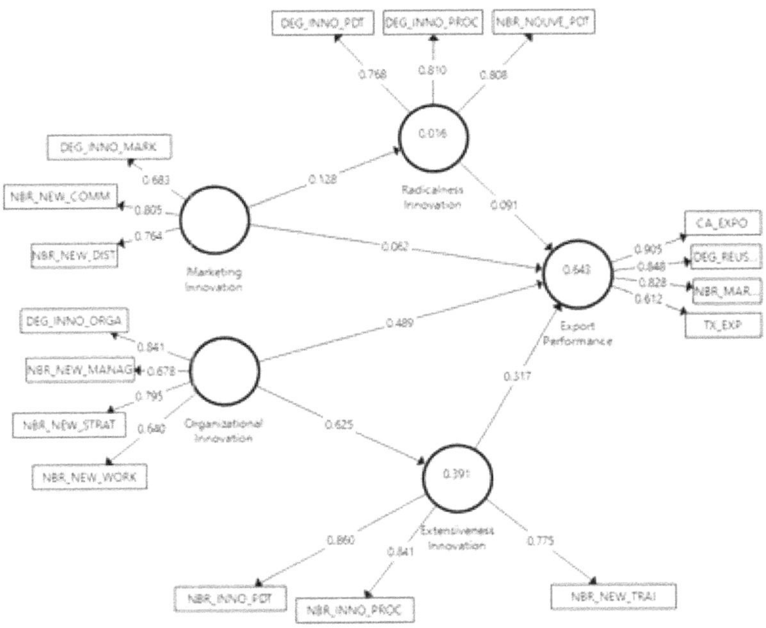

Source: Own elaboration (data from own survey)

Discussion

The results underscore the importance of innovation in enhancing the export performance of SMEs in the Marrakech-Safi region. The number of innovations adopted by SMEs has a positive and significant impact on export performance, confirming that the greater the variety of innovations integrated by an SME, the better its export performance. This finding aligns with Porter's (1985) theory of competitiveness through innovation.

Non-technological innovations also play a critical role, particularly in sectors where competitive advantage is driven by differentiation strategies, such as tourism, crafts, and agriculture. Organizational innovations, including improvements in managerial practices, better internal structures, and enhanced communication, create a conducive environment for experimentation. These innovations enable companies to adapt their products and services to meet the demands of foreign markets, thereby facilitating their international integration and expansion.

However, in the case of Moroccan SMEs in the Marrakech-Safi region, marketing innovation strategies do not directly impact export performance. This could be attributed to the budgetary constraints faced by SMEs, as the effects of new brand images or digital strategies are not immediate. Such effects may take several years to manifest and have a significant impact on export sales.

IMPLICATIONS AND RECOMMENDATIONS

Theoretical Implications

The results confirm several theoretical frameworks and align with previous studies:

First, we found that organizational innovation has a positive and strong effect on export performance. According to the Resource-Based View (Barney, 1991), organizational innovation allows firms to optimize internal processes, improve coordination, and better utilize resources, leading to enhanced export performance. Our findings are consistent with studies demonstrating that innovative SMEs are better equipped to meet the demands of international markets.

Second, we identified the mediating role of the extensiveness of innovation in the relationship between organizational innovation and export performance. The mediating role of innovation extensiveness supports theories on innovation dynamics (Schumpeter, 1934). The concept suggests that the accumulation of various types of innovations (product, process, organizational, etc.) strengthens competitive advantages and enhances performance. This mediation effect has also been observed in empirical studies of exporting SMEs (e.g., Love & Roper, 2015).

Practical Implications

Our findings emphasize the importance for business managers to invest in innovative organizational practices and diversify the types of innovations they adopt to maximize their impact on export performance. Policymakers should also support initiatives that promote organizational innovation and innovation diversification, such as offering subsidies, training programs, or fostering partnerships.

Recommendations

Based on the results, this study offers several valuable recommendations for business leaders, policymakers, and researchers.

First, SMEs should adopt a balanced approach between technological and non-technological innovation to optimize their export performance. A strategic investment in improving organizational and managerial practices can be particularly beneficial for enhancing export success.

Second, public policies should support exporting SMEs by exempting innovation-related services from taxes and offering incentive grants aimed at boosting their international competitiveness.

Finally, it is recommended that researchers expand this study to include other sectors and contexts, exploring the dynamics of innovation and its impact on export performance in additional emerging countries.

Future Directions

Our study lays the foundation for several promising future research avenues. For example, it would be valuable to explore the differentiated impact of technological and non-technological innovations across various sectors. Additionally, investigating the effects of digitalization would provide insights into how the digital transformation of SMEs influences their access to international markets. Moreover, comparing these results with data from other regions in Morocco or similar emerging countries could offer a broader understanding of innovation's role in export performance.

CONCLUSION

The purpose of this research was to examine the impact of innovation on the export performance of SMEs in the Marrakech region, Morocco. The study had two primary objectives: first, to test the relationship between technological and non-technological innovation and export performance, and second, to explore the mediating role of innovation extensiveness in the relationship between organizational innovation and export performance.

The findings highlight the significance of non-technological innovations, particularly organizational innovation, and underscore the importance of diversifying the types of innovations adopted by SMEs. The results suggest that the greater the variety of innovations integrated by an SME, the better its export performance. This supports the theory of competitiveness through innovation (Porter, 1985), particularly in the context of SMEs in emerging economies.

While this research provides valuable theoretical and practical insights, it also has some limitations. The sample size of 92 SMEs, although providing a useful overview, may not fully represent SMEs in other regions or sectors. Additionally, the specific characteristics of the Marrakech-Safi region limit the generalizability of the results to other areas in Morocco or to different developing countries.

Future Research Directions

Future research could explore the role of contextual variables such as competitive environment, business maturity, and government support in shaping the innovation-export performance relationship. Longitudinal studies would provide valuable insights into the temporal dynamics of innovation's impact on export performance. Expanding the study to other regions within Morocco, or even incorporating a larger, more diverse sample of SMEs, would enhance the generalizability and robustness of the findings. Furthermore, examining the impact of digitalization and technological innovation across various sectors would provide a more comprehensive understanding of the drivers of export performance in the modern global marketplace.

REFERENCES

Armbruster, H., Bikfalvi, A., Kinkel, S., & Lay, G. (2008). Organizational innovation: The challenge of measuring non-technical innovation in large-scale surveys. technovation, 28(10), 644-657.

Azar, G., & Ciabuschi, F. (2017). Organizational innovation, technological innovation, and export performance: The effects of innovation radicalness and extensiveness. *International Business Review*, *26*(2), 324–336. DOI: 10.1016/j.ibusrev.2016.09.002

Azar, G., & Drogendijk, R. (2016). Cultural distance, innovation and export performance: An examination of perceived and objective cultural distance. *European Business Review*, *28*(2), 176–207. DOI: 10.1108/EBR-06-2015-0065

Bao, Y., Chen, X., & Zhou, K. Z. (2012). External learning, market dynamics, and radical innovation: Evidence from China's high-tech firms. *Journal of Business Research*, *65*(8), 1226–1233. DOI: 10.1016/j.jbusres.2011.06.036

Barasa, L., Knoben, J., Vermeulen, P., Kimuyu, P., & Kinyanjui, B. (2017). Institutions, resources and innovation in East Africa: A firm level approach. *Research Policy*, *46*(1), 280–291. DOI: 10.1016/j.respol.2016.11.008

Barney, J. B. (2000). Firm resources and sustained competitive advantage. In *Economics meets sociology in strategic management* (pp. 203–227). Emerald Group Publishing Limited. DOI: 10.1016/S0742-3322(00)17018-4

Becker, S. O., & Egger, P. H. (2013). Endogenous product versus process innovation and a firm's propensity to export. *Empirical Economics*, *44*(1), 329–354. DOI: 10.1007/s00181-009-0322-6

Bodlaj, M., Kadic-Maglajlic, S., & Vida, I. (2020). Disentangling the impact of different innovation types, financial constraints and geographic diversification on SMEs' export growth. *Journal of Business Research*, *108*, 466–475. DOI: 10.1016/j.jbusres.2018.10.043

Boso, N., Story, V. M., Cadogan, J. W., Micevski, M., & Kadić-Maglajlić, S. (2013). Firm Innovativeness and Export Performance: Environmental, Networking, and Structural Contingencies. *Journal of International Marketing*, *21*(4), 62–87. DOI: 10.1509/jim.13.0052

Caldera, A. (2010). Innovation and exporting: Evidence from Spanish manufacturing firms. *Review of World Economics*, *146*(4), 657–689. DOI: 10.1007/s10290-010-0065-7

Cassiman, B., & Golovko, E. (2011). Innovation and internationalization through exports. *Journal of International Business Studies*, *42*(1), 56–75. DOI: 10.1057/jibs.2010.36

Cavusgil, S. T., & Zou, S. (1994). Marketing strategy-performance relationship: An investigation of the empirical link in export market ventures. *Journal of Marketing*, *58*(1), 1–21. DOI: 10.1177/002224299405800101

Churchill, G. A.Jr. (1979). A paradigm for developing better measures of marketing constructs. *JMR, Journal of Marketing Research*, *16*(1), 64–73. DOI: 10.1177/002224377901600110

Cieślik, A., Michałek, A., Michałek, J. J., & Mycielski, J. (2015). Determinants of export performance: Comparison of central European and Baltic firms. *Finance a Uver. Czech Journal of Economics & Finance*, *65*(3).

Cieślik, A., & Michałek, J. J. (2017). Innovation Forms and Firm Export Performance: Empirical Evidence from ECA Countries. *Entrepreneurial Business and Economics Review*, *5*(2), 85–99. DOI: 10.15678/EBER.2017.050205

Damanpour, F., & Aravind, D. (2011). Managerial innovation: Conceptions, processes and antecedents. *Management and Organization Review*, *8*(2), 423–454. DOI: 10.1111/j.1740-8784.2011.00233.x

Damanpour, F., & Evan, W. M. (1984). Organizational innovation and performance: The problem of" organizational lag. *Administrative Science Quarterly*, *29*(3), 392–409. DOI: 10.2307/2393031

Damanpour, F., Szabat, K. A., & Evan, W. M. (1989). The relationship between types of innovation and organizational performance. *Journal of Management Studies*, *26*(6), 587–602. DOI: 10.1111/j.1467-6486.1989.tb00746.x

Damanpour, F., & Wischnevsky, J. D. (2006). Research on innovation in organizations: Distinguishing innovation-generating from innovation-adopting organizations. *Journal of Engineering and Technology Management*, *23*(4), 269–291. DOI: 10.1016/j.jengtecman.2006.08.002

Dupuis, M. (1998). L'innovation dans la distribution, son implication dans les relations industrie-commerce. *Décisions marketing*, (3), 29-41.

Edeh, J. N., Obodoechi, D. N., & Ramos-Hidalgo, E. (2020). Effects of innovation strategies on export performance: New empirical evidence from developing market firms. *Technological Forecasting and Social Change*, *158*, 120167. DOI: 10.1016/j.techfore.2020.120167

El Chaarani, H., Vrontis, P. D., El Nemar, S., & El Abiad, Z. (2022). The impact of strategic competitive innovation on the financial performance of SMEs during COVID-19 pandemic period. *Competitiveness Review*, *32*(3), 282–301. DOI: 10.1108/CR-02-2021-0024

Evans, J., Mavondo, F. T., & Bridson, K. (2008). Psychic distance: Antecedents, retail strategy implications, and performance outcomes. *Journal of International Marketing*, *16*(2), 32–63. DOI: 10.1509/jimk.16.2.32

Exchange Office. (2022). Multidimensional analysis of the evolution of the profile of Moroccan exporters.

Exposito, A., & Sanchis-Llopis, J. A. (2020). The effects of innovation on the decisions of exporting and/or importing in SMEs: Empirical evidence in the case of Spain. *Small Business Economics*, *55*(3), 813–829. DOI: 10.1007/s11187-019-00184-3

Favre-Bonté, V. (2015). *Les stratégies internationales et d'innovation comme leviers de développement des entreprises* (Doctoral dissertation, université de Savoie Mont lanc).

Fonchamnyo, D. C., & Wujung, V. A. (2016). Innovation and export performance: An empirical insight on the effect of innovation on manufacturing firms in Cameroon. *Applied Economics and Finance*, *3*(4), 123–133. DOI: 10.11114/aef.v3i4.1682

Freeman, S., & Cavusgil, S. T. (2007). Toward a typology of commitment states among managers of born-global firms: A study of accelerated internationalization. *Journal of International Marketing*, *15*(4), 1–40. DOI: 10.1509/jimk.15.4.1

Freixanet, J., & Rialp, J. (2022). Disentangling the relationship between internationalization, incremental and radical innovation, and firm performance. *Global Strategy Journal*, *12*(1), 57–81. DOI: 10.1002/gsj.1412

Grant, R. M. (1991). The resource-based theory of competitive advantage: Implications for strategy formulation. *California Management Review*, *33*(3), 114–135. DOI: 10.2307/41166664

Gunday, G., Ulusoy, G., Kilic, K., & Alpkan, L. (2011). Effects of innovation types on firm performance. *International Journal of Production Economics*, *133*(2), 662–676. DOI: 10.1016/j.ijpe.2011.05.014

Hall, B. H., & Mairesse, J. (1995). Exploring the relationship between R&D and productivity in French manufacturing firms. *Journal of Econometrics*, *65*(1), 263–293. DOI: 10.1016/0304-4076(94)01604-X

Hashmi, H. B. A., Voinea, C. L., Ooms, W., & Caniëls, M. C. (2022). The impact of breakthrough innovations on the export performance of SMEs in developing countries: The moderating role of institutional factors. *Frontiers in Psychology*, *13*, 888697. DOI: 10.3389/fpsyg.2022.888697 PMID: 35645878

Heredia, J., Flores, A., Heredia, W., Arango, R., & Medina, L. (2019). How innovation influences on export performance: A configuration approach for emerging economies. *Journal of Technology Management & Innovation*, *14*(4), 54–65. DOI: 10.4067/S0718-27242019000400054

Hervas-Oliver, J. L., Sempere-Ripoll, F., & Boronat-Moll, C. (2014). Process innovation strategy in SMEs, organizational innovation and performance: A misleading debate? *Small Business Economics*, *43*(4), 873–886. DOI: 10.1007/s11187-014-9567-3

Hultman, M., Katsikeas, C. S., & Robson, M. J. (2011). Export promotion strategy and performance: The role of international experience. *Journal of International Marketing*, *19*(4), 17–39. DOI: 10.1509/jim.11.0022

Johanson, J., & Vahlne, J.-E. (1977). The internationalization process of the firm. *Journal of International Business Studies*, *8*(1), 23–32. DOI: 10.1057/palgrave.jibs.8490676

Katsikeas, C. S., Leonidou, L. C., & Morgan, N. A. (2000). Firm-level export performance assessment: Review, evaluation, and development. *Journal of the Academy of Marketing Science*, *28*(4), 493–511. DOI: 10.1177/0092070300284003

Laufs, K., & Schwens, C. (2014). Foreign market entry mode choice of small and medium-sized enterprises: A systematic review and future research agenda. *International Business Review*, *23*(6), 1109–1126. DOI: 10.1016/j.ibusrev.2014.03.006

Laursen, K., & Salter, A. (2006). Open for innovation: The role of openness in explaining innovation performance among UK manufacturing firms. *Strategic Management Journal*, *27*(2), 131–150. DOI: 10.1002/smj.507

Leonidou, L. C., Katsikeas, C. S., & Samiee, S. (2002). Marketing Strategy Determinants of Export Performance: A Meta-Analysis. *Journal of Business Research*, *55*(1), 51–67. DOI: 10.1016/S0148-2963(00)00133-8

Lewandowska, M. S., Szymura-Tyc, M., & Gołębiowski, T. (2015). Innovation complementarity, cooperation partners, and new product export: Evidence from Poland. *Journal of Business Research*, *69*(9), 3673–3681. DOI: 10.1016/j.jbusres.2016.03.028

Lo Turco, A., & Maggioni, D. (2015). Dissecting the impact of innovation on exporting in Turkey. *Economics of Innovation and New Technology*, *24*(4), 309–338. DOI: 10.1080/10438599.2014.946311

Love, J. H., & Roper, S. (2015). SME innovation, exporting and growth: A review of existing evidence. *International Small Business Journal*, *33*(1), 28–48. DOI: 10.1177/0266242614550190

Majocchi, A., Bacchiocchi, E., & Mayrhofer, U. (2005). Firm size, business experience and export intensity in SMEs: A longitudinal approach to complex relationships. *International Business Review*, *14*(6), 719–738. DOI: 10.1016/j.ibusrev.2005.07.004

Manual, O. (2005). *Proposed guidelines for collecting and interpreting technological innovation data*. OECD: Statistical Office of the European Communities.

Martínez-Román, J. A., Gamero, J., de Loreto Delgado-González, M., & Tamayo, J. A. (2019). Innovativeness and internationalization in SMEs: An empirical analysis in European countries. *Technological Forecasting and Social Change*, *148*, 119716. DOI: 10.1016/j.techfore.2019.119716

Mol, M. J., & Birkinshaw, J. (2009). The sources of management innovation: When firms introduce new management practices. *Journal of Business Research*, *62*(12), 1269–1280. DOI: 10.1016/j.jbusres.2009.01.001

Muhammad, S. A., & Chelliah, S. (2024). Organisational innovation, competitive advantage, and export performance. *International Journal of Innovation and Learning*, *35*(2), 161–176. DOI: 10.1504/IJIL.2024.136471

Naidoo, V. (2010). Firm Survival Through a Crisis: The Influence of Market Orientation, Marketing Innovation, and Business Strategy. *Industrial Marketing Management*, *39*(8), 1311–1320. DOI: 10.1016/j.indmarman.2010.02.005

Ortigueira-Sánchez, L. C., Welsh, D. H., & Stein, W. C. (2022). Innovation drivers for export performance. *Sustainable Technology and Entrepreneurship*, *1*(2), 100013. DOI: 10.1016/j.stae.2022.100013

Pla-Barber, J., & Alegre, J. (2007). Analysing the link between export intensity, innovation and firm size in a science-based industry. *International Business Review*, *16*(3), 275–293. DOI: 10.1016/j.ibusrev.2007.02.005

Porter, M. E. (1990). *The Competitive Advantage of Nations*. Free Press, MacMillan. DOI: 10.1007/978-1-349-11336-1

Porter, M. E. (2008). *Competitive advantage: Creating and sustaining superior performance*.

Raymond, L., & St-Pierre, J. (2010). R&D as a determinant of innovation in manufacturing SMEs: An attempt at empirical clarification. *Technovation*, *30*(1), 48–56. DOI: 10.1016/j.technovation.2009.05.005

Rehman, N. U. (2017). Self-selection and learning-by-exporting hypotheses: Micro-level evidence. *Eurasian Economic Review*, *7*(1), 133–160. DOI: 10.1007/s40822-016-0063-8

Ribau, C. P., Moreira, A. C., & Raposo, M. (2019). The role of exploitative and exploratory innovation in export performance: An analysis of plastics industry SMEs. *European Journal of International Management*, *13*(2), 224–246. DOI: 10.1504/EJIM.2019.098149

Rodil, Ó., Vence, X., & del Carmen Sánchez, M. (2016). The relationship between innovation and export behaviour: The case of Galician firms. *Technological Forecasting and Social Change*, *113*, 248–265. DOI: 10.1016/j.techfore.2015.09.002

Rothwell, R., & Dodgson, M. (1991). External linkages and innovation in small and medium-sized enterprises. *Research Management*, *21*(2), 125–138.

Sarstedt, M., Ringle, C. M., & Hair, J. F. (2021). Partial least squares structural equation modeling. In *Handbook of market research* (pp. 587–632). Springer International Publishing. DOI: 10.1007/978-3-319-05542-8_15-2

Schumpeter, J. (1934). *The Theory of Economic Development*. Harvard University Press.

Schumpeter, J. A. (2008). The theory of economic development.

Tandrayen-Ragoobur, V. (2022). The innovation and exports interplay across Africa: Does business environment matter? *The Journal of International Trade & Economic Development*, *31*(7), 1041–1071. DOI: 10.1080/09638199.2022.2051588

Tavassoli, S. (2018). The role of product innovation on export behavior of firms: Is it innovation input or innovation output that matters? *European Journal of Innovation Management*, *21*(2), 294–314. DOI: 10.1108/EJIM-12-2016-0124

Teece, D. J. (2010). Business models, business strategy and innovation. *Long Range Planning*, *43*(2-3), 172–194. DOI: 10.1016/j.lrp.2009.07.003

Tellis, G. J., Prabhu, J. C., & Chandy, R. K. (2009). Radical innovation across nations: The preeminence of corporate culture. *Journal of Marketing*, *73*(1), 3–23. DOI: 10.1509/jmkg.73.1.003

Tushman, M., & Nadler, D. (1986). Organizing for innovation. *California Management Review*, *28*(3), 74–92. DOI: 10.2307/41165203

Van Beveren, I., & Vandenbussche, H. (2010). Product and process innovation and firms' decision to export. *Journal of Economic Policy Reform*, *13*(1), 3–24. DOI: 10.1080/17487870903546267

Vernon, R. (1992). International investment and international trade in the product cycle. In *International economic policies and their theoretical foundations* (pp. 415–435). Academic Press. DOI: 10.1016/B978-0-12-444281-8.50024-6

Wadho, W., & Chaudhry, A. (2018). Innovation and firm performance in developing countries: The case of Pakistani textile and apparel manufacturers. *Research Policy*, *47*(7), 1283–1294. DOI: 10.1016/j.respol.2018.04.007

Wakasugi, R. (1992). Why are Japanese firms so innovative in engineering technology? *Research Policy*, *21*(1), 1–12. DOI: 10.1016/0048-7333(92)90023-W

Wernerfelt, B. (1984). A resource-based view of the firm. *Strategic Management Journal*, *5*(2), 171–180. DOI: 10.1002/smj.4250050207

Zahra, S. A., & Covin, J. G. (1995). Contextual influences on the corporate entrepreneurship-performance relationship: A longitudinal analysis. *Journal of Business Venturing*, *10*(1), 43–58. DOI: 10.1016/0883-9026(94)00004-E

Zou, S., & Cavusgil, S. T. (2002). The GMS: A broad conceptualization of global marketing strategy and its effect on firm performance. *Journal of Marketing*, *66*(4), 40–56. DOI: 10.1509/jmkg.66.4.40.18519

Zou, S., & Stan, S. (1998). The determinants of export performance: A review of the empirical literature between 1987 and 1997. *International Marketing Review*, *15*(5), 333–356. DOI: 10.1108/02651339810236290

Chapter 16
Leveraging AI for Disruptive Green Human Resource Management:
Challenges and Opportunities to Strategic People Management

Mariana Carôla
ISCTE, Instituto Universitário de Lisboa, Portugal

Rosa Isabel Rodrigues
https://orcid.org/0000-0001-6548-418X
Instituto Superior de Gestão, Portugal

Paula C. N. Figueiredo
https://orcid.org/0000-0002-5267-546X
Universidade Lusófona, Portugal

ABSTRACT

Artificial intelligence (AI) is an emerging technology that has significant impacts on the management of organizations, especially green human resource management (GHRM), which aims to promote sustainable and responsible human resource practices. The aim of this chapter is to explore the impacts of AI on GHRM and how AI can boost green practices in organizations. The results of the literature review showed that AI has an important role in promoting the performance, development, and retention of green employees, as well as in improving organizational culture and corporate social responsibility. However, AI also brings some challenges for GHRM, such as ethical, legal, and social issues. The chapter concludes that AI and HRM should be integrated strategically and responsibly in organizations, aiming at their sustainable development.

DOI: 10.4018/979-8-3693-4030-1.ch016

INTRODUCTION

There is a great deal of interest from researchers regarding the impacts on human resources management (HRM) of the digital age we are living in (Khurana et al., 2025), more specifically how HRM strategies will evolve with the digital transformation phenomenon (Husen et al., 2024). Still about transformation, but in this case green transformation, some researchers have linked green HRM practices to sustainability issues (Cesário et al., 2025), particularly employees' perceptions of companies' green initiatives (Garavan et al., 2023). It therefore seems obvious to us to understand whether Artificial Intelligence (AI) can contribute to and even drive the transformation and implementation of green human resource management (GHRM) practices.

The evolution of AI and its application in the field of management, and more specifically in HRM, has been the subject of study by several researchers in recent years (*e.g.*, Arslan et al., 2022; Budhwar et al., 2022; Dima et al., 2024; Husen et al., 2024; Jiaping, 2022; Johnson et al., 2022; Kambur & Akar, 2022; Menaka, 2023; Palos-Sánchez et al., 2022; Pan & Froese, 2023; Prikshat et al., 2023a; Prikshat et al., 2023b; Sari et al., 2020; Umrao et al., 2024; Vrontis et al., 2022). Due to the difficulties of attracting and retaining talent, some researchers have also focused on an approach centred on employee experience, but always supported by the technological factor (Figueiredo, 2025; Jayalakshmi & Jayanthi, 2024; Thorat & Saste, 2024).

However, with the need to adapt organizations to digital transformation and the green transition, these aspects need to be aligned with business strategy and management practices, particularly HRM. Organizations need a culture based on the assumptions of the digital transformation brought about by technological advances and the green transformation brought about by the pressing challenges of climate change (Roscoe et al., 2019). They also need leaders who are deeply committed and involved in these issues, both now and in the future (Robertson & Carleton, 2018).

The role of leadership is fundamental in this process of aligning business strategy with sustainability issues. In this sense, digital transformation and AI are the leader's allies in promoting change in the organization (Kollmann et al., 2023; Quaquebeke & Gerpott, 2023). In this context, interest is beginning to emerge in understanding how AI can boost the strategy for GHRM practices and environmental sustainability (Alvi, 2021; Garg et al., 2018; Ogbeibu et al., 2023). These practices include five dimensions: green recruitment and selection, green training, green performance management, green pay and reward, and green involvement (Jabbour & Renwick, 2018). However, GHRM practices are presently confined to a few activities: hiring, training, and reward (Khateeb & Nabi, 2023).

Therefore, the research question is whether the impacts of IA on GHRM contribute to strategic people management in line with the principles of sustainability. This approach adopts a qualitative methodology based on a review of the existing literature.

RESEARCH METHODOLOGY

The aim of this research is to explore the impacts of AI on GHRM and how AI can boost GHRM practices in organizations through scientific studies carried out on these topics. So, the methodological essay is carried out through a literature review, which consists of the search, selection, analysis, and synthesis of recent and relevant secondary sources on the topic. This exploratory methodology is the most appropriate because it is a current and very recent topic. The theoretical framework is based initially on the theoretical concepts of AI AND GHRM and their relevance. Next, studies are presented that demonstrate the relationship between the concepts and dimensions of GHRM that are influenced by AI, as well as the models and applications of AI in GHRM. Finally, the benefits and challenges of AI for GHRM.

THEORETICAL FRAMEWORK

AI and GHRM: Definitions and Relevance

The ability of robots to simulate or improve human intelligence, including reasoning and experience-based learning, is known as artificial intelligence (AI) (Afzal et al., 2023). AI is defined as "a system's ability to correctly interpret external data, to learn from such data, and to use those learnings to achieve specific goals and tasks through flexible adaptation" by (Kaplan & Haenlein, 2019). According to Morandini et al., (2023), these skills are changing the character of work and organizational models by posing challenges to management politics, procedures, and practices, which in turn affect the current and future skills required to meet the issues associated with them (Ekuma, 2023; Hancock et al., 2020; Jaiswal et al., 2022; Morandini et al., 2023).

A new chapter in information technology management is brought in by the management of artificial intelligence (AI). A constantly evolving frontier of computational advances that relate to human intelligence in handling progressively complicated decision-making problems is what managing artificial intelligence (AI) entails leading, coordinating, controlling, and communicating (Berente et al., 2021). There are many definitions of IA in the literature, and some scholars have even suggested a single definition after examining the benefits and drawbacks of IA for HRM (Pan & Froese, 2023). Therefore, it is assumed in this research that artificial intelligence (IA) refers to robots' ability to learn, reason, comprehend, and behave rationally and intelligently, much like people do (Sousa et al., 2023). These authors claim that the modern definition of artificial intelligence (AI) is a machine's capacity to comprehend its surroundings and make decisions that optimize the likelihood that it will accomplish its objectives.

Algorithms are gradually taking over management roles in the AI and automation era (Alahmad & Robert, 2020). The disintegration of the interpersonal, often identification-based, relationship between managers and staff is one major obstacle. It has been demonstrated that building rapport and trust is particularly influenced by an employee's level of identification with their management, or how much the manager is a part of their identity. Additional topics of importance to academics and subjects of study include organizational fairness (Robert et al., 2020) and the place of people in leadership in the future (Quaquebeke & Gerpott, 2023). The latter has significant ramifications for training and developing leaders (Figueiredo et al., 2024). In this way, leaders are the main ones trying to figure out how to link the technical side, made up of technology, with the social side, made up of the organization and the people who are part of it. This link is presented in a conceptual framework (Tursunbayeva & Chaluz-Ben Gal, 2024).

Organizational leaders are under pressure to retain talented and well-managed staff members considering the global challenges to combat global warming in the context of advancements in smart technology, AI, robotics, algorithms (STARA), and digitization (Ogbeibu et al., 2022). The academic literature is currently evolving the idea of STARA competences (Yudiatmaja et al., 2021). As a result, businesses require leaders with a certain set of abilities to act and compete in the new environment. The social orientation of SMEs and the significance of training in competences for sustainable development are significantly influenced by the competences of their leaders (Álvarez-García et al., 2022), which in turn affects the performance of green innovation.

In addition to the importance of AI at the management level (Figueiredo, 2024), there is a critical component of management that deals with an organization's most valuable asset - its people - HRM. Researchers, professionals, and entrepreneurs are interested in HRM because the field is currently being examined considering sustainability (Massood, 2018).

Different stakeholders, including management researchers, practitioners, governments, the corporate world, including business executives, and consumers, have a remarkably increased awareness of GHRM practices (Khan et al., 2017; Xie & Lau, 2023; Kant et al., 2025). Due to its new approach to environmental sustainability, GHRM is the most important component of sustainability (Ren et al., 2018; Sharma & Gupta, 2015). Humans are the 'instruments' that make management systems and goals focused on sustainability possible, especially environmental sustainability.

GHRM is a new subject in today's world (Agyabeng-Mensah et al., 2020). The centrality of sustainability in society in recent years has resulted in increased interest and focus on GHRM, leading to growth in research and the application of practices associated with the concept in organizations (Yong et al., 2020). According to Benevene and Buonomo (2020), the majority of articles published on GHRM began to emerge around 2018. However, the integration of the "green" aspect into HRM originated with Wehrmeyer (1996). Subsequently, the scientific community adopted the GHRM concept, leading to various studies over the years.

For Renwick et al. (2013), GHRM is a concept that encompasses all organizational activities related to the development, maintenance, and implementation of a system that strives to encourage and develop "green" employees. Muisyo et al. (2022) define the concept as a set of practices aimed at the sustainable growth of companies, influencing the culture, strategy, structure, and policy development of companies. The concept has evolved and now includes not only awareness of environmental issues but also the social and economic well-being of organizations and workers (Ahmad, 2015). Bose and Gupta (2017) argue that the GHRM concept emerged through green practices implemented within businesses. According to Ren et al. (2018), GHRM is an end that emphasizes the human resources (HR) of organizations and the ability of GHRM to promote behavioral and attitudinal changes in workers.

GHRM practices depend on three components: developing green skills, motivating employees for sustainable practices, and facilitating green opportunities for employees (Renwick et al., 2013). However, GHRM practices are defined in different ways by different researchers. Jamal et al. (2021) and Renwick et al. (2013) studied green recruitment and selection practices, green training, green rewards and benefits, and green employee involvement. Ahmad (2015) states that green performance management, green relationships, and other initiatives such as green construction, paperless offices, energy conservation, and recycling.

In this context, green recruitment practices include green hiring criteria, green employer branding, and candidates' green awareness as three characteristics of green recruitment and selection (Renwick et al., 2013). Green training consists of activities that encourage workers to have concerns and skills related to environmental issues (Jabbour, 2011). For Tang et al. (2018), this practice involves three elements: knowledge management, green awareness, and environmental protection activities. According to Renwick et al. (2013), training should be part of comprehensive programs and linked to performance management to create a green work environment. Green reward systems include financial and non-financial benefits that attract, retain, and motivate workers to contribute to the organization's green objectives (Muisyo et al., 2022). Finally, another perspective on these practices is presented by Massood (2018), who refers to

GHRM practices as paperless offices, energy conservation, green printing, green construction, water-saving, recycling and waste disposal, green rewards, and electricity savings from a "go green" perspective.

Relationship between AI and GHRM

The intersection between AI and GHRM represents a significant milestone for organizations seeking to implement more sustainable practices (Garg et al., 2018). The application of advanced algorithms offers a unique opportunity to optimize processes, reduce waste, and simultaneously promote environmental responsibility (Malik et al., 2023). Consequently, AI plays a crucial role in analyzing large datasets related to HRM practices, as it enables the identification of inefficiencies and opportunities for sustainable improvement (Yong et al., 2023). Through predictive analytics, organizations can anticipate consumption patterns and reduce the carbon footprint associated with HRM activities (Aykan, 2017). The automation provided by AI in GHRM contributes to increased operational efficiency by automating routine and bureaucratic tasks (Prikshat et al., 2023a), allowing HR professionals to focus on the implementation of more sustainable strategies (John & Pramila, 2023). These procedures not only streamline internal processes but also promote an organizational culture oriented towards environmental awareness (Mabungela, 2023).

The personalization of development and training strategies can also be positively enhanced by AI (Murugesan et al., 2023). The use of algorithms allows for the analysis of individual performance and the identification of skill gaps, contributing to the development of a more qualified workforce that is conscious of environmental issues (Malik et al., 2023). However, it is important to ensure transparency in the implementation of AI in GHRM, as it is crucial to balance operational efficiency with the preservation of ethical values and equity in decision-making when adopting this technology (Rodgers et al., 2023). The integration of AI in GHRM represents a significant advancement in the implementation of more sustainable corporate practices, promoting environmental responsibility, and optimizing organizational efficiency (Ranjbar, 2023).

Dimensions of GHRM influenced by AI

According to Gupta (2021), the existence of a diverse workforce in terms of generations (Baby Boomers, Generation X, Generation Y, and Generation Z) requires organizations to use AI tools and techniques to adopt GHRM practices. The author also argues that AI tools and techniques can be employed in HR processes, engaging and motivating workers to voluntarily participate in sustainable activities. GHRM with AI provides employees with job satisfaction, well-being, and people. Policies included in GHRM are compensation and management reward, security of employment, and teamwork for better productivity (Sabale & Gomathi, 2022).

Garg et al. (2028) found that an IT company in Bengaluru had an application for managing its employees, including a chatbot in the HR area. The chatbot resolves employee queries related to leave and company policies. It fields technical support queries and provides simple ways of expressing employee appreciation across all levels of corporate hierarchy. It specializes in giving an overview of openings and jobs that will be created in the future, along with the preferred skills for each. This author developed a model and argues that AI assists in various areas of GHRM. In hiring, "artificial intelligence helps parse and screen CVs on the basis of the profile required better than humans and saves time as well." In performance appraisal, advanced software aids in continuous assessment, in operations, for example, creating chat rooms and chat boards that facilitate problem-solving and addressing workers' doubts, and in the workforce (Garg et al., 2018, p. 115).

AI helps organizations adopt GHRM services by offering various applications and smart technologies that simplify work and increase employee productivity (Gupta, 2021). According to Garg et al. (2018), automation has reduced traditional recruiters's jobs by 55% to 60%, making the process more efficient and cost-effective.

Models and Applications of AI in GHRM

As previously mentioned, AI plays a significant role in GHRM because it enables the development of more sustainable and efficient practices in various HR areas (Yong et al., 2023). According to Rathore (2023), AI can be utilized to analyze CVs and identify candidates with the necessary skills for the roles the organization needs to fill. Additionally, AI-based tools used in virtual interviews provide insights that allow the evaluation of candidates' body language and verbal language (Thakur et al., 2023). The use of AI makes the recruitment and selection process faster and more effective (Ore & Sposato, 2022).

AI-driven training platforms offer personalized programs for each employee as they focus on their specific development needs (Sabil et al., 2023). Performance management and evaluation are also positively impacted by AI models, as they can predict employee performance based on their professional history (Anayat, 2023). By identifying areas for improvement, AI suggests measures to optimize individual and collective performance (Yadav et al., 2023).

AI systems also help analyze employee engagement, significantly contributing to talent retention and reducing the organization's turnover rate (Alzyoud, 2022). AI algorithms facilitate the management of work schedules and logistics related to energy consumption, thus helping to minimize the carbon footprint associated with the movement of human and material resources within facilities (Bhuiyan, 2024).

Reddy et al. (2024) add that by identifying sustainable HRM practices, AI encourages environmentally friendly behaviors among employees and provides valuable information for making strategic decisions related to GHRM. AI-based chatbots provide information on sustainable HR practices, answer employee queries, and promote awareness of the importance of environmental management (Bijoria, 2024).

By integrating AI into HRM, organizations promote more sustainable practices, reduce operational costs, and enhance their reputation as environmentally responsible leaders (Bhardwaj, 2023). This integrated approach strengthens corporate social responsibility and contributes to creating more conscious and engaged work environments (Ghosal & Ganguly, 2024). Moreover, the data-driven approach offers a significant competitive advantage to organizations seeking to align their HRM practices with sustainable development goals (Moorthi, 2024).

Benefits and Challenges of AI for GHRM

The application of AI in GHRM offers numerous benefits for organizations (Chowdhury et al., 2023). Beyond optimizing recruitment and selection processes, performance management and evaluation, and training, as mentioned previously, it enhances organizational efficiency (Gupta, 2021). By analyzing large datasets, AI identifies relevant patterns and trends to formulate sustainable strategies for GHRM (Hossin et al., 2021). AI-based systems assist in monitoring and evaluating employee performance concerning sustainable practices, facilitating the implementation of corrective measures or the recognition of eco-friendly behaviors (John & Pramila, 2023).

The automation of routine tasks through AI allows HR professionals to devote more time to developing strategic sustainable activities (Singh & Pandey, 2024). Despite its benefits, AI also presents challenges, primarily related to privacy and ethics (Prikshat et al., 2022). Therefore, it is crucial to ensure that data collection and usage comply with current data protection regulations.

The implementation of AI technologies in GHRM sometimes faces resistance from employees who fear job loss and distrust new technologies (Park et al., 2021). Additionally, AI systems can be quite costly, especially for small and medium-sized enterprises, which may hinder adoption (Chatterjee et al., 2024). Ravesangar et al. (2024) note that AI in GHRM may require professionals to acquire specific skills to interact effectively with these technologies. To maximize the benefits and mitigate the challenges of AI implementation in GHRM, it is essential for organizations to adopt an approach that considers both operational efficiency and ethical and social impacts (Sachan et al., 2024).

CONCLUSIONS AND IMPLICATIONS

The study concludes that the integration of AI in GHRM represents an innovative and strategic approach to fostering the sustainable development of organizations, particularly when aiming to implement a sustainable HRM system. AI enables a deeper understanding of employee needs, identifying opportunities for the implementation of sustainable practices. By automating routine tasks, AI frees up time and resources, allowing HR professionals to focus on the development and implementation of more humanized and sustainable strategies. Furthermore, AI can be used to predict employee behavior trends, aiding in the anticipation of challenges and the implementation of proactive GHRM initiatives.

Therefore, according to the literature review conducted in this study, the impacts of AI on GHRM do contribute to strategic people management in line with the principles of sustainability. AI plays an important role in promoting the performance, development, and retention of green employees, as well as in improving organizational culture and corporate social responsibility, which are key aspects of sustainability. The strategic and responsible integration of AI and GHRM in organizations is seen as essential for their sustainable development.

Theoretical Implications

- This study contributes to the growing body of literature exploring the intersection of AI and HRM, specifically focusing on the emerging field of Green Human Resource Management (GHRM). It reinforces the theoretical relevance of both AI in management and the increasing importance of GHRM in contemporary organizations facing the pressures of digital and green transitions.
- The research highlights the potential of AI as an enabler of GHRM practices, expanding the theoretical understanding of how technology can be leveraged to promote environmental sustainability within HRM functions. It connects the capabilities of AI (such as data analysis, automation, and predictive analytics) to the different dimensions of GHRM, including recruitment, training, performance management, and employee engagement.
- The study also contributes to the theoretical understanding of strategic people management by demonstrating how the integration of AI in GHRM can align HR practices with broader organi-

zational sustainability goals. This emphasizes the strategic role of HRM in achieving sustainable development.

Practical Implications

- Organizations should strategically and responsibly integrate AI into their HRM practices to boost green initiatives and contribute to sustainable development. This involves recognizing AI as a valuable tool for optimizing HR processes and promoting environmental responsibility.
- HR professionals can leverage AI tools for various GHRM dimensions, such as using AI for efficient and effective green recruitment and selection by analyzing CVs and conducting virtual interviews. AI can also personalize green training programs and improve green performance management by identifying areas for improvement.
- AI can be used to enhance employee engagement in sustainability initiatives by analyzing engagement levels, facilitating communication through chatbots that provide information on sustainable HR practices, and promoting environmental awareness.
- Organizations need to be mindful of the challenges associated with AI implementation in GHRM, including ethical, legal, and social issues, particularly concerning data privacy, potential for bias, and employee resistance. Therefore, ensuring transparency, fairness, and addressing employee concerns are crucial for successful adoption.
- Investing in reskilling and upskilling HR professionals is necessary to enable them to effectively interact with AI technologies and leverage their capabilities for GHRM.
- Organizations, especially SMEs, should consider the cost implications of implementing AI systems and explore solutions that are both effective and financially feasible.
- Leadership plays a fundamental role in aligning business strategy with sustainability issues, and digital transformation and AI can be valuable allies for leaders in promoting change towards GHRM within the organization. Leaders should be deeply committed to these issues and understand how to link the technical aspects of AI with the social side of the organization and its people.

RECOMMENDATIONS AND FUTURE DIRECTIONS

While this approach is suitable for exploring a current and recent topic, it inherently carries the limitations associated with literature reviews. These limitations might include:

- a lack of primary data collection and empirical validation of the relationships and impacts discussed. The findings are based on the synthesis of existing research, rather than new empirical evidence;
- the possibility that the literature reviewed may not be exhaustive of all perspectives and studies on the intersection of AI and GHRM; and,
- the rapidly evolving nature of AI might mean that the reviewed literature could become quickly outdated.

In the field of research, it is necessary to continue promoting more studies on AI, particularly its impact on management. AI evolves every day, so it is crucial to prepare organizations and their people for constant learning in order to be able to take advantage of technological advances in more sustainable organizational practices. Some recommendations for future research might include:

- Given that AI evolves every day, it is crucial to prepare organizations and their people for constant learning to take advantage of technological advances in more sustainable organizational practices. This implies that future research should focus on understanding and facilitating this continuous learning process within organizations.
- Future research has to go beyond literature reviews and explore the topic through empirical studies to validate the findings and provide deeper insights into the practical application and impact of AI on GHRM. This could involve quantitative research, case studies, or mixed-methods approaches to gather primary data and assess the real-world outcomes.
- Further research could explore the specific ethical, legal, and social issues associated with the integration of AI in GHRM in more detail, providing frameworks and guidelines for responsible implementation.
- Studies could also investigate the challenges faced by different types of organizations, such as SMEs, in adopting AI for GHRM and explore potential solutions or tailored approaches.
- Future research could delve into the specific skills and competencies that HR professionals need to acquire to effectively interact with AI technologies in the context of GHRM.
- The recommendation to align people management with societal demands and contribute to global sustainable development suggests that future research could explore the broader societal and environmental impacts of AI-driven GHRM practices.

REFERENCES

Afzal, M. N. I., Shohan, A. H. N., Siddiqui, S., & Tasnim, N. (2023). Application of AI on Human Resource Management: A Review. *Journal of Human Resource Management - HR Advances and Developments, 2023*(1), 1–11. DOI: 10.46287/FHEV4889

Agyabeng-Mensah, Y., Ahenkorah, E., Afum, E., Agyemang, A., Agnikpe, C., & Rogers, F. (2020). Examining the influence of internal green supply chain practices, green human resource management and supply chain environmental cooperation on firm performance. *Supply Chain Management, 25*(5), 585–599. DOI: 10.1108/SCM-11-2019-0405

Ahmad, H., & Ibrahim, B. (2015). Leadership and the Characteristic of Different Generational Cohort towards Job Satisfaction. *Procedia: Social and Behavioral Sciences, 204*, 14–18. DOI: 10.1016/j.sbspro.2015.08.104

Ahmad, S. (2015). Green Human Resource Management: Policies and practices. *Cogent Business & Management, 2*(1), 1–13. DOI: 10.1080/23311975.2015.1030817

Alahmad, R., & Robert, L. (2020). Artificial Intelligence (AI) and IT identity: Antecedents identifying with AI applications. *26th Americas Conference on Information Systems, AMCIS 2020*, 1–10.

Álvarez-García, J., Hormiga-Pérez, E., Sarango-Lalangui, P., & Río-Rama, M. (2022). Leaders' sustainability competences and small and medium-sized enterprises outcomes: The role of social entrepreneurial orientation. *Sustainable Development (Bradford), 30*(5), 927–943. DOI: 10.1002/sd.2291

Alvi, F. (2021). A New Technological Dimension In The Hrm - Finding The Correlation Between Artificial Intelligence Applications and Green Human Resource Management. *Turkish Online Journal of Qualitative Inquiry, 12*(7), 5483–5494.

Alzyoud, A. (2022). Artificial intelligence for sustaining green human resource management: A literature review. In *2022 ICETSIS - International Conference in Emerging Technologies for Sustainability and Intelligent Systems* (pp. 321-326). IEEE.

Anayat, S. (2023). Human resources management after industry 4.0: Blending AI and HRM. In *Strategic Human Resource Management in the Hospitality Industry: A Digitalized Economic Paradigm* (pp. 97–115). IGI Global. DOI: 10.4018/978-1-6684-7494-5.ch005

Arslan, A., Cooper, C., Khan, Z., Golgeci, I., & Ali, I. (2022). Artificial intelligence and human workers interaction at team level: A conceptual assessment of the challenges and potential HRM strategies. *International Journal of Manpower, 43*(1), 75–88. DOI: 10.1108/IJM-01-2021-0052

Aykan, E. (2017). Gaining a competitive advantage through green human resource management. In Corporate governance and strategic decision making. In *Corporate Governance and Strategic Decision Making* (pp. 159- 175). DOI: 10.5772/intechopen.69703

Benevene, P., & Buonomo, I. (2020). Green Human Resources Management: An evidence-based systematic literature review. *Sustainability (Basel), 12*(15), 1–25. DOI: 10.3390/su12155974

Berente, N., Gu, B., Recker, J., & Santhanam, R. (2021). Managing artificial intelligence. *Management Information Systems Quarterly*, *45*(3), 1433–1450. DOI: 10.25300/MISQ/2021/16274

Bhardwaj, B. (2023). Leveraging AI for the Reinforcement of GHRM. In *AI and Emotional Intelligence for Modern Business Management* (pp. 64–76). IGI Global. DOI: 10.4018/979-8-3693-0418-1

Bhuiyan, M. (2024). Carbon Footprint Measurement and Mitigation Using AI. Available at *SSRN*: https://ssrn.com/abstract=4746446 DOI: 10.2139/ssrn.4746446

Bijoria, S. (2024). A New Revolution in Green Human Resource Management (GHRM) Using Artificial Intelligence (AI). *International Journal of Innovative Research in Technology and Science*, *12*(2), 93–100.

Bose, I., & Gupta, V. (2017). Green HRM Practices in Private HealthCare & Banking Sectors in India. *Indian Journal of Industrial Relations*, *53*(1), 48–58. https://www.jstor.org/stable/26536436

Budhwar, P., Malik, A., Silva, M., & Thevisuthan, P. (2022). Artificial intelligence–challenges and opportunities for international HRM: A review and research agenda. *International Journal of Human Resource Management*, *33*(6), 1065–1097. DOI: 10.1080/09585192.2022.2035161

Cesário, F., Sabino, A., Palma-Moreira, A., & Pinto-Coelho, M. (2025). Exploring Links Between Green HRM, Greenwashing, and Sustainability: The Role of Individual and Professional Traits. *Sustainability (Basel)*, *17*(4), 1764. DOI: 10.3390/su17041764

Chatterjee, S., Jemima, A., Ray, S., Kumar, M., & Ahluwalia, G. (2024). A Systematic Review of Artificial Intelligence (AI) And Impact on Human Resource Management (HRM): Challenges, Risks and Opportunities. *Naturalista Campano*, *28*(1), 558–591.

Chowdhury, S., Dey, P., Joel-Edgar, S., Bhattacharya, S., Rodriguez-Espindola, O., Abadie, A., & Truong, L. (2023). Unlocking the value of artificial intelligence in human resource management through AI capability framework. *Human Resource Management Review*, *33*(1), 100899. DOI: 10.1016/j.hrmr.2022.100899

Dima, J., Gilbert, M. H., Dextras-Gauthier, J., & Giraud, L. (2024). The effects of artificial intelligence on human resource activities and the roles of the human resource triad: opportunities and challenges. In *Frontiers in Psychology* (Vol. 15). Frontiers Media SA. DOI: 10.3389/fpsyg.2024.1360401

Ekuma, K. (2023). *Rethinking Upskilling and Reskilling in the Age of AI and Automation: A fsQCA Approach*. DOI: 10.20944/preprints202309.0055.v1

Figueiredo, P. (2024). *Complex AI Dynamics and Interactions in Management*. IGI Global., DOI: 10.4018/979-8-3693-0712-0

Figueiredo, P. (2025). *Approaching Employee Experience Management With Data Science*. IGI Global., DOI: 10.4018/979-8-3693-7848-9

Figueiredo, P. C. N., Alves, M. J. J., & Rouco, J. C. (2024). Reskilling in the artificial intelligence era. *Complex AI Dynamics and Interactions in Management*, 48–64. https://doi.org/DOI: 10.4018/979-8-3693-0712-0.ch003

Garavan, T., Ullah, I., O'Brien, F., Darcy, C., Wisetsri, W., Afshan, G., & Mughal, Y. H. (2023). Employee perceptions of individual green HRM practices and voluntary green work behaviour: A signalling theory perspective. *Asia Pacific Journal of Human Resources, 61*(1), 32–56. DOI: 10.1111/1744-7941.12342

Garg, V., Srivastav, S., & Gupta, A. (2018). Application of Artificial Intelligence for Sustaining Green Human Resource Management. In *ICACE 2018 - International Conference on Automation and Computational Engineering* (pp. 113-116). IEEE. https://doi.org/DOI: 10.1109/ICACE.2018.8686988

Ghosal, A., & Ganguly, C. (2024). Unleashing Work Engagement: The Power Of Green Human Resource Management Practices In Socially Responsible Companies. *Educational Administration: Theory and Practice, 30*(4), 8452–8460. DOI: 10.53555/kuey.v30i4.2744

Gupta, M. (2021). The role of artificial intelligence in adopting green HRM practices. In Rana, G., Khang, A., Sharma, R., Goel, A., & Dubey, A. (Eds.), *Reinventing Manufacturing and Business Processes Through Artificial Intelligence* (pp. 1–18). CRC Press. DOI: 10.1201/9781003145011-1

Hancock, J. T., Naaman, M., & Levy, K. (2020). AI-Mediated Communication: Definition, Research Agenda, and Ethical Considerations. *Journal of Computer-Mediated Communication, 25*(1), 89–100. DOI: 10.1093/jcmc/zmz022

Hossin, M., Ulfy, M., & Karim, M. (2021). Challenges in adopting artificial intelligence (AI) in HRM practices: A study on Bangladesh perspective. *International Fellowship Journal of Interdisciplinary Research, 1*(1), 66–73. DOI: 10.5281/zenodo.4480245

Husen, S., Wahidah, R. N., & Mustajab, D. (2024). Evolution of HRM Strategies in the Digital Age: A Comprehensive Review. [AMAR]. *Amkop Management Accounting Review, 4*(1), 42–58. DOI: 10.37531/amar.v4i1.1535

Jabbour, C. (2011). How green are HRM practices, organizational culture, learning and teamwork? A Brazilian study. *Industrial and Commercial Training, 43*(2), 98–105. DOI: 10.1108/00197851111108926

Jabbour, C., & Renwick, D. (2018). The soft side of environmentally-sustainable organizations. *RAUSP Management Journal, 53*(4), 622–627. DOI: 10.1108/RAUSP-07-2018-0044

Jaiswal, A., Arun, C. J., & Varma, A. (2022). Rebooting employees: Upskilling for artificial intelligence in multinational corporations. *International Journal of Human Resource Management, 33*(6), 1179–1208. DOI: 10.1080/09585192.2021.1891114

Jamal, T., Zahid, M., Martins, J., Mata, M., Rahman, H., & Mata, P. (2021). Perceived green human resource management practices and corporate sustainability: Multigroup analysis and major industries perspectives. *Sustainability (Basel), 13*(6), 1–17. DOI: 10.3390/su13063045

Jayalakshmi, K., & Jayanthi, N. (2024). AI-Powered HR: Enhancing Employee Experience and Engagement. *Proceedings of the International Conference on Digital Transformation in Business: Navigating the New Frontiers Beyond Boundaries (DTBNNF 2024), Advances in Economics, Business and Management Research*, 281–294. DOI: 10.2991/978-94-6463-433-4_20

Jiaping, Y. (2022). Enterprise Human Resource Management Model by Artificial Intelligence Digital Technology. *Computational Intelligence and Neuroscience, 2022*, 1–9. DOI: 10.1155/2022/6186811 PMID: 36479021

John, J., & Pramila, S. (2023). Integrating AI Tools into HRM to Promote Green HRM Practices. In *International Conference on Information and Communication Technology for Competitive Strategies* (pp. 249-259). Springer Nature Singapore.

Johnson, B., Coggburn, J., & Llorens, J. (2022). Artificial Intelligence and Public Human Resource Management: Questions for Research and Practice. *Public Personnel Management, 51*(4), 538–562. DOI: 10.1177/00910260221126498

Kambur, E., & Akar, C. (2022). Human resource developments with the touch of artificial intelligence: A scale development study. *International Journal of Manpower, 43*(1), 168–205. DOI: 10.1108/IJM-04-2021-0216

Kant, S., Adula, M., Kumari, M., & Nimbrain, N. (2025). Role of SMEs in Greening the Economy by Mediation of Eco-Innovation in the Horn of Africa. In *Examining Green Human Resources Management and Nascent Entrepreneurship* (pp. 173–200). IGI Global., DOI: 10.4018/979-8-3693-7046-9.ch007

Kaplan, A., & Haenlein, M. (2019). Siri, Siri, in my hand: Who's the fairest in the land? On the interpretations, illustrations, and implications of artificial intelligence. In *Business Horizons* (Vol. 62, Issue 1, pp. 15–25). Elsevier Ltd. DOI: 10.1016/j.bushor.2018.08.004

Khan, N., Rasli, A., & Qureshi, M. (2017). Greening human resource management: A review policies and practices. *Advanced Science Letters, 23*(9), 8934–8938. DOI: 10.1166/asl.2017.9998

Khateeb, F., & Nabi, T. (2023). Green Human Resource Management: A review of two decades of research. *Management Research and Practice, 15*(2), 43–64.

Khurana, P., Thakral, S., Grover, D., & Kapoor, R. (2025). Human Resource Management in the Era of AI. In Jain, M. K., Goel, S. K., & Verma, Y. (Eds.), *Transformative Role of Commerce Management and Technology* (pp. 21–36). MGM Publishing House., DOI: 10.62823/MGM/2025/9788198191335

Kollmann, T., Kollmann, K., & Kollmann, N. (2023). Artificial Leadership: Digital Transformation as a Leadership Task between the Chief Digital Officer and Artificial Intelligence. *International Journal of Business Science and Applied Management, 18*(1), 76–95. DOI: 10.69864/ijbsam.18-1.172

Mabungela, M. (2023). Artificial Intelligence (AI) and Automation in the World of Work: A Threat to Employees? *Research in Social Sciences and Technology, 8*(4), 135–146. DOI: 10.46303/ressat.2023.37

Malik, A., Budhwar, P., & Kazmi, B. (2023). Artificial intelligence (AI)-assisted HRM: Towards an extended strategic framework. *Human Resource Management Review, 33*(1), 606–659. DOI: 10.1016/j.hrmr.2022.100940

Massood, Z. (2018). Green HRM: A need for 21st Century. *Journal of Emerging Technologies and Innovative Research, 5*(6), 356–359.

Menaka, R. (2023). Role of Artificial Intelligence (AI) in Human Resource Management (HRM) in Recent Era. *Shanlax International Journal of Management*, *11*(2), 32–38. DOI: 10.34293/management.v11i2.6664

Moorthi, V. (2024). Effective Contribution of Green Human Resources Practices on Environmental Sustainability in the Era of Industry 4.0: Evidence from India. In *Renewable Energy and AI for Sustainable Development* (pp. 197-214). CRC Press.

Morandini, S., Fraboni, F., De Angelis, M., Puzzo, G., Giusino, D., & Pietrantoni, L. (2023). The Impact of Artificial Intelligence on Workers' Skills: Upskilling and Reskilling in Organisations. *Informing Science*, *26*, 39–68. DOI: 10.28945/5078

Muisyo, K., Qin, S., Julius, M., Ho, H., & Huong, T. (2022). Green HRM and employer branding: The role of collective affective commitment to environmental management change and environmental reputation. *Journal of Sustainable Tourism*, *30*(8), 1897–1914. DOI: 10.1080/09669582.2021.1988621

Murugesan, U., Subramanian, P., Srivastava, S., & Dwivedi, A. (2023). A study of Artificial Intelligence *impacts on Human Resource Digitalization in Industry 4.0.Decision Analytics Journal*, *7*, 1–9. DOI: 10.1016/j.dajour.2023.100249

Ogbeibu, S., Chiappetta Jabbour, C., Burgess, J., Gaskin, J., & Renwick, D. (2022). Green talent management and turnover intention: The roles of leader STARA competence and digital task interdependence. *Journal of Intellectual Capital*, *23*(1), 27–55. DOI: 10.1108/JIC-01-2021-0016

Ogbeibu, S., Emelifeonwu, J., Pereira, V., Oseghale, R., Gaskin, J., Sivarajah, U., & Gunasekaran, A. (2023). Demystifying the roles of organisational smart technology, artificial intelligence, robotics and algorithms capability: A strategy for green human resource management and environmental sustainability. *Business Strategy and the Environment*, *33*(3), 1–20. DOI: 10.1002/bse.3495

Ore, O., & Sposato, M. (2022). Opportunities and risks of artificial intelligence in recruitment and selection. *The International Journal of Organizational Analysis*, *30*(6), 1771–1782. DOI: 10.1108/IJOA-07-2020-2291

Palos-Sánchez, P., Baena-Luna, P., Badicu, A., & Infante-Moro, J. (2022). Artificial Intelligence and Human Resources Management: A Bibliometric Analysis. *Applied Artificial Intelligence*, *36*(1), 1–28. DOI: 10.1080/08839514.2022.2145631

Pan, Y., & Froese, F. (2023). An interdisciplinary review of AI and HRM: Challenges and future directions. *Human Resource Management Review*, *33*(1), 1–22. DOI: 10.1016/j.hrmr.2022.100924

Park, H., Ahn, D., Hosanagar, K., & Lee, J. (2021). Human-AI interaction in human resource management: Understanding why employees resist algorithmic evaluation at workplaces and how to mitigate burdens. In *Proceedings of the 2021 CHI conference on human factors in computing systems* (pp. 1-15). ACM. DOI: 10.1145/3411764.3445304

Prikshat, V., Islam, M., Patel, P., Malik, A., Budhwar, P., & Gupta, S. (2023a). AI-Augmented HRM: Literature review and a proposed multilevel framework for future research. *Technological Forecasting and Social Change*, *193*, 1–19. DOI: 10.1016/j.techfore.2023.122645

Prikshat, V., Malik, A., & Budhwar, P. (2023b). AI-augmented HRM: Antecedents, assimilation and multi-level consequences. *Human Resource Management Review*, *33*(1), 1–18. DOI: 10.1016/j.hrmr.2021.100860

Prikshat, V., Patel, P., Varma, A., & Ishizaka, A. (2022). A multi-stakeholder ethical framework for AI-augmented HRM. *International Journal of Manpower*, *43*(1), 226–250. DOI: 10.1108/IJM-03-2021-0118

Quaquebeke, N., & Gerpott, F. (2023). The Now, New, and Next of Digital Leadership: How Artificial Intelligence (AI) Will Take Over and Change Leadership as We Know It. *Journal of Leadership & Organizational Studies*, *30*(8), 1–12. DOI: 10.1177/15480518231181731

Ranjbar, M. (2023). Optimizing the human resource management process with artificial intelligence algorithmic approach. *Journal of System Management*, *9*(3), 285–298. DOI: 10.30495/jsm.2023.1978493.1756

Rathore, S. (2023). The Impact of AI on Recruitment and Selection Processes: Analysing the role of AI in automating and enhancing recruitment and selection procedures. *International Journal for Global Academic & Scientific Research*, *2*(2), 51–63. DOI: 10.55938/ijgasr.v2i2.50

Ravesangar, K., Halid, H., Mahadzir, S., & Singh, R. (2024). The Reinvention of HRM Practices Through Artificial Intelligence: Opportunities and Challenges in the Digital World of Work. *Artificial Intelligence Enabled Management: An Emerging Economy Perspective*, *87*, 87–104. Advance online publication. DOI: 10.1515/9783111172408-006

Reddy, M., Deepthi, S., Bhattaru, S., Srilakshmi, V., & Singh, H. (2024). Harmony in HR: Exploring the Synergy of Artificial Intelligence and Green Practices for Sustainable Workplaces. In *MATEC Web of Conferences* (pp. 10-39). EDP Sciences.

Ren, S., Tang, G., & Jackson, S. (2018). Green human resource management research in emergence: A review and future directions. *Asia Pacific Journal of Management*, *35*(3), 769–803. DOI: 10.1007/s10490-017-9532-1

Renwick, W., Redman, T., & Maguire, S. (2013). Green Human Management: A Review and Research Agenda. *International Journal of Management Reviews*, *15*(1), 1–14. DOI: 10.1111/j.1468-2370.2011.00328.x

Robertson, J., & Carleton, E. (2018). Uncovering How and When Environmental Leadership Affects Employees' Voluntary Pro-Environmental Behavior. *Journal of Leadership & Organizational Studies*, *25*(2), 197–210. DOI: 10.1177/1548051817738940

Rodgers, W., Murray, J., Stefanidis, A., Degbey, W., & Tarba, S. (2023). An artificial intelligence algorithmic approach to ethical decision-making in human resource management processes. *Human Resource Management Review*, *33*(1), 1–20. DOI: 10.1016/j.hrmr.2022.100925

Roscoe, S., Subramanian, N., Jabbour, C., & Chong, T. (2019). Green human resource management and the enablers of green organisational culture: Enhancing a firm's environmental performance for sustainable development. *Business Strategy and the Environment*, *28*(5), 737–749. DOI: 10.1002/bse.2277

Sabale, A., & Gomathi, S. (2022). AI-supported green hrm practices for socially responsible and environmental sustainability. *Manager: The British Journal of Administrative Management*, *58*(152), 115–123.

Sabil, S., Bangkara, B., Mogea, T., Niswan, E., & Timotius, E. (2023). Identification of HRM Improvement Strategy Using Artificial Intelligence in Modern Economic Development. *International Journal of Professional Business Review*, *8*(6), 1–14. DOI: 10.26668/businessreview/2023.v8i6.1835

Sachan, V., Katiyar, A., Somashekher, C., Chauhan, A., & Bhima, C. (2024). The Role of Artificial Intelligence In HRM: Opportunities, Challenges, And Ethical Considerations. *Educational Administration: Theory and Practice*, *30*(4), 7427–7435. DOI: 10.53555/kuey.v30i4.2588

Sari, R., Min, S., Purwoko, H., Furinto, A., & Tamara, D. (2020). Artificial Intelligence for a Better Employee Engagement. *International Research Journal of Business Studies*, *13*(2), 173–188. DOI: 10.21632/irjbs.13.2.173-188

Sharma, R., & Gupta, N. (2015). Green HRM: An Innovative Approach to Environmental Sustainability. *Twelfth AIMS International Conference on Management*, 825-830.

Singh, A., & Pandey, J. (2024). AI Driven HR Automation-A Boon for Industrial Relations: Empirical Evidence. *Indian Journal of Industrial Relations*, *59*(3), 411–425.

Sousa, M. J., Sousa, M., Rocha, Á., & Di Virgilio, F. (2023). Scoping Review on AI as a Driver for Industry. In R. Pereira, I. Bianchi, & Á. Rocha (Eds.), *Digital Technologies and Transformation in Business, Industry and Organizations* (Springer, Vol. 2, pp. 235–243). DOI: 10.1007/978-3-031-40710-9_13

Tang, G., Chen, Y., Jiang, Y., Paillé, P., & Jia, J. (2018). Green human resource management practices: Scale development and validity. *Asia Pacific Journal of Human Resources*, *56*(1), 31–55. DOI: 10.1111/1744-7941.12147

Thakur, A., Hinge, P., & Adhegaonkar, V. (2023). Use of Artificial Intelligence (AI) in recruitment and selection. In *ICAMIDA - International Conference on Applications of Machine Intelligence and Data Analytics* (pp. 632-640). Atlantis Press.

Thorat, T. D., & Saste, S. P. (2024). The Influence of Technology on HR Practices and Employee Experience. *International Journal of Research Publication and Reviews*, *5*(3), 7466–7470. www.ijrpr.com

Tursunbayeva, A., & Chaluz-Ben Gal, H. (2024). Adoption of artificial intelligence: A TOP framework-based checklist for digital leaders. *Business Horizons*, *67*(4), 1–12. DOI: 10.1016/j.bushor.2024.04.006

Umrao, S., Dron, S., & Saxena, R. (2024). An Examination of the Impact of Artificial Intelligence on Human Resource Management: Improving Efficiency and Employee Experience. *International Journal of Innovations In Science Engineering And Management*, *3*(2), 87–94. DOI: 10.69968/ijisem.2024v3si287-94

Vrontis, D., Christofi, M., Pereira, V., Tarba, S., Makrides, A., & Trichina, E. (2022). Artificial intelligence, robotics, advanced technologies and human resource management: A systematic review. *International Journal of Human Resource Management*, *33*(6), 1237–1266. DOI: 10.1080/09585192.2020.1871398

Xie, H., & Lau, T. (2023). Evidence-Based Green Human Resource Management: A Systematic Literature Review. *Sustainability*, *15*(14), 1–23. DOI: 10.3390/su16010001

Yadav, P., Kollimath, U., Giramkar, S., Pisal, D., Badave, S., & Swamy, S. (2023). Impact of Artificial Intelligence (AI) on Various Aspects of Human Resources Management (HRM) in India. In *3rd International Conference on Emerging Smart Technologies and Applications* (pp. 1-6). IEEE. DOI: 10.1109/eSmarTA59349.2023.10293625

Yong, J., Yusliza, M., Ramayah, T., Farooq, K., & Tanveer, M. (2023). Accentuating the interconnection between green intellectual capital, green human resource management and sustainability. *Benchmarking*, *30*(8), 2783–2808. DOI: 10.1108/BIJ-11-2021-0641

Yudiatmaja, W., Salomo, R., & Prasojo, E. (2021). Smart Technologies, Artificial Intelligence, Robotics, and Algorithms (STARA) competencies during COVID-19: A confirmatory factor analysis using SEM approach. *Journal of Physics: Conference Series*, *2049*(1), 1–10. DOI: 10.1088/1742-6596/2049/1/012014

KEY TERMS AND DEFINITIONS

Artificial Intelligence: refers to a robot's ability to learn, reason, comprehend, and behave intelligently and rationally, much like humans.

Human Resource Management: It is the practice of managing people within an organization. It involves finding, hiring, integrating, training, developing and rewarding employees so that the company can achieve its objectives. It means looking after the company's most important asset: its employees.

Green Human Resource Management: It's an approach to HRM that considers the environmental impact of human resources practices. This can include things like online recruitment to save paper, offering carpooling programs to reduce carbon emissions and training employees on sustainability practices. It's like traditional HRM, but with an ecological footprint.

Strategic People Management: It's the process of aligning human resources goals with the company's overall objectives. Strategic people management ensures that everyone is working on the right things to achieve the company's success.

Sustainable Development: It's development that meets the needs of the present without compromising the ability of future generations to meet their own needs. It involves practices that protect the environment, but also consider present and future social and economic well-being.

Chapter 17

Innovation and Its Relation to Strategic Positioning With Mediation of Market Competition in the Horn of Africa

Shashi Kant
https://orcid.org/0000-0003-4722-5736
Bule Hora University, Ethiopia

Wako Jio
https://orcid.org/0009-0003-6192-459X
Bule Hora University, Ethiopia

ABSTRACT

This investigation manifests the association among Innovation, competition in market, and Strategic Positioning, highlighting their influence on structural success. It recognizes innovation as a decisive driver of strategic edge, emphasizing the mediating function of competition in market in this dynamic. The investigation demonstrates that organizations that integrate innovation within a competitive framework are better equipped to respond to market changes and achieve sustainable progression. Key recommendations include nurturing an ethos of innovation, conducting regular market analyses, and promoting collaborative influences across teams. Also, the investigation advocates for active methodologies to adapt swiftly to evolving market circumstances. By providing insights into the back-and-forth of these elements, this investigation contributes to the strategic management literature and offers Applied guidance for managers navigating complex business environments.

DOI: 10.4018/979-8-3693-4030-1.ch017

INTRODUCTION

Innovation stands as a cornerstone of fiscal enlargement and strategic edge in today's rapidly evolving global landscape. In the Eastern Africa, a section categorized by diverse ethos, economies, and radical structures, innovation presents both a challenge and an opportunity (Azhdary Moghadam et al., 2024). With global importance, the Eastern Africa is increasingly recognized for its strategic position in international trade routes, resource endowments, and its potential as a burgeoning market. The integration of innovation into strategic positioning can not only enhance the competitiveness of local firms but also contribute to broader fiscal resilience and stability (Zuhroh & Rini, 2024). The notion of innovation has evolved expressively since its inception. Initially rooted in technological progressions, the term now encompasses a wide array of processes, including structural, communal, and business model innovations (Pundziene et al., 2022). The empirical literature highlights various dimensions of innovation, illustrating its critical function in driving productivity and fiscal progression. Studies indicate that innovative firms are better equipped to navigate competition in market, adapt to changing consumer demands, and leverage technological progressions (Elgarhy & Abou-Shouk, 2023). Also, the existing body of investigation often overlooks the unique contextual influences that influence innovation in the Eastern Africa, particularly the back-and-forth among competition in market and strategic positioning (Alsafadi & Aljuhmani, 2024).

Despite the recognized importance of innovation, several applied encounters persist in the Eastern Africa. These encounters include inadequate infrastructure, limited Contact to financing, and a lack of skilled labor (Klein et al., 2021). Also, radical instability and regulatory blockades further complicate the innovation landscape. These issues not only hinder the ability of firms to innovate but also stifle competitive dynamics within the section. The gaps in empirical literature reflect a need for more localized studies that address these specific encounters and contextual influences (Hayat & Qingyu, 2024). This chapter aims to fill these gaps by providing a comprehensive analysis of how innovation can mediate the influences of competition in market on strategic positioning in the Eastern Africa. By synthesizing existing literature and incorporating case studies from the section, this investigation will shed light on influence strategies for nurturing innovation amidst the unique encounters faced by local firms. It will also offer insights for policymakers aiming to create an enabling environment for innovation and competitiveness.

The central question guiding this investigation is: ***"How can innovations serve as a mediator among competition in market and strategic positioning in the Eastern Africa, and what strategies can be employed to overcome existing blockades?"*** This question underscores the need for a focused exploration of the association among these variables, highlighting the importance of innovation in shaping the competitive landscape.

BACKGROUND OF INVESTIGATION

The Eastern Africa is a section marked by a complex back-and-forth of socio-fiscal, radical, and ethnic dynamics. Comprising economies, this area is categorized by a rich history yet faces substantial encounters that impede its enlargement (Hayat & Qingyu, 2024). Innovation, while recognized as decisive for fiscal progression, is often stymied by a variety of obstacles unique to this section. Considerate the background of these encounters is indispensable for framing the function of innovation in augmenting strategic positioning and nurturing competitive markets (Klein et al., 2021). One of the primary encounters in implementing innovative practices in the Eastern Africa is inadequate infrastructure (Pundziene et al., 2022). Likewise, Ethiopia has made strides in sectors like agriculture and know-how, yet the lack of reliable transportation and communication networks hampers the distribution of innovative offerings and amenities. A notable case is the Ethiopian startup, ZayRide, which intended to provide ride-hailing amenities similar to Uber. Despite initial success, the company struggled with limited internet Contact and inconsistent electricity supply, influencing its operations and scalability. This highlights that without a robust infrastructure; even innovative ideas can falter in execution (Zuhroh & Rini, 2024).

Contact to finance is another critical blockade that many businesspersons face. In economies such as Somalia, where official banking systems are underdeveloped, businesses often rely on in official lending sources (Alateeg & Alhammadi, 2024).. A pertinent example is the mobile money platform, EVC Plus, which emerged in Somalia to facilitate transactions. Although it has gained popularity, its progression has been curtailed by regulatory uncertainties and the lack of comprehensive financial amenities, limiting the potential for innovation in other sectors. This case underscores the necessity for a stable financial ecosystem that supports innovative ventures (Elgarhy & Abou-Shouk, 2023). Also, radical instability and insecurity have profound implications for innovation in the section. Likewise, in South Sudan, ongoing conflict has resulted in a fractive fiscal environment where businesspersons are hesitant to invest in new technologies. The case of the South Sudanese agribusiness, AgriTech, illustrates this challenge (Alsafadi & Aljuhmani, 2024). Although the company sought to introduce modern farming techniques, the volatile security situation and erratic government policies created an atmosphere of uncertainty, leading to halted projects and diminished investor confidence. Such instability not only influences individual enterprises but also stifles overall competition in market and strategic positioning (Klein et al., 2021).

Also, the skills gap in the workforce poses a substantial challenge for innovation. Many economies in the Eastern Africa contend with a lack of trained professionals who can drive technological progressions and innovative practices (Pundziene et al., 2022). In Kenya, while the tech industry has seen rapid progression with hubs like Silicon Savannah, the shortage of skilled labor in specialized fields, such as data science and software engineering, remains a hurdle. The case of M-Pesa, a mobile banking service, demonstrates both innovation and the necessity for ongoing training and enlargement. While M-Pesa has revolutionized financial transactions in Kenya, the need for continuous workforce enlargement is decisive to sustain and expand such innovations (Kahpi et al., 2024). Therefore, the Eastern Africa presents a rich yet challenging environment for nurturing innovation. The back-and-forth of inadequate infrastructure, limited Contact to finance, radical instability, and workforce skill gaps creates a complex landscape where innovative ideas struggle to materialize into viable business ventures. Addressing these encounters is indispensable for leveraging innovation as a mediator among competition in market and strategic positioning, paving the way for sustainable fiscal progression in the section.

STATEMENT OF PROBLEM

The back-and-forth among innovation, strategic positioning, and competition in market in the Eastern Africa presents a multifaceted problem that requires comprehensive analysis and considerate (Hayat & Qingyu, 2024). Ideally, innovation should serve as a driving force that empowers local enterprises to enhance their competitive edge and navigate the complexities of the market landscape. Also, the current reality in the section starkly contrasts this ideal state (Velyako & Musa, 2024). Despite the recognition of innovation as a critical component for fiscal progression, numerous contradictory evidences suggest that local firms struggle to leverage innovative practices influence due to a myriad of blockades, including inadequate infrastructure, limited Contact to finance, and a lack of skilled labor (Alsafadi & Aljuhmani, 2024). One substantial gap identified in the literature is the inconclusive results surrounding the influence of innovation in mediating competition in market. While some studies highlight positive associations among innovation and strategic edge, others present mixed findings, particularly in the context of the Eastern Africa (Kakeesh et al., 2024). This inconsistency raises questions about the circumstances under which innovation can influence enhance strategic positioning, suggesting a need for more nuanced investigation tailored to the section's unique circumstances. Also, theoretical gaps exist in Considerate the specific mechanisms by which innovation interacts with competition in market and strategic positioning, especially in environments categorized by volatility and uncertainty (Pundziene et al., 2022).

Knowledge gaps further complicate the landscape, as much of the existing investigation tends to focus on broader African contexts or developed nations, leaving the Eastern Africa underexplored. This oversight results in a lack of relevant frameworks and models that can accurately represent the section's fiscal dynamics (Moqaddamerad & Ali, 2024). Contextual gaps also emerge, as many studies fail to account for the socio-radical intricacies that shape business operations in the Eastern Africa. For example, the influence of sectional conflicts, governance issues, and ethnic influences on innovation practices remains inadequately addressed (Hayat & Qingyu, 2024). Applied encounters compound these theoretical and contextual gaps, as businesspersons in the Eastern Africa face real-world obstacles that hinder their ability to innovate. Issues such as radical instability, regulatory uncertainty, and an underdeveloped financial ecosystem create an environment where innovative ideas struggle to take root. These Applied encounters not only influence individual firms but also stifle overall competition in market, limiting the potential for strategic positioning on a broader scale (Sudirjo et al., 2024).

This investigation aims to fill these gaps by providing a comprehensive framework that elucidates the association among innovation, competition in market, and strategic positioning in the Eastern Africa. By conducting localized case studies and empirical analyses, the investigation will explore the specific circumstances under which innovation can thrive in this unique context. Also, the investigation will develop Applied recommendations for policymakers, focusing on creating an enabling environment that fosters innovation and enhances competitive dynamics. Ultimately, this investigation seeks to provide a clearer direction for future scholars and practitioners, equipping them with the insights necessary to navigate the complexities of innovation in the Eastern Africa and contribute to sustainable fiscal enlargement.

THEORETICAL FOUNDATION

Schumpeterian Innovation Theory: Joseph Schumpeter's innovation theory emphasizes the function of the entrepreneur as a key driver of fiscal enlargement through innovation. Schumpeter posited that innovation occurs through "creative destruction," where new innovations disrupt existing market structures, leading to fiscal progression (Hayat & Qingyu, 2024). This theory highlights the importance of technological progressions, new product enlargement, and improved processes as indispensable components of strategic edge. In the context of the Eastern Africa, Schumpeterian theory can help explain how innovative firms may challenge established competitors and reshape market dynamics. Also, the

Applied encounters of implementing this theory, such as the lack of infrastructure and financial support in the section, necessitate a deeper examination of its applicability (Pundziene et al., 2022).

Resource-Based View (RBV): The Resource-Based View maintains that resources and capabilities are strategic for any firm and form the foundation for a strategic advantage. Relational theory pays attention to internal resources to which innovations correspond, including skilled labour, non-physical assets, and technology. In the eastern Africa region the following are some of the challenges that firms encounter when trying to Contact these resources The constraint presented by Contacting these resources can therefore limit their innovation abilities. The RBV insights enable Considerate how local firms can effectively harness their contextual resources to drive innovation even under external encounters. Secondly, the theory frequently supposes a stable environment across the organization and it is in contrast with the dynamic environment in the section (Zuhroh & Rini, 2024).

Open Innovation Theory: Open innovation theory, introduced by Henry Chesbrough, suggests that firms can benefit from external ideas and technologies, thus augmenting their innovation capabilities. This approach encourages collaboration among organizations, universities, and investigation institutions to foster innovation. In the Eastern Africa, where local firms may lack the resources to innovate independently, open innovation can serve as a viable strategy for overcoming resource constraints. Likewise, partnerships with international organizations or universities can provide Contact to new technologies and expertise. Also, the influenceness of open innovation in this context can be limited by influences such as regulatory blockades and ethnic differences in collaboration practices (Alsafadi & Aljuhmani, 2024).

Innovation Systems Theory: Innovation systems theory emphasizes the interconnectedness of various actors—such as firms, government agencies, and investigation institutions—in driving innovation within a specific context. This approach recognizes that innovation does not occur in isolation but is influenced by the broader socio-fiscal environment, including policies, institutions, and ethnic influences. In the Eastern Africa, this theory highlights the importance of creating supportive ecosystems that facilitate knowledge exchange and collaboration among stakeholders. Also, the theory often lacks specificity regarding how these systems can be influence developed in sections with substantial socio-radical encounters.

For this investigation, ***"Innovation Systems Theory"*** serves as the main theoretical lens. This choice is rooted in the need to understand the complex interactions among various stakeholders involved in the innovation process within the Eastern Africa. Given the section's unique socio-radical context and the multifaceted encounters faced by local firms, Innovation Systems Theory provides a comprehensive framework for analyzing how different actors—such as government agencies, private enterprises, and educational institutions—can collaborate to foster an environment conducive to innovation. By focusing on these interactions, the investigation aims to uncover mechanisms that can enhance the innovative capacity of firms while addressing the contextual blockades inherent in the Eastern Africa. This theoretical lens will guide the analysis of empirical data and case studies, ultimately contributing to a deeper Considerate of how innovation can mediate the association among competition in market and strategic positioning in this section.

DEFINITION AND ORIGIN OF CONCEPTS

Innovation: Innovation is therefore defined as creating and implementing new products, services, assets, facilities, systems or methods that deliver added utility or productivity. It is a broad concept that covers four forms – technological, business model, and structural advancements. Innovation can thus be of two types: continuous innovation where innovations are small improvements to existing offers; and disruptive innovation that bring radical changes to a market (Alsafadi & Aljuhmani, 2024). The word innovation comes from the Latin term "innovare "which means to change or to renew. It in fact rest on the understanding of the modern economy and was recognized in the early part of the twentieth century with theorists like Joseph Schumpeter placing the role of business people at the center of fiscal activity through innovation. In the present time fiscals, business studies and enlargement theory have specific overture to innovation as it plays an imperative role in enhancing strategic stance and promoting fiscal development.

Strategic Positioning: Strategic positioning can be defined as the place of the company or organization based on the competitive field. It includes identification of markets, identification and distinction of goods and services or facilities, and differentiation of the value proposal. Importance of influence strategic positioning also helps the firms to enhance their competitive capabilities, serve the needs of customers, and create competitive advantage in their sector (Klein et al., 2021). Strategic positioning as a notion belongs, therefore, to the field of strategic management and was developed in the 1980s. Mis beforehand put in place by influential texts like Michael Porter's "Competitive Strategy," where features like the Generic Strategies model for strategies to gain strategic position were developed. Strategic positioning has gradually evolved into an essential component of the strategic management process and serves as a framework for firms to apply within a particular marketplace (Alsafadi & Aljuhmani, 2024).

Market Competition: Market competition therefore means the fighting that businesses subject themselves to, as they vie for the patronage of clients and the corresponding market percentage. It expands into different types of competition such as BN UD and product differentiation and service quality competition of price. Competition in market is a basic component of free-market economies; it sustains invention, effectiveness and customers 'options (Rua & Santos, 2022). The idea of competition in market can be dated back to Classical fiscals most probably as represented by Adam Smith in the eighteenth century. Describing Smith's simplification of the competitive nature into the "invisible hand" concept as displaying how individual self- interest results into efficient working of the market. However, over time it has changed and with the help of other Economists such as Alfred Marshall and Joseph Schumpeter the idea of Considerate of competition in market has elaborated including its function of innovation and business in the market competition (Zuhroh & Rini, 2024).

Innovation Systems: Innovation systems can be described as the arrangement of institutions, related organizations, and performers, who come together to support the innovation process. This covers the central and local government, investigative organizations, universities and private companies, all of which expand and disseminate new technologies and thoughts. It is the case that innovation systems are concerned with the sharing of ideas and ideas generation to foster the right conditions for innovation (Zuhroh & Rini, 2024). It has however been since the late 20th century those scholars realized the inadequacies of the linear models of integration and thus developed the concept of innovation systems. Innovative studies like Bengt-Åke Lundvall and Christopher Freeman were done in focusing the function of the communal and institutional environment in the process of innovation. Currently, for the analysis of sectional and national innovation policies, the concept of Innovation Systems is applied outlining the interactions between various key players, regarding the fiscal enlargement.

Collectively, these terms provide a foundational Considerate of the themes central to this investigation, highlighting the interconnectedness of innovation, strategic positioning, and competition in market within the broader context of innovation systems.

EMPIRICAL LITERATURE REVIEW

Innovation Association To Strategic Positioning

Still, innovation goes hand in hand with strategic positioning since it becomes one of the main sources that help firms to achieve competitive advantage in the market. In its simplest terms strategic positioning entails drawing a clear and unique value proposition, which separates a firm from its competitors (Rua & Santos, 2022).. Being a central function of this process, innovation enables firms to create new products and service offerings, physical facilities or even new service delivery models that reflect the changing needs and aspirations of their customers. The fact remains that innovation is possible and ongoing, which means firms are able to vary their products to better meet the dynamics of the marketplace and therefore enhance their appeal to the customer (Alsafadi & Aljuhmani, 2024). In addition, there is an addition to the scale of unusual competencies and assets that support a firm's strategic stance (Zuhroh & Rini, 2024). As an example, a firm may use advanced working knowledge towards enhancing productive efficacy as well as towards establishing a competitive advantage based on quality or client relations. This association is particularly decisive in competitive scenarios where firms are not only expected to respond to change but also drive it. Therefore, innovation is not merely a solitary action; it is a core aspect of influence strategic position; through which firms are armed to deal with competition, market consumers' pressures, to mention but few with a view of aspiring to long-term success.

Innovation Association To Competition In Market

Competition in market is addressed by the aspect of innovation because it provides firms with the opportunity to obtain strategic positioning. In the competitive market there is always immense pressure on the organization to improve the existing products and services, to attract more and more customers and gain more and more market share. Change is assumed to be the facilitation of the above influences through the introduction of new products; services, or a new flow that may disorient the market (Klein et al., 2021). Similarly, a firm ends up creating a new know-how that revolutionizes the industry standard to force an incumbent to shift or fall out of the market. This constant cycle of new formulations creates a culture that encourages companies to partake in investigation and enlargement to create benefits for the consumer due to improved product options and quality (Elgarhy & Abou-Shouk, 2023).

Also, innovation helps move the competitive structure of a market by changing the barriers to entry within the space. The formulation could be the advent of new technologies or business models in which costs may come down and gates are open for Nan players to potentially establish their business in such markets that are already occupied by large players (Alsafadi & Aljuhmani, 2024). Similar to the current use of digital platforms in organizing Contact, the entry of startups has been able to level the playing ground for influence against large corporations. This disruption does not only increase competition but also shifts the rate at which innovation is expected to occur since companies set in a disrupted environment are forced to innovate with a high frequency in order to retain their competitive advantage. In this case, it can also be concluded that the use of innovation initiatives plays the role of defense and attack in fighting in the competitive environment of the business competition (Rua & Santos, 2022).

Furthermore, innovation is identified as supporting competition dynamism in a market since it fosters the culture for creativity and trial. By nature, innovative organizations seek to promote the work culture that creates both the need to take calculated risks and fosters cooperation, therefore creating new ideas and solutions. It also pays for individual firms since those who discover useful practices that form the basis of new technologies get to make substantial profits while at the same time raising the standards in the industry. That is why, along with the best examples and their successful experience, competition becomes more diverse and meets the consumer needs (Zuhroh & Rini, 2024).

Therefore, relationship between innovation and competition in market can be described as a cycle in which innovation promotes competitive advantage, redraws market dynamics and creates a culture of contingency. Creative business organizations not only promote their own needs but also affect the competitive market environment and make positive changes in the offerings and services available to consumers (Hayat & Qingyu, 2024).

Competition in Market Influence On Strategic Positioning

With regards to the market, the concept of competition expressively applies to strategic positioning since business has to constantly adjust to change. Where competition is intense, often firms are forced to articulate their strategies of differentiation more clearly than they would have wished. This necessity makes organizations to understand their counterpart strengths and weaknesses, the consumers and the trends in the markets, outcomes to more informed strategic plans (Alsafadi & Aljuhmani, 2024). Similarly a firm that identifies a particular market field, for instance an unfulfilled consumer want can align its products and services to satisfy this need, position itself as a market champion. Besides increasing the overall visibility of the firm, this proactive approach is usually followed by a firm to develop a good brand image that will be appealing to the target market (Elgarhy & Abou-Shouk, 2023).

That is, the level of competition determines which resources and capabilities of the firms' strategic positioning influence. Various types of market structure also give some basic strategies where the company can fight back, which are innovations, better customer service or cost leadership. It can allow companies that invest in investigation and enlargement to create new adaptations either in features or offerings that will differentiate them or use exaggerated customer service to lock in revenues and re-orders. On the other hand, firms that seem not to change their strategic positioning owing to pressures from competitors may find themselves outcompeted as consumers shift towards new strategic positions that are more innovative or fully-serving their needs (Rua & Santos, 2022).

At the same time, competition in market requires constant process of re-orientation of strategic positioning. It is again important for companies to continue their activities or change strategies as competitors start coming up with new products or shifting their market position. This keeps the strategic market analysis and competitive benchmarking of firms ongoing, in order to maintain relevance and influence of positioning strategies. Thus, organizations that manage such encounters effectively can not only defend their market share but also build on competitive pressures enhancing their overall influence and robustness in the market place even further (Zuhroh & Rini, 2024).

In conclusion, issues relating to competition in market have a huge impact to strategic positioning because it forces firms to right-position themselves, identify important capabilities, and align them innovatively to ever existing market conditions. By influence responding to competitive dynamics, organizations can effectively establish their competitive platform, create and maintain customer loyalty and subsequently achieve progressive advance (Hayat & Qingyu, 2024).

Innovation, Strategic Positioning, Mediation Of Competition In Market

Innovation and strategic positioning are seen as being linked by competition in market, yielding interplay between the two forces that leads to the manner in which firms operate within their competitive framings. Innovation is thus invariably used as a foundational platform on which business can distinguish their goods and services, and thereby improve on the value additions that accompany such goods and services (Alsafadi & Aljuhmani, 2024). However in a competitive environment the need to innovate becomes even more acute as firms seek to adapt to new consumer needs and or re-strategies to counteract what their competitors have to offer the market. This necessity arises due to the fact the companies need to invest in investigation and enlargement, adapt new technologies, and adopt new organizational business models, which are imperative for firms seeking to create a strategic and competitive niche market in the demand (Rua & Santos, 2022).

Competition in market moderates this association in a way that determines the level and type of innovation within a given market. More specifically, in highly competitive industries, the forces of change are not simply progressive but are survival imperatives. Similarly, the shift by one player to a disruptive know-how implies that other players will be forced to act fast in response and develop new innovations as part of the competitive environment (Cahyadi et al., 2024). This competitive pressure may help to spur the cycle, under which firms may be prompted to upgrade not only the offerings or amenities they provide but also their strategic position. Hence, organizational innovation enables companies to respond to competitive forces more effectively than they can potentially change their competitive standing and increase customer satisfaction (Elgarhy & Abou-Shouk, 2023). Also, the mediation function of competition in market speaks of the right timing and responsive methods in carrying out innovations. It may be added that those companies which are able to predict the market shifts and act fast due to the threats from competitors will be able to use innovation for establishing better competitive positions. On the other hand, organizations at the back end of innovation influences might discover they are unable to effectively contest influence with more active competitors. This responsiveness also serves to strengthen the strategic context of the firm/venture but also unknowingly builds an image of the firm as an industry innovator (Zuhroh & Rini, 2024).

Altogether, there is the mutual connection between the concepts of innovation, strategic positioning and competition in market suggesting that innovation is a critical factor that cannot be spared for strategic positioning. Competition in market conditions our association by influencing the nature of innovation in the markets compelling firms to strategically pursue the relevant positions. Therefore, if an organization is able to identify and capitalize on this feature, then it is assured of increasing its competitiveness and can, therefore, transform into a sustainable organization as it operates in the market that is volatile (Hayat & Qingyu, 2024).

CONCEPTUAL FRAMEWORK

Figure 1. Conceptual framework

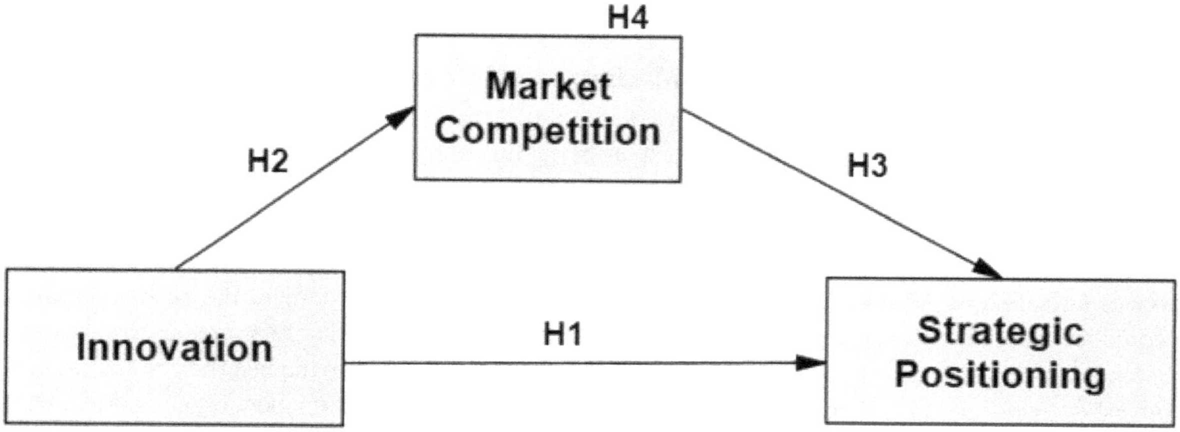

INVESTIGATION METHODOLOGY

The investigation methodology employed in this investigation utilized a quantitative approach to analyze the association among innovation, strategic positioning, and competition in market. A sample size of 390 respondents was selected through a stratified random sampling technique to ensure a representative distribution across various sectors relevant to the investigation. The primary data collection method involved the use of structured questionnaires, which were designed to measure the constructs of interest based on established scales from prior investigation. This approach allowed for the systematic collection of data that could be quantitatively analyzed. To assess the validity and reliability of the measurement scales, the investigations conducted a preliminary analysis using the Kaiser-Meyer-Olkin (KMO) test. The KMO statistic indicated the adequacy of sampling for factor analysis, with values above 0.7 considered acceptable. Following the KMO assessment, Exploratory Factor Analysis (EFA) was performed to identify the underlying structure of the data and to confirm the factor loadings of the items associated with each construct. The EFA results revealed distinct influences corresponding to innovation, strategic positioning, and competition in market, thus supporting the dimensionality of the scales.

Subsequently, the investigations employed Structural Equation Modeling (SEM) using AMOS software to test the hypothesized association among the constructs. SEM allowed for the evaluation of both the direct and indirect influences of innovation on strategic positioning, mediated by competition in market. The model fit indices were assessed to ensure a good fit among the proposed model and the observed data. Criteria such as the Chi-square statistic, Comparative Fit Index (CFI), and Root Mean Square Error of Approximation (RMSEA) were utilized to determine the adequacy of the model fit. Therefore, the investigation methodology integrated a quantitative approach with a comprehensive analysis involving KMO, EFA, and SEM. The use of AMOS software facilitated a robust examination of the association among the constructs, while the scale enlargement process ensured that the measures used were both valid and reliable. This methodological framework provided a solid foundation for Considerate the back-and-forth among innovation, strategic positioning, and competition in market in the context of the investigation.

Scale Synthesis

The scale enlargement for this investigation was grounded in a thorough review of related literature, ensuring that the constructs of innovation, strategic positioning, and competition in market were accurately measured and reflected in the survey instrument. The initial step involved identifying established scales from previous investigation that had demonstrated reliability and validity in similar contexts. For the innovation construct, items were adapted from the work of O'Sullivan and Dooley (2009), which highlighted dimensions such as product innovation, process innovation, and structural innovation. These dimensions were decisive for capturing the multifaceted nature of innovation as it relates to strategic edge.

For the strategic positioning construct, items were derived from Porter's (1985) framework on competitive strategies, emphasizing differentiation and cost leadership. The scale included items that assessed how firms position themselves in the market, the uniqueness of their offerings, and their approach to meeting customer needs. This linkage to established theoretical frameworks ensured that the scale was robust and contextually relevant.

Competition in market was measured using scales developed by Dess and Davis (1984), which focused on aspects such as competitive intensity, market share, and the degree of rivalry among industry players. These items were indispensable for Considerate how competitive dynamics influence a firm's strategic positioning and innovation influences. Once the initial items were drafted, the investigations conducted a pre-test with a small sample to refine the wording and ensure clarity. Feedback from this pilot investigation informed adjustments to the scale, improving item clarity and relevance. Subsequently, the finalized scale included multiple-item measures for each construct, allowing for a comprehensive assessment of the association under investigation.

The final scale was structured to facilitate quantitative analysis, with responses measured on a Likert scale ranging from 1 (strongly disagree) to 5 (strongly agree). This approach not only provided a nuanced Considerate of respondents' perceptions but also enabled the application of advanced statistical techniques in the subsequent analysis. Overall, the scale enlargement process, rooted in related literature, ensured that the measures employed were both theoretically sound and empirically validated, laying a solid foundation for the investigation's findings.

DATA ANALYSIS

Table 1. KMO and Bartlett's Test

Kaiser-Meyer-Olkin Measure of Sampling Adequacy.		**.950**
Bartlett's Test of Sphericity	Approx. Chi-Square	1419.838
	df	91
	Sig.	.000

Source: Authors

Table 1 presents the results of the Kaiser-Meyer-Olkin (KMO) Measure of Sampling Adequacy and Bartlett's Test of Sphericity, which were conducted to evaluate the suitability of the data for factor analysis. The KMO statistic was found to be 0.950, indicating a high level of sampling adequacy and suggesting that the associations among variables were suitable for performing factor analysis. According to Kaiser (1974), KMO values above 0.6 are considered acceptable, while values above 0.9 are considered excellent. Therefore, the obtained KMO value demonstrates that the sample size was adequate for the analysis. Also, Bartlett's Test of Sphericity yielded an approximate Chi-Square value of 1419.838 with 91 degrees of freedom and a significance level (Sig.) of 0.000. This result indicates that the association matrix is expressively different from an identity matrix, implying that there are likely association among the variables that warrant further exploration through factor analysis. The significance of Bartlett's Test reinforces the appropriateness of proceeding with factor analysis on the data set. Collectively, these findings provide strong evidence supporting the validity of the constructs and the overall robustness of the dataset for the subsequent analysis.

Table 2. Total variance explained

Component	**Initial Eigenvalues**			**Extraction Sums of Squared Loadings**			**Rotation Sums of Squared Loadings**		
	Total	**% of Variance**	**Cumulative %**	**Total**	**% of Variance**	**Cumulative %**	**Total**	**% of Variance**	**Cumulative %**
1	4.51	32.235	32.235	4.51	32.235	32.235	3.00	21.451	21.451
2	1.37	9.887	42.022	1.37	9.887	42.022	2.26	15.491	36.942
3	1.27	9.087	51.210	1.27	9.087	51.210	1.98	14.268	51.210
4	.949	6.882	57.992						
5	.971	6.223	64.215						
6	.873	5.519	69.734						

Extraction Method: Principal Component Analysis.
Source: Authors

Table 2 summarizes the results of the Total Variance Explained from the factor analysis, employing Principal Component Analysis as the extraction method. The table presents the initial eigenvalues, the extraction sums of squared loadings, and the rotation sums of squared loadings for each component. The analysis revealed that the first component had an initial eigenvalue of 4.51, accounting for 32.235%

of the total variance. This indicates that the first factor captures a substantial portion of the variability in the dataset. The second component showed an eigenvalue of 1.37, contributing an additional 9.887% to the total variance, bringing the cumulative percentage to 42.022%. The third component similarly demonstrated an eigenvalue of 1.27, explaining 9.087% of the variance and cumulative variance of 51.210%. Together, these three components accounted for over half of the total variance, highlighting their importance in representing the underlying data structure. Further components with initial eigenvalues below 1.0, specifically the fourth (0.949), fifth (0.971), and sixth (0.873) components, indicated diminishing returns in variance explained, reinforcing the decision to retain only the first three components for further analysis. The extraction sums of squared loadings mirrored these findings, illustrating that the first three influences were sufficient for capturing the primary dimensions of the constructs under investigation. Overall, the results from Table 2 validate the appropriateness of the influences identified and underscore their relevance in the context of the investigation.

Confirmatory Principle Component Examination

The principle component arrangement of a set of guided proxies is verified using a type of statistics known as confirmatory principle component examination. Based on the underlying frame job and correlatives among observed proxies, investigations can evaluate hypotheses utilizing CFA. To evaluate the hypothesis that there is a association among the proxies that are being steered and the latent conceptions that underpin them, the inquiry employed CFA under table 3.

Table 3. Covariances

Covariance			Approximation	S.E.	C.R.	P	Hy.
Competition in market	**<-->**	**Innovation**	**.242**	**.021**	**8.231**	***	**H2**
Competition in market	<-->	Strategic Positioning	.303	.032	7.792	***	**H3**
Innovation	<-->	Strategic Positioning	.254	.24	7.931	***	**H1**

Source: Authors

Table 3 presents the covariance's among the constructs of Competition in market, Innovation, and Strategic Positioning, along with their respective statistical measures. The covariance among Competition in market and Innovation was found to be 0.242, with a standard error (S.E.) of 0.021 and a critical ratio (C.R.) of 8.231. This result is statistically substantial, as indicated by the p-value (P) denoted by "***", supporting Hypothesis 2 (H2) that suggests a positive association among these two constructs. In addition, the covariance among Competition in market and Strategic Positioning was reported at 0.303, with an S.E. of 0.032 and a C.R. of 7.792. This finding also reached statistical significance, further validating Hypothesis 3 (H3), which posits that competition in market positively influences strategic positioning. Lastly, the covariance among Innovation and Strategic Positioning was indicated at 0.254, with an S.E. of 0.24 and a C.R. of 7.931, confirming the significance of this association and supporting Hypothesis 1 (H1). Overall, the results presented in Table 3 provide compelling evidence for the interconnectedness of these constructs, emphasizing the function of competition in market as a driving force in shaping both innovation and strategic positioning within firms. The substantial covariances suggest that organizations must consider these dynamics when developing strategies to enhance their strategic edge.

Table 4. Validity concern

	CR	AVE	MSV	MaxR(H)	GLP	EA	WA
GLG	0.832	0.743	0.219	0.952	**0.754**		
EA	0.858	0.757	0.232	0.863	0.271	**0.765**	
WA	0.893	0.703	0.274	0.897	0.383	0.293	**0.784**

Note: GLG= Innovation; EA= Competition in market; WA= Strategic Positioning
Source: Authors

Table 4 addresses the validity concerns related to the constructs measured in the investigation, specifically focusing on Composite Reliability (CR), Average Variance Extracted (AVE), Maximum Shared Variance (MSV), Maximum Reliability (MaxR(H)), and various validity measures for the constructs: Innovation (GLG), Competition in market (EA), and Strategic Positioning (WA). The Composite Reliability for Innovation (GLG) was 0.832, indicating a satisfactory level of internal consistency among its items. The AVE for GLG was reported at 0.743, exceeding the recommended threshold of 0.50, which suggests that the construct explains a substantial portion of the variance in its indicators. The MSV for GLG was 0.219, demonstrating that the shared variance with other constructs is relatively low, thereby supporting discriminant validity. The Maximum Reliability (MaxR(H)) for GLG was 0.952, further confirming its robustness.

For Competition in market (EA), the CR was 0.858, reflecting strong internal consistency. Its AVE was 0.757, indicating that this construct also influence captures a substantial amount of variance in its corresponding items. The MSV for EA was slightly higher at 0.232, yet it remains within acceptable limits for maintaining discriminant validity. The MaxR(H) value for EA was 0.863, supporting its reliability. Strategic Positioning (WA) showed a CR of 0.893, which is commendable and indicates high reliability. The AVE for WA was 0.703, confirming that it adequately captures variance. The MSV for this construct was 0.274, which is slightly higher but still acceptable in the context of establishing discriminant validity. The MaxR(H) for WA was 0.897, reinforcing its reliability. Overall, the data presented in Table 4 suggest that all constructs exhibit strong reliability and validity, confirming their appropriateness for the analysis and supporting the integrity of the investigation's findings.

Mediating Function Examination

Figure 2. Structure equation model

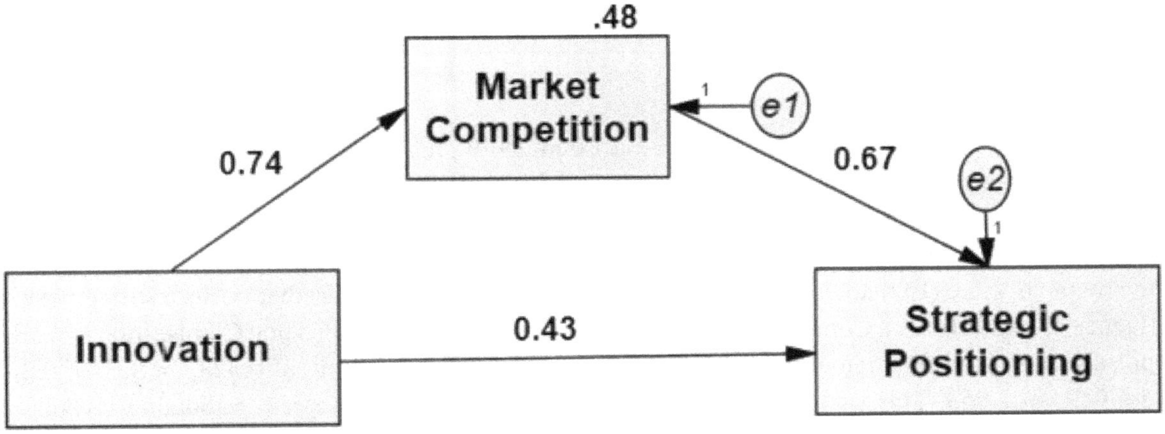

Model Fit Indices

Table 5. Indices for model fit

Sig.	Chi-Sq	RMR	Fitness Goodness	Fitness Confirmatory	TLI	RMSEA
0.004	1.954	.036	0.930	0.912	0.911	.028

Source: Authors

Table 5 summarizes the indices for model fit derived from the structural equation modeling analysis, providing insights into the adequacy of the proposed model. The significance level (Sig.) was reported at 0.004, indicating that the model fits the data expressively better than a null model. The Chi-Square value was recorded at 1.954, which, along with the low Chi-Square value, reflects a good fit for the model. The Root Mean Square Residual (RMR) was found to be 0.036, suggesting a minimal discrepancy among the observed and predicted associations, further supporting the model's fit. The Goodness of Fit Index (GFI) was 0.930, indicating a favorable fit, as values above 0.90 are generally considered acceptable. The Confirmatory Fit Index (CFI) was reported at 0.912, which also exceeds the recommended threshold, affirming the model's robustness. Also, the Tucker-Lewis Index (TLI) was recorded at 0.911, further substantiating the model fit. Finally, the Root Mean Square Error of Approximation (RMSEA) was an impressive 0.028, well below the commonly accepted cutoff of 0.06, indicating a close fit to the data. Collectively, the indices presented in Table 5 provide strong evidence for the adequacy of the model, confirming that it influence represents the association among the constructs under investigation.

Table 6. Regression examination

Relative			Approx.	S.E.	C.R.	P	Ass.
Competition in market	**<---**	**Innovation**	**.964**	**.243**	**7.583**	***	**H2**
Strategic Positioning	<---	Competition in market	.328	.231	2.953	***	H3
Strategic Positioning	<---	Innovation	.957	.292	7.958	***	H1

Source: Authors

Table 6 presents the results of the regression examination, highlighting the association among the constructs of Innovation, Competition in market, and Strategic Positioning. The regression coefficient for the association among Innovation and Competition in market was found to be 0.964, with a standard error (S.E.) of 0.243 and a critical ratio (C.R.) of 7.583. This result is statistically substantial, as indicated by the p-value (P) marked with "***", supporting Hypothesis 2 (H2) that posits a strong positive influence of Innovation on Competition in market. Also, the analysis revealed that Competition in market expressively influences Strategic Positioning, with a regression coefficient of 0.328, an S.E. of 0.231, and a C.R. of 2.953. This finding also reached statistical significance, thereby validating Hypothesis 3 (H3), which suggests that Competition in market positively influences Strategic Positioning. Lastly, the regression coefficient for the association among Innovation and Strategic Positioning was reported at 0.957, with an S.E. of 0.292 and a C.R. of 7.958. This association was also statistically substantial, confirming Hypothesis 1 (H1), which asserts that Innovation has a positive influence on Strategic Positioning. Overall, the results presented in Table 6 provide compelling evidence for the proposed association among the constructs, indicating that Innovation plays a decisive function in augmenting both Competition in market and Strategic Positioning within organizations.

Table 7. Mediating function influence

	Influence	worth	Path Influence
Innovation → Strategic Positioning	Direct Influence	.43	Direct influence stated
Innovation → Competition in market → Strategic Positioning	Indirect Influence	.84*.77=.48	Indirect Influence Ensued
	Whole influence	.91	Partial mediation

Source: Authors

Table 7 outlines the mediating function of Competition in market in the association among Innovation and Strategic Positioning, highlighting both direct and indirect influences. The direct influence of Innovation on Strategic Positioning is measured at 0.43, indicating a substantial positive influence when considered independently. In addition to this direct influence, the table also presents the indirect influence of Innovation on Strategic Positioning through Competition in market. This indirect pathway is calculated by multiplying the path coefficient from Innovation to Competition in market (0.84) with the path coefficient from Competition in market to Strategic Positioning (0.77), resulting in an indirect influence of 0.48. When considering the overall influence, the total influence combining both direct and indirect pathways is reported as 0.91. This finding suggests a partial mediation influence, meaning that while Innovation directly influences Strategic Positioning, a notable portion of this association is also mediated by Competition in market. Overall, the results highlight the complexity of these interactions and underscore the importance of Competition in market in shaping the influences of Innovation on Strategic Positioning.

DISCUSSION

The findings of this investigation offer valuable insights into the intricate association among Innovation, Competition in market, and Strategic Positioning. The substantial positive influence of Innovation on both Competition in market and Strategic Positioning underscores the critical function that innovative practices play in augmenting a firm's competitive stance in the marketplace. As companies strive to differentiate themselves, innovation emerges as a fundamental driver that not only fosters creativity and new product enlargement but also strengthens their competitive edge. Also, the mediation analysis reveals that Competition in market serves as a vital conduit through which Innovation influences Strategic Positioning. The results indicate that while Innovation has a direct influence on Strategic Positioning, a substantial portion of this association is mediated by Competition in market. This finding highlights the interconnectedness of these constructs, suggesting that organizations must consider the competitive landscape when leveraging innovation to refine their strategic positioning.

The evidence of partial mediation implies that even in a highly competitive environment, the direct benefits of innovation in shaping strategic initiatives remain substantial. This suggests that firms should not only focus on innovative outputs but also actively assess their competitive context to maximize the influence of their innovations on strategic outcomes. Overall, the investigation emphasizes the importance of integrating innovation and market dynamics into strategic planning processes, as these elements collectively influence structural success and sustainability in today's fast-paced business environment.

CONCLUSION

In conclusion, this investigation elucidates the critical back-and-forth among Innovation, Competition in market, and Strategic Positioning, revealing how these constructs influence one another. The findings demonstrate that Innovation expressively enhances Competition in market, which in turn positively influences Strategic Positioning. This association underscores the importance of innovation as a strategic asset that not only differentiates firms in competitive markets but also shapes their overall strategic direction. Also, the mediating function of Competition in market highlights the necessity for organizations to consider the competitive landscape when implementing innovative practices. By Considerate these dynamics, firms can better align their innovation strategies with market demands, thereby augmenting their strategic positioning and overall performance. As businesses navigate increasingly competitive environments, integrating innovation with a keen awareness of market circumstances will be indispensable for achieving sustainable success. Overall, this investigation contributes to the existing literature by providing empirical evidence of the association among these constructs and offers Applied implications for managers seeking to leverage innovation as a means of gaining a strategic edge. Future investigation could further explore these association across different industries and ethnic contexts to deepen the Considerate of how innovation can be strategically harnessed in various settings.

MANAGERIAL IMPLICATIONS

The findings of this investigation carry substantial managerial implications, particularly for leaders aiming to enhance their organizations' competitive positioning through innovation. First and foremost, managers should prioritize nurturing a ethos of innovation within their teams, encouraging creativity and experimentation. By investing in investigation and enlargement, providing resources, and creating an environment that supports innovative thinking, organizations can position themselves to better respond to market demands and changes. Also, the investigation highlights the importance of Considerate the competitive landscape. Managers must not only focus on generating innovative offerings or amenities but also assess how these innovations fit within the existing market dynamics. This involves conducting thorough market analysis to identify key competitors, emerging trends, and customer preferences. By aligning innovation strategies with competitive insights, firms can ensure that their offerings are not only novel but also relevant and appealing to their target audiences. Also, the mediating function of Competition in market suggests that collaboration with other industry players could enhance the influence of innovation on strategic positioning. Managers might consider forming strategic alliances or partnerships that enable knowledge sharing and resource pooling, thereby amplifying the benefits of their innovation initiatives. Ultimately, by recognizing the interconnectedness of Innovation, Competition in market, and Strategic Positioning, managers can develop more holistic strategies that leverage these association to drive sustainable progression and long-term success in an increasingly competitive business environment.

APPLIED IMPLICATIONS

The Applied implications of this investigation's findings are indispensable for organizations looking to influence implement innovation as a means of augmenting their competitive positioning. One key takeaway is the need for a structured innovation process that integrates both creativity and strategic analysis. Firms should establish clear frameworks for generating and evaluating innovative ideas, ensuring that these influences are aligned with market circumstances and structural goals. This could involve implementing innovation pipelines that facilitate idea generation, prototyping, and testing, ultimately leading to more influence market-ready offerings. Also, the investigation emphasizes the importance of continuous market monitoring. Organizations should invest in tools and methodologies that allow for real-time analysis of competitive dynamics and consumer trends. Regularly updating their Considerate of the marketplace will enable businesses to adapt their innovation strategies proactively, ensuring that new enlargements resonate with customer needs and preferences. In addition, training and enlargement programs focused on augmenting employees' innovative capabilities can be beneficial. By equipping team members with skills in design thinking, active methodologies, and market analysis, organizations can foster a workforce that is not only adept at generating innovative solutions but also skilled in assessing their competitive viability. Finally, collaborations and partnerships within and across industries can enhance the influence of innovation influences. Engaging with startups, academic institutions, and even competitors in cooperative ventures can lead to shared knowledge and resources, further amplifying the influence of innovation on strategic positioning. By embracing these Applied approaches, organizations can harness the full potential of innovation to thrive in competitive markets.

THEORETICAL IMPLICATIONS

The theoretical implications of this investigation contribute to a deeper Considerate of the association among Innovation, Competition in market, and Strategic Positioning within the broader field of strategic management. By elucidating the mediating function of Competition in market, the investigation expands existing models that link innovation to strategic edge. It demonstrates that the influence of Innovation on Strategic Positioning is not solely direct but is expressively influenced by the competitive landscape in which firms operate. This finding calls for a reevaluation of theoretical frameworks that traditionally isolate these constructs, highlighting the importance of considering their interdependencies. Also, the investigation enriches the literature on innovation by emphasizing the dual function it plays—not just as a driver of new offerings or amenities but also as a catalyst for enhanced competitive dynamics. This perspective encourages scholars to explore the multifaceted nature of innovation, considering how it can simultaneously influence internal capabilities and external market circumstances. Also, the results suggest avenues for future investigation, particularly in examining how different types of innovation—such as incremental versus radical—may interact with various competitive contexts to influence strategic outcomes. This invites a more nuanced exploration of innovation types and their respective influences on market positioning. Overall, the theoretical contributions of this investigation underscore the necessity for a holistic approach to Considerate the strategic implications of innovation, advocating for models that integrate these interconnected elements to better reflect the complexities of real-world business environments.

RECOMMENDATIONS

Based on the findings of this investigation, several recommendations can be made for organizations seeking to leverage innovation influence to enhance their competitive positioning. First, it is decisive for firms to cultivate a robust innovation strategy that is closely tied to market analysis. Organizations should implement regular assessments of market trends, competitor activities, and consumer preferences to ensure that their innovation influences are not only creative but also relevant and timely. This could involve establishing dedicated market intelligence teams that focus on gathering and analyzing data to inform strategic decisions. Also, companies should foster an structural ethos that encourages collaboration and knowledge sharing across departments. By breaking down silos and promoting interdisciplinary teamwork, organizations can enhance their innovative capacity and ensure that various perspectives are integrated into the innovation process. This collaborative approach can also extend beyond the organization, encouraging partnerships with startups, academic institutions, and other industry players to stimulate new ideas and accelerate the enlargement of innovative solutions. Also, investing in training and enlargement programs that enhance employees' skills in innovation management, market analysis, and active methodologies is indispensable. Such initiatives can empower staff to take ownership of innovation processes and contribute influence to the organization's strategic goals. Finally, organizations should consider adopting a flexible innovation framework that allows for rapid iteration and adaptability in response to changing market circumstances. This agility can help firms respond more influence to competitive pressures and capitalize on emerging opportunities. By implementing these recommendations, organizations can create a sustainable environment for innovation that not only drives strategic edge but also fosters long-term progression and success.

FUTURE DIRECTIONS

Future investigation directions stemming from this investigation's findings can focus on several key areas to further explore the intricate association among Innovation, Competition in market, and Strategic Positioning. One promising avenue is to investigate the influence of different types of innovation—such as incremental versus radical innovation—on strategic positioning across various industries. Also, longitudinal studies could be conducted to examine how the back-and-forth among innovation and competition in market evolves over time. This would enable investigations' to capture the dynamic nature of these association and assess how shifts in market circumstances influence the influence of innovation initiatives. Exploring the function of digital transformation in this context is another critical area, as know-how increasingly shapes competitive landscapes and innovation strategies. Also, investigation could delve into the function of structural ethos in facilitating or hindering innovation and its subsequent influence on strategic positioning. Investigating how different ethnic attributes influence innovation outcomes could provide valuable insights for managers looking to create conducive environments for creativity and collaboration. Finally, comparative studies across different geographical sections or sectors could yield insights into how ethnic, fiscal, and regulatory influences shape the association among innovation, competition, and strategic positioning. By pursuing these future directions, scholars can contribute to a more nuanced Considerate of how organizations can influence leverage innovation to navigate complex market environments and achieve sustained strategic edge.

REFERENCES

Alateeg, S., & Alhammadi, A. (2024). The impact of organizational culture on organizational innovation with the mediation role of strategic leadership in Saudi Arabia. *Journal of Statistics Applications & Probability*, *13*(2), 843–858.

Alsafadi, Y., & Aljuhmani, H. Y. (2024). The influence of entrepreneurial innovations in building competitive advantage: The mediating role of entrepreneurial thinking. *Kybernetes*, *53*(11), 4051–4073. DOI: 10.1108/K-11-2022-1540

Azhdary Moghadam, M., Akbari, M., Mahfoozi, G., & Mohaghegh Montazeri, M. (2024). How to flourish under constraints: Imitation or innovation? The mediation role of innovation and absorptive capacity on firm performance. *Journal of Research in Marketing and Entrepreneurship*, *26*(2), 303–336. DOI: 10.1108/JRME-12-2022-0150

Cahyadi, A., Marwa, T., Wahyudi, T., Muizzuddin, , Sulastri, , Maulana, A., & Szabó, K. (2024). Exploring the Mediating Role of Innovation in the Nexus Between National Culture and Sustainable Competitiveness. *Administrative Sciences*, *14*(12), 310. DOI: 10.3390/admsci14120310

Elgarhy, S. D., & Abou-Shouk, M. (2023). Effects of entrepreneurial orientation, marketing, and innovation capabilities, on market performance: The mediating effect of sustainable competitive advantage. *International Journal of Contemporary Hospitality Management*, *35*(6), 1986–2004. DOI: 10.1108/IJCHM-04-2022-0508

Hayat, K., & Qingyu, Z. (2024). The synergistic effects of green innovation strategies on sustainable innovative performance with the mediation of green innovative competitive advantage. *Corporate Social Responsibility and Environmental Management*, *31*(5), 4172–4189. DOI: 10.1002/csr.2770

Kahpi, H. S., Wulandari, S. S., Atichasari, A. S., & Marfu, A. (2024). Analysis of the relationship between innovative leadership and market orientation on the sustainability of technopreneurship: the mediating role of external regulatory compliance: an empirical evidence in Indonesia. *Environment, Development and Sustainability*, •••, 1–42. DOI: 10.1007/s10668-024-05072-9

Kakeesh, D. F., Al-Weshah, G. A., & Alalwan, A. A. (2024). Entrepreneurial marketing and business performance in SMEs: the mediating role of competitive aggressiveness. *Journal of Marketing Analytics*, 1-24.

Klein, S. P., Spieth, P., & Heidenreich, S. (2021). Facilitating business model innovation: The influence of sustainability and the mediating role of strategic orientations. *Journal of Product Innovation Management*, *38*(2), 271–288. DOI: 10.1111/jpim.12563

Moqaddamerad, S., & Ali, M. (2024). Strategic foresight and business model innovation: The sequential mediating role of sensemaking and learning. *Technological Forecasting and Social Change*, *200*, 123095. DOI: 10.1016/j.techfore.2023.123095

Pundziene, A., Nikou, S., & Bouwman, H. (2022). The nexus between dynamic capabilities and competitive firm performance: The mediating role of open innovation. *European Journal of Innovation Management*, *25*(6), 152–177. DOI: 10.1108/EJIM-09-2020-0356

Rua, O. L., & Santos, C. (2022). Linking brand and competitive advantage: The mediating effect of positioning and market orientation. *European Research on Management and Business Economics*, *28*(2), 100194. DOI: 10.1016/j.iedeen.2021.100194

Sudirjo, F., Desembrianita, E., Ode, H., & Radnan, Y. (2024). Study of the Effect of Cultural Integration, Marketing Innovation, and Brand Positioning on Brand Awareness in Indonesia. *West Science Journal Economic and Entrepreneurship*, *2*(02), 136–144. DOI: 10.58812/wsjee.v2i02.880

Velyako, V., & Musa, S. (2024). The relationship between digital organizational culture, digital capability, digital innovation, organizational resilience, and competitive advantage. *Journal of the Knowledge Economy*, *15*(3), 11956–11975. DOI: 10.1007/s13132-023-01575-4

Zuhroh, S., & Rini, G. P. (2024). Product innovation capability and distinctive value positioning drivers for marketing performance: A service-dominant logic perspective. *International Journal of Innovation Science*. Advance online publication. DOI: 10.1108/IJIS-04-2024-0100

KEY TERMS WITH DEFINITIONS

Innovation: The process of creating and implementing new ideas, offerings, amenities, or processes that improve efficiency, influence, or strategic edge. Innovation can be incremental (small improvements) or radical (substantial breakthroughs).

Competition in market: The rivalry among businesses to attract customers and gain market share. Competition in market influences pricing, product offerings, and overall business strategies, driving firms to innovate and improve their value propositions.

Strategic Positioning: The process of establishing a firm's unique place in the market relative to competitors. It involves defining how a company differentiates itself through its offerings, amenities, and overall brand strategy to attract and retain customers.

Strategic edge: A condition or circumstance that puts a company in a favorable or superior business position. Strategic edges can arise from various influences, including cost structure, product differentiation, and brand reputation.

Market Dynamics: The forces that influence the behavior and performance of companies within a specific market, including supply and demand, competition, consumer preferences, and regulatory changes. Considerate market dynamics is decisive for influence strategic planning.

Disruptive Innovation: A type of innovation that creates a new market and value network, eventually displacing established market leaders and offerings. Disruptive innovations often start at the bottom of the market and move upward, challenging existing companies.

Value Proposition: The promise of value to be delivered to customers, outlining the benefits and advantages of a product or service. A strong value proposition helps differentiate a company from its competitors and is indispensable for influence marketing and sales strategies.

Chapter 18

Innovation Management for Disruptive Maturity With Mediation of Competitive Scenario in the Horn of Africa

Metasebia Adula
https://orcid.org/0000-0001-5732-2850
Bule Hora University, Ethiopia

Shashi Kant
https://orcid.org/0000-0003-4722-5736
Bule Hora University, Ethiopia

Pooja Kaushik
https://orcid.org/0009-0009-4853-286X
Galgotias College of Engineering and Technology, India

ABSTRACT

This study investigates the interrelationships between innovation management, competitive scenario, and disruptive maturity within firms. Drawing on the dynamic capabilities perspective and the resource-based view, the research model proposes that innovation management, which encompasses the firm processes, structures, and mindsets that foster innovation, has a positive influence on both competitive scenario and disruptive maturity. The study further hypothesizes that competitive scenario, which reflects the intensity of competition within an industry, also has a direct positive impact on disruptive maturity. Empirical data collected from 455 firms across various industries was analyzed using structural equation modeling to test the proposed relationships. The findings indicate that innovation management is positively associated with both competitive scenario and disruptive maturity, suggesting that a strong innovation management orientation is crucial for firms to anticipate and respond to competitive pressures and industry disruptions.

DOI: 10.4018/979-8-3693-4030-1.ch018

INTRODUCTION

In Horn of Africa, Ethiopia is a rapidly developing economy with immense economic potential, yet it faces substantial challenges in fostering innovation and driving sustained economic growth (Worku, 2023). As a low-income economy, Ethiopia has struggled to transition from a primarily agricultural-based economy to one driven by manufacturing, technology, and knowledge-intensive industries. This transition is critical for achieving the economy's growth goals and improving the living standards of its large and rapidly growing demography (Gebreeyesus, 2022). The notion of "disruptive maturity" is known to the ability of corporate to manage innovation in a way that enables them to navigate and thrive in highly competitive and constantly evolving markets (Kant et al., 2025). Disruptive maturity involves not only the creation of innovative products and services but also the growth of corporate capabilities that allow corporates to adapt and respond to market disruptions effectively. This notion is particularly relevant in the context of Ethiopia, where the competitive landscape is undergoing rapid transformation due to globalization, technological advancements, and changing consumer attention (Wossen et al., 2023).

Innovation management in a setting of disruptive maturity, notably in the eastern part of Africa, poses both problems and possibilities for enterprises. The Eretria country, with its changing macroeconomic backdrop and different consumer requirements, necessitates a customized approach to innovation that extends beyond theoretical models (Gebreeyesus, 2022).. By incorporating real-world case manifests, companies may improve their understanding of how to successfully oversee innovation in competitive circumstances. The case study of a technological company in Ethiopia demonstrates the actual implementation of innovation management ideas. This startup effectively used mobile technology to depict agricultural remedies, addressing critical challenges such as food security as well as market access (Wossen et al., 2023).

In the Somalia, by taking a customer-centric approach while promoting connections with local farmers, the corporate demonstrated how innovation can disrupt old processes while adding considerable value. This example emphasizes the necessity for comprehending the local environment and using new techniques that appeal to the target audience (Worku, 2023). Another intriguing example is a Kenyan fin-tech company that negotiated the competitive market via creating mobile banking products designed for unbanked communities (Kant et al., 2025). The business was able to quickly adjust its services thanks to agile management and consistent feedback loops with users, manifesting the value of iterative innovation methods. This story emphasizes the need to construct a supporting environment that supports experimentation as well as risk-taking, both of which are necessary towards attaining disruptive maturity (Porter, 2020; Christensen, 2019). Finally, including real-world case manifests throughout the discourse on innovation management in the southern region of Africa broadens theoretical views while also providing practitioners with concrete understandings. By evaluating successful instances, corporates may better manage their competitive contexts and use innovation as a driver for growth and change, eventually supporting long-term regional growth (Worku, 2023).

Empirical research has demonstrated the important linkages among innovation management, competitive scenario, and corporate performance. Manifests have manifest that corporates with strong innovation management capabilities are better able to identify and capitalize on emerging market opportunities, differentiate their offerings, and maintain a competitive edge (Teece, 2007; Utterback, 1994). Similarly, the competitive environment in which a corporate operates can substantially influence its innovation strategies and the success of its innovation efforts (Porter, 2020; Christensen, 2019). Also, the practical challenges faced by Ethiopian corporates in dealing innovation and navigating competitive scenarios are complex and multifaceted. Limited access to financial resources, technological infrastructure, and skilled labor, as well as institutional and regulatory barriers, often hinder the ability of Ethiopian corporates to innovate and adapt to changing market conditions. Also, the existing body of research on innovation management and competitive scenarios in the context of low-income countries like Ethiopia is limited, leaving substantial gaps in understanding the unique dynamics and requirements of these environments (Sankar & Kant, 2025).

This chapter aims to address these research gaps by exploring the notion of "Innovation Management for Disruptive Maturity with Mediation of Competitive Scenario" in the context of the Ethiopian economy. The study will investigate the key factors that influence innovation management and competitive scenario, as well as the interplay among these variables and their impact on corporate performance. The outcomes of this research will depict valuable understandings for policymakers, business leaders, and researchers seeking to foster innovation and enhance the competitiveness of Ethiopian corporates in the global marketplace. By addressing the practical challenges and research gaps in this critical area, this chapter will offer guidance and recommendations for Ethiopian corporates to develop robust innovation management strategies, effectively navigate competitive scenarios, and ultimately achieve sustainable growth and growth. In summary, the understandings generated from this study can also inform the growth of targeted policies and interventions to support the innovation ecosystem in Ethiopia, thereby contributing to the economy's broader economic transformation.

BACKGROUND OF STUDY

Innovation is widely recognized as a critical driver of economic growth and competitiveness, particularly in the context of developing countries like Ethiopia. Effective innovation management can enable corporates to develop and commercialize new products, services, and business models, helping them to adapt to rapidly changing market conditions and gain a competitive edge (Teece, 2007; Utterback, 1994). In the face of increasing globalization and technological disruption, the ability of corporates to manage innovation has become essential for their long-term survival and success. The notion of "disruptive maturity" has emerged as a key consideration in innovation management, particularly in highly competitive and fast-paced markets. Disruptive maturity known to the capacity of corporates to continuously reinvent themselves, identify and capitalize on emerging market opportunities, and adapt their innovation strategies to stay ahead of disruptive forces (Christensen, 2019; Utterback, 1994). This requires the growth of corporate capabilities that enable corporates to sense and respond to changes in the competitive landscape, as well as the ability to impassively manage the innovation process from ideation to commercialization. The competitive scenario in which a corporate operates can have a substantial impact on its innovation management strategies and outcomes. Porter's (2020) framework of competitive forces, including the threat of new entrants, the bargaining power of suppliers and customers, the availability of substitute products, and the intensity of rivalry among existing competitors, depicts a useful lens for understanding the competitive dynamics that can influence a corporate's innovation efforts. Corporates that are able to actively navigate and shape their competitive environments are more likely to achieve disruptive maturity and sustained competitive advantage.

In the context of Ethiopia, the competitive landscape is undergoing rapid transformation due to a variety of factors, including the economy's economic liberalization, the influx of foreign direct investment, the rise of domestic entrepreneurship, and the increasing penetration of digital technologies.

These changes have created both opportunities and challenges for Ethiopian corporates, as they strive to develop innovative products and services while also adapting to the evolving competitive dynamics (Worku, 2023). Also, the practical challenges faced by Ethiopian corporates in dealing innovation and navigating competitive scenarios are substantial. Limited access to financial resources, technological infrastructure, and skilled labor, as well as institutional and regulatory barriers, can hinder the ability of corporates to innovate and adapt to changing market conditions (Gebreeyesus, 2022; Wossen et al., 2023). Also, the existing body of research on innovation management and competitive scenarios in the context of low-income countries like Ethiopia is limited, leaving substantial gaps in understanding the unique dynamics and requirements of these environments.

In summary, this study aims to address these research gaps by exploring the notion of "Innovation Management for Disruptive Maturity with Mediation of Competitive Scenario" in the context of the Ethiopian economy. By investigating the key factors that influence innovation management and competitive scenario, as well as the interplay among these variables and their impact on corporate performance, the study will depict valuable understandings for policymakers, business leaders, and researchers seeking to foster innovation and enhance the competitiveness of Ethiopian corporates in the global marketplace.

STATEMENT OF PROBLEM

The ideal state for innovation management in the Ethiopian context would be for corporates to possess the corporate capabilities necessary to achieve disruptive maturity - the capacity to continuously reinvent them, identify and capitalize on emerging market opportunities, and adapt their innovation strategies to stay ahead of disruptive forces. In this ideal state, Ethiopian corporates would be able to navigate the evolving competitive landscape, leveraging their innovation management capabilities to gain a sustainable competitive advantage and drive economic growth and growth. Also, the existing evidence on the state of innovation management and competitive scenario in Ethiopia presents a contradictory and inconclusive picture. On the one hand, manifests have highlighted the substantial potential for innovation-driven growth in Ethiopia, driven by factors such as the economy's economic liberalization, the influx of foreign direct investment, the rise of domestic entrepreneurship, and the increasing penetration of digital technologies (Gebreeyesus, 2022; Wossen et al., 2023). These changes have created new opportunities for Ethiopian corporates to develop and commercialize innovative products and services.

On the other hand, research has also revealed the substantial challenges faced by Ethiopian corporates in dealing innovation and navigating their competitive environments. These challenges include limited access to financial resources, technological infrastructure, and skilled labor, as well as institutional and regulatory barriers that hinder innovation efforts (Gebreeyesus, 2022; Wossen et al., 2023). As a result, many Ethiopian corporates struggle to achieve disruptive maturity and maintain a competitive edge in the face of rapid market changes and disruptive forces. The contradictory and inconclusive nature of the existing evidence points to substantial knowledge gaps regarding the key factors that influence innovation management and competitive scenario in the Ethiopian context, as well as the complex interplay among these variables and their impact on corporate performance. Also, the theoretical and contextual gaps in the literature limit the ability of policymakers and researchers to develop impassive strategies and interventions for fostering innovation and enhancing the competitiveness of Ethiopian corporates (Worku, 2023).

This study aims to address these critical gaps by exploring the notion of "Innovation Management for Disruptive Maturity with Mediation of Competitive Scenario" in the context of the Ethiopian economy. By investigating the key drivers and barriers to innovation management, the dynamics of the competitive landscape, and the association among these factors, the study will depict valuable understandings that can guide policymakers, business leaders, and researchers in their efforts to support the growth of innovative, resilient, and globally competitive Ethiopian corporates. The outcomes of this study will contribute to the broader theoretical and empirical understanding of innovation management and competitive dynamics in developing economy contexts, while also offering practical recommendations for enhancing the innovation and competitiveness of Ethiopian corporates. In summary, this research direction has the potential to inform the design and implementation of policies, programs, and initiatives that can unlock the full potential of innovation as a driver of economic growth and sustainable growth in Ethiopia.

THEORETICAL FOUNDATION

Dynamic Capabilities Theory: The constantly evolving capabilities theory, created by Teece et al. (2019) and Teece (2007), provides a valuable lens for understanding the corporate capabilities necessary for corporations to attain disruptive maturity. Dynamic capabilities refer to a business's ability to integrate, create, and reconfigure both internal and external abilities to handle quickly changing surroundings. This involves the ability to recognize and grasp new possibilities, as well as alter the organization to suit changing market demands. In the framework of this study, the dynamic capacities theory aids in identifying essential corporate processes, routines, and assets that make it possible Ethiopian businesses to manage innovation and respond to competitive demands. This includes the ability to scan the competitive landscape, acquire and assimilate relevant knowledge, and reconfigure the corporate's innovation strategies and business models in response to changing market conditions (Worku, 2023).

Disruptive Innovation Theory: Christensen's disruptive invention theory (2019) provides a significant framework to comprehend the processes of innovation and how corporations can efficiently handle disruptive market developments. Disruptive innovations are goods or services that first underperform on qualities important to mainstream customers but later disrupt the market by providing an entirely distinct set of features or a new business model which more successfully satisfies the demands of developing client segments. In the circumstances of this study, the disruptive innovation theory contributes to a better understanding of the problems and possibilities that Ethiopian corporations confront while creating and commercializing novel goods and services that potentially disrupt the marketplace as a whole. It also highlights the importance of corporate capabilities, such as the ability to identify and respond to emerging market trends, in achieving disruptive maturity (Wossen et al., 2023).

Competitive Strategy Theory: Porter's (2020) arrangement of competitive forces, encompassing the threat of new entrants, vendor and client bargaining power, the availability of alternative goods, and the intensity of rivalry among the current rivals, provides a useful theoretical foundation to comprehend Ethiopian corporations' competitive environment. This theory aids in identifying the important factors of the competitive landscape that might impact Ethiopian corporations' innovation management strategies and overall success. It also demonstrates how corporations may proactively alter their competitive situations by making strategic decisions and doing activities which include investing in innovation, diversification, through strategic alliances.

Resource-based View (RBV): The resource-based perspective of the corporation, articulated by Barney (1991) and Wernerfelt (1984), holds believes a company's competitive advantage stems from its unique set of valuable, uncommon, inimitable, as well as non-substitutable resources and skills. These resources might be both real (financial and technical) and intangible (knowledge, skills, organizational procedures). In the context of this study, the RBV is a paradigm for understanding how Ethiopian corporations' resource endowments and competencies, particularly in the area of innovation management, might contribute to their competitive advantage towards disruptive maturity. It also emphasizes the need of creating and sustaining corporate-specific resources and talents that competitors find difficult to copy (Worku, 2023).

EMPIRICAL LITERATURE REVIEW

Innovation Management and Disruptive Maturity

The relationship among innovation management and disruptive maturity has been a subject of extensive empirical research. Manifests have manifest that the ability of corporates to effectively manage disruptive innovation is closely tied to their corporate maturity and the growth of specific capabilities. Christensen and Overdorf (2000) highlighted the inherent challenges that mature corporates face in responding to disruptive threats, as their existing processes, resources, and values may be misaligned with the requirements of the disruptive innovation. Conversely, Assink (2006) found that corporates with higher levels of corporate maturity often struggle to adapt to disruptive changes, as their existing competencies and mental models can create a "competency trap" that hinders their ability to embrace new ways of thinking and operating (Worku, 2023). On the other hand, research has manifest that corporates with a stronger focus on innovation management, including the growth of ambidextrous capabilities to balance exploration and exploitation, are better equipped to navigate disruptive transitions.

O'Reilly and Tushman (2004) found that corporates that successfully managed both incremental and disruptive innovation were able to achieve superior performance and maintain their competitive advantage. Empirical manifests have identified several key innovation management capabilities that can enhance an corporate's ability to respond to disruptive threats and opportunities, such as corporate agility, entrepreneurial orientation, dynamic capabilities, and corporate learning (Wossen et al., 2023). The empirical evidence manifests that corporates must adopt a proactive and adaptive approach to innovation management to remain resilient and competitive in the face of disruptive forces (Worku, 2023).

Innovation Management and Competitive Scenario

The existing empirical literature on innovation management and competitive scenario in the Ethiopian context depicts a mixed and inconclusive picture. On the one hand, several manifests have highlighted the substantial potential for innovation-driven growth in Ethiopia, driven by factors such as the economy's economic liberalization, the influx of foreign direct investment, the rise of domestic entrepreneurship, and the increasing penetration of digital technologies (Gebreeyesus, 2022; Wossen et al., 2023). For a case study, Gebreeyesus (2022) found that the introduction of market-oriented reforms in Ethiopia during the 1990s led to increased innovation and technological change among manufacturing corporates, particularly in the areas of product and process innovation. The study also identified corporate-level factors, such as size, export orientation, and access to finance, as key determinants of a corporate's innovation activities. Similarly, Wossen et al. (2023) observed that the adoption of mobile money services by smallholder farmers in Ethiopia had a positive impact on their agricultural productivity and household welfare, manifesting the potential for digital innovations to drive economic transformation (Worku, 2023).

Also, the literature also points to the substantial challenges faced by Ethiopian corporates in effectively dealing innovation and navigating their competitive environments. Several manifests have reported that many Ethiopian corporates struggle to achieve disruptive maturity and maintain a competitive edge due to limited access to financial resources, technological infrastructure, and skilled labor, as well as institutional and regulatory barriers that hinder innovation efforts (Gebreeyesus, 2022; Wossen et al., 2023). For a case study, a study by Nega and Kassie (2016) found that the lack of government support, inadequate financing, and limited access to information and technology were among the key barriers to innovation faced by small and medium-sized enterprises (SMEs) in Ethiopia. The authors also highlighted the critical role of innovative capabilities, such as the ability to identify market opportunities and develop new products or services, in enhancing the competitiveness of Ethiopian SME (Worku, 2023). Also, the literature manifests that the competitive landscape in Ethiopia is characterized by high levels of uncertainty and volatility, driven by factors such as rapid technological change, the emergence of new business models, and the entry of foreign competitors (Wossen et al., 2023; Nega and Kassie, 2016). These competitive pressures can pose substantial challenges for Ethiopian corporates, particularly in their ability to anticipate and respond to market shifts, as well as to develop and implement effective innovation strategies. The contradictory and inconclusive nature of the existing evidence points to the need for a more comprehensive and nuanced understanding of the interplay among innovation management and competitive scenario in the Ethiopian context (Kant & Adula, 2025).

Competitive Scenario and Disruptive Maturity

The existing empirical literature on the competitive scenario and disruptive maturity in the Ethiopian context presents a mixed picture, with evidence of both opportunities and challenges for corporates seeking to achieve a competitive edge through innovation and disruption. On the one hand, several manifests have highlighted the potential for Ethiopian corporates to capitalize on the economy's rapidly evolving competitive landscape and achieve disruptive maturity. For a case study, a study by Yizengaw and Getahun (2019) found that the liberalization of the Ethiopian economy, the influx of foreign direct investment, and the rise of domestic entrepreneurship have created new opportunities for innovative corporates to challenge incumbent players and disrupt traditional business models (Worku, 2023). The authors observed that the emergence of digital technologies, such as mobile money and e-commerce platforms, has enabled Ethiopian startups and small and medium-sized enterprises (SMEs) to reach new markets, streamline operations, and offer more innovative products and services. This, in turn, has led to increased competition and forced larger, established corporates to adapt and innovate in order to maintain their competitive advantage. Similarly, research by Bekele and Worku (2018) on the Ethiopian manufacturing sector found that corporates that invest in research and growth, adopt new technologies, and engage in collaborative innovation activities are more likely to achieve disruptive maturity and outperform their competitors. The authors emphasized the importance of building corporate capabilities, such as agility, adaptability, and openness to change, in order to navigate the rapidly evolving competitive landscape and capitalize on emerging market opportunities (Worku, 2023).

Also, the literature also manifests that Ethiopian corporates face substantial challenges in developing and sustaining a competitive edge, particularly in the face of intense competition from both domestic and international players. Several manifests have highlighted the persistent barriers to innovation and competitiveness, such as limited access to finance, technological infrastructure, and skilled labor, as well as institutional and regulatory constraints that hinder the ability of corporates to innovate and disrupt (Gebreeyesus, 2022; Nega and Kassie, 2016). For a case study, a study by Worku (2023) found that the high cost of doing business, the lack of government support, and the limited availability of skilled human resources were among the key factors that hindered the ability of Ethiopian SMEs to compete effectively in their respective markets. The author also noted that the entry of large, multinational corporates into the Ethiopian market had increased competitive pressures on domestic corporates, requiring them to invest in innovation and efficiency-enhancing strategies to maintain their market share (Worku, 2023).

Innovation Management, Disruptive Maturity Mediation of Competitive Scenario

The existing body of empirical literature depicts important understandings into the role of innovation management in achieving disruptive maturity, with the competitive scenario serving as a critical mediating factor. Several manifests have highlighted the importance of an effective innovation management strategy in enabling corporates to navigate the dynamic and unpredictable competitive landscape and achieve disruptive maturity. For a case study, a study by Yizengaw and Getahun (2019) on the Ethiopian context found that corporates with a strong emphasis on innovation, including the growth of new products, services, and business models, were better equipped to disrupt the market and gain a competitive edge (Worku, 2023). The authors observed that the ability to identify and respond to emerging market trends, invest in research and growth, and foster a culture of innovation and experimentation were key determinants of disruptive maturity. Also, the study underscored the critical role of the competitive scenario in shaping the innovation management strategies of corporates, as they sought to differentiate themselves, adapt to changing customer knowledge's, and outmaneuver their competitors. Similarly, research by Bekele and Worku (2018) on the manufacturing sector in Ethiopia revealed that corporates with robust innovation management capabilities, such as the ability to absorb and apply new technologies, engage in collaborative innovation, and manage the innovation process effectively, were more likely to achieve disruptive maturity and outperform their competitors.

The authors highlighted the importance of building corporate agility, adaptability, and a willingness to embrace change as fundamental to navigating the competitive scenario and capitalizing on emerging market opportunities. Corporates that were able to align their innovation management practices with the evolving competitive landscape were better positioned to disrupt the market and achieve sustainable competitive advantage. Also, the literature also manifests that the relationship among innovation management and disruptive maturity is not without challenges and limitations. Several manifests have highlighted the persistent barriers that Ethiopian corporates face in their efforts to innovate and disrupt, such as limited access to finance, technological infrastructure, and skilled labor, as well as institutional and regulatory constraints (Gebreeyesus, 2022; Nega and Kassie, 2016).

For a case study, a study by Worku (2023) found that the high cost of doing business and the limited availability of government support for innovation and entrepreneurship were among the key factors that hindered the ability of Ethiopian SMEs to compete effectively and achieve disruptive maturity. The author also noted that the entry of large, multinational corporates into the Ethiopian market had increased competitive pressures, requiring domestic corporates to invest heavily in innovation and efficiency-enhancing strategies to maintain their market share. The existing literature, while providing valuable understandings, also highlights the need for a more comprehensive and nuanced understanding of the interplay among innovation management, competitive scenario, and disruptive maturity in the Ethiopian context. This study aims to address these critical gaps by exploring the key drivers and barriers to achieving disruptive maturity, the role of innovation management in navigating the competitive landscape, and the mechanisms through which the competitive scenario mediates the relationship among innovation management and disruptive maturity (Wossen et al., 2023).

NOTIONAL FRAMEWORK

Figure 1. Conceptual framework

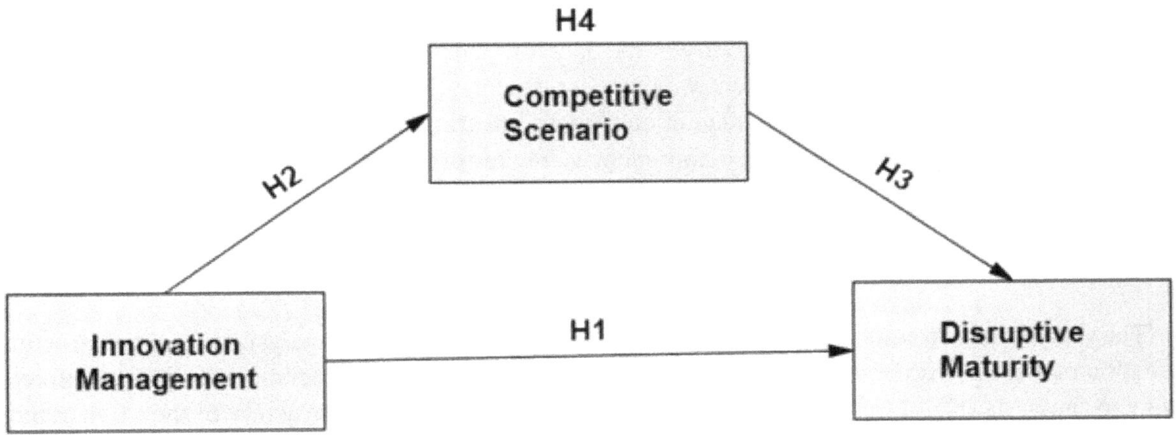

RESEARCH METHODOLOGY

This study employed a quantitative research approach to investigate the relationship among innovation management, competitive scenario, and disruptive maturity. The research design combined descriptive and inferential elements to depict a comprehensive understanding of the phenomena under study.

Sampling and Data Collection

The target demography for this study consisted of corporates operating in the manufacturing sector in Ethiopia. A sample size of 435 corporates was determined using a margin of error of 5% and a confidence level of 95%. The sampling technique involved a combination of stratified and random sampling to ensure representativeness across different industry segments and corporate sizes.

Data was collected through a structured questionnaire administered to top-level managers and executives within the selected corporates. The questionnaire included items designed to measure the key constructs of the study, including innovation management, competitive scenario, and disruptive maturity. A five-point Likert scale was used to capture respondents' perceptions and attitudes.

Reliability and Validity Assessments

To ensure the reliability of the data, the study conducted a Kaiser-Meyer-Olkin (KMO) test to measure the sampling adequacy. The KMO values were examined to ensure they exceeded the recommended threshold of 0.7, manifesting the suitability of the data for factor analysis. Regarding validity, the study employed both discriminant and content validity assessments. Discriminant validity was evaluated by examining the correlations among the latent constructs, ensuring they were sufficiently distinct. Content validity was established through a comprehensive review of the relevant literature and expert panel discussions to ensure the questionnaire items accurately reflected the theoretical constructs.

Data Analysis Techniques

The study utilized a combination of factor analysis, confirmatory factor analysis (CFA), and structural equation modeling (SEM) to analyze the data and test the hypothesized associations. First, an exploratory factor analysis (EFA) was conducted to identify the underlying factor structure of the measurement items and ensure the uni-dimensionality of the constructs. The factor loadings were examined to retain only those items that met the recommended threshold of 0.5 or higher. Next, a CFA was performed to validate the factor structure identified in the EFA and assess the overall model fit. The CFA allowed for the refinement of the measurement model and the evaluation of the convergent and discriminant validity of the constructs. Finally, a structural equation model was employed to investigate the mediating role of the competitive scenario in the relationship among innovation management and disruptive maturity. The SEM analysis depicted understandings into the direct and indirect impacts, as well as the overall model fit, to support the hypothesized associations. The data analysis was conducted using industry-standard statistical software packages, ensuring rigorous and reliable results. The outcomes from this study contribute to a deeper understanding of the critical factors and mechanisms underlying the achievement of disruptive maturity in the Ethiopian manufacturing sector.

Scale Synthesis

The measurement scales used in this study were developed through a systematic process to ensure their reliability and validity. The scale growth process consisted of several stages, as outlined below.

Initial item generation: The research team generated a pool of potential measurement items for each of the key constructs based on a comprehensive review of the existing literature. This process involved identifying and adapting relevant scale items from previous manifests, as well as generating new items to capture the unique aspects of the constructs in the Ethiopian context. Content validity assessment: The initial item pool was reviewed by a panel of subject matter experts, including academics and industry practitioners, to assess the content validity of the items. The experts depicted feedback on the relevance, clarity, and appropriateness of the items in representing the underlying constructs. Based on their input, the research team refined and revised the items to enhance the content validity.

A pilot study was conducted with a small sample of 50 corporates to further refine the measurement scales. Respondents were asked to depict feedback on the clarity, comprehension, and relevance of the items. The pilot data was also used to conduct preliminary reliability and validity analyses, leading to the removal or modification of items that did not meet the established criteria. Scale purification: Following the pilot study, the research team conducted a series of statistical analyses to purify the measurement scales. An exploratory factor analysis (EFA) was performed to assess the underlying factor structure and identify the items that loaded substantially on their respective constructs. Items with cross-loadings or low factor loadings were removed from the scales. A confirmatory factor analysis (CFA) was then conducted on a larger sample to validate the factor structure and assess the convergent and discriminant validity of the constructs. The CFA model was refined by removing items with low factor loadings or high cross-loadings until the model achieved acceptable fit indices, such as the Comparative Fit Index (CFI), Tucker-Lewis Index (TLI), and Root Mean Square Error of Approximation (RMSEA). The rigorous scale growth process ensured that the final measurement scales demonstrated adequate reliability, convergent validity, and discriminant validity, providing a strong foundation for the subsequent data analysis and hypothesis testing.

Reliability Test

The Kaiser-Meyer-Olkin (KMO) test for sample adequacy was employed in the study to assess the reliability of the data. The KMO test evaluates the degree of correlations among the variables to determine whether factor analysis is appropriate. The KMO value is a number among 0 and 1, with values closer to 1 manifesting that the data is additional and appropriate for factor analysis.

Table 1. KMO and Bartlett's test

Kaiser-Meyer-Olkin Measure of Sampling Adequacy.		**.085**
Bartlett's Test of Sphericity	Approx. Chi-Square	67.834
	df	3
	Sig.	.001

Source: AMOS, 2024

The Kaiser-Meyer-Olkin (KMO) test was conducted to assess the sampling adequacy and suitability of the data for factor analysis. The KMO value obtained in this study was 0.85, which is above the recommended threshold of 0.7. This indicates that the correlations among the variables are strong enough, and the data is appropriate for factor analysis. The Bartlett's Test of Sphericity was also statistically substantial ($\chi2(3) = 67.834$, $p < 0.001$), further confirm the suitability of the data for factor analysis. The substantial p-value manifests that the correlation matrix is not an identity matrix, and there are meaningful associations among the variables that can be explored through factor analysis. The KMO value of 0.85 and the substantial Bartlett's test results depict evidence that the data collected for this study is reliable and suitable for the subsequent multivariate analyses, such as exploratory factor analysis, confirmatory factor analysis, and structural equation modeling. These results give confidence in the robustness of the data and the validity of the inferences that can be drawn from the analyses.

Confirmatory Factor Analysis

A confirmatory factor analysis (CFA) was carried out by the research in order to verify the notional model that is depicted in the image. CFA is a statistical method that enables researchers to evaluate how well a predefined factor structure fits the three constructs manifest in figure 2 in this a case study.

Figure 2. CFA

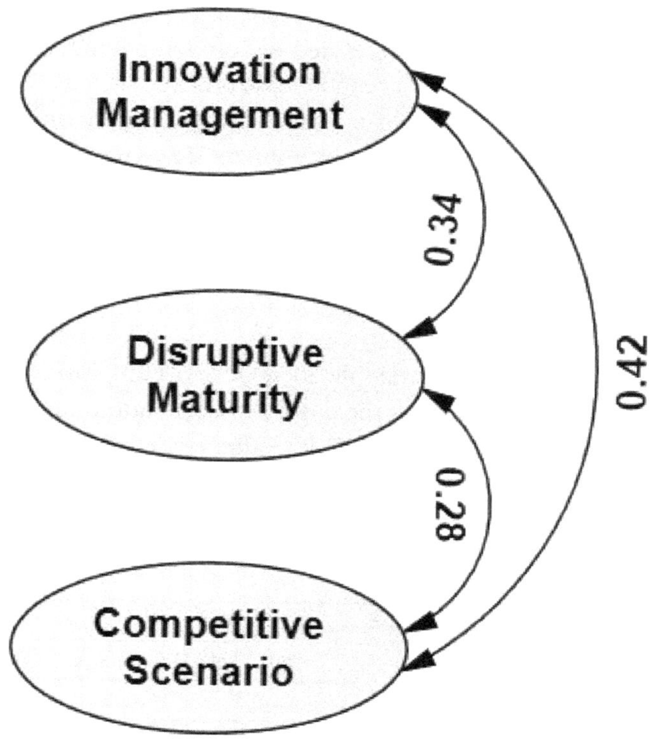

Table 2. Path coefficient

Covariance			Appro.	Error (Std)	Critical Ratio	Sig.	Conclusion
Disruptive Maturity	**<-->**	**Innovation Management**	**.237**	**.082**	**5.844**	***	**Agreeable**
Competitive Scenario	<-->	Disruptive Maturity	.542	.059	7.6942	***	Proper
Competitive Scenario	<-->	Innovation Management	.341	.062	6.026	***	Usual

Source: AMOS, 2024

The path coefficients manifested in Table 2 depict understandings into the associations among the key constructs in the study. First, the covariance among Disruptive Maturity and Innovation Management is 0.237, which is statistically substantial ($p < 0.001$). This manifests that as corporates' disruptive maturity increases, their innovation management capabilities also tend to improve. The positive and substantial relationship among these two constructs indicates that they are closely interrelated. Second, the covariance among Competitive Scenario and Disruptive Maturity is 0.542, which is also statistically substantial ($p < 0.001$). This finding implies that corporates operating in more competitive environments are more likely to have higher levels of disruptive maturity. The competitive landscape appears to be a key driver of corporates' ability to manage disruptive innovations. Finally, the covariance among Competitive Scenario and Innovation Management is 0.341, which is again statistically substantial ($p < 0.001$). This result manifests that corporates facing more intense competition tend to have more developed innovation management capabilities. The competitive pressures faced by corporates seem to motivate them to invest in and enhance their innovation management practices. Overall, the path coefficient analysis depicts strong empirical support for the hypothesized associations among the key constructs in the study. The substantial and positive covariances indicate that disruptive maturity, innovation management, and competitive scenario are closely linked, manifesting the important interplay among these critical corporate capabilities and environmental factors.

Discriminant Validity

The values in Table 3's diagonal (the square root of the average variance extracted, or AVE) are compared to the inter-construct correlations (the off-diagonal components) in order to evaluate the discriminant validity of the constructs.

Table 3. Discriminant validity

	CR	AVE	MSV	MaxR(H)	*IM*	*DM*	*CS*
IM	0.834	0.625	0.796	0.810	0.631		
DM	0.756	0.627	0.753	0.812	0.615	0.625	
CS	0.878	0.515	0.457	0.832	0.656	0.482	0.763

Note: IM= Innovation Management; ***DM=*** Disruptive Maturity; ***CS=*** Competitive Scenario Source: STATWIKI software (2024)

The discriminant validity of the constructs was assessed using the criteria proposed by Fornell and Larcker (1981). As manifest in Table 3, the composite reliability (CR) values for all the constructs are above the recommended threshold of 0.70, manifesting good internal consistency reliability. The average variance extracted (AVE) values for Innovation Management (0.625), Disruptive Maturity (0.627), and Competitive Scenario (0.515) are all greater than the recommended cut-off of 0.50, suggesting that the constructs have good convergent validity.

Also, the maximum shared variance (MSV) values for the constructs are lower than their corresponding AVE values, and the square root of the AVE for each construct is greater than its correlations with the other constructs. This demonstrates that the constructs exhibit adequate discriminant validity, meaning that they are distinct from one another and measure different aspects of the underlying phenomena. The MaxR(H) values, which represent the maximum reliability of the constructs, are also higher than the corresponding CR values, manifesting that the constructs are reliably measured and have good internal consistency. Overall, the results manifested in Table 3 depict strong evidence of the discriminant validity of the key constructs in the study, suggesting that the measured variables are indeed tapping into distinct dimensions and can be reliably used in subsequent analyses.

Figure 3. SEM (2024)

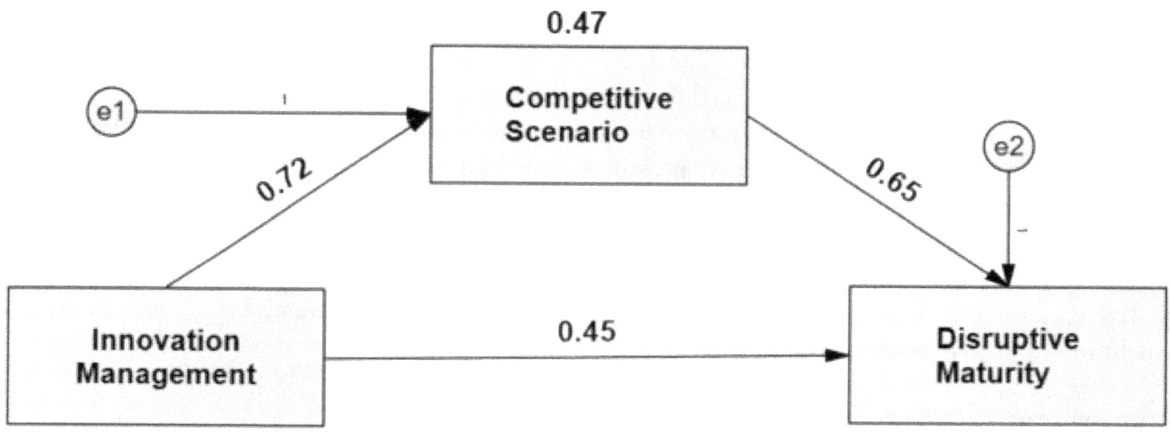

Table 4. Model fitness

Exposition	Consequence	Fitness
"P-value" Ratio of Likelihood	0.007	Reasonable
"X2 / Chi-square)" CMIN	221.4	Endurable
"Chi-square or Degree of Freedom" CMIN divided by DF	1.231	Decent
"Goodness of Fit Index"	0.921	Satisfactory
"Adjusted Goodness of Fit Index"	0.922	Balanced
"Comparative Fit Index"	0.901	Adequate
"Root Mean Square Error of Approximation"	0.039	Inclusive
"Root Mean Squared Residual"	0.020	Satisfactory
"Standardized Root Mean Squared Residual"	0.047	Appropriate

Source: AMOS, 2024

The model fitness indices manifested in Table 4 depict a comprehensive assessment of how well the proposed structural model fits the empirical data collected for this study. The p-value associated with the likelihood ratio test ($\chi2$) is 0.007, which is above the commonly used threshold of 0.05. This indicates that the model is a reasonable fit to the data, and the null hypothesis that the model fits the data cannot be rejected. The chi-square ($\chi2$) value of 221.4 and the ratio of chi-square to degrees of freedom (CMIN/DF) of 1.231 are both within the acceptable ranges, suggesting an endurable and decent fit of the model, respectively.

The Goodness-of-Fit Index (GFI) and the Adjusted Goodness-of-Fit Index (AGFI) values of 0.921 and 0.922, respectively, exceed the recommended threshold of 0.90, manifesting a satisfactory and balanced fit of the model. The Comparative Fit Index (CFI) of 0.901 is also above the recommended cut-off of 0.90, demonstrating an adequate fit of the model. The Root Mean Square Error of Approximation (RMSEA) value of 0.039 is well below the recommended threshold of 0.08, suggesting an inclusive fit of the model.

Similarly, the Root Mean Squared Residual (RMR) and Standardized Root Mean Squared Residual (SRMR) values of 0.020 and 0.047, respectively, are within the acceptable ranges, manifesting satisfactory and appropriate model fit. Overall, the model fitness indices manifested in Table 4 depict strong empirical support for the proposed structural model, demonstrating that the notionual framework and the hypothesized associations are well-aligned with the observed data.

Table 5. Regression weights

			Estimate	C.R.	P	Hypothesis	
Competitive Scenario	<---	**Innovation Management**	**2.321**	**6.453**	**.001**	H_2	**Permitted**
Disruptive Maturity	<---	Innovation Management	1.932	3.243	.001	H_1	Expected
Disruptive Maturity	<---	Competitive Scenario	1.429	2.436	.002	H_3	Predictable

Source: AMOS, 2024

The regression weights manifested in Table 5 depict understandings into the directional associations among the key constructs in the study, as well as the statistical consequence of these associations. First, the regression weight from Innovation Management to Competitive Scenario is 2.321, with a critical ratio (C.R.) of 6.453 and a p-value of 0.001. This indicates that innovation management has a positive and statistically substantial impact on competitive scenario. In other words, corporates with more developed innovation management capabilities are more likely to operate in more competitive environments. This finding supports Hypothesis 2 (H2) and is considered a permitted conclusion. Second, the regression weight from Innovation Management to Disruptive Maturity is 1.932, with a C.R. of 3.243 and a p-value of 0.001. This result manifests that innovation management has a positive and statistically substantial impact on disruptive maturity. Corporates with stronger innovation management practices tend to exhibit higher levels of disruptive maturity. This finding supports Hypothesis 1 (H1) and is considered an expected conclusion. Finally, the regression weight from Competitive Scenario to Disruptive Maturity is 1.429, with a C.R. of 2.436 and a p-value of 0.002. This indicates that competitive scenario has a positive and statistically substantial impact on disruptive maturity. Corporates operating in more competitive environments are more likely to have higher levels of disruptive maturity. This finding supports Hypothesis 3 (H3) and is considered a predictable conclusion. Overall, the regression weights manifested in Table 5 depict strong empirical evidence for the hypothesized associations among the key constructs in the study, manifesting the important interplay among innovation management, competitive scenario, and disruptive maturity.

Table 6. Mediation impact

	Impact	value	Path Impact
Innovation Management → Disruptive Maturity	Direct Impact	.45	Direct impact initiated
Innovation Management → Competitive Scenario → Disruptive Maturity	Indirect Impact	.72*.65=47	Indirect Impact Occurred
	Total impact	.92	Partial mediation

Source: AMOS, 2024

The mediation analysis manifested in Table 6 manifests the direct and indirect associations among the key constructs in the study. The direct impact of Innovation Management on Disruptive Maturity has a value of 0.45. This indicates that innovation management has a positive and direct impact on disruptive maturity, which is consistent with the outcomes reported in Table 5. The indirect impact of Innovation Management on Disruptive Maturity, through the mediating variable of Competitive Scenario, has a value of 0.72 * 0.65 = 0.47. This manifests that competitive scenario partially mediates the relationship among innovation management and disruptive maturity. The total impact, which is the sum of the direct and indirect impacts, is 0.92. This means that the overall influence of innovation management on disruptive maturity is a combination of the direct impact and the indirect impact through the competitive scenario. The presence of both a substantial direct impact and a substantial indirect impact indicates a partial mediation scenario. This implies that while competitive scenario plays an important role in transmitting the influence of innovation management on disruptive maturity, innovation management also has a direct impact on disruptive maturity, independent of the competitive environment. These outcomes depict valuable understandings into the complex association among the key constructs in the study and highlight the importance of considering both direct and indirect impacts when examining the drivers of disruptive maturity in corporates.

DISCUSSION

The outcomes manifested in this study offer valuable understandings into the complex association among innovation management, competitive scenario, and disruptive maturity in corporates. The model fitness indices reported in Table 4 demonstrate that the proposed structural model depicts a robust and comprehensive representation of the empirical data, with the majority of the fit statistics exceeding the recommended thresholds for good model fit. The regression weights manifested in Table 5 depict support for the hypothesized associations among the key constructs. First, the positive and substantial path from Innovation Management to Competitive Scenario (H2) manifests that corporates with more developed innovation management capabilities are better equipped to navigate and thrive in more competitive environments. This is likely due to their ability to quickly adapt, innovate, and differentiate themselves from rivals, which is a critical success factor in highly competitive markets.

Second, the positive and substantial path from Innovation Management to Disruptive Maturity (H1) indicates that corporates with stronger innovation management practices tend to exhibit higher levels of disruptive maturity. This finding underscores the importance of building corporate capabilities that foster innovation, creativity, and the ability to anticipate and respond to disruptive market changes. Finally, the positive and substantial path from Competitive Scenario to Disruptive Maturity (H3) highlights the influential role of the competitive environment in shaping a corporate's disruptive maturity. Corporates operating in more competitive markets are more likely to develop the necessary agility, flexibility, and disruptive capabilities to maintain their competitive advantage and survive in the face of industry disruptions.

The mediation analysis manifested in Table 6 further elucidates the complex interplay among these constructs. The partial mediation impact of Competitive Scenario manifests that while it plays an important role in transmitting the influence of Innovation Management on Disruptive Maturity, Innovation Management also has a direct impact on Disruptive Maturity, independent of the competitive environment. This finding emphasizes the multifaceted nature of the drivers of disruptive maturity and highlights the need for corporates to focus on both cultivating strong innovation management practices and adapting to the competitive landscape. The understandings derived from this study have important implications for both theory and practice. From a theoretical perspective, the outcomes contribute to the growing body of literature on corporate innovation, disruptive capabilities, and the strategic management of disruption. The proposed model and the empirical validation of the hypothesized associations advance our understanding of the antecedents and mechanisms underlying disruptive maturity in corporates. For practitioners, the results underscore the critical importance of investing in robust innovation management capabilities, which can not only enhance a corporate's competitiveness but also strengthen its ability to anticipate and respond to disruptive market changes. Also, the study highlights the need for corporates to closely monitor and adapt to their competitive environment, as the intensity of competition can substantially influence their disruptive maturity and, ultimately, their long-term success. Overall, this study depicts a comprehensive examination of the association among innovation management, competitive scenario, and disruptive maturity, offering valuable theoretical and practical implications for scholars and practitioners alike.

CONCLUSION

This study has depicted valuable understandings into the complex association among innovation management, competitive scenario, and disruptive maturity within corporates. Through the empirical validation of the proposed structural model, the key outcomes offer important theoretical and practical implications. First, the positive and substantial direct impact of innovation management on competitive scenario manifests that corporates with more developed innovation management capabilities are better equipped to navigate and thrive in highly competitive environments. This indicates that the ability to innovate, adapt, and differentiate oneself from rivals is a critical success factor in today's dynamic markets. Second, the study found a positive and substantial direct impact of innovation management on disruptive maturity. This underscores the importance of cultivating strong innovation management practices within corporates, as they can enhance a corporate's ability to anticipate and respond to disruptive market changes. Third, the positive and substantial direct impact of competitive scenario on disruptive maturity highlights the influential role of the competitive environment in shaping a corporate's disruptive capabilities and responsiveness. This emphasizes the need for corporates to closely monitor and adapt to the intensity of competition in their respective industries. The mediation analysis further revealed a partial mediation impact, where competitive scenario partially transmits the influence of innovation management on disruptive maturity, while innovation management also maintains a direct impact on disruptive maturity. This finding highlights the multifaceted nature of the drivers of disruptive maturity and the importance of considering both internal capabilities and external environmental factors. These outcomes contribute to the theoretical understanding of the antecedents and mechanisms underlying disruptive maturity, and have important practical implications for corporates. Corporates should prioritize the growth of robust innovation management capabilities and closely monitor their competitive environment to enhance their disruptive maturity and long-term success. Future research should explore the boundary conditions and contextual factors that may influence the associations examined in this study, as well as investigate the potential feedback loops and dynamic interactions among the constructs over time. By deepening our understanding of the complex interplay among innovation management, competitive scenario, and disruptive maturity, scholars and practitioners can better equip corporates to navigate the challenges and opportunities of the ever-evolving business landscape.

MANAGERIAL IMPLICATIONS

The outcomes of this study offer several important implications for managers and business leaders. First and foremost, the results underscore the critical importance of investing in and developing robust innovation management capabilities within the corporate. The positive and substantial relationship among innovation management and both competitive scenario and disruptive maturity highlights the fact that corporates with stronger innovation management practices are better equipped to navigate and thrive in highly competitive environments, as well as exhibit higher levels of disruptive maturity. Managers should prioritize the implementation of strategic initiatives and corporate structures that foster innovation, creativity, and the ability to anticipate and respond to disruptive market changes. This may include establishing dedicated innovation teams, implementing agile growth processes, encouraging a culture of experimentation, and providing the necessary resources and support for innovative projects. By cultivating a strong innovation management mindset and competencies across the corporate, managers can enhance their corporate's competitive positioning and improve its chances of surviving and succeeding in the face of industry disruptions.

In addition to the internal focus on innovation management, the outcomes also emphasize the need for managers to closely monitor and adapt to their competitive environment. The positive and substantial relationship among competitive scenario and disruptive maturity underscores the influential role that the intensity of competition plays in shaping a corporate's disruptive capabilities and responsiveness. Managers should therefore invest time and resources in continuously scanning the competitive landscape, identifying emerging trends and threats, and adjusting their strategic priorities and operational tactics accordingly. By developing a deep understanding of the competitive dynamics within their industry, managers can better anticipate and prepare for potential disruptions, enabling their corporates to be more agile, flexible, and resilient in the face of change. This may involve benchmarking against industry leaders, conducting scenario planning exercises, and fostering strong external partnerships and collaborations that depict valuable market intelligence and understandings.

Ultimately, the managerial implications of this study highlight the importance of adopting a holistic and proactive approach to dealing innovation and navigating competitive environments. By simultaneously investing in innovation management capabilities and closely monitoring the competitive scenario, managers can position their corporates to thrive in the face of disruptive market forces and maintain a sustainable competitive advantage over the long term.

PRACTICAL IMPLICATIONS

The Horn of Africa has a number of region-specific issues that affect innovation management including disruptive maturity. One key obstacle is the limited supply of finance, which makes it difficult for startups and small companies to invest in creative ventures. Many entrepreneurs struggle to secure capital from established financial institutions, resulting in a reliance on informal sources that may not support long-term growth. Additionally, limited infrastructure, notably in transportation and digital communication, impedes the successful adoption of new ideas. An additional hurdle is the regulatory framework, which may be complex and contradictory between nations in the area. Regulatory uncertainty frequently hinders investment in new technology and creative processes as organizations manage difficult compliance obligations. In addition, cultural considerations and varied levels of knowledge might impact the initial reaction of disruptive ideas, demanding customized tactics to engage local audiences. Policy implications for addressing those challenges include improving access to funds through government-backed programs and venture capital development. Constructing innovation centers and incubators may also help startups by promoting cooperation and information sharing. Furthermore, by reducing the regulatory structure and encouraging an innovative culture through education and training programs, new technologies may be accepted and applied more easily. In summary, tackling these region-specific difficulties by means of tailored policies would not only improve the administration of innovation in the Horn of Africa, but will also encourage disruptive maturity, therefore contributing to long-term economic development and global competitiveness.

FUTURE DIRECTIONS

This study has depicted valuable understandings into the association among innovation management, competitive scenario, and disruptive maturity, but there are several avenues for future research that could further enhance our understanding of these important constructs. One potential area for future exploration is the investigation of boundary conditions and contextual factors that may influence the associations examined in this study. Researchers could explore how industry characteristics, corporate size, or geographical location might shape the dynamics among innovation management, competitive scenario, and disruptive maturity. Examining these contingencies could help identify the specific circumstances under which the observed associations are strengthened or weakened, providing a more nuanced understanding of the phenomenon. Additionally, future manifests could delve into the potential feedback loops and dynamic interactions among the constructs over time. The associations examined in this cross-sectional research may evolve as corporates navigate changing market conditions and competitive pressures. Longitudinal research designs could shed light on the temporal dynamics and potential recursive impacts, offering a more comprehensive perspective on the interplay among these key variables. Another promising avenue for future research is the investigation of the specific mechanisms and processes through which innovation management and competitive scenario influence disruptive maturity. While this study has established the direct and indirect associations, a deeper exploration of the underlying factors, such as corporate ambidexterity, dynamic capabilities, or risk-taking propensity, could further elucidate the causal pathways and depict a more granular understanding of the drivers of disruptive maturity.

Moreover, future manifests could expand the scope of the investigation by incorporating additional relevant constructs into the research model. For a case study, exploring the role of corporate culture, leadership styles, or technological capabilities in shaping the associations among the focal variables could yield valuable understandings and contribute to a more holistic understanding of the antecedents and consequences of disruptive maturity. Finally, researchers could explore the generalizability of the outcomes by conducting cross-cultural or cross-industry comparisons. Examining the proposed associations in different contextual settings could depict valuable understandings into the boundary conditions and enhance the external validity of the outcomes, ultimately leading to a more comprehensive understanding of the complex interplay among innovation management, competitive scenario, and disruptive maturity.

By addressing these future research directions, scholars can continue to build upon the foundations established in this study and deepen our knowledge of the drivers and implications of disruptive maturity within corporates. Such advancements can inform more impassive managerial practices and help corporates navigate the challenges and opportunities of the ever-evolving business landscape.

RECOMMENDATIONS

Based on the key understandings gleaned from this research, several recommendations can be made to help corporates enhance their disruptive maturity and navigate the competitive business environment more impassively. The first and foremost recommendation is for corporates to prioritize the growth of robust innovation management capabilities within the corporate. The positive and substantial relationship among innovation management and both competitive scenario and disruptive maturity underscores the critical importance of investing in initiatives and structures that foster innovation, creativity, and the ability to anticipate and respond to disruptive market changes. This may involve establishing dedicated innovation teams, implementing agile growth processes, encouraging a culture of experimentation, and providing the necessary resources and support for innovative projects. By cultivating a strong innovation management mindset and competencies across the corporate, corporates can improve their competitive positioning and increase their chances of surviving and succeeding in the face of industry disruptions. Managers should therefore make innovation management a strategic priority, ensuring that it is corporately embedded in the corporate's culture, processes, and resource allocation decisions. In addition to the internal focus on innovation management, corporates should also closely monitor and adapt to their competitive environment. The positive and substantial relationship among competitive scenario and disruptive maturity underscores the influential role that the intensity of competition plays in shaping a corporate's disruptive capabilities and responsiveness. Corporates should invest time and resources in continuously scanning the competitive landscape, identifying emerging trends and threats, and adjusting their strategic priorities and operational tactics accordingly.

This may involve benchmarking against industry leaders, conducting scenario planning exercises, and fostering strong external partnerships and collaborations that depict valuable market intelligence and understandings. By developing a deep understanding of the competitive dynamics within their industry, corporates can better anticipate and prepare for potential disruptions, enabling them to be more agile, flexible, and resilient in the face of change. Also, corporates should consider adopting a more holistic and proactive approach to dealing innovation and navigating competitive environments. By simultaneously investing in innovation management capabilities and closely monitoring the competitive scenario, corporates can position themselves to thrive in the face of disruptive market forces and maintain a sustainable competitive advantage over the long term. Ultimately, the recommendations derived from this study emphasize the importance of cultivating a culture of innovation, adaptability, and strategic foresight within corporates. By prioritizing these critical capabilities, corporates can enhance their disruptive maturity and better navigate the challenges and opportunities manifested by the ever-evolving business landscape.

REFERENCES

Barney, J. (1991). Corporate resources and sustained competitive advantage. *Journal of Management, 17*(1), 99–120. DOI: 10.1177/014920639101700108

Bekele, A., & Worku, Z. (2018). Determinants of innovation capability in the manufacturing sector of Ethiopia. *Journal of Innovation and Entrepreneurship*, *7*(1), 1–16.

Christensen, C. M. (2019). *The innovator's dilemma: When new technologies cause great corporates to fail*. Harvard Business Review Press.

Gebreeyesus, M. (2022). Innovation and microenterprises growth in Ethiopia. In Szirmai, A., Naudé, W., & Goedhuys, M. (Eds.), Entrepreneurship, innovation, and economic growth (pp. 122-146). Oxford University Press.

Kant, S., & Adula, M. (2025). Integrating Gender and Indigenous Knowledge in Digital Innovations for Ethiopia's Renewable Energy Transition. In *Digital Innovations for Renewable Energy and Conservation* (pp. 167–194). IGI Global.

Kant, S., Adula, M., Kumari, M., & Nimbrain, N. (2025). Role of SMEs in Greening the Economy by Mediation of Eco-Innovation in the Horn of Africa. In *Examining Green Human Resources Management and Nascent Entrepreneurship* (pp. 173–200). IGI Global., DOI: 10.4018/979-8-3693-7046-9.ch007

Kant, S., Niguse, T., Terefe, A., & Adula, M. (2025). Emerging Trends and Innovations: Effect on Social Sustainability in Manufacturing Supply Chains by Shifts in Consumer Behavior in the Horn of Africa. *Enhancing Social Sustainability in Manufacturing Supply Chains*, 135-162. http://dx.doi.org/ DOI: 10.4018/979-8-3693-9740-4.ch005

Nega, F., & Kassie, G. T. (2016). Small and medium enterprise's access to finance in Ethiopia: Synthesis of demand and supply. *Cogent Business & Management*, *3*(1), 1210439.

Porter, M. E. (2020). *Competitive strategy: Techniques for analyzing industries and competitors*. Free Press.

Sankar, J. G., & Kant, S. (2025). Assessing Consumer Trust and Perception in Online Food Marketplaces. In *Innovative Trends Shaping Food Marketing and Consumption* (pp. 375–394). IGI Global Scientific Publishing. DOI: 10.4018/979-8-3693-8542-5.ch015

Teece, D. J. (2007). Explicating dynamic capabilities: The nature and microfoundations of (sustainable) enterprise performance. *Strategic Management Journal*, *28*(13), 1319–1350. DOI: 10.1002/smj.640

Teece, D. J., Pisano, G., & Shuen, A. (2019). Dynamic capabilities and strategic management. *Strategic Management Journal*, *18*(7), 509–533. DOI: 10.1002/(SICI)1097-0266(199708)18:7<509::AID-SMJ882>3.0.CO;2-Z

Utterback, J. M. (1994). *Mastering the dynamics of innovation*. Harvard Business Press.

Wernerfelt, B. (1984). A resource-based view of the corporate. *Strategic Management Journal*, *5*(2), 171–180. DOI: 10.1002/smj.4250050207

Worku, Z. (2023). Challenges and prospects of business growth in Ethiopia. *Journal of Small Business and Entrepreneurship*, *29*(1), 11–30.

Wossen, T., Abdoulaye, T., Alene, A., Haile, M. G., Feleke, S., Olanrewaju, A., & Manyong, V. (2023). Impacts of extension access and cooperative membership on technology adoption and household welfare. *Journal of Rural Manifests*, *54*, 223–233. PMID: 28989229

Yizengaw, Z. A., & Getahun, T. (2019). The influence of disruptive digital technologies on the competitiveness of SMEs in Ethiopia. *International Journal of Entrepreneurship and Small Business*, *37*(1), 1–18.

Chapter 19
Variables That Explain the Way of Innovating in Colombian Companies:
A Perspective Under the Rise Model

Rafael Ignacio Pérez-Uribe
https://orcid.org/0000-0001-9924-6657
Independent Researcher, Colombia

Carlos Salcedo-Pérez
https://orcid.org/0000-0002-4433-5537
Politécnico Grancolombiano, Colombia

Carlos Mario Muñoz-Maya
https://orcid.org/0000-0003-1451-1648
Universidad de la Salle, Colombia

Maria Del Pilar Ramirez-Salazar
Fundación Universitaria Juan N. Corpas, Colombia

ABSTRACT

To contribute to the development of the concept of innovation for sustainability in organizations, a group of professors from the EAN University in Bogotá (Colombia) "created the Route for Innovative and Sustainable Enterprises (RISE) model" that serves for diagnosis, evaluation and organizational intervention has served to highlight and close technological gaps in innovation. The objective of this chapter is to present the analysis of the variables in the light of the RISE model, which best explain the way of innovating in Colombian companies. The data from the 148 organizations as a convenience sample "were processed with the Statgraphic centurion XVI statistical package, using multiple regression analysis" and three (3) variables were obtained (Ecodesign, model that includes circular economy and work environment) of the 48 that make up the model, which explain between 42,8 and 44% the managerial decision to improve the way of innovating that facilitates the achievement of organizational goals.

DOI: 10.4018/979-8-3693-4030-1.ch019

INTRODUCTION

Innovation for organizations means creating and developing new organizational methods in business practices, workplace or in external relationships. The objectives of innovation in organizations may be aimed at improving business results by reducing costs, expenses, and transactions, improving customer satisfaction (more income) and workers (improving productivity), and ensuring the creation and use of knowledge (Correa Tipán et al., 2019).

"Innovation for sustainability not only enables organizations to adapt to environmental regulations but also generates economic and competitive value by creating greener and more socially responsible products and services." (Palacios Cardona et al., 2025. p. 14). In this sense, the top management of organizations must provide workers with the necessary working conditions to motivate them. "to be more proactive, aggressive, and daring in exploring opportunities and experimenting with innovative approaches before competitors. Considering that, organizations have begun to emphasize reorienting culture and motivation toward more environmentally and socially sustainable behaviors" (Rodriguez, p. 26).

For the above reason, the RISE model (Route of innovation for sustainable enterprises), was created in 2016 to address the needs of assessing their status in innovation management and defining best practices focused on the sustainability of companies established in Colombia. Up to the year 2023 had been applied to 148 companies from varied sizes and industries in Colombia. Each organization has developed its own paths of innovation for sustainability based on the results obtained by the application of the RISE Model.

The fundamental problem to be solved is in the central hypothesis of this work: What are the variables that best explain innovation in Colombian companies considering the RISE model?

Thus, this chapter presents the variables that explain in higher degree the development of the concept of the way to innovate in the Colombian context. This research aims at improving the knowledge of the academic, governmental, and business communities, related to innovation to strengthen the workers' capabilities to add value to their life and organizations. Also, the present work presents the theoretical foundations about innovation, from different theoretical approaches, and the methodology applied in the research aimed at determining the factors and dimensions of the RISE Model, finding the most relevant variables to explain innovation ways in Colombia. In addition, the conclusions and recommendations proposed will allow the academic, business, and governmental communities to have aid in developing activities to strengthen innovation in their organizations.

BACKGROUND

RISE Model

Since 2016, the RISE Model has been conceptually developed, considering 4 managerial models: The Model of Modernization for Management of Organizations -MMGO- ((Perez-Uribe et al, 2013); The Open Collaborative Innovation Model (Ramírez-Salazar, 2016; Ramírez-Salazar et al., 2021); the Design Thinking Model (Razzouk & Shute, 2012); and the Global Reporting Initiative (GRI, 2016; PNUMA, 2016). The general structure of the model is presented below:

General Structure

Based on concepts and methodologies from the already mentioned models, the RISE Model is composed of nine (9) factors, four (4) dimensions, and thirty-six (36) descriptors (Perez-Uribe, et al., 2023, p. 383) (Table 1).

Factors

1. **Innovation:** Innovation is the creation or transformation of a product, service, process, market strategy, and business model. An innovative company is one that changes, evolves, becomes dynamic, offers new products, and adopts or implements new manufacturing processes (Escorsa Castells & Jaume Valls, 2003). For true innovation to exist, an entrepreneurial action must occur, that is, acting to make things happen (González Candia et al., 2014). According to Miller and Morris (2008), innovation is said to be the process of transformation and invention into something that is commercially useful and valuable. In this way, the final consumer will be the one who validates that usefulness by satisfying a need. According to Ramírez-Salazar (2016) the concept of Collaborative Open Innovation is added to the concept of Innovation: "a collaborative open innovation model is one in which several entities or people modify or change an unresolved issue that then serves others. That is, it's about creating solutions to business challenges, using the organization's external knowledge, which can then serve as a reference for others" (p. 43).
2. **Sustainable production:** Sustainable production involves promoting the efficient use of resources to improve the quality of life for current and future generations. One definition of sustainable production is that it seeks to "improve the quality of human life without exceeding the carrying capacity of the ecosystems that sustain it" (Fundación Vida Silvestre, 2019). Sustainable consumption and production involve promoting the efficient use of resources and energy, building environmentally friendly infrastructure, improving access to basic services, and creating green jobs with fair pay and good working conditions. All of this translates into a better quality of life for all and helps achieve comprehensive development plans that reduce economic, environmental, and social costs, increase competitiveness, and reduce poverty (UNDP, 2018).
3. **Leadership and Strategic Direction:** Flight log of any type of organization where the directional parameters towards which the company must go are defined and on which all collaborators must work. Strategic direction is the potential for the future achievement of an organization's objectives, Ansoff (1997) as stated by Scholes & Johnson (2001), the scope of strategic direction is greater than that of any of the operational management areas. According to Camacho (2002), strategic direction "is a managerial approach that allows senior management to determine a clear course and promote the necessary activities so that the entire organization works in the same direction" (p. 2). According to Pérez-Uribe (2018), strategic direction as the flight log of any type of organization where the directional parameters towards which the company must go are defined and on which all collaborators must work.
4. **Organizational Culture:** For Schein (1988) culture is the basis of basic assumptions and beliefs shared by the members of a company and says that culture can be intervened by management, planned and with mechanisms that generate organizational change. Culture is understood as a system of symbols and shared meanings, product of negotiations of people in their interaction (Calderón Hernández et

al., 2003). The culture of a company is made up of the tradition, conditions and values that provide guidelines for a pattern of activities, opinions, and actions (Ouchi, 1982).

5. **Recognition:** A social habit of valuing and recognizing people's contributions through symbolic and material manifestations and accrediting actions. Recognition should not only be given in words or symbolic manifestations but must be accompanied by accrediting actions (Pérez-Uribe & Ramirez-Salazar, 2021). These authors also claim that recognition is conceived in various ways, but its position is close to that every time there is an affirmation about a person, this affirmation must be positive and in such a way that it allows the recipient to identify with their qualities and thus achieve greater autonomy. According to Lovibond (2002) human recognition has become a social habit in which the reasons for verifying the value of other people have been linked in the process of socialization between people.
6. **Collaborative processes:** These are all those actions in which people or entities participate in associations, guilds, cooperatives, or interest groups to develop projects with a common objective and obtain mutual benefits (Pérez and Ramírez, 2021). For several people of different identities and companies to participate in a project, it is necessary to cement the principle of trust (Ramírez-Salazar, 2016; Cheikhrouhou et al., 2013). Trust makes people feel safe interacting with each other and thus sharing their knowledge (Lozano-Correa et al., 2020). Collaborative processes become successful to the extent that networks, associations, and alliances are formed. The most pertinent aspect of collaborative processes is to be clear about the scope, goals, and intellectual property of each of the people and entities participating (Ramírez-Salazar, 2016).
7. **New Markets:** A group of consumers who can purchase a product or service from a company entering a new market to grow its sales (Pérez and Ramírez, 2021). Underserved markets worldwide are those at the base of the pyramid (International Finance Corporation, 2007). A series of standards and requirements must be met to enter new markets, and qualitative and quantitative variables must be evaluated to make an effective selection (Peralta, 2010, cited by Pérez and Ramírez, 2021):

> Qualitative variables include product suitability, the search for a market with potential consumption, and geographic proximity. Quantitative variables include Analyze market size, GDP per capita, population characteristics, national production of the product, imports of the product, prices, tariffs, additional taxes, tariff preferences, current destination of exports from the country of origin of the product, availability of international transportation, language skills (p. 18).

8. **Technology:** Set of theories and techniques that allow the practical use of scientific knowledge for the solution of specific problems in an effective way and is the means through which scientific knowledge is transferred to the solution of specific problems in an effective way (Tula and Giuliano, 2015). Van Wyk (2004) states that from the previous concept of technology there is a tendency to value science in terms of what contributes to society and to create competencies expressed in technological entities that consist of devices, procedures, and skills. For his part, Benavides (1998) defines technology as "the system of knowledge and information derived from research, experimentation or experience and that, together with the methods of production, marketing and management that are specific to it, allows the creation of a reproducible form or the generation of new or improved products, processes or services" (p. 3).

9. **Financial indicators:** According to Ortiz (2021), these are metrics that show the financial health of the company in terms of its Liquidity, Efficiency, Performance, Productivity, Indebtedness, and Financial Diagnosis. The main financial indicators to consider in companies are (Carrión et al., 2019): 1) Liquidity, corresponding to the ability to pay off short-term debts, 2) Efficiency, which measure the costs associated with production, 3) Performance, related to the fulfillment of objectives, 4) Productivity. Related to the highest production per person in the company, 5) Indebtedness, related to the degree to which creditors participate in the financing of the company, 6) Financial Diagnosis, which are obtained from the balance sheet, cash flow, and the profitability of the company is measured in a more dynamic way.

Table 1. Factors, dimensions, and descriptors of the RISE Model

No	FACTORS	Dimension	DESCRIPTOR
1	**Innovation**	Social	Innovation ways
		Environmental	Eco-design-Circular Economy
		Managerial	Innovation type
		Economic	Value creation
2	**Sustainable Production**	Social	Suppliers- Raw materials or supplies for operations
		Environmental	Efficient use of water- Sewage Water-Energy-Air Emissions-Solid waste
		Managerial	Plans, environmental certifications
		Economic	Budget for environmental management – Business model of Circular Economy
3	**Leadership and Strategic Direction**	Social	Social trends, capability to mobilize, ethics, values, and anti-corruption policies
		Environmental	Accountability- Value of Sustainability
		Managerial	Corporate Government-Knowledge Management-Corporate Strategy
		Economic	Analysis of the environment-Decision-making
4	**Organizational Culture**	Social	Work environment
		Environmental	Change of paradigms
		Managerial	Communication
		Economic	Shared value
5	**Recognition**	Social	Happiness at work
		Environmental	Environment aware worker
		Managerial	Motivation practices
		Economic	Recognition at work
6	**Collaborative Processes**	Social	Associativeness
		Environmental	Safety at work
		Managerial	Agreements -Consensus
		Economic	Corporate volunteering

continued on following page

Table 1. Continued

No	FACTORS	Dimension	DESCRIPTOR
7	**New Markets**	Social	Degree of influence of the products in other markets
		Environmental	Green markets
		Managerial	Strategic plan to enter into new markets
		Economic	Commercial strategy
8	**Technology**	Social	Democratization of technology
		Environmental	Clean technologies
		Managerial	Prospective
		Economic	Technology transfer
9	**Financial Indicators**	Social	Impact on stakeholders
		Environmental	Environment protection or recovery
		Managerial	ROA= Return on Assets
		Economic	EBITDA

Source: Perez-Uribe, et al., (2023)

Dimensions

The RISE Model studies four dimensions for each factor: social, environmental, managerial, and economic, which are described in tables 2-5:

Table 2. Social dimension

In relation to this dimension, companies must conceive of the human factor that surrounds them in terms of society. In this sense, employees, shareholders, suppliers, competitors, the community, employees' families, government entities, and customers will be the primary stakeholders that the company must benefit directly and indirectly.	
FACTORS	**ASPECTS ANALYZED (descriptors)**
INNOVATION	All stakeholders are considered to answer: Who do we innovate with? Are all members or just the managerial team of the company considered?
SUSTAINABLE PRODUCTION	Inventory, raw material, supplies, and suppliers, are periodically analyzed.
STRATEGIC LEADERSHIP AND DIRECTION	The morals of the organization, considering social trends, are permanently analyzed.
ORGANIZATIONAL CULTURE	The work environment is analyzed. Is it toxic? Does it promote trust and camaraderie? Do workers feel a sense of belonging? Are workers proud to work in the company?
RECOGNITION	How do workers feel? Are they happy? Is there a balance between work and personal life? How can that be measured?
COLLABORATIVE PROCESSES	Are managers in favor of belonging to unions and associations? Does the company belong to a union or association? Are experts from other areas or companies consulted to develop new projects?
NEW MARKETS	The company has clearly defined one or more market segments. If so, does it have a real influence on them? Does it know them well? Does the company know the culture of the communities where it has any presence?

continued on following page

Table 2. Continued

In relation to this dimension, companies must conceive of the human factor that surrounds them in terms of society. In this sense, employees, shareholders, suppliers, competitors, the community, employees' families, government entities, and customers will be the primary stakeholders that the company must benefit directly and indirectly.	
FACTORS	**ASPECTS ANALYZED (descriptors)**
TECHNOLOGY	Do all members of the company have easy access to technology? Is technology available just to a few privileged members?
FINANCIAL INDICATORS	Does the company have a budget aimed at creating a positive impact on the communities where it has a presence?

Source: Perez-Uribe et al., (2023)

Table 3. Environmental dimension

This dimension analyzes how companies are affecting the environment (in terms of physical, social and economic conditions) due to their atmospheric emissions, water treatment, energy management and solid waste management.	
FACTORS	**ASPECTS ANALYZED (descriptors)**
INNOVATION	Does the management team take steps to fulfill the strategies of ecological design? Is the company promoting and fulfilling circular economy policies?
SUSTAINABLE PRODUCTION	What policies and practices are the company implementing related to use of water, sewage water, hazardous materials, solid waste, energy and air pollution.
STRATEGIC LEADERSHIP AND DIRECTION	Has the situation regarding transparency, accountability, and circular economy improved?
ORGANIZATIONAL CULTURE	Are the company's stakeholders changing paradigms? Has the company updated the ways to approach customers, potential and current customers, by using innovative technologies and social media?
RECOGNITION	Has the company made efforts to prepare and train the staff in matters of environmental awareness?
COLLABORATIVE PROCESSES	Are staff security and well-being a priority for the company? The information about industrial security and technical specifications is periodically updated, therefore reducing risks, dangers, and accidents?
NEW MARKETS	Is the company selling in ecological markets? Where? What percentages of the demand are supplied by green suppliers?
TECHNOLOGY	Does the company use clean energy? Which type? Has this lead to improving the processes and measurement of its impact on the environment?
FINANCIAL INDICATORS	Is the company investing in recovering its environment, mitigating the effects of its presence? Has the company invested in recovering/building sidewalks, parks, or others? Is it participating on reforesting initiatives?

Source: Perez-Uribe et al., (2023)

Table 4. Managerial dimension

In the context of the RISE model, the managerial dimension relates to the way the company is led to achieve its objectives related to business innovation and sustainability.	
FACTORS	**ASPECTS ANALYZED (descriptors)**
INNOVATION	What are the current innovation processes? Do they include marketing? Is the company using state-of-the-art marketing strategies, organizational structures, production processes and business models? Does the company promote social innovation to favor and benefit the most vulnerable members?
SUSTAINABLE PRODUCTION	Are the sustainable production and processes certified by a proper environmental authority?
STRATEGIC LEADERSHIP AND DIRECTION	Are the handbooks of procedures and functions available for corporate management, knowledge management and corporate strategy? How frequently are they used?
ORGANIZATIONAL CULTURE	Is the communication horizontal or vertical? Is it assertive, transparent, and open?
RECOGNITION	Is the staff motivated based on incentives? Is it motivated by recognition and based on merit? Is it motivated by a combination of the previous factors?
COLLABORATIVE PROCESSES	How are agreements reached? Does the company promote coexistence and community committees? Are protocols and agreements available so the parties have an adequate follow-up during negotiations?
NEW MARKETS	Entering a new market is the consequence of: a strategic process and exhaustive research? demand satisfaction? or intuition?
TECHNOLOGY	Is prospective used to plan the company's future?
FINANCIAL INDICATORS	Does the company fulfill its promise to generate added value to its stakeholders?

Source: Perez-Uribe et al., (2023)

Table 5. Economic dimension

This dimension should show the investments that companies make each year to benefit their stakeholders and how this reflects the creation of value for all the people interested in the company.	
FACTORS	**ASPECTS ANALYZED (descriptors)**
INNOVATION	Does the company make effors to create value for its stakeholders? Is the added value the result of innovation processes?
SUSTAINABLE PRODUCTION	Does the company have clear environmental governance? If so, does it allocate a specific budget that considers the circular economy principles?
STRATEGIC LEADERSHIP AND DIRECTION	Is top management committed with circular economic principles? Is decision-making based on exhaustive environmental analysis performed periodically?
ORGANIZATIONAL CULTURE	What is the shared value of the company with its stakeholders?
RECOGNITION	Does the staff receive fair rewards and recognition for their performance? Is recognition above or below the market average?
COLLABORATIVE PROCESSES	Are company members and stakeholders committed to corporate volunteering?
NEW MARKETS	Is there a solid, coherent, and clear market strategy?
TECHNOLOGY	Does the company usually participate in congresses, debates, or scientific events related to technology transfer? Is the company engaged in a process of bidirectional technology transfer with stakeholders, such as suppliers and companies that belong to its industry?
FINANCIAL INDICATORS	Is the company in a good financial situation? Are liquidity and profitability indicators, and debt at long-term indicators coherent with a good financial situation?

Source: Perez-Uribe et al., (2023)

Levels of Analysis

The RISE Model uses a percentual scale ranging from 0 to 100 to measure the innovation level for sustainability; when applying it to organizations, it can be classified in 5 levels, as shown in Table 6 (Perez-Uribe et al., 2021):

Table 6. Levels of organizations in innovation for sustainability

LEVEL	DESCRIPTION
One: Emerging (0-20%)	It corresponds to an organization on its first stages of administrative development. All processes are incipient or are just being created. Times are uncertain since the founders/managers feel comfortable and rarely feel the need to find advisors in managerial topics.
Two: Survival (21-40%)	Organizations with more than 4 years of existence. Colombian MSMEs endure demanding conditions such as precarious government support and inexistent or insufficient support from financial banks. Entrepreneurs struggle to survive. During this time, it is necessary to get familiar with the environment of the company, start to become part of innovation systems and regional clusters.
Three: Development (41-60%)	Organizations experience sales and market share growth, buy technology, test new ways to produce goods and services, search for external financing sources, and eventually, import and export. During this stage entrepreneurs test trial and error. Financing is the key, and it is essential to know how much the company can grow and the cost of external financing.
Four: Outstanding (61-80%)	The organization has clear differentiation from other companies in the market. It is a consolidation stage, with a differentiated market segment. The company is financially stable, and it is recognized for the high quality of its products. However, uncertainty comes since success leads to complacency which could make the organization vulnerable. This is the time to plan the succession to keep the organization on the path to success, satisfying the needs and expectations of the stakeholders.
Five: World Class (81-100%)	World class organizations, recognized by the quality of their products and for their corporate management and policies. They are considered as referents in the country and overseas. They are praised by the media because of their achievements and performance; their directors/founders are praised by society, academics, businesspeople, and government officers. Their social labor is recognized. During this stage, the entrepreneur must get ahead of the competitors, trends, and the big companies that try to imitate or acquire the organization. This is a moment of deep introspection. It is difficult to resist the possibility to sell the company to a bigger organization.

Source: Perez-Uribe et al., (2021)

Way to Innovate

When if companies always intent to stand out in their industries because of their innovations, it is essential to understand that innovation is performed by people, so the human talent is the most important resource to become a world class company (Álvarez-Aros & Bernal-Torres, 2017).

Companies have diverse ways to innovate. In some cases, workers develop only what the general managers order; in other cases, the management team relies on the different divisions of the company or on external experts. The following are the several types of innovation models:

Technology Push Model

The company innovates due because technological innovations make them do it. When Internet appeared, companies had to innovate changing the ways in which they sold their products; Internet connected everybody online all over the world (Ramírez-Salazar, 2016; Ramirez-Salazar et al., 2019; Martínez Vázquez & Espinosa Yañes, 2017).

Pull Market

Innovation occurs due to market needs. For example, innovation in the sport footwear market: nowadays companies must change their designs since people of all ages prefer to walk, go to work, or to the gym wearing sport shoes with exclusive designs (Atarama-Rojas et al., 2022; García-Legaz, 2021).

Model by Stages

Innovation occurs in the departments of innovation and development of the company. Such department explores technology and market needs (Barreto Fereira & Petit Torres, 2017).

Integrated Models

Innovation performed by multidisciplinary teams of experts. This shows the importance of collaborative innovation (Landazury-Villalba et al., 2018).

Network Models

Under these models, when innovation occurs, the company's systems and subsystems are considered. Networking is necessary; knowledge networks become valuable (Terán-Bustamante et al., 2019).

Open Innovation Models

Innovation projects involve talented people who have the solution to the challenges of the organization. This model is widely used nowadays, however, it is necessary to create teams that include people from inside and outside of the organization to keep the know-how inside the company (Ramírez-Salazar, 2016; Álvarez-Aros & Bernal-Torres, 2017). The way to innovate is about how to involve people in the process of creating or improving products, management models, or market strategies; however, due to globalization, demographic changes and technological developments, companies can find different ways to innovate using management models that help managers to propose strategies and new ways to implement practices inside the companies to create a real innovation culture (Ramírez-Salazar et al., 2021).

Variables that Explain the Way to Innovate in the RISE Model

The variables included in the RISE Model that mostly explain the innovation ways are, as proven in this research and shown later: Ecodesign and circular economy, business model that includes circular economy, and work environment (as will be shown in the methodology and findings).

Ecodesign and Circular Economy

They characterized the environmental dimension of the innovation factor, stating that, first, any innovation is accepted as long as it considers a sustainable production and a more rational use of resources (Ecodesign) (Rodríguez Viñas, 2019) and that the organization is different from others in its industry because it creates products under the principles of circular economy.

The ecodesign, is playing a relevant role in the design or redesign of products and processes, since it allows the inclusion of ecological aspects that reflect the mutual dependence between the environment, humans, and sustainable development (Predassi, 2023).

The concept of eco-design, understood as the type of design of products that minimizes the consumption of energy and maximizes the benefits for users during their lifecycle, started in the Netherlands in the 90´s, and from there, initially it spread to Germany, Belgium, the United Kingdon, and Australia. From the perspective of sustainability, eco-design has consolidated as a key tool for a strategy of Producers Extended Responsibility (Predassi, 2023).

Clearly, the traditional design model, which uses practices of linear economies, must move towards a closed and cyclical system, better than the current existing model in which waste generation is considered normal; it must do so to use less resources and have a system in which nature plays the main role (Cabello Cervantes, et al., 2020). An industrial ecosystem that incorporates waste as input to new fabrication processes saves resources and reduces the generation of waste. That is why it is important to take measures, so waste and the energetic content of manufactured products is not just dumped in a garbage filling or loses its value in recycling processes; it is then that the principle of circular economy appears (Cerdá & Khalilova, 2016).

The origin of the circular economy dates back to the 90's; back then, McDonough (American architect) and Braungart (German chemist) developed a design philosophy called "from cradle to cradle", which considers the materials used in processes as nutrients, classifying them into technical and biological, focusing on the positive impact generated by the flow of products, different from the traditional approach that focuses on the reduction of negative impacts (WIPO, 2007; Mcdonough & Braungart, 2003). Thus, an innovative way of industrial design and production is proposed, based on the closing of the lifecycle of products, just as it happens in the nature (Balboa & Somonte, 2014).

The application of circular economy in productive processes must consider certain implementation processes (Prieto-Sandoval, et al., 2017; Perez-Uribe & Ramírez- Salazar, 2020): (i) the concept of garbage or waste disappears because the products are unassembled once they are not useful anymore, and their biological components become part of the natural cycles or industrial (technical) with a minimum energy consumption; (ii) diversity strengthens the most resistant and resilient natural systems and therefore, a similar approach must be applied in economic environments, to improve the capacity to respond to economic and productive crisis; (iii) the elements of the system are highly connected; (iv) the property model must be reconceptualized, since the producing company can rent technology towards the user; (v) fossil and nuclear energy must be substituted by renewable energy; (vi) prices must reflect the real cost of the product to favor a rational consumption.

Business Model that Included Circular Economy

According to Perez-Uribe et al., (2020) the principles of circular economy are 1) the projection of products to generate zero waste, emphasizing on the optimization of processes in different scenarios; 2) products are designed considering their reuse; 3) use of renewable energy.

There is a long path to create a better environment through environmental management and sustainability in organizations. It is necessary to move from planning to implementation, and this requires resources. That is why companies must think about the need to use the enough number of tools to positively impact the quality of life, the reduction of the impact on the environment, and the sustainable competitiveness of companies (Carrillo & Pomar, 2021). That is why, when reviewing the contributions of different professions to the topic of sustainability, and the implementation of a cleaner production and circular economy, it is necessary to consider the active participation of administrative and accounting professionals, since, an amount of money allocated to a project is a good indicator of the interest of the company regarding projects related to such topics (other way is by measuring the impact obtained, that could be economic or financial (Perez-Uribe, 2020).

Many times, companies do not plan their environmental management, neither their sustainability: therefore, they do not identify the impact of their operations, cannot define a clear environmental policy and cannot establish objectives about it; the lack of planning causes that most companies do not have an adequate control of their processes, thus limiting the use of indicators for control, follow-up, investments, and results (Martín Toquero, 2020).

The economic dimension of the factor of sustainable production of a company can be characterized by the follow-up and control made through indicators of actions in which there is investment to control and mitigate negative environmental impacts, and by the budget allocated to programs related to the environment and sustainability (and evidence of results obtained such as seals, certifications, and savings) (Barrera Castro & Bossa Eljaiek, 2021). The only certification recognized for circular economy is Cradle to Cradle® (Braungart & McDonough, 2009).

Work Environment

In the RISE Model, the work environment as descriptor of organizational culture in the social dimension refers to the actions developed by the organization to promote respect and dignity in an excellent way, providing concrete answers to the feelings of alienation, anguish, and stress caused by the environment. Organizations that promote such actions are great places to work (Perez-Uribe, 2012).

Of course, that is related to the managerial commitment to building an adequate environment, so such actions spread all over the organization. When conditions that focus on dignity and respect are created, workers will be more committed to implement, follow-up and evaluate their performance and the use of resources (Machado et al., 2023). Thus, the execution of strategies and action or operational plans required to reach the expected objectives will be easier (Perez-Uribe, 2018). This concept includes looking for workers' motivation, which is directly in the hands of business leaders by means of three factors: key people, transparency, and atomization (Kim and Mauborgne, 2005).

Key People

It is necessary to mobilize workers from all organizational levels to have a strategic change. Efforts must concentrate on key people or on those that are more influential in the organization, identifying natural born leaders, who are usually respected and persuasive people, who can help to block or unblock the events who could harm the processes. Getting those leaders' support creates a domino effect in workers, who then will be more prone to support the strategy.

Transparency

It is essential to communicate the acts of key people broadly and repeatedly to significantly motivate them. To do so, management must be transparent, inclusive, and equal. This implies reviewing (in a periodical, mandatory, participative, and visible meetings with those leaders selected, and they should also meet with their workers) the progress and objectives of the strategy. The visual and explicit presentation allows workers to better understand the performance of the execution of the strategy. Each leader must explain the activities, describe how his/her workers correct problems, and explain why results increase or decrease. This demanding work that leaders do generates a high-performance culture in a matter of weeks. In this framework, the concept of equity refers to the need to include all those affected, so they can know firsthand the base of the decisions and criteria to be promoted or to leave behind a worker in the future.

Atomization

It refers to the way in which the strategic challenge is presented. Change is unlikely to be successful if people do not believe they are going to overcome the challenge. It is necessary to have an attainable challenge and divide it into parts that have a meaning for workers at diverse levels of the organization. This means that each person must be in charge to improve and reaching the results in his/her process, area, or department.

Those three factors must go along with the creation of a culture of trust and commitment that motivates people to execute the agreed strategy. Hearts and minds must be aligned with the strategy in a way that each person voluntarily makes it his/hers, to apply it by his/her own initiative and not by force. Under these conditions a process called equitable by these authors arises; such a process allows people to accept people from the beginning, starting with the design of the strategy. When this occurs, workers perceive that the playfield is leveled, which brings voluntary cooperation when strategic decisions are executed (Kim and Mauborgne, 2005, cited by Perez-Uribe, 2018).

Kim and Mauborgne (2005) consider that there are three linked principles of the equitable process, necessary to understand the concept of equity: 1) participation; 2) explanation; and 3) clear expectations.

1) Participation

It is the involvement of people in strategic decisions that affect them by asking them for their ideas, allowing them to refute other's ideas and assumptions. Whenever participation and controversy are permitted, there are positive effects on people, such as: the management team shows respect for the individual and his/her ideas, the reflection of people sharpens, there is a better dynamic of the collective knowledge, and people are better committed regarding such decisions.

2) Explanation

All those involved and affected understand the reason strategic decisions are taken. When managers explain the reasons of such decisions, people consider that their opinions have been taken into account, that decisions are neutral, and are taken for the benefit of the company. These managerial actions allow workers to trust managers' intentions and create an environment of feedback that strengthens the learning processes.

3) Clear Expectations

It is the communication of the working parameters for employees, including goals, objectives, stages, and those responsible, in a way that all those involved know what is expected of them, so they can concentrate their efforts on the quick development of the strategy.

MAIN FOCUS OF THE CHAPTER

Methodology

This study is based on the application of the RISE Mode located in Colombia; the sample of 148 companies, as source of the study, was chosen using a convenience sampling (Otzen, & Manterola, 2017), of which 34.8% were micro-enterprises, 24.42% were small, 17.44% medium and 23.26% large (table 7 and figure 1).

Table 7. Company size.

COMPANIES SIZE	%
LARGE	23,26%
MEDIUM	17,44%
SMALL	24,42%
MICRO	34,88%
TOTALES	**100,00%**

Source. Own elaboration

Figure 1. Companies size.

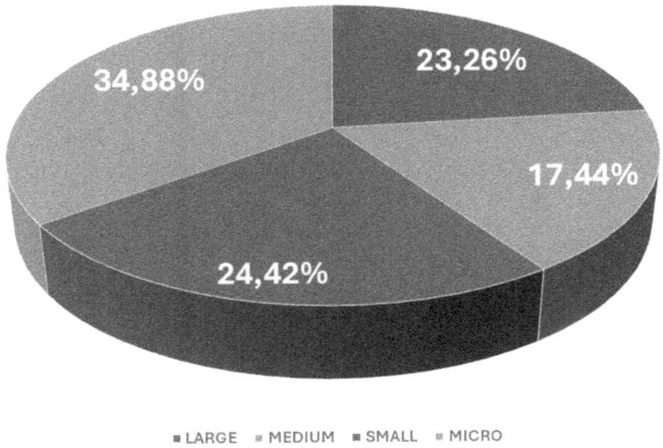

Source. Own Elaboration

Of this convenience sample of 148 companies, 63% correspond to the services sector, 24% to the manufacturing industry, 7% to the trade sector and 6% to the construction sector (table 8 and figure 2).

Table 8. Economics sector of companies

ECONOMIC SECTORS	TOTAL
SERVICES	63%
MANUFACTURING INDUSTRY	24%
TRADE	7%
CONSTRUCTION	6%
TOTAL	100%

Source. Own elaboration

Figure 2. Economics sector of companies

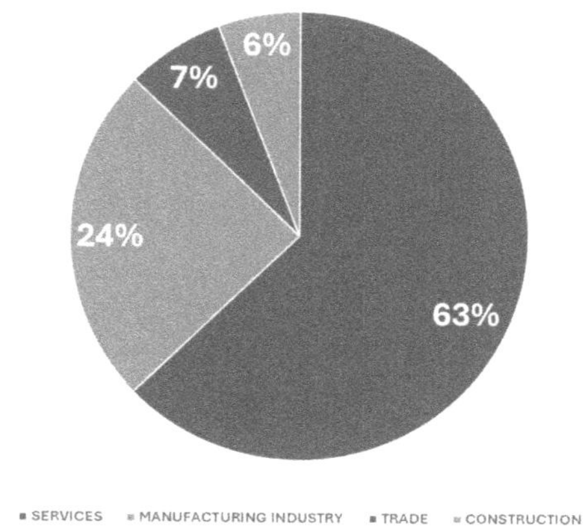

Source. Own Elaboration

A multiple regression analysis was performed to determine the relation or association degree (Hernández-Sampieri, *et al.*, 2014) between the dependent variable (way to innovate) and the thirty-five (35) descriptors or variables of the RISE Model. The statistical software Statgraphics Centurion XVI was used to perform the analysis (Perez-Uribe and Ramírez-Garzón, 2020); thirty (30) analyses were performed given automatically by the statistical package used; the last one (the thirtieth analysis) showed the adjusted equation for those variables that better explain the way to innovate. Managers should work on those variables to develop an innovative organizational culture in their organizations, applying the best practices, facilitating the achievement of organizational objectives.

Findings

The 30th analysis showed the results of the adjustment of the linear multiple regression model to describe the relation between the way to innovate (dependent variable) and the three (3) independent variables that mostly explain it. The equation of the adjusted model is:

WAY TO INNOVATE= 6,11411 + 0,385728* **Ecodesign and Circular Economy** + 0,300827* **Business Model that Includes Circular Economy** + 0,196812* **Work Environment**

The R-Squared indicates that the adjusted model explains 44.0117% of the variability of the way to innovate. The Adjusted R-Squared, which is more appropriate to compare models with different numbers of independent variables, is 42.8533% (Table 9). The standard error of the estimate shows that the standard deviation of the residuals is 17.5453 (this value can be used to calculate limits for future observations), and the median absolute error, with a result of 13.8206, is the average value of the residuals (Table 9). The Durbin-Watson examines the residuals to determine if there is a significant correlation based on the order in which data is presented in the datafile; in this case, it is lower than two (2), so the values analyzed are considered correct.

Table 9. Summary of statistical results of the multiple regression analysis

Statistics	Values
R-squared =	44,0117%
R-squared (adjusted for g.l.) =	42,8533%
Standard error of the estimate=	17,5453
Mean absolut error =	13,8206
Durbin-Watson =	1,79449 (P=0,1054)
Autocorrelation of residuals in 1 =	0,0908964

Source. The authors using Statgraphics Centurion XVI

To determine if the model could be simplified, note that the highest P-value of the independent variables was 0.0008, corresponding the variable of work environment. Since the P-value was lower than 0.05, it is considered that this is statistically significant with a confidence level of 95.0%. Therefore, no variable was eliminated from the model (Tables 10 and 11).

Table 10. Data obtained in the 30th analysis.

Parameter	*Estimation*	*Standard Error*	*T Statistics*	*P Value*
CONSTANTE	6,11411	5,14382	1,18863	0,2365
Ecodesign and circular economy	0,385728	0,0774334	4,98141	0,0000
Business model that included circular economy	0,300827	0,0769789	3,90792	0,0001
Work Environment	0,196812	0,0577175	3,40991	0,0008

Source. The authors using Statgraphics Centurion XVI

Table 11. Variance analysis

Source	*Sum of squares*	*Gl*	*Half square*	*F test*	*P value*
Model	35088,0	3	11696,0	37,99	0,0000
Residual	44636,3	145	307,836		
Total (Corr.)	79724,3	148			

Source. The authors using Statgraphics Centurion XVI

SOLUTIONS AND RECOMMENDATIONS

This part presents the recommendations arising from some of the activities considered as best practices, considering the characteristics of the companies analyzed using the RISE Model, for the following descriptors: 1) way to innovate (dependent variable) in the social dimension of the innovation factor; 2) ecodesign and circular economy (independent variable) in the environmental dimension of the innovation factor; 3) business model that included circular economy (independent variable) in the economic dimension of the sustainable production factor; 4) work environment (independent variable) in the social dimension of the organizational culture factor.

Way to Innovate

Development of techniques of design thinking or similar with the support of experts to solve problems, needs, or innovation opportunities (Ramírez-Salazar et al., 2023).

Integration of internal and external innovation processes, through design thinking models to generate added value to its stakeholders (Amador et al., 2022).

Ecodesign and Circular Economy

Implementation of an innovation process in projects to improve its processes, products, the environment, the use of alternative energies, and the development of sustainable production with the rational consumption of resources (Da Silva et al., 2016).

Development of the concept of circular economy to create new processes and products, validated by national and international certifications (Prieto-Sandoval et al., 2017).

Business Model that Includes Circular Economy

To lead a group, area, or department that periodically performs the analysis of the lifecycle of the product, and to work in the redesign of products, focusing on circular economy (Rozo Doncel, 2019; Almeida-Guzmán & Díaz-Guevara, 2020; Carrillo González & Pomar Fernández, 2021).

Work Environment

Permanent development of actions (a combination of best practices described in the RISE Model in the levels 4 and 5 of each descriptor) to promote the respect and dignity of the human being, to concretely respond to the feelings of alienation, anguish, and stress of workers, which are generated by the environment (Perez-Uribe et al., 2023; Fisher, 2003; Furstenberg, 2015; Lozano-Correa et al., 2020).

Application of a policy based on equity to set salaries and recognition for all workers, permanently adjusting based on best practices (Rodríguez, 2015; López Chanes et al., 2019; Fernández, 2015).

Application of a program to keep the ethical code and anticorruption policy as part of the strategic direction and the organizational culture of all stakeholders; this includes performance indicators and comparison with international standards as input for strategic projects (Montenegro, 2021).

Implementation of processes to answer to demands and expectations of the society, exceeding the international standards of corporate governance, in a dynamic process and with an adequate equilibrium between the organizational culture and structure (Flores Konja & Rozas, 2008; Danish & Usman, 2010).

FUTURE RESEARCH

The RISE Model will be applied in more companies in the future, therefore providing more data to keep performing further analyses like the one presented in this chapter.

The authors invite colleagues, businesspeople, academics, researchers, and consultants in innovation focused on organizational sustainability, to validate the resulting equation shown in this chapter, considering the recommendations presented. In the development of the activities for each variable (way to innovate, eco-design, circular economy, and work environment), it is important to evidence the continuous improvement of the way in which companies develop innovations.

Thus, and after compiling the information of the application of such activities, it is important to publish cases, and, if there is a representative number of companies, additional statistical analyses can be performed to have evidence of support to enrich the theory about the "way to innovate".

Based on the results of this research, the following aspects could also be investigated: 1) In "Variables that Explain the Way to Innovate in the RISE Model", it would be interesting to investigate experiences of companies around the world and how Colombian companies are in this scenario, 2) In "Work Environment": what impacts do the three factors addressed: key people, transparency and atomization have on companies around the world and in Colombia and, 3) It would be interesting to investigate the contextualization of the subject by addressing the SDGs (Sustainable Development Goals) and how companies around the world are adapting to achieve the same sustainable target according to RISE MODEL.

CONCLUSION

It was proven that the way to innovate is explain between 42.85% and 44.01% by the eco-design, model of circular economy, and work environment (based on the RISE Model). It is important to note that those are the most representative variables based on the analyses performed for purposes of this chapter. The authors consider that strengthening these variables is a priority, through the application of the activities mentioned in the recommendations of this chapter, which will help organizations to have better results in their ways to innovate.

However, organizations and their members are complex, and this must be considered when managers try to create the conditions necessary to implement the best practices to innovate (Muñoz-La Rivera et al., 2021). Besides the mentioned variables, the following aspects, shown in Figure 3, must be considered) Perez-Uribe et al., (2023):

- The type of technology to develop core business.
- External variables that affect an organization (political, economic, social, technological, ecological, and legal).
- The level of development of innovation and knowledge in an organization regarding its core business.
- Everything related to culture and work environment.
- The relational capital of the company with its stakeholders.
- The geographic and strategic location of the organization (the country if it is a public organization).
- The type of company (public or private).
- The type of activities performed by the staff.
- The type of leadership of the top management; the working capital.

The size of the company.
The management of information used for decision-making.
The type of strategy.
The infrastructure; and
The MYU- μ (unexpected aspects that may occur such as pandemics, acts of God, death of a member, internal or external acts of corruption that affect the organization, among others).

Figure 3. Aspects affecting organization's complexity.

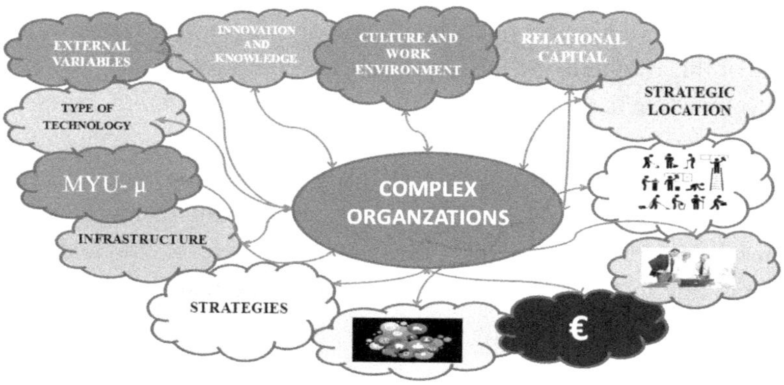

Source. Perez-Uribe et al., (2023)

REFERENCES

Almeida-Guzmán, M., & Díaz-Guevara, C. (2020). Circular economy, a strategy for sustainable development. Advances in Ecuador. *Management Studies: International Journal of Management*, (8), 34–56.

Álvarez-Aros, E. L., & Bernal-Torres, C. A. (2017). Open Innovation Model: Focus on Human Potential. *Technology Information*, *28*(1), 65–76.

Amador, J. A. L., Espinoza, Á. A. E., & Payares, A. M. C. (2022). Internal or external innovation? An organizational dilemma. *Economic Outlook*, *30*(1), 25–40.

Atarama-Rojas, T., Robledo-Dioses, K., & Kam Arteaga, L. (2022). Branded content y marcas de moda de lujo: Un estudio sobre la composición de los fashion films. *Correspondence & Analysis*, (15), 57–82.

Barrera Castro, L. A., & Bossa Eljaiek, A. C. (2021). *Business Model: Guilt-Free Fashion, a Circular Economy Approach.* Master's Degree in Business Innovation Management. Center for Innovation and Sustainability. Faculty of Business Administration. Universidad Externado de Colombia Bogotá. https://bdigital.uexternado.edu.co/bitstreams/c8d6a38d-021f-447a-929c-0132f892f1a8/download

Barreto Fereira, J. R., & Petit Torres, E. E. (2017). Explanatory models of the technological innovation process in organizations. [Universidad del Zulia, Venezuela. https://www.redalyc.org/articulo.oa?id=29055964004]. *Venezuelan Journal of Management*, *22*(79), 2017.

Benavides, C. A. (1998). *Technology, Innovation and Business*. Pyramid Editions.

Braungart, M., & McDonough, W. (2009). *Cradle to cradle*. Random House. https://books.google.com.co/books?hl=es&lr=&id=13hfHzBstcEC&oi=fnd&pg=PT2&dq=Cradle+to+cradle+&ots=IenVdJBSTw&sig=ZhkBVi35M27Rou0YrsJcXBs3aso&redir_esc=y#v=onepage&q=Cradle%20to%20cradle&f=false

Cabello Cervantes, L. M., Estrada López, H. H., Cáceres Martelo, A., & Fontalvo Torres, F. (2020). Circular economy and ecodesign in the furniture industry. Querétaro, Universidad Autónoma de Querétaro, 2020. https://rd.udb.edu.sv/server/api/core/bitstreams/80b1f06e-1edf-4e07-98c7-0cc078261317/content

Calderón Hernández, G., Murillo Galvis, S., & Torres Narváez, K. (2003). Organizational Culture and Workplace Well-being. *Administration Notebooks*, *16*(25), 109–137.

Camacho, M. M. (2002). Strategic direction: Analysis of a powerful tool. *Via Salud Journal*, (21), 2.

Carrillo González, G., & Pomar Fernández, S. (2021). The circular economy in new business models. *Entreciencias: dialogues in the knowledge society, 9*(23).

Carrión, R. D. I., Huanca, O. A. C., & Tinitana, K. M. L. (2019). Financial indicators and their impact on corporate insolvency. *Latin American Economic Observatory*, (11), 54.

Cerdá, E., & Khalilova, A. (2016). Circular economy. *Economía Industrial*, *401*(3), 11–20. https://www.mintur.gob.es/Publicaciones/Publicacionesperiodicas/EconomiaIndustrial/RevistaEconomiaIndustrial/401/CERD%C3%81%20y%20KHALILOVA.pdf

Cheikhrouhou, N., Pouly, M., & Madinabeitia, G. (2013). Trust categories and their impacts on information exchange processes in vertical collaborative networked organisations. *International Journal of Computer Integrated Manufacturing*, *26*(1-2), 87–100. DOI: 10.1080/0951192X.2012.681913

Correa Tipán, J. L. (+), Ledesma Cervantes, J.S., y Peñaherrera Larenas, F. (2019). Importance of innovation and creativity in product development. *Pro-Sciences: Revista de Producción, Ciencias E Investigación*, E-ISSN: 2588-1000, vol. 3, N 22, mayo 2019, PP. 31-37

Da Silva, D. N., Vieira, R. K., Vieira, A. K., & de Santiago, M. (2016). Optimizing the innovation process for internal projects in companies. *Technology Information*, *27*(3), 119–130.

Danish, Q. D., & Usman, A. (2010). Impact of reward and recognition on job satisfaction and motivation: An empirical study from Pakistan. *International Journal of Business and Management*, *5*(2), 159–167. DOI: 10.5539/ijbm.v5n2p159

Escorsa Castells, P., & Jaume Valls, P. (2003). Technology and Innovation in Business. Volume 148 of *POLitext Business Organization,* Polytechnic University of Cataluña, ISBN 8483017067, 9788483017067

Fernández, I. (2015). *Organizational happiness. How to build happiness at work.* 1 edition. Ediciones B. Chile S.A. https://books.google.es/books?hl=es&lr=&id=zlQsDwAAQBAJ&oi=fnd&pg=PT5&dq=felicidad+en+el+trabajo&ots=CfkE_X1KXd&sig=L3Eq_Hg6ZWqp8f01BpoMVPEqCCY#v=onepage&q=felicidad%20en%20el%20trabajo&f=false

Fisher, C. (2003). Why do lay people believe that satisfaction and performance are correlated? Sources of a commonsense theory. *Journal of Organizational Behavior*, *24*(6), 753–777. DOI: 10.1002/job.219

Flores Konja, J. V., & Rozas Flores, A. E. (2008). Corporate governance: A modern approach. *Quipukamayoc*, *15*(29), 7–22. DOI: 10.15381/quipu.v15i29.2065

Fundación Vida Silvestre. (2019). *Sustainable production.* https://www.vidasilvestre.org.ar/nuestro_trabajo/concientizacion_y_educacion/campanas/dia_del_medio_ambiente/dia_mundial_del_medio_ambiente/cinco_propuestas_para_una_argentina_sustentable/produccion_sostenible/

Furstenberg, C. (2015). The inherent dignity of man must be respected, protected, and promoted in every human being: A worthwhile struggle for the protection of humanity. *Nursing Horizon*, *26*(2), 47–56.

García-Legaz, A. C. (2021). Innovation has a history. *Puente de Hierro,* V1, n.1. https://www.puentedehierro.org/ojs/index.php/pdh/article/view/13/7

González Candía, J., García Coliñanco, L., Caro, C. L., & Romero Hernández, N. (2014). Innovation strategy and culture, resource management, and idea generation: practices for managing innovation in companies. *Scientific journal Pensamiento y Gestión, 36.*

GRI. (2016). *Global Reporting Iniciative.* GRI 103: Management Approach 2016. https://www.globalreporting.org/standards/gri-standards-download-center/?g=30f6860d-7c3e-4d29-b3ae-98549f76774f

Hernández-Sampieri, R., Fernández-Collado, C., y Baptista-Licio, MdeP. (2014). *Research methodology.* McGrawHill. https://docs.google.com/viewer?a=v&pid=sites&srcid=ZGVmYXVsdGRvbWFpbnxjb250YWR1cmlhcHVibGljYTk5MDUxMHxneDo0NmMxMTY0NzkxNzliZmYw

International Finance Corporation (IFC). (2007). *Practical Guide: Strategies for Serving New Markets through Inclusive and Sustainable Business Models.* https://www.icex.es/icex/wcm/idc/groups/public/documents/documento_anexo/mde3/nzm1/~edisp/dax2017735778.pdf

Kim, W. C., & Mauborgne, R. (2005). *Blue Ocean Strategy*. Harvard Business School Press.

Landazury-Villalba, L. F., Jaafar-Orfale, H., Cristofani, M. A., & Canales-Cuba, R. (2018). Innovation and management models: their transformative reflection from the human and knowledge perspectives. *Revista Espacios*. Vol. 39 (Nº 13) Year 2018 • Pag. 20.

López Chanes, F., Casique Guerrero, A., & Ferrer Guevara, J. (2019). Job Satisfaction: An Analysis Based on Expectancy and Equity Theories. *Administration and Organizations, 9*(17), 117-134. https://rayo.xoc.uam.mx/index.php/Rayc/article/view/249

Lovibond, E. (2002). *Ethical Formation*. Harvard University Press.

Lozano-Correa, L. J., Perez-Uribe, R. I., & Ocampo-Guzman, D. (2020). Trusthfulness as a key factor in competitiveness in medium sized companies in Colombia. 19 p. 280-298. DOI. .DOI: 10.4018/978-1-7998-2019-2.ch015

Machado, J. A. L., Escobar, M. M., Valentón, E. M., Ramírez, P. E. G., & Escobar, A. L. (2023). Perception of respect for the right to education and decent work among medical interns in a secondary-level hospital. *Medicina Interna de México*, *39*(4), 602–609.

Martín Toquero, G. (2020). Circular economy business models. University of Valladolid School of Industrial Engineering Degree in Industrial Organization Engineering. Valladolid, July 2020. https://uvadoc.uva.es/bitstream/handle/10324/41750/TFG-I-1601.pdf?sequence=1&isAllowed=y

Martínez Vázquez, G., & Espinosa Yañes, A. (2017, July-December). Building business and technological capabilities. The case of a small Mexican technology-push company. [Universidad de La Habana La Habana, Cuba]. *Economy and Development*, *158*(2), 117–128.

Mcdonough, W., & Braungart, M. (2003). *Cradle to Cradle*. Editorial McGraw-Hill.

Miller, W. L., & Morris, L. (2008). *Fourth generation R&D: Managing knowledge, technology, and innovation*. John Wiley & Sons.

Montenegro, S. L. F. (2021). *The code of ethics and its influence on the anti-corruption policy of a public entity in eastern Peru.* Graduate School. Academic Program of master's in public management. Universidad Cesar Vallejo. https://hdl.handle.net/20.500.12692/73000

Muñoz-La Rivera, F., Hermosilla, P., Delgadillo, J. & Echeverría, D. (2021). Proposal for the construction of innovation skills in engineering education in the context of industry 4.0 and sustainable development goals (SDG). *University education*, *14*(2), 75-84.

Ortiz, H. (2021). Financial indicators. *Applied Financial Analysis. Bajo NIIF*, *16*, 191–202.

Otzen, T., & Manterola, C. (2017). Sampling Techniques for a Study Population. *International Journal of Morphology*, *35*(1), 227–232. DOI: 10.4067/S0717-95022017000100037

Ouchi, W. (1982). *Theory Z: How companies can meet the Japanese challenge.* Bogotá, Colombia. Ed.Norma.

Palacios Cardona, A. I., Mejía Mazo, L. F., & Pérez Rengifo, C. M. (2025). *Innovation for sustainability in six large Colombian companies belonging to the G12.* https://repository.eafit.edu.co/items/79bae7e8-76dd-49d2-9bb1-c01aecfa514d

Perez –Uribe. R. (2018). *Corporate Strategic Management.* Ediciones Ecoe Ltda. First Edition. ISBN: 978-958-771-630-6. eISBN: 978-958-771-631-3. 259 p. April. https://www.ecoeediciones.com/libros/libros-de-administracion-ecoe/gerencia-estrategica-corporativa/

Perez-Uribe, R. (2020). Innovation and sustainability in the company L. *Revista Citas, Universidad Santo Tomas.* Vol. 6 Núm. 1 (2020), Innovación. January 1. Pages. 1-16. DOI:

Perez-Uribe, R., Ocampo-Guzmán, D., Moscoso-Duran, F., & Ramírez-Salazar, M. P. (2021). Innovation and Sustainability Management as a Key Factor in Global Purpose-Driven Micro, Small, and Medium-Sized Bogotanas Companies: Its Impact on Financial Results in MSMEs. *Advances in Business Strategy and Competitive Advantage*, *26*, 354–379. DOI: 10.4018/978-1-7998-4909-4.ch019

Perez-Uribe, R., Ramírez-Garzón, M. T., Ramírez-Salazar, M. D. P., & Salcedo-Perez, C. (2020). Transitioning From Medium to Large Companies in the Circular Economy. *Advances in Business Strategy and Competitive Advantage*, *21*, 323–344. DOI: 10.4018/978-1-7998-5116-5.ch017

Pérez-Uribe, R. I., & Ramírez-Salazar, M. D. (2021). Guide for managing the Route for innovation and sustainability for Enterprises model (RISE). DOI: DOI: 10.13140/RG.2.2.13026.45762

Perez-Uribe, R. I., Ramírez-Salazar, M. D., Ocampo-Guzmán, D., & Saiz-Álvarez, J. M. (2021). Organizational Sustainability and Value Creation by Means of Innovation in Chaotic Environments. In R. Perez-Uribe, D. Ocampo-Guzman, N. Moreno-Monsalve, & W. Fajardo-Moreno (Ed.), *Handbook of Research on Management Techniques and Sustainability Strategies for Handling Disruptive Situations in Corporate Settings.*17/06/2021. (pp. 1-29). 29 pages. IGI Global. http://doi:10.4018/978-1-7998-8185-8.ch001 DOI: 10.4018/978-1-7998-8185-8.ch001

PNUMA. (2016). *Manual of the Vienna Convention for the Protection of the Ozone Layer.* Tenth edition of 2016. https://observatoriop10.cepal.org/sites/default/files/documents/treaties/vc-handbook-2016-spanish.pdf

Predassi, S. (2023). Circular Future: Ecodesign and Circular Economy Strategies. *Proyectual-D*, (1), e004-e004. https://papelcosido.fba.unlp.edu.ar/ojs/index.php/proyectual-d/article/view/1847

Prieto-Sandoval, V., Jaca-García, C., & Ormazabal-Goenaga, M. (2017). Circular economy: Relationship with the evolution of the concept of sustainability and strategies for its implementation. *Engineering Research Report*, num. 15 (2017) 85 ISSN 2301-1092 https://dadun.unav.edu/bitstream/10171/53653/1/Economia_Circular.pdf

Prieto-Sandoval, V., Jaca-García, C., & Ormazabal-Goenaga, M. (2017). *Circular Economy: Relationship with the Evolution of the Concept of Sustainability and Strategies for its Implementation. Engineering Research Report,* num. 15 (2017) ISSN 2301-1092 • ISSN (online) 2301-1106.

Ramírez- Salazar. MP. (2016). *Collaborative Open Innovation Model for Development Banking: The Bancóldex Case*. PhD. dissertation. PhD, in Management. Universidad EAN. 378 P. ISBN: 978-958-756-384-9.

Ramírez-Salazar, M. D. P., Salcedo-Perez, C., & Castañeda, S. E. (2021). Open Collaborative Innovation at Colombia's National Spectrum Agency. *Advances in Business Strategy and Competitive Advantage*, *24*, 169–192. Advance online publication. DOI: 10.4018/978-1-7998-3568-4.ch009

Razzouk, R., & Shute, V. (2012, September). What Is Design Thinking and Why Is It Important? *Review of Educational Research*, *82*(3), 330–348. DOI: 10.3102/0034654312457429

Perez-Uribe, RI., Ocampo-Guzmán, D., Moscoso-Duran, FF., Salcedo-Perez, C. & Muñoz Maya, CM. (2023). Happiness at Work in Colombian Companies: A Perspective Under the Rise Model. Source Title: *Transformational Leadership Styles, Management Strategies, and Communication for Global Leaders*. Pages: 22. (pp. 380-401). DOI: DOI: 10.4018/978-1-6684-8822-5.ch019

Rodríguez, A. (2015). People, a source of innovation and value creation in the company.*Publicaciones de la Real Academia de Ciencias Económicas y Financieras*. https://www.researchgate.net/publication/288499382_Las_personas_fuente_de_innovacion_y_de_creacion_de_valor_en_la_empresa

Rodríguez-Peña, A. (2025). Corporate Entrepreneurship and the sustainable development goals. exploring the connection. *Journal of Innovation and Entrepreneurship*, *14*(1), 33. Advance online publication. DOI: 10.1186/s13731-025-00481-0

Rodríguez Viñas, C. (2019). Ecodesign: Concept and Applications. CIC: 5th Bulletin of the Creativity Research Center *UCAL*. Vol. 1 (nov. 2019). https://repositorio.ucal.edu.pe/bitstream/handle/20.500.12637/288/Arti%cc%81culo_1.pdf?sequence=1&isAllowed=y

Rozo Doncel, G. (2019). *State of the art of the circular economy in Colombia*. Universidad Cooperativa de Colombia. Faculty of Administrative, Economic, and Related Sciences. Business Administration Program. Bogotá. https://repository.ucc.edu.co/server/api/core/bitstreams/52ae3557-008c-46e0-b1b8-9e7dfa222483/content

Schein, E. (1988). *Corporate culture and leadership: a dynamic view*. Plaza y Janés.

Scholes, K., & Johnson, G. (Eds.). (2001). *Exploring public sector strategy*. Pearson Education.

Terán Bustamante, A., Dávila Aragón, G., & Castañón Ibarra, R. (2019). Technology and innovation management: A Bayesian network model. *Economía: Teoría y Practica*, (50), 63–100.

Van Wyk, R. J. (2004). A template for graduate programs in management of technology (MOT), Report to the Education Committee. *International Association for Management of Technology (IAMOT)*. Retrieved from: https://www.iamot.org/homepage/2004-MOTTemplateE

WIPO. (2007). *"Cradle to Cradle" - Fully "green" life cycle*. Abril. https://www.wipo.int/wipo_magazine/es/2007/02/article_0002.html

KEY TERMS AND DEFINITIONS

Core Business: it is also called the core competence, essential competence, or key competence of an organization. In managerial terms, it refers to the activity that generates value and that is necessary to create a beneficial competitive advantage for an organization. The distinctive competences are determined by an analysis of processes of chain value that allows us to know the activities necessary to satisfy the customer's needs, costs, qualities, and corresponding efficiency.

Cradle To Cradle: this philosophy proposes to face the roots of the problems: instead of reducing energy consumption, from the idea and design of any product, strategy, or policy all stages of the products must be considered (extraction, process, use, reuse, recycling), in a way that energy uses are not even necessary, even having positive balances of expenses and contributions. The key concepts of cradle to cradle are intuitive and based on the nature and the connection with it: 1) use of the energy that reaches the planet (sun) instead of fossil fuels; 2) the full closing of cycles of materials: in the ecosystems of the planet the concept of waste does not exist. Our societies can do the same by designing all products in a way that materials are recycled while they are used, or in a way that the subsequent use will have a higher value than the initial one.

GRI: The Global Reporting Initiative Standards are the best international practices designed to inform the public about a variety of economic, environmental, and social impacts. The presentation of reports of sustainability from these standards provide information about contributions (positive or negative) of organizations to sustainable development.

UNEP: The UN Environment Programme is the main global environmental authority, in charge of evaluating the state of the world environment and the causes of environmental impacts.

RISE: The RISE Model (Route of Innovation for Sustainability to Enterprises) allows companies to perform an assessment based on five levels to manage innovation for sustainability. The results obtained allow organizations to develop their own paths with the best practices of innovation for sustainability.

WIPO: The World Intellectual Property Organization is the global forum for intellectual property policy, services, information, and cooperation.

Chapter 20
Efficacy of Deep Neural Embeddings-Based Approach vs. Traditional Machine Learning:
An Experimental CASE Study

Kaneeka Vidanage
https://orcid.org/0000-0002-1923-9084
General Sir John Kotelawala Defense University, Sri Lanka

M. W. P. Maduranga
Sri Lanka Technology Campus, Sri Lanka

ABSTRACT

Building robust machine learning models requires large, annotated datasets often unavailable for specific business domains. In fields such as medicine, agriculture, etc, the annotation process adds to the overhead, requiring the involvement of domain specialists. This is time-consuming and costly. Additionally, data scientists must perform exploratory data analysis to address dataset imbalances, adding another layer of complexity. Data collection from diverse sources requires careful exploration to avoid compromising important features. Even after overcoming these challenges, model training and hyperparameter tuning remain resource-intensive tasks. Furthermore, introducing new data necessitates retraining, often requiring the entire process to be repeated. In such contexts, this research explores the potential advantages of embedding techniques as an alternative to traditional dataset-centric machine learning via an experimental case study. The experimental case study yielded satisfactory outcomes for the embedding approach on par with traditional machine learning techniques.

DOI: 10.4018/979-8-3693-4030-1.ch020

INTRODUCTION

To build mature and robust machine learning models, large, annotated datasets are required. However, given the real business use cases, it is not possible to locate appropriately sized datasets for specific business domains (Mascolini et al., 2021). For example, the COVID-19 pandemic was a new experience for the entire world. Due to the lack of suitably sized annotated datasets, the use of machine learning techniques to derive the computational boost for COVID-19-related experimental investigation purposes was very primitive. This was mainly due to the non-availability of the appropriately sized annotated datasets linked with the COVID-19 domain (Fritz et al., 2022).

Though the dataset has been identified, adequate annotation and data exploration workflows cause additional overheads (Guerra et al., 2022). In fields like medicine, agriculture, construction, etc., the annotation process requires the cooperation of domain specialists as well. Obtaining the support of domain professionals over time until the annotation work is accomplished will be financially costly (Mende et al., 2023). Additionally, data scientists must undertake exploratory data analysis to resolve the imbalances and discrepancies in the dataset. This will likewise be a time-critical operation with the involved costs (Batzolis et al., 2023). In addition, data collection from various sources must be done with extreme caution ensuring the important features are not compromised (Howe & Papa, 2023). For example, to train a machine learning model for a supermarket`s item detection, each individual item's required image counts must be obtained in a good setting with a non-disturbing background. These images obtained are used for the training of the respective machine learning models. Therefore, establishing and executing a comprehensive image acquisition pipeline is a significant overhead for the stakeholders.

However, after successfully overcoming all of the aforementioned problems, a machine-learning model can be trained. Henceforth, again a significant amount of work must be put into the hyper-parameter tuning procedure to achieve consistently high accuracy (Saikia et al., 2023). Aside from that, when new data points are introduced to the dataset, the models must be re-trained (Raymaekers et al., 2024). This is another cumbersome procedure, as adding new data can occasionally reduce accuracy compared to the prior edition. As a result, hyper-parameter tuning operations need to be repeated from the beginning. Therefore, the development teams have a significant challenge in scaling up the existing machine-learning models to meet the increased business expectations (Martineau et al., 2023).

Embeddings-based techniques align well with established innovation management frameworks such as Diffusion of Innovations (Raman et al., 2024) and Open Innovation (Li et al., 2024). By enabling efficient model development with limited labeled data, embeddings foster rapid adoption of AI-driven solutions across organizations, accelerating the diffusion of machine learning innovations. Moreover, these techniques support open innovation by facilitating collaboration with external partners, such as research institutions and industry stakeholders, through shared pre-trained models and transfer learning. However, challenges such as data privacy concerns and integration complexities must be addressed to fully leverage embeddings within an organization's innovation ecosystem.

As previously stated, several bottlenecks negatively impact on the training of machine learning models to solve business challenges. Hence, this research investigates the advantages of the probable use of embedding techniques over pure dataset-based machine learning processes.

Contribution to the paper is as follows.

- This paper triggers the awareness and insights that leverages the use of embedding techniques over the traditional model training amidst machine learning solutions

- Emphasizes on the potential advantages over the tradeoffs associated with traditional model training approach via an experimental case study with quantitative resolutions.
- Justify on the efficiency gains and cost reductions associated with
    - Data annotation procedures and complexities.
    - Hyper-parameter tuning costs avoid overfitting / underfitting challenges.
    - Model`s scalability elasticity on new data elements.
    - Overall variations related to administration costs in using the embeddings technique over the traditional model training.

LITERATURE REVIEW

Traditional Machine Learning Techniques

The use of Machine learning mechanisms has become a very popular approach in developing intelligent applications (Li et al., 2022). However, machine learning is just another technique residing under the Artificial Intelligence umbrella (Sigatapu et al., 2023). Machine learning mechanisms can be further segmented into subcategories such as supervised learning, unsupervised learning and reinforcement learning (Unhelkar & Gonsalves, 2021). In supervised learning, it's mandatory to have significantly sized large, annotated datasets. Therefore, once the required datasets have been acquired, annotation procedures need to commence with the involvement of domain specialists as needed (Shetty et al., 2022).

Subsequently, machine learning engineers need to play a pivotal role in properly fine-tuning the model and overcoming the overfitting / underfitting issues. Artificial Neural Networks also belong to the segment of supervised learning. Once neural network architecture is designed with multiple hidden layers it`s referred to as a deep learning model (Fink et al., 2024). In the process of training deep neural network architectures also, a large size of pre-processed and feature-engineered datasets is a must. Afterwards, numerous parameters such as kernel size, batch size, learning rate, epoch counts, activation function types, and optimizer adjustments can be done as hyper-parameter tuning and refinements to ensure proper accuracy (Saleh & Issa, 2020). This is mostly a trial-and-error type of domain-specific experimental attempt.

In circumstances where the dataset is extensively large, and handling is very complex, unsupervised learning can be introduced as a remedy (Zhang & Peng, 2024). Additionally, unsupervised learning is an ideal strategy for dimensionality reduction as well (Roh et al., 2025). Unsupervised learning itself can be utilized as a strategy in data pre-processing and exploratory analysis. Because, once the required cluster counts are provided, the dataset is segmented into different clusters of the provided count. Inside each cluster, the residing elements are homogeneous. Therefore, once the unsupervised learning has been applied to cleanse the dataset, afterwards supervised learning can be applied to each cluster as required. Unsupervised learning does not require data annotation (Roh et al., 2025).

However, when providing an optimal business resolution, it's recommended to use unsupervised learning first to derive the homogenous clusters and then apply supervised learning to each cluster with appropriate data annotations (Roh et al., 2025; Zhang & Peng, 2024).

The next widely used mechanism is re-enforcement learning. In this mechanism, models learn based on punishments and rewards (Dawud, 2020). This strategy is ideal for interaction-based dynamic environments. Unlike previously discussed machine learning strategies, re-enforcement learning is not that greedy, as it absorbs its learning mainly from the interactions and the feedback (Zhang, 2024). However, the rudimentary data needs to be supplied for the re-enforcement learning process to be begun effectively.

Machine Learning-Based Embeddings

The use of machine learning embeddings is another alternative approach that could be utilized for intelligent application development (Wang, Hohman, & Chau, 2023). Intelligence derived from traditional machine learning strategies is quite different compared to the intelligence developed via embedding-based approaches. Traditional machine learning techniques focused on learning the underlying patterns rooted in the datasets used (Wu & He, 2023). In the embeddings-based approach, the learning aspect is not present. However, it allows for a more detailed and fine-grained comparison. This allows to perform precise segregation sometimes even better than the traditional machine learning techniques (Abbas et al., 2024).

The significant usage of the embeddings-based mechanism is not prominent with all machine learning techniques (i.e. SVM, Naïve Bayes, Regression, Trees). However, it can be more accurately utilized with deep neural network architectures (López & Gosztolya, 2021). When the target and the source images, text or audio files are sent through the specified deep neural network architecture, it will outcome specific tensor sequences. These tensor sequences derived from both the source and target files can be compared against each other using rational variation assessment techniques like cosine similarity (Wu et al., 2023), Jacquard co-efficient (Chowdhury & Sinha, 2024) or vector space (Buatoom, 2024) techniques. This similarity/dissimilarity analysis allows us to recognize the matching percentages between the source and the target inputs. This percentage output can be utilized for rational segregation leading to an intelligent outcome. This cannot be recognized as real intelligence as in traditional machine learning techniques; hence the model will not learn anything from the dataset provided. However, the fine-grained granular feature-based tensor`s output analysis allows to yield of intelligent alike accurate outcomes.

Nevertheless, traditional machine learning techniques are much more stable to be applied to dynamically varying problem contexts (Jui & Rivas, 2024), such as the use of computer vision in an autonomous vehicle, analyzing of vocals of an apprentice singer etc. Because all such problem domains are dynamic and could vary over time. However, the traditional machine learning mechanisms would perform reasonably well in such contexts as they are tuned to learn the underlying patterns associated with the problem of concern and adjust accordingly. The embeddings-based technique relies on fine-grained feature-based tensor comparison (Wang, Hohman, & Chau, 2023). In this mechanism learning of the underlying patterns will not take place as in traditional machine learning approaches, Therefore, embeddings-based approaches are not suitable for the dynamically changing atmospheres as it can perform well on feature-based static use cases only. However, one of the significant advantages of embedding-based mechanisms is the ability to function with a stable high accuracy even during the absence of large, annotated datasets (Pattun & Kumar, 2023). Hence, the embedding technique can be introduced as a rational alternative for developing intelligent solutions in a much more cost-effective manner due to the non-requirement of the huge datasets. Additionally, this saves computational resources as well, by eliminating the need for long hours of rigorous model training (Truică & Apostol, 2023).

Embeddings Applications

1. Sentiment Analysis – Word2Vec (Zhu & Ren, 2023), Glove (Putra & Setiawan, 2023), and fastText (Ghosal & Jain, 2023) are some of the widely used words embedding generation techniques. These tools contain different deep neural network architectures. Based on the inputs provided, numerous tensor-based word embeddings are generated.

The generated word embeddings can be used for text comparison, sentiment analysis, semantic similarity assessment, information retrieval needs and document classification purposes.

2. Computer Vision – Convolutional Neural Network (CNN) architectures can be used to generate image embeddings (Li et al., 2020). These image embeddings generated can be used for purposes such as image captioning, image matching and searching, object detection, facial recognition, facial attribute analysis, facial expressions and emotion analysis (Sbrolli et al., 2024) and many more.
3. Audio Analysis – Via transferring the respective audio clips through the specified deep neural network architectures, audio embeddings can be generated (Koh & Dubnov, 2021). These could be used for purposes like audio analysis, audio recommendations, tune pitch assessment and numerous other applications.

The aforementioned strategies are implemented in the form of recommendation systems, profiling systems, filtering systems, diagnostics systems, and optimization systems in abundant domains (Ettorre et al., 2021). The unique advantage of embedding-based applications across the domains is the ability to operate with very limited or almost no training datasets (Wilkinghoff & Fritz, 2023). This eliminates several rigorous procedures associated with dataset preparations contributing to an effective cost-advantage.

METHODOLOGY

This research conducts a comparative experimental study on the utilization of traditional deep learning architecture training versus the application of deep learning-based image embedding analysis over pneumonia detection through chest X-ray images of the lungs as depicted in Figure 01.

Figure 1. Chest X-ray image of the lung

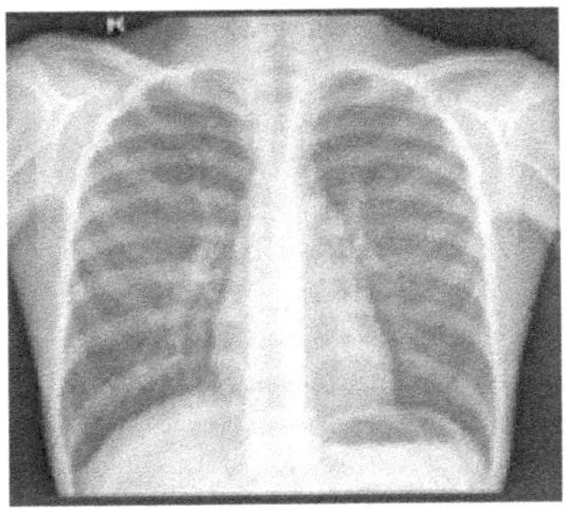

A pneumonia chest X-ray image dataset derived from the Kaggle was selected for this experiment. It comprises a total of 5216 images covering both pneumonia-affected and normal chest x-ray images.

Experiment 01 is conducted with the traditional training approach, VGG-16 deep neural network architecture (Gupta & Patel, 2023) has been utilized for it. To reduce the training efforts, transfer learning has been applied with the top FALSE of the VGG-16 architecture. After flattening the layer outputs derived from the trained ImageNet weights, a trainable dense layer has been introduced as the custom layer with SoftMax activation to determine the final decision as pneumonia positive or negative.

As in experiment 02, a source image of pneumonia-positive chest X-ray has been sent through the identical VGG-16 architecture with transfer learning which was used in experiment 01. The produced image embeddings were saved inside the list object as the first element. Subsequently, the target pneumonia-positive chest X-ray image was sent across the VGG-16 architecture and generated embeddings were stored in the same list object as the second element.

Consequently, cosine similarity was conducted across the two embedded results stored in the list object. A similar workflow has been followed for a non-pneumonia-based chest X-ray image as well. Multiple trials were conducted in both experiment options 1 and 2. The same test images were provided for the comparison amidst both experiments along with the same deep neural network architecture. The complete experimental pipelines utilized in experiments 01 and 02 have been depicted in Figure 02.

Figure 2. Experimental pipelines

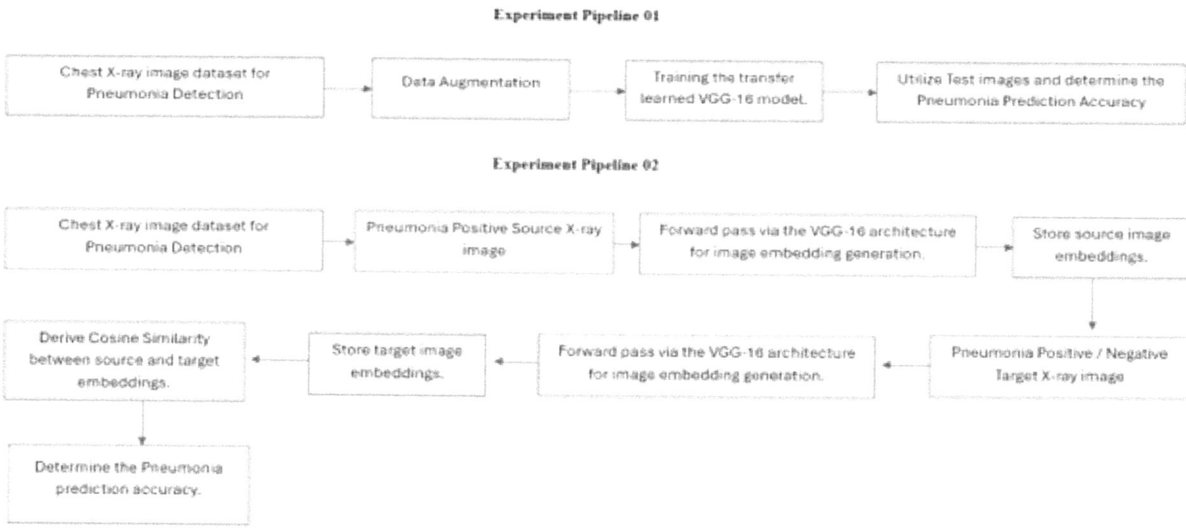

RESULTS & DISCUSSIONS

In experiment-01, prediction confidence was utilized as the judging parameter and in experiment-02 similarity metric derived was used as the judging parameter. In both cases, a judging metric value greater than 0.80 is determined as a successful classification. In cases with a judging parameter less than 0.80, it's recognized as weak predictions are decided as True Negatives (TN).

Fifty experimental trials were conducted using experiment pipelines 1 and 2. Accurate classifications derived were identified as True Positives (TP) and wrong classifications as False Positives (FP). Henceforth, using confusion matrix calculations, matrices like Sensitivity, Specificity, Precision, Accuracy and F-Measure were derived and logged in Table 1.

Table 1. Log of matrix calculations

Pneumonia detection using Chest X-ray images of the lung	Metric	Experiment – 01 (Training-based)	Experiment-02 (Embedding-based)
	Sensitivity	0.96	0.89
	Specificity	0.97	0.88
	Precision	0.95	0.84
	F – Measure	0.96	0.92

Figure 3. Confusion matrix for experiment 1

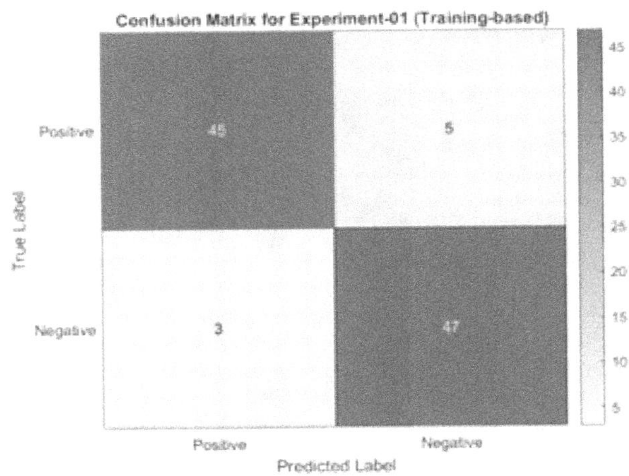

Figure 4. Confusion matrix for experiment 2

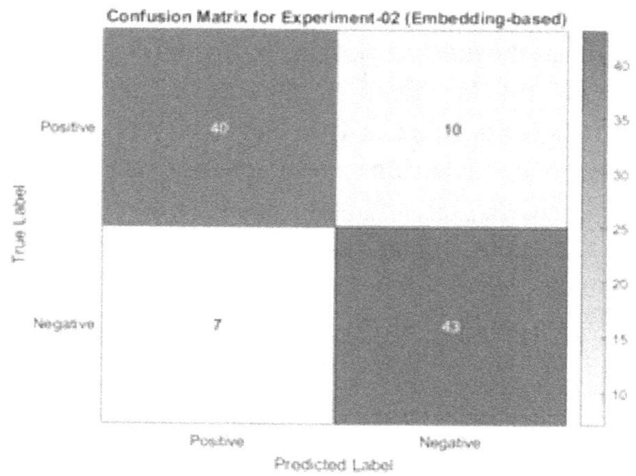

Figure 3 shows that the traditional training-based approach (Experiment-01) performs well, with a high number of True Positives (TP) and True Negatives (TN) and minimal False Positives (FP) and False Negatives (FN), indicating strong sensitivity and specificity. In contrast, Figure 4 reveals that while the embedding-based approach (Experiment-02) also performs well, it has a slightly higher rate of FP and FN, resulting in lower classification accuracy compared to Experiment-01. The increased misclassification affects both sensitivity and specificity.

Figure 5. Performance matrix comparison

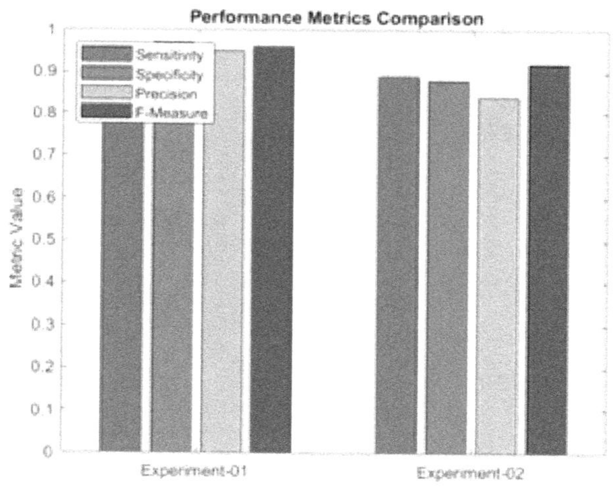

Figure 6. ROC curve comparison

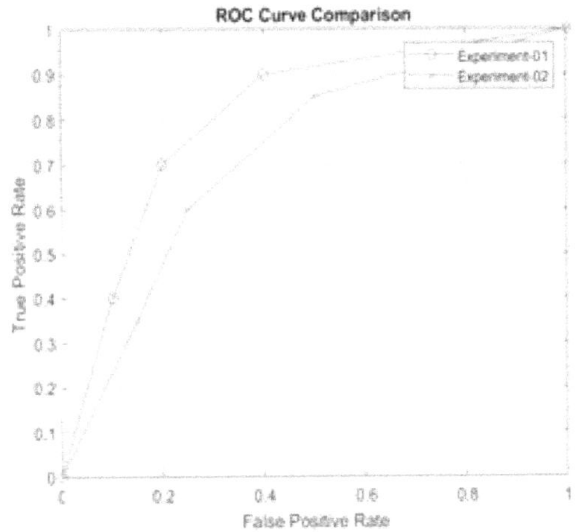

Figure 5 compares the key performance metrics—Sensitivity, Specificity, Precision, and F-Measure—between the two experimental approaches. It is evident that the traditional training-based method (Experiment-01) outperforms the embedding-based method (Experiment-02) across all metrics. The sensitivity and specificity of Experiment-01 are higher, which indicates that the model can better detect true pneumonia cases while minimizing false alarms. Although Experiment-02 shows relatively high Precision and F-Measure, its performance is still very slightly inferior to the traditional approach. This suggests that while the embedding-based method may be computationally more efficient, it sacrifices little amount of accuracy.

The ROC curves in Figure 6 display the trade-off between the true positive rate (Sensitivity) and the false positive rate (1 - Specificity) for both Experiment-01 and Experiment-02. The curve for Experiment-01 (traditional training) is closer to the top-left corner, indicating a superior balance between Sensitivity and Specificity. In contrast, the ROC curve for Experiment-02 (embedding-based) is slightly below that of Experiment-01, suggesting a slightly lower ability to distinguish between pneumonia-positive and pneumonia-negative cases. This analysis reinforces the earlier conclusion that the traditional training-based approach yields better classification results. However, even with small amount of data embeddings based technique is also not that inferior against the traditional model training approach.

Figure 7. Cosine similarity distribution

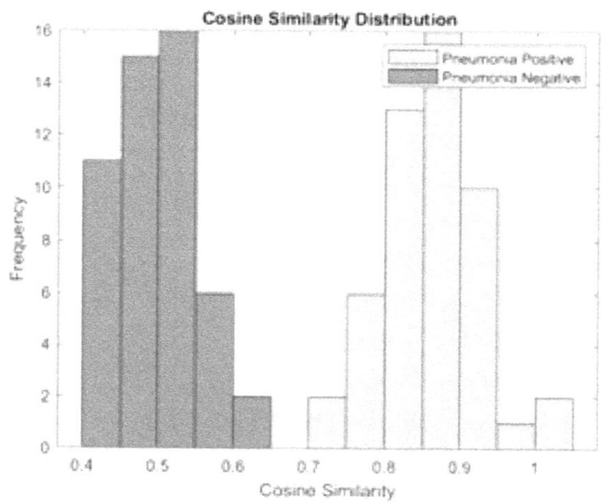

Figure 8. Accuracy over 50 trails

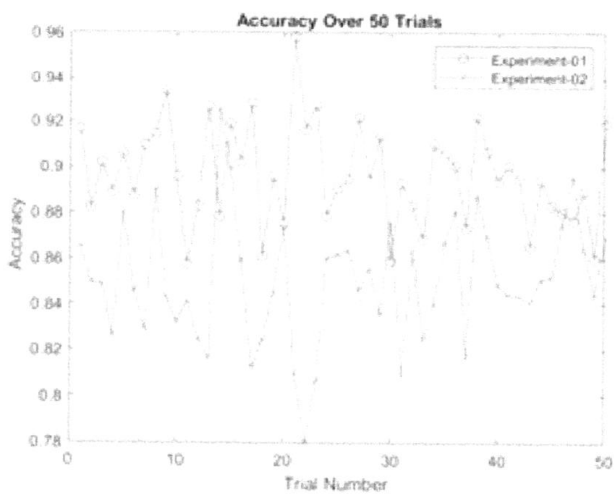

Figure 7 depicts the distribution of cosine similarity values for pneumonia-positive and pneumonia-negative chest X-ray images in Experiment 02. The results show a clear distinction between the cosine similarity values of positive and negative cases. Pneumonia-positive cases tend to cluster around higher similarity values (closer to 1), whereas pneumonia-negative cases have lower similarity values. However, there is some overlap between the two distributions, indicating that the embedding-based method occasionally struggles to differentiate between the two classes. This overlap contributes to occasionally a little higher number of false positives and false negatives observed in the confusion matrix of Experiment-02.

Figure 8 illustrates the accuracy of both Experiment-01 and Experiment-02 over 50 trials. The traditional training-based approach (Experiment-01) demonstrates consistently higher accuracy across trials, with minimal fluctuations. On the other hand, the embedding-based approach

(Experiment-02) exhibits a bit of variability, with little lesser average accuracy. These results confirm that while the embedding-based approach can achieve reasonable accuracy, it lacks the stability and consistency observed in the traditional training approach. The traditional method is more reliable for pneumonia detection based on chest X-ray images.

Figure 9. Precision-recall curve

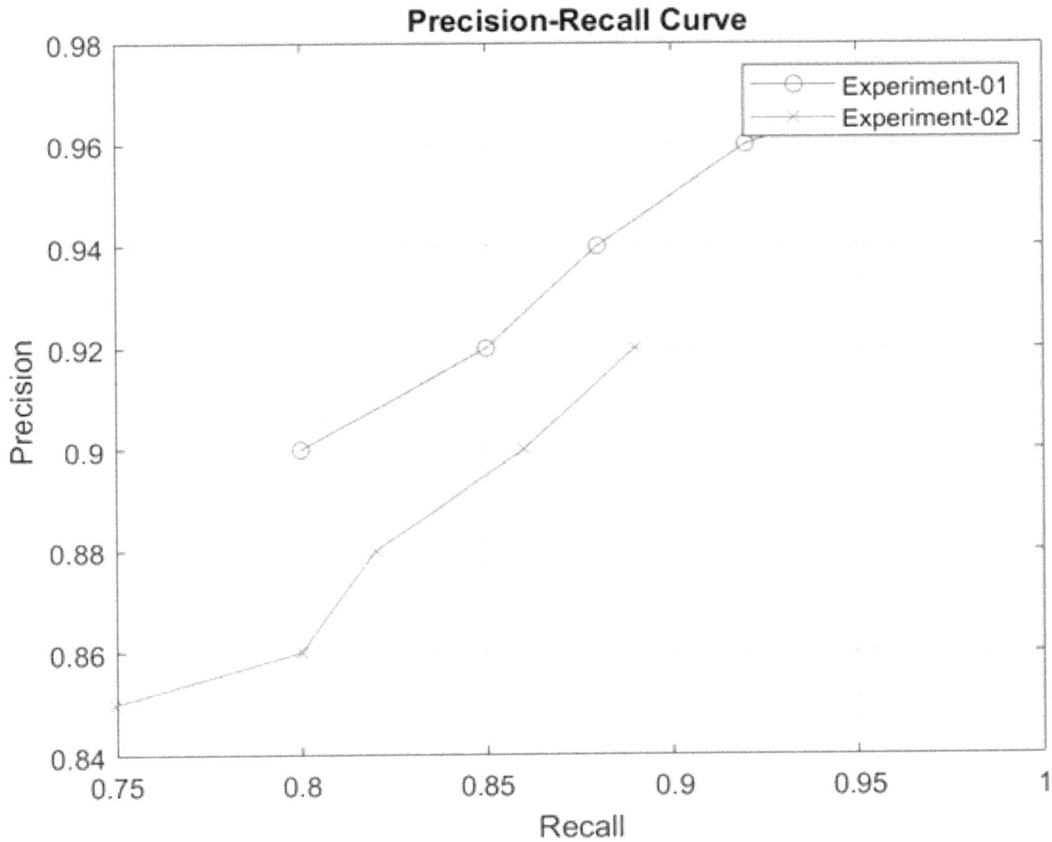

The Precision-Recall curve in Figure 8 highlights the relationship between Precision and Recall (Sensitivity) for both experiments. The curve for Experiment-01 lies above that of Experiment-02, indicating better precision and recall trade-off. Experiment-01 is more capable of maintaining high precision while also achieving strong recall, meaning it can accurately classify pneumonia cases without compromising the detection of positive cases. Experiment 02, while still achieving reasonable performance, shows a slight drop in precision as recall increases, indicating a greater likelihood of false positives in the embedding-based approach.

Embeddings-based techniques offer significant advantages beyond classification accuracy, particularly in resource-constrained environments where computational efficiency is crucial. By leveraging pre-trained embeddings, organizations can minimize reliance on extensive labeled datasets and high-end computational resources, accelerating model deployment and adaptation to new scenarios. This efficiency fosters a culture of rapid innovation, enabling businesses to experiment with novel AI-driven applications at lower costs. Beyond healthcare, embeddings-based methods have demonstrated impact across diverse industries. In finance, they enhance fraud detection by capturing complex transactional patterns with minimal labeled data. In retail, embeddings improve personalized recommendations by understanding customer preferences based on purchase history. In cybersecurity, anomaly detection models leverage embeddings to identify emerging threats with greater accuracy. Furthermore, these techniques facilitate seamless knowledge transfer across domains, allowing organizations to repurpose existing models for different tasks, improving scalability, and maintaining a competitive edge in dynamic, AI-driven industries.

While embeddings-based approaches offer numerous advantages, they also come with limitations, particularly in dynamic environments where data distributions evolve over time. Since embeddings are often pre-trained on static datasets, their representations may become less effective when dealing with concept drift, requiring periodic fine-tuning or retraining. Additionally, embeddings-based models can be prone to overfitting if trained on limited or biased data, leading to poor generalization in real-world applications. Conversely, underfitting may occur when embeddings fail to capture domain-specific nuances, reducing model effectiveness. A promising solution lies in hybrid approaches that integrate embeddings with traditional ML techniques, such as combining embeddings with adaptive learning algorithms or reinforcement learning to dynamically update representations. By blending embeddings-based methods with domain-specific feature engineering and fine-tuned classifiers, organizations can achieve greater robustness, improving adaptability and long-term model performance.

CONCLUSION

By analyzing the results presented in Table 1 and the accompanying visualizations, it is evident that the training-based technique has achieved consistently high accuracy across all cases. However, the embedding-based approach has also demonstrated stable and reliable performance, despite requiring minimal dataset utilization and no epoch-based training cycles. This presents significant advantages, such as drastically reducing dataset acquisition, annotation, and preprocessing costs while minimizing computational resource requirements for extensive training. Since obtaining large, high-quality datasets for all problem domains is often impractical, embedding-based techniques offer a viable alternative, enabling machine learning adoption in data-scarce environments.

To effectively implement embedding-based techniques, organizations should consider several key steps. First, they must assess their domain-specific constraints, ensuring that embeddings capture the necessary contextual information. Second, they should explore transfer learning strategies to leverage pre-trained models while fine-tuning them for specialized tasks. Additionally, hybrid approaches combining embeddings with traditional ML techniques can help improve adaptability in dynamic environments. For resource-constrained settings, organizations should prioritize computationally efficient embeddings and optimize deployment strategies, such as on-device inference or cloud-based solutions.

Long-term, the adoption of embedding-based approaches can drive innovation by lowering entry barriers for AI adoption, fostering rapid prototyping, and enabling cross-domain knowledge transfer. By reducing dependency on large, annotated datasets and extensive model training cycles, businesses can accelerate AI-driven solutions, enhance decision-making, and gain a competitive edge in fast-evolving industries. Thus, embeddings-based strategies represent a transformative shift in machine learning adoption, offering both technical and economic benefits for organizations navigating data-limited environments.

CONFLICT OF INTEREST

Authors have no conflict of interest to be declared

REFERENCES

Abbas, F., Malah, M., & Agaba, R. (2024). Image rendering with generative adversarial networks. In *Applications of Generative AI* (pp. 117–135). Springer International Publishing. DOI: 10.1007/978-3-031-46238-2_6

Batzolis, E., Vrochidou, E., & Papakostas, G. A. (2023). Machine Learning in Embedded Systems: Limitations, Solutions and Future Challenges. 2023 IEEE 13th Annual Computing and Communication Workshop and Conference (CCWC), 345-350. DOI: 10.1109/CCWC57344.2023.10099348

Buatoom, U. (2024). Improving Term Weight with Normalized Class Mutual Information for Sentiment Classification. *2024 9th International Conference on Business and Industrial Research (ICBIR)*. IEEE. DOI: 10.1109/ICBIR61386.2024.10875727

Chowdhury, P., & Sinha, B. B. (2024). Evaluating the Effectiveness of Collaborative Filtering Similarity Measures: A Comprehensive Review. *Procedia Computer Science*, *235*, 2641–2650. DOI: 10.1016/j.procs.2024.04.249

Dawud, J. (2020). A Novel Link Prediction Model Based on Deep Reinforcement Learning. *Solid State Technology*, 7190–7203.

Ettorre, A., Bobasheva, A., Faron-Zucker, C., & Michel, F. (2021). A systematic approach to identify the information captured by Knowledge Graph Embeddings. *IEEE/WIC/ACM International Conference on Web Intelligence and Intelligent Agent Technology*. DOI: 10.1145/3486622.3494027

Fink, Z., Parasyris, K., Rathi, P., Georgakoudis, G., Menon, H., & Bremer, P.-T. (2024). HPAC-ML: A Programming Model for Embedding ML Surrogates in Scientific Applications. *SC24: International Conference for High Performance Computing, Networking, Storage and Analysis*, 1-16. DOI: 10.1109/SC41406.2024.00078

Fritz, H., Tang, M., Kinney, K., & Nagy, Z. (2022). Evaluating machine learning models to classify occupants' perceptions of their indoor environment and sleep quality from indoor air quality. *Journal of the Air & Waste Management Association*, *72*(12), 1381–1397. DOI: 10.1080/10962247.2022.2105439 PMID: 35939653

Ghosal, S., & Jain, A. (2023). Depression and Suicide Risk Detection on Social Media using fastText Embedding and XGBoost Classifier. *Procedia Computer Science*, *218*, 1631–1639. DOI: 10.1016/j.procs.2023.01.141

Guerra, E., Wilhelmi, F., Miozzo, M., & Dini, P. (2022). The Cost of Training Machine Learning Models Over Distributed Data Sources. *IEEE Open Journal of the Communications Society*, *4*, 1111–1126. DOI: 10.1109/OJCOMS.2023.3274394

Gupta, V., & Patel, R. (2023). Lungs Disease Classification using VGG-16 architecture with PCA. *2023 International Conference on Advancement in Computation & Computer Technologies (InCACCT)*, 495-500. DOI: 10.1109/InCACCT57535.2023.10141690

Howe, A., & Papa, M. (2023). Feature Engineering in Machine Learning-Based Intrusion Detection Systems for OT Networks. *2023 IEEE International Conference on Smart Computing (SMARTCOMP)*, 361-366. DOI: 10.1109/SMARTCOMP58114.2023.00086

Jui, T. D., & Rivas, P. (2024). Fairness issues, current approaches, and challenges in machine learning models. *International Journal of Machine Learning and Cybernetics*, *15*(8), 3095–3125. DOI: 10.1007/s13042-023-02083-2

Koh, E.S., & Dubnov, S. (2021). Comparison and Analysis of Deep Audio Embeddings for Music Emotion Recognition. ArXiv, abs/2104.06517.

Li, J., Cai, K., Chen, H., Xu, L., Lin, Q., & Xu, F. (2022). Machine Learning Framework for Intelligent Detection of Wastewater Pollution by IoT-Based Spectral Technology. *Wireless Communications and Mobile Computing*, *2022*, 1–10. DOI: 10.1155/2022/9203335

Li, L., Yu, H., & Kunc, M. (2024). The impact of forum content on data science open innovation performance: A system dynamics-based causal machine learning approach. *Technological Forecasting and Social Change*, *198*, 122936. DOI: 10.1016/j.techfore.2023.122936

Li, Y., Li, X., Sun, Q., & Dong, Q. (2020). SAR Image Classification Using CNN Embeddings and Metric Learning. *IEEE Geoscience and Remote Sensing Letters*, *19*, 1–5.

López, J. V., & Gosztolya, G. (2021). Deep Neural Network Embeddings for the Estimation of the Degree of Sleepiness. ICASSP 2021 - 2021 IEEE International Conference on Acoustics, Speech and Signal Processing (ICASSP), 7288-7292.

Martineau, T., He, S., Vaidyanathan, R., & Tan, H. (2023). Hyper-parameter tuning and feature extraction for asynchronous action detection from sub-thalamic nucleus local field potentials. *Frontiers in Human Neuroscience*, *17*, 17. DOI: 10.3389/fnhum.2023.1111590 PMID: 37292583

Mascolini, A., Cardamone, D., Ponzio, F., Cataldo, S. D., & Ficarra, E. (2021). Exploiting generative self-supervised learning for the assessment of biological images with lack of annotations. *BMC Bioinformatics*, 23. PMID: 35871688

Mende, H., Frye, M., Vogel, P. A., Kiroriwal, S., Schmitt, R. H., & Bergs, T. (2023). On the importance of domain expertise in feature engineering for predictive product quality in production. *Procedia CIRP*, *118*, 1096–1101. DOI: 10.1016/j.procir.2023.06.188

Pattun, G., & Kumar, P. (2023). Emotion Classification using Generative Pre-trained Embedding and Machine Learning. *2023 IEEE International Conference on Machine Learning and Applied Network Technologies (ICMLANT)*, 1-6. DOI: 10.1109/ICMLANT59547.2023.10372980

Putra, D.T., & Setiawan, E.B. (2023). Sentiment Analysis on Social Media with Glove Using Combination CNN and RoBERTa. Jurnal RESTI (Rekayasa Sistem dan Teknologi Informasi).

Raman, R., Mandal, S., Das, P., Kaur, T., Sanjanasri, J. P., & Nedungadi, P. (2024). Exploring university students' adoption of ChatGPT using the diffusion of innovation theory and sentiment analysis with gender dimension. *Human Behavior and Emerging Technologies*, *2024*(1), 3085910.

Raymaekers, J., Rousseeuw, P. J., Verdonck, T., & Yao, R. (2024). Fast linear model trees by PILOT. *Machine Learning, 113*(9), 6561–6610. DOI: 10.1007/s10994-024-06590-3

Roh, H. R., Kim, C. S., Lee, Y., & Lee, J. M. (2025). Dimensionality Reduction for Clustering of Nonlinear Industrial Data: A Tutorial. *Korean Journal of Chemical Engineering, 42*(5), 1–15. DOI: 10.1007/s11814-025-00402-7

Saikia, P., Mangal, A., Jain, H., & Singh, A. (2023). Cuisine Prediction from Ingredients using Hyper Parameter Tuning on Machine Learning Algorithms. *2023 IEEE Silchar Subsection Conference (SILCON)*, 1-6. DOI: 10.1109/SILCON59133.2023.10404585

Saleh, Y., & Issa, G. F. (2020). Arabic Sign Language Recognition through Deep Neural Networks Fine-Tuning. *Int. J. Online Biomed. Eng., 16*(5), 71–83. DOI: 10.3991/ijoe.v16i05.13087

Sbrolli, C., Cudrano, P., & Matteucci, M. (2024). Can Shape-Infused Joint Embeddings Improve Image-Conditioned 3D Diffusion? *2024 International Joint Conference on Neural Networks (IJCNN).* DOI: 10.1109/IJCNN60899.2024.10649907

Shetty, S., Shetty, S., Singh, C. S., & Rao, A. (2022). *Supervised Machine Learning: Algorithms and Applications.* Fundamentals and Methods of Machine and Deep Learning.

Sigatapu, L., Sundar, S., Padmalatha, K., K, S., Ooha, D., & Devi, P.U. (2023). Artificial Intelligence in Healthcare- An Overview. Asian Journal of Pharmacy and Technology.

Truică, C., & Apostol, E.S. (2023). It's All in the Embedding! Fake News Detection Using Document Embeddings. ArXiv, abs/2304.07781.

Unhelkar, B., & Gonsalves, T. (2021). *Machine learning types.* Artificial Intelligence for Business Optimization. DOI: 10.1201/9781003120926-4

Wang, Z.J., Hohman, F., & Chau, D. (2023). WizMap: Scalable Interactive Visualization for Exploring Large Machine Learning Embeddings. ArXiv, abs/2306.09328.

Wilkinghoff, K., & Fritz, F. (2023). On Using Pre-Trained Embeddings for Detecting Anomalous Sounds with Limited Training Data. 2023 31st European Signal Processing Conference (EUSIPCO), 186-190.

Wu, S., Lu, F., Raff, E., & Holt, J. (2023). Exploring the Sharpened Cosine Similarity. ArXiv, abs/2307.13855.

Wu, Z., & He, H. (2023). *Traditional Machine Learning Models for Building Energy Performance Prediction: A Comparative Research.* Machine Learning Research.

Zhang, C. (2024). Advancing DRL agents in commercial fighting games: Training, integration, and agent-human alignment. *arXiv preprint arXiv:2406.01103.*

Zhang, H., & Peng, Y. (2024). Image clustering: An unsupervised approach to categorize visual data in social science research. *Sociological Methods & Research, 53*(3), 1534–1587. DOI: 10.1177/00491241221082603

Zhu, J., & Ren, Z. J. (2023). The Evolution of Research in Resources, Conservation & Recycling Revealed by Word2vec-Enhanced Data Mining. SSRN Electronic Journal. DOI: 10.1016/j.resconrec.2023.106876

Chapter 21
Innovation Management in Health Technology Scenarios: A Reflection on Immunosuppressant Drugs for Organ Transplants

Renata Fagundes
https://orcid.org/0000-0002-1665-5867
Oswaldo Cruz Foundation, Brazil

Carla Silveira
Oswaldo Cruz Foundation, Brazil

Henrique Koch Chaves
https://orcid.org/0000-0003-3035-6799
Oswaldo Cruz Foundation, Brazil

Jorge Lima de Magalhães
https://orcid.org/0000-0003-2219-5446
Oswaldo Cruz Foundation, Brazil

ABSTRACT

According to United Nations the number of inhabitants on the planet could reach around 8.5 billion in 2030 and 9.7 billion in 2050. In 50 years, organ, tissue, and cell transplantation has become a successful worldwide practice that can prolong and improve the quality of life for many patients. Advances in harvesting, conservation, surgery, and immunosuppression techniques have improved the effectiveness, cost-benefit, and cultural acceptability of transplants, with a consequent increase in demand for these practices. However, the capacity to perform transplants in different health systems is still heterogeneous. The constant evolution of immunosuppressive therapy, with research and clinical trials, has allowed the creation of specific protocols for each type of transplant, with a diverse range of medications available, always in search of an excellent immunosuppressive response, combined with a low incidence of adverse effects. Therefore, the contribution of this chapter aims to present an overview of technological management to support decision makers in health systems

DOI: 10.4018/979-8-3693-4030-1.ch021

1. INTRODUCTION:

In 2011, the world population reached the level of 7 billion inhabitants and, just eleven years later, on November 15, 2022, it reached the level of 8 billion people. United Nations (UN) projections, published on World Population Day on July 11, 2022, suggest that the number of inhabitants on the planet could reach around 8.5 billion in 2030 and 9.7 billion in 2050 Furthermore, it is estimated that the peak will reach around 10.4 billion people during the 2080s, remaining at this level until 2100. Faced with this constant growth, the world has always faced issues that transcend national borders, making it difficult to achieve success. economic and social progress of the population, such as climate change, human rights, food security and health (NU, 2022).

According to the WHO, health corresponds to being healthy in a complete state of physical, mental, and social well-being, and not just the absence of pathologies or illnesses. Therefore, it is necessary to think about treatment actions in a holistic concept, which provides the individual with not only their treatment and/or diagnosis of the disease, but a life of well-being (WHO, 2023).

In the health area, it is estimated that the global pharmaceutical sector will spend around US$2.3 trillion in 2028, on medicines alone. (IQVIA, 2024). However, the World Health Organization (WHO) declares, among several critical issues, that the transplantation of human cells, tissues and organs is one of the topics in the 21st century, due to its complexity and importance for humanity. According to the WHO, transplantation is a form of intervention and treatment that increases human longevity and restores essential functions, as there are no alternatives of comparable effectiveness (WHO, 2022).

Regarding chronic kidney disease and its main risk factors, diabetes, and hypertension, they have been constantly increasing and today it is a global public health problem due to its medical, social, and economic consequences for individuals, families, communities, and health systems. Chronic kidney disease stands out as a silent epidemic and highlights the neglect of prevention and treatment in all countries. It is estimated that 1 in 10 people worldwide have chronic kidney disease (SIN, 2017).

Organ, tissue, and cell transplantation have become a successful global practice that can prolong and improve the quality of life of many patients. Advances in harvesting, conservation, surgery, and immunosuppression techniques have improved the effectiveness, cost-benefit ratio, and cultural acceptability of transplants, with a consequent increase in demand for this practice. However, the capacity to perform transplants in different health systems is still heterogeneous. Despite the notable increase in organ donation by deceased people and even living donors in recent years, the supply and availability of cells, tissues and, mainly, organs are much lower than demand (OPAS, 2009).

Worldwide, an increase in the prevalence and incidence of chronic non-communicable diseases has been observed, some of which can be treated by organ, tissue and cell transplantation (Malta et al., 2017; OPAS, 2019b).

Furthermore, the lack of consolidated national programs, the scarcity of competent human resources and the high cost of transplants and maintenance therapies, together with insufficient coverage and financial protection, represent barriers to equitable access for the entire population in need of these procedures (OPAS, 2019a).

Given that, this chapter contributes an overview of immunosuppressants in organ transplantation, as well as technological management data to support managers in decision making.

2. ORGAN TRANSPLANT SCENARIO

The term transplant is defined in the World Health Organization (WHO) Global Glossary as the transfer (graft) of human cells, tissues or organs from a living or deceased donor to a recipient with the aim of restoring functions in the body and prolong its life (WHO, 2009).

The surgical procedure for transplanting human organs began to develop in fact at the beginning of the 20th century, with a series of experimental studies, which in general, seek advances in immunosuppressive therapy, better surgical methods, organ procurement, preservation techniques and post-operative care protocols (WHO, 1986).

One of the first major milestones in the history of transplants was the improvement of organ revascularization techniques developed by doctor Alexis Carrel (Carrel & Guthrie, 1906). Between 1901 and 1910, Carrel (1963) recognized that allografts of transplanted organs did not success, although at the time he did not know why. He used experimental animals, in addition to the technical contributions that underpinned conventional vascular surgery. Considering the pioneering work in this area of activity, in 1912 Carrel received the Nobel Prize in Medicine (Carrel, 1963; Carrel & Guthrie, 1905).

The first successful transplant was performed in Boston, in 1954, between identical twins, still without any form of immunosuppression (Noronha et al., 1997). The recipient lived for eight years, having died as a result of ischemic heart disease, and the donor died of a stroke in December 2011, 57 years after the donation (Merrill et al., 1956; Schatzki, 2003).

It is worth noting that only after years of research and discoveries, the clinical application of transplants became viable and practical when long-term allograft survival with success in recipient patients treated with immunosuppressive agents. The most notable development in this area was Jean Borel's discovery of an immunosuppressive drug in the mid-1970s (Borel & Kis, 1991). Cyclosporine was approved for commercial use in November 1983 (Garcia et al., 2004; Stähelin, 1996).

Advances in immunosuppressive therapy, improved surgical methods including microvascular anastomosis, organ harvesting, preservation techniques, and postoperative surgical care regimens have had a major impact on the safety and efficacy of organ transplantation. In the 21st century, kidney transplantation has become a standard treatment for irreversible kidney failure and heart and liver transplantation with successful methods for treating end-stage diseases of these organs. However, many questions regarding the appropriate targeting and prioritization of specific areas in clinical organ transplantation remain unanswered. The "experimental treatment" label having removed from heart, liver and pancreas transplants can no longer be used as a reason to retain these techniques. However, ethical and economic issues related to this type of advanced technology cannot be ignored (WHO, 1986).

Medical history shows that once a new effective method of treatment is developed, it is difficult to deny its use when necessary; Effective methods to prevent premature death and improve quality of life, such as organ transplantation, are used until better technology can replace them, confronting society with a range of social, economic, medical, political and ethical issues (WHO, 1986).

Organ transplantation offers effective treatment for patients with end-stage vital organ diseases, measured in years of survival and rehabilitation. There are still many unresolved issues and room for improvement. Rejection is still one of the most serious complications. It seems reasonable, given the prominence of molecular biology, to anticipate that patients will have access to specific immunosuppression with few or no side effects. It seems appropriate to predict that future advances in transplantation will result from a joint attack by immunology and surgery, as well as other scientific disciplines (DiMartini et al., 2005; Fishman, 2017).

The dream of artificial organs to replace the heart, liver or kidney seems doubtfully achievable in the near future, given the unique metabolic functions of these organs (Duy Nguyen et al., 2021). To exemplify its complexity, recent research shows that the heart is the source of a peptide that is vasoactive and has systemic regulatory actions. Most likely, metabolic feedback functions between organs cannot be realized in the current state of knowledge by any man-made construct (Martyniak & Tomasik, 2023). If this is true, organ transplantation will likely remain an evolving technology for many years to come (Edgar et al., 2020).

As an example of costs in organ transplants, about the stages of Chronic Kidney Disease (CKD), the annual cost of treatment increases substantially. Patients with stage 2 CKD, on average, require double the treatment costs of patients without CKD. For patients with End Stage Renal Disease (ESRD) without dialysis, the cost of treatment is more than 10 times higher than the cost for those with CKD. Early diagnosis and treatment not only leads to a better quality of life for patients, but also reduces the overall cost of treatment by slowing disease progression (Figure 1) (IQVIA, 2022).

Figure 1. Annual costs of chronic kidney disease by stage (2007 - 2012)

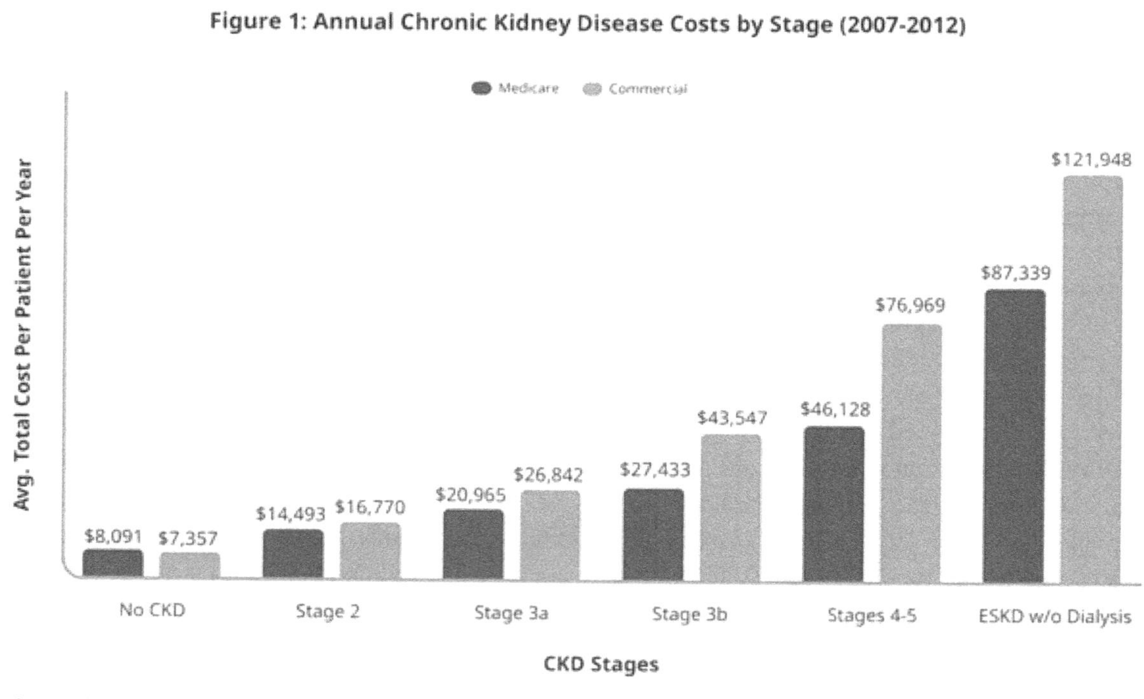

Source: IQVIA, 2022.

According to the Centers for Disease Control and Prevention[1] (CDC), nine out of 10 adults with CKD do not know they have CKD. Inequalities and disparities in awareness, testing, diagnosis, and intervention may be more prominent at certain stages throughout a patient's life (IQVIA, 2022).

Likewise, liver diseases cause approximately 2 million deaths a year worldwide; Cirrhosis is the eleventh leading cause of death and liver cancer is the sixteenth (Asrani et al., 2019; Marcellin & Kutala, 2018). In Latin America, hepatitis C is the most common cause of liver disease and the main indication for liver transplantation (Salvalaggio et al., 2014). In 2019, the WHO estimated that more than 100,000 people die daily due to lack of an organ transplant (Søborg et al., 2022).

Kidney and corneal transplants are two examples in which transplantation offers better results in terms of survival, quality of life and cost-effectiveness compared to other therapies (Gain et al., 2016; Rosselli et al., 2015).

According to the activity report of the Global Observatory on Donation and Transplantation (2020), 129,681 patients from 89 countries underwent an organ transplant in 2020, of which 54,084 occurred in the Americas region, showing an increase in organ transplantation activity. transplants worldwide (GODT, 2020).

On the other hand, considering only the United States of America (USA), there are more than 106 thousand candidates on the waiting list for solid organ transplants. The Organ Procurement and Transplantation Network (OPTN), a non-profit organization responsible for evaluating and sourcing organs from deceased donors for transplantation, periodically publishes the waiting list for donor organs (Table 1) (OPTN, 2022).

Table 1. Waitlisted candidates in the US in 2022

Solid organ	Patients
Kidney	89.958
Liver	11.096
Heart	3.363
Kidney/Pancreas	1.927
Lung	1.002
Pancreas	863
Intestine	198
Heart/Lung	30
Kidney	**105.852***

Source: adapted OPTN (2022).
* Number of candidates is less than the total sum due to candidates waiting for various bodies

According to information from the European Directorate for the Quality of Medicines and Healthcare, in 2022, on average, 39 thousand patients received a transplant, and 48 thousand new patients are waiting lists in the Member States of the Council of Europe. And, as a consequence of the imbalance between the active and passive demand for transplants, the result is an average of 19 patients dying every day while waiting for an organ transplant in Europe (EDQM, 2023).

According to Abreu (2017), the evolution and success of organ transplant procedures is directly related to innovation and access to effective, safe immunosuppressive medications, capable of preventing graft rejection, thus, access to immunosuppressive medications was crucial for the survival of organ recipients (Abreu, 2017).

3. OVERVIEW OF IMMUNOSUPPRESSANT DRUGS

Given the technological development of immunosuppressive medications and advances in solid organ transplantation, in the long term, transplant patients have achieved a significant increase in graft survival. However, acute organ rejection, chronic adverse effects and toxicity caused by medications continue to be a major concern in transplant surgery, were advanced immunosuppressive procedures. Given the technological development of immunosuppressive medications and advances in solid organ transplantation, in the long term, transplant patients have achieved a significant increase in graft survival. However, acute organ rejection, chronic adverse effects and toxicity caused by medications continue to be a major concern in transplant surgery, where advanced immunosuppressive procedures are used (Castelli et al., 2012; Correa et al., 2022; Matos, 2024; L. S. da S. Soares et al., 2019).

Although the use of multidrug therapy reduces toxicity by reducing the target doses of each medication, it is a fact that this technique still has a major impact on patients' quality of life. The aim of treatment is to prevent graft rejection and improve tolerance induction but, unfortunately, a perfect immunosuppressive agent has not yet been available (Parlakpinar & and Gunata, 2021; Pastural et al., 2002; Tönshoff, 2019).

The development timeline of the main immunosuppressive drugs related to organ transplantation (Figure 2).

Figure 2. Transplant-related immunosuppressive drug development timeline

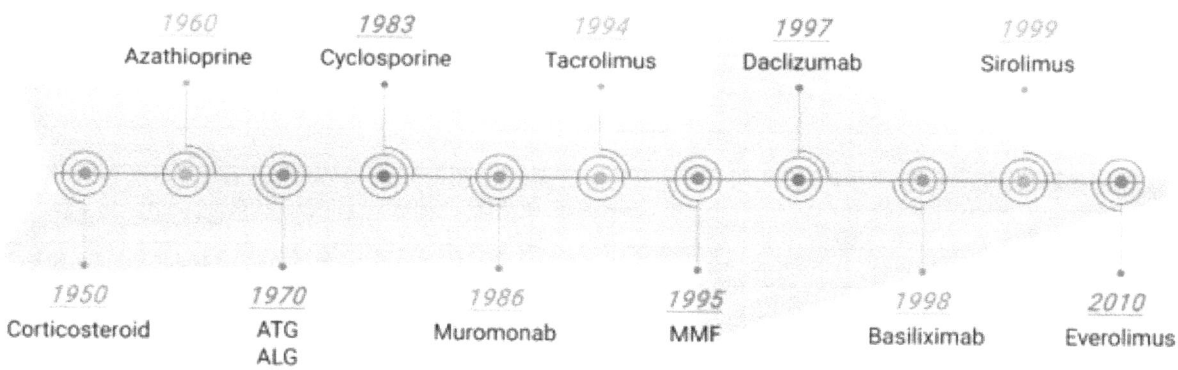

Source: PARLAKPINAR; GUNATA, 2021.

4. AN OVERVIEW OF INFORMATION MANAGEMENT, KNOWLEDGE MANAGEMENT AND COMPETITIVE INTELLIGENCE IN MEDICINES

According to Tarapanoff, building a society in which everyone can create, access, use and share information and knowledge is the challenge facing all nations and corporations in the world, intensely based on information and knowledge technologies, in which intangible assets[2] acquire increasing importance. For this, the following bases can be distinguished related to the creation, access, use and sharing of information and knowledge in organizations, which are the areas of information management, knowledge management and competitive intelligence (Tarapanoff, 2006).

Information management is defined as the application of administrative principles to the acquisition, organization, control, dissemination and use of information for the effective operationalization of organizations of all types (WILSON, 1988), or as the management of an organization's entire informational environment. Information is a determining factor in improving processes, products and services, having strategic value in organizations (Davenport, 1994).

Knowledge management systematically works with information and knowledge aiming to increase the company's ability to respond to the environment with innovation and competence, developing effectiveness and corporate knowledge. It also allows the organization to expand its skills and creates capabilities for the future (Tarapanoff, 2006).

The knowledge management process is an independent activity, but, when associated with the decision-making process, it is strongly related to the information management process and the work and analysis of information. Competitive (strategic) intelligence can be considered a synthesis of the information and knowledge work process, capable of indicating new paths for innovation (Tarapanoff, 2006).

Regarding its functions and characteristics, competitive intelligence serves as the basis for four main functions: mastery of scientific and technological know-how; detection of risks and opportunities in the domestic and foreign markets; ability to define appropriate individual and collective strategies; and define influence strategies in support of actions (UNESCO, 1997).

In the modern business context, competitive intelligence based on technological surveillance plays a strategic role in decision-making and investment allocation, increasing the chance of business success (Dou & Hassanaly, 1993; Quoniam & Lucien, 2009).

It is important to emphasize that intelligence obtained through information and knowledge management is only possible when it becomes an asset, that is, when it takes on a concrete and coherent form, such as a strategic information system or a consultation base., and also when captured in a way that is described, shared and explored by the organization (STEWART, 1998).

The organization of information allows managers and employees of institutions to view and analyze data quickly and safely, allowing them to act strategically in the face of different situations and scenarios (Magalhães et al., 2014).

5. OVERVIEW OF SCIENTIFIC AND TECHNOLOGICAL PROSPECTION

5.1. Scientific

With the evolution of information systems, the use of databases has become increasingly greater, especially in the scientific world, which has made it possible to index several types of documents (books, theses, dissertations, periodicals, event annals, etc.). Therefore, it is essential to use technological tools for managing and accessing this data, such as document management software, or even cloud storage. In this context, bibliometrics presents itself as a facilitating practice in the prospecting and analysis of information stored in these databases, making it possible to obtain a panoramic view of the scientific and technological level of a country or area of knowledge (P. B. Soares et al., 2016).

Bibliometrics is an area of information science that uses statistical and mathematical methods to analyze scientific production, identifying patterns of publication, impact and evolution of research in different fields of knowledge. The term was introduced by Alan Pritchard in 1969, who defined bibliometrics as "the application of mathematical and statistical methods to books and other media" (Egghe & Rousseau, 1990; Glänzel, 2003; Pritchard, 1969).

Widely used in the field of applied social sciences, the use of bibliometric techniques facilitates the understanding of new themes, helping to identify trends for future research (Quevedo-Silva et al., 2016). Aiming to explore global scientific production, its metrics and indicators allow measuring aspects of academic production, making it possible to calculate the production and dissemination of knowledge that contribute to the development of science (Guedes, 2005). These indicators can be both the text of the publication and the elements that compose it, such as title, source, authors, language, keywords, classification and citations (Silva, 1987; Zhu et al., 1999).

Bibliometric indexes prove to be strong indicators of scientific evaluation, as well as one of the main means of objectively measuring the scientific impact of a researcher, institution or journal Bibliometric indexes prove to be strong indicators of scientific evaluation, as well as one of the main means of objectively measuring the scientific impact of a researcher, institution or journal (Barreto, 2013). These indicators generated by the analysis make it possible to point out immediate results to support the fields of science and technology (S&T), helping, for example, in the creation of public policies in a country. These indicators generated by the analyzes make it possible to point out immediate results to support the fields of science and technology (S&T), helping, for example, in the creation of public policies in a country (Avena & Barbosa, 2017; Giménez-Espert & Prado-Gascó, 2019).

It is worth highlighting that science repositories provide access to theses, dissertations, and journals free of charge, such as Google Scholar, Scielo, Scopus (Elsevier), Pubmed (Medline) and Web of Science (ISI - Institute for Scientific Information).

In this context, it is relevant to study existing and available data on immunosuppressive medications, within the universe of scientific knowledge available within a specific repository. In this way, we intend to demonstrate how bibliometrics can help in the management of knowledge of technologies for the area of Public Health, based on the study, identifying the global scenario in research and scientific development in immunosuppressive medicines, considering the time span of 20 years It is worth highlighting that science repositories provide access to theses, dissertations and journals free of charge, such as Google Scholar, Scielo, Scopus (Elsevier), Pubmed (Medline) and Web of Science (ISI - Institute for Scientific Information).

In this context, it is relevant to study existing and available data on immunosuppressive medications, within the universe of scientific knowledge available within a specific repository. In this way, we intend to demonstrate how bibliometrics can help in the management of knowledge of technologies for the area of Public Health, based on the study, identifying the global scenario in research and scientific development in immunosuppressive medicines, considering the time span of 20 years.

5.2. Technological

In order to incorporate information into the management process, technological prospecting tries to predict possible future states of technologies or conditions that affect their contribution to established goals and involves actors around this issue. (Canongia et al., 2006; Gariboldi et al., 2021).

As a prospecting practice in science and technology, the data mining method, also known as Tech mining, consists of analyzing information from patents and other repositories (Porter & Cunningham, 2004). However, when mining uses only patent documents, it is called Patent informatics (Trippe, 2003).

In terms of information sources, patent registrations stand out, organized in large-volume databases. Recent data indicates that there are approximately 3.5 million such valid registrations, excluding those whose exclusivity rights have expired, in a sample of one hundred and fifty-seven patent authorities around the world (WIPO, 2023b). These records aggregate relevant information on technological, economic and legal aspects (Mailänder, 2013).

In the technological context, patents cover the description of technology and its main aspects. This is the first widely accessible documentary record made available to society about emerging technology. In the economic context, patent registrations also relate to technologies with commercialization or innovation potential, being used as an asset to evaluate the performance and competitiveness of companies, countries, and regions. Finally, from a legal point of view, they provide information on technological opportunities by allowing the identification of technologies that are subject to exclusivity through industrial property rights and their respective duration. (Mailänder, 2013).

Most prospective studies that are based on information from patent documents focus on the following objectives: trend analysis, technological prospecting, strategic planning, detection of violations of third-party rights, assessment of the quality of patents, analysis of the chances of obtaining patents, mapping technological trajectories, identifying technological opportunities and competitors to support decision-making processes (Abbas et al., 2014; Satos et al., 2004).

This method of patent analysis requires automated and computational resources to facilitate its implementation, which involves text mining and visualization of results, which are carried out by specific software. (Porter & Cunningham, 2004).

In this context, patents represent the right of ownership of an invention and, consequently, constitute a repository of technological and scientific knowledge. Patent databases represent a vast source of Big Data information that offers a valuable source of technological knowledge. Patents have a wide range of applications and are constantly expanding to support public policies and reveal trends in science and technology. Technical-scientific mapping techniques help to obtain essential information for managers' decision-making, since prospective studies provide results to anticipate trends and manage changes (Antunes, 2013).

Due to these drug life cycle management strategies, a single drug on the market may be the subject of more than one patent, with several types of claims and filed at various stages of the drug's development. Each of these patents may have different protection objectives (for example defensive or offensive blocking of competitors) and aim to extend as much as possible the period in which the medicine is exclusive on the market. (INPI, 2021; Sternitzke, 2012).

6. EXAMPLES OF TECHNOLOGICAL MANAGEMENT IN PROSPECTIVE SCENARIOS FOR IMMUNOSUPPRESSANTS

6.1. Scientific Prospective Scenario

Initially, we build search strings to be able to perform analyses in the databases. In order to identify the descriptors, the Descriptors in Science and Health (DeCS)/Medical Subject Headings (MeSH) portal was used (https://decs.bvsalud.org/) (ALVES, 2024). The word "immunosuppressant" is as a descriptor in Portuguese (only in first time). Likewise, the query was conducted using the terms "kidney transplant," "liver transplant" and "heart transplant." Finally, the search strings consider the solid organs kidney, liver and heart and used in the "Title" and "Summary" filters in the search fields and the Boolean operator "OR" linking these fields (Table 2).

Table 2. Searches for strings

Solid human organ	*String*	
Kidney	("imunossupressores" OR "Agente Imunossupressor" OR "Agentes Imunossupressores" OR "Droga Imunossupressora" OR "Drogas Imunossupressoras" OR "Fármaco Imunossupressor" OR "Fármacos Imunossupressores" OR "Imunossupressor" OR "Medicamento Imunossupressor" OR "Medicamentos Imunossupressores" OR "Substância Imunossupressora" OR "Substâncias Imunossupressoras" OR Immunosuppressant OR Immunosuppressants OR "Immunosuppressive Agent")	AND ("Transplante de rim" OR "Enxerto de Rim" OR "Transplantação Renal" OR "Transplantação de Rim" OR "Transplante Renal" OR "Kidney Transplantation" OR "Kidney Grafting" OR "Kidney Transplantations" OR "Renal Transplantation" OR "Renal Transplantations")
Liver	("imunossupressores" OR "Agente Imunossupressor" OR "Agentes Imunossupressores" OR "Droga Imunossupressora" OR "Drogas Imunossupressoras" OR "Fármaco Imunossupressor" OR "Fármacos Imunossupressores" OR "Imunossupressor" OR "Medicamento Imunossupressor" OR "Medicamentos Imunossupressores" OR "Substância Imunossupressora" OR "Substâncias Imunossupressoras" OR Immunosuppressant OR Immunosuppressants OR "Immunosuppressive Agent") AND	AND ("Transplante de Fígado" OR "Enxerto de Fígado" OR "Transplantação de Fígado" OR "Transplante Hepático" OR "Transplantação Hepática" OR "Liver Transplantation" OR "Hepatic Transplantation" OR "Hepatic Transplantations" OR "Liver Grafting" OR "Liver Transplant" OR "Liver Transplantations" OR "Liver Transplants")
Heart	("imunossupressores" OR "Agente Imunossupressor" OR "Agentes Imunossupressores" OR "Droga Imunossupressora" OR "Drogas Imunossupressoras" OR "Fármaco Imunossupressor" OR "Fármacos Imunossupressores" OR "Imunossupressor" OR "Medicamento Imunossupressor" OR "Medicamentos Imunossupressores" OR "Substância Imunossupressora" OR "Substâncias Imunossupressoras" OR Immunosuppressant OR Immunosuppressants OR "Immunosuppressive Agent")	AND ("Transplante de coração" OR "Enxerto Cardíaco" OR "Enxerto de Coração" OR "Transplantação Cardíaco" OR "Transplantação de Coração" OR "Transplante Cardíaco" OR "Heart Transplantation" OR "Cardiac Transplantation" OR "Cardiac Transplantations" OR "Heart Grafting" OR "Heart Graftings" OR "Heart Transplantations")

Source: created by the authors, 2023.

To conduct the searches, extraction, and retrieval of national scientific productions in Brazil and internationally, the Web of Science, Scopus and Pubmed databases were used, on August 8, 2023, the three strings expressed in table 2. Regarding the inclusion criteria, scientific production data refers to articles and review articles published in indexed journals, between the years 2003 and 2022 (Figures 3, 4 and 5). Then, the "Types of documents" filter was applied, choosing the options "articles" and "review articles," and finally the "Country" filter, with the option "Brazil." It is observed that in the

Pubmed database the third filter is described as "classic articles and review," in addition to not having the "Country" filter.

Figure 3. Manuscripts found in the Web of Science, Scopus and Pubmed databases for the solid organ kidney

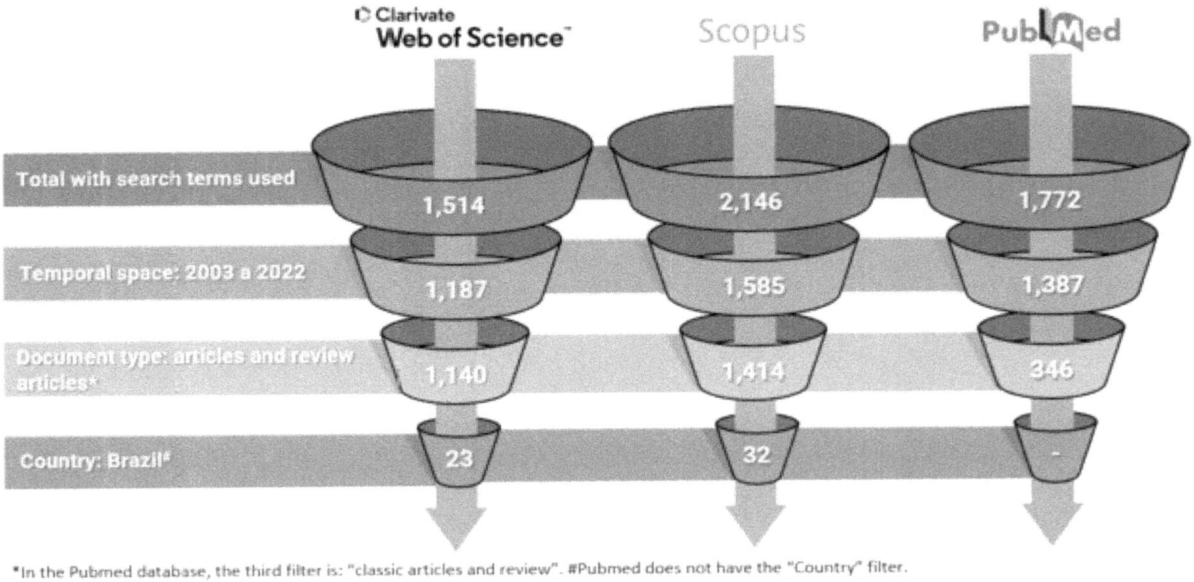

Source: created by the authors with Power Point, 2023.

Figure 4. Manuscripts found in the Web of Science, Scopus and Pubmed databases for the solid organ liver

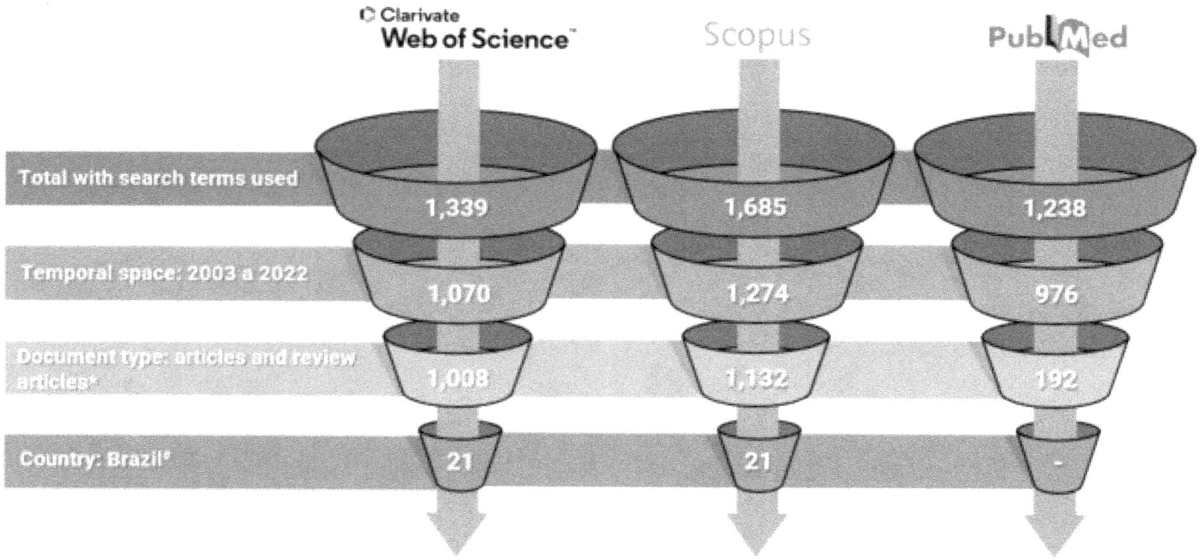

*In the Pubmed database, the third filter is: "classic articles and review". #Pubmed does not have the "Country" filter.

Source: created by the authors with Power Point, 2023.

Figure 5. Manuscripts found in the Web of Science, Scopus and Pubmed databases for the solid organ heart

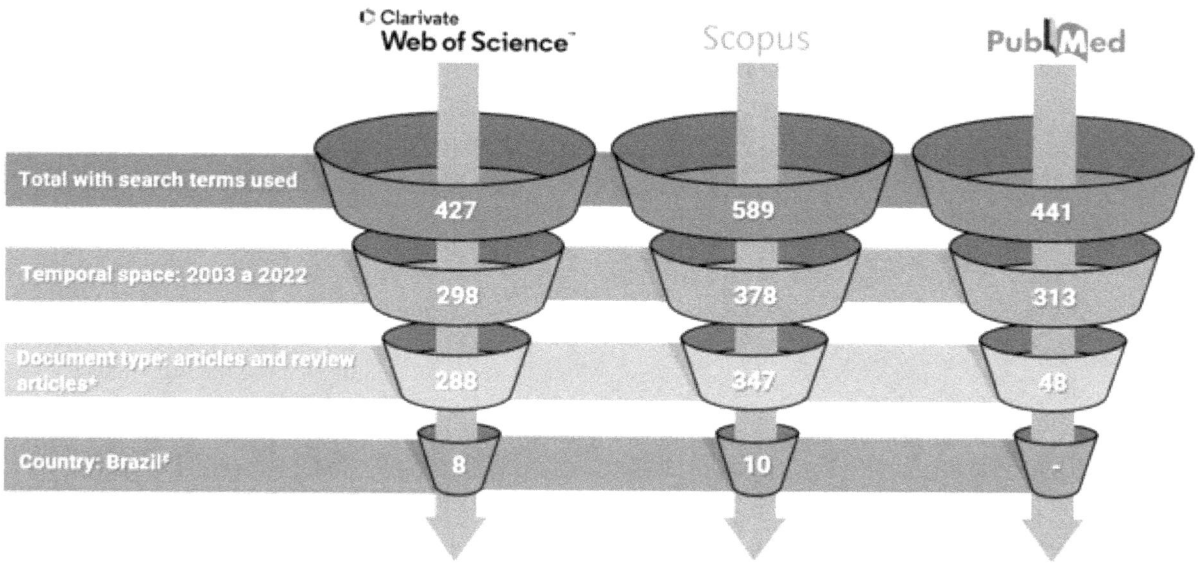

*In the Pubmed database, the third filter is: "classic articles and review". #Pubmed does not have the "Country" filter.

Source: created by the authors with Power Point, 2023.

Table 3 shows the temporal distribution of article publications in periodicals over the years 2003 to 2022 of the three bodies considered in this phase of the research. A relative stability in the number of total publications can be observed in the timeline, with few increases and decreases over the years, reaching a maximum number of publications in the year 2022 for kidney (98), 2021 for liver (79) and 2004 for heart (30); and a minimum in 2012 for kidney (54), 2005 for liver (34) and 2016 for heart (6).

Table 3. Publications between 2003 and 2022 in the Scopus database

Year	Kidney	Liver	Heart	Total
2022	98	72	17	**187**
2021	84	79	22	**185**
2020	72	69	15	**156**
2019	65	63	11	**139**
2018	80	51	19	**150**
2017	68	48	10	**126**
2016	64	60	6	**130**
2015	63	71	13	**147**
2014	70	55	16	**141**
2013	68	53	12	**133**
2012	54	57	18	**129**
2011	66	41	13	**120**
2010	71	56	17	**144**
2009	86	75	19	**180**
2008	62	57	22	**141**
2007	59	58	18	**135**
2006	84	59	27	**170**
2005	74	34	21	**129**
2004	60	39	30	**129**
2003	66	35	21	**122**
Total	**1.414**	**1.132**	**347**	**2.893**

Table 4 shows the published articles in the English language (2,510) followed by Chinese (191) and German (46). The predominance of English in scientific publications is due to several factors such as: a globally dominant language, a long tradition of use in academia and scientific publishing, easy communication and efficient dissemination and impact.

As for Chinese, the growth of scientific research in China in recent decades has led to a significant increase in the number of articles published in this language. This reflects the massive investment in research and development in the country, as well as the scientific community's desire to publish in their native language. The predominance of English followed by Chinese and then German in the publication of scientific articles and reviews reflects the distribution of global scientific research and linguistic trends within the academic community.

Table 4. Publications by language between 2003 and 2022 in the Scopus database

Language	Kidney	Liver	Heart	Total
English	1.197	1.001	312	**2.510**
Chinese	111	70	10	**191**
German	25	13	8	**46**
Spanish	16	16	5	**37**
Japanese	22	9	3	**34**
French	10	6	2	**18**
Russian	3	10	1	**14**
Portuguese	4	2	5	**11**
Korean	4	5	1	**10**
Czech	4	2	3	**9**
Italian	6	1	1	**8**
Polish	7		1	**8**
Persian	4			**4**
Turkish	2		2	**4**
Bulgarian	3			**3**
Dutch	2	1		**3**
Hungarian	2	1		**3**
Croatian	2			**2**
Lithuanian	2			**2**
Ukrainian	1		1	**2**
Bosnian	**1**			**1**
Moldova	1			**1**
Moldovan	1			**1**
Romanian	1			**1**
Slovak	1			**1**
Total	**1.432**	**1.137**	**355**	**2.924**

Source: created by the authors, 2023.

With remarkably similar numbers of articles published, we see Japan and Germany occupying third and fourth place.

Table 5 indicates that medicine is the dominant area of study within the scope of this search. Also noteworthy are the areas of Biochemistry, Genetics and Molecular Biology, Pharmacology, Toxicology and Pharmaceuticals and Immunology and Microbiology as these are immunosuppressive medications which are pharmacological substances used to suppress or reduce the body's immune response and in the context of blood transplants. organs, these drugs administered to prevent rejection of the transplanted organ by the recipient's immune system.

Table 5. Publications of articles in the Scopus database by area of study between the years 2003 and 2022.

Study area	Kidney	Liver	Heart	Total
Medicine	1.286	1.050	325	**2.661**
Biochemistry, Genetics and Molecular Biology	157	105	21	**283**
Pharmacology, Toxicology and Pharmaceutics	149	79	32	**260**
Immunology and Microbiology	81	64	28	**173**
Engineering	56	22	3	**81**
Nursing	23	14	2	**39**
Health Professions	12	8	3	**23**
Neuroscience	9	10	3	**22**
Chemical Engineering	14	4	2	**20**
Multidisciplinary	10	9	1	**20**
Agricultural and Biological Sciences	7	9	2	**18**
Chemistry	14	4		**18**
Social Sciences	12	4	2	**18**
Computer Science	10	2	1	**13**
Dentistry	8	2	2	**12**
Decision Sciences	5	3	1	**9**
Materials Science	4	3	2	**9**
Environmental Science	6	1	1	**8**
Psychology	3	2		**5**
Veterinary	3	2		**5**
Physics and Astronomy	1	1	1	**3**
Arts and Humanities	1	1		**2**
Economics, Econometrics and Finance	1			**1**
Mathematics	1			**1**

Source: created by the authors, 2023. Source: created by the authors, 2023.

Considering the central objective of this study, the metadata of the selected "core" documents are from the Scopus database, in CSV (Microsoft Excel) format. Then, the data obtained from the Scopus database conduct the treatment using the VOSviewer® program. In this way, it allowed the creation of bibliometric maps based on network data (Van Eck; Waltman, 2010).

Regarding co-citation analysis of references and sources, these are present in the articles retrieved. In this sense, to exemplify the technique, the main choice is solid organ "kidney" due to substantial number of articles found, volume of information and, also, it is the most performed organ transplant surgery in Brazil (ABTO, 2023).

Using the VOSviewer® software, co-citation analyzes of references, authors, the co-occurrence of keywords (All Keywords) and, finally with the bibliographic coupling of the articles analyzed.

In Figure 6, one can observe a co-citation network of references extracted from the 1,414 Scopus documents, which highlights the references shared by the authors of the documents. References in the figure are joints based on similarity. The assorted colors present in the network indicate the groupings in clusters and the lines show the interrelationship of these references. Clusters are groups that form due to affinity or proximity, identifying the strongest relationships within the group. Clusters are exclusive, not allowing a node to be present in two clusters. There are still nodes that do not cluster (Mira et al., 2022). Nodes that have the same color are more similar than nodes that have distinct colors. Nodes that have greater similarity are closer to each other. The greater the weight of a bibliographic reference in the network (based on the total link strength between them, the Total Link Strength), the larger the label and the circle. Thus, the more lines between the nodes, the greater the interaction between them. Regarding the thickness of the lines, the thicker they are, the more frequent the nodes are connected to each other (Codato, 2019).

In the network formed, the nodes and names of the largest references show their greatest occurrence (Figure 6). Of the 46,409 references cited, 118 have at least 5 citations, forming a total of 8 clusters. In the fourth largest cluster (17 items), identified in the figure in yellow, the reference with the highest number of co-citations (24), link strength (60) and total link strength (85), which was: Halloran, P. F. Immunosuppressive drugs for kidney transplantation. The New England Journal of Medicine, v. 351, n. 26, p. 2715–2729, 2004.

Figure 6. Co-citation network of 118 references with at least 5 citations

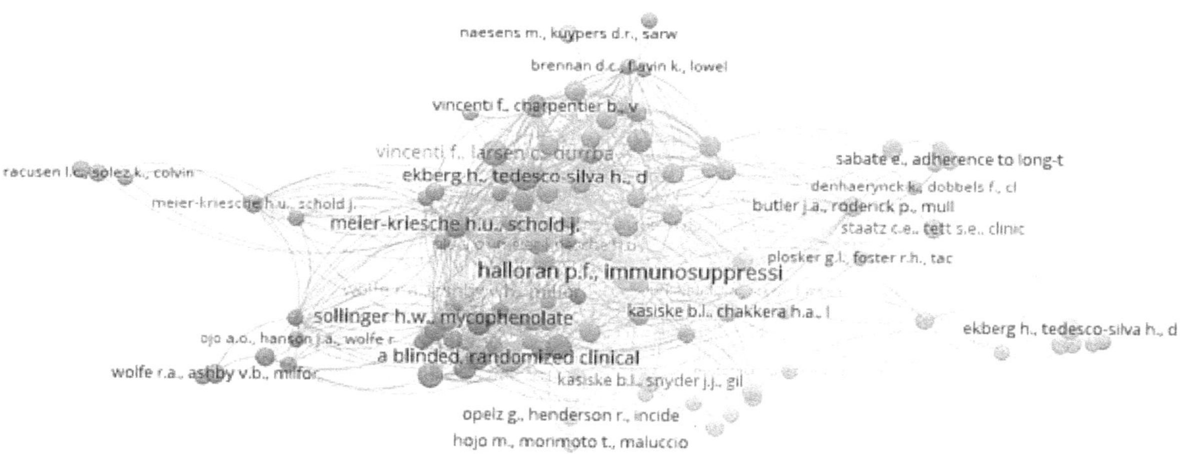

Source: Created by the authors with VOSViewer® software, 2023.

Figure 7 presents the author co-citation analysis. This analysis measures the strength of attraction between cited authors according to the understanding of the community of citing authors. The proximity between co-cited authors shows the intensity of publications in which two or more authors appear together in bibliographic references. In this way, it can be inferred that the strength of co-citation will be determined by the researchers' reaction to the published articles (Clemente et al., 2022; Grácio, 2016).

Of the 69,398 authors cited, 1,004 have at least 20 citations. This figure shows that there are 7 groups of authors conducting research in the field of study in question. It is possible to verify that the network is quite overlapping and branched. The most cited author is Flávio Vicenti (vicenti f.) (462 citations; link strength: 976; total link strength: 49,414; cluster 1 – 206 authors - red). The author is a professor at the University of California (San Francisco – USA) and is a specialist in kidney and pancreas transplantation. In 1978, Vicenti co-authored an article in the New England Journal of Medicine on the role of blood transfusions in improving kidney transplants. He has also written a series of articles on the complications of transplants and has been a leading researcher and writer on the use of anti-interleukin 2 receptor monoclonal antibodies, used to reduce the chance of organ rejection. Additionally, he is an advocate for minimizing immunosuppressive drugs after transplants. Vincenti is a member of many professional organizations, including *Transplantation Society*, *American Society of Nephrology*, *American Society of Transplantation* e *International Society of Nephrology*. He is co-lead of the kidney section of the Immune Tolerance Network. Additionally, he serves on the editorial boards of Graft magazine and the American Journal of Kidney Diseases (UCSF HEALTH, 2024).

Figure 7. Co-citation network of 1,004 authors with at least 20 citations

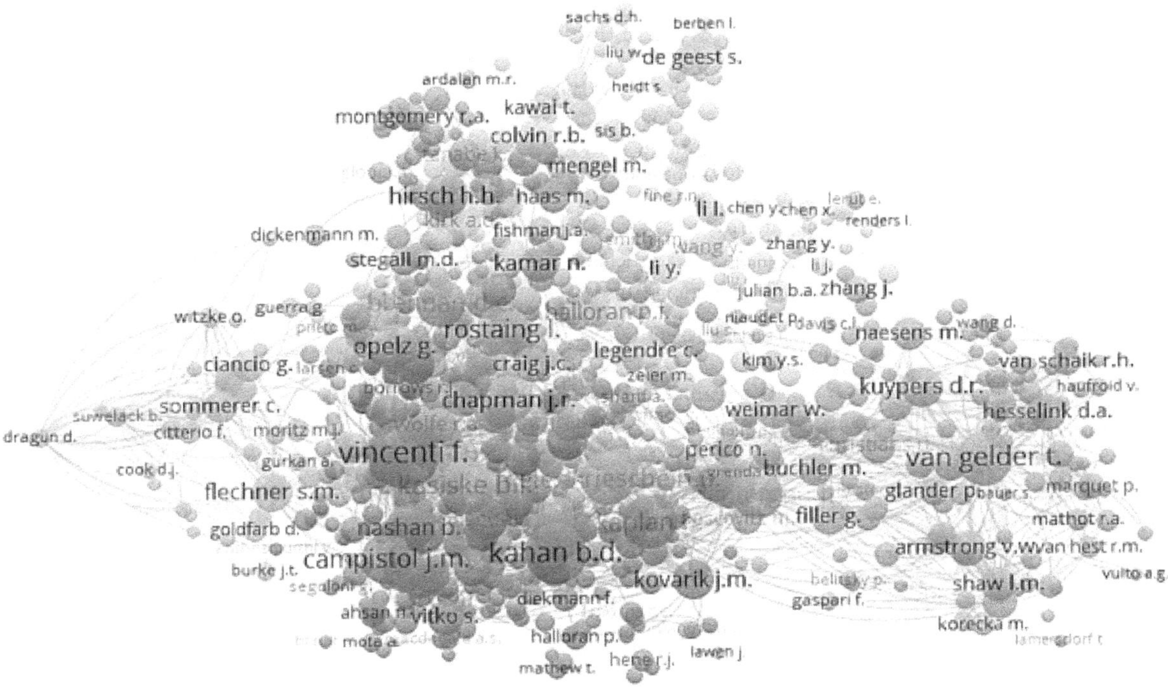

Source: Created by the authors with VOSViewer® software, 2023.

6.2. Technological Prospective Scenario

With the same keywords used in item 5.1, a total of 95 patent documents are technological (see figure 8). A time span of 20 years (2004 to 2023) was also chosen due to the term of protection for invention patents (INPI, 2023). After reading the 67 documents found, note that some of them did not fit the scope of the research, leaving a total of 47 patent documents.

Figure 8. Patent documents found on the Orbit Intelligence platform®

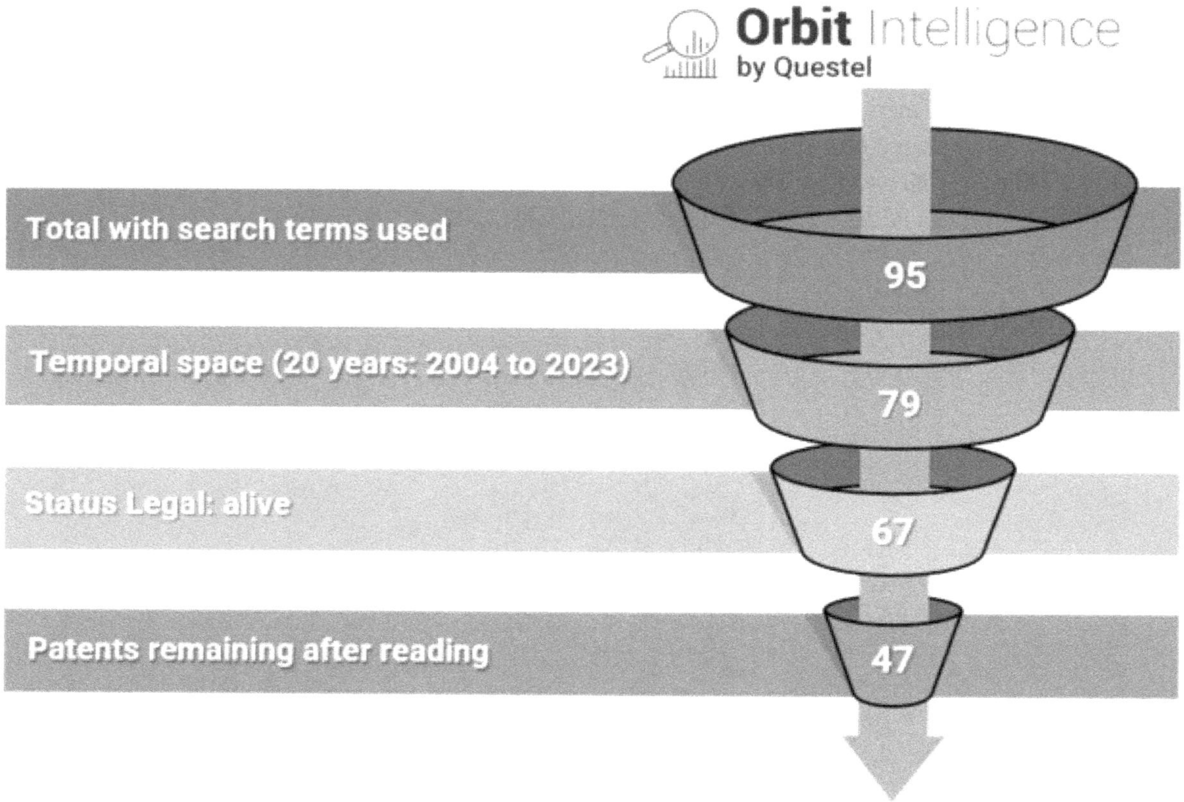

Source: Created by the authors with Orbit Intelligence®, 2024.

Although expired patent documents have not been analyzed, it is worth noting that they are in the public domain and can be exploited for the development of new drugs, medicines or therapies, including by other countries that do not have the financial resources to work with RD&I (Research, Development and Innovation). Due to the confidentiality period of the patent application (an 18-month interval that exists between the filing of the application and its publication), it is worth highlighting that the data presented in this work present a moment, due to the gap that exists between the updated information and the number of patents filed (INPI, 2023). Private investment in RD&I for the development of new medicines requires a high degree of protection of the knowledge created due to the high costs, duration and associated risks (Hasenclever et al., 2008).

Figure 9 shows the distribution of patent document filings across countries in the period analyzed. The United States leads in the number of patent documents filed, with a total of 1,058 documents (73.73%). In second place was China with a total of 258 documents (17.98%). Next, Australia stands out with 149 documents (10.38%), South Korea with 81 (5.64%), Japan with 63 (4.39%), the United Kingdom with 60 (4.18%), Canada with 48 (3.34%), Italy with 44 (3.07%), New Zealand with 38 (2.65%) and France with 18 (1.25%). As previously discussed, the results of the present study indicate the concentration of patent documents in developed countries. This point is due to greater investments in the areas of research and development (R&D), with emphasis once again on the United States.

Figure 9. Worldwide distribution of patent document filings between 2004 and 2023

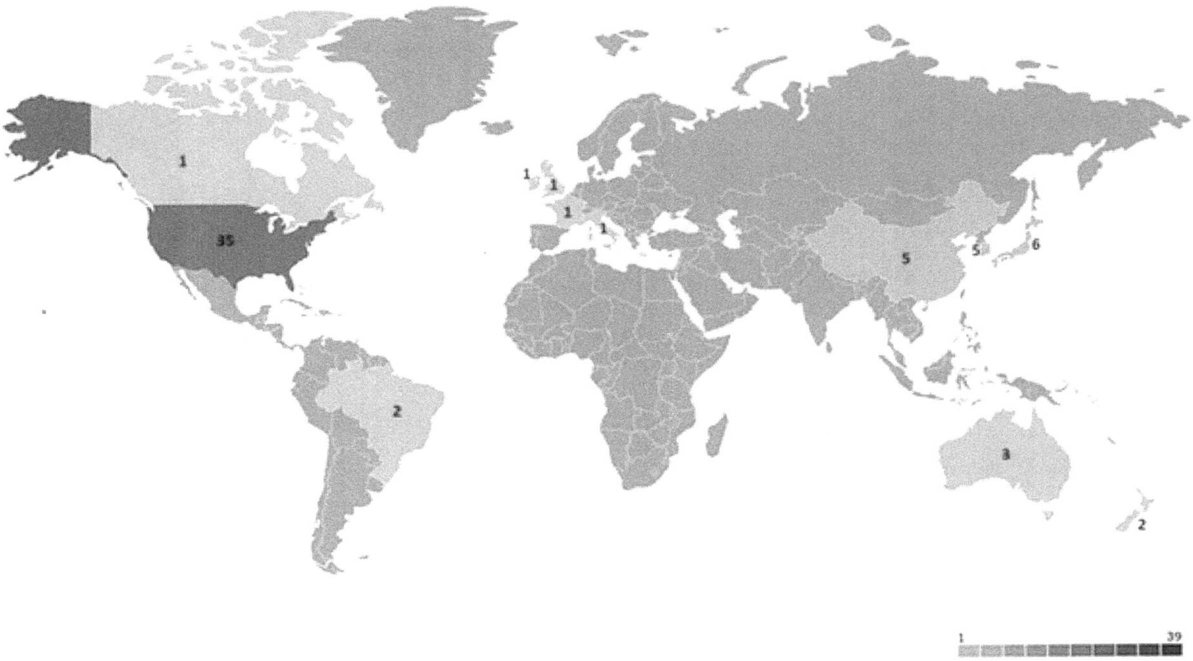

Source: Created by the authors with Orbit Intelligence® (2024).

Figure 10, generated by Orbit Intelligence®, groups patent documents according to their IPC (International Patent Classification) codes or, in Portuguese, CIP (International Patent Classification). The body responsible for publication assigns the IPC symbols valid at the time of publication of the invention application (Espacenet, 2017). The IPC classification, in addition to facilitating access to technological information, also serves as an instrument for the orderly arrangement of patent documents, facilitating research into the state of the art in a given technological field and serves as a basis for preparing industrial property statistics, enabling the evaluation technological development in various areas of knowledge (WIPO, 2023a). Most patent databases use this classification as a filter to locate patents in the technological area of interest. It is important to highlight, it is important to highlight, that it is possible to identify other patent applications that have the same purpose into the group(s) to which the patent application refers. This happens because the chart uses a text mining process, resulting in a visually informative map, in which the larger the size of the foam is the more patent families by their code.

Figure 10. IPC codes present in patent family documents

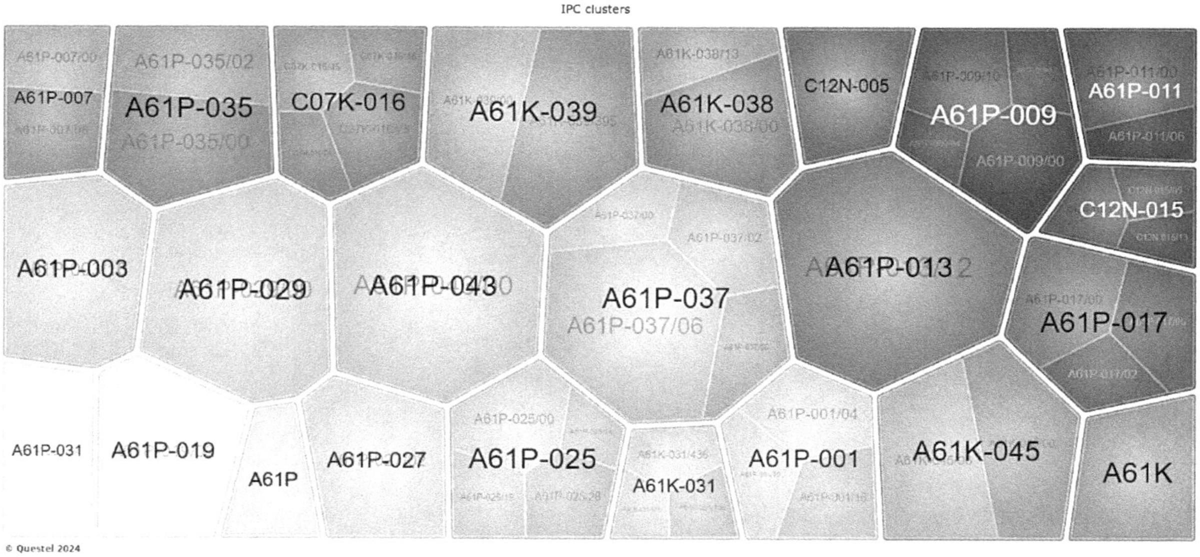

Source: created by the author with Orbit Intelligence® (2024).

It is possible to see that most patents belong to the A61P subclass. This subclass code is used to identify patents with "specific therapeutic activity of chemical compounds or medicinal preparations."

6.3. Final Considerations

The pharmaceutical market is a research-intensive sector and is one of the highest-earning sectors globally. Estimates state that by the year 2028 a record of US$ 2.3 trillion in medicines. In this scenario, the WHO highlights the problem of treatments for transplantation of human cells, tissues, and organs in the 21st century. It is worth noting that transplantation is a form of intervention and treatment that increases human longevity and restores the individual's essential functions and well-being. In this scenario, chronic kidney disease associated with the main risk factors, such as diabetes and hypertension, is an example of a global public health problem. Estimates are that 1 in 10 people worldwide have chronic kidney disease.

In the context of human organs, tissue, and cell transplants, it has become a successful worldwide practice, prolonging the quality of life of many patients.

Information management of all knowledge generated in the health sector, specifically scientific and technological knowledge (patents) is a source of essential information for decision-making by managers of public or private institutions in the health sector. The data generated in all databases, after process and treatment by management software, constitutes a strong ally to identify scientific and technological trends.

It was possible to observe through the results obtained for scientific trends that research on immunosuppressants for the three organs remained constant in the time analyzed, with increases and decreases in the number of articles published over the years, with emphasis on articles on immunosuppressants for kidney transplants. This fact can be caused by the much higher number of kidney transplants performed annually compared to heart and liver transplants. Using the VOSViewer® software, it was possible to perform data mining, with subsequent recompositing of information to stratify authors, sources, keywords, and references linked to publications.

The visualization of the processed data, generating information on the analysis results, provided important graphical representations, as they enable both academics and professionals from various areas of science, such as Medicine, Pharmacy, Knowledge Management, Information Science, among others, to evaluate the its level of development, main trends, as well as pointing out possible gaps still present in the field of this study.

Concerning technological trends, it was possible to observe the leading role of the United States in the development of modern technologies in immunosuppressants for kidney transplantation, justified by the substantial number of patent documents in relation to other countries. The scientific and technological data available in a patent document are a rich repository of the state of the art in the area analyzed, providing a better understanding of technological trends, new market opportunities and new scientific directions. Since society wins with the acquisition of new knowledge and new inventions, patent documents can function as "innovation engines" for economies, enabling not only incremental innovations, but also radical innovations through the creation of new pharmaceutical principles.

REFERENCES

Abreu, J. C. D. (2017). *Technological prospection applied in the optimization of patent granting in Brazil: a case study in patents for immunosuppressive drugs. 2017. 342 f* (Doctoral dissertation, Thesis (Doctorate)-Graduate Program in Public Policies, Strategies and Development of the Institute of Economics of the Federal University of Rio de Janeiro, Federal University of Rio de Janeiro, Rio de Janeiro).

Asrani, S. K., Devarbhavi, H., Eaton, J., & Kamath, P. S. (2019). Burden of liver diseases in the world. *Journal of Hepatology*, *70*(1), 151–171. DOI: 10.1016/j.jhep.2018.09.014 PMID: 30266282

Carrel, A. (1963). 1 The Operative Technique of Vascular Anastomoses and the Transplantation of Viscera. *Clinical Orthopaedics and Related Research (1976-2007), 29*, 3-6.

Carrel, A., & Guthrie, C. C. (1905). Functions of a transplanted kidney. *Science*, *22*(563), 473–473. DOI: 10.1126/science.22.563.473.a PMID: 17753234

Carrel, A., & Guthrie, C. C. (1906). A new method for the homoplastic transplantation of the ovary. *Science*, *23*(589), 591–591. DOI: 10.1126/science.23.589.591 PMID: 17797348

Gain, P., Jullienne, R., He, Z., Aldossary, M., Acquart, S., Cognasse, F., & Thuret, G. (2016). Global survey of corneal transplantation and eye banking. *JAMA Ophthalmology*, *134*(2), 167–173. DOI: 10.1001/jamaophthalmol.2015.4776 PMID: 26633035

Garcia, S. C., Lopes, L. D. S., Schott, K. L., Beck, S. T., & Pomblum, V. J. (2004). Cyclosporine A and tacrolimus: A review. *Jornal Brasileiro de Patologia e Medicina Laboratorial*, *40*(6), 393–401. DOI: 10.1590/S1676-24442004000600007

Global Observatory on Donation and Transplantation. (2020). *Reports*. https://www.transplant-observatory.org/wp-content/uploads/2022/07/2020-Global-report-para-web.pdf

Malta, D. C., Bernal, R. T. I., Lima, M. G., Araújo, S. S. C. D., Silva, M. M. A. D., Freitas, M. I. D. F., & Barros, M. B. D. A. (2017). Chronic non-communicable diseases and the use of health services: Analysis of the National Health Survey in Brazil. *Revista de Saude Publica*, *51*(suppl 1), 4s. DOI: 10.1590/s1518-8787.2017051000090 PMID: 28591353

Marcellin, P., & Kutala, B. K. (2018). Liver diseases: A major, neglected global public health problem requiring urgent actions and large-scale screening. *Liver International*, *38*(S1), 2–6. DOI: 10.1111/liv.13682 PMID: 29427496

Merrill, J. P., Murray, J. E., Harrison, J. H., & Guild, W. R. (1956). Successful homotransplantation of the human kidney between identical twins. *Journal of the American Medical Association*, *160*(4), 277–282. DOI: 10.1001/jama.1956.02960390027008 PMID: 13278189

Organ Procurement and Transplantation Network. (2022a). *Data - OPTN*. https://optn.transplant.hrsa.gov/data/

Organ Procurement and Transplantation Network. (2022b). *Learn about the history of transplant*. https://optn.transplant.hrsa.gov/patients/about-transplantation/history/

Rosselli, D., Rueda, J.-D., & Diaz, C. E. (2015). Cost-effectiveness of kidney transplantation compared with chronic dialysis in end-stage renal disease. *Saudi Journal of Kidney Diseases and Transplantation*, *26*(4), 733. DOI: 10.4103/1319-2442.160175 PMID: 26178546

Salvalaggio, P. R., de Almeida, M. A., de Lima, L. R., & Silva, E. M. (2014). Liver transplantation in Latin America: The state-of-the-art and future trends. *Transplantation*, *98*(3), 241–246. DOI: 10.1097/TP.0000000000000198 PMID: 25093292

Schatzki, S. (2003). The first kidney transplantation. *AJR. American Journal of Roentgenology*, *181*(1), 190. DOI: 10.2214/ajr.181.1.1810190 PMID: 12818857

World Health Organization. (1986). *Human organ transplantation: Report by the Director-General.* https://apps.who.int/iris/handle/10665/161923

World Health Organization. (2022). *Transplantation*. https://www.who.int/health-topics/transplantation

ENDNOTES

1 *Center for Disease Control and Prevention* – CDC (https://www.cdc.gov/index.html) is the nation's leading science-based, data-driven service organization protecting public health is the nation's leading science-based, data-driven service organization protecting public health.

2 The competence of its employees, processes and relationship networks are the intangible assets of organizations, their intellectual capital The competence of its employees, processes and relationship networks are the intangible assets of organizations, their intellectual capital (www.sveiby.com.au/article/Brennan/ic_defined.htm).

Chapter 22

Innovation Positioning Panorama Influence on Innovation Source Diversity in the Healthcare Sector Through R&D Investment in Ethiopia

Tamire Ashuro
Bule Hora University, Ethiopia

Shashi Kant
https://orcid.org/0000-0003-4722-5736
Bule Hora University, Ethiopia

ABSTRACT

In the healthcare industry, this investigation explores the complex connections among innovation source variety, R&D spending, and positioning of innovation. Given the importance of innovation in raising firm's competitiveness, the investigation uses exploratory factor analysis to pinpoint the main variables influencing innovation dynamics. Organizations that strategically connect their innovation activities with sufficient money are better positioned to investigate a variety of sources of innovation, according to the findings, which show a high association among successful positioning of innovation and greater R&D expenditure. The investigation also emphasizes the mediating function of R&D expenditure, showing that it plays a critical function in converting innovative positioning into observable innovation results. These revelations highlight how crucial it is to have a strong R&D ecosystem and use an organized approach to innovation strategy. In the conclusion, this investigation advances the theoretical knowledge of innovation management and provides useful advice for healthcare.

DOI: 10.4018/979-8-3693-4030-1.ch022

INTRODUCTION

Investigating positioning of innovation in the healthcare industry has become crucial, especially when considering emerging countries like Ethiopia. Since healthcare innovation has a direct influence on patient outcomes, system efficiency, and public health in general, its significance on a worldwide scale cannot be understated (Pak et al., 2024). Understanding how positioning of innovation influences the variety of innovation sources is crucial in a time when healthcare requirements are dictated by rapidly changing demographic trends and technology breakthroughs (Niguse et al., 2025). This is particularly relevant in Ethiopia, where local settings present healthcare systems with particular opportunities and problems (Pak et al., 2024). The notion of positioning of innovation has its roots in the more general domains of innovation theory and strategic management. It discusses how businesses may set themselves up to take use of a variety of innovation sources, such as investments in R&D and partnerships with academic institutions and the commercial sector (Dereso et al., 2023). Though it frequently ignores the unique dynamics at work in low-resource contexts, empirical investigation emphasizes the critical function that R&D plays in promoting innovative diversity. Investigation shows that although R&D spending is substantial, it might be difficult to turn these expenditures into real discoveries, especially in areas with inadequate financing and infrastructure (Simon Kaggwa et al., 2023).

A shortage of qualified personnel, disjointed healthcare systems that make collaboration difficult, and insufficient financing for healthcare investigation are some of Ethiopia's real-world problems. These problems limit the range of innovation sources available, which reduces the possibility of ground-breaking discoveries (Kant et al., 2025). Also, there are a lot of unanswered questions in the literature, especially when it comes to how R&D expenditures might be positioned strategically to increase the variety of innovation sources in the healthcare industry (Birhane et al., 2024). There is a knowledge gap about the particular elements that influence innovation in Ethiopian healthcare as most investigation focuses on high-income nations (Simon Kaggwa et al., 2023). By examining the association among positioning of innovation and Investment in R&D in the Ethiopian healthcare industry, this chapter seeks to close these gaps. It will give future scholars and policymakers useful insights by looking at case studies and actual data. This investigation's main inquiry is: ***"How does positioning of innovation influence the variety of innovation sources in Ethiopia's healthcare industry through Investment in R&D?"*** In addition to shedding light on the intricacies of healthcare innovation in Ethiopia, answering this question will provide a path for improving R&D tactics and building a more resilient and varied healthcare innovation ecosystem.

STATEMENT OF PROBLEM

The notional state would have a strong system with a variety of innovative sources propelled by substantial R&D expenditures. Such a system should notionally increase the overall influence of healthcare delivery, promote sustainable healthcare practices, and improve patient outcomes. But this notional is far different from reality (Simon Kaggwa et al., 2023). The efficiency of R&D spending in promoting innovation variety is the subject of conflicting empirical investigation, which frequently shows a lack of association among increasing funding and observable inventive results (Adula & Kant, 2025). The methods by which R&D spending influence positioning of innovation and the resulting variety of innovation sources are called into question by this discrepancy (Kapoor, 2024). Also, there are knowledge gaps in the literature about the particular contextual elements that support or impede innovation in the Ethiopian healthcare industry (Adula et al., 2025). According to some investigation, external partnerships and collaborations can foster innovation, but structural inefficiencies and capacity constraints frequently hinder these initiatives. Numerous studies' findings are still equivocal and do not provide a clear framework among increased innovation diversity with R&D spending(Fernandes, 2023; Gomes, 2024; Xiong, 2022). Theoretical gaps in our knowledge of how innovative positioning interacts with local settings, especially in low-resource countries, result from this ambiguity.

Also, little is known about the particular difficulties Ethiopian investigations and healthcare professionals have while pursuing innovation. The situation is further complicated by real-world issues including poor infrastructure, a lack of funds, and a lack of qualified workers. The absence of efficient policy frameworks that may leverage R&D spending to strengthen innovation sources exacerbates these problems. These practical issues are frequently ignored by the corpus of current knowledge, which results in recommendations that are not based on the real-world experiences of those involved in the Ethiopian healthcare system (Pak et al., 2024). By offering a thorough examination of how positioning of innovation may be used to increase the variety of innovation sources in Ethiopia's healthcare industry through R&D spending, this investigation seeks to close these gaps. This investigation will provide light on the manner in which R&D may be used efficiently by combining the body of current literature, carrying out empirical investigations, and interacting with regional stakeholders. Also, it will investigate the contextual elements that influence innovation dynamics, providing a sophisticated knowledge that may direct future scholars and decision-makers. The ultimate goal of our effort is to help create a more knowledgeable and useful framework for boosting healthcare innovation in Ethiopia, opening the door for long-lasting developments that are substantial and pertinent.

THEORETICAL FRAMEWORK

Open Innovation Theory: The associations among positioning of innovation, R&D expenditure, and the variety of innovation sources in the healthcare industry are fundamentally understood by a number of theories. One relevant theory is Open Innovation Theory, which holds that businesses may and ought to employ both internal and external notions and routes to market in order to develop their inventions and technologies. This notion, which is especially pertinent in the Ethiopian setting, highlights the value of cooperation amongst different stakeholders, including academics, business, and the government. The diversity and influence of healthcare innovations may be greatly increased by utilizing external collaborations, especially in light of the restricted resources available (Kapoor, 2024).

Diffusion of Innovations Theory: Another relevant theory is Everett Rogers' Diffusion of Innovations Theory that provides a framework for explaining how and why and at what rate, new ideas and technology attributes are taken up by adopters. According to this theory, communication channels and social networks that facilitate the diffusion of innovations are especially important for the acceptance and encouragement of an environment that supports the diffusion of creative activity and products. Knowledge of factors that hinder dissemination of healthcare progresses in the nation might help show how R&D costs can be better aligned with regional strengths and requirements in Ethiopia (Aggarwal et al., 2024).

Resource-Based View: Also, the Resource-Based View (RBV) provides insightful information about how businesses may use their own internal resources and competencies to obtain a competitive edge in innovation. According to this notion, promoting creativity requires special resources, such as trained workers and technological prowess. Using the RBV in Ethiopia can assist in pinpointing the precise resources that can be used to improve innovation diversity (Kapoor, 2024; Pesqueira et al., 2023).

EMPIRICAL LITERATURE REVIEW

Positioning of Innovation Panorama and Innovation Source Diversity

Numerous empirical studies have examined the association among positioning of innovation and source diversity in the healthcare industry, especially as firms look to improve their inventive capabilities through strategic R&D expenditures. Influence positioning of innovation has been manifest to have a major influence on the variety of innovation sources, which is essential for developing long-term healthcare solutions (Gokhberg et al., 2023). Investigation has demonstrated, for example, that healthcare companies that embrace an open innovation strategy—actively pursuing outside partnerships and using a variety of knowledge bases—generally demonstrate a higher degree of innovative diversity. This is especially true in high-income nations, where collaborations among government organizations, tech companies, and academic institutions have successfully produced cutting-edge medical procedures and technologies (Pesqueira et al., 2023).

Empirical investigation indicates a more nuanced link in the situation of emerging countries like Ethiopia. Increased R&D expenditures may improve innovation outcomes, according to certain investigation, although systemic obstacles frequently impede such advancement. For instance, investigation by Chataway et al. (2014) highlights how local factors, such as infrastructure problems and regulatory frameworks, may have a big influence on how well Investment in R&Ds drive innovation. In Ethiopia, healthcare institutions' capacity to use R&D for a variety of innovations is sometimes hampered by a lack of resources and qualified staff (Gu et al., 2023). Additional empirical data emphasizes how institutional support promotes innovation variety. Healthcare systems with strong regulatory frameworks and government support are better able to foster a variety of innovation sources, according to a investigation by Tzeng et al. (2016). These frameworks frequently promote cooperation among the public and private sectors, creating an atmosphere that is conducive to the development of creative notions. However, Ethiopia frequently experiences fragmented attempts due to a lack of integrated rules and incentives, which limits the potential for diversified innovation (Gokhberg et al., 2023).

Also, the investigation emphasizes how crucial cultural elements are in influencing innovative methods. According to investigation by (Aggarwal et al., 2024), cultural perspectives on risk and teamwork may either support or undermine innovative posture. Traditional procedures and hierarchical structures in Ethiopian healthcare may make it difficult to implement more open and collaborative innovation strategies, which might have an influence on the variety of innovation sources. Empirical investigation has also looked into how innovation diversity is influenced by technology developments. Investigation by (Pak et al., 2024), for example, shows how new technologies like digital health and telemedicine may greatly broaden the range of innovation sources in the healthcare industry. However, the degree of digital literacy among patients and healthcare practitioners as well as the infrastructure that is currently in place frequently determine how widely these technologies are used in Ethiopia.

H1: *Positioning of innovation panorama has a substantial influence on Innovation Source Diversity in Healthcare Sector*

Positioning of Innovation Panorama and Investment in R&D

In the empirical investigation, the connection among R&D expenditure and positioning of innovation has received a lot of attention, especially as businesses look to improve their competitiveness and inventive skills. Influence positioning of innovation has been repeatedly manifest to result in more strategic and substantial R&D expenditures, which in turn may alter the broader innovation landscape across a range of industries, including healthcare Pak et al., 2024).

The importance of strategic alignment among R&D spending and innovative positioning is a recurring issue in the literature. According to studies, companies are more likely to produce substantial results when their R&D efforts are in line with their overall innovation strategy. For instance, a investigation by (Simon Kaggwa et al., 2023) shows that companies that focus on market trends and consumer wants, or market-oriented positioning of innovation, tend to deploy their R&D resources more efficiently, producing innovations that meet market expectations. In the healthcare industry, where knowledge about patient requirements and regulatory ecosystems may greatly influence R&D goals, this alignment is essential.

The influence of outside partnerships on the efficacy of Investment in R&Ds is another substantial discovery. Chesbrough's (2003) investigation highlights the notion of open innovation, which supports the notion that businesses may improve their R&D processes by utilizing outside notions and technology. According to empirical investigation, healthcare organizations may benefit from a variety of knowledge and experience by forming alliances with academic institutions, investigation groups, and business leaders. This collaboration can lead to more creative solutions. According to a investigation by (Simon Kaggwa et al., 2023), for example, companies that actively collaborate with outside partners tend to have higher R&D productivity, which results in a wider range of successful discoveries. Also, investigation emphasizes the significance of contextual elements that influence R&D expenditures in connection to innovative positioning. The success of Investment in R&Ds may be greatly influenced by institutional support, leadership, and firms culture, as demonstrated by investigation by (Fernandes, 2023). Successful R&D results are more likely to occur in ecosystems where leadership actively fosters creativity and encourages teamwork. This discovery is especially pertinent to Ethiopia, where institutional frameworks and cultural attitudes may influence the ability to position R&D influence.

Also, the influence of funding mechanisms on Investment in R&D strategies has been a focal point in the investigation of empirical data. According to a investigation by (Xiong, 2022), the ability to make Investment in R&Ds may be improved by having access to a variety of financing sources, including public-private partnerships, private investments, and government subsidies. Understanding how to optimize available resources through strategic positioning becomes crucial for promoting innovation in nations like Ethiopia where financing for healthcare may be scarce. The investigation also emphasizes how substantial it is to have a thorough grasp of measuring and evaluation frameworks in order to evaluate how R&D spending influence innovation results. According to investigation by (Gomes, 2024), a lack of reliable measurements makes it difficult for firms to assess the success of their R&D initiatives. This evaluation gap may make it more difficult to modify R&D plans in response to performance, which might eventually influence innovative positioning.

H2: *Positioning of innovation panorama influence has a substantial influence on Investment in R&D*

Mediation Function of R&D Investment

Recent empirical investigation has manifest that R&D expenditure plays a critical function in mediating the association among innovative positioning and innovation source diversity, especially in the healthcare industry. In terms of various sources of innovation, this connection emphasizes how smart R&D expenditures may ease the shift from creative positioning to observable results. Understanding how R&D functions as a mediator that not only improves but also changes the dynamics among these two constructs has been a growing area of investigation (Pesqueira et al., 2023). The Resource-Based View (RBV), which highlights the significance of internal resources, particularly R&D skills, in gaining competitive advantage, is one of the fundamental notions in this field. According to investigation by (Gu et al., 2023), companies that have strong R&D departments are better able to use their innovation strategy. In particular, focused Investment in R&Ds may assist in converting proactive positioning of innovation by organizations—for example, a focus on customer-centric healthcare solutions—into a variety of useful inventions. An investigation by Bessant and Tidd (2011), for example, shows that coordinating R&D expenditures with creative positioning can result in notable improvements in medical technology, increasing source variety.

The literature also emphasizes the importance of Investment in R&D in creating collaborative networks, which is especially crucial in the healthcare industry. According to (Aggarwal et al., 2024), open innovation frameworks show how R&D may mediate the link among innovative positioning and source diversity by fostering collaborations with academic institutions, investigation centers, and business stakeholders. Investigation indicates that healthcare companies that make strategic investments in investigation and development are more likely to form partnerships that expand their opportunities for innovation. According to a investigation by (Simon Kaggwa et al., 2023), for instance, businesses who use R&D spending to support external partnerships have better access to a wider range of innovation sources, which improves overall performance. Also, the literature highlights how crucial contextual elements like institutional support and firms culture are in determining how Investment in R&D functions as a mediator. A culture that encourages creativity and risk-taking can greatly increase the efficacy of R&D spending as a mediator, according to investigation by (Kapoor, 2024). Knowing how to establish an atmosphere that supports investigation and development might be crucial in the Ethiopian healthcare sector, where structural issues frequently impede innovation. The link among positioning and source diversity is successfully mediated by empirical data showing that firms with strong leadership support for R&D are better positioned to align their innovation strategy with varied sources of innovation.

Also, investigation has looked into how financing structures influence Investment in R&D's mediating function. According to investigation by (Fernandes, 2023), having access to a variety of financing sources can increase the ability to invest in R&D, which in turn makes it easier to create a variety of innovation sources. This is especially true in Ethiopia, where a lack of finance may make it more difficult for healthcare institutions to carry out R&D projects that promote diversity and creativity. Organizations can better position themselves to investigate a greater range of creative notions by carefully utilizing the R&D money that is available.

H3: R& D Investment play substantial Mediation function among Positioning of innovation panorama and Innovation Source Diversity in Healthcare Sector

CONCEPTUAL FRAMEWORK

Figure 1. Conceptual framework

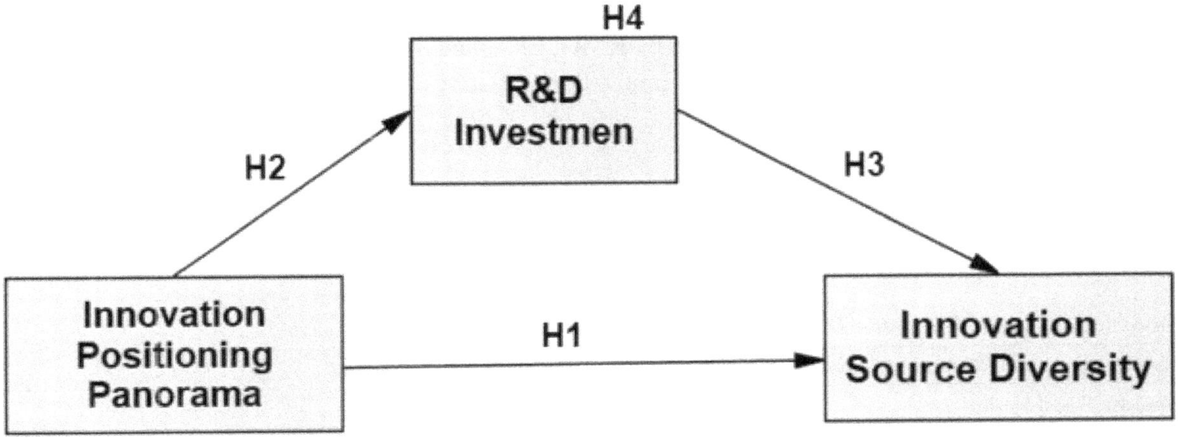

METHODOLOGY

The purpose of this investigation's investigation technique was to examine how Investment in R&D functions as a mediator among innovation source variety and positioning of innovation in the healthcare industry. To make it easier to analyze the correlations among the variables under examination in a methodical manner, a quantitative investigation methodology was used. With a total sample size of 390 individuals, a wide range of stakeholders in Ethiopia's healthcare system, including investigations, policymakers, and healthcare practitioners, were represented.

A thorough scale development procedure was carried out to guarantee the measuring tools' validity and reliability. In order to find pertinent concepts and objects pertaining to positioning of innovation, Investment in R&D, and innovation source variety, the body of current literature was first evaluated. A preliminary collection of items was developed as a result of this investigation, and its content validity was confirmed by expert review. After that, a smaller sample was used in a pilot investigation to improve the items in response to comments and statistical analysis.

The Kaiser-Meyer-Olkin (KMO) measure of sampling adequacy was the first of several statistical analyses performed on the data gathered from the primary investigation. To determine if the data was suitable for factor analysis, the KMO statistic was computed; values greater than 0.70 indicate sufficient sampling. Exploratory Factor Analysis (EFA) was then used to establish the factor structure of the constructs and uncover underlying dimensions. The EFA results made it easier to identify the substantial elements that contributed to each construct and to reduce the number of items. Following the establishment of the factor structure, the interactions among the constructs in the manifested model were investigated using Structural Equation Modeling (SEM) with AMOS software. SEM made it possible to evaluate both direct and indirect influences inside the model and test several associations at once. To ascertain if the model adequately represented the data, the analysis also evaluated model fit indices, including the Chi-square statistic, Comparative Fit Index (CFI), and Root Mean Square Error of Approximation (RMSEA). By using this thorough approach, the investigation sought to offer solid insights into how innovation source variety, Investment in R&D, and positioning of innovation interact within the Ethiopian healthcare industry. The findings were statistically valid and dependable thanks to the quantitative method, which was bolstered by KMO, EFA, and SEM analyses. This helped to produce useful suggestions for improving healthcare innovation initiatives in Ethiopia.

DATA ANALYSIS

Table 1. KMO and Bartlett's Test

Kaiser-Meyer-Olkin Measure of Sampling Adequacy.		**.850**
Bartlett's Test of Sphericity	Approx. Chi-Square	1419.738
	df	91
	Sig.	.000

Source: Authors

As manifest in Table 1, the Kaiser-Meyer-Olkin (KMO) Measure of Sampling Adequacy and Bartlett's Test of Sphericity were evaluated before the data analysis began. A high degree of sample adequacy for factor analysis was indicated by the KMO statistic, which was computed at 0.850. KMO levels near 1.0 indicate that the data are suitable for factor analysis, whereas values over 0.70 are often regarded as acceptable. The variables chosen for the investigation exhibit enough correlations to support the use of component analysis, as indicated by this strong KMO result, suggesting that the underlying dimensions among the constructs are probably substantial.

The dataset's eligibility for factor analysis was further supported by Bartlett's Test of Sphericity, which produced an estimated Chi-Square value of 1419.738 with 91 degrees of freedom and a significance level (Sig.) of 0.000. Substantial correlations among the variables are manifested by the statistically substantial result ($p < 0.001$), which shows that the correlation matrix is not an identity matrix. This result validates the choice to do exploratory factor analysis (EFA) by confirming that the variables are connected and appropriate for additional investigation.

Table 2. Total variance explained

Component	**Initial Eigenvalues**			**Extraction Sums of Squared Loadings**			**Rotation Sums of Squared Loadings**		
	Total	**% of Variance**	**Cumulative %**	**Total**	**% of Variance**	**Cumulative %**	**Total**	**% of Variance**	**Cumulative %**
1	4.51	32.235	32.235	4.51	32.235	32.235	3.00	21.451	21.451
2	1.37	9.787	42.022	1.37	9.787	42.022	2.16	15.491	36.942
3	1.27	9.087	51.110	1.27	9.087	51.110	1.98	14.168	51.110

continued on following page

Table 2. Continued

Component	Initial Eigenvalues			Extraction Sums of Squared Loadings			Rotation Sums of Squared Loadings		
	Total	**% of Variance**	**Cumulative %**	**Total**	**% of Variance**	**Cumulative %**	**Total**	**% of Variance**	**Cumulative %**
4	.949	6.782	57.892						
5	.871	6.223	64.115						
6	.773	5.519	69.634						

Extraction Method: Principal Component Analysis.
Source: Authors

The Total Variance Explained findings, obtained from the Exploratory Factor Analysis (EFA) with Principal Component Analysis (PCA) as the extraction technique, are manifest in Table 2. The amount of variation explained by each component is manifest by the first eigenvalues. With an eigenvalue of 4.51, the first component explains 32.235% of the variance. This high proportion implies that the first element is crucial for comprehending the links among the constructs as it captures a sizable amount of the data's underlying structure. With an eigenvalue of 1.37, the second component accounts for 9.787% of the variance, increasing the total variance explained by the first two components to 42.022%. With an eigenvalue of 1.27, the third component contributes an additional 9.087%, bringing the cumulative variance to 51 110%. These three elements together explain more than half of the variation, indicating that they cover the main facets of the data. The explanatory power of the following components decreases, even if they still contribute to the overall variation. With an eigenvalue of 0.949, the fourth component, for instance, explains 6.782% of the variance, but the fifth and sixth components, respectively, explain 6.223% and 5.519%. These subsequent components' declining eigenvalues suggest that their overall contribution to the data's variability is less substantial.

Confirmatory Principle Component Examination

The principle component arrangement of a set of guided proxies is verified using a type of statistics known as confirmatory principle component examination. Based on the underlying frame job and correlatives among observed proxies, investigations can evaluate hypotheses utilizing CFA. To evaluate the hypothesis that there is a association among the proxies that are being steered and the latent conceptions that underpin them, the inquiry employed CFA under table 3.

Table 3. Covariances

Covariance			Approximation	S.E.	C.R.	P	Hy.
Positioning of innovation panorama	<-->	**R& D Investment**	**.242**	**.021**	**8.231**	***	**H2**
R& D Investment	<-->	Innovation Source Diversity	.303	.032	7.692	***	**H3**
Positioning of innovation panorama	<-->	Positioning of innovation panorama	.254	.24	7.831	***	**H1**

Source: Authors

The covariance's among the main constructs in the manifested model are manifest in Table 3, which sheds light on the connections among R&D expenditure, innovation source diversity, and positioning of innovation panorama. The covariance's are presented together with their significance levels (P), critical ratios (C.R.), and standard errors (S.E.), which together show how strong and substantial the connections are.With a standard error of 0.021, the correlation among Investment in R&D and positioning of innovation panorama is determined to be 0.242. With a significance level of *** ($p < 0.001$) and a critical ratio of 8.231, there is a highly substantial positive correlation among these two constructs. This investigation supports the hypothesis (H2) that positioning of innovation has a positive influence on Investment in R&D by indicating that firms' Investment in R&Ds are likely to rise as they improve their positioning of innovation strategies.

Also, Investment in R&D and innovation source diversification have a correlation of 0.303 and a standard error of 0.032. A substantial positive correlation among these variables is further supported by the crucial ratio of 7.692 and the significance level of ***. This finding demonstrates the mediating function of R&D in creating a rich pool of creative notions and solutions within the healthcare industry and supports hypothesis H3, which states that higher R&D spending are linked to greater variety in innovation sources. Also, the correlation among the positioning of innovation panorama and itself is revealed to be 0.254, with a significance level of *** and a critical ratio of 7.831. This self-covariance supports the internal consistency and dependability of the model (H1) by demonstrating that the construct is accurately measuring its intended variable.

Table 4. Validity concern

	CR	AVE	MSV	MaxR(H)	IMS	TQMS	ECE
IPP	0.732	0.643	0.219	0.852	**0.654**		
RDI	0.758	0.657	0.132	0.763	0.171	**0.665**	
ISD	0.793	0.603	0.274	0.797	0.383	0.193	**0.684**

Note: IPP= Positioning of innovation panorama*; RDI*= Investigation and Development Investment*; ISD*= Innovation Source Diversity
Source: Authors

With an emphasis on Composite Reliability (CR), Average Variance Extracted (AVE), Maximum Shared Variance (MSV), and other pertinent metrics for each of the constructs—Positioning of innovation Panorama (IPP), Investigation and Development Investment (RDI), and Innovation Source Diversity (ISD)—Table 4 manifests the validity concerns for the constructs measured in the investigation. These indicators are crucial for assessing the measurement model's validity and dependability. All of the constructs' Composite Reliability (CR) values—IPP at 0.732, RDI at 0.758, and ISD at 0.793—are over the permissible cutoff of 0.7. These values suggest that the items are successfully measuring the same underlying construct since they show a high degree of internal consistency among the items within each construct. The constructs' validity is further supported by the Average Variance Extracted (AVE) values, which show scores of 0.643 for IPP, 0.657 for RDI, and 0.603 for ISD. In comparison to the measurement error, all of these numbers are over the 0.5 cutoff, suggesting that the constructs capture a substantial amount of variation. This implies that there is excellent convergent validity in the constructs.

Also, there is little overlap across the constructs, as indicated by the very low Maximum Shared Variance (MSV) values (0.219 for IPP, 0.132 for RDI, and 0.274 for ISD). Because the components measure different aspects of the healthcare innovation ecosystem, this supports the model's discriminant validity. The measurement model's dependability is further supported by the noteworthy Maximum dependability (MaxR(H)) values, which include IPP at 0.852, RDI at 0.763, and ISD at 0.797. Although not discussed in detail in this investigation, the presence of IMS (Intermediary Management System) and TQMS (Total Quality Management System) values may point to other constructs that should be investigated in connection with the primary variables.

Mediating Function Examination

Figure 2. Structure equation model

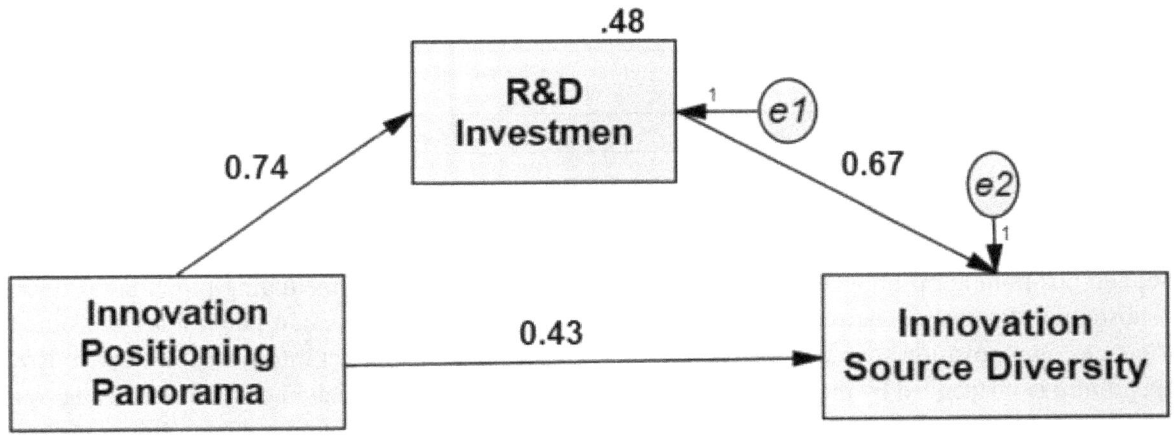

Model fit indices

Table 5. Indices for model fit

Sig.	Chi-Sq	RMR	Fitness Goodness	Fitness Confirmatory	TLI	RMSEA
0.004	1.954	.036	0.930	0.912	0.911	.028

Source: Authors

In order to assess the overall suitability of the manifested model that looks at the mediating function of R&D expenditure among positioning of innovation and innovation source diversity, Table 5 provides the model fit indices for the structural equation model. Although this is typically expected in social science investigation, the chi-square test is statistically substantial with a significance level (Sig.) of 0.004, indicating that the model does not fully match the data.

When combined with the degrees of freedom, the reported chi-square value of 1.954 results in a comparatively low ratio, indicating a strong degree of fit among the observed and predicted covariance matrices. The model's adequacy is further supported by the Root Mean Square Residual (RMR) value of 0.036, as lower values denote better fit. The values for the Adjusted Goodness-of-Fit Index (AGFI) and Goodness-of-Fit Index (GFI) are displayed as 0.912 and 0.930, respectively. With values above 0.90 suggesting a strong fit, these indices imply that the model explains a substantial percentage of the variation in the data. This notion is further supported by the Tucker-Lewis Index (TLI), which is 0.911; values greater than 0.90 are often regarded as indicating a well-fitting model. Finally, a very excellent match is manifest by the reported Root Mean Square Error of Approximation (RMSEA), which is very low at 0.028. While values up to 0.08 reflect a decent fit to the data, numbers below 0.05 are frequently viewed as showing a close fit.

Table 6. Mediating function influence

	Influence	worth	Path Influence
Positioning of innovation panorama → Innovation Source Diversity	Direct Influence	.43	Direct influence stated
Positioning of innovation panorama → R& D Investment → Innovation Source Diversity	Indirect Influence	.74*.67=.48	Indirect Influence Ensued
	Whole influence	.91	Partial mediation

Source: Authors

Substantial new information on the connections among innovation source variety, R&D expenditure, and positioning of innovation is revealed by the examination of the mediating function influence. A substantial direct association is indicated by the positioning of innovation panorama's 0.43 direct influences on innovation source diversity. The data also reveals an indirect influence, too, where R&D expenditure is influenced by the positioning of innovation landscape, which in turn influences the variety of innovation sources. This indirect pathway shows how R&D expenditure mediates the link among innovative positioning and source diversity, yielding a total influence of 0.48 (calculated as 0.74 multiplied by 0.67). A substantial overall link is manifest by the cumulative influence, which is computed at 0.91 when taking the full model into account. The outcome shows partial mediation, which means that although innovation source diversity is directly influenced by positioning of innovation; this connection is substantially enhanced by the mediation provided by R&D spending. In the end, this enriches the innovation ecosystem by highlighting the significance of Investment in R&D as a critical intermediate component that amplifies the influence of innovative positioning on diversifying innovation sources.

DISCUSSION

The data analysis's conclusions provide substantial new information on the complex connections among innovation source variety, R&D spending, and positioning of innovation. A solid foundation for factor analysis is established by the substantial findings from Bartlett's Test of Sphericity and the strong Kaiser-Meyer-Olkin (KMO) score of 0.850. This implies that there is sufficient correlation among the variables that were chosen, supporting the notion that the underlying aspects of these conceptions are substantial and deserving of more investigation. Such a strong basis is necessary to guarantee that the outcomes of the ensuing analysis are trustworthy and understandable. The first three components together account for more than half of the variation, according to the results of the Exploratory Factor Analysis (EFA), highlighting their significance in capturing the key dimensions that establish the connections among the constructs. This conclusion is substantial because it implies that these crucial characteristics may be used to understand positioning of innovation, R&D spending, and innovation source diversity. It also offers a structured lens through which to examine the associations among these aspects.

Also, the analysis's covariance's demonstrate the constructive connections among the structures. Improving positioning of innovation methods is likely to result in higher R&D expenditure, as seen by the substantial correlation among positioning of innovation and Investment in R&D. This link supports the notion that successful positioning of innovation has a favorable influence on Investment in R&D by highlighting the strategic significance of matching innovation initiatives with sufficient finance. Also, the notion that more R&D spending leads to a wider variety of innovation sources is supported by the large correlation among Investment in R&D and innovation source diversity, underscoring the crucial mediating function that R&D plays in promoting inventive results. The validity measures also support the dependability of the constructs, with Composite dependability (CR) values surpassing the acceptable threshold and Average Variance Extracted (AVE) scores suggesting high convergent validity. The constructs' uniqueness is further supported by the low Maximum Shared Variance (MSV) values, which indicate little overlap among them. This resilience improves the measuring model's legitimacy and ability to capture the intricacies of the healthcare innovation ecosystem. With a noteworthy chi-square value and positive goodness-of-fit indices, including an outstanding Root Mean Square Error of Approximation (RMSEA) value of 0.028, the model fit indices also show a solid overall fit. These outcomes support the conclusions about the mediating function of R&D spending by showing that the model correctly depicts the underlying connections in the data.

CONCLUSION

The research emphasizes the significance of R&D expenditures in determining innovative positioning in Ethiopia's healthcare industry, notably through its impact on innovation source variety. The results of Bartlett's Test for Sphericity and the Kaiser-Meyer-Olkin (KMO) Measure prove that the dataset is robust enough for factor analysis, demonstrating notable patterns throughout the association between these variables. Exploratory Factor Analysis (EFA) demonstrated that the top three components account for a significant amount of the variance, highlighting necessary factors that promote creativity. Positive covariances amongst these dimensions show whether R&D investment influences the hyperlink amongst source variety as well as innovative positioning. This form of mediation is further supported by an overall score of 0.91 through the impact study, demonstrating the relevance of both direct and indirect channels in increasing innovation source variety. The measuring approach precisely reflects the intricacies of the healthcare innovation ecosystem, and its accuracy is backed by validity measures that demonstrate the constructs' originality as well as dependability. Positive model fit metrics indicate the model's ability to successfully convey the underlying connection among variables. In the end, our findings underscore the strategic importance of R&D investments in stimulating creative ventures as well as cultivating a varied innovation environment. Organizations which successfully combine these features have a stronger chance to negotiate the difficulties of the health care industry, resulting in long-term innovation results that optimize their competitive advantage along with operational efficiency. This complete understanding calls for a holistic approach that places emphasis on the interaction of these essential aspects, which will guide Ethiopia's processes for innovation forward.

MANAGERIAL IMPLICATIONS

The investigation's conclusions have substantial management ramifications for companies looking to improve their innovation tactics in the healthcare industry. First and foremost, managers should give top priority to aligning their innovation plans with sufficient money, as there is a strong correlation among innovation posture and R&D expenditure. Organizations may encourage more Investment in R&Ds, which are necessary to cultivate a wide range of innovation sources, by proactively bolstering their positioning of innovation initiatives. Also, the analysis's emphasis on the mediating function of Investment in R&D suggests that managers should prioritize building a strong R&D ecosystem in addition to direct innovation activities. This entails cultivating an innovative culture that promotes teamwork and the distribution of resources toward investigation and development. By doing this, businesses may use their R&D resources to investigate a wider variety of creative concepts and solutions, which will ultimately increase their total production of inventive notions. Also, managers should use an organized approach while examining the elements driving innovation, as indicated by the Exploratory Factor Analysis's emphasis on the first three components. Organizations may more influence negotiate the complexity of the innovation ecosystem and make sure that their strategies are both successful and flexible enough to adjust to shifting market conditions by comprehending and concentrating on these substantial elements. Finally, because the healthcare industry is dynamic and demands flexibility in response to new trends and problems, the findings support ongoing monitoring and assessment of innovation initiatives. Managers will be better able to make decisions that promote competitive advantage and sustainable growth if they take the initiative to evaluate how positioning of innovation, Investment in R&D, and source diversity interact. All things considered, a comprehensive strategy that incorporates these components will enable businesses to prosper in a setting that is becoming more complicated and competitive.

PRACTICAL IMPLICATIONS

The investigation's conclusions have a number of applications for businesses in the healthcare industry. First, the obvious connection among Investment in R&D and positioning of innovation emphasizes the need for useful frameworks that help managers match their financial resources with their innovation strategy. To guarantee that innovation projects receive sufficient funding, organizations should set up explicit rules and procedures. This may entail giving priority to projects that show a high potential for influence. The survey also emphasizes how crucial it is for businesses to have a strong R&D culture. In practice, this entails funding educational initiatives that provide staff members the know-how to spur creativity. R&D efforts may also be improved by promoting cross-functional cooperation since diverse teams can provide more innovative notions because they bring a range of viewpoints and experience to the table. Companies should think about putting in place collaborative platforms and technologies that encourage staff members to share notions and information. Also, identifying the critical elements that influence innovation gives businesses a useful road map for assessing and improving their innovation initiatives. To stay competitive, managers may utilize this information to compare their procedures to industry norms. Organizations may discover opportunities for development and modify their strategy to suit changing market demands by regularly evaluating positioning of innovation, R&D spending, and source diversity. Lastly, companies should use data analytics to regularly assess how well their innovation efforts are working. Managers can decide on resource allocation and strategic direction by using metrics that monitor R&D results and innovation success. In addition to increasing the chances of successful innovation, this proactive strategy puts businesses in a position to react quickly to shifts in the healthcare industry, which eventually promotes long-term sustainability and success.

THEORETICAL IMPLICATIONS

The investigation's conclusions have substantial theoretical ramifications that add to the corpus of investigation on innovation management in the healthcare industry. First, the theoretical framework surrounding innovation dynamics is expanded by the proven linkages among innovation source variety, Investment in R&D, and positioning of innovation. The intricacy of these associations is highlighted by this investigation's identification of Investment in R&D as a crucial mediator. It is recommended that future investigation examine more mediating and moderating factors in order to better understand the ways in which positioning of innovation influences innovation outcomes. Also, the usefulness of employing empirical methods to examine abstract notions in innovation investigation is reinforced by the investigation's reliance on component analysis to uncover fundamental dimensions. Theoretical knowledge of the associations among these constructs is enhanced by the identification of the main factors that contribute substantially to the variation. Future investigation in this area will be more rigorous and reliable as a result of this strategy. which motivates other investigations to use comparable approaches.

Also, the results encourage a reassessment of current notions about innovation strategy, indicating that conventional models may profit from including a more sophisticated comprehension of the association among R&D expenditure and positioning of innovation. The analysis's positive covariance's cast doubt on oversimplified conceptions of innovation as a linear process and support a more comprehensive viewpoint that takes into account how dynamic and interconnected invention activities are. Lastly, new theoretical models that take into consideration the particular possibilities and difficulties faced by the healthcare industry might be developed using the knowledge gathered from this investigation. Scholars can further theoretical discussions on innovation management by incorporating the complex associations found in this investigation, which will ultimately lead to a more thorough understanding of how organizations can successfully navigate the complexities of innovation in a rapidly changing ecosystem.

RECOMMENDATIONS

The investigation's conclusions lead to a number of substantial suggestions for businesses looking to improve their innovation tactics in the healthcare industry. First and foremost, managers must give top priority to matching R&D expenditure with innovative positioning. Companies should set up specialized teams to evaluate and carefully distribute funds to innovative projects that have the potential to have a big influence. This may be accomplished by routinely evaluating project proposals and performance indicators to make sure that funds are allocated to the most creative and feasible concepts. Also, it is crucial to cultivate a culture of cooperation and ongoing education in R&D teams. Mentorship programs and interdisciplinary workshops that promote information exchange and creative thinking should be implemented by organizations. Organizations may increase their capacity for creativity and guarantee that a variety of viewpoints are taken into account when developing new solutions by fostering an atmosphere where staff members feel encouraged to share notions and work together on projects.

Also, firms should employ technology and data analytics to assess the performance of their innovation efforts regularly. Implementing technologies that track R&D outcomes, innovation development, and market trends can give useful insights into the influence of various efforts. Managers can make well-informed decisions about resource allocation and strategic adjustments by routinely evaluating this data, which will keep the company flexible and responsive to shifts in the healthcare industry.

Lastly, it is recommended that organizations invest in building strategic partnerships and networks with other healthcare entities, investigation institutions, and technology firms. Such collaborations can enhance access to diverse innovation sources and foster the exchange of notions and resources. By actively engaging in partnerships, organizations can broaden their innovation horizons and strengthen their competitive position in the market. Overall, these recommendations aim to create a robust framework for driving influence innovation in a rapidly evolving healthcare ecosystem.

FUTURE DIRECTIONS

This investigation opens up a number of investigation directions for the future that will help us better understand the dynamics of innovation in the healthcare industry. Investigating new mediating and moderating factors that could influence the connections among innovation source variety, Investment in R&D, and positioning of innovation is one encouraging avenue. A more sophisticated knowledge of the ways in which these components interact and influence innovation results may be obtained by looking into elements like company culture, leadership philosophies, and outside ecosystem influences. Longitudinal studies that monitor the influences of R&D spending over time on innovative positioning and source diversity are another substantial subject for future investigation. These kinds of investigation might assist firms improve their innovation strategy by offering empirically supported insights into the long-term influences of strategic choices and resource allocation. Also, taking into account various geographic settings and investigating the ways in which cultural variations influence innovative processes may provide insightful information that improves the generalizability of current notions. Future studies might also examine how new technologies, including big data and artificial intelligence, are influencing innovation tactics in healthcare institutions. Navigating the complexity of today's healthcare concerns will need an understanding of how these technologies may support R&D initiatives and innovate positioning. Lastly, another crucial topic for investigation is how laws and regulatory frameworks influence innovative behaviors. Examining how various regulatory frameworks influence R&D spending and the variety of innovation sources may yield insightful information for both healthcare executives and legislators. Scholars may make a substantial contribution to the developing conversation on innovation management by tackling these future directions, which will eventually promote more efficient and long-lasting procedures in the healthcare industry.

REFERENCES

Adula, M., & Kant, S. (2025). Impact of Smart Entrepreneurship and Territorial Marketing on Regional Branding: Case of Ethiopia. In *Sustainable and Intelligent Territorial Marketing and Entrepreneurship* (pp. 229–256). IGI Global., DOI: 10.4018/979-8-3693-9775-6.ch008

Adula, M., Kant, S., & Geda, H. D. (2025). Innovation Management Landscape for Digital Economy Transformation Through Innovation Resilience in Ethiopia. In *Innovation Management for a Resilient Digital Economy* (pp. 177–204). IGI Global Scientific Publishing., DOI: 10.4018/979-8-3693-8357-5.ch007

Aggarwal, A., Baker, H. K., & Joshi, N. A. (2024). Firms Innovation as Business Strategy: A Review and Bibliometric Analysis. *Journal of the Knowledge Economy, 0123456789*. Advance online publication. DOI: 10.1007/s13132-024-01830-2

Birhane, M., Amentie, C., Borji, B., & Kant, S. (2024). Are Ethiopian Coffee Farmers Proficient to Predict Capital Structure with Mediation of AI Based Knowledge Systems and Asset Liquidity? In AI in Agriculture for Sustainable and Economic Management (pp. 199-215). CRC Press.

Dereso, C. W., Kant, S., Muthuraman, M., & Tufa, G. (2023). Effect of Point of Service on Health Department Student's Creativity in Comprehensive Universities of Ethiopia: Moderating Role of Public-Private Partnership and Mediating Role of Work Place Learning. In: Jain, S., Groppe, S., Mihindukulasooriya, N. (eds) Proceedings of the International Health Informatics Conference. Lecture Notes in Electrical Engineering, vol 990. Springer, Singapore. https://doi.org/DOI: 10.1007/978-981-19-9090-8_13

Fernandes, M. A. (2023). *Investigation of influence frameworks to support medical technology innovation in the hospital setting.*

Gokhberg, L., Meissner, D., & Kuzminov, I. (2023). What semantic analysis can tell us about long term trends in the global STI policy agenda. *The Journal of Technology Transfer*, *48*(6), 2249–2277. DOI: 10.1007/s10961-022-09959-5 PMID: 35967600

Gomes, J. S. (2024). *Journal of Management & Technology e-ISSN: 2177-6652. 24*, 149–166.

Gu, F., Gao, J., Zhu, X., & Ye, J. (2023). The influence of digital inclusive finance on SMEs' technological innovation activities—Empirical analysis based on the data of new third board enterprises. *PLoS ONE, 18*(11 NOVEMBER), 1–21. DOI: 10.1371/journal.pone.0293500

Kaggwa, S., Akinoso, A., Dawodu, S. O., Uwaoma, P. U., Akindote, O. J., & Eloghosa, S. O. (2023). Entrepreneurial Strategies for Ai Startups: Navigating Market and Investment Challenges. *International Journal of Management & Entrepreneurship Investigation*, *5*(12), 1085–1108. DOI: 10.51594/ijmer.v5i12.662

Kant, S., & Adula, M. (2025). Integrating Gender and Indigenous Knowledge in Digital Innovations for Ethiopia's Renewable Energy Transition. In *Digital Innovations for Renewable Energy and Conservation* (pp. 167–194). IGI Global.

Kant, S., Adula, M., Kumari, M., & Nimbrain, N. (2025). Role of SMEs in Greening the Economy by Mediation of Eco-Innovation in the Horn of Africa. In *Examining Green Human Resources Management and Nascent Entrepreneurship* (pp. 173–200). IGI Global., DOI: 10.4018/979-8-3693-7046-9.ch007

Kant, S., Niguse, T., Terefe, A., & Adula, M. (2025). Emerging Trends and Innovations: Effect on Social Sustainability in Manufacturing Supply Chains by Shifts in Consumer Behavior in the Horn of Africa. *Enhancing Social Sustainability in Manufacturing Supply Chains*, 135-162. http://dx.doi.org/ DOI: 10.4018/979-8-3693-9740-4.ch005

Kapoor, R. (2024). Strategic Human Resource Management Practices and Firms Innovation: Evidence from Small and Medium Enterprises in India. *International Journal of Strategic Management*, *3*(1), 30–40. DOI: 10.47604/ijsm.2481

Niguse, T., Kant, S., & Adula, M. (2025). Mediation Effect of Customer Loyalty in Relationship Between Market Orientation, Entrepreneurial Orientation, and Firm Performance in Ethiopia. In *Leveraging AI for Effective Digital Relationship Marketing* (pp. 419–458). IGI Global.,

Pak, A., Seo, D. J., & Roh, T. (2024). The influence of intellectual property rights on firm performance in service firms: The function of process and firms innovation. *Cross Cultural & Strategic Management*, (August). Advance online publication. DOI: 10.1108/CCSM-08-2023-0163

Pesqueira, A., Sousa, M. J., & Pereira, R. (2023). Individual Dynamic Capabilities Influential Factors in Blockchain Technology Innovation from Hospital Settings. *WSEAS Transactions on Biology and Biomedicine*, *20*, 275–294. DOI: 10.37394/23208.2023.20.29

Xiong, W. (2022). *Insta360 Catches the Wave of Industry Growth*. Atlantis Press International BV., DOI: 10.2991/978-94-6463-052-7_46

KEY TERMS WITH DEFINITIONS

Positioning of innovation: The strategic placement of an organization’s innovations in the market, focusing on how these innovations are perceived relative to competitors. It involves defining the unique value proposition and differentiating features of an organization's products or services.

Investment in R&D: Financial and resource allocation dedicated to investigation and development activities aimed at creating new products, services, or processes. This investment is crucial for fostering innovation and maintaining competitive advantage.

Innovation Source Diversity: The variety of origins from which an organization draws notions and inputs for innovation. This can include internal sources (employees, departments) and external sources (customers, suppliers, investigation institutions) that contribute to the innovation process.

Chapter 23
Leveraging Blockchain in High Frequency Trading for Shaping Market Behavior and Enhancing Portfolio Returns

Rohit Sood
https://orcid.org/0000-0002-4127-7781
Lovely Professional University, India

Ashwani Panesar
https://orcid.org/0000-0003-0605-8951
Lovely Professional University, India

ABSTRACT

Capital markets have been consistently evolving in the last decade since the global financial crises struck. The growth of Groww and Zerodha along with other fintech companies who have made investing in capital markets more accessible to the people as a result millions of investors have thronged the markets either directly or indirectly. Mutual funds have received record number of folios; capital markets have seen highest number of retail participation. This has created lot of volatility in the trades as the share prices rise and fall under the emotions of the investors even if macro indicators are good. But with the industry adopting AI, ML and Blockchain the results have been more consistent as machines have been programmed to perform the human tasks. With human emotions coming into play, HFT has been designed to ensure that human interference comes to the lowest. In this chapter the entire thrust has been made to understand the HFT and the integration of blockchain and the emotionality in having better portfolio returns.

INTRODUCTION

Capital markets have been consistently evolving in the last decade since the global financial crisis struck. The stakeholders of the capital markets have managed their revenue models efficiently to survive in the market. In the last decade or so, there has been a paradigm shift in the investor's geography &

DOI: 10.4018/979-8-3693-4030-1.ch023

demography as more and more investors are coming from the towns and rural areas. With the growth of Groww & Zerodha along with other fintech companies, the investment in capital markets has been made more accessible resulting in millions of investors thronging the markets either directly or indirectly. Mutual funds have received record numbers of folios; capital markets have seen the highest number of retail participation. This has created a lot of volatility in the trades as the share prices rise and fall under the emotions of the investors even if macro indicators are good. But with the industry adopting AI, ML and Blockchain the results have been more consistent as machines have been programmed to perform human tasks. With human emotions coming into play, HFT has been designed to ensure that human interference comes to the lowest.

The concept of blockchain was first traced in Bitcoin's paper published by Satoshi Nakamoto in 2008 (Nakamoto, 2008), which was primarily designed and developed for cryptocurrency, but blockchain's decentralized ledger mechanism quickly gained attention for its potent applications in financial markets. The challenges associated with algorithmic trading like latency, security, and regulatory concerns paved the way for HFT in the late 20th century (Aldridge & Krawciw, 2017) for better data integrity and smooth transactions. Studies of early 2010s explored the application of blockchain in trading systems which laid the foundation of its integration into HFT strategies (Yermack, 2017).

In this chapter the entire thrust has been made to understand the HFT and the integration of blockchain and the emotionality in the light of financial analysis to augment better portfolio returns.

The global high frequency trading servers' market which was valued at USD 604.63 million in 2023 and is expected to grow at a CAGR of 6.4% in the next 8 years (Polaris Market research). The reason for the growth of this industry is largely driven by better level of infrastructure development, accessibility to electricity, accessibility to banking, digitalization, digitization and internet penetration. Also, the millennials inclination towards adoption of technology further helped the sector for trusting the trading platforms for enhanced usage frequency.

The graph below shows that the growth in the Asia-Pacific is growing in double digits largely attributed to the above-mentioned factors. The rise of fintech industry is seen as the major reason in bringing paradigm shift to convert non-investors to investors to regular investors.

Figure 1. High Frequency Trading growth projections across the World

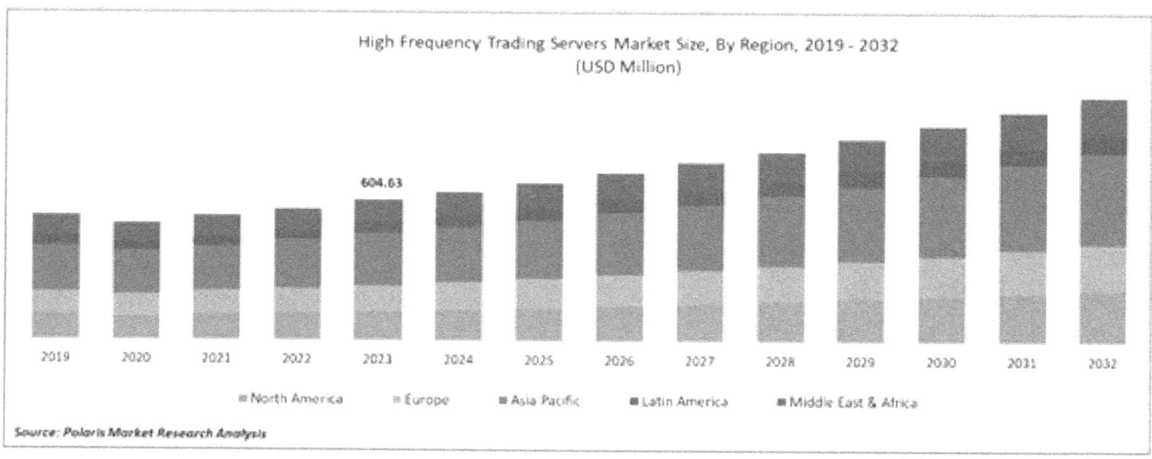

Source: Polaris Market Research Analysis

During the entire process of investment for trading purposes, traders have used manual interventions to buy/sell stocks. But sometimes in the event of heavy selling like the one in 2009-financial crises, the manual intervention does not work. Taking into cognizance all these problems, the concept of HFT came to existence which supports traders who do heavy frequency of trade without much self-intervention. some call this as improved version of algorithm trading.

Blockchain technology has made massive transformation in global financial markets thus impacting on high-frequency trading (HFT). Usage of blockchain in trading brings far more transparency, security, and efficiency in its operations to create faith among investors. Global financial hubs like USA, China, and the EU, have explored blockchain applications in trading to minimize market manipulation, abide regulatory compliance (Gai et al., 2019). Globally, millions of orders are executed in stock markets, its integration will help in real-time settlement, reducing latency issues and improving trading efficiency (Nash et al., 2021). Lately, traders, brokers and investors are recognizing the role of blockchain in HFT to influence market behavior, investors' confidence, and portfolio returns.

The rise of the HFT is largely attributed due to the penetration of high internet bandwidth. High-frequency trading (HFT) refers to trading in financial instruments, such as securities and derivatives, transacted through supercomputers executing trades within microseconds or milliseconds (or, in the technical jargon, with extremely low latency). Normally, traders or Investors with heavy volumes of trade use HFT strategies in order to earn a small amounts of profit per trade whereby manual intervention will not help in desired results.

HFT consists of automated trading that encompasses:

(a) algorithms for decision making, order initiation, generation, routing, or execution, for each individual transaction without human direction.
(b) low-latency technology that is designed to minimize response times, including proximity and co-location services.
(c) high speed connections to markets for order entry; and
(d) high message rates (orders, quotes or cancellations).

Traders use strategies which can be categorized as passive or aggressive strategies to stay afloat in the competitive trading market.

Figure 2. Application of Aggressive and Passive Strategies through High Frequency Trading

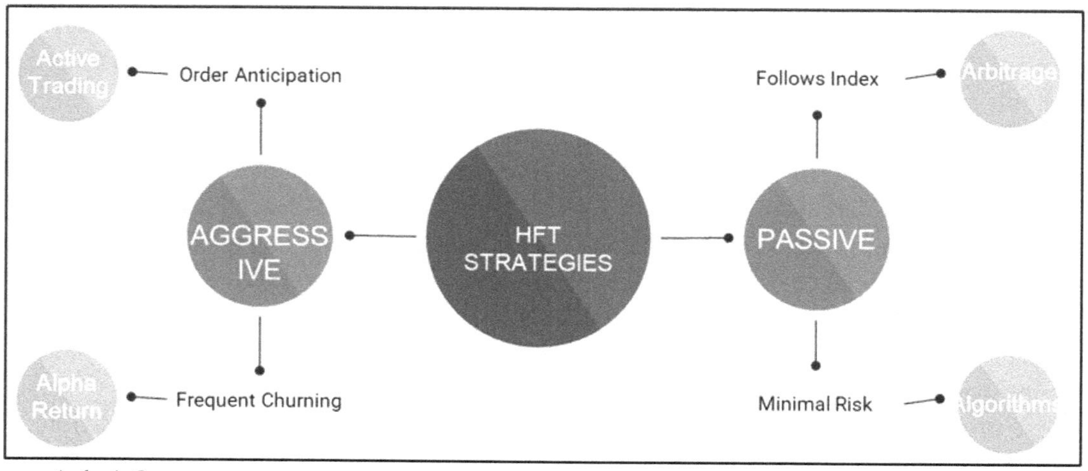

Source: Author's Own

Passive strategies include arbitrage trading which attempt to profit from price differentials for the same stocks or their derivatives traded on different trading venues and passive market making, in which profits are generated by spreads between bid and ask prices. On the other hand, passive trading aims to match the returns of an index. Passive investors can leverage HFT to precisely replicate the index's components within their portfolio, achieving maximum returns with minimal risk associated with the index.

Aggressive strategies include those known as order anticipation or momentum ignition strategies. This may be because such strategies can share some similarities to practices such as front-running and spoofing, which are generally illegal as they are associated with increased market fragility and volatility. Active trading allows investors to generate alpha returns by capitalizing on market fluctuations. High-frequency trading (HFT) also contributes to alpha generation by efficiently executing trades, ultimately enhancing portfolio performance.

Moreover, the demand for Ultra-Low Latency (ULL) in trading systems, coupled with advancements in quantum computing will support the growth in the coming years. Additionally, the increasing requirement for intent-based networking to support high-frequency trading environments is expected to open opportunities for future expansion. These servers, driven by mathematical algorithms, enable lightning-fast trading transactions in mere nanoseconds. Notably, these high-speed executions leverage robust computational analysis, surpassing the capabilities of traditional stock trading servers.

Empirical Literature on Blockchain and High-Frequency Trading

Numerous studies have investigated the impact of blockchain on financial markets, in the light of HFT whereby it has been proved that usage of blockchain drastically reduced transaction time by eliminating intermediaries. Xu et al. (2020) reported that blockchain-based HFT reduce latency in system to make price adjustments more efficient. Li and Zhang (2021) empirically reported that decentralized trading platforms incorporating blockchain improved liquidity and reduced arbitrage opportunities. Moreover, some comparative studies among traditional electronic trading platforms and blockchain-powered exchanges found that blockchain integration enhanced efficiency while mitigating risks associated with market crashes and price manipulation (Biais et al., 2019).

Blockchain in HFT for Shaping Market Behavior

Blockchain integration has transformed HFT by addressing incompetence in market operations. Traditional HFT systems operate on centralized infrastructures, leading to challenges such as latency manipulation and a lack of transparency in trade execution (Aldridge, 2009; Brogaard, 2010). The integration of blockchain provides a decentralized ledger to improve trade validation, minimize settlement risks, by addressing the inefficiencies due to delays in order execution (Biais et al., 2019). By implementing the smart contracts, refine the entire HFT process by automated trade execution thus reducing the reliance of intermediaries (Buterin, 2016) for better operational and transactional efficiency. Moreover, blockchain's timestamping mechanism plays a critical role in order sequencing to reduce practices like front-running and order spoofing to maintain market stability (Bonneau et al., 2015; Nakamoto, 2008). Blockchain usage also provides a comprehensive and transparent trading environment by reducing fragmentation across multiple platforms (Li & Zhang, 2021) for seamless clearing and settlement among the traders. Blockchain helps decentralized exchanges to reduce transactional risks among parties by providing participants with more reliable and direct trading opportunities with no reliance on intermediaries (Avgouleas & Kiayias, 2019). But blockchain's adoption in HFT pose a challenge due to its computational requirements (Chen et al., 2019). On the one side it strengthens security and transparency, but the transaction execution speed remains a concern compared to existing HFT systems can execute millions of trades in a faction of time (Xu et al., 2020). With better bandwidth, off-chain scaling and hybrid blockchain models this gap between speed and security is set to be narrowed (Gai et al., 2019; Noble & Patil, 2021).

Blockchain in HFT for Enhancing Portfolio Returns

Integration of blockchain with HFT will help to augment portfolio management by improving trade efficiency and execution. Traditional HFTs' encountered with problems like liquidity fragmentation leading to inefficient capital deployment due to market scenario and latency disparities (Avellaneda & Stoikov, 2008). Blockchain technology has helped to addresses these inefficiencies by introducing decentralized trade execution to better market depth and reduce the price slippage leading to improved portfolio performance (Li & Zhang, 2021), thus eliminating the need of traditional clearinghouses and intermediaries (Avgouleas & Kiayias, 2019). This structural shift targeted at optimizing capital allocation helps traders to execute high-speed strategies with more confidence on settlement. Smart contracts have further streamlined the post-trade processes which have helped for maximizing returns in dynamic environment (Buterin, 2016). Blockchain-driven HFT, used for portfolio management, helps in risk mitigation. Participants have complained that traditional HFT systems have caused order mismatches and execution delays leading to erosion in profitability (Biais et al., 2019). Blockchain's characteristics like real-time reconciliation and transparent ledger help to reduce disparity in trade execution and arbitrage inefficiencies to enhance returns vs market fluctuations (Nakamoto, 2008; Bonneau et al., 2015). The application of blockchain in trade data analytics also augment predictive modeling in HFT.ML-driven strategies integrated with blockchain's immutable transaction history allow for more precise recognition of patterns in trading behaviors (Treleaven et al., 2017), to result in improved risk-adjusted returns and portfolio performance (Harvey et al., 2018). Blockchain provides lot of benefits, but its scalability remains a concern in context of HFT, as the need for ultra-fast order execution is inhibited by consensus mechanisms (Chen et al., 2019). By integrating off-chain scaling solutions and hybrid blockchain frameworks, this gap can be bridged to leverage blockchain's security while maintaining execution speeds comparable to traditional ones (Xu et al., 2020). As technology continues to evolve, optimization of various technologies and their platforms will help for better portfolio returns as penetration of trading continues to rise. For better risk mitigation (Nair & Bhagat, 2020).

RESEARCH GAPS IN THE EXISTING LITERATURE

Several research gaps exist in adoption of blockchain in HFT after exploring various studies which are being discussed in the forthcoming sections. Firstly, there are limited empirical studies which have analyzed the usage effect of blockchain's adoption on market stability and investor behavior (Harvey et al., 2018). This calls for framing new theoretical models and for more in-depth longitudinal studies. Secondly, the interactive studies involving the impact of hybrid blockchain models on improving trading efficiency remains unexplored (Rauchs et al., 2018). Thirdly, research on the regulatory implications of blockchain-powered HFT being infancy stage and call for more development for framing comprehensive legal structure (Zohar, 2015). Lastly, more research is desired to comprehend the role of AI and ML with its integration to blockchain for better analytical and automated trading strategies in HFT environment (Treleaven et al., 2017).

Table 1. Research Gap and contribution of study to uncover research Gap

Research Gap	Contribution
Blockchain adoption for efficient and sound capital markets.	- Innovative financial models and empirical studies to analyze investor behavior, uncover psychological traits, and explore regulatory perspectives.
Blockchain models for enhanced trading efficiency	- Conceptual frameworks for the unexplored impacts of blockchain hybrid blockchain model integrating AI and ML for improved trading efficiency
Indecisive regulatory research for blockchain empowered HFT	- A standardized legal framework to enhance blockchain-enabled high-frequency trading (HFT) -Pilot sandbox initiatives to debunk myths surrounding blockchain applications in HFT
Role of Emotional Baises in Capital market investments	- Insight of various investor bias in irrational market movements. - Blockchain as a mitigator of emotional bias powered for optimized trading efficiency

Source: Author's Own

In this chapter, an effort has been made to find the various theoretical frameworks and find as to how by adoption of blockchain practices, comprehensive model can be designed keeping in context the regulatory frameworks and developing conceptual models for pilot testing. Its usage to provide better efficiency with a seamless integration of AI and ML with blockchain technology.

THEORETICAL UNDERPINNINGS

Table 2. Theoretical Underpinnings and contribution of study to respective Gaps

Gaps	Details	Description	Contribution
Contradictory Evidence Gaps	AAA Model • A- Accountability Gaps • A- Accessibility Gaps • A- Adaptability Gaps	Blockchain enhances trading efficiency with transparent and secure transactions.	Blockchain-AI-ML models to enhance trading efficiency gradually.
Theoretical Gaps	• Limited Theoretical Insights • Hybrid Blockchain Models • Blockchain-AI-ML Integration	Blockchain enhances trading with zero-knowledge proof (ZKPs).	Innovative conceptual frameworks managing theoretical Gaps.
Knowledge Gaps	Market Dynamics and Regulation	Need for dynamic and updated regulatory frameworks.	Strategic Solutions to uncover knowledge gaps.

Source: Author's Own

Contradictory Evidence Gaps

Accountability remains an area of concern in context to scalability for broader adoption which is uncertain and being debated due to the ULL and high-speed execution required in HFT trading. Proof-of-stake (PoS) and sharding techniques could further accelerate order execution, but there remains a lot of potential for further improvement to meet the HFT demands.

Transparent practices enabled by immutable transactions and smart contracts can enhance operational efficiency, but the concerns related to the risk of publicly accessible transaction data being manipulated for individual advantage in HFT trading. Blockchain has huge potential in SME financing even if it encounters hurdles such as regulatory approvals and operational complexities. Accessibility to information, cost and vulnerabilities in the prices of contracts might become a bottleneck in blockchain-based trading. These issues collectively have raised concerns as to how a decentralized model could help the rapid execution of HFT.

Contribution: More empirical research is needed to compare PoW and PoS blockchain systems as to how they impact market stability. Integrated blockchain-AI-ML model helps to test the efficiency of HFT trading which is aimed to improve speed and optimize trade execution.

Theoretical Gaps: Studies conducted on date lack deeper theoretical insights for optimization of financial analysis of the executed trades. Innovative blockchain-based models incorporating zero-knowledge proof (ZKPs) and confidential transactions can further help to improve scalability and security. Predictive analytics models backed blockchain with AI and ML further optimize HFT algorithmic trading.

Contribution: This study tries to propose an integrated conceptual model ideal for hybrid blockchains whereby AI/ML could be integrated for better application of theoretical concepts.

Knowledge Gaps: The impact of blockchain on market dynamics and regulatory frameworks need more requires updated research so as to optimize blockchain-empowered HFT trading.

Contribution: The study proposes strategic solutions to reduce investor behavior biases and cognitive limitations so that gaps can be addressed effectively.

Evolution and Interaction of High-Frequency Trading with Emotional Dynamics

HFT being a complex and advanced form of algorithmic trading that has dramatically shaped the pillars of capital markets in recent years. Essentially, HFT being precisely engineered, executes large number of orders at incredibly high speeds, by integrating cutting-edge technology and computational algorithms.it rose to limelight with its ability to find minute price differences due to market inefficiencies within fractions of a second. This quantum of speed is only possible with the help of direct market access (DMA) and co-location, whereby HFT firms try to position their servers as close to the exchange servers to reduce latency. HFT executes large number of trades in a very short time period, and this contributes significantly to turnover of markets. These strategies used by HFT are possible only with the help of advanced algorithms for analytics, detect share movement patterns, and automatically execute trades based on predetermined criteria. Market manipulators use this HFT strategy as they consistently provide liquidity to the market by placing large numbers of buy and sell orders. This is possible by narrowing the bid-ask spread whereby HFT market makers help facilitate smoother transactions for other market participants, but the provision of this liquidity is only limited and can be withdrawn as and when the market conditions shift their pattern. Arbitrage is another strategy used by HFT whereby it exploits the price differences between various trading platforms and makes profit from these discrepancies. *Like, an HFT algorithm might simultaneously buy a security in one market across the globe where the prices are lower and tend to sell it in another market where the price is higher by taking advantage of the price gap.* Another feature of HFT is its ability to hold securities for short periods, which being apart from traditional investment methods. HFT platforms primarily focuses on profiting from minute price movements in a very short time window, maintaining positions for only for fraction of a second. This rapid turnover helps the firms to quickly adapt to dynamic market conditions which helps them to execute large number of trades in a limited timeframe. Although HFT has paved the way for high speed coupled with better efficient operations, it is marked by controversy and regulatory scrutiny. Market watch dogs have conveyed that HFT improves liquidity but narrows bid-ask spreads, and aids in price discovery. It has raised concerns about the possible market manipulation, systemic risks, and the perceived advantage over the traditional investment platforms. This entire transition to the HFT has been shaped by regulatory shifts, technological interventions and some prominent market events. With the financial markets consistently evolving, HFT is set to play a crucial role in shaping the future of financial markets. In the present technological era amidst debates of whether its usage is good or bad regulatory adjustments will likely continue to influence the direction and development of HFT.

Evolution of HFT over the years:

Early 17th Century: The 17th century witnessed significant advancements in traditional trading methods, particularly with the establishment of the East India Company (VOC). The introduction of transferable securities and active share trading in Amsterdam revolutionized trading practices. This transformation turned Amsterdam into a major financial hub, increasing trading volumes and modernizing stock trading with both cash and derivatives. By 1680, Amsterdam had achieved exceptional trading techniques comparable to contemporary financial standards, leading to innovative market development strategies and global expansion. These innovations paved the way for the adoption of standardized global market practices, which enhanced stock trading penetration worldwide.

Late 17th and 18th Century: This period marked the rapid dissemination of information, overcoming traditional bookkeeping barriers. It was the era when High-Frequency Trading (HFT) began to emerge. With HFT, traders aimed to execute trades with precision and certainty, leveraging speed and accuracy. The development of centralized marketplaces in the 18th century, including the establishment of the London Stock Exchange, further evolved HFT. Traders sought to outperform market returns by capitalizing on the speed and efficiency of HFT.

1900 – 1970: Despite economic downturns affecting sectors like railways and automobiles from 1900 to 1929, financial markets continued to grow with improved trading infrastructure and innovative techniques. The advent of computerization led to market making in stock markets. The New York Stock Exchange introduced systems like Designated Order Turnaround (DOT) and SuperDOT, which revolutionized electronic trading by replacing manual order execution. The 19th century was transformative for information dissemination, facilitating swift communication and trade execution, laying groundwork for high-frequency trading.

1980 – 2000: The period from 1982 to 2007 saw significant growth in U.S. financial markets, driven by modern trading equipment and software. The financial sector experienced rapid expansion from low to high volumes, thanks to technological advancements, efficient regulatory practices, and innovative financial products. This era also saw the emergence of algorithmic trading, facilitated by widespread internet access. The development of electronic communication networks (ECNs) and Market By Watch (MBW) reduced bid-ask spreads and increased trading volumes globally. The SEC played a crucial role in high-frequency trading by authorizing electronic exchanges in 1998. HFT leverages advanced technology to execute trades at speeds that are thousands of times faster than human capability, leading to its widespread adoption and causing a substantial impact on financial markets. HFT leverages advanced technology to execute trades at speeds that are thousands of times faster than human capability, leading to its widespread adoption and causing a substantial impact on financial markets.

2000 – 2011: During the early 2000s, high-frequency trading constituted less than 10% of equity orders, but its adoption escalated rapidly in subsequent years. As per the New York Stock Exchange (NYSE), the period from 2005 to 2009 witnessed an extraordinary 164% surge in high-frequency trading volume. The year 2011 marked a significant milestone with the introduction of Nano trading technology. Fixnetix, a firm, developed a microchip capable of executing trades in nanoseconds, equivalent to one billionth of a second. To put this into perspective, 1 Nanosecond equals 0.000000001 seconds. The concept of co-location involves situating the computers owned by high-frequency trading (HFT) firms and proprietary traders on the same premises where the exchange's computer servers are located. The prominence of co-location began to rise in the early 2000s, coinciding with exchanges initiating the offering of co-location services. In the pursuit of enhanced speed, CoreSite, a Denver-based data center company, played a significant role. They operate a facility allowing traders to install "co-located" computers directly within the heart of Washington.

2011- 2023: The emergence of the internet in the late 20th century had a profound influence on news-based trading, revolutionizing the accessibility of news globally and fostering an equitable environment for traders. Online news platforms, financial websites, and social media channels assumed central roles as primary sources of information for investors, offering real-time updates and analyses. In September 2012, Dataminr introduced a groundbreaking service, converting social media streams into actionable trading signals with substantial backing of $30 million. This innovative service provided business news to clients up to 54 minutes faster than conventional sources. By scrutinizing on-the-ground conversations, consumer product responses, online community discussions, and public attention patterns, Dataminr identified micro-trends, showcasing the increasing influence of artificial intelligence (AI) and machine learning in the financial landscape. During 2012, real-time analytics engines examined the vast volume of daily tweets to identify linguistic and propagation patterns. This period coincided with the dominance of High-Frequency Trading (HFT) in stock markets, accounting for 70% of US equity trade.

In recent years, there has been a significant improvement in trading precision, with stock prices transitioning from fractions to pennies. High-frequency trading (HFT) has played a pivotal role in injecting liquidity into the market and narrowing bid-ask spreads. Information technology (IT) companies have made substantial investments in HFT technology, encompassing ultra-fast computer chips capable of executing trades in microseconds. Furthermore, there was a proposal for a $300 million transatlantic cable aimed at reducing transaction times between New York City and London by a fraction of a second.

The monitoring of social media by the FBI marked the inception of the instant impact of social media on securities. On April 2nd, 2013, the Securities and Exchange Commission (SEC) and the Commodity Futures Trading Commission (CFTC) imposed restrictions on public company announcements through social media. In response, just two days later on April 4th, 2015, Bloomberg Terminals integrated live Tweets into its economic data service. Bloomberg Social Velocity tracked unusual spikes in discussions about specific companies.

Emotional Dynamics refers to the interplay of emotions that influence human behavior and decision-making processes in various contexts. In financial markets, emotional dynamics can significantly impact trading behaviors, risk assessments, and market reactions. Emotions such as fear, greed, and anxiety can drive volatility and decision-making, often leading to irrational actions that deviate from purely logical analysis.

For example, behavioral finance acknowledges that emotions like fear and overconfidence can drive market participants to buy or sell assets in ways that might contradict the fundamentals of those assets (Lo, 2012). Emotional dynamics play a critical role in shaping market behavior by influencing traders' perceptions and reactions, thus contributing to market cycles of booms and busts (Shiller, 2017).

(Zaloom and Knorr Cetina and Bruegger), demonstrated as to how technology influences, shapes, and reconfigures trader subjectivity and prompt new ways of acting and thinking in markets.

The findings of Zaloom further demonstrated that pit traders and click/screen traders alike are preoccupied with transforming and conducting themselves to have a more disciplined approach to trading. Also, certain suggestions are being made to reduce the impact of emotionality on trades among the traders like

To have a separate personal and professional life.

They seek to control the impact of losses on trading behavior, so that one or more losses will not lead them to make excessively risky trades to regain the losses.

They attempt to treat each trade in isolation regardless of previous success or failure.
They learn to focus minutely about the market and get involved

In the context of HFT, emotional dynamics are less overt than in traditional trading due to the automated and algorithmic nature of decision-making.

Figure 3. Interrelation of HFT with Volatility, Efficiency and Liquidity

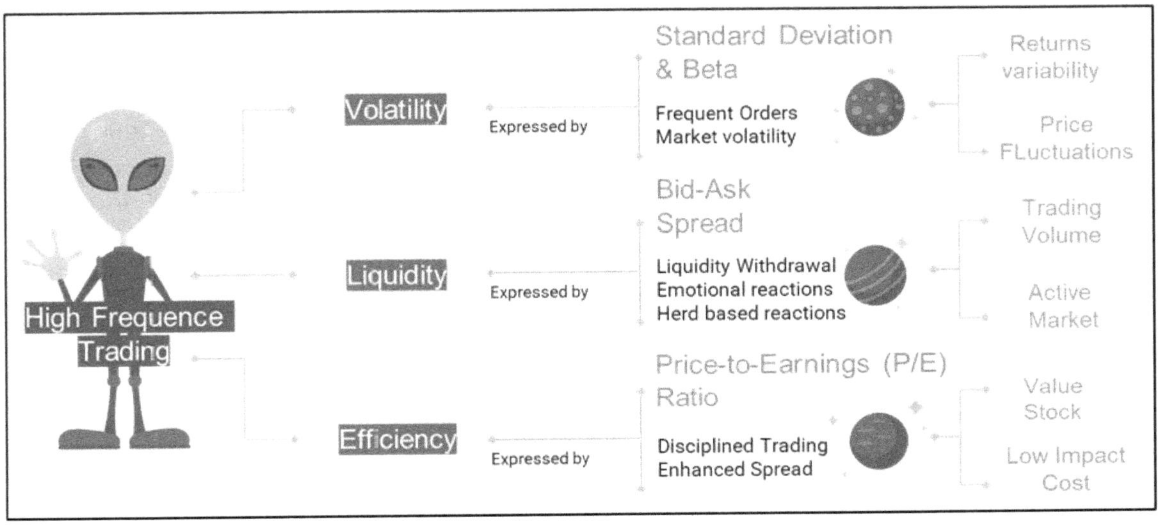

Source: Author's Own

However, emotional dynamics still play an indirect yet important role in influencing market behavior, especially in the design, monitoring, and reactions to HFT strategies.

1. ***Algorithm Design and Trader Emotions:*** Human beings have created algorithms which being the base of HFT and logically their personal experience and bias will come into play for fine-tuning it. Like a risk-averse trader will design an algorithm in such a way that are overly conservative, while a more aggressive trader may create algorithms that chase high-risk opportunities (Hendershott et al., 2011). (Knorr Cetina, K. and Bruegger, U., 2000) emphasized as to how traders' emotional attachment tc markets in light of the development in markets and their personal thoughts, which has lately become problematic for the traders, even they are being part of the market (Zaloom, 2006) but hey need to be avoided to become successful (Smith, 1999). Also, traders' interference of emotions on trading sessions also plays an important role (Zaloom).
2. ***Market Sentiment and HFT Reaction:*** High-frequency traders often respond to market sentiment, which are driven by the emotional reactions of other stakeholders. HFT algorithms capitalize on short-term market inefficiencies often driven by emotional trading behaviors, such as panic selling or irrational exuberance. Studies have shown that tone being labelled as tenor, direction or sentiment in context of economic news as to whether the market is bullish or bearish also affects the customer sentiments with a delay of at least one month (Alsem et al., 2008; Hollanders and Vliegenthart, 2011; Wu et al., 2002). Political development variables also affect consumer confidence (De Boef and

Kellstedt, 2004) in the short and the long run coupled with news on government popularity (Shah et al., 2009). While HFT reduces human emotional responses in trading decisions, it can exacerbate the effects of market emotions by amplifying price movements triggered by these emotional dynamics (Kirilenko et al., 2017).

3. ***Volatility and Emotional Feedback Loops:*** Market volatility which gets triggered by various factors, ranging from economic events and geopolitical tensions to bring changes to investor behavior. HFT can also contribute to feedback loops in volatile markets, where sharp price movements driven by emotional responses (like fear or greed) trigger algorithmic trades that intensify those movements. In all, navigating emotional responses to market changes helps in being more successful in volatility like developing a long-term strategy, staying informed, diversifying your portfolio, and having a support system in place, you can navigate market volatility with confidence and improve your chances of long-term success (Faster-capital). For instance, during flash crashes, HFT algorithms react to market signals caused by panic selling, often intensifying the downward spiral of prices (Aldridge, 2013).

Thus, while HFT aims to reduce the direct impact of human emotions, emotional dynamics still permeate the market through their influence on market sentiment, the behavior of other traders, and even the algorithms themselves.

COMPARISON OF VARIOUS THEORIES IN THE HFT

Table 3. Comparative analysis of HFT Theories

Theory	Year	Key Concept	Key Points on Blockchain & HFT	Real-World Example
			Blockchain enhances **market transparency** and reduces **information asymmetry**.	
Efficient Market Hypothesis (EMH)	1970 (Fama)	Financial markets reflect all available information, making it difficult to consistently outperform the market.	Real-time settlement limits **arbitrage opportunities** for HFT firms.	**DTCC's Project Ion** improves settlement speed, reducing inefficiencies in stock trading.
			Price efficiency improves, reducing chances of **market manipulation**.	
			- Blockchain-driven **liquidity pools** allow markets to adapt dynamically.	
Adaptive Market Hypothesis (AMH)	2004 (Lo)	Markets evolve based on behavioral changes, investor adaptation, and external conditions.	- Decentralized finance (DeFi) platforms introduce **new market structures**.	**Uniswap's Automated Market Makers (AMMs)** adapt pricing based on liquidity supply and demand.

continued on following page

Table 3. Continued

Theory	Year	Key Concept	Key Points on Blockchain & HFT	Real-World Example
			- AI-powered HFT models continuously adjust to **blockchain-based trading patterns**.	
			Blockchain reduces **latency in trade settlements**, impacting **HFT arbitrage**.	
Market Microstructure Theory	1985 (O'Hara)	Analyzes how market structure, trading mechanisms, and price discovery processes impact financial markets.	Transparency in **order book dynamics** lowers the risk of **front-running**.	**Ethereum-based dYdX** executes HFT strategies with smart contracts for instant trade settlement.
			Smart contracts automate trade execution, reducing human intervention.	
			Blockchain-based **smart contracts** reduce reliance on intermediaries.	**Curve Finance (DEXs)** removes broker dependency, ensuring fair transactions without intermediaries.
Principal-Agent Theory	1976 (Jensen & Meckling)	Focuses on conflicts of interest between investors (principals) and brokers/ traders (agents).	Transparency minimizes **agency conflicts** between investors and brokers.	**Curve Finance (DEXs)** removes broker dependency, ensuring fair transactions without intermediaries.
			HFT traders benefit from **automated, trustless execution** in decentralized environments.	
			Blockchain facilitates **fractional ownership**, making assets more liquid.	
Liquidity Preference Theory	1936 (Keynes)	Investors prefer highly liquid assets and demand higher premiums for holding illiquid ones.	24/7 trading ensures **continuous liquidity**, unlike traditional markets.	**Security tokens on tZERO and Securitize** allow investors to trade tokenized assets like stocks.
			Tokenization converts **illiquid assets** (real estate, private equity) into **tradable tokens**.	
			On-chain transparency reduces uncertainty and influences investor sentiment.	

continued on following page

Table 3. Continued

Theory	Year	Key Concept	Key Points on Blockchain & HFT	Real-World Example
Prospect Theory (Behavioral Finance)	1979 (Kahneman & Tversky)	Investors make decisions based on perceived gains/ losses rather than rational calculations.	Behavioral patterns in HFT can be analyzed using **blockchain transaction data**.	**Glass node and Nansen** provide insights into blockchain transaction behavior for better risk management.
			Risk-adjusted decision-making improves with **real-time blockchain analytics**.	

Source: Author's Own

Among the various theories listed, Market Microstructure Theory (1985, O'Hara) being the most relevant and valid for analyzing the impact of blockchain in HFT. It focuses on trading mechanisms & price discovery; explains Liquidity, Transparency & Efficiency and has a strong empirical relevance.

Figure 4. Conceptual Framework (Blockchain and sound financial markets)

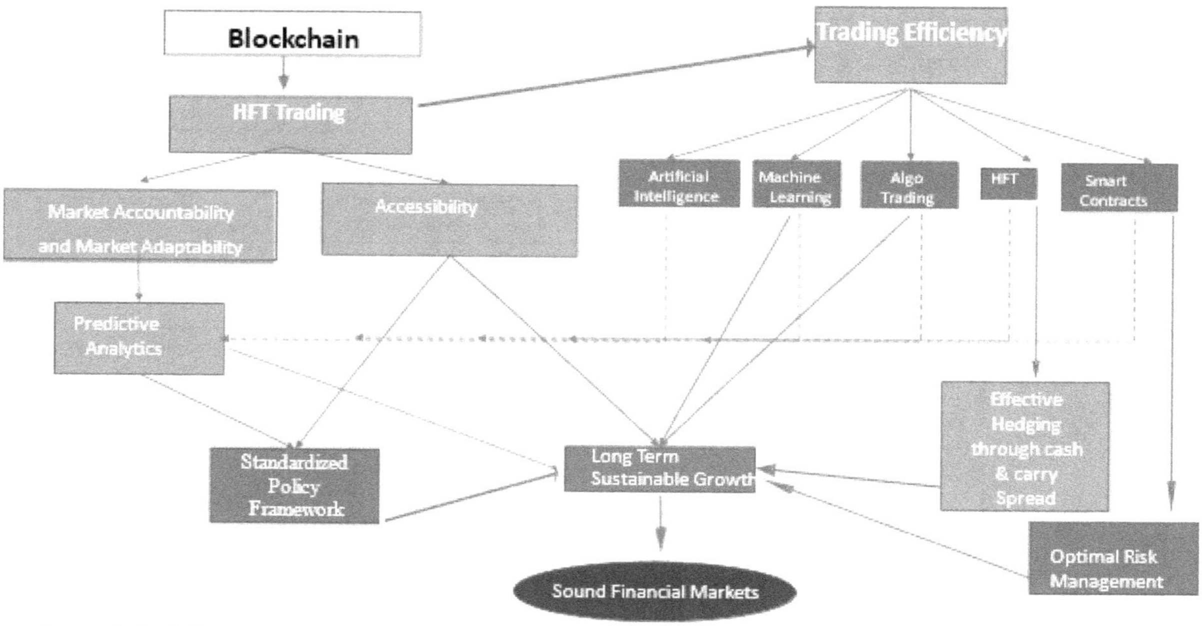

Source: Author's Own

CONCEPTUAL FRAMEWORK

Blockchain technology can optimize high-frequency trading, leading to increased trading efficiency. This, in turn, enhances accessibility, adaptability, and accountability for stakeholders, enabling them to act more efficiently and effectively. The improved effectiveness facilitates near-precise predictive analysis for standardized policy behavior and long-term sustainable growth. By leveraging blockchain, trading efficiency is enhanced, contributing to the creation of robust financial markets. These markets utilize machine learning, smart contracts, artificial intelligence, and logical algorithms to achieve optimal returns and effective risk management. Ultimately, this conceptual framework can pave the way for the evolution and prosperity of sound financial markets, aligning with economic advancements and development.

Blockchain in Financial Markets

With the advent of industry 4.0 and its integration in various industries have brought a paradigm shift in the entire operations of industry. From mass marketing to differentiation to customization to personalization are some examples of change. But with more technology intervention like Blockchain, an era of digitalization has resulted in sweeping changes to the industry mindset. This has helped capital markets firms to address all the challenges faced with the traditional trading methods. *Nakamoto, S. (2008)defined,*" Blockchain as a distributed ledger technology that enables the secure, transparent, and immutable recording of transactions across a network of computers." Blockchain has far reaching impact on the capital markets across the buy side, sell side, and market infrastructure with the promise of reducing the role of intermediaries.

Blockchain technology adoption in securities helps to offer number of advantages like:

i. Help to track their holdings and predict risk
ii. Help to streamline KYC/AML processes for more efficiency
iii. Help in faster and more transparent settlement and clearing
iv. Help to address the risk of liquidity by programable nature of digital assets

Moreover, Blockchain technology has fundamentally transformed how financial transactions are processed, providing several benefits that mitigate emotional bias in investments, which are as follows:

a) ***Transparency and Security:*** Systems built on the Blockchain might potentially assist in complete transparency in stock trading, enabling all participants to access the same information at the same time. Traditional trade finance techniques have been a significant source of annoyance for companies because the drawn-out procedures usually cause operations to be disrupted and make managing liquidity challenging. Blockchain increases transparency in stock trading since it offers a decentralized platform by ending convoluted multi-person engagement process. Blockchain has the potential to ease cross-border procedures and trade finance transactions. It makes it easier for firms to transact with one another beyond regional or geographic boundaries.
b) ***Efficiency and Automation through Smart Contracts:*** Smart contracts being a self-existing program that automates activities in blockchain transactions. Its incorporation helps to remove the intermediaries. These self-executing contracts are coded to be executed when certain conditions are met. This reduces transaction costs and speeds up settlement times, reducing uncertainties that typically

trigger emotional reactions like fear and greed. Faster, more reliable execution of trade helps to alleviate emotional volatility during periods of market uncertainty.

c) ***Trust through Decentralization:*** Blockchain operates without central authorities, making it less prone to manipulation and fraud. The decentralized nature of the ledger creates trust among investors by providing a permanent, tamper-proof record of transactions. With a higher level of trust in the system, investors are less likely to experience overconfidence or irrational exuberance, as there is less opportunity for human error or manipulation.

Emotional Bias in financial Market Investment

Emotional factors in investors could have a negative impact on the capital market, causing unusual market movements. This is due to lack of interpretation of information among investors due to cognitive and emotional factors that affect them to act rationally or irrationally.

Figure 5. HFT impact on Investors' Cognitive behavior

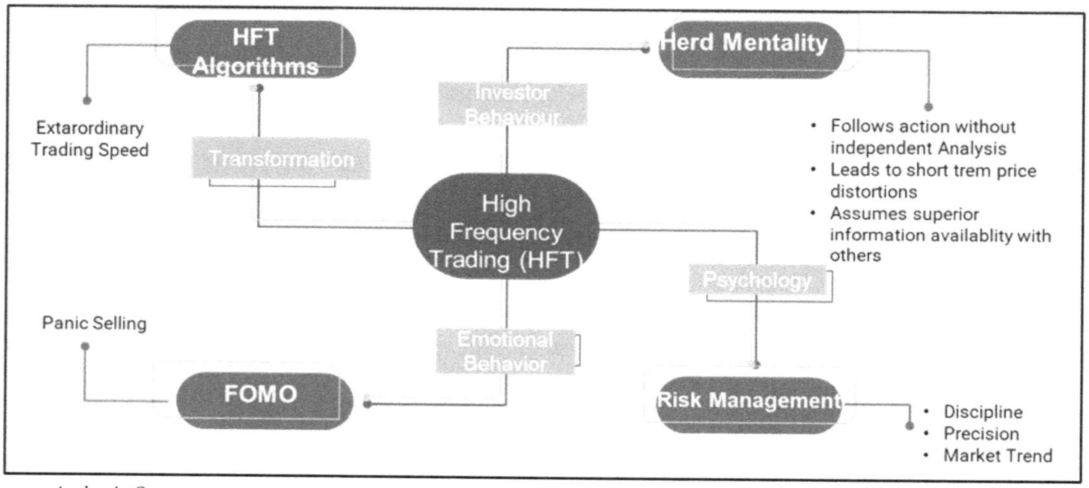

Source: Author's Own

Cognitive bias is a deviation from an investor in understanding, processing, and finally making a decision on an information or fact, and information processing bias falls into this category, whereas emotional bias is a deviation caused by feelings and spontaneity rather on facts. Studies have revealed that investors' psychology can influence investment decision making or their behavior. Studies also found that overconfidence, endowment bias, confirmation bias, and recency bias all have a significant impact on investment decisions (Tjandrasa & Tjandraningtyas, 2018; Qasim et al., 2019; Rudiawarni et al., 2020; Arman syah, 2021; Gavrilakis & Floros, 2022).

Like, if an investor is looking to buy a stock at $20 apiece, he will focus more on current price and ignore the other factors, which can be due to fact that stock once was trading at $42 apiece and now he thinks it is cheap.

Overconfidence refers to an investor's excessive belief in something. Overconfidence leads to overestimation of knowledge and underestimation of predictions made due to their superior abilities (Nofsinger, 2016). Overconfidence is the behavior of someone who is overconfident in his abilities and predictive skills and believes he will always succeed. Overconfidence can be beneficial because it allows investors to make decisions without relying on the choices of other investors or market news, but it can also be detrimental if the decision is solely based on investor confidence without considering in-depth analysis of market conditions. Malik et al. (2019) shows that overconfidence bias has a positive relation with investment decisions.

Like, younger generations are more confident of making money from the capital markets as compared to their parents.

Figure 6. Investors' behavior biases and solutions

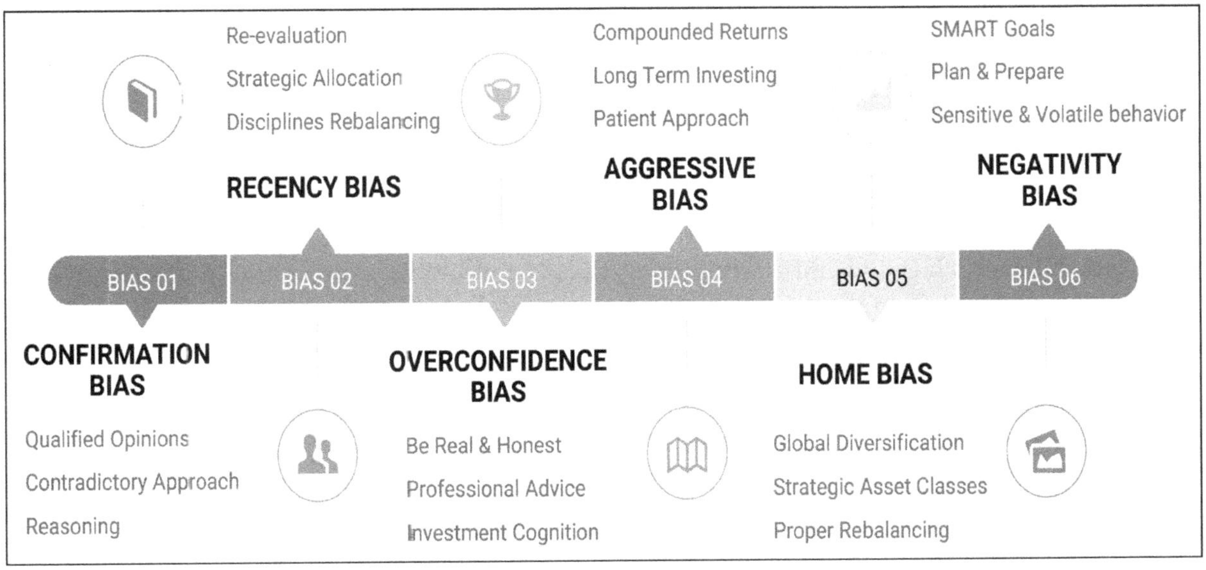

Source: Author's Own

Herding bias is a type of investing behavior in which an investor tends to follow the decisions of other investors without first conducting personal analysis such as fundamentals or techniques. Armansyah (2021) explained that herding is a phenomenon that occurs to investors or a group of investors because of the desire of investors to achieve the same profit by mimicking the behavior of other investors. Gavrilakis and Floros (2022) shows that investors exhibit irrational behavior, such as making decisions based on the decisions of other investors, investors tend to react quickly when other investors' decisions change in investing especially when constructing a portfolio.

Endowment bias is a condition where a person adds value to his or her goods, because it is considered as their own property which has a substantial influence on the economy (Ericson & Fuster, 2014). One of these behaviors can also be seen when someone appreciates an asset that is already his property. Endowment bias itself shapes individual behavior in making investment decisions which results in lower investment decision making.

Likewise, an investor bought a share at $20 of a company but with the market fluctuating the price falls to $15. The investor refuses to sell the stock even if there are other investment opportunities as he is attached and has a sense of ownership to it. As per them, their stocks are more profitable as they own them.

Confirmation bias is a term that describes a person's unwillingness to change previously made beliefs (Cheng, 2018). While selecting stocks, investors consider the timeframe for the investment either for the short term or for the long term to meet the needs of the future. When investors are looking for a compatible model, they will join the community. The investor is most likely to exhibit confirmation bias by joining virtual communities to seek information that confirms their previous beliefs and opinions (Trehan & Sinha, 2021). Kurniawan and Murhadi (2018) found that confirmation bias has no impact on investment decisions, particularly when purchasing life insurance.

Think of an investor who buys stocks of a company based upon his research that this company is bound to grow based upon its future business pipeline, but the market started giving missed reviews and the price becomes stagnant. Investor now looks for only positive news and looks to join some groups which speak good about this stock's selection and future.

Recency bias is influenced by the recency effect that occurs in a person's tendency to make a judgment that is more influenced by the information that investors last saw or heard (Ahlawat, 1999). The recency effect itself is a final assessment that has a big influence on a decision that will be taken by investors (Almilia et al. 2013). It can be said that recency bias is a behavior bias that occurs in an individual because of remembering or based on the latest information they have just obtained (Patel, 2005).

Mohini did research and identified three companies she would like to invest; last 10 years data of these companies showed their average annual returns were 15%, 20% and 30% respectively. She would like to invest in funds with 50% return, however she decides to avoid as some bad news came of the company business methodology. But there being no inter-connection between methodology and stock. In other words, investors can develop negative associations with a company, irrespective of its past record, as the vent being a recent one.

Looking into the above discussion, investors often struggle with various forms of emotional biases that mainly drive irrational decisions, leading to their suboptimal portfolio performance. These biases are more evident during the phases of market volatility. Some of the reactions of investors are:

a) *Fear and Panic Selling*: During downturns, many investors react emotionally and sell assets to minimize losses, even when the fundamentals remain strong.
b) *Greed and Overconfidence:* In rising markets, investors often overestimate their ability to time the market, leading to overexposure in risky assets and ignoring diversification.
c) *Loss Aversion:* Many investors hold onto losing positions longer than they should, hoping to recoup losses rather than cutting their losses early. This is a classic example of emotional bias interfering with rational decision-making.

Moreover, emotional biases cause investors to overreact to short-term market fluctuations, relying on incomplete or outdated information rather than making decisions based on long-term fundamentals. These emotional responses can be mitigated through mechanisms that promote transparency, such as blockchain. By incorporating the Blockchain, the impact of these biases can be reduced to the lowest level.

INDUSTRY INSIGHTS ON PRACTICAL IMPLICATIONS

The section below gives some theoretical case studies to have better practical learning as to how some companies have encashed or crashed with the incorporation of HFT.

1) From Manual Trading to AI-Driven HFT

The transition from manual trading to algorithmic execution marked the rise of HFT, enabling firms to execute thousands of trades per second. Virtu Financial, an early adopters leveraged the automated systems to improve the liquidity and reduce the bid-ask spreads across the securities. But as the competition became intense, firms were forced to move beyond basic algorithms and incorporate AI and ML, as seen in Renaissance Technologies' Medallion Fund. These AI-driven models analyze large amounts of data to predict price movements, and redefine trading strategies being suited for the market conditions allowing firms to provide a cutting edge over their rivals.

2) Blockchain's Entry into HFT and Real-Time Settlements

Traditional HFT systems used to rely largely on decisions from centralized clearinghouses but, blockchain technology usage helped to reduce settlement delays and risks among the traders. Nasdaq Linq and t ZERO pioneered blockchain-based settlement solutions, showcasing how distributed ledgers could further augment transparency and trade finality. HFT integration with the blockchain has helped the firms to dive for real-time transaction processing while curtailing market inefficiencies. This entire transition seems easy but the scalability of operations to perform this function at a high speed remains a challenge.

3) Regulatory Adjustments and the Evolution of Market Controls

The rapid expansion of HFT too had led to some market disruptions with most notably being the *2010 Flash Crash*, which exposed large scale discrepancies in automated trading. To improve its work, regulators introduced measures such as **MiFID II** and mandatory risk controls for providing transparent and accountable algorithmic trading systems. This has forced the HFT operators to provide real-time risk monitoring and AI-driven compliance systems to reduce and prevent market instability. These adjustments were aimed at bringing a shift from an unregulated, profit-driven HFT landscape to a more meticulous, competent, and resilient trading platform.

4) Citadel Securities: Overcoming Latency Issues with Co-Location

Citadel Securities, one of the largest HFT firms, faced challenges in execution of orders of large volumes owing to speed due to latency issues, where milliseconds of delay meant losing trading opportunities to more efficient competitors. The firm needed to find a way which could pave the way to reduce time lags in accessing and analyzing market data to execute trades. Citadel was able to arrest this problem by investing money in co-location services, whereby its servers were placed inside exchange data centers which drastically reduced the time required to access order books and execute trades. To support this process and be competitive, the company further implemented microwave communication networks to transmit data faster than the traditional fiber-optic cables. By optimizing all these innovations Citadel was able to improve trading efficiency and regain their dominance in global financial markets.

5) Knight Capital: From a $440M Trading Disaster to Risk-Control Enhancements

In 2012, Knight Capital experienced a disaster in the use of HFT when a faulty algorithm caused the firm to mistakenly buy and sell millions of shares, resulting in a loss of $440 million in about 45 minutes. The issue arose from outdated software that malfunctioned upon deployment, exposing a lack of proper risk controls. After the near-collapse, Knight Capital implemented rigorous pre-trade risk checks and fail-safe mechanisms to prevent errant trades. They also adopted kill switches, which automatically halt trading activities if anomalies are detected. The firm's risk management overhaul became an industry benchmark, influencing regulators to enforce stricter compliance measures for HFT firms to prevent similar incidents.

Blockchain Technology as a Mitigator of Emotional Bias

The intervention of technology has helped the capital market investors overcome the fear of heavy losses due to an unprecedented rise or fall of the stock prices, this is largely possible due to the role of AI, ML and blockchain which has ensured that investors are able to do multiple trades and harvest the profits and losses as per the algorithm written for the stock. Moreover, Blockchain's key features—transparency, immutability, and automation—can help reduce the impact of emotional biases in investment decisions. The figure below represents as to how the transformation has taken place in stock markets in the light of blockchain and being impacted by the emotionality of an investor.

Figure 7. Role of Blockchain and emotions in Capital Markets

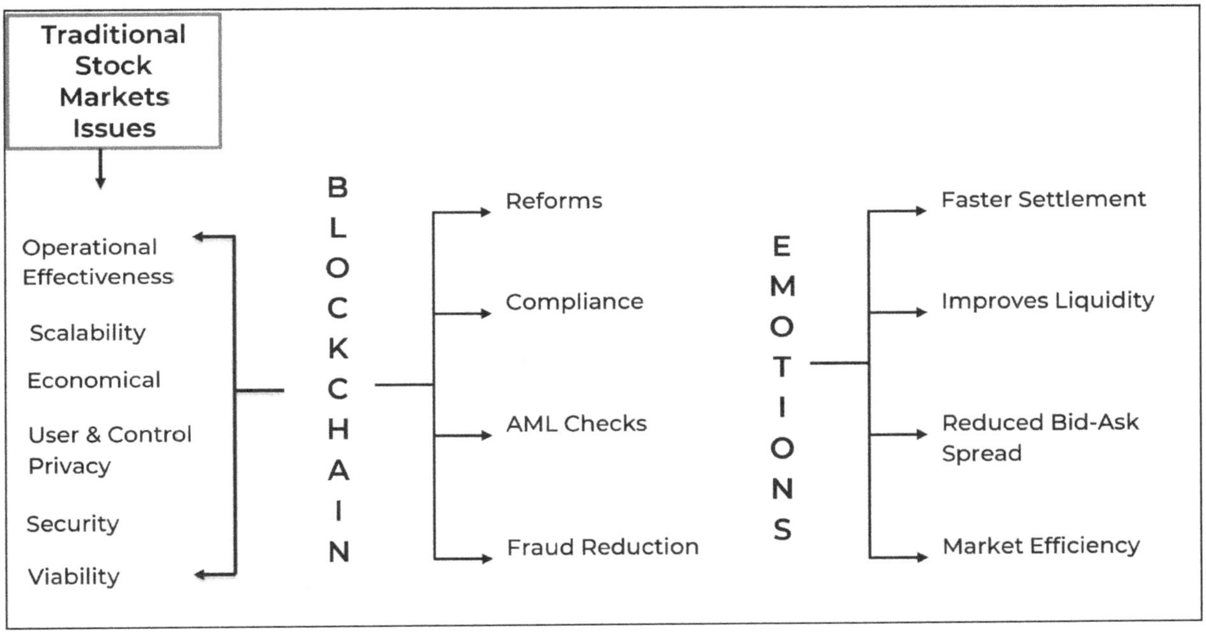

Source: Author's Own

The figure above depicts that apart from the issues being reported in the traditional stock markets which have been addressed by the blockchain to ensure that brokers and the traders have sight of relief and government can prevent financial trading frauds. Also, the same in the light of the emotionality have transformed the operationality of the markets. Some of the advantages of this integration is discussed below:

a) ***Real-Time Data and Transparency***: Blockchain's decentralized ledger provides real-time access to data for all market participants, ensuring that decision-making is based on the most accurate and up-to-date information. When investors have access to reliable sources of data which allows them to make more informed and rational decisions to minimize the risks caused by panic selling.
b) ***Immutable Record of Transactions***: Blockchain maintains a complete record of all the transactions undertaken, which boosts the investors' confidence in the context of the market. This imbibed confidence reduces their fear and uncertainty which could otherwise trigger an emotional response. Integrating blockchain in trading helps investors to verify ownership and historical data, providing them with more clarity and help them to counter loss aversion.
c) ***Lessening of Overconfidence and Herd Behavior:*** The decentralized nature of blockchain removes the need for intermediaries, thereby minimizing potential biases that could influence market outcomes. This reduction in third-party involvement helps curb herd behavior and investor overconfidence. Additionally, by executing trades based on predefined conditions, smart contracts ensure that investors adhere to their strategies, limiting the impact of emotional decision-making.

The incorporation of blockchain technology into financial markets transforms transaction processes, playing a crucial role in minimizing emotional biases in several ways.

1. ***Blockchain boosts trust and reduces emotional volatility:*** In highly volatile markets, blockchain enhances stability by offering a transparent and secure framework. Access to verifiable transaction data helps investors make more rational decisions, minimizing emotional reactions during periods of market uncertainty. *Like, in a scenario of market correction, blockchain's immutable records and real-time data can alleviate investor panic and provide a more stable environment for decision-making.*
2. ***Smart contracts to automate decision-making:*** Blockchain-backed smart contracts in automated trading help mitigate cognitive biases by ensuring trades are executed according to predefined criteria. This structured approach minimizes the likelihood of overconfidence and emotional decision-making in trading. *Like, a trading investor might set a smart contract to automatically sell a stock if it drops by 5%, reducing the chance of holding onto a losing position due to emotional attachment.*

Blockchain technology serves as an effective mechanism for minimizing emotional biases in stock market investments by enhancing transparency, trust, and automation. With real-time data availability and the removal of intermediaries, it fosters a more stable and open trading environment. This stability helps reduce emotional fluctuations that often result in irrational choices, enabling investors to rely on data-driven, rational decision-making. As blockchain technology continues to develop, it may become an essential tool in promoting sound investment practices and mitigating the effects of emotional bias on financial transactions.

Interrelationship of Emotional and Financial analysis in defining investor behavior

Financial analysis traditionally relies on historical data and market trends. The entire methodology of selection of stocks has undergone a paradigm shift whereby previously the book value, EPS etc. were the benchmark of selection but now it's the future business standings, projected revenues which play a bigger role in shaping their investment behavior. However, with the rise of big data and AI and ML financial analysis is increasingly factoring in behavioral patterns, including emotional dynamics. The diagram below depicts the association of financial analysis, emotional analysis and AI driving into the new age of financial analysis.

Figure 8. New Age Financial Analysis through co-integration of Financial Analysis, Emotional Analysis and Artificial Intelligence

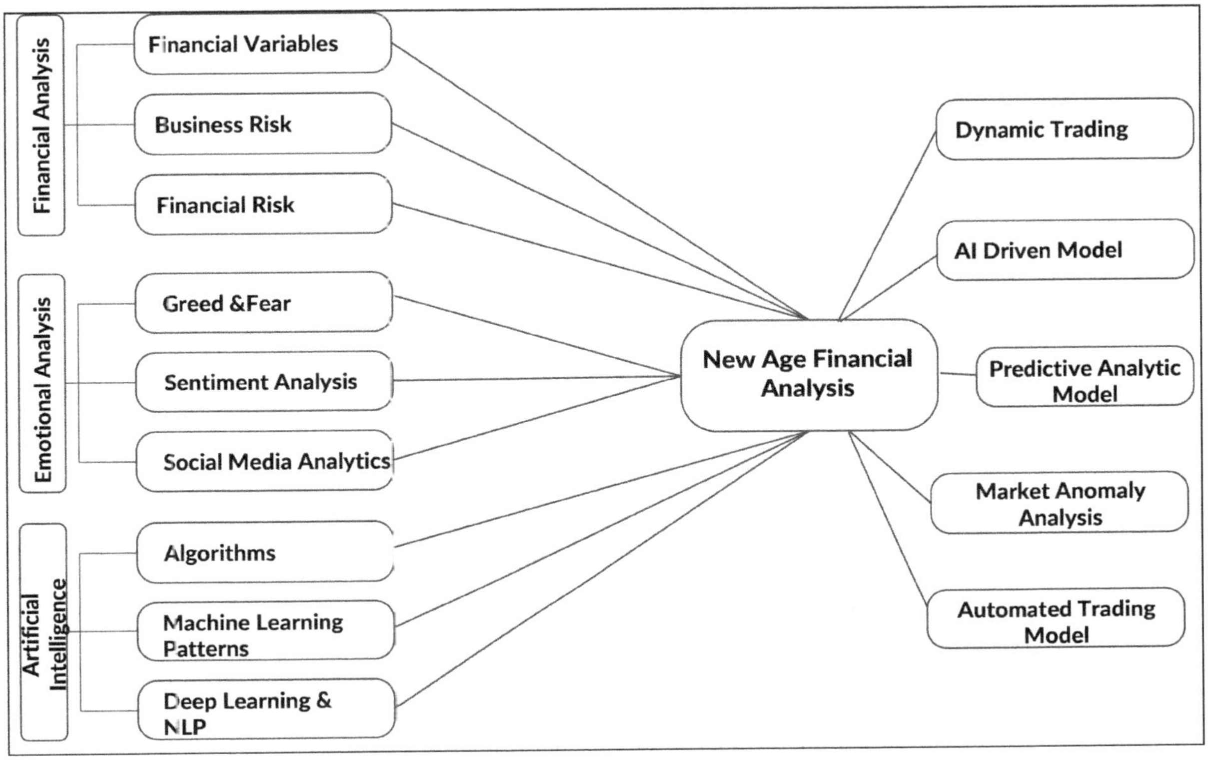

Source: Author's Own

Modern financial analysis can also incorporate sentiment analysis, which uses tools like AI to gauge the emotional tone of news, posts on social media, and investor and trader reactions on the market. By understanding the emotional state of the market participants, traders and analysts can make more informed predictions. From the above diagram we would like to discuss the interrelationship of all the factors with the new age financial analysis like

1. Financial analysis and New-age financial analysis (NAFA):

Financial analysis provides in-depth data analysis based on the financial statements to help make decisions. NAFA helps to integrate novel technologies like AI, ML, analytics and blockchain for deeper insights and by making algorithms in trading helps to ensure that with evolving financial performance, HFT also needs to be augmented and provide better returns to investors. But with NAFA building its foundation on the analysis of traditional methods to improve efficiency, accuracy, and depth of insights with the use of technology the results have proved to be remarkable. NAFA moves one step ahead then tis peers of the old regime as it considers the data of both the financial and non-financial data, like ESG factors, investor sentiment, and global macroeconomic indicators to help the investors to make money in the fall and rise of the markets.

For instance, due to some news of international crises like war or rate cut by any other bank, a negative sentiment will be built, the algorithms supporting HFT will accommodate this information in their models and help to redefine the position of investors.

2. Emotional analysis and New-age financial analysis

Emotional analysis and NAFA have lately become the pillars of HFT to bridge the gap between capital markets & trader psychology. NAFA comprises of AI, ML and analytics which allow investors/traders to interpret big data along with their emotional analysis. A holistic view could be developed to drive better profits and returns. Also, emotional analysis helps HFT systems to formulate sentiment-driven trading strategies. It also helps to predict market volatility raised by the greed or fear of investors; external events like elections, natural disasters etc.; panic selling.

Likewise, USA's Federal bank is proposed to give the commentary about the status of interest rates, so the market will remain unstable. But instead of proposing, the bank cut the rates by 0.25 percent to increase the liquidity in the market and promote consumerism, the HFT will tap the news and shift the approach to sell now as money will start moving from developing to developed nations.

3. Artificial Intelligence and New-age financial analysis

AI based HFT helps to consistently learn and upgrade the algorithm to keep it in line with any future trends in the markets. This change is based on volatility, only buyers/ sellers in stock, unrealistic price movements and some remarks on the social media posts about the present/ future business. Modern forms of technology help to measure sentiments to identify potential service gaps. This will help the companies to make their algorithms more interactive and meaningful. Also, AI-supported voice and speech analysis can further predict customer moods and help to suggest a appropriate solution. AI in NAFA integrates the elements of technology like ML, data science, NLP, and algorithmic trading to create a more efficient, responsive, analytical and predictive financial system.

Like, a company has announced its yearly results, and the market starts reacting with its price sliding, but NAFA will analyze the key ratio and the management commentary as to how the results are like this and the future prices of stock will bound to rise.

CONCLUSION

The above discussion helps to bring deeper insights into traders' subjectivity, but our analysis raises more pertinent questions in the light of achieving their financial goals. Finance plays a key role in transforming society and ensures that it helps to improve the standard of living of people. The rise of investment in capital markets in recent years has proved that people look forward to different avenues for investment. The unprecedented rise in global capital markets has made all investors switch towards equity as an option but the scope of their earnings is limited. HFT gave an option to all the traders to trade with more volumes and make profits. Emotionality played a key role in formulating algorithms behind this trading with biases being never removed. Blockchain further ensures that investors make money on the back of technology and HFT. Reports published by various agencies have reported that usage of HFT may lead to unfair practices due to insider information and price manipulation claims.

Practical Challenges in Implementing Blockchain for HFT

Blockchain integration with HFT has posed unique challenges. Firstly, scalability as most blockchain networks struggle with handling the vast number of transactions required for HFT (Buterin, 2016). Also, the inherent delay in consensus mechanisms, such as proof-of-work and proof-of-stake, might hamper real-time execution at speeds required for adoption of HFT strategies (Bonneau et al., 2015). With entire financial markets becoming integrated, the regulatory bodies operating across the diverse geographies pose significant obstacles to develop such frameworks which could govern blockchain-based trading (OECD, 2020). Secondly, being the interoperability as the existing HFT infrastructure primarily relies on centralized systems, making its operations seamless in cohesion with decentralized blockchain complex and costly networks (Gandal et al., 2018). With securities expected to be more vulnerable, exploitation through smart contract remains a cause of concern for institutional investors keeping into context the adoption of blockchain in HFT (Chen et al., 2019).

Managerial Implications

Managers need to prioritize the adoption and implementation of integrated models by leveraging AI, ML, and blockchain to optimize the operations so that volatility could be reduced, and operational efficiency could be boosted (Aldridge & Krawciw, 2017). Regulators need to also incorporate those practices which could be updated to build robust risk mitigation strategies, ensuring better compliance in line with the internal, external, and regulatory standards (Serbera & Paumard, 2016). The structure of complex financial markets needs to be made more accessible and understandable for all the stakeholders to facilitate the integration of blockchain based technologies with traditional market practices which will enable them to adapt to current market nuances effectively (Fry & Serbera, 2017).

Theoretical Implications

The optimization of HFT with blockchain in capital markets could result in enhanced and transparent price discovery, although it might increase price volatility (Benos & Sagade, 2012). To promote the mental well-being among investors and reduce the impact of emotional trading, rational behavioral responses also need to be prioritized (Yermack, 2017). With technology going for major upswing, the regulatory authorities need to be more realistic to frame the guidelines which could foster the investor sentiments (Boehm et al., 2015), to provide a much informed and stable market operation.

Practical Implications

The practical application in context of integration of blockchain and HFT in light of traditional capital market practices need a very unrealistic high network speed with modern infrastructure, which are critical for their successful implementation (Fry & Serbera, 2017). Blockchain technology will also empower stakeholders through better learning of advanced algorithms and automated processes, resulting in higher precision in trading (Serbera & Paumard, 2016). Furthermore, the use of blockchain combined with smart contracts in capital markets can help mitigate flash crashes reported by exchanges, thereby preventing market inefficiencies and instabilities (Hudson et al., 2015). This integration can enhance market stability and efficiency by addressing these challenges effectively.

REFERENCES

Aldridge, I. (2009). *High-frequency trading: a practical guide to algorithmic strategies and trading systems* (Vol. 459). John Wiley & Sons.

Aldridge, I., & Krawciw, N. (2017). *Real-Time Risk: What Investors Should Know About Fintech, High-Frequency Trading, and Flash Crashes*. Wiley. DOI: 10.1002/9781119319030

Alsem, K. J., Brakman, S., Hoogduin, L., & Kuper, G. (2008). The impact of newspapers on consumer confidence: Does spin bias exist? *Applied Economics*, *40*(5), 531–539. DOI: 10.1080/00036840600707100

Avellaneda, M., & Stoikov, S. (2008). High-frequency trading in a limit order book. *Quantitative Finance*, *8*(3), 217224. DOI: 10.1080/14697680701381228

Avgouleas, E., & Kiayias, A. (2019). The promise of blockchain technology for global securities and derivatives markets: The new financial ecosystem and the 'holy grail'of systemic risk containment. *European Business Organization Law Review*, *20*(1), 81–110. DOI: 10.1007/s40804-019-00133-3

Benos, E., & Sagade, S. (2012). High-frequency trading: A literature review. *Journal of Financial Markets*, *15*(3), 278–304.

Boehm, F., et al. (2015). *Blockchain and Financial Securities/Markets*. [Unpublished manuscript].

Brogaard, J. (2010). High frequency trading and its impact on market quality, Northwestern University Kellogg School of Management Working Paper 66.

De Boef, S., & Kellstedt, P. M. (2004). The political (and economic) origins of consumer confidence. *American Journal of Political Science*, *48*(4), 633–649. DOI: 10.1111/j.0092-5853.2004.00092.x

Fry, J., & Serbera, J. P. (2017). *Modelling and Mitigation of Flash Crashes*. MPRA Paper No. 82457.

Harvey, C. R., Ramachandran, A., & Santoro, J. (2018). The Adoption of Blockchain Technology: Implications for Financial Markets. *Journal of Financial Perspectives*, *5*(3), 1–20.

Hollanders, D., & Vliegenthart, R. (2011). The influence of negative newspaper coverage on consumer confidence: The Dutch case. *Journal of Economic Psychology*, *32*(3), 367–373. DOI: 10.1016/j.joep.2011.01.003

Hudson, R., et al. (2015). *Flash Crashes and High-Frequency Trading*. [Unpublished manuscript].

Knorr Cetina, K. and Bruegger, U. (2000) 'The Market as an Object of Attachment: Exploring

Nair, R., & Bhagat, A. (2020). An application of blockchain in stock market. In *Transforming businesses with bitcoin mining and blockchain applications* (pp. 103–118). IGI Global. DOI: 10.4018/978-1-7998-0186-3.ch006

Nakamoto, S. (2008). Bitcoin: A Peer-to-Peer Electronic Cash System.

Noble, D., & Patil, K. (2021). Blockchain in stock market transformation: A systematic literature review. REVISTA GEINTEC-GESTAO INOVACAO E TECNOLOGIAS, 11(4), 5088-5111.

Serbera, J. P., & Paumard, P. (2016). *Algorithmic Trading and Market Microstructure*. [Unpublished manuscript].

Shah, D. V., Watts, M., Domke, D., Fan, D., & Fibison, M. (2009). News coverage, economic cues, and the public's presidential preferences, 1984–1996. *The Journal of Politics*, *61*(4), 914–943. DOI: 10.2307/2647548

Treleaven, P., Gendal Brown, R., & Yang, D. (2017). *Blockchain Technology in Finance*. IEEE Computer Society Journal. DOI: 10.1109/MC.2017.3571047

Wu, H. D., Stevenson, R. L., Chen, H. C., & Guner, Z. N. (2002). The conditioned impact of recession news: A time-series analysis of economic communication in the United States, 1987–1996. *International Journal for Quality in Health Care : Journal of the International Society for Quality in Health Care*, *14*(1), 19–36. DOI: 10.1093/ijpor/14.1.19

Yermack, D. (2017). Corporate governance and blockchains. *Review of Finance*, *21*(1), 7–31.

Zohar, A. (2015). Regulating Blockchain Technologies: Challenges and Opportunities. *Communications of the ACM*, *58*(9), 38–40.

KEY TERMS AND DEFINITIONS

HFT (High-Frequency Trading): A trading strategy that combines the advantages of speed and voluminous trades for enhanced returns.

ULL (Ultra-Low Latency): The instantaneous execution of trade at thunderous speed.

ECN (Electronic Communication Networks): An automated software/algorithm to match bid and asks for financial instruments.

AML (Anti-Money Laundering): Regulations and procedures designed to prevent illegal activities such as money laundering and terrorist financing.

ZPK (Zero-Knowledge Proof): A cryptographic method that allows one party to prove they know a value without revealing the value itself.

Decentralized Exchanges: Peer-to-peer trading platforms that operate without a central authority, using blockchain technology to facilitate transactions.

Algorithmic Trading: The use of computer algorithms to execute trades automatically based on predefined criteria.

Chapter 24
Strategic Innovation Management Using Emerging Technologies for Building Resilience in Competitive Scenarios

Nidhi Suthar
https://orcid.org/0009-0000-9791-7349
Woxsen University, India

Dinesh Kumar
https://orcid.org/0000-0001-5943-1444
Woxsen University, India

Akashdeep Joshi
https://orcid.org/0000-0002-6445-6651
Lovely Professional University, India

ABSTRACT

Organizations desirous of competing in the markets and building resilience in a dynamic environment must adopt strategic innovation management. The integration of innovation in general, and of emerging technologies such as artificial intelligence, blockchain and the Internet of Things, in particular, is explored in this chapter and the role played by these technologies in the context of long-term strategy is emphasized. Key components of this chapter include promotion of organizational adaptability, address barriers to innovation and leverage technology for sustainable value creation. The chapter also examines policy and managerial implications of the emerging technologies for building resilience for strategic innovation management. Future directions in the chapter emphasize ethical innovation, collaboration and sustainability. This work provides actionable insights for policymakers, managers and researchers and it offer a roadmap for harnessing innovation to achieve growth and resilience in an evolving global landscape.

DOI: 10.4018/979-8-3693-4030-1.ch024

1. INTRODUCTION

Organizations' competitive global market demands that they take up strategic innovation management as it is now a highly considered strategy. Due to this evolution in the industry, organizations struggle with the issues of technological disruption, economic turbulence and consumer preference shifts. These factors have signaled the requirement for building resilience. Depending on the authors, resilience is defined as the capacity to adjust, recover and flourish in the face of adversity and to foster long term innovation (Lengnick-Hall et al. 2011). In the quest for long term innovation, various strategies are needed. However, if these strategies are related to organizational resilience then organizational risks can be mitigated and competitive advantage obtained (Hamel & Prahalad, 1994). In this chapter, the capacity of strategic innovation management aligned with the rise of new technologies is explored to boost organizational resilience to competitive situations.

There is more to strategic innovation than the development of new products, services, processes, structures, or business models. In fact, it is strongly associated with the planning over a long period and bringing innovation efforts in line with corporate goals. Strategic innovation recasts the ways in which businesses operate, compete and create value (Hamel & Prahalad, 1994; Teece, 2010) unlike operational innovation. For complex challenges in competitive markets, strategic innovation is a vital tool to transform both the most fervently competitive markets and to a lesser extent less competitive ones (Christensen, 1997).

Resilience underpins the success of strategic innovation initiatives in terms of the organizations' ability to plan for disruptions. Flexibility, adaptability and robustness (Lengnick-Hall et al. (2011)) are promoted through resilience. This enables organizations to continue operations through a crisis and to flourish after. To achieve adaptive and sustainable growth organizations can embed resilience into their innovation strategies.

Artificial intelligence (AI), blockchain, the Internet of Things (IoT) and advanced robotics, all emerging technologies, have completely redefined the innovation tableau. These are catalyst technologies for strategic innovation. Gartner (2020) notes that these thus present unprecedented opportunities to improve operational efficiency, decision making and in the process redefine value creation. Let's take it one step further: AI allows companies to handle massive amounts of data, extract insights uncover underlying patterns and make data driven decisions which are crisply reactive and agile. It also makes the IoT seamless, bringing physical and digital systems together. These technologies are driving innovation, but in addition it enables organizations to be resilient by the ability to predict, adapt and recover from disruptions.

For emerging technologies to be used in management of corporate strategic innovation, a very comprehensive understanding of their potentials and potentials is needed (Susskind & Susskind, 2022). Challenges ahead are navigating such organizations to unlock the full potential of emerging technologies to build organizational resilience. For example, blockchain can heighten external entities' transparency and security within supply chains, which depends on an investment of considerable resources and expertise. Just as advanced robotics can increase productivity, yielding concerns over workforce displacement and ethical issues (European Commission, 2021), advanced robotics can also add to training costs.

In particular, its application to competitive scenarios is of interest for the interplay between strategic innovation and resilience. Today's markets are characterized by fast and inconstant change, market disruptions and increased competition. Such environments force organizations to innovate or grow resilience or risk obsolescence. On the other hand, if those that integrate innovation and resilience do so successfully, they thus obtain sustainable competitive advantages (McGrath, 2013). The focus on this interplay emphasizes the importance of fitting into a wider strategic innovation management by considering technological and operational aspects as well as the cultural and organizational ones.

Organizations need to adopt a forward looking approach if they want to use strategic innovation building resilience. It entails forecasting potential disruptions, highlighting the chances of innovation and securing the company's internal resources for successful tackling challenges. For instance, using AI based predictive analytics, the organizations can predict market trends and take decisive decisions (Gartner, 2020). Like agile methodologies, flexibility and responsiveness are shared.

Overall, this chapter intends to concentrate about a comprehensive perspective for strategic innovation management and the role that it plays in the construction of resilience using emerging technologies. The conceptual framework, which starts with defining the key constructs and examining their interrelations, provides a basis for this work. Subsequent sections delve into the role of emerging technologies as catalysts for innovation. It also explores the processes and challenges involved in building organizational resilience. The chapter also incorporates case studies to illustrate successful implementations and concludes with policy and managerial implications, future research directions and a synthesis of key insights.

2. AN OVERVIEW OF STRATEGIC INNOVATION MANAGEMENT FRAMEWORK

A strategic innovation management is a cornerstone that organizations look to use in order to navigate through complex and volatile competitive spaces. The aim of this chapter is to align innovation with long term strategic objectives so that firms can support resilience and adaptability while exploiting emerging technologies (Hamel and Prahalad, 1989; Teece, 2010). This framework rests on the assumption that innovation must become a core element of an organization's strategic decision making, providing it with a proactive engagement with rapidly changing market conditions, rather than viewing innovation as something that the organization may sometimes be lucky enough to stumble across passively. This is composed of a collective network effect among strategic intents, innovation capabilities, organizational agility and resilience to enable the competitive advantage of the business over time (Doz & Kosonen, 2008; Sutcliffe & Vogus, 2003). Table 1 depict components of strategic innovation management.

Table 1. Key components of strategic innovation management

Component	Definition	Role in Strategic Innovation Management
Strategic Intent	The organization's long-term vision and objectives guiding innovation efforts.	Provides a clear direction, aligns resources and ensures innovation supports overarching goals (Hamel & Prahalad, 1989).
Innovation Capability	The organization's ability to generate, implement and scale innovative ideas.	Drives the development of new solutions, fosters creativity and leverages resources for impactful innovation (Lengnick-Hall et al., 2011).
Organizational Agility	The capacity to adapt quickly and effectively to changes in the external environment.	Enhances responsiveness to disruptions, allows pivoting strategies and capitalizes on emerging opportunities (Doz & Kosonen, 2008).
Resilience	The ability to withstand, recover and thrive amid disruptions or uncertainties.	Ensures sustainability of innovation efforts, mitigates risks and builds long-term competitive advantage (Sutcliffe & Vogus, 2003).

Strategic innovation is very different from operational innovation in terms of focus and role. Different from each other, operative and strategic innovation both emphasize incremental efficiency or productivity improvements, or radical markets, industries change and value propositions. Innovation is strategic, not limited to technology, it also includes new ways to model business, processes and structures so as to address the long term goals (Hamel & Prahalad, 1989). Let's take the story of Netflix, from a simple DVD rental business to a world leader in streaming and content creation, as an example, which illustrates the ability of strategic innovation to recast an entire industry and maintain its growth.

Strategic intent is the first critical component of strategic innovation management and guides the innovation efforts. Strategic intent is that statement of purpose that articulates what the organization intends to achieve in the long run and how innovation will help to accomplish them (Hamel & Prahalad, 1989). Organizations with a strong strategic intent are more likely to synchronize its innovation initiatives with its bigger picture, such that resources are allocated well and innovation efforts are targeted at important areas. As an example, a company looking to be a leading light in renewable energy innovation might put its money towards creating its own breakthrough solar technologies and sustainable energy storage solutions. Strategic intent not only drives innovation but also fosters a shared sense of purpose among employees, motivating them to contribute to the organization's innovation agenda.

Another essential component is innovation capability, which refers to the organization's ability to generate, implement and scale innovative ideas. Innovation capability spans multiple factors, including leadership, culture, resources and processes, that enable the organization to drive an idea to result. Fostering of innovation capability is largely the responsibility of leadership through the establishment of an environment that supports experimentation, taking a risk and working with others (Lengnick-Hall et al., 2011). An environment where innovation is celebrated and which provides for ongoing learning enhances employee's ability to implement new ideas, to engage in the organization's innovation efforts. Furthermore, investing in research and development (R&D), knowledge management systems and other emerging technologies strengthens an organizations' capability to innovate so it can stay competitive in fast changing markets.

The second part of the conceptual framework is organizational agility. Agility describes the capacity of an organization to quickly and successfully respond to changes in the external environment, including market disruption, technological innovation and new consumer phenomenon (Doz & Kosonen, 2008). Organizations in an agile context are expected to be flexible in their decision making processes, be responsive in their operations and be able to pivot at a moment's notice. Agility is important in general especially for industries where technology disruptions are happening more and more frequently like information technology (IT) and telecommunications. Companies like Amazon have shown huge agility by reconfiguring their business models all the time to take on new technologies like AI and IoT. Not only does agility promote innovation, but it has also been proven to support reliability that helps in resilience and gives competitive advantage.

From the definition, resilience is the propensity of an organization to resist and rehabilitate from interruptions to work into a form and having long run enduring and fruitful future. Resilience in strategic innovation management can be seen as the ability of developing robust systems, processes and strategy that will help an organization to navigate through uncertainties thereby increasing strength of resiliency (Sutcliffe & Vogus, 2003). Proactive resilient organizations that understand the value of anticipating potential risks and disruptions, are proactive in integrating these considerations in their innovation strategy. For example, organizations that had invested in digital transformation were able to flexibly respond to the sudden adoption of remote work and shifting consumer behavior, even in the midst of the COVID-19 pandemic, where there was some conceptual and logistical pushback against such innovation.

Strategic intent, innovation capability, organizational agility and resilience are mutually dependent and mutually reinforcing components. In terms of innovation capability, strategic intent acts as foundation to guide the innovation path (Hamel & Prahalad, 1989; Sun & Abdullahi Usman, 2025). Organizations with innovation capability help them develop and implement solutions to increase agility and resilience (Teece, 2010). Supporting the early execution of innovation initiatives, agility enables organizations to handle changes and take advantage of opportunities, quickly (Doz & Kosonen, 2008). This will ensure that all innovation efforts are sustainable that serve the Organization to recover from disruption thereby continuing to pursue its Strategic Objectives (Sutcliffe & Vogus, 2003).

Strategic innovation management needs emerging technologies to enable it. Organizations have different such tools available such as AI, blockchain, IoT and advanced robotics to drive innovation, become more agile and build resilience (Brynjolfsson & McAfee, 2014; Iansiti & Lakhani, 2020). AI powered analytics can be used by organizations to detect market trends, predict customer needs and improve decision making processes. Supply chains with blockchain technology are more transparent and secure and lower the risks and increase trust among stakeholders. With IoT devices, organizations get real time performance data and can monitor for performance then respond proactively to potential disruptions. Repetitive tasks are assumed by advanced robotics, which excel at running operations and provide more efficiency to free up more resources for strategic operations. Organizations can gain competitive advantages from integrating these technologies to their innovation frameworks and have resilience into future challenges.

The conceptual framework also suggests the importance to create an organization's culture of innovation and resilience. To meaningfully drive strategic innovation and build resilience a culture that values collaboration, experimentation and adaptability is essential (Goleman et al., 2002; Lengnick-Hall et al., 2011). It's time to encourage employees to take calculated risks, offers ideas and contribute to the innovation technology initiatives. This culture is shaped by leadership whose vision is set, who provides support and who proves its commitment to innovation and the ability to bounce back. For instance, Google, as well as other companies have been building environments where people are willing to sacrifice time for innovative projects that lead to something like Gmail and Google Maps. Strategic innovation management not only is supported by a strong innovation culture but is also enhanced by it and the resulting increased employee engagement and organizational performance.

One of the emphasis areas of the framework is alignment between innovation strategy and sustainability goals. This has been accompanied by increasing pressure by organizations to address environmental and social challenges, a role that has become a prominent driver of organizational innovation (Elkington, 1998; Porter & Kramer, 2011). Principles of sustainability, like minimizing carbon footprint, taking a circular economy approach and promoting ecofriendly products and services, ought to be integrated into strategic innovation management. Take Unilever, for example, that has added sustainability into its innovation strategy, which focused on inventing products that address environmental and consumer requirements simultaneously. Stakeholder value and longevity are created not only through aligning innovation to sustainability, but through building resilience.

The conceptual framework concludes with an understanding that the heart and soul of strategic innovation management is dealing with challenges and barriers. Resistance to change, resource constraints or the complexity of integrating the emerging technology with existing system can be stumbling blocks for organizations faced with implementing new technologies. To succeed in this, leadership, collaboration and targeted investments (Christensen et al., 2013; McGrath, 2013) are all required. That is why leaders need to communicate the value of innovation, ask and involve employees in the decision making process and create opportunities for training to develop the needed skills and competencies. Access to needed resources and expertise may also be provided by organizations by collaboration with other external partners like technology providers, research institutions or industry peers. More importantly, you should also take a phased approach to technology implementation, beginning with testing in pilot projects and only taking the most effective efforts to scale.

3. STRATEGIC INNOVATION: HOW EMERGING TECHNOLOGIES ARE CATALYSTS

These technologies are changing industries and changing the way organizations view innovation. These technologies are enablers for a firm's strategic initiatives – creation of new value propositions, operating efficiency & resilience. These technologies are leveraged by strategic innovation management to address complex challenges facing organizational partners, as well as the dynamic competitive environments in which those companies operate. Beyond the better known technologies such as AI, blockchain and the IoT, a number of other innovations including advanced robotics, quantum computing, augmented reality (AR), 5G, big data analytics, additive manufacturing (3D printing), biotechnology, nanotechnology, edge computing, digital twins, autonomous vehicles, wearable devices, cloud computing and green energy technologies are powerful tools to drive strategic innovation (Brynjolfsson & McAfee, 2014). Table 2 describes emerging technologies and their application strategic innovation management.

Table 2. Emerging technologies and their application in strategic innovation management

Technology	Description	Applications in Strategic Innovation
AI	Enables data-driven decision-making through advanced analytics, machine learning and automation.	Predictive analytics for market trends, personalized customer experiences and operational efficiency.
Blockchain	A decentralized ledger technology providing secure and transparent transactions.	Enhancing supply chain transparency, reducing fraud and enabling peer-to-peer energy trading.
IoT	Connects devices to collect and exchange real-time data.	Real-time monitoring in smart factories, predictive maintenance and energy optimization in smart cities.
Advanced Robotics	Autonomous systems and robots capable of performing complex tasks with precision.	Warehouse automation, robotic-assisted surgeries and manufacturing process optimization.
Quantum Computing	A computational technology solving problems beyond the capacity of classical computers.	Supply chain optimization, drug discovery and financial modeling.
Augmented Reality (AR)	Enhances physical environments with digital overlays for interactive experiences.	Retail product visualization, immersive employee training and customer engagement.
Virtual Reality (VR)	Creates immersive digital environments for simulations and experiences.	Training in high-risk industries, virtual collaboration and enhanced entertainment experiences.
5G Connectivity	High-speed and low-latency networks	Autonomous vehicle operations, IoT device integration and enhanced telemedicine applications.
Big Data Analytics	Processes large datasets to generate actionable insights.	Fraud detection, consumer behavior analysis and predictive maintenance.

continued on following page

Table 2. Continued

Technology	Description	Applications in Strategic Innovation
3D Printing	Additive manufacturing enabling the creation of complex designs with minimal waste.	Rapid prototyping, custom medical devices and lightweight components for aerospace.
Biotechnology	Advances in biological processes and genetic engineering.	Precision medicine, sustainable agriculture and bio-based materials.
Nanotechnology	Manipulation of materials at the atomic and molecular scale.	Drug delivery systems, advanced materials for energy storage and improved manufacturing processes.
Edge Computing	Processes data closer to its source, reducing latency and improving efficiency.	Real-time analytics in manufacturing, autonomous systems and energy management.
Digital Twins	Virtual replicas of physical systems for simulation and monitoring.	Production process optimization, scenario testing and predictive maintenance.
Autonomous Vehicles	Vehicles operating without human intervention using AI and sensors.	Logistics optimization, last-mile delivery and sustainable transportation solutions.
Wearable Devices	Smart devices worn on the body to monitor health or enhance user capabilities.	Real-time health monitoring, employee safety tracking and fitness management.
Green Energy Technologies	Renewable energy solutions reducing environmental impact.	Solar energy systems, wind power and sustainable energy storage innovations.

AI remains a cornerstone technology for strategic innovation to enable organizations to analyze vast amounts of data, uncover trends and automate decision-making processes (Brynjolfsson & McAfee, 2017). It has applications in both the retail markets, such as personalization and also the manufacturing markets, especially predictive maintenance. Optimal supply chains, their enhanced customer experiences and their forecasting of market shifts need their own AI-powered tools. AI is being used to advance precision agriculture where data from drones' sensors are used by AI to optimize crop yield and resource use (Russell & Norvig, 2021). How will strategic innovation expand exponentially as AI capabilities continue to grow?

Secure and tamperproof recordkeeping of transactions is granted by blockchain technology which changes the notion of trust and transparency, like Nakamoto (2008). Implications of this technology have been identified in the areas of finance, supply chain management and healthcare. The facility of Blockchain to be used in creating an innovative business model has made it important in terms of creating decentralized systems. It is used in the energy sector to facilitate peer to peer energy trading by consumers buying and selling renewable energy directly (Tapscott & Tapscott, 2016). Blockchain is changing the way industries that heavily rely on trust and data integrity approach their strategic innovation by enhancing transparency and removing inefficiencies.

The IoT connects devices and systems, enabling seamless data exchange and real-time monitoring (Evans, 2011). IoT applications are prevalent in industries such as manufacturing, logistics and health-care. IoT sensors in smart factories monitor health of equipment, allowing prediction maintenance and minimization of down time (Manyika et al., 2015). Wearable devices with IoT technology capable of tracking patient health metrics in real time feed medical professional with real time data. By integrating IoT with other emerging technologies – i.e., AI and Blockchain – IoT will be able to further facilitate strategic innovation by creating intelligent and safe connected ecosystems.

The future of quantum computing is poised to entirely revolutionize industries by tackling problems classical computers cannot and rendering insight into otherwise challenging problems. The implications, however, are significant across the areas of supply chain optimization, drug discovery and financial modeling. For example, quantum algorithms can develop logistics company routes that will minimize expenses and increase efficiency (Preskill, 2018). The concept of Quantum Computing is still in its infancy, but it is a strategic innovation force with great potential to deliver unparalleled computational power to assess the more deeply complex of business problems.

Customer Engagement and Operational Processes are undergoing a technological Reformation via AR and VR (Milgram & Kishino, 1994). AR puts digital information on top of the physical world, creating new experiences in industries like retail, real estate and health care. For instance, AR is being used by retailers who provide their customers with a chance to see the products in their own home before they make the purchase as occurred (Carmigniani et al., 2011). In contrast VR offers immersive environments for training, simulations and entertainment (Slater & Sanchez-Vives, 2016). The products and services sustained by these technologies are leading innovation, inspired by the ability to find new ways to interact with products and services and are thereby increasing customer satisfaction and operational efficiency.

IoT devices, autonomous vehicles, smart cities (Ford, 2017) and the like, are now running due to 5G technology that offers a high speed connectivity and low latency. That's because it allows real time transmission of data, which is key for allowed for real time communication and hence also to the original innovation that requires instant communication. For instance, the networks of autonomous transportation systems (Campolo et al., 2017), in which different vehicles must exchange information with one another and with infrastructure in real time, represent an example of this. The innovation of 5G will mean innovations in so many different sectors like logistics, healthcare ...

Another technology driving strategic innovation in transforming raw data into actioned insights is big data analytics (Davenport, 2014). Big data is used by organizations to comprehend how customers behave, improve performance and spot new fashions. Big data analytics is used in financial sector to detect fraud and evaluate credit risk and in marketing to support targeted campaign and personalization (Chen et al., 2012). With big data, organizations can get to the bottom of who's saying what, who deserves what and what additional resources need to be allocated to support those individuals as they deserve.

Additive manufacturing, or 3D printing, has revolutionized production processes by enabling the creation of complex and customized products with minimal waste (Lipson & Kurman, 2013). Industries such as aerospace, healthcare and automotive have embraced this technology for rapid prototyping and manufacturing. For instance, 3D printing is used to produce lightweight components for airplanes and custom prosthetics for patients. Its ability to reduce production costs and lead times makes it a key driver of strategic innovation, particularly in industries that prioritize customization and sustainability.

Biotechnology is another emerging technology with profound implications for healthcare, agriculture and environmental sustainability (National Research Council, 2009). Advances in genetic engineering, synthetic biology and biopharmaceuticals have enabled breakthroughs in precision medicine, crop enhancement and biofuels. For example, CRISPR technology allows for precise gene editing, leading to the development of targeted therapies and disease-resistant crops (Doudna & Charpentier, 2014). Biotechnology's ability to address global challenges positions it as a crucial tool for innovation (Wheeler, 2025).

Nanotechnology, the manipulation of matter at the atomic and molecular scale, is driving innovation in materials science, healthcare and energy (Bhushan, 2017). It enables the creation of materials with unique properties, such as increased strength, reduced weight and enhanced conductivity. In healthcare, nanotechnology is used to develop drug delivery systems that target specific cells, improving treatment efficacy. Its applications in energy include the development of more efficient solar panels and energy storage solutions.

Processing data closer to its source (Shi et al., 2016) has caused edge computing to be a critical technology. Specifically, this is important for industries including manufacturing and logistics, which rely on real time decision making. Especially in its application in digital twins, virtual replicas of physical systems, edge computing enables faster data processing and shortens dependence on centralized systems, enhancing organizational agility and resilience even for strategic innovation. Manufacturing, construction and in Energy use digital twins to automate performance measurement and anticipate problems. It comes particularly useful to manufacturers enabling a digital twin of the process of production and allowing it to simulate the processes to identify bottlenecks and to optimize the use of resources (Negri et al., 2017). With autonomous vehicles — self driving cars and drones — transforming the transportation and logistics, innovation brings new insights to guide decision making and optimize operations efficiency. Using AI, sensors and real time data, these technologies navigate without human intervention as they make decisions. In the logistics context, drones facilitate last mile delivery which would reduce costs and improve efficiency (Goodchild & Toy, 2018). As a quantum leap in innovation, autonomous vehicles serve to create safer, more efficient and sustainable transportation systems (Litman, 2024).

Increasingly, healthcare innovation has become dependent on wearable devices — from fitness trackers to medical monitors — that provide additional information to patients and their doctors. Real time health data are captured in these devices that help the individuals manage their wellbeing and medical professionals personalize the care. For instance, smartwatches with ECG sensors can track irregular heart rhythms and perform early intervention (Liu et al., 2020). Health outcomes are being improved which is driving strategic innovation in wearables as well as opening new opportunities for preventive healthcare.

A number of these emerging technologies are enabled by our ability to build scalable and flexible data storage, processing and analytics infrastructure with cloud computing. Organizations use cloud platforms to create innovative solutions, without the need to maintain heavy on premises as intensive infrastructure (Armbrust et al., 2010). Collaboration, cost reduction and time to market for new technologies are facilitated by cloud computing and therefore it constitutes part of strategic innovation management practice.

It's no secret that environmental challenges are driving innovation in green energy technologies like solar, wind and energy storage systems. These technologies are adopted by organizations to decrease their carbon footprint, comply with the regulations, as well as match consumer preferences for sustainability (IRENA, 2019). That includes companies working to free themselves from the dependency and vulnerability of energy supply through the adoption of renewable energy solutions in operations.

4. STRATEGIC INNOVATION MANAGEMENT AS A BUILDING RESILIENCE STRATEGY

In the light of environmental volatility and competitiveness, resilience, the ability of the organization to withstand, adjust and thrive in the face of disruptions, has started to receive increased interest (Bhamra et al., 2011). The role of strategic innovation management, explicitly, in building resilience through embedding innovation in organizational architecture to deal with risks, emerges to adapt to challenges and support long term growth is played (Teece et al., 2016). While resilience is certainly about surviving disruptions, it is also about using disruption as an opportunity for transformation and strategic advantage (Lengnick-Hall et al., 2011). This section discusses how, through strategic innovation management, organizational resilience is built by discussing key principles, strategies and practical applications. Figure 1 depict stretegies of building organizational resilience through innovation management.

Figure 1. Building organizational resilience through innovation management

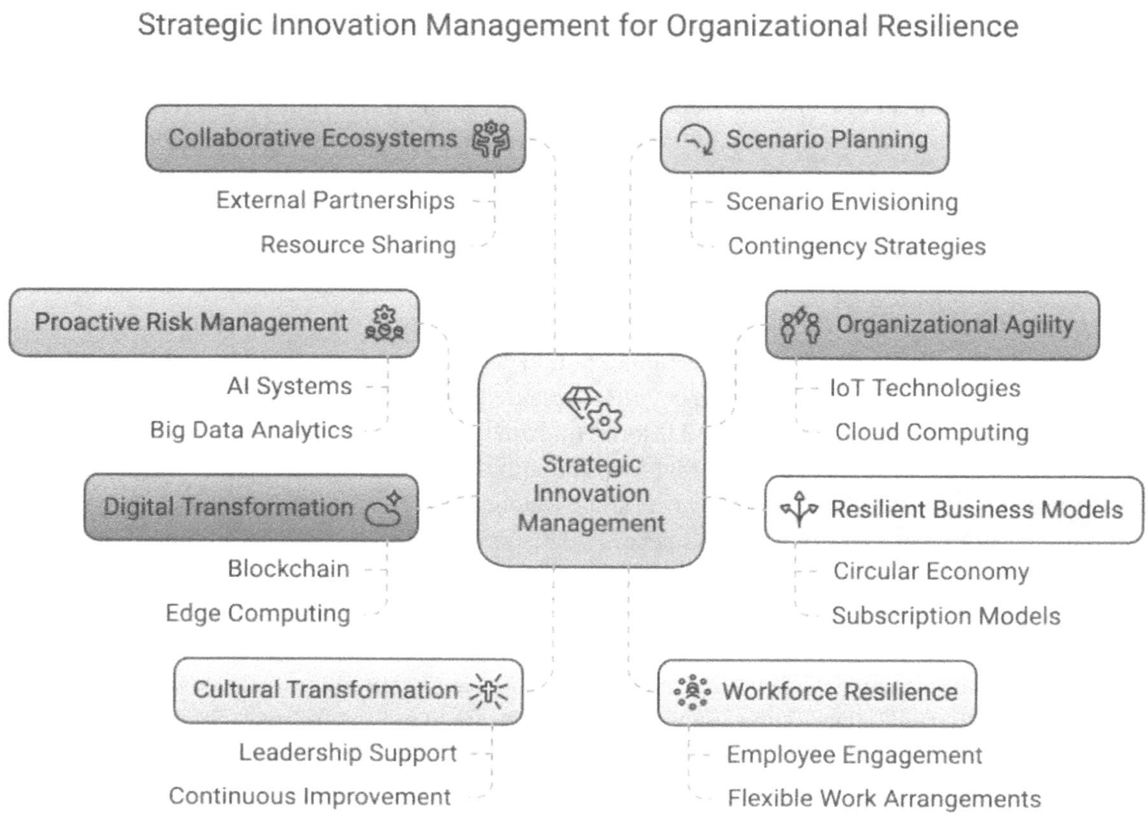

Source: Created by authors

In the context of strategic innovation management, resilience means the creation of systems, processes and cultures capable to know and respond to new challenges whilst maintaining operational continuity (Hamel & Välikangas, 2003). Given the technological disruptions, economic uncertainties, environmental crises and changing market dynamics, businesses are faced with the interplay of innovation and the resilience necessary to meet that challenge (Teece et al., 2016). Those organizations that build resilience into their innovation strategy are more likely to be able to adapt to changing environments and recover from setbacks to earn sustainable competitive advantages (Pisano 2015).

The strategic innovation adds to the fundamental aspect of building resilience, that of proactive risk management. Proactive risk management consists in identifying potential disruptions, estimating their impacts and developing innovative solution to mitigate them (Kaplan & Mikes, 2012). In this process, it is made clear emerging technologies like AI, big data analytics etc. which plays an important role by giving the tools for modeling and scenario planning. The following is an example of how using AI powered systems, such as AI systems that have a capability to analyze historical data as well as market trends to predict potential risks that could cause impacts to businesses to enable organizations to act well ahead of time through mechanisms that minimize or prevent risks (Brynjolfsson & McAfee, 2014). Likewise, big data analytics reveal supply chain vulnerabilities to enable a business to optimize inventory management and to minimize the risk of disruptions (Waller & Fawcett, 2013).

Another important facet of resilience building is organizational agility, i.e. the capacity to rapidly and properly respond to the modifications in the external market. Strategic innovation management rests on agility, which enables organizations to reorient their strategy, redeploy resources and discover new opportunities to response to changing challenges (Doz & Kosonen, 2010). With the IoT and cloud computing technologies, agile methodologies are becoming a means by which organizations improve their responsiveness. For example, prior to the COVID-19 pandemic, organizations have become agile through rapid transition to remote work modes using cloud based collaborative tools to continue normal business operations (Soto-Acosta, 2020). Such adaptability speaks to the relevance of creating a gap that innovation strategies can plug to increase resilience.

Strategic innovation management cannot happen without the existence of resilient business models. In such organizations, models must be designed such that they would be robust and at the same time flexible so as to absorb disruptions and take advantage of the emerging trends (Osterwalder, & Pigneur 2010). For instance, companies who have embraced the philosophy of circular economy have shown resilience – by reducing dependence on finite resources and building up sustainable value chains (MacArthur, E., 2013). Resilience is boosted by circular business models, fueled by mechanisms of recycling, reusing and repurposing materials (Geissdoerfer et al. 2017). Additionally, subscription-based revenue models, such as those used by software-as-a-service (SaaS) companies, provide predictable income streams, reducing financial volatility and enhancing stability (Lindström et al., 2024).

Digital transformation is a powerful enabler of resilience through strategic innovation. The integration of digital technologies into core business processes enhances an organization's ability to adapt to changing environments and deliver value (Bharadwaj et al., 2013). Building resilience, technologies like blockchain, edge computing, digital twins have come to the fore. For instance, blockchain guarantees data integrity and transparency in supply chains, mitigate risks as a result of fraud and inefficiency (Dobrovnik et al., 2018). Edge computing makes real time decision-making feasible with the data being processed closer to its source, which in turn reduces the time taken by organizations to respond to disruptions (Shi et al., 2016). On the other hand, digital twins of physical system create virtual representation of physical system, which enables organizations to simulate, test and understand how to respond to possible challenges and be very prepared and at the minimum reduce downtime (Tao et al., 2018).

In contrast, organizations are able to evaluate potential solutions in digital twin environment without incurring the associated risk of replacing hardware in production systems (Tao et al., 2018). As the recognition of sustainability driven innovation as a critical driver of resilience rises, a pathologic review is needed. Organizations that engage in Environmental and Social responsibility innovations are in better position to tackle global challenges such as climate change and resource shortages (Porter & Kramer, 2011). For instance, the investment in renewable energy technologies, the energy efficient operation and the green supply chain already help companies adapt to regulatory and market pressures, in addition to reducing its environmental impact (Geissdoerfer et al., 2017). Sustainability-driven innovation drives organizational goals to embody broader societal needs and creates shared value as well as strengthens stakeholder relationships (Bocken et al., 2014).

Constructing resilience through strategic innovation management also comprises another important aspect of cultural transformation. An organization with such a resilient culture is one that is open to change, collaborative and that has the commitment for continuous improvement (Schein, 2010). In this culture, leadership is crucial as it leads the promotion of innovation, the encouragement of experimentation and aid to employees through uncertainties (Edmondson, 1999). For instance, the headline company 3M and Google strive for something similar; companies that supports innovation and encourage employees to have the guts to take calculated risks (Amabile, 1998). Such cultural resilience enhances an organization's capacity to innovate and adapt in dynamic environments.

Workforce resilience is an often overlooked but critical component of building organizational resilience. Strategic innovation management must include initiatives to upskill and reskill employees, equipping them with the knowledge and capabilities required to adapt to new technologies and changing roles (Bhamra et al., 2011). For instance, Frey and Osborne (2017) solve that organizations that devoted resources in workforce development were capable to combat job displacement risks and keep employee engagement. It also helps to promote employee wellbeing in flexible work arrangements, mental health support and development programs through initiatives that will promote the workforce resilience and will allow employees to participate in the efforts on innovation (Soto-Acosta, 2020).

Further, the integration of collaborative ecosystems into strategic innovation management further expands resilience. Organizations collaborate with external partners like technology providers, academic institutions and industry peers; the purpose is to share resource, expertise and network with external organization to enhance innovation (Powell et al., 1996). Consider for example how collaborative ecosystems sped the development of COVID 19 vaccines when pharmaceutical companies worked alongside governments, research institutions and nonprofit agencies to create and manufacture innovation (Ball & Maxmen, 2020). Organizations build collaborative relationships that work better together, pool resources and solve complex problems.

Scenario planning and contingency strategies are powerful resilience building mechanisms through strategic innovation. For example, scenario planning is a process of envisioning several future scenarios and developing strategies for addressing one or more of them, all to be prepared for similar outcomes (Schoemaker, 1995). An example of this is in the energy sector and scenario planning used by such organizations to anticipate changes in oil prices and alternative strategies developed to maintain profitability (Van der Heijden, 1996). Contingency strategies, for instance, making a backup supply chain or being diversified in revenue streams, give to organizations flexibility to react to disruptions without creating a conflict between core operation (Kaplan & Mikes, 2012).

There are practical applications of strategic innovation management about industries that are building resilience. Companies in manufacturing have adopted IoT enabled smart factories to make the manufacturing process more efficient and less vulnerable. IoT sensors monitor equipment health allowing for predictive maintenance keeping downtime at minimum (Shi et al., 2016). Blockchain technology has been applied in the financial sector, making transaction transparency and security plus fostering trust and lowering fraud risks (Dobrovnik et al., 2018). Arguably, AI-driven analytics are most implemented where it has expedited inventory management and personalized customer experience while making operations more agile to respond to market changes (Brynjolfsson & McAfee, 2014). The examples presented here provide a snapshot of how organizations apply strategic innovation to leverage resilience and long term success.

Although it is a functional process with many benefits, constructing resilience through strategic innovation management couldn't be easy. As there are existing barriers including resource constraints, resistance to change and integrating more complex technologies (Hamel & Välikangas, 2003), organizations often need to overcome a number of barriers the technologies have to overcome. An approach that takes a proactive leadership, collaboration and targeted investments (Teece et al., 2016) approach, addressing these challenges, is required. Resilience must be communicated at leadership level, employees must be involved in the innovation process and proper support for the navigation of transitions must be provided (Kotter, 1995). Organizations are sometimes constrained for resources and sometimes lack certain specialized expertise; thus, collaboration with external partners can aid organizations in overcoming resource limitations to achieve collaboration with those with specific expertise (Powell et al., 1996). The adoption of a phased approach to technology implementation, for example, begins with pilot projects and those that emerge successful, are scaled up, diminishes the risks and assures technology integration (Brown & Eisenhardt, 1997).

Strategic innovation management in support of resilience is not an oxymoron. By offering incentives for innovation, supporting workforce development and creating enabling environments through the creation of frameworks for ethical and sustainable practices, governments and regulatory bodies can create enabling environments for that innovation (Mazzucato, 2013). For example, the OECD (2015) mentions those policies which would grant tax credits for R&R investments, or funding for renewable energy projects that would motivate organizations to make resilience building efforts a priority. Effective collaboration of policymakers and of industry leaders is necessary to provide the innovation goals aligned with the regulatory requirements and to tackle social problems properly (Porter & Kramer, 2011).

5. COMPETITIVE SCENARIOS AND CHALLENGES OF STRATEGIC INNOVATION MANAGEMENT

Strategic innovation management plays a very important part of the organizations as it is a dynamic and competitive environment in which it is operating. Global markets of today require facing rapid technological advancements; an ever changing landscape of consumer behavior; regulatory changes; geopolitical uncertainties and so on (Drucker, 1993; Tidd & Bessant, 2020). As a result, these factors make the business scenarios competitive and thus creates scenarios demanding constant innovation for business to live and stay competitive. However, the strategic innovation poses many problems in implementation. The competitive scenarios that organizations experience, the inherent challenge in bringing innovation line with strategic objectives, are explored in this section (See Figure 2).

Technological disruptions are a prominent competitive scenario in today's markets. AI, blockchain and IoT have disrupted traditional business models completely forcing businesses to switch or have themselves wiped out from the market (Brynjolfsson & McAfee, 2017). As a case study, the rise of AI powered platforms has ushered in the exception with total efficiency and personalization across industries such as healthcare, retail, logistics (Manyika et al. 2017). Those companies that don't embed these technologies into their strategies risk being left behind by competitors. For example, the many years that traditional taxi services would take to ready themselves from ride-hailing platforms like Uber and Lyft, leveraging mobile app technology and data analytics would lead to better customer experiences (Cramer & Krueger, 2016). In such circumstances, strategic innovation management must concentrate on integrating disruptive technologies in the organization's value chain.

Globalization and market expansion does not end competition pressure. There are significant cultural differences and variations in regulatory and here comes the biggie - diverse consumer expectations (Friedman, 2005). Globalization allows the organization to expand its market; but at the same time, globalization brings the organization into the competition market with the international players. For example, the Indian e-commerce space present the regional platform like Alibaba in China and Flipkart, both face stiff competition from the e-commerce giants like Amazon. In such scenarios, strategic innovation management involves product and service tailoring of products and services for local market and using global resources to achieve efficiency.

Another competitive scene is developing consumer preferences. Today, consumers expect personalized experience, an ethical business practice and innovative products that resonate with their value (Prahalad & Ramaswamy, 2004). These expectations are demanding organizations must innovate constantly to differentiate in order to meet them. Similarly, take the beauty and personal care industry — Fenty Beauty is a good example, one that has disrupted the market by providing inclusive product lines that respond to the lack of inclusive products in this market. Competing companies that do not respond with a new type of innovation to meet consumer demands lose relevance.

Completive scenarios are shaped by environmental and regulatory pressures. Continued tightening of environmental regulations, along with increasing expectations of organizations to align their operations with sustainability goals all place great pressure on companies to meet constant demand for greater efficiency in resource use (Porter & Kramer, 2011). As an example, automotive manufacturers are spending on electric vehicles (EVs) so that they will comply with regulatory mandates and cater to rising consumers' demand for platform ecofriendly products. This area represents Tesla's success in this space, demonstrating potential for innovation to create competitive advantage in an environmentally conscious market (Stringham et al., 2015).

For competitive scenarios, geopolitical uncertainties and supply chain disruptions make matters harder. Organizations' need to innovate to address such vulnerabilities (exposure to events like trade wars, pandemics and natural disasters) always expose vulnerabilities in global supply chains (Christopher & Peck, 2004). As an example, companies that had adopted digital supply chain technologies (e.g., blockchain and IoT) greatly benefited from being able to navigate disruptions over the course of the COVID-19 pandemic (Ivanov & Dolgui, 2020).

These competitive scenarios require innovation but challenging strategic innovation management. The major challenge is balance between short term performance and long term innovation goals. Within tight margin industries, this becomes especially common for organizations that choose to prioritize immediate financial results over investments in innovation. This short termism could prevent the emergence of game changing innovations that demand intensive attention (Chistensen, 1997; Govindarajan, & Trimble, 2010). Imagine a company in the pharmaceutical industry, where the speed with which to deliver the quarterly results is high, yet they are investing in R&D for what will likely be years of drugs to commercialize (Pisano, 2006). To overcome this challenge, a combination of incremental and radical innovation projects should be funded.

Another big barrier to strategic innovation is resistance to change. New technologies, processes, or business models may scare employees, customers, or even stakeholders because of fear of disruption, or fear of an unknown outcome (Kotter, 1996). For example, one of the reasons adoption of automation technology in manufacturing is met with pushback from the workers, worried about job displacement (Brynjolfsson & McAfee, 2017). Effective change management strategies like transparent communication, employee involvement and training program to compliment the new industry must be used to overcome resistance to change (Hiatt, 2006).

Organizations and critical among them SMEs, continue to face persistent resource constraints. That puts innovation initiatives in an often substantial investment in research and development (R&D), for talent acquisition and technology infrastructure (Tidd & Bessant, 2020). Both big and small organizations may offer their products or services at smaller costs compared to SMEs. For example, a one who has started up a new one Ai led solution gets into challenges in scaling its product as a result of not much funding (Ries, 2011). In resource constrained environments, Strategic innovation management involves identifying high impact projects, pursuing partnerships and taking advantage of government incentives to support innovation effort.

Organizations must adopt the latest technologies including AI, Blockchain or IoT and both the technological complexity and integration challenges increase. These technologies aren't easy to integrate into existing systems (Westerman et al., 2014). For example, deploying blockchain in the operations of a supply chain has to entail reconfiguring processes, training personnel and making the system work on legacy systems (Tapscott & Tapscott, 2016). Organizations must adopt phased implementation strategies, starting with pilot projects to test feasibility and scalability before full-scale deployment.

Fragmented organizational structures and silos can impede collaboration and knowledge sharing, which are critical for innovation. Departments operating in isolation may pursue conflicting priorities, limiting the organization's ability to execute cohesive innovation strategies (Senge, 2006). Imagine, for example, that the company's R&D team develops a new product with great promise, but if marketing and sales are not aligned to commercialize the product, then it will fail (Chesbrough, 2003). These silos must be addressed using strategic innovation management, such as boosting cross functional collaboration, developing said innovation governance structures and promoting said organization vision.

In addition, ethical and regulatory challenges make strategic innovation efforts more complicated. For example, when data privacy issues associated with AI applications like data mining on blockchain raises ethical issues with their emergence technologies e.g. Fldomi, 2013). Slow innovation is also navigating complex regulatory landscapes. As fintech companies developing blockchain-based payment system must show compliance with financial regulations and take care of data security worries (Nakamoto, 2008). Ethical frameworks must be adopted by organizations and regulators need to be negotiated by them in order to meet the innovation with the prevailing societal and legal expectations.

Like any new initiative, innovation is filled with uncertainty and risk that could manifest from an initiative whose outputs are not what was expected, or that is not so easily understood or solved as other projects more new ideas that have a more solid run rate. To manage these risks, the organizations have to have flexible strategies and a culture of experimenting (Thomke, 2003). For instance, such as minimum viable products (MVPs) are used to test innovations in the market before investing on large scale (Ries, 2011). The strategic innovation management environment characterizes an organizational capacity to perceive failures as learning opportunities to improve on future offerings and to refine the approach to their environments.

Organizations must take targeted steps to overcome these challenges which have to be aligned to the corresponding competitive scenario. There is no way around it: You need to build a culture of innovation if you want to overcome resistance to change and promote collaboration. In the end, Leaders have a critical role to portray this culture by promoting innovation as a core value, highlighting contributions, do the experiments (Tidd & Bessant, 2020). For instance, 3M has long incubated innovation cultures, such that employees had time (and sometimes money) to devote to exploratory projects that often culminated in products that have radically changed the nature of how societies function (Govindarajan & Trimble, 2010).

Figure 2. Competitive scenarios and challenges of strategic innovation management

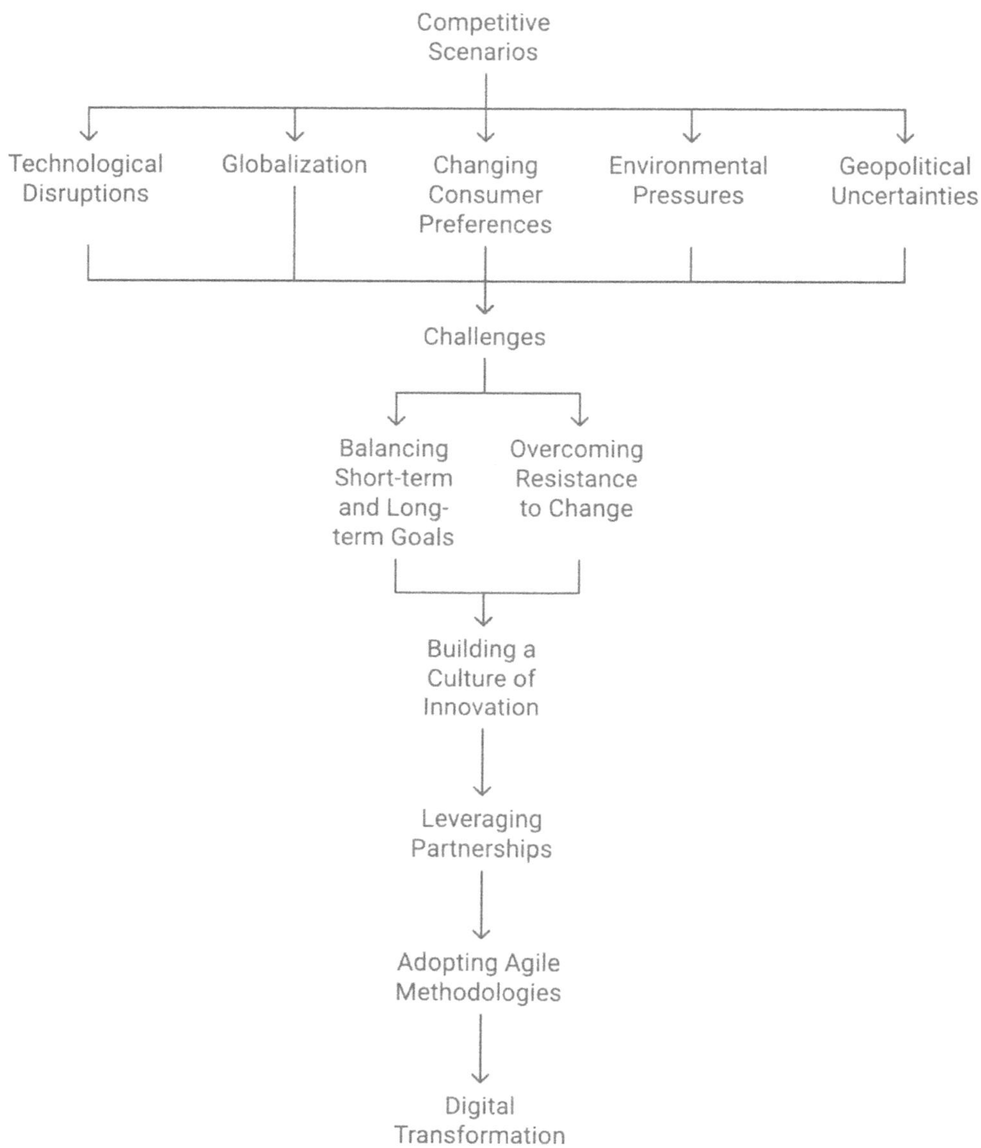

Source: Created by authors.

A second, powerful strategy to manage resource constraints and increase the innovation capabilities of an organization is to leverage partnerships and ecosystems. Organizations collaborate with startups, academic institutions and technology peers to leverage whomever and whatever they have because it would be hard to create this type of resources from within. For instance, pharmaceutical companies may team up with research universities to quicken drug discovery. However, these partnerships produce a synergy that helps fuel innovation and for companies to respond more with the same competitive pressures.

Adopting agile methodologies are a requirement of organizational agility in fast changing markets. Organizations applying Agile frameworks are allowed to respond quickly to changes, this has been shown by Rigby et al. (2016), with iterative development and rapid prototyping, taking the feedback Series. As an example, software companies have embraced agile practices to speed up product development and to keep up with changing customer needs. Applied to broader innovation initiatives, these methodologies create greater responsiveness and mitigate the associated risks of uncertainty.

Digital transformation requires investing to navigate technological complexity and integrate newly emerging technologies within the core operations. For digital initiatives (Westerman et al., 2014), organizations must invest in infrastructure, talent and training. An example of this is that to leverage the power of AI in analytics a company needs to ensure its employees are trained to handle data and apply insights. Not only does it support innovation, but digital transformation also provides organizations with resilience by which they can adjust to disruption and take advantage of new potential.

6. CASE STUDIES OF SUCCESSFUL IMPLEMENTATION OF STRATEGIC INNOVATION

Changes in strategic innovation management have favored various organizations in gaining competitive advantage. These organizations have aligned their innovation initiatives with long-term goals (Tidd & Bessant, 2020; Christensen, 1997). This part includes cases of successful strategic innovation management that highlight the opportunities an innovative approach might provide for diverse organizations (See Table 3).

Table 3. Case studies overview in strategic innovation management

Organization	Industry	Innovation Implemented	Key Outcomes
Amazon	E-Commerce	AI-powered recommendation system for personalized shopping experiences.	Enhanced customer satisfaction, increased sales and improved customer retention (Kantor & Streitfeld, 2015).
Tesla	Automotive	Electric vehicles (EVs) with advanced battery technology and autonomous driving.	Market leadership in EVs, increased brand loyalty and global recognition for sustainability (Stringham et al., 2015).
Unilever	Consumer Goods	Sustainability-driven innovations like eco-friendly packaging and products.	Reduced environmental impact, enhanced brand reputation and attraction of eco-conscious consumers.
IBM	Technology	Blockchain-based "Food Trust" platform for supply chain transparency.	Improved food safety, reduced waste and increased consumer trust (Kamath, 2018).
Siemens	Manufacturing	IoT-enabled smart factories with predictive maintenance capabilities.	Reduced downtime, enhanced operational efficiency and improved decision-making (Cotler & Sniderman, 2017).

continued on following page

Table 3. Continued

Organization	Industry	Innovation Implemented	Key Outcomes
Mastercard	Financial Services	Contactless payment systems and biometric authentication.	Faster transactions, improved security and greater customer convenience (Accenture, 2022).
Netflix	Entertainment	Data-driven content creation and subscription-based streaming model.	Global market leadership, increased customer retention and expansion of original content (Küng, 2016).
Johnson & Johnson	Healthcare	Open innovation through partnerships in JLABS incubators for healthcare solutions.	Accelerated vaccine development, increased collaboration and breakthroughs in drug discovery (Lazonick & Tulum, 2011).
DHL	Logistics	Autonomous vehicles and drones for last-mile delivery.	Reduced costs, improved delivery speed and minimized environmental impact.
Adobe	Technology	Shift to subscription-based "Creative Cloud" platform.	Stabilized revenue streams, increased customer retention and continuous product innovation (Cusumano et al., 2015).

6.1 Case Study 1: Amazon – AI-Powered Personalization

In many respects, Amazon proves the application of AI for conducting strategic innovation to enhance the customer experience. The recommendation system of the company analyzes large datasets, including customers' browsing history, purchase tendencies and interests to recommend products that cater to a particular individual (Smith & Linden, 2017). This AI-led innovation not only helped improve customer satisfaction but also contributed to increased sales and enhanced customer retention (Kantor & Streitfeld, 2015). With AI embedded in its business strategy, Amazon has put itself ahead in the competitive e-commerce market. The good thing taken from the way Amazon has used data-driven innovation to make customers engage and drive revenue.

6.2 Case Study 2: Tesla – Disruptive Innovation In Automotive Industry

Disruptive strategic innovation has led to radical changes in the automotive industry by focusing on a combination of sustainability and advanced technology initiatives. The EVs produced by Tesla represent the newest breed of automakers, providing high-performance electrified options, powered by state-of-the-art battery technology, also set for autonomous driving technology that redefines future transportation (Stringham et al., 2015). Differentiating the company, innovative enhancements, such as over-the-air software updates, keep vehicles within their maximum functional capacity by enhancing software after purchase. Melding its innovation strategy with the global paradigm shift for sustainability has given Tesla the position of market leader among electric vehicle manufacturers. This case illustrates the significance of disruptive innovation in creating markets and success over the long haul.

6.3 Case Study 3: Unilever – Sustainability-Driven Innovation

Strategic Innovation driven by Sustainability; Unilever has been able to successfully embed sustainability in its business processes. Through the Sustainable Living Plan, the company brings together environmental and social goals into its innovation strategy to develop eco-friendly products and packaging. For instance, it has introduced compressed deodorant cans that use less aluminum and emit fewer greenhouse gases when produced. Aligning sustainability goals with innovation has allowed Unilever to boost its brand reputation, minimize environmental impact and attract green customers. The case shows how sustainability can further innovation while solving global challenges.

6.4 Case Study 4: IBM – Blockchain For Supply Chain Transparency

IBM leveraged the blockchain technology to improve the traceability and efficiency of the supply chain. Through Food Trust, IBM partnered with retailers like Walmart and created a safe and traceable system to track food products from the farm on the shelf (Kamath, 2018). This leads to increased food safety, less waste and more consumer trust based on the guaranteed authenticity of the product. IBM's strategic use of blockchain does exhibit the promising potential of emerging technologies to solve the complex challenges in the supply chain. It does constitute an evident case for collaboration and technology adoption to further strategic innovation.

6.5 Case Study 5: Siemens – IoT-Enabled Smart Factories

Siemens has implemented IoT into the manufacturing process to create smart factories to optimize productivity and cut down downtime. IoT sensors embedded in the machinery monitor real-time equipment performance, permits predictive maintenance and transitions to prevent disruptive production. This innovation has not only enhanced productivity but also added resilience in the face of unforeseen breakdowns. The case of Siemens demonstrates how solutions driven by IoT will change traditional industries by improving decision-making and allocation of resources.

6.6 Case Study 6: Mastercard – Agile Innovation In Financial Services

Mastercard's adoption of agile methodologies has enabled it to adapt quickly to changing market and technological trends. New payment systems on handheld devices and biometric authentication greatly enhanced speed, security and customer convenience for Mastercard transactions (Accenture, 2022). Such innovations proved invaluable during the COVID-19 pandemic, as buyers looked for touch-free solutions. The strategic focus of Mastercard on agile innovation shows the necessity of adaptability and customer-centricity in the financial service sector. This case emphasizes the importance of iterative development and continuous improvement in strategic innovation.

6.7 Case Study 7: Netflix – Reinventing Entertainment

The transformation of Netflix from a DVD rental shop to a global leader in streaming and content production marks one very significant story of strategic innovation. The company employed data analytics to ascertain viewer preferences from which it then generated original content that suited audience interests (Küng, 2016). The shift to a subscription-based revenue model and investment in proprietary technology allowed Netflix to disrupt traditional media. This case also emphasizes the significance of data-driven decision-making and business model innovation in achieving sustained competitive advantages.

6.8 Case Study 8: Johnson & Johnson - Open Innovation in Healthcare Solutions

Johnson & Johnson has committed itself to open innovation so as to expedite healthcare innovations. In its incubators, Johnson & Johnson collaborates with start-up companies, academic institutions and researchers towards advancing breakthroughs in developing pharmaceutical products, medical devices and consumer health products. For example, in the fight against and development of COVID-19 vaccines, J&J partnered with many corporations by sharing resources and knowledge to speed up innovation for a greater impact. This illustrates how effective collaborative ecosystems can be for urgent health challenges. The case illustrates the important role that partnerships can play in enabling a strategic innovation agenda with significant social impact.

6.9 Case Study 9: DHL - Logistics Innovation with Autonomous Vehicles

DHL has infused autonomous vehicle technology in order to take logistics operations to a different level of delivery efficiency. Costs have gone down; determination of delivery has been shortened and environmental burden greatly reduced because DHL utilizes drones and self-driving vehicles for last-mile delivery. The strategic application of emerging technologies is synonymous with DHL's greater commitment to innovation in logistics. The case highlights the opportunities available for autonomous systems to complement or change traditional operations and attend to the strategic objectives that accompany their needs.

6.10 Case Study 10. Adobe – Transformation Through Subscription Models

Adobe's movement from a one-off software purchase to a subscription-based service has redefined the company's value proposition and streams of revenue. With the introduction of the Creative Cloud platform, Adobe provided customers access to continuous updates, cloud storage and collaboration tools (Cusumano & others, 2015). This strategic innovation not only led to enhanced customer retention but also ensured revenue stability through continuous streams of income. The transformation of Adobe is illustrative of the need for business models to align with ever-changing market dynamics to ensure continued growth and competitiveness.

6.11 Insight from the Case Studies

The lessons drawn from these success stories provide valuable insights into the strategy management of innovation. First, customer-centricity is a red thread since when organizations focus on understanding and meeting customer needs, they usually stand on the road to sustained success. Transformation is competitive, whether via personalization, sustainability, or convenience, depending on how far innovation meets the expectations of customers (Smith & Linden, 2017). Second, new technology complements innovation. Technological advancements such as AI, IoT, blockchain and autonomous systems enable organizations to overcome hurdles, improve efficiencies and create new propositions for customers. Such technologies are enablers of strategic innovation and influence transformation across industries (Kamath, 2018). Third, collaborative ecosystems amplify innovation efforts. Partnering with startups, academic institutions and industry peers grants access to key resources, expertise and networks. This accelerates innovation. The case studies of J&J and IBM present mixed incentives on the collaboration that is crucial to solving complex challenges (Lazonick & Tulum, 2011; Kamath, 2018). Fourth, agility and adaptability are imperative to deal with dynamically changing market conditions. In light of uncertainties, iterating through development and adjusting strategies have advantageously played important roles, as demonstrated by organizations like Mastercard and Netflix in seizing opportunities accordingly (Küng, 2016; Accenture, 2018). Lastly, aligning innovation with sustainability and social goals further enhances resilience and builds stakeholder trust. Unilever and Tesla exemplify that innovation to address environmental and social challenges can create sustaining value to the company and society both (Stringham, et al.).

7. POLICY AND MANAGERIAL IMPLICATIONS

Strategic innovation management is one of those processes or actions critical to the success or survival of an organization in a dynamic and competitive environment, but effective implementation requires cooperation among policymakers and managers to establish conditions for innovations to germinate, ensuring sustainability, resilience and ethical responsibility. In this light, this section will discuss policy and managerial implications derived from strategic innovation management and provide actionable suggestions for stakeholders seeking to tackle existing problems, seize opportunities and dovetail innovation with social aims.

7.1 Policy Implications

Governments play a pivotal role in promoting innovation ecosystems by enabling collaboration among businesses, academia and technology providers. Innovation hubs and incubators like Silicon Valley in the U.S. and Bengaluru in India enable sharing of resources, cross-pollination of ideas and acceleration of technological advancements by providing infrastructure, funding and mentorship (Audretsch et al., 2015). Financial incentives like tax credits, grants and subsidies encourage entities to invest in R&D. This scenario is particularly beneficial for SMEs, which have scarce resources. R&D tax credits have driven innovation in clean energy and biotechnology in the EU (d'Andria et al., 2018). Emerging technologies from blockchain to AI rely heavily on digital and technological literacy, wherein governments will cooperate within education and training programs. Industry 4.0 in Germany provides support for policy alignment with industry needs focusing on reskilling automation and smart manufacturing (Veile et al., 2020). Policymakers ought to tackle fundamental ethical challenges in data privacy, algorithmic biases and workforce displacement with regulations establishing transparency and fairness for all stakeholders. The EU's GDPR has established global standards for data privacy and has a clear impact on organizational practices (Voigt & von dem Bussche, 2017). Offering incentives for green innovation and circular economy practices to promote sustainability aligns innovation with climate goals, such as the electric vehicle manufacturing subsidies provided by China (Ma et al., 2017). Finally, policies to improve access to capital and lower bureaucratic barriers, along with mentorship, will have a direct impact on the innovative capacity of startups and SMEs. In most countries, government-backed venture capital funds and public-private partnerships prioritize support for relatively new sectors, such as fintech and cleantech (OECD, 2020).

7.2 Managerial Implications

Based on this, managers play a major part in realizing a policy direction that will lead to effective strategies that spur creativity and are useful to organizational objectives. Thus, managing strategic innovation requires ensuring a creative culture, promoting technologies and proactively initiating solutions to organizational or social problems (Dodgson et al., 2008; Tidd & Bessant, 2020). They should go on to fully integrate innovation into the organization's strategic vision by setting strategic goals, allocating suitable resources and balancing innovation performance using performance ma- management e.g., balanced scorecard (Kaplan & Norton, 1996; Anthony, 2014). Valuing creativity, experimentation and collaboration through initiatives such as hackathons and cross-functional teams is necessary in building a culture that supports it (Amabile, 1998). Recognizing and rewarding innovative endeavours through legitimate leadership reinforces the values (Christensen et al., 2013). For maximizing AI and IoT efficiencies, phased implementation and goal alignment with strategic goals are critical (Brynjolfsson & McAfee, 2014; Chui et al., 2018). Equally, managers should address antagonism towards change by leaning towards good communication, engaging stakeholders and providing training to ensure that these values are realized (Kotter, 1996; Hiatt, 2006). They should ensure that a balance is struck between short-term and long-term goals so as to better sustain competitiveness through a portfolio approach (Anthony, 2014; Christensen et al., 2013). With external collaboration, including universities and technology providers, ecosystem-based innovation builds a culture of innovation while cooperating over allocation of risks and resources (Chesbrough, 2003; Pisano & Verganti, 2008). Innovation practices must ensure that ethical and sustainable innovation policies prioritize data privacy, inclusiveness and environmental care, while being transparent in order to reinforce stakeholder trust (Elkington, 1997; Freeman, 2010). Resilience in the workforce-even with a solid basis on upskilling, flexible working arrangements and mental health options-will be key to ease the working of technological disruptions (World Economic Forum, 2020; Spreitzer & Porath, 2012). Finally, measurement of innovation through indicators such as ROII and customer satisfaction is vital for continuous improvement to stay aligned with existing organizational goals (Tidd & Bessant, 2020).

7.3 Implications for Collaboration Between Policymakers and Managers

The overlap between the realm of policy and managerial action is vital for extracting maximum value from strategic innovation management. Policymakers and managers need to work collectively to bring innovation response to societal needs in accordance with an organization's objectives (Chesbrough, 2003; Dodgson et al., 2008). Many programs could be initiated—such as public-private partnerships in infrastructure, renewable energy and healthcare—that can aid in solving social challenges while also advancing innovation (Hartley et al., 2013). Advocacy and engagement help organizations model the regulatory space in such a way that it reflects the realities of the industry (Freeman, 2010; Tidd & Bessant, 2020). Shared investments in education and training guarantee that an appropriately skilled workforce will emerge in the innovation economy (World Economic Forum, 2020).

8. FUTURE DIRECTIONS IN STRATEGIC INNOVATION MANAGEMENT

Strategic innovation management continues to evolve and develop under the influences of technological advances, market dynamics and global challenges. Emerging trends will accentuate advanced technologies, sustainability, collaboration, inclusiveness and ethics (Tidd & Bessant, 2020). The opportunity will be to integrate AI and ML into automatic decisions and prediction of trends toward personalized customer experience (Brynjolfsson & McAfee, 2014). Sustainability focuses on innovations in green technologies and circular models, gaining more and more weight (Elkington, 1997; Hart, 1995). Collaborative strategies across ecosystems create disruptive solutions through open innovation (Chesbrough, 2003; Pisano & Verganti, 2008). Ethics and inclusivity will ensure fairness and transparency (Floridi et al., 2021). Resilience-based decision-making strategies meet uncertainties through adaptation processes (Taleb, 2014). Finally, the digital transformation utilizing blockchain and quantum computing welcomes the reshaping of strategic innovation management (Brynjolfsson & McAfee, 2014).

9. CONCLUSION

Strategic innovation management is undoubtedly essential for organizations operating in complex and competitive dynamic environments. Aligning innovation with long-term objectives assists organizations through challenges, capturing opportunities and enhancing resilience. New-age technologies such as AI, blockchain and the IoT act as catalysts for change, improving efficiency, decision-making and value creation. Important themes include aligning innovation with goals, supporting creativity and adaptability and preventing barriers such as resistance to change or conflicts with resource constraints. Practical examples show the impact across industries, emphasizing the benefits of technological integration, collaboration and sustainability. Moving forward, a deepening of technological integration, sustainability and inclusivity will mark its evolution, making real changes and generating meaningful value.

REFERENCES

Accenture. (2022). *Growing payments to new heights: The value propositions that pay*. https://www.accenture.com/content/dam/accenture/final/industry/banking/document/Accenture-Growing-Payments-New-Heights.pdf

Amabile, T. M. (1998). How to kill creativity. *Harvard Business Review*, *76*(5), 76–87. PMID: 10185433

Anthony, S. D. (2014). *The first mile: a launch manual for getting great ideas into the market*. Harvard Business Review Press.

Armbrust, M., Fox, A., Griffith, R., Joseph, A. D., Katz, R. H., Konwinski, A., Lee, G., Patterson, D., Rabkin, A., Stoica, I., & Zaharia, M. (2010). A view of cloud computing. *Communications of the ACM*, *53*(4), 50–58. DOI: 10.1145/1721654.1721672

Audretsch, D. B., Belitski, M., & Desai, S. (2015). Entrepreneurship and economic development in cities. *The Annals of Regional Science*, *55*(1), 33–60. DOI: 10.1007/s00168-015-0685-x

Ball, P., & Maxmen, A. (2020). The epic battle against coronavirus misinformation. *Nature*, *581*(7809), 371–374. DOI: 10.1038/d41586-020-01452-z PMID: 32461658

Bhamra, R., Dani, S., & Burnard, K. (2011). Resilience: The concept, a literature review and future directions. *International Journal of Production Research*, *49*(18), 5375–5393. DOI: 10.1080/00207543.2011.563826

Bharadwaj, A., El Sawy, O. A., Pavlou, P. A., & Venkatraman, N. (2013). Digital business strategy: Toward a next generation of insights. *Management Information Systems Quarterly*, *37*(2), 471–482. DOI: 10.25300/MISQ/2013/37:2.3

Bhushan, B. (2017). *Springer handbook of nanotechnology*. Springer. DOI: 10.1007/978-3-662-54357-3

Bocken, N. M. P., Short, S. W., Rana, P., & Evans, S. (2014). A literature and practice review to develop sustainable business model archetypes. *Journal of Cleaner Production*, *65*, 42–56. DOI: 10.1016/j.jclepro.2013.11.039

Brown, S. L., & Eisenhardt, K. M. (1997). The art of continuous change: Linking complexity theory and time-paced evolution in relentlessly shifting organizations. *Administrative Science Quarterly*, *42*(1), 1–34. DOI: 10.2307/2393807

Brynjolfsson, E., & McAfee, A. (2014). *The second machine age: Work, progress and prosperity in a time of brilliant technologies*. W.W. Norton & Company.

Campolo, C., Molinaro, A., Araniti, G., & Berthet, A. O. (2017). Better platooning control toward autonomous driving: An LTE device-to-device communications strategy that meets ultralow latency requirements. *IEEE Vehicular Technology Magazine*, *12*(1), 30–38. DOI: 10.1109/MVT.2016.2632418

Carmigniani, J., Furht, B., Anisetti, M., Ceravolo, P., Damiani, E., & Ivkovic, M. (2011). Augmented reality technologies, systems and applications. *Multimedia Tools and Applications*, *51*(1), 341–377. DOI: 10.1007/s11042-010-0660-6

Chen, H., Chiang, R. H., & Storey, V. C. (2012). Business intelligence and analytics: From big data to big impact. *Management Information Systems Quarterly*, *36*(4), 1165–1188. DOI: 10.2307/41703503

Chesbrough, H. (2003). *Open innovation: The new imperative for creating and profiting from technology*. Harvard Business Review Press.

Christensen, C., Raynor, M. E., & McDonald, R. (2013). *Disruptive innovation*. Harvard Business Review.

Christensen, C. M. (1997). *The innovator's dilemma: When new technologies cause great firms to fail*. Harvard Business Review Press.

Christopher, M., & Peck, H. (2004). Building the resilient supply chain. *International Journal of Logistics Management*, *15*(2), 1–14. DOI: 10.1108/09574090410700275

Chui, M., Manyika, J., & Miremadi, M. (2018). What AI can and can't do (yet) for your business. *The McKinsey Quarterly*.

Cramer, J., & Krueger, A. B. (2016). Disruptive innovation and taxi services: The case of Uber. *The American Economic Review*, *106*(5), 177–182. DOI: 10.1257/aer.p20161002

Cusumano, M. A., Kahl, S. J., & Suarez, F. F. (2015). Services, industry evolution and the competitive strategies of product firms. *Strategic Management Journal*, *36*(4), 559–575. DOI: 10.1002/smj.2235

d'Andria, D., Pontikakis, D., & Skonieczna, A. (2018). Towards a European R&D incentive? An assessment of R&D provisions under a common corporate tax base. *Economics of Innovation and New Technology*, *27*(5–6), 531–550. DOI: 10.1080/10438599.2017.1376168

Davenport, T. H. (2014). *Big data at work: Dispelling the myths, uncovering the opportunities*. Harvard Business Review Press. DOI: 10.15358/9783800648153

Dobrovnik, M., Herold, D. M., Fürst, E., & Kummer, S. (2018). Blockchain for and in logistics: What to adopt and where to start. *Logistics (Basel)*, *2*(3), 18. DOI: 10.3390/logistics2030018

Dodgson, M., Gann, D. M., & Salter, A. (2008). The management of technological innovation: strategy and practice. oxford university press.

Doudna, J. A., & Charpentier, E. (2014). The new frontier of genome engineering with CRISPR-Cas9. *Science*, *346*(6213), 1258096. DOI: 10.1126/science.1258096 PMID: 25430774

Doz, Y. L., & Kosonen, M. (2010). Embedding strategic agility: A leadership agenda for accelerating business model renewal. *Long Range Planning*, *43*(2-3), 370–382. DOI: 10.1016/j.lrp.2009.07.006

Drucker, P. F. (1993). *Innovation and entrepreneurship: Practice and principles*. HarperBusiness.

Edmondson, A. (1999). Psychological safety and learning behavior in work teams. *Administrative Science Quarterly*, *44*(2), 350–383. DOI: 10.2307/2666999

Elkington, J. (1998). *Cannibals with forks: The triple bottom line of 21st-century business*. New Society Publishers.

European Commission. (2021). *Shaping Europe's digital future: Advancing resilience and sustainability*. Retrieved from https://digital-strategy.ec.europa.eu/en

Evans, D. (2011). The internet of things. *How the Next Evolution of the Internet is Changing Everything, Whitepaper* [IBSG]. *Cisco Internet Business Solutions Group, 1*, 1–12.

Floridi, L., Cowls, J., King, T. C., & Taddeo, M. (2021). How to design AI for social good: Seven essential factors. *Ethics, Governance and Policies in Artificial Intelligence*, 125-151.

Ford, R. D. (2017). *Low latency fifth-generation cellular networks* (Doctoral dissertation, Polytechnic Institute of New York University).

Freeman, R. E. (2010). *Strategic management: A stakeholder approach*. Cambridge University Press. DOI: 10.1017/CBO9781139192675

Frey, C. B., & Osborne, M. A. (2017). The future of employment: How susceptible are jobs to computerization? *Technological Forecasting and Social Change, 114*, 254–280. DOI: 10.1016/j.techfore.2016.08.019

Friedman, T. L. (2005). *The world is flat: A brief history of the twenty-first century*. Macmillan.

Gartner. (2020). *Resilience and innovation: Key capabilities for organizational sustainability*. Retrieved from https://www.gartner.com/smarterwithgartner/build-organizational-resilience-for-today-and-tomorrow

Geissdoerfer, M., Savaget, P., Bocken, N. M. P., & Hultink, E. J. (2017). The circular economy–A new sustainability paradigm? *Journal of Cleaner Production, 143*, 757–768. DOI: 10.1016/j.jclepro.2016.12.048

Goleman, D., Boyatzis, R., & McKee, A. (2002). *Primal leadership: Realizing the power of emotional intelligence*. Harvard Business School Press.

Goodchild, A., & Toy, J. (2018). Delivery by drone: An evaluation of unmanned aerial vehicle technology in reducing CO2 emissions in the delivery service industry. *Transportation Research Part D, Transport and Environment, 61*, 58–67. DOI: 10.1016/j.trd.2017.02.017

Govindarajan, V., & Trimble, C. (2010). *The other side of innovation: Solving the execution challenge*. Harvard Business Review Press.

Hamel, G., & Prahalad, C. K. (1989). Strategic intent. *Harvard Business Review, 67*(3), 63–76. PMID: 10303477

Hamel, G., & Välikangas, L. (2003). The quest for resilience. *Harvard Business Review, 81*(9), 52–63. PMID: 12964393

Hart, S. L. (1995). A natural-resource-based view of the firm. *Academy of Management Review, 20*(4), 986–1014. DOI: 10.2307/258963

Hartley, J., Sørensen, E., & Torfing, J. (2013). Collaborative innovation: A viable alternative to market competition and hierarchical public sector organizations? *Public Administration Review, 73*(6), 821–830. DOI: 10.1111/puar.12136

Hiatt, J. M. (2006). *ADKAR: A model for change in business, government and our community*. Prosci.

Iansiti, M., & Lakhani, K. R. (2020). *Competing in the age of AI: Strategy and leadership when algorithms and networks run the world*. Harvard Business Review Press.

IRENA. (2019), Climate Change and Renewable Energy: National policies and the role of communities, cities and regions (Report to the G20 Climate Sustainability Working Group (CSWG)), International Renewable Energy Agency, Abu Dhabi. Retrieved from https://www.irena.org

Ivanov, D., & Dolgui, A. (2020). Viability of intertwined supply networks: Extending the supply chain resilience angles towards survivability. A position paper motivated by COVID-19 outbreak. *International Journal of Production Research*, *58*(10), 2904–2915. DOI: 10.1080/00207543.2020.1750727

Kamath, R. (2018). Food Traceability on Blockchain: Walmart's Pork and Mango Pilots with IBM. *The Journal of the British Blockchain Association*, *1*(1), 1–12. DOI: 10.31585/jbba-1-1-(10)2018

Kantor, J., & Streitfeld, D. (2015, August 15). Inside Amazon: Wrestling Big Ideas in a Bruising Workplace. *The New York Times*. https://www.nytimes.com/2015/08/16/technology/inside-amazon-wrestling-big-ideas-in-a-bruising-workplace.html

Kaplan, R. S., & Mikes, A. (2012). Managing risks: A new framework. *Harvard Business Review*, *90*(6), 48–60.

Kaplan, R. S., & Norton, D. P. (1996). *The balanced scorecard: Translating strategy into action*. Harvard Business Review Press.

Kotter, J. P. (1995, May 1). Leading Change: Why Transformation Efforts Fail. *Harvard Business Review*. https://hbr.org/1995/05/leading-change-why-transformation-efforts-fail-2

Küng, L. (2016). *Strategic management in the media: From theory to practice*. SAGE Publications.

Lazonick, W., & Tulum, O. (2011). US biopharmaceutical finance and the sustainability of the biotech business model. *Research Policy*, *40*(9), 1170–1187. DOI: 10.1016/j.respol.2011.05.021

Lengnick-Hall, C. A., Beck, T. E., & Lengnick-Hall, M. L. (2011). Developing a capacity for organizational resilience through strategic human resource management. *Human Resource Management Review*, *21*(3), 243–255. DOI: 10.1016/j.hrmr.2010.07.001

Lindström, C. W. J., Maleki Vishkaei, B., & De Giovanni, P. (2024). Subscription-based business models in the context of tech firms: Theory and applications. *International Journal of Industrial Engineering and Operations Management*, *6*(3), 256–274. DOI: 10.1108/IJIEOM-06-2023-0054

Lipson, H., & Kurman, M. (2013). *Fabricated: The new world of 3D printing*. John Wiley & Sons.

Litman, T. (2024). Autonomous vehicle implementation predictions: Implications for transport planning. *Victoria Transport Policy Institute*. Retrieved from https://www.vtpi.org/avip.pdf

Liu, J., Liu, M., Bai, Y., Zhang, J., Liu, H., & Zhu, W. (2020). Recent progress in flexible wearable sensors for vital sign monitoring. *Sensors (Basel)*, *20*(14), 4009. DOI: 10.3390/s20144009 PMID: 32707637

Ma, L., Fan, Y., & Feng, L. (2017). An evaluation of government incentives for new energy vehicles in China focusing on vehicle purchasing restrictions. *Energy Policy*, *110*, 609–618. DOI: 10.1016/j.enpol.2017.07.057

MacArthur, E. (2013). Towards the circular economy. *Journal of Industrial Ecology*, *2*(1), 23–44.

Manyika, J.. (2015). *The internet of things: Mapping the value beyond the hype*. McKinsey Global Institute.

Manyika, J., Chui, M., Miremadi, M., Bughin, J., George, K., Willmott, P., & Dewhurst, M. (2017). A future that works: AI, automation, employment and productivity. *McKinsey Global Institute Research. Tech. Rep*, *60*, 1–135.

Mazzucato, M. (2013). *The entrepreneurial state: Debunking public vs. private sector myths*. Anthem Press.

McGrath, R. G. (2013). *The end of competitive advantage: How to keep your strategy moving as fast as your business*. Harvard Business Review Press.

Milgram, P., & Kishino, F. (1994). A taxonomy of mixed reality visual displays. *IEICE Transactions on Information and Systems*, *77*(12), 1321–1329.

Nakamoto, S. (2008). Bitcoin: A peer-to-peer electronic cash system. Retrieved from https://bitcoin.org/bitcoin.pdf

National Research Council. (2009). *A new biology for the 21st century: Ensuring the United States leads the coming biology revolution*. National Academies Press.

Negri, E., Fumagalli, L., & Macchi, M. (2017). A review of the roles of digital twin in CPS-based production systems. *Procedia Manufacturing*, *11*, 939–948. DOI: 10.1016/j.promfg.2017.07.198

OECD. (2015). *Frascati manual 2015: Guidelines for collecting and reporting data on research and experimental development*. OECD Publishing.

OECD. (2020). *Financing SMEs and Entrepreneurs 2020: An OECD Scoreboard*. OECD., DOI: 10.1787/061fe03d-

Osterwalder, A., & Pigneur, Y. (2010). *Business model generation: A handbook for visionaries, game changers and challengers*. Wiley.

Pisano, G. P. (2006). *Science business: The promise, the reality and the future of biotech*. Harvard Business Review Press.

Pisano, G. P. (2015). You need an innovation strategy. *Harvard Business Review*, *93*(6), 44–54.

Pisano, G. P., & Verganti, R. (2008). Which kind of collaboration is right for you? *Harvard Business Review*, *86*(12), 78–86.

Porter, M. E., & Kramer, M. R. (2011). Creating shared value. *Harvard Business Review*, *89*(1/2), 62–77.

Powell, W. W., Koput, K. W., & Smith-Doerr, L. (1996). Interorganizational collaboration and the locus of innovation: Networks of learning in biotechnology. *Administrative Science Quarterly*, *41*(1), 116–145. DOI: 10.2307/2393988

Prahalad, C. K., & Ramaswamy, V. (2004). Co-creating unique value with customers. *Strategy and Leadership*, *32*(3), 4–9. DOI: 10.1108/10878570410699249

Preskill, J. (2018). Quantum computing in the NISQ era and beyond. *Quantum : the Open Journal for Quantum Science*, *2*, 79. DOI: 10.22331/q-2018-08-06-79

Ries, E. (2011). *The lean startup: How today's entrepreneurs use continuous innovation to create radically successful businesses*. Crown Publishing.

Rigby, D. K., Sutherland, J., & Takeuchi, H. (2016). Embracing agile. *Harvard Business Review, 94*(5), 40–48.

Russell, S. J., & Norvig, P. (2021). *Artificial intelligence: A modern approach* (4th ed.). Pearson.

Schein, E. H. (2010). *Organizational culture and leadership* (4th ed.). Jossey-Bass.

Schoemaker, P. J. H. (1995). Scenario planning: A tool for strategic thinking. *Sloan Management Review, 36*(2), 25–40.

Senge, P. M. (2006). *The fifth discipline: The art and practice of the learning organization*. Broadway Business.

Shi, W., Cao, J., Zhang, Q., Li, Y., & Xu, L. (2016). Edge computing: Vision and challenges. *IEEE Internet of Things Journal, 3*(5), 637–646. DOI: 10.1109/JIOT.2016.2579198

Slater, M., & Sanchez-Vives, M. V. (2016). Enhancing our lives with immersive virtual reality. *Frontiers in Robotics and AI, 3*, 74. DOI: 10.3389/frobt.2016.00074

Smith, B., & Linden, G. (2017). Two decades of recommender systems at Amazon.com. *IEEE Internet Computing, 21*(3), 12–18. DOI: 10.1109/MIC.2017.72

Soto-Acosta, P. (2020). COVID-19 pandemic: Shifting digital transformation to a high-speed gear. *Information Systems Management, 37*(4), 260–266. DOI: 10.1080/10580530.2020.1814461

Spreitzer, G., & Porath, C. (2012). Creating sustainable performance. *Harvard Business Review, 90*(1/2), 92–99. PMID: 22299508

Stringham, E. P., Miller, J. K., & Clark, J. R. (2015). Overcoming barriers to entry in an established industry: Tesla Motors. *California Management Review, 57*(4), 85–103. DOI: 10.1525/cmr.2015.57.4.85

Sun, X., & Abdullahi Usman, M. (2025). Drivers of platform ecosystem adoption: Does innovation capability translate these drivers into improved firm performance. *Business Process Management Journal, 31*(1), 118–145. DOI: 10.1108/BPMJ-12-2023-0972

Susskind, R., & Susskind, D. (2022). *The future of the professions: How technology will transform the work of human experts*. Oxford University Press.

Sutcliffe, K. M., & Vogus, T. J. (2003). Organizing for resilience. In Cameron, K. S., Dutton, J. E., & Quinn, R. E. (Eds.), *Positive organizational scholarship: Foundations of a new discipline* (pp. 94–110). Berrett-Koehler.

Taleb, N. N. (2014). *Antifragile: Things that gain from disorder* (Vol. 3). Random House Trade Paperbacks.

Tao, F., Zhang, H., Liu, A., & Nee, A. Y. C. (2018). Digital twin in industry: State-of-the-art. *IEEE Transactions on Industrial Informatics, 15*(4), 2405–2415. DOI: 10.1109/TII.2018.2873186

Tapscott, D., & Tapscott, A. (2016). *Blockchain revolution: How the technology behind bitcoin is changing money, business and the world*. Portfolio.

Teece, D. J. (2010). Business models, business strategy and innovation. *Long Range Planning*, *43*(2–3), 172–194. DOI: 10.1016/j.lrp.2009.07.003

Teece, D. J., Peteraf, M., & Leih, S. (2016). Dynamic capabilities and organizational agility: Risk, uncertainty and strategy in the innovation economy. *California Management Review*, *58*(4), 13–35. DOI: 10.1525/cmr.2016.58.4.13

Thomke, S. (2003). *Experimentation matters: Unlocking the potential of new technologies for innovation*. Harvard Business Review Press.

Tidd, J., & Bessant, J. R. (2020). *Managing innovation: integrating technological, market and organizational change*. John Wiley & Sons.

Van der Heijden, K. (1996). *Scenarios: The art of strategic conversation*. Wiley.

Veile, J. W., Kiel, D., Müller, J. M., & Voigt, K. I. (2020). Lessons learned from Industry 4.0 implementation in the German manufacturing industry. *Journal of Manufacturing Technology Management*, *31*(5), 977–997. DOI: 10.1108/JMTM-08-2018-0270

Voigt, P., & Von dem Bussche, A. (2017). The eu general data protection regulation (GDPR). *A Practical Guide, 1st Ed., Cham: Springer International Publishing, 10*(3152676), 10-5555.

Waller, M. A., & Fawcett, S. E. (2013). Data science, predictive analytics and big data: A revolution that will transform supply chain design and management. *Journal of Business Logistics*, *34*(2), 77–84. DOI: 10.1111/jbl.12010

Westerman, G., Bonnet, D., & McAfee, A. (2014). *Leading digital: Turning technology into business transformation*. Harvard Business Review Press.

Wheeler, N. E. (2025). Responsible AI in biotechnology: Balancing discovery, innovation and biosecurity risks. *Frontiers in Bioengineering and Biotechnology*, *13*, 1537471. DOI: 10.3389/fbioe.2025.1537471 PMID: 39974189

World Economic Forum. (2020). *The future of jobs report 2020*. Retrieved from https://www.weforum.org

Chapter 25
The Dynamic Model for Management Improvement Projects

Heitor Leopoldo Nogueira Coutinho
https://orcid.org/0009-0004-5814-9614
FDC, Brazil

ABSTRACT

This chapter presents a model for dealing with management improvement projects in a turbulent world. The dynamic model addresses the main variations of a management improvement project in a company. The basis of this model is the star diagram, a powerful mechanism for defining paths based on the biggest challenges of projects. The challenges are complexity, singularity, speed, and technical and opposition. Based on the challenges, the methods to be used are defined. The model aims to ensure that projects are successful through the necessary procedural dynamics. The success of a project can be established from four perspectives: performance, impact, result, and sustainability. Based on the dynamic model, companies can develop their methodologies. The process described can also be used without customization. The flexibility it provides means that projects of different categories can be developed effectively.

INTRODUCTION

The Discourse on Method

Achieving strategic advantages in a turbulent world imposes new ways of thinking and acting on organizations to achieve their results. There is always a pertinent strategy in response to the competitive market and the increasingly present uncertainties. Thinking and formulating strategies will need to be comprehensively active to be effective. On the other hand, living harmoniously and productively in dynamic and unpredictable environments will only happen if the formulated strategy is successfully implemented (Coutinho, 2025). Successful implementation is fundamentally achieved through adopting selectively advanced project management practices combined with innovation and change management. The purpose of this chapter is to present, in a structured way, methods and instruments for carrying out management improvement projects that guarantee that the organizational objectives are achieved in full.

DOI: 10.4018/979-8-3693-4030-1.ch025

Nowadays, the great challenge for institutions of any kind is to know which factors determine successful execution and how to deal with them despite the uncertainty and current challenges.

Projects to solve an organization's strategic or real problem are mechanisms for improving management and developing talent. When applied correctly, they promote a rapid increase in capacity and a lasting dialog between fundamentals and practice. The model we describe below is based on analyzing driving challenges, project development methods, and human behavior in predictive or uncertain situations.

THE DRIVING CHALLENGES

Competitive rivalry is present in almost every business. And the business world has become more unpredictable and dynamic, making competitive advantages that were once sustainable in the long term only temporary (Reeves & Whitaker, 2022). These are strategic obstacles that affect many sectors of the economy in some way. However, some companies manage to identify opportunities and position themselves uniquely. Most, however, find themselves dispersed in this new environment. There is little agility and capacity to turn ideas into reality due to a lack of awareness about what matters, despite progress in knowledge and the recent efforts made by the design community. Significant advances in research and development have occurred over the last fifty years. Nonetheless, failure rates in project management remain alarming, regardless of the market, region, or sector in which companies operate (Coutinho, 2019).

Business schools, universities, research centers, institutes, managers, and specialists have long sought to understand the possible causes of failures in strategic projects. The difficulty of this analysis and understanding the origin of errors arises because it is a complex multidisciplinary environment that is naturally ambiguous. It is also a fact that most organizations that have been researched have used rigorously structured project management processes and techniques, and their teams have been trained to perform their duties. So why do failures prevail and, in certain conditions, such as turbulent markets, even increase?

Over the years, we have conducted research and monitored strategic moves in public and private organizations, and over the last decade, we have dedicated ourselves to finding the causes of these failures. It has been a lengthy investigation process in search of information to identify the reasons that led to unsatisfactory results. Many projects were thoroughly analyzed, leaving us no doubt about the findings. However, we were surprised about aspects in which there was already a good level of maturity in the organizations surveyed.

The results of this study confirmed the hypothesis that there were peculiar situations that were difficult to deal with and which, when combined, increased the possibility of failure. Based on these findings, the first case studies were produced, and we developed the methodology for assessing the challenge level of project portfolios. This fact gave rise to the Star Diagram, a simple, high-value tool capable of guiding the execution of strategies uniquely toward success. The Star Diagram analyzes the biggest challenges in implementing strategies through programs and projects. It is a progressive resource for implementing classic and adaptive strategies capable of breaking paradigms and increasing the agility and effectiveness of organizations in their development (Coutinho, 2019).

Methodology and Its Application

The Star Diagram represents five vertices containing the biggest challenges for executing organizational strategies, as shown in Figure 1. These challenges are Complexity, Singularity, Speed, Technical Challenge, and Opposition Challenge.

Even before discussing complexity, we thought it deserved special attention because it is a growing phenomenon. To better understand this variable, seminars entitled “Complexity & Management” were held, involving professionals from administration and other sciences. The multidisciplinary vision could bring new insights, as indeed it did. We also followed and participated in initiatives by the ICCPM (International Center for Complex Project Management) in Australia, which brings together research in this area.

Uncertainty was the second object of study identified, restricted to the product scope in R&D, software development, and innovation projects. The results of innovation projects pointed to high failure rates due to the uncertainty inherent in these categories. This situation led to the understanding that the analysis would focus on the highest singularity. In the case of organizational change projects, the reasons for failure are associated with stakeholder behavior: resistance due to fear of the unknown, the possibility of failure, and the feeling of loss of power, control, status, or security. Technical-operational and technological challenges, such as the lack of specialized knowledge in developing certain products and services, were also identified as obstacles to the success of projects and programs. Finally, the last variable found was related to the time available for the development and execution of strategic initiatives on their scale. If an organization needs to make large-scale changes or even create something new quickly, it must deal with high speed, which can be risky.

Based on these observations and analyses, the model was created so that, right at the beginning of each initiative, the main challenges for its execution could be measured and evaluated. In this way, the challenges to its implementation were found to maximize the chances of success. For example, the overall level of risk identified is very high. In that case, a company may choose not to invest or, having decided to implement it, must improve its effectiveness by acting in line with the situation's requirements. Thus, some challenges, such as complexity and velocity, can be practically avoided if the situation allows, reducing the scope into smaller parts and increasing deadlines. On the other hand, risks also bring opportunities, and taking on high challenges to capture more value in the market can be advantageous.

So, let's understand what each challenge means.

Figure 1. The Star Diagram

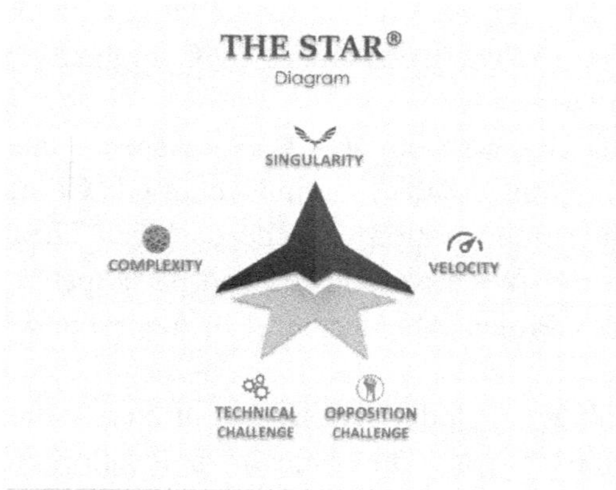

For a long time, experts in complexity have been dedicated to studying the phenomenon in management and engineering, physics, humanities, social, and biological sciences. From an organizational perspective, there is a consensus that complexity has progressively increased managerial action and risk exposure.

The Complexity

Although complexity is widely referenced for its relevance, there is no single and convergent understanding of its meaning and definition. It is common, for example, to be imprecise about what is complex and what is complicated. Complicated situations are challenging and have many parts. However, they have a certain linearity, and it is possible to understand their correlations and causalities more easily. On the other hand, complexity is related to the unpredictability and ambiguity of the results due to the large number of variables and their elusive interrelationships.

One mistake the managers sometimes make is thinking that large projects are always complex. A project's size strongly correlates with complexity, but this information is not decisive (Yadav & Paul, 2024). In other words, large projects tend to be complex, but there are large projects with moderate complexity. To illustrate, a project complexity construction of many standardized buildings may be less complex than developing a new product in the medical field. Unlike complexity, the size of a project is determined directly, without obstacles, and can therefore be easily measured (Davis, 2017).

Therefore, the first attribute of complexity is the number of variables involved. The second refers to the quantity of interdependencies between the variables. These two attributes analyzed together make it possible to analyze complexity by the level of uncertainty and ambiguity generated (Botchkarev & Finnigan, 2015). Let's consider a project to implement a new integrated management system (or ERP - Enterprise Resource Planning), which impacts all parts of the company and thus interferes with many variables. It will undoubtedly be complex because, in this situation, many stakeholders are involved, processes and areas are affected. There are many interdependencies between the areas, and such a complex intervention generates uncertain consequences that are difficult to interpret and prevent.

By understanding what characterizes complexity and using a measurement instrument, the manager will be able to assess a project's complexity level, which varies from primary, at level 1, where complexity is low or negligible, to mega-complex, at level 4. As mega-complex projects are likely to fail, it is vital to manage the difficulties and ambiguities common to them appropriately. Among the various actions recommended for dealing with complexity are methodological rigor, effective stakeholder management, and critical evaluation of integration processes (Denney & Watkins, 2023).

The Singularity

Singularity in projects has a direct relationship with the degree of novelty of the products generated and a direct relationship with innovation. Innovation is defined as an idea, practice, or object new to the entity adopting it and, sometimes, to the market. Transformational, radical, or disruptive innovation projects are the most challenging. They involve the need to make decisions and act under adverse conditions (Mignenan, 2023).

Innovation is directly related to uncertainty, which results from the fact that, on the one hand, events in the future do not follow the course of past events, and, on the other, knowledge about the future is always incomplete (Kotter et al., 2021). Some authors classify innovation uncertainty into market, organizational, and technological uncertainty. Market and organizational uncertainties are considered in the Singularity challenge of the Star Diagram; technological uncertainty is part of the Technical Challenge. Singularity, then, in the managerial and market spheres, is restricted to the level of understanding about the scope. In the early stages of innovation projects, goals are often nothing more than an expression of a desired end state, with no certainty that it can be achieved. The solution, then, has not yet been explored.

Uniqueness is present in all projects by definition, but each project has different levels, ranging from incremental to breakthrough. At the incremental level, a project has little novelty and aims to improve the performance of a product, service, or process known to the company (Brix & Peters, 2015). At the level of more outstanding singularity, creating something innovative is sometimes related to significant societal transformations, as it disrupts conventional models.

Flexible, iterative methods are necessary to deal with singularity. These methods allow the production of a minimum viable product that can be evolved incrementally.

The Velocity

Competitive organizations are the ones that most impose the need to conceive, develop, and implement projects quickly. Record-time developments can provide competitive advantages in the market through pioneering spirit, for example. However, the pressure imposed can have harmful or unsustainable consequences, such as discouraging creativity and even reducing overall productivity by affecting team motivation in some cases.

The challenge of speed is the relationship between time and effort, the latter being the project's number of deliverables (products and/or services). The greater the speed required to make many changes, the more critical factors must be managed simultaneously, and the greater the costs and difficulties of carrying them out. The key, then, is to know how to scale the velocity when it is possible to make time and/or effort more flexible. If this is not possible, decisions must be made to deal with this challenge successfully (Lister, 2018).

The velocity challenge can also be assessed on a four-level scale, from low to extreme speed. The first level usually includes initiatives focused on the quality of the product to be developed when the final deadline is not critical. Projects in extreme situations are typically set up in reaction to crises, i.e., when an organization's survival is compromised, or severe losses will occur if changes are not implemented quickly.

Dealing with velocity requires, first and foremost, autonomy to make decisions that will have to be supported unconditionally by senior management. Success in this condition will come from the indispensable monitoring of deadlines in line with effective communication and unique risk management processes. Velocity is different from agility. Velocity occurs when the deadline to deliver products is short, and agility is the best deadline for meeting specific situations. Generally, when making cultural changes, agility occurs when the deadlines are long.

The Technical Challenge

The project's Technical Challenges are related to the absence of the organization's prior knowledge, availability, and difficulty of acquiring the required technical capacity. They refer to the barriers imposed by the scope of the product to be developed because the organization does not have the necessary technical competence, either due to the degree of topicality of the subject and/or process, the level of expertise required, or the uncertainty about some technology to be used (Mohammad et al., 2016).

The scale of the Technical Challenge starts at the basic level and rises to the deep level, with the regular and broad levels in between. The first level is characterized by the basic need for knowledge to develop the project. If there is a technology to be used in the project, it is mature, has been widely used in the market, and presents no difficulty in being understood and applied. At the deep level, the organization has no systemic knowledge or understanding of the functionalities or technologies that will be needed. If there is technology external to the project, it is very advanced, unused in the market and susceptible to many technical uncertainties and failures. This type of technology is called hi-tech. If the subject of the project is unknown, there is also a high technical challenge. Developing a project on logistics without mastering the subject is very risky.

Dealing with technical challenges means knowing how to manage technical competence. Management must simultaneously consider alternatives for training and acquiring specialists. In this situation, teamwork must be valued to open all possibilities for sharing knowledge and practices during the project life cycle process. As far as the management process is concerned, periodic reviews and, usually, late definitions of the design are very welcome.

The Opposition Challenge

Opposition in the workplace is more common than we think, especially in organizational change projects. Discord is usually veiled and not always perceived. Changes in organizations threaten or alter power structures. The consequence is the natural reaction of those who feel affected. When not handled properly, the reaction can turn into resistance and characterize real opposition to the design, development, and implementation of necessary changes. Opposition generates rumors that can be stronger than any fact. Opposition is negative resistance. Resistance can also be beneficial when it alerts the team to find a better solution (Gibbons & Kennedy, 2023).

Thus, analyzing and measuring the Opposition Challenge in the Star Diagram will make it possible to understand the degree of opposition better. The spectrum of opposition starts at a low level of resistance to change. It goes up to large movements opposing specific initiatives or even the entire strategy implementation process. The level of minimal opposition occurs when the stakeholders reacting to the changes have low power or influence over organizational decisions, and there is no opposing coalition, i.e., these stakeholders are not grouped to demonstrate against the change. The evolution of resistance levels by stakeholders with more significant power raises the opposition challenge. At an extensive level, opinion formers can lead these movements. Thus, in addition to the importance of the extent of resistance and the collective organization of resistant groups, there is another situation. Consider the level of power/influence of the resistors. The degree of opposition is high when there is authority.

Managing opposition is the art of balance, which requires political and relational skills (Hubbs et al., 2021). Knowing how to negotiate, achieve harmony between interests and general expectations, and engage everyone through dialog is good. It is an adaptive leadership that can make profound changes. It helps individuals and organizations adapt and thrive in challenging environments, but fundamentally, it exists to innovate and create the resources that match the aspirations of winning organizations.

The Star Diagram is the key instrument for dealing with the significant challenges of strategy execution and organizational change processes. By assessing the level of challenge of strategic initiatives, it is possible to make better decisions about the degree of risk a company is willing to take or adjust them. However, the Star Diagram technique also looks at the requirements, practices, and competencies for dealing with each challenge, thus significantly increasing the likelihood of ultimate success. Project and change management models are essential for implementing initiatives, but they are not enough to deal with the plurality and dynamics of contemporary organizations. The ability to deal with the five challenges will make the difference. It is from this understanding that a plausible path will be formed. That's why you must know how much weight to give standardization, recognize that each situation requires specific treatment, and appreciate the results that are sure to come.

PROJECT DEVELOPMENT METHODS

The development method is a process adopted for creating, elaborating, and specifying an idea, product, or service to ensure the achievement of objectives. Based on your prior knowledge about the products you intend to deliver, you can determine the most appropriate method for a project's life cycle. Four methods use different approaches to creation and development: predictive, agile, hybrid, and innovative (Wysocki, 2019).

The predictive method, widely used in engineering and organizational projects, is also known as the traditional or waterfall method. The predictive method is used when there is a good initial understanding of the scope of the product. This scope does not include detailed information but is limited to a sense of its scope at the outset. In this way, it is possible to plan the schedule more precisely on what will be specified in time, making the cost plan feasible. The process is sequential because the more incredible foresight into the future facilitates linear, multi-process work.

The agile method, applied to information systems development projects and more often to product and service development projects, is also known as the adaptive method. This approach is suitable when there is little knowledge about the scope of the products and services to be delivered. The development of innovation projects has this characteristic. As it is impossible to grasp the scope beforehand clearly, it isn't easy to describe a precise timetable. Therefore, development is adapted in cycles of interaction with the active participation of stakeholders. The process used is iterative and progressive to achieve the necessary products.

The hybrid method combines some elements of the predictive and agile approaches to meet organizations' specific needs. Its use has been growing significantly worldwide due to its adaptability to different required situations. Due to their inherent adaptability through customization, hybrid project management methods can provide excellent project performance and customization. It is essential to be aware of the benefits of methodological approaches to meet the changing needs in business reality. This evolution underscores the importance of aligning management practices with the complexity and dynamics of particular transformations.

The innovative method, most used in research and development projects, is also known as the extreme method. This type of process is used in situations of high future uncertainty, requiring studies and research to ensure the project's central objective. At first, it is consequently difficult to understand what the products will be. That's why adopting a flexible approach, subject to many changes, is necessary. Innovative methods are essential for anticipating market possibilities, strategic pioneering, and higher levels of innovation (Shenhar at all, 2016).

Other methods must be combined with development methods to meet the challenges of one project. The challenge levels defined in the Star Diagram, shown in Figure 2, are a valuable diagnostic for deciding on the appropriate combination of responses to some challenges.

Figure 2. The indication of methods by the challenges

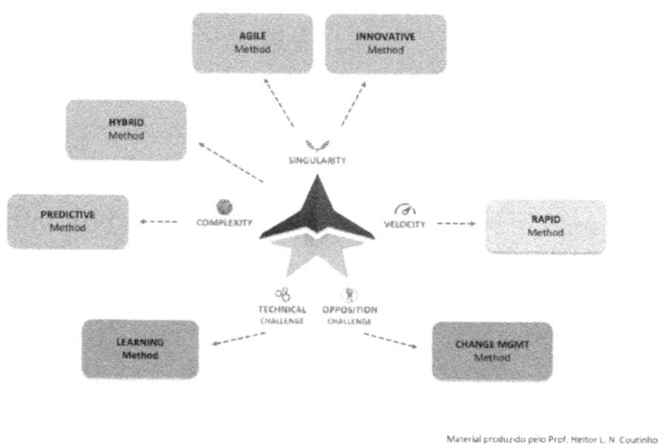

The rapid method simplifies one of the four development methods we have discussed. It uses resources to reduce the total project lifecycle time by streamlining documentation recording, reducing analysis, fast tracking, crashing, the RAD (Rapid Application Development) approach, and focusing on criticality. In decisive circumstances, this method postpones work that can be improved later so that the final deadline is not compromised. Crises, aggressive competition, and implementing an organizational survival strategy are among the most common reasons for using rapid methods.

The learning method raises the quality of the knowledge to be acquired for absorption on a specific subject. A qualified and competent team is key to transforming education into results. Corporate education methodologies based on andragogy, cognitive thinking, coaching, leadership techniques, and experiential learning, for example, speed up the assimilation of knowledge, practices, and applications. Learning is a new strategic mentality that will change the rules of competition. And technical challenges are prevalent in projects. That's why teaching-learning strategies are relevant for management professionals.

The Change Management method focuses on organizational behavior and culture due to transformations. It is a preventative method to deal with the potential for future opposition to changes arising from projects. The initial phases of this method are usually recommended, despite the degree of challenge, so that negative resistance does not occur. A good management improvement project should not be started without human attention, even in an initial situation of low opposition. A change leader helps to engage stakeholders and provide an environment for collaboration and growth (White et al., 2022).

You understood the differences between the methods and the demands of business development. It is inevitable for organizations to have comprehensive models. The levels of challenge shown in the Star Diagram and the need for change supported by different methods demonstrate that methodologies using a single path are ineffective. Therefore, it is necessary to assume that the heterogeneity present in organizational project management requires solutions that use paths coherent with the composition of the change situations. This knowledge will be demonstrated below.

THE PROJECT MODEL FOR ORGANIZATIONAL IMPROVEMENT

Improvement through adequate progress at all levels is the current aspiration of most organizations. Ensuring a distinctive and prosperous future is what the execution of a good strategy through projects must achieve. The mechanisms to be adopted must not shy away from the success that investments aim to achieve despite the unpredictability of the business world. Unfortunately, current market practice is still far from being able to address the impacting variables we have mentioned so far. However, the correct responses to challenges and situations are the primary factors determining change's success. For decades, the failure rate of projects worldwide has not improved considerably.

The key to this challenge lies in systematizing the project development process to ensure repeatability. This affirmation is the right way to achieve maturity without restricting paths. Each project is unique, and its difficulties are equally original. There is usually a combination of challenges that require a coordinated approach. Consequently, it is necessary to have a clear awareness of the situation, the variables to be analyzed, and the journeys leading to the best proposals. The solution for each particularity must meet the primary objective that is to be achieved.

The Success of a Project

Evaluating a project's success depends on its performance and the relative importance of certain factors. For a practical judgment, consider four levels of achievement in a cumulative and interdependent manner. An entirely successful project would achieve the following levels, which will be determined over time in this order: performance, impact, result, and longevity (Coutinho, 2019).

The first level, performance, describes the project's efficiency during its life cycle. It is equivalent to saying that its conception, planning, development, execution, and completion were successful if we consider that the scope, cost, deadline, and quality targets (in terms of compliance) were met. Was it executed according to the specified scope, and were the phases approved by the client? Have the deadlines been met, and have the costs not exceeded what was planned? In addition, have the quality requirements been executed by what was envisioned and agreed upon? (Gošnik et al., 2024). Estimating the targets of the performance plans with tolerance percentages based on the uncertainty found is essential. After all, projects are not repetitive activities and include a distinct lack of precision in their development. The result of calculating each factor will take into account the variation stipulated for its target. In this way, performance can be measured.

The second level of evaluating success concerns the impact generated on stakeholders. A project needs to be efficient and generate appropriate implications according to assessing potentially related people and entities. The project leader's job is to manage expectations, negotiate, and adjust the interests of the various stakeholders throughout the life cycle and after the project has been completed. The client and users must be satisfied with the products and services implemented. More specifically, the principles of usefulness and value perceived by customers and users need to be met if impact is ensured.

However, other stakeholders also play important roles and can be impacted differently. The experience should be unique, highly motivating, challenging, and full of learning and professional growth for the project team. It will become even more prosperous and appropriate for the diversity of thought, gender, culture, and skills. Diversity in a project team is an important differentiating factor and will generate positive effects if it is welcomed. However, the conflicts generated throughout the project's life cycle need to be well managed if the intended benefits are to be achieved. The team's goal will be achieved if the morale of its members increases or remains high.

Still, on the subject of stakeholders, the project's sponsors, those who have the essential role of providing the necessary resources, giving direction, assisting in the communication process, and approving semi-finished or final products, have the understanding that the process has been well conducted, and that success will result. Suppliers and other entities outside the organization that somehow interact with the project and/or can be involved must also be considered in the impact assessment. The degree of engagement of different stakeholders is a trend indicator. It is, therefore, a criterion to be considered.

Although the impact level is after the performance level related to time, where it is better ascertained, stakeholder perceptions also occur throughout the project's development and can change as results are obtained. Therefore, the impact target refers to the various stakeholders and will be achieved if they have confidence during the performance and operation of the products and services.

The third level of determining success is the result. The result is recorded in the target of the project's primary and/or secondary objectives. It can be measured in different ways. It can be expressed as a financial result, such as revenue, cost savings, profit, or return on investment, or it can express a particular benefit of the project. A project can have the primary objective of improving urban mobility or the perception of the sustainable value of an offer. It can also contain secondary objectives and benefit other stakeholders. For example, the delivery of software (product) can increase sales (result) and benefit internal productivity within the company (secondary outcome).

The perenniality level represents the project's ability to generate lasting results over time. In this sense, it is essential to look at factors that make organizations sustainable and not just focus on short-term results. To this end, the GPM Global standard can be used as a reference, a tool developed to measure projects in terms of their sustainability-related impacts. GPM Global, Green Project Management, is a social certification organization focused on sustainability. Its mission is to decouple socio-environmental degradation from economic growth by infusing sustainable practices into managing projects, programs, and portfolios. It was the first project organization to join the United Nations Global Compact.

The GPM P5 standard takes into account five categories for measuring sustainability: the maturity and efficiency of the project development process, following ISO 21.500; the life cycle of the product generated from a social, environmental, and economic perspective; the social base (people), which includes recommendations on labor practices and decent work, relations with society and clients, human rights and ethical behavior; the financial base (profit), which includes recommendations on return on investment, business agility, and economic simulation; and the environment (planet), which includes recommendations on transport, energy, water, and waste.

In addition to these universal sustainability principles for organizations, other elements contribute to achieving a certain level of sustainability. A management system's capacity perspective generally contains objectives to achieve long-term results. For example, investments in skills development and a culture of innovation and technology can contribute to organizations' longevity.

On the other hand, it must also be understood that not all projects need to include sustainability levels to succeed. It is the only level that is optional if it is not advantageous. Some projects are created for immediate purposes, such as solving problems or preparing the environment for other projects. However, no one should consume resources that could compromise the organization's future. In this sense, special care should be taken with cost-cutting initiatives. They can bring results in the short term, increasing the possibility of generating higher incentive pay for employees, which is a good attraction. However, they can occasionally compromise the organization in the future, for example, if the costs of maintaining facilities and developing people are reduced too much.

The Organization of the Project Model

There are different ways of structuring models for application in organizations. Some are more abstract, others have restricted use, and everything will depend on their purpose. In the case of projects, the main recommendation is to optimize their applicability to suit different situations. An uncomplicated model that can be used more often is more likely to be practiced as a routine. The dynamic model was designed with an organization based on principles, a fundamental process, roles, and attributes. These four parts include definitions that will guide each company's specific methodologies. This model can also support a methodology for use in different businesses. A multi-applicable guiding model as a foundation will ensure that indispensable analyses are carried out for projects to succeed.

The Principles

The primary principle of the dynamic model is to ensure the definition of success in projects. There are models in the literature that focus on performance because this is the part that covers the life cycle of a project, from conception to closure. However, considering only the project life cycle may not be enough to express the full value expected from a given investment. To certify this conclusion, the project's performance does not show the desired result. Therefore, high performance does not guarantee that the results will happen. Good performance will only allow products to be delivered. A consistent model considers the project and operation cycles of the project's products and services. Organization project methodologies must meet the circumstances we have presented for success to be fully achieved.

Another principle is the flexibility of the process to be used. A procedural system that certifies the quality of development is indispensable. Similarly, the model must provide flexibility to strengthen reflection and innovation in ambiguous and uncertain environments. This fact is the structural paradox of modeling, which encompasses systematization and flexibility simultaneously. It doesn't allow for extremes such as rigidity on the one hand, which stems from systematization, or carelessness on the other, which stems from flexibility. Both aspects are relevant, and their weight will depend on the composition indicated in the star diagram.

A third principle refers to the use of methods. An organization with a single method for doing projects is making a mistake that could make results impossible. It may seem unbelievable to some, but some companies and professionals adopt a single method as their preference This situation may have origins in the past when the stability of the entrepreneurial world and macroenvironment limited a single method. For a model to be widely followed, it must at least address the four methods mentioned in the previous sections. Only in this way will the different natures be successfully incorporated.

The fourth principle deals with specific attributes particular to a given situation. Decisions and actions specific to each type of challenge and intensity are needed to deal well with project challenges. When dealing with different difficulties in combination, activities and actions become even more unique. The star diagram is a highly guiding resource in the models. Based on the composition of the resulting challenges, the specific attributions for that situation will be included in the methodology. This flexibility is possible by the technique that gives the model the logic of a Lego, providing unique and necessary paths.

The model that considers these principles has consistency, starting with its design stage, where its requirements are. The requirements have to be structured to underpin development and execution. This fact explains why the first stage of a project development process is crucial. Categorical definitions occur in the first moments and depend on strategic choices for modeling the scope. Let's see how the principles are present in the process at every stage.

The Process

The dynamic model describes a standard process for developing projects. Because it is the essence of systematization in project management, it is called the base process. The base process addresses the knowledge demonstrated in this chapter practically. In this way, it aspires to meet the variety of paths and analysis content required for project development. Its structure can be used in organizations' methodologies to cover the particular demands of each area. It is also possible without adapting it to market segments, making development and execution more agile.

The process is structured in phases and stages that detail its characteristics. The phases are conception, development, execution, and completion. They are common in business practice. In this article, we will focus on the first two phases.

The Conception Phase

The project begins when the interest in what is to be worked on has already been discussed, whether it be the formulation of a strategy, usefulness, or importance for the organization. The initial phase allows the idea to be conceived, the methods to be chosen, and the structuring definitions to be established. These steps simplify classifying and determining the likely products and services. The most significant decisions of the project take place in the conception phase. For this reason, it is recommended that this phase be developed with the participation of the sponsor and clients. The reason for creating the project, the benefits to be captured, and the investment required for implementation are all matters that should be discussed at the outset, even if their accuracy is still presumed. The stages of the phase can be seen in Figure 3.

Figure 3. The stages of the conception phase

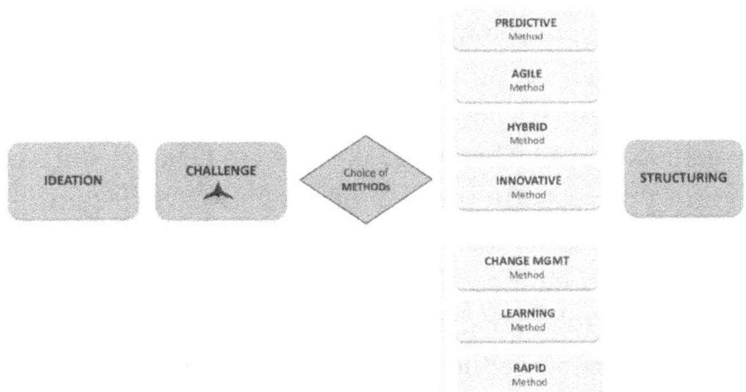

There are four steps to conclude the initial phase:

- The ideation to ensure that the project team with stakeholders thinks through the ideas;
- The challenge is to classify estimated levels of stimulus;
- The definition of the methods that will be used from this stage onwards; and
- Structuring is needed so that scope-directing decisions can be made.

Many techniques help at this stage (ideation step), such as traditional brainstorming, brainwriting, storyboarding, mind mapping, or part of design thinking. These techniques allow you to think more freely, creatively, and without judgment. Articulated discussions should be recorded for use in the process. For ideas to emerge more selectively, it is necessary to understand the problem or opportunity, its causes or benefits, the stakeholders affected, and any barriers.

When measuring the challenges, the level of each factor in the star diagram is indicated on a scale, as we mentioned earlier. The complexity challenge ranges from the primary level, where the number of variables is limited to a few occurrences, to the mega level, where many interdependencies occur, causing the project to face adverse, ambiguous, and highly challenging conditions. The singularity spectrum begins at the progression level, characterized by scope risks that occur more frequently in the life cycle. The highest level is that of imagination, where expressive innovation happens, and there is no initial clarity about the objective and its meaning. Pioneering in the sector market is the most characteristic type, representing transformations, disruptions, or radical changes.

Regarding velocity, the scale starts at a low level, representing projects where the deadline is not critical, and there is plenty of time for product development. The last level is intense, with a tight deadline for reacting to crises and business-critical situations. In the technical challenge, the first level, the knowledge needed to develop the project, is essential, generating a small and easily achievable challenge. The knowledge required to create the project is specialized at a deep level, and there is limited literature on the subject. The opposition challenge starts at a minimal level, ideal in highly cooperative organizations, where stakeholders are partners in the proposed improvement or have little influence or power over organizational decisions. The opposite extreme is extensive, with collective oppositional movements and significant influence from key stakeholders who resist change.

The next stage is to define the methods using the available conception information. The first part involves choosing one of the four project management methods. This decision uses the definition of the method's application in conjunction with the diagram results. When the method is chosen, the critical success factors in its management must be incorporated. It is desirable to anticipate reflections on objectives and products to support this decision. Subsequently, new methods are added to the path to be followed. Change management is always necessary, and its scope of use will depend on the scale of the challenge. The learning method is used when there is a need to develop skills of any origin. Knowledge development is usually part of projects. In the case of a higher-level technical challenge, this resource becomes indispensable. The rapid method is proper when speed is of the essence. This method option aims to optimize processes and resource allocation, provide autonomy for the team, strengthen monitoring, and improve relationships with stakeholders. The pragmatic and rational use of information technology enhances the effectiveness of the rapid method.

In structuring, decisions are made to develop the project's products. This stage focuses on defining the scope and the information that may affect it. The attributes to be analyzed will be different for each project management method. Defining the primary objective and the products to be delivered using the predictive method is essential. Being thorough in resolving these attributes will make a difference in future definitions as they mark the project's performance, impact, and outcome. In an agile method, the content to be delivered is smaller. A backlog of products is determined, although it is still subject to subsequent confirmation. The process itself will confirm that the product list is still valid. Depending on the uncertainty inherent in the method used, the scope structuring will have its corresponding precision. However, it's good to focus on structuring so that the phase is a guiding light for the project in all circumstances. Structuring can also include prototyping. Prototypes are valuable resources for dealing with uncertainty. It helps people imagine, exchange ideas, and shape future states. Prototyping is also intended to increase learning about the scope of the product, enrich communication with stakeholders, certify the quality of integrations between components and systems, present the progress of specifications, validate with users, and be part of a progressive training process.

The Development phase

The development phase will have a process according to the chosen project management method. The four methods have both a work cycle and specific stages. That's why the development phase will be particular to each of them. Management improvement projects usually use predictive methods, except in a few cases. The main exception is a new product development project which uses the agile method. However, the agile method should be evaluated when there is uncertainty about products for management improvement. Management improvement projects depend on information technology for implementation, which is usually developed using agile methods. Look at the generic proposal for development paths in predictive and agile methods. Hybrid methods have a flexible development mix. Innovative methods adopt the agile method in continuity with a research method.

Developing predictive methods

In the predictive option, development follows the traditional project life cycle guidelines with sequential activities per stage. Recurrences can be used casually, maintaining the prerogative of the linear process. As there is also a good product vision, some management aspects must be planned in parallel. Development in the life cycle can be seen in Figure 4.

Figure 4. The development of a predictive method

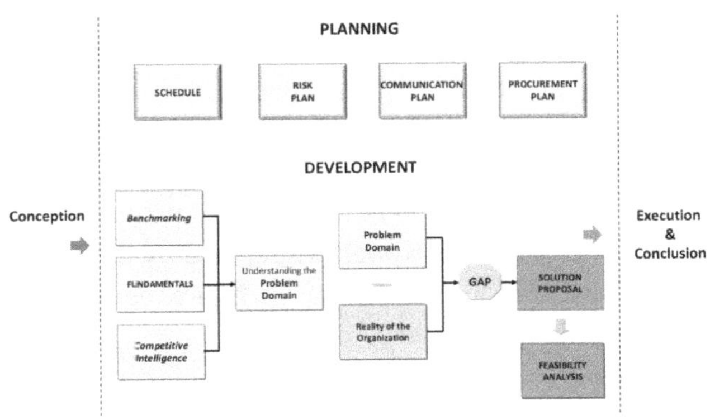

The planning stages are represented by the schedule, which contains information on activities, deadlines, and costs; the risk plan; the communication plan, drawn up following the mapping of stakeholders; and the procurement plan, when necessary. The planning stages are updated according to the results achieved, keeping management active.

Development begins by exploring external factors: fundamentals, benchmarking, and competitive intelligence. The fundamentals are developed by researching the project's themes in books, technical, academic, and consultancy articles, websites, and published reports. Benchmarking is carried out in organizations that have already applied similar knowledge and have become a reference in the field. Competitive intelligence is used to assimilate knowledge about competitive practice and is obtained through various regular means. Benchmarking and competitive intelligence are not always possible or relevant, but the essential support given by fundamentals must not be lacking. These elements lend credibility to the project and should be given priority.

The fundamentals establish the perspective from which the problem is situated and the references to guide possible data collection and analysis. They must provide a basis for the topics dealt with. Their function is to allow familiarization with the subject that interests the work and to enable access to what has already been researched and validated about the knowledge and practices. The fundamentals can also broaden the view of the study topic since links can be discovered between the subject and other problems and management areas. It is impossible to interpret and understand reality without a basis for explaining the knowledge developed about the facts. They can provide a reflection and understanding of the state of the art about the project's themes.

A second stage of seeking knowledge from an external perspective is equally crucial. The Japanese have the word dantotsu, which means striving to become the "best of the best," based on self-improvement, which consists of looking for, finding, and overcoming competitors' strengths. This concept has taken root in approaches to executing strategy through projects, has produced impressive results in many companies, and is known as benchmarking. Benchmarking is an ongoing process of comparing products, services, and business practices among the strongest competitors or companies recognized as leaders. It is a research process that allows comparisons of processes and practices "company by company" to identify the best of the best and achieve a level of superiority or competitive advantage. However, developing projects does not necessarily require looking for a leader but for a company that serves as a reference for the topic being studied. Most of the time, the companies that were researched are not competitors and are in another industry. Only in an environment where there is no competition do they feel open to providing information that can be of great value in transposing to the reality of your project. Benchmarking emerged as a need for information and a company's desire to learn how to deal with a business issue faster or generate results in less time. Knowing how to make and adapt benchmarks in the organization's process can allow opportunities to be exploited, constituting a safe shortcut to excellence since it uses all the intellectual and empirical work accumulated by other organizations, thus avoiding mistakes and pitfalls. They are based on learning from the best experiences of similar companies and help explain the whole organizational performance process.

Competitive intelligence is acquiring and organizing relevant information about competitors' behavior, customers, and the market. This external look may be necessary; it increases your strategic capacity to understand the need for investment, allowing for a better short- and long-term decision-making process. The business strategy itself contains information about the market and the competitive environment. However, it is natural that the strategy does not include some aspects that you need. If this is the case, you should analyze sector trends, uncertainties, and trends relating to your demand in the global and national macro-environment.

External information provides an understanding of the problem or opportunity to be exploited, which is valuable knowledge for organizational development. The next step is to critically assess the organization's current state. Its reality concerning the issues being dealt with reveals shortcomings that need to be evaluated to make decisions.

Introspective analysis allows for a greater understanding of your practices and can be worked on even before you appreciate how other organizations live. To do this, you must look at the information systems, current processes, various indicators, historical data, management models, tools, and company plans. Be exhaustive in your search for internal information. In the same way as planning benchmarking, it is also advisable to take the time to establish scripts and instruments for internal research. You must mark the research universe once you have defined the data collection instruments. Evaluate the stakeholders identified in the design phase and analyze the possibility of making a "vertical cut" in the organization to include the various hierarchical levels. The sample of people to be interviewed must be representative. Understanding current projects and potential interdependencies with your project is also part of the internal perspective. The history of the project portfolio can also provide insight into the organizational culture and its constraints on change.

The difference between the problem domain and the organization's reality will demonstrate the gap for analysis. It is essential to consider which aspects should be included in the current project, which should be discarded, and which can be implemented in the future. Estimating maturity levels can help with this decision. It is not advisable to jump ahead in any aspect of the maturity level next to your reality. This problem is an inconsistency in organizational evolution that can lead to damage if making coherent progress.

With the scope defined, the next step is to propose a solution. The solution proposal is the central part of the paper and occupies a separate space. It aims to develop the central answer, set it in context, and highlight the relevant details and inherent benefits. The specific project solution requires an examination of what has already been defined. Developing the solution requires creativity, reasoning, and strategic vision. It also involves aligning with the project's primary stakeholders, especially the sponsor, throughout the development process. Solutions developed with the close involvement of clients, sponsors, and users are more likely to succeed. They minimize the possibility of future reactions to change and create a sense of ownership and commitment to the implementation and the generation of lasting results. In many cases, it is interesting to demonstrate solution options, considering evaluation items and weights to determine the advantages and disadvantages of each possibility. That is not only a way of dealing with the problem from a broader perspective but also a way of providing greater flexibility and involvement and empowering decision stakeholders.

The proposed solution must be analyzed for feasibility, with or without options. Some aspects of feasibility are certainly developed during the solution stage, such as technical feasibility. However, ensuring implementation is necessary to confirm what has been analyzed. Many aspects need to be checked. The most applied is economic feasibility, which considers the costs of results. Operational viability checks how well a product will meet needs and whether it will be used and maintained after implementation. In certain situations, market viability examines whether the project meets the demands and expectations of the target market. Depending on the type of project, cultural, socio-environmental, and legal feasibility are also required to determine the practicality of a proposed project or business venture.

Developing agile methods

In the agile option, development takes place iteratively and incrementally. It is oriented towards disciplined project management processes with regular inspection and adaptation. In this approach, plans and results are constantly monitored. This type of method accelerates response to change through the adaptive approach. In a linear method like the predictive method, adaptations are challenging and unnatural as in the iterative process. This approach is based on self-organization, teamwork, and collaboration with stakeholders. Thus, project development is carried out in stages that deliver by-products. A completed cycle is a sprint that provides part of the project's product. There is a lot of learning during each sprint because feedback from stakeholders is obtained regularly. That is how consistent value is increased.

Figure 4 shows the development phase, which starts with the product backlog from the conception phase. The creation of this stage is the responsibility of the Product Owner, a person dedicated to maximizing a product's value. Based on this scope, a planning meeting is held to find the best structure for defining by-products to form a sprint backlog. The sprint backlog contains the items selected by the development teams to work on in the next sprint. The sprint backlog represents the primary output of the recursive activities planning meeting.

Sprint cycles then begin for the delivery of by-products. During these times, it is good to identify and integrate the missing parts of the solution while maintaining extensive engagement. Iterative processes are ideal for creating new products and services because they meet customers' needs in a participatory way. Intermediate versions give the customer something to work on so they can learn from it and discover additional features needed. In this way, the scope and functioning of the future solution are better understood and will be incorporated through feedback. The missing parts will also come to light through prototyping or other resources that broaden the view of reality. The method also has spontaneous mechanisms to encourage feedback and revisions. For this reason, costs and deadlines are lower than for other approaches. The Agile method has a philosophy and principles that guide its implementation. However, there are variations in the techniques derived from it on the market and additions that can help organizations create their methodologies.

Project management is continually adapted to the project's changing situation and environment. Its efficiency depends on an organizational culture based on internal and external collaboration so that decisions are informed agilely. New interactions naturally happen incrementally, and plans are reworked, executed, and completed to add new requirements in a new cycle. A baseline schedule should also be updated regularly according to circumstances, the ideas that emerge from iterations, and additional requirements. That is not a conventional schedule but a guiding and dynamic roadmap representing the sprint plan's backlog.

Figure 5. The agile process

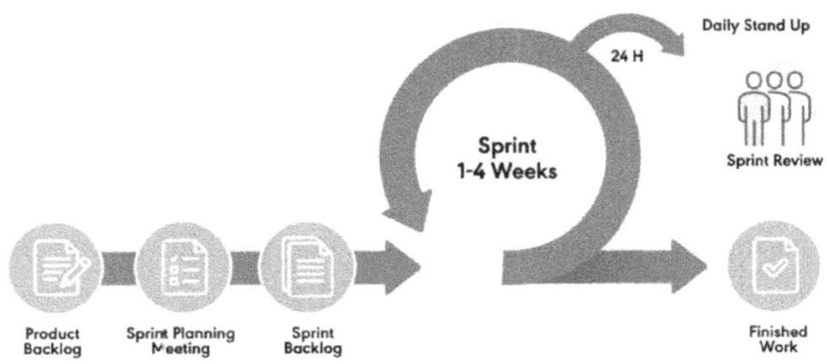

The roles

The essential remit of the core project team should be defined at the conception structuring stage. Other stakeholders can also be included in the initial phase because of their relevance and contribution. A stakeholder is any person, organization, or entity involved and/or interested in your project, may be affected by your products/services, or has power or influence over the results. The project manager should feed into the communication plan with each group of stakeholders, taking into account their levels of power or influence, the degree to which the change will impact them, or even the reaction they will have to the project. Therefore, a stakeholder is someone who is potentially related, a fact that is characterized by systemic attention and proximity.

The core team should include the project leader/manager, the sponsor, the business analyst, and the project team. The project manager ensures that the project results are produced within the agreed scope, timeframe, and cost and with the quality required to deliver the defined benefits. He is also responsible for communicating with everyone. The business analyst, the project's knowledge manager, is tasked with understanding the business needs for change, assessing the business impact of these changes, capturing, analyzing, and documenting the requirements, and supporting the communication and delivery of the requirements with the relevant stakeholders. Business analysts are at the heart of knowledge management - they identify knowledge sources, learn from them, and influence how knowledge flows within the organization to deliver successful results. The sponsor ensures that the project is focused on achieving strategic objectives. They take responsibility for the value of the investment and ensure the adoption of a cost-benefit ratio that balances the demands of the business, users, suppliers, and customers. They can analyze projects and their impact on the organization as an entity. The other participants in the team will have roles defined according to their attributions during the life cycle. Additional roles can be established to provide critical capabilities. A change manager may be appointed when behavioral and cultural aspects require concentrated effort. A technical leader may be needed to meet the demand for specialized knowledge. They provide guidance and advice on technical issues, clarify issues for business partners, and support products and services to ensure the planned quality.

Agile methods have different roles, which are not directly related to classic roles that have been described. A project manager performs additional tasks in an agile process. There are also different roles for agile methods in the market. For example, Scrum, Kanban, Lean Development, and XP have their roles. In Scrum, the most widely used on the market, a project manager has several of the Scrum Master's attributions and is responsible for facilitating the team's work but also for the Product Owner's responsibilities, who is responsible for products.

Own attributes for dealing with challenges

As we have seen, a dynamic model to meet the diverse instigations of change projects can bring benefits and help increase success rates. Determining the methods at the conception stage is critical but not enough. Each challenge assessed has specific responses to deal with its intensity. We call these responses our own attributes, characterized by intrinsic actions and decisions. In addition to the responses to each challenge in its intensity, there is an essential coherence to deal well with the combination of challenges. It is common to have projects with several challenges simultaneously. Attributes specific to the combinations are necessary to complete the path with the most constructive responses.

In certain situations, you have to work hard to combine the answers. That kind of work is because there may be paradoxes to deal with. For example, a project with high complexity and singularity coexists with dualities that are essential for success. It is necessary to determine, for example, attributes that can coexist with opposing decisions, such as methodological rigor to deal with complexity and flexibility to deal with singularity. In this case, you can see the evidence of taking on your attributes in a model. There is no standard answer for all projects, but the existence of a dynamic model is edifying. Project development depends on analytical thinking combined with an innovative mindset. The capacity of the project professional must support this way of dealing with organizational stimuli for organizational evolution. Thinking is everyone's responsibility, but it must not exist without direction. A practical model provides direction if it allows us to respond with the attributes that each situation calls for.

Active Change Management

Organizational improvement is achieved by actively managing organizational behavior and culture. Traditional approaches on the market describe change processes with an initial focus on overcoming resistance. The active approach considers resistance a point of attention throughout the life cycle. It focuses on encouraging change in the early stages and treating resistance only when necessary. Change is a strategic opportunity to learn and grow. People have a natural propensity to evolve in life. Agile organizations, essential for today's times, have dynamic structures and high levels of trust and collaboration. They are not environments compatible with regular resistance. The orientation of change management must be positive (Coutinho, 2025).

The change cycle is subject to variations arising from human interference, which are actual and likely to occur. That's why every cycle should be indicative and used as a reference (Gibbons et al., 2024). Methods are generally beneficial in different contexts if they remain flexible to the environment. In the preferred path, the elements for intervention action and the exercise of leadership are the valuable aspects. Four stages should be considered in the change process: prevention through specific perspectives, continuous reinforcement through recognizing and encouraging people, eliminating barriers to change, and overcoming remaining resistance or resistance that arises on the transformation journey.

Prevention can rely on the acronym LEAD2C. Each letter represents a perspective to be considered. The first is leadership (L) for change, which proposes collaboration, service through facilitation, and the desire to transform the organization by creating an environment conducive to value creation and personal growth. Leadership remains active by practicing engagement through dialogue (letters E and D of the acronym). Dialogue encourages cooperation and potential development and is an opportunity to propose a constructive direction for making changes. The determination for the conversation must happen with alignment (A). Alignment with strategic choice and purpose must prevail. Finally, the two Cs refer to the capacity (C) needed for everyone to get along and contribute to the process and organizational culture (C), which must be respected and modified according to the project. In this way, prevention promotes inclusive participation and ownership. The prevailing mindset is one of readiness for change (Patel, 2022).

Reinforcement, the second referential stage, includes maintaining all forms of communication with stakeholders. These moments should include celebrating partial achievements, recognizing team performance, and rewarding accomplishments. The third stage excludes common barriers to change processes, and new habits should be encouraged. In this sense, leadership reduces any unnecessary energy expended. It is done by strengthening freedom and control over decisions, promoting the renunciation of old procedures, reducing the distance of change by proposing negotiation in a mobile environment, relieving uncertainty whenever possible, and demonstrating corroborating evidence to resolve issues (Berger, 2020). When it occurs, the final stage may have been mitigated by the previous stages. However, it is essential to consider that resistance to change needs to be carefully analyzed for its multiple faces and origins. They can be positive and be important warnings for improving the project or the change process. In adverse situations, the feelings inherent in the emotional journey must be dealt with appropriately.

Excellence in a management improvement project is achieved through the model described. In this way, the relevant path is created by the intelligible composition of the challenges. The Star Diagram combines advanced attributes of project management and change management. So, this is the uncomplicated response to achieving success at all levels. Lasting results will be achieved by fully understanding that each project is unique. Each path is particular to meet unique needs. The dynamic model has the consistency to meet multiple real-world specificities.

REFERENCES

Berger, J. (2020). *The catalyst: How to change anyone's mind*. Simon & Schuster.

Botchkarev, A., & Finnigan, P. (2015). Complexity in the context of systems approach to project management. *Organisational Project Management*, *2*(1), 15–34. DOI: 10.5130/opm.v2i1.4272

Brix, J., & Peters, L. S. (2015). Exploring an innovation project as a source of change in organization. *Journal of Organizational Design*, *4*(1), 29–43. DOI: 10.7146/jod.17574

Coutinho, H. (2019). *Da estratégia ágil aos resultados: Uma combinação de abordagens adaptativas, mudanças dialógicas e gestão avançada de projetos*. Editora Saraiva.

Coutinho, H. (2025). *Vantagem estratégica na incerteza: A agilidade na governança corporativa, na estratégia e na gestão de mudanças em um mundo dinâmico e imprevisível*. Editora Altabooks.

Denney, V. P., & Watkins, D. V. (2023). Using phenomenological inquiry to understand stakeholder management success criteria on projects. *Global Journal of Business Disciplines*, *7*(1).

Gibbons, P., & Kennedy, T. (2023). *Change myths: The professional's guide to separating sense from nonsense*. Phronesis Media.

Gibbons, P., & Kennedy, T. (Eds.), & Ulrich, D. (Foreword). (2024). *The future of change management: Collected essays from leading thinkers and practitioners*. Phronesis Media Press.

Gošnik, D., Janjušić, D., & Nikolić, G. (2024, August). Analysis of project success factors and project success criteria in micro and small firms: Evidence from Slovenia. *Sciendo*, (3).

Green Project Management. (2021). *The GPM P5 standard for sustainability in project management* (Version 3.1). GPM Global. https://www.greenprojectmanagement.org/

Hubbs, G., O'Rourke, M., & Orzack, S. H. (2021). *The Toolbox Dialogue Initiative: The power of cross-disciplinary practice*. CRC Press.

Kotter, J. P., Akhtar, V., & Gupta, G. (2021). *Change: How organizations achieve hard-to-imagine results in uncertain and volatile times*. John Wiley & Sons.

Kotter, J. P., Akhtar, V., & Gupta, G. (2021, August 13). Is your organization surviving change — or thriving in it? *Harvard Business Review*. https://hbr.org/2021/08/is-your-organization-surviving-change-or-thriving-in-it

Lister, J. (2018, April 24). How can fast tracking affect a project? *Chron*. https://smallbusiness.chron.com/can-fast-tracking-affect-project-32112.html

Mignenan, V. (2023). Management practices and success of innovation projects. *American Journal of Management*, *23*(3), 79. DOI: 10.33423/ajm.v23i3.6361

Mohammad, H., ... (2016). Determining the technical competencies of construction managers in Malaysia's construction industry. In *3rd International Conference on Civil and Environmental Engineering for Sustainability* (pp. 1–10). MATEC Web Conference. DOI: 10.1051/matecconf/20164704021

Patel, K. (2022). Change management needs a prevention strategy. *Journal of Business Agility*.

Reeves, M., Whitaker, K., & BCG Henderson Institute. (2022). *Inspiring the next game: Strategy ideas for forward-looking leaders*. BCG Henderson Institute.

Shenhar, A. J., Dvir, D., Levy, O., & Maltz, A. C. (2016, April/May). The challenge of innovation in highly complex projects: What can we learn from Boeing's Dreamliner experience? *Project Management Journal*, *47*(2), 62–78. DOI: 10.1002/pmj.21579

White, A., Smets, M., & Canwell, A. (2022, July 18). Organizational transformation is an emotional journey. *Harvard Business Review*. https://hbr.org/2022/07/organizational-transformation-is-an-emotional-journey

Wysocki, R. K. (2019). *Effective project management: Traditional, agile, extreme, hybrid* (8th ed.). Wiley. [Kindle edition] DOI: 10.1002/9781119562757

Yadav, P. S., & Paul, V. K. (2024). Investigating the determinants of construction project complexity impacting project success: An India perspective. *International Journal of Construction Management*, *24*(16), 1760–1770. DOI: 10.1080/15623599.2023.2295609

Zucker, R., & Rowell, D. (2021, April 26). 6 strategies for leading through uncertainty. *Harvard Business Review*. https://hbr.org/2021/04/6-strategies-for-leading-through-uncertainty

About the Contributors

Nuno Geada Master degree in Systems Information Management by Polytechnic Institute of Setúbal - School of Business Sciences and Management -Setúbal, Degree in Industrial Management and Technology by Polytechnic Institute of Setúbal - School of Technology of Setubal. He has written chapters, and papers to journals about topics regarding information technology management and strategic change management. He is from the Editorial Board - Associate Editor from International Journal of Business Strategy and Automation (IJBSA). He is the Editor of the book Reviving Businesses with New Organizational Change Management Strategies. His main research interests in information systems management, strategy, knowledge management, change management, and information technology management adoption in business contexts throw models and frameworks. He works as a Professor and a Researcher.

Liliane Jamil Graduation in Medicine from the Federal University of Minas Gerais (2007-2013). Specialization in Basic Care and Family Health by the Federal University of Minas Gerais (2013-2014). Residence in General Surgery at Santa Casa de Misericórdia in Belo Horizonte (2014-2016). Specialization in Plastic Surgery at Hospital Felício Rocho (2016- ongoing). Master's degree in Science, Technology and Management applied to Tissue Regeneration by the Federal University of São Paulo (2018- ongoing).

***

Felipe Alcantara is an engineer with extensive experience in business management, strategic planning, and development. He holds an MBA in Corporate Finance from Fundação Getúlio Vargas and completed the Executive Development Program at Fundação Dom Cabral. As a former executive in large corporations and currently a senior consultant at Veritus Consulting, he has led transformative projects that integrate technology, processes, and people to deliver exceptional results. Passionate about management and innovation, he strongly believes that management and leadership are distinct yet complementary skills, critical to driving business success. His expertise in strategic planning allows him to connect vision with execution, ensuring sustainable growth and value creation.

Thierry Burger-Helmchen is a professor at the Faculty of Economics and Management, University of Strasbourg, specializing in innovation management. His research at the BETA lab examines how organizations craft innovation strategies, focusing on emerging technologies like AI. He teaches at MBA, DBA and PhD levels and regularly delivers conferences for public and private organizations.

Mariana Carôla has a Master's degree in Human Resources Management and Organisational Consulting and a degree in Sociology. Her research focuses on Green HRM practices, Employer Attractiveness and Corporate Social Responsibility. She currently works in Talent Acquisition at Deloitte.

Marcos Corrêa PhD and Master's degree in Business Administration, with a focus on Organizational Studies, Work, and People, from the Federal University of Minas Gerais (UFMG). University professor in Human Resources Management, General Administration, Project Management, Leadership Skills and CSR/ESG at SKEMA Business School and the Federal University of Minas Gerais (UFMG).

Younesse El Menyari, Department of Economics and Management, Research laboratory in Business Intelligence, Governance of Organizations, Finance and Economic policies (BIGOFE), Faculty of Law, Economics and Social Sciences, Aïn Chock, Hassan II University – Casablanca- Morocco. email: youness.elmenyari@gmail.com

Renata Fagundes PhD candidate at Professional Postgraduate Program in Management, Research and Development in Pharmaceutical Industry, Oswaldo Cruz Foundation (FIOCRUZ), Brazil; Public Health Analyst at Centre for Technological Innovation - NIT FAR/FIOCRUZ

Paula Figueiredo has a degree in Economics with a specialization in Human Resources Economics, a Master in Human Resources Management and a PhD in Management. She is currently a professor at higher education institutions, and she has also been the Director of an institution since 2019. Paula has 20 years of experience in the business area with functions of direction, coordination, and consultancy. Throughout her professional experience, she has collaborated on leadership development projects. At the research level, she has developed research on Human Resource Development, more specifically Leadership Development in organizations, with some publications in scientific journals and conferences.

Xiaomei He received her bachelor's degree in International Economics and Management and a master's degree in Economics and Management of Innovation from the Faculty of economics and Management - University of Strasbourg, France, in 2021 and 2023, respectively. She is currently pursuing a Ph.D. in Innovation Management at the research lab BETA-CNRS. Her research focuses on technology-driven innovation, knowledge management, new product development, and the applications of artificial intelligence in innovation processes.

Akashdeep Joshi is an Assistant Professor at Mittal School of Business, Lovely Professional University.

Dinesh Kumar is an accomplished professional with extensive expertise spanning the military, academia, business, and charity sectors. He holds a PhD from the prestigious Indian Institute of Technology (IIT) Roorkee and received admission offers for the Fellow Program in Management from the Indian Institute of Management (IIM) Ranchi that made him the first Indian soldier to secure doctoral admission offers from both IIT and IIM. He is a former Chief Executive Officer of Pomento and the propounder of the concept of 'Lifestyle Social Work.' Currently, he serves as a faculty member in Business Analytics at the School of Business, Woxsen University.

MWP Maduranga (Senior Member, IEEE) is currently a Senior Lecturer at the Faculty of Computing & IT, Sri Lanka Technology Campus (SLTC), Sri Lanka, and also serves as a Co-Principal Investigator at the Centre for Telecommunication Research at the same institution. He received his B.Sc. Eng. in Electronic Engineering from the School of Engineering, Asian Institute of Technology (AIT), Thailand, in 2013, his M.Sc. Eng. in Electrical and Electronic Engineering from the University of Peradeniya, Sri Lanka, in 2017, and his Ph.D. from the IIC University of Technology, Kingdom of Cambodia, in 2022.

Frederico Cesar Mafra Pereira PhD and Master in Information Science from the School of Information Science (ECI) at UFMG, Bachelor in Economics from FACE-UFMG, Specialist in Strategic Marketing Management from CEPEAD-UFMG and Specialist in Innovation in Information Units from ITI-UFSCar. Studying for a Post-Doctorate in Information Science at UFSCar. An educator for 22 years, currently as an Adjunct Professor at the Federal University of Minas Gerais, in the School of Information Science, Department of Information Theory and Management (DTGI), and as a Permanent Professor of the Postgraduate Program in Knowledge Organization Management at the Federal University of Minas Gerais, in the research line Information and Communication Technology Management (GETIC). Researcher in the topics of Information and Knowledge Management, Competitive Intelligence, Digital Transformation, Innovation Management and Strategy. Leader of the Research Group "IKI - Information, Knowledge Innovation", and member of the Research Group "Núcleo de Informação, Tecnologia e Inovação - ITI UFSCar", both at CNPq. Mentor and speaker, and Member of the Editorial Review Board of The Bottom Line Journal of Emerald.

Jorge Magalhães Postgraduated in Competitive Intelligence for Public Health. Doctor and Master in Sciences in Management and Technological Innovation. Has over 20 years of experience in strategic management in the pharmaceutical Industrial Operations and in the last 14 years to act in R, D & I for Public Health area at FIOCRUZ. Published 03 books, 9 book chapters and several articles in journals indexed. Actually work in Technology Innovation Center (NIT-Far) at FIOCRUZ. It is leader of the CNPq Research Group Knowledge Management and Prospecting Health. The emphasis of their research permeate the identification, extraction and analysis of essential information within the "Big data" for Health, regarding the Management and Technological Innovation. The topics covered are inherent in Global Health, in which involves the pharmaceutical industry, pharmochemical and public health. Investigations are carried out through the development of prospective and technological scenarios of information science tools, Competitive Intelligence and Knowledge Management. Included in this context the analysis of BIG DATA, Web 2.0, Health 2.0 Technological Trends, market, Patents and Knowledge Translation.

Heloisa Menezes Economist at Pontifícia Universidade Católica de Minas Gerais, master in Sciences in Agricultural Development at Universidade Federal Rural do Rio de Janeiro, PhD Candidate at Universidad de Ciencias Empresariales y Sociales UCES de Buenos Aires. Invited profesor at Fundação Dom Cabral, Brazil. Consultant in innovation, digital economy and digital risks.

Fábio Luiz Daudt Morais Morais Public Health Technologist at the Institute of Science and Technology in Biomodels (ICTB) of the Oswaldo Cruz Foundation (FIOCRUZ/RJ). She holds a master's degree in Computer Science with an emphasis on artificial intelligence from the Federal University of the State of Rio de Janeiro (2020). She currently develops innovative technological solutions as an alternative to the use of animals in research at the ICTB/Fiocruz Innovation Laboratory, research projects that apply the 3Rs Principle and projects to apply digital medicine to public health. She has experience in the area of information technology, working mainly on the following topics: artificial intelligence, machine learning, data science, data engineering, analytical systems, business intelligence, Big Data, database, data integration, internet of things and innovation.

Carlos Muñoz-Maya PhD in Administration from the University of Celaya (Mexico), master's degree in teaching from the University of La Salle, specialist in Logistics from the EAN University, specialist in Finance Management from the Grancolombiano Polytechnic, Business Administrator from the Javeriana University, with extensive experience in teaching at the national level. Experience as a researcher and co-investigator on issues such as organizational management, competitiveness, sustainable management and corruption in micro, small and medium-sized companies and as an advisor and consultant on issues related to financial management and the formulation and evaluation of projects in both public and private companies. Director in national companies with responsibilities in planning, direction and management. Experience in the financial, logistics and commercial area in the financial and manufacturing sector.

Marion Neukam is a professor of management sciences at the BETA laboratory at the University of Strasbourg. With experience in the industrial sector, she specializes in innovation management in international firms. Her research focuses on knowledge management and the creative capacity of international teams, with a particular emphasis on the impact of sustainability on corporate innovation strategies. She teaches at various academic levels and contributes to conferences and industry discussions on innovation, sustainability, and business model transformation.

Karine Panerai Specialist in Competition Law from Fundação Getulio Vargas (FGV), with a Bachelor's degree in Business Administration from the University of Brasília (UnB). Currently serving as the General Coordinator of Strategic People Management at the Administrative Council for Economic Defense (CADE).

Ashwani Panesar An Associate Professor in Mittal School of Business, Lovely Professional University, Punjab (India); Dr. Ashwani Panesar has a rich teaching experience of over 13 years, taking up various business management subjects at university level. Having worked closely with the sales aspect of the pharmaceutical industry (8 years), he had pursued his Doctorate in pharmaceutical marketing strategies from Jaipur National University, Rajasthan (India).

Rafael Pérez-Uribe Doctor in Business Sciences Nebrija University. Diploma of Advanced Studies in Applied Economics, Nebrija University. Master in Organization Management EAN University. Maître es Sciences Université du Québec a Chicoutimi. Specialist in Evaluation and Construction of Management Indicators for Higher Education, School of Business Administration. Postgraduate studies such as Kenkyusei (Researcher attached) in Total Quality Control and Quality Circles, Fukushima University, Japan. Business Administrator, Jorge Tadeo Lozano University. Former Full professor assigned to the Research Department and director of the management research group of large, small,l and medium-sized companies (G3PYMES-A1 in Minciencias) of the EAN University. At the present time, Senior Researcher categorized by Minciencias and Independent Management Professor. Modeling expert. I have participated in the structuring in many management models that include diagnosis, evaluation, and organizational intervention.

Rogério Ramalho Professor and Researcher linked to the Department of Information Science at the Federal University of São Carlos. CNPq Research Productivity Scholarship holder, he is dedicated to the interrelation and balance between applied research, directed at the development and use of digital technologies, and research that addresses the theoretical dimension that underpins the representation and organization of information. President of the Institute of Information for Innovation - i2i (Mandate 2023 - 2025). Advisor to the Brazilian Chapter of the International Society for Knowledge Organization - ISKO (Mandate 2022 2025) and member of several scientific associations, such as: International Association of Artificial Intelligence (I2AI), National Association for Research and Postgraduate Studies in Information Science (ANCIB), Brazilian Society for the Advancement of Science (SBPC) and Brazilian Association for Education in Information Science (ABECIN). Leader of the Center for Information, Technology and Innovation (ITI UFSCar). Permanent professor of the Postgraduate Programs in Information Science at UFSCar and UEL and coordinator of the MBA in Information, Technology and Innovation for Business, and of the Specialization course in Innovation in Information Units, linked to the Open University of Brazil Program (UAB).

Pilar Ramírez Ph.D. in Management at Universidad EAN. Ph.D. in Entrepreneurial Sciences at Universidad Nebrija- España. Master in Business Management at Universitée du Quebec and Universidad EAN. Graduate degree in Management Indicators to Evaluate Higher Education Institutions) at Universidad EAN. BA in Business Administration at Universidad EAN. BA in Education at Universidad Pedagógica Nacional.Expert at online postgraduate education. Senior consultant in open collaborative innonovation prpjects.

Oscar Rete Universidad Tecnológica Nacional. Research Professor at UTN and UCES in Argentina. Guest lecturer at HS Mainz, Germany. Consultant in business strategy, technology applications, digital transformation and innovation.

Rosa Rodrigues PhD in Management, specialization in Human Resources and Organizational Behavior from ISCTE-IUL. Professor at the Instituto Superior de Gestão – Economics & Business School where teaches in the Bachelor and Master degrees. Research areas include organizational behaviour, human resource management, work and organizational psychology, psychological evaluation, soft skills, and test construction and validation.

Carlos Salcedo-Pérez Business Administrator with a M.S. in Economic Development and a PhD in Entrepreneurial Sciences. Experience working in economic development research. Academic experience in research focused on MSMEs. Teaching of classes in Management, Strategy among other subjects.

Aline Suave She holds a degree in Philosophy from the State University of Northern Paraná (2011) and Pedagogy from the Higher Institute of Education Alvorada (2013); Specialization in Modern and Contemporary Philosophy from the State University of Londrina (2013) and Specialization in Inclusive Special Education from the Faculty of Technology Paulista (2014). She attended the subjects of Special Topics in Philosophy: Ethics and Political Philosophy, Ethics and Political Philosophy II and Special

Topics in Philosophy: Aesthetics and Subjectivity II at the State University of Londrina, as a special student (2013). She is currently a professor - Municipality of Londrina and professor of the specialization course Innovation in Information Units ITI UFSCar / CAPES. She has experience in the area of Philosophy, Philosophy and Information Science, Philosophy of Information. Master in Information Science from the State University of Londrina. PhD in Information Science from the State University of Paulista - Campus Marília.

Nidhi Suthar holds a PhD from MPUAT, Rajasthan. She has worked for more than decade in academia and industry. She is a management faculty member at the School of Business, Woxsen University.

Index

A

B

C

D

E

F

L

M

N

O

P

Q

R

S

T

V

W

www.ingramcontent.com/pod-product-compliance
Lightning Source LLC
LaVergne TN
LVHW080618300926
849162LV00005B/1199